THE ELECTROMAGNETICS PROBLEM SOLVER®

REGISTERED TRADEMARK

A Complete Solution Guide to Any Textbook

Staff of Research and Education Association
Dr. M. Fogiel, Chief Editor

Research and Education Association
61 Ethel Road West
Piscataway, New Jersey 08854

THE ELECTROMAGNETICS
PROBLEM SOLVER®

Printed in the United States of America

Library of Congress Catalog Card Number 99-74250

International Standard Book Number 0-87891-550-8

III-1

WHAT THIS BOOK IS FOR

Students have generally found electromagnetics a difficult subject to understand and learn. Despite the publication of hundreds of textbooks in this field, each one intended to provide an improvement over previous textbooks, students continue to remain perplexed as a result of the numerous conditions that must often be remembered and correlated in solving a problem. Various possible interpretations of terms used in electromagnetics have also contributed to much of the difficulties experienced by students.

In a study of the problem, REA found the following basic reasons underlying students' difficulties with electromagnetics taught in schools:

(a) No systematic rules of analysis have been developed which students may follow in a step-by-step manner to solve the usual problems encountered. This results from the fact that the numerous different conditions and principles which may be involved in a problem, lead to many possible different methods of solution. To prescribe a set of rules to be followed for each of the possible variations, would involve an enormous number of rules and steps to be searched through by students, and this task would perhaps be more burdensome than solving the problem directly with some accompanying trial and error to find the correct solution route.

(b) Textbooks currently available will usually explain a given principle in a few pages written by a professional who has an insight in the subject matter that is not shared by students. The explanations are often written in an abstract manner which leaves the students confused as to the application of the principle. The explanations given are not sufficiently detailed and extensive to make the student aware of the wide range of applications and different aspects of the principle being studied. The numerous possible variations of principles and their applications are usually not discussed, and it is left for the students to discover these for themselves while doing exercises. Accordingly, the average student is expected to rediscover that which has been long known and practiced, but not published or explained extensively.

(c) The examples usually following the explanation of a topic are too few in number and too simple to enable the student to obtain a thorough grasp of the principles involved. The explanations do not provide sufficient basis to enable a student to solve problems that may be subsequently assigned for homework or given on examinations.

The examples are presented in abbreviated form which leaves out much material between steps, and requires that students derive the omitted material themselves. As a result, students find the examples difficult to understand--contrary to the purpose of the examples.

Examples are, furthermore, often worded in a confusing manner. They do not state the problem and then present the solution. Instead, they pass through a general discussion, never revealing what is to be solved for.

Examples, also, do not always include diagrams/graphs, wherever appropriate, and students do not obtain the training to draw diagrams or graphs to simplify and organize their thinking.

(d) Students can learn the subject only by doing the exercises themselves and reviewing them in class, to obtain experience in applying the principles with their different ramifications.

In doing the exercises by themselves, students find that they are required to devote considerably more time to electromagnetics than to other subjects of comparable credits, because they are uncertain with regard to the selection and application of the theorems and principles involved. It is also often necessary for students to discover those "tricks" not revealed in their texts (or review books), that make it possible to solve problems easily. Students must usually resort to methods of trial-and-error to discover these "tricks," and as a result they find that they may sometimes spend several hours to solve a single problem.

(e) When reviewing the exercises in classrooms, instructors usually request students to take turns in writing solutions on the boards and explaining them to the class. Students often find it difficult to explain in a manner that holds the interest of the class, and enables the remaining students to follow the material written on the boards. The remaining students seated in the class are, furthermore, too occupied with copying the material from the boards, to listen to the oral explanations and concentrate on the methods of solution.

This book is intended to aid students in electromagnetics to overcome the difficulties described, by supplying detailed illustrations of the solution methods which are usually not apparent to students. The solution methods are illustrated by problems selected from those that are most often assigned for class work and given on examinations. The problems are arranged in order of complexity to enable students to learn and understand a particular topic by reviewing the problems in sequence. The problems are illustrated with detailed step-by-step explanations, to save the student the large amount of time that is often needed to fill in the gaps that are usually found between steps of illustrations in the textbooks or review/outline books.

The staff of REA considers electromagnetics a subject that is best learned by allowing students to view the methods of analysis and solution techniques themselves. This approach to learning the subject matter is similar to that practiced in various scientific laboratories, particularly in the medical fields.

In using this book, students may review and study the illustrated problems at their own pace; they are not limited to the time allowed for explaining problems on the board in class.

When students want to look up a particular type of problem and solution, they can readily locate it in the book by referring to the index which has been extensively prepared. It is also possible to locate a particular type of problem by glancing at just the material within the boxed portions. To facilitate rapid scanning of the problems, each problem has a heavy border around it. Furthermore, each problem is identified with a number immediately above the problem at the right-hand margin.

To obtain maximum benefit from the book, students should familiarize themselves with the section, "How To Use This Book," located in the front pages.

To meet the objectives of this book, staff members of REA have selected problems usually encountered in assignments and examinations, and have solved each problem meticulously to illustrate the steps which are usually difficult for students to comprehend. Special gratitude is expressed to them for their efforts in this area, as well as to the numerous contributors who devoted brief periods of time to this work.

Gratitude is also expressed to the many persons involved in the difficult task of typing the manuscript with its endless changes, and to the REA art staff who prepared the numerous detailed illustrations together with the layout and physical features of the book.

The difficult task of coordinating the efforts of all persons was carried out by Carl Fuchs. His conscientious work deserves much appreciation. He also trained and supervised art and production personnel in the preparation of the book for printing.

Finally, special thanks are due to Helen Kaufmann for her unique talents to render those difficult border-line decisions and constructive suggestions related to the design and organization of the book.

Max Fogiel, Ph. D.
Program Director

HOW TO USE THIS BOOK

This book can be an invaluable aid to students in electromagnetics as a supplement to their textbooks. The book is subdivided into 14 chapters, each dealing with a separate topic. The subject matter is developed beginning with vector analysis, electric charges, field intensity, dielectrics, capacitance, magnetic fields and circuits, and Maxwell's equations. Also included are waves, transmission lines, and antennas. An extensive number of applications have been included, since these appear to be most troublesome to students. A special Section II has been included as a summary of electromagnetic propagation in conducting media. This section reviews basic equations and theorems, plane waves, dipole sources, long line sources, and antennas. Section II serves as a thorough review through which students can acquire a deeper understanding of electromagnetics.

HOW TO LEARN AND UNDERSTAND A TOPIC THOROUGHLY

1. Refer to your class text and read the section pertaining to the topic. You should become acquainted with the principles discussed there. These principles, however, may not be clear to you at the time.

2. Then locate the topic you are looking for by referring to the "Table of Contents" in the front of this book.

3. Turn to the page where the topic begins and review the problems under each topic, in the order given. For each topic, the problems are arranged in order of complexity, from the simplest to the more difficult. Some problems may appear similar to others, but each problem has been selected to illustrate a different point or solution method.

To learn and understand a topic thoroughly and retain its contents, it will be generally necessary for students to review the problems several times. Repeated review is essential in order to gain experience in recognizing the principles that should be applied, and to select the best solution technique.

HOW TO FIND A PARTICULAR PROBLEM

To locate one or more problems related to particular subject matter, refer to the index. In using the index, be certain to note that the numbers given there refer to problem numbers, not to page numbers. This arrangement of the index is intended to facilitate finding a problem more rapidly, since two or more problems may appear on a page.

If a particular type of problem cannot be found readily, it is recommended that the student refer to the "Table of Contents," and then turn to the chapter which is applicable to the problem being sought. By scanning or glancing at the material that is boxed, it will generally be possible to find problems related to the one being sought, without consuming considerable time. After the problems have been located, the solutions can be reviewed and studied in detail.

For the purpose of locating problems rapidly, students should acquaint themselves with the organization of the book as found in the "Table of Contents."

In preparing for an exam, it is useful to find the topics to be covered in the exam from the "Table of Contents," and then review the problems under those topics several times. This should equip the student with what might be needed for the exam.

CONTENTS

Chapter No. **Page No.**

1 VECTOR ANALYSIS 1

Scalars and Vectors 1
Gradient, Divergence and Curl 26
Line, Surface and Volume Integrals 39
Stoke's Theorem 48

2 ELECTRIC CHARGES 55

Charge Densities and Distributions 55
Coulomb's Law 59
Electric Field 66

3 ELECTRIC FIELD INTENSITY 86

Electric Flux 86
Gauss's Law 98
Charges 119

4 POTENTIAL 125

Work 125
Potential 136
Potential and Gradient 170
Motion in Electric Field 181
Energy 187

5 DIELECTRICS 197

Current Density 197
Resistance 208
Polarization 212
Boundary Conditions 224
Dielectrics 239

6 CAPACITANCE 260

Capacitance 260
Parallel Plate Capacitors 261
Coaxial and Concentric Capacitors 277
Multiple Dielectric Capacitors, Series and Parallel Combinations 285
Potential 297
Stored Energy and Force in Capacitors 303

7 POISSON'S AND LAPLACE'S EQUATIONS 316

Laplace's Equation 316
Poisson's Equation 355
Iteration Method 372
Images 375

8 STEADY MAGNETIC FIELDS 389

Biot-Savart's Law 389
Ampere's Law 392
Magnetic Flux and Flux Density 401
Vector Magnetic Potential 434
H-Field 445

9 FORCES IN STEADY MAGNETIC FIELDS 457

Forces on Moving Charges 457
Forces on Differential Current Elements 464
Forces on Conductors Carrying Currents 468
Magnetization 475
Magnetic Boundary Conditions 488
Potential Energy of Magnetic Fields 495

10 MAGNETIC CIRCUITS 498

Reluctance and Permeance 498
Determination of Ampere-Turns 506
Flux produced by a given mmf 521
Self and Mutual Inductance 538
Force and Torque in Magnetic Circuits 558

11 TIME - VARYING FIELDS AND MAXWELL'S EQUATIONS 574

Faraday's Law 574
Maxwell's Equations 604
Displacement Current 611
Generators 614

12 PLANE WAVES 624

Energy and the Poynting Vector 624
Normal Incidence 636
Boundary Conditions 649
Plane Waves in Conducting Dielectric Media 658
Plane Waves in Free Space 690
Plane Waves and Current Density 697

13 TRANSMISSION LINES 702

Equations of Transmission Lines 702
Input Impedances 722
Smith Chart 728
Matching 748
Reflection Coefficient 756

14 WAVE GUIDES AND ANTENNAS 763

Cutoff Frequencies for TE and TM Modes 763
Propagation and Attenuation Constants 777
Field Components in Wave-Guides 791
Absorbed and Transmitted Power 805
Characteristics of Antennas 814
Radiated and Absorbed Power of Antennas 822

SECTION II

SUMMARY OF ELECTROMAGNETIC PROPAGATION IN CONDUCTING MEDIA

II-1 BASIC EQUATIONS AND THEOREMS 1-1

Maxwell's Equations 1-1
Auxiliary Potentials 1-2
Harmonic Time Variation 1-4
Particular Solutions for an Unbounded Homogeneous Region with Sources 1-6
Poynting Vector 1-6
Reciprocity Theorem 1-7
Boundary Conditions 1-8
Uniqueness Theorems 1-9
TM and TE Field Analysis 1-10

II-2 PLANE WAVES 2-1

Uniform Plane Waves 2-1
Nonuniform Plane Waves 2-3
Reflection and Refraction at a Plane Surface 2-4
Refraction in a Conducting Medium 2-8
Surface Waves 2-10
Plane Waves in Layered Media 2-12
Impedance Boundary Conditions 2-15
Propagation Into a Conductor With a Rough Surface 2-17

II-3 ELECTROMAGNETIC FIELD OF DIPOLE SOURCES 3-1

Infinite Homogeneous Conducting Medium 3-1
Semi-Infinite Homogeneous Conducting Medium 3-2
Static Electric Dipole 3-3
Harmonic Dipole Sources 3-5
Far field 3-6
Near field 3-9
Quasi-static field 3-15
Layered Conducting Half Space 3-27

II-4 ELECTROMAGNETIC FIELD OF LONG LINE SOURCES AND FINITE LENGTH ELECTRIC ANTENNAS 4-1

Infinite Homogeneous Conducting Medium 4-1
Long Line Source 4-1
Finite Length Electric Antenna 4-2
Semi-Infinite Homogeneous Conducting Medium 4-3
Long Line Source 4-3
Finite Length Electric Antenna 4-8
Layered Conducting Half Space 4-12
Long Line Source 4-12
Finite Length Electric Antenna 4-16

APPENDIX A-1

Parameters of Conducting Media A-1
Dipole Approximation Scattering B-1
Antenna Impedance C-1
ELF and VLF Atmospheric Noise D-1

INDEX 940

CHAPTER 1

VECTOR ANALYSIS

SCALARS AND VECTORS

• PROBLEM 1-1

If for each value of the real variable u in an interval $u_1 \le u \le u_2$ there corresponds a vector $\bar{r}$ in the vector space then $\bar{r}$ is a vector function of u over that interval. We write

$$\bar{r} = \bar{r}(u) \qquad \text{for} \qquad u_1 \le u \le u_2$$

to indicate that $\bar{r}$ is a function of u.

Let $\bar{r}(t)$ be a vector function given by

$$\bar{r}(t) = t\bar{a} + (1 - t)\bar{b}\ ,\ 0 \le t \le 1 \qquad (1)$$

where

$$\bar{a} = (a_1, a_2, a_3)$$

$$\bar{b} = (b_1, b_2, b_3) \qquad (2)$$

are two given vectors.

Express $\bar{r}(t)$
1) in the component form
2) in terms of the base vectors.

Solution: 1) In the n dimensional vector space a vector $\bar{a}$ is expressed in its component form, when all its components $a_1, a_2, \ldots, a_n$ are given. We write

$$\bar{a} = (a_1, a_2, \ldots, a_n)$$

The maximum number of components of a vector in the n dimensional space is n. To express vector $\bar{r}(t)$ in the component form $r_1(t)$, $r_2(t)$, $r_3(t)$ should be found such that

$$\bar{r}(t) = [r_1(t), r_2(t), r_3(t)]$$

Substituting Eq. (2) into Eq. (1),

$$\bar{r}(t) = t\bar{a} + (1 - t)\bar{b} = t(a_1,a_2,a_3) + (1 - t)(b_1,b_2,b_3) \quad (3)$$

Multiplying and adding the corresponding components in Eq. (3)

$$\bar{r}(t) = (ta_1,ta_2,ta_3) + [(1-t)b_1,(1-t)b_2,(1-t)b_3]$$
$$= [ta_1 + (1-t)b_1, ta_2 + (1-t)b_2, ta_3 + (1-t)b_3] \quad (4)$$

Eq. (4) expresses the vector $\bar{r}(t)$ in terms of its components.

2) Let $\bar{a}$ be any vector in the n dimensional vector space and $\bar{i}_1,\bar{i}_2,\ldots,\bar{i}_n$ be the base vectors of this space. Note that n dimensional vector space has n base vectors. Vector $\bar{a}$ can be expressed uniquely as a linear combination of the base vectors

$$\bar{a} = a_1\bar{i}_1 + a_2\bar{i}_2 + \ldots + a_n\bar{i}_n$$

In the present case the vector space is three dimensional therefore there are three base vectors. The simplest choice is

$$\bar{i} = (1,0,0)$$
$$\bar{j} = (0,1,0) \quad (5)$$
$$\bar{k} = (0,0,1)$$

First express vectors $\bar{a}$ and $\bar{b}$ in terms of the base vectors. Then

$$\bar{a} = a_1\bar{i} + a_2\bar{j} + a_3\bar{k}$$
$$\bar{b} = b_1\bar{i} + b_2\bar{j} + b_3\bar{k} \quad (6)$$

Substituting Eq. (6) into Eq. (1) obtain

$$\bar{r}(t) = t\bar{a} + (1-t)\bar{b} = t(a_1\bar{i} + a_2\bar{j} + a_3\bar{k}) + (1-t)(b_1\bar{i} + b_2\bar{j} + b_3\bar{k})$$
$$= [ta_1 + (1-t)b_1]\bar{i} + [ta_2 + (1-t)b_2]\bar{j} + [ta_3 + (1-t)b_3]\bar{k} \quad (7)$$

Eq. (7) expresses the vector $\bar{r}(t)$ as a linear combination of the base vectors $\bar{i},\bar{j},\bar{k}$.

The expression

$$\bar{a} \times (\bar{b}\times\bar{c})$$

is called the vector triple product. Prove the following identities:

$$\bar{a} \times (\bar{b}\times\bar{c}) = (\bar{a}\cdot\bar{c})\bar{b} - (\bar{a}\cdot\bar{b})\bar{c} \tag{1}$$

$$(\bar{a}\times\bar{b}) \times \bar{c} = (\bar{c}\cdot\bar{a})\bar{b} - (\bar{c}\cdot\bar{b})\bar{a} \tag{2}$$

Solution: To prove Eq. (1) compute the components of both sides of Eq. (1). From the definition of the vector product, obtain

$$\begin{aligned}
\bar{a} \times (\bar{b}\times\bar{c}) &= (a_1,a_2,a_3) \times (b_2c_3-b_3c_2,\ b_3c_1-b_1c_3,\\
&\quad b_1c_2-b_2c_1)\\
&= [a_2(b_1c_2-b_2c_1) - a_3(b_3c_1-b_1c_3),\\
&\quad a_3(b_2c_3-b_3c_2) - a_1(b_1c_2-b_2c_1),\\
&\quad a_1(b_3c_1-b_1c_3) - a_2(b_2c_3-b_3c_2)] \qquad (3)\\
&= [b_1(a_2c_2+a_3c_3) - c_1(a_2b_2+a_3b_3),\\
&\quad b_2(a_1c_1+a_3c_3) - c_2(a_1b_1+a_3b_3),\\
&\quad b_3(a_1c_1+a_2c_2) - c_3(a_1b_1+a_2b_2)]\\
&= [b_1(a_1c_1+a_2c_2+a_3c_3) - c_1(a_1b_1+a_2b_2\\
&\quad + a_3b_3),\ b_2(a_1c_1+a_2c_2+a_3c_3)\\
&\quad - c_2(a_1b_1+a_2b_2+a_3b_3),\\
&\quad b_3(a_1c_1+a_2c_2+a_3c_3) - c_3(a_1b_1+ a_2b_2\\
&\quad +a_3b_3)]\\
&= [b_1(\bar{a}\cdot\bar{c}),\ b_2(\bar{a}\cdot\bar{c}),\ b_3(\bar{a}\cdot\bar{c})]\\
&\quad - [c_1(\bar{a}\cdot\bar{b}),\ c_2(\bar{a}\cdot\bar{b}),\ c_3(\bar{a}\cdot\bar{b})]\\
&= (\bar{a}\cdot\bar{c})\bar{b} - (\bar{a}\cdot\bar{b})\bar{c}
\end{aligned}$$

That completes the proof. The same method can be used to prove Eq. (2).

The shorter method is to rewrite Eq. (1) in the form

$$(\bar{b}\times\bar{c}) \times \bar{a} = (\bar{a}\cdot\bar{b})\bar{c} - (\bar{a}\cdot\bar{c})\bar{b} \tag{4}$$

Replace

vector $\bar{b}$ by $\bar{a}$

$$\bar{b} \to \bar{a}$$

vector $\bar{c}$ by $\bar{b}$

$$\bar{c} \to \bar{b}$$

vector $\bar{a}$ by $\bar{c}$

$$\bar{a} \to \bar{c}$$

and obtain

$$\begin{aligned} (\bar{b}\times\bar{c}) \times \bar{a} &= (\bar{a}\cdot\bar{b})\bar{c} - (\bar{a}\cdot\bar{c})\bar{b} \\ (\bar{a}\times\bar{b}) \times \bar{c} &= (\bar{c}\cdot\bar{a})\bar{b} - (\bar{c}\cdot\bar{b})\bar{a} \end{aligned} \tag{5}$$

which proves identity (2).

• PROBLEM 1-3

1) Express the derivative of a vector function $\bar{r}(t)$ in terms of the derivatives of the components of $\bar{r}(t)$.

2) Illustrate graphically the construction of $\frac{d\bar{r}}{dt}$.

Solution: 1) The derivative of a vector function $\bar{r}(t)$ is

$$\frac{d\bar{r}}{dt} = \lim_{\Delta t \to 0} \frac{\bar{r}(t + \Delta t) - \bar{r}(t)}{\Delta t} \tag{1}$$

Since the vector $\bar{r}(t)$ is a function of t, each of its components is a function of t, thus

$$\bar{r}(t) = (x_1(t), x_2(t), x_3(t)) \tag{2}$$

For $\bar{r}(t + \Delta t)$ it can be written that

$$\bar{r}(t + \Delta t) = (x_1(t+\Delta t), x_2(t+\Delta t), x_3(t+\Delta t)) \tag{3}$$

Substituting Eq. (2) and Eq. (3) into Eq. (1) obtain

$$\frac{d\bar{r}}{dt} = \lim_{\Delta t \to 0} \frac{\bar{r}(t + \Delta t) - \bar{r}(t)}{\Delta t} \tag{4}$$

$$= \lim_{\Delta t \to 0} \frac{(x_1(t+\Delta t), x_2(t+\Delta t), x_3(t+\Delta t)) - (x_1(t), x_2(t), x_3(t))}{\Delta t}$$

$$= \lim_{\Delta t \to 0} \frac{(x_1(t+\Delta t) - x_1(t), x_2(t+\Delta t) - x_2(t), x_3(t+\Delta t) - x_3(t))}{\Delta t}$$

$$= \left(\frac{dx_1}{dt}, \frac{dx_2}{dt}, \frac{dx_3}{dt}\right)$$

In Eq. (4) the definition of derivative of an ordinary function is used, that is

$$\lim_{\Delta t \to 0} \frac{x_1(t + \Delta t) - x_1(t)}{\Delta t} = \frac{dx_1}{dt} \tag{5}$$

The same equation can be written for $x_2(t)$ and $x_3(t)$.

From Eq. (4) it can be concluded that to differentiate a vector function first differentiate each component separately. In the general case where

$$\bar{r}(t) = (x_1(t), x_2(t), \ldots, x_n(t)) \tag{6}$$

is an n-dimensional vector, its derivative is given by

$$\frac{d\bar{r}}{dt} = \frac{dx_1}{dt}, \frac{dx_2}{dt}, \ldots, \frac{dx_n}{dt} \tag{7}$$

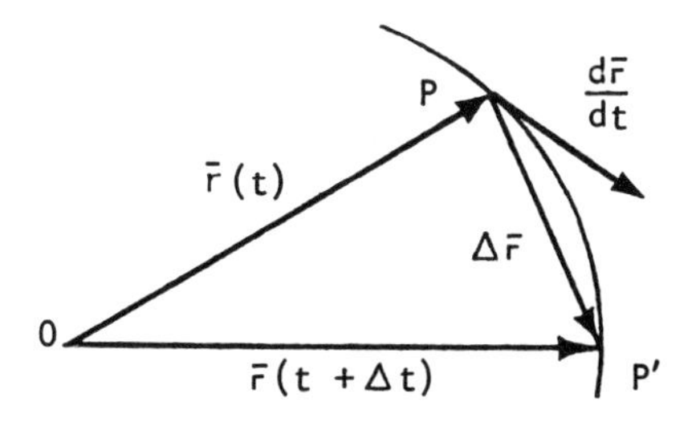

2) The construction of $\frac{d\bar{r}}{dt}$ is shown in the figure: Vector $\overline{OP}$, with 0 fixed represents $\bar{r}(t)$ while vector $\overline{OP'}$ represents $\bar{r}(t') = \bar{r}(t + \Delta t)$.

From the figure we have

$$\bar{r}(t) + \Delta\bar{r} = \bar{r}(t + \Delta t) \tag{8}$$

or

$$\Delta\bar{r} = \bar{r}(t + \Delta t) - \bar{r}(t) \tag{9}$$

$\Delta\bar{r}$ represents vector $\overline{PP'}$ which is the displacement of the moving point P in the interval t to t + Δt. Of course $\Delta\bar{r}$ is a vector and $\frac{\Delta\bar{r}}{\Delta t}$ is a vector, since it is a vector $\Delta\bar{r}$ multiplied by a scalar $\frac{1}{\Delta t}$. The limit

$$\lim_{\Delta t \to 0} \frac{\Delta\bar{r}}{\Delta t} = \frac{d\bar{r}}{dt} \quad \text{is also a vector.}$$

When $\frac{d\bar{r}}{dt} \neq \bar{0}$ the vector $\frac{d\bar{r}}{dt}$ represents the tangent to the curve at the point P. The tangent to a curve is often defined as the limiting position (if the limit exists) of a secant.

• PROBLEM 1-4

For $\bar{r}(t) = [\ln(t^3+1),\ e^{-2t},\ t^2]$, find

1) $\dfrac{d\bar{r}}{dt}$

2) $\left|\dfrac{d\bar{r}}{dt}\right|$

3) $\dfrac{d^2\bar{r}}{dt^2}$

4) $\left|\dfrac{d^2\bar{r}}{dt^2}\right|$

at $t = 0$.

Solution: 1) Differentiate each component separately and obtain

$$\frac{d\bar{r}}{dt} = \frac{d}{dt}[\ln(t^3+1),\ e^{-2t},\ t^2] \tag{1}$$

$$= \left[\frac{d}{dt}\ln(t^3+1),\ \frac{d}{dt}e^{-2t},\ \frac{d}{dt}t^2\right]$$

$$= \left[\frac{3t^2}{t^3+1},\ -2e^{-2t},\ 2t\right]$$

At $t = 0$, Eq. (1) becomes

$$\left.\frac{d\bar{r}}{dt}\right|_{t=0} = [0,-2,0] \tag{2}$$

2) The magnitude of $\dfrac{d\bar{r}}{dt}$ at $t = 0$ is

$$\left.\left|\frac{d\bar{r}}{dt}\right|\right|_{t=0} = \sqrt{4} = 2 \tag{3}$$

3) Using the definition of the second derivative

$$\frac{d^2\bar{r}}{dt^2} = \frac{d}{dt}\left(\frac{d\bar{r}}{dt}\right)$$

and Eq. (1) we obtain

$$\frac{d^2\bar{r}}{dt^2} = \frac{d}{dt}\left(\frac{3t^2}{t^3+1}, -2e^{-2t}, 2t\right) \tag{4}$$

$$= \left(\frac{-3t^4 + 6t}{(t^3+1)^2}, 4e^{-2t}, 2\right)$$

$$\left.\frac{d^2\bar{r}}{dt^2}\right|_{t=0} = (0,4,2) \tag{5}$$

4) At t = 0 the magnitude of

$\frac{d^2\bar{r}}{dt^2}$ is

$$\left|\left.\frac{d^2\bar{r}}{dt^2}\right|_{t=0}\right| = \sqrt{16 + 4} = \sqrt{20} \tag{6}$$

• PROBLEM 1-5

Find the general solution of the differential equation

$$\frac{d^2\bar{x}}{dt^2} - 2\frac{d\bar{x}}{dt} - 3\bar{x} = 10\bar{a}\sin t + \bar{b}(2t+1) \tag{1}$$

where $\bar{a}$ and $\bar{b}$ are constant vectors.

Solution: Use the following theorem to solve the problem.

Theorem. The general solution of the linear vector differential equation

$$\left(f_n\frac{d^n}{dt^n} + f_{n-1}\frac{d^{n-1}}{dt^{n-1}} + \ldots + f_1\frac{d}{dt} + f_0\right)\bar{x} = \bar{a} \tag{2}$$

denoted $G[\bar{x}] = \bar{a}$

where $\bar{a}$ and $f_0, f_1, \ldots, f_n$ are given functions of the scalar t and $\bar{x}$ is an unknown vector, is given by

$$\bar{x} = \bar{Y} + \bar{A} \tag{3}$$

where $\bar{A}$ is a particular solution of this differential equation, and

$$\bar{Y} = \bar{c}_1 y_1 + \bar{c}_2 y_2 + \ldots + \bar{c}_n y_n \tag{4}$$

$y_1, y_2, \ldots, y_n$ are n linearly independent solutions of the homogeneous scalar differential equation $G[y] = 0$ and $\bar{c}_1, \bar{c}_2, \ldots, \bar{c}_n$ are arbitrary constant vectors.

To solve Eq. (1) find two linearly independent solutions of

$$\frac{d^2y}{dt^2} - 2\frac{dy}{dt} - 3y = 0 \tag{5}$$

The auxiliary equation is

$$m^2 - 2m - 3 = 0$$

with the roots -1,3. We obtain

$$\bar{Y} = \bar{c}_1 e^{-t} + \bar{c}_2 e^{3t} \tag{6}$$

To find a particular solution $\bar{A}$ of Eq. (1) use the method of indetermined coefficients. Since the function on the right hand side of Eq. (1) is

$$10\bar{a} \sin t + \bar{b}(2t + 1) \tag{7}$$

look for a particular solution $\bar{A}$ in the form

$$\bar{A} = \bar{c} \sin t + \bar{d} \cos t + \bar{e}t + \bar{f} \tag{8}$$

where $\bar{c}$, $\bar{d}$, $\bar{e}$, $\bar{f}$ are constant vectors.

Substituting Eq. (8) into Eq. (1),

$$\begin{aligned} -\bar{c} + 2\bar{d} - 3\bar{c} &= 10\bar{a} \\ -\bar{d} - 2\bar{c} - 3\bar{d} &= \bar{0} \\ -3\bar{e} &= 2\bar{b} \\ -2\bar{e} - 3\bar{f} &= \bar{b} \end{aligned} \tag{9}$$

Solving the system (9)

$$\begin{aligned} \bar{e} &= -\frac{2}{3}\bar{b} \qquad & \bar{c} &= -2\bar{a} \\ \bar{f} &= \frac{1}{9}\bar{b} \qquad & \bar{d} &= \bar{a} \end{aligned} \tag{10}$$

Thus the solution of the equation is

$$\bar{x} = \bar{Y} + \bar{A} = \bar{c}_1 e^{-t} + \bar{c}_2 e^{3t} - 2\bar{a} \sin t + \bar{a} \cos t$$

$$- \frac{2}{3} \bar{b}t + \frac{1}{9} \bar{b}$$

$$= \bar{c}_1 e^{-t} + \bar{c}_2 e^{3t} + \bar{a}(\cos t - 2 \sin t) + \bar{b}(- \frac{2}{3} t + \frac{1}{9}) \quad (11)$$

• PROBLEM 1-6

The position vector of the point P is given by

$$\bar{u} = (A + \alpha t, \; B + \beta t, \; C + \gamma t) \quad (1)$$

where t is the time and $A,B,C,\alpha,\beta,\gamma$ are constants. Find the trajectory of the point P, its velocity vector and its speed.

Solution: From Eq. (1) it can be concluded that the point P moves according to the equations

$$x_1 = A + \alpha t$$
$$x_2 = B + \beta t \quad (2)$$
$$x_3 = C + \gamma t$$

Eqs. (2) are parametric equations of a straight line in space through the point (A,B,C) and with direction numbers α,β,γ.

The trajectory of the point P is a straight line.

The velocity of a point is defined as the first derivative of the position vector with respect to the time

$$\bar{v} = \frac{d\bar{u}}{dt} \quad (3)$$

From Eq. (1)

$$\bar{v} = \frac{d\bar{u}}{dt} = (\alpha,\beta,\gamma) \quad (4)$$

The speed of the point is defined as the magnitude of its velocity vector.

$$v = |\bar{v}| = \frac{ds}{dt} \quad (5)$$

Substituting Eq. (4) into Eq. (5)

$$v = |\bar{v}| = \sqrt{\alpha^2 + \beta^2 + \gamma^2} \tag{6}$$

Now investigate the special case when the velocity vector is a unit vector, $|\bar{v}| = 1$. From Eq. (6)

$$v = |\bar{v}| = \sqrt{\alpha^2 + \beta^2 + \gamma^2} = 1 \tag{7}$$

and α, β, γ are direction cosines, due to the condition

$$\sqrt{\alpha^2 + \beta^2 + \gamma^2} = 1 \tag{8}$$

Let s be the distance traversed by the point from time t_0 to time t. Then

$$\frac{ds}{dt} = \sqrt{\left(\frac{dx_1}{dt}\right)^2 + \left(\frac{dx_2}{dt}\right)^2 + \left(\frac{dx_3}{dt}\right)^2}$$

and (9)

$$v = |\bar{v}| = \frac{ds}{dt}$$

For $|\bar{v}| = 1$ Eq. (9) gives

$$\frac{ds}{dt} = 1 \tag{10}$$

and both parameters can be identified within the accuracy of an additive constant f, thus

$$s = t + f \tag{11}$$

If s is measured from the position $t = 0$ and in the direction of increasing t it can be written that

$$s = t \tag{12}$$

Eqs. (2) become

$$x_1 = A + \alpha s$$

$$x_2 = B + \beta s \tag{13}$$

$$x_3 = C + \gamma s$$

which are the parametric equations of the straight line in terms of the parameter s.

• PROBLEM 1-7

A particle moves according to the equations

$$x = \sin 2t$$
$$y = \cos 2t \qquad (1)$$
$$z = e^{-t}$$

where t is the time.

1) Determine the velocity of the particle.
2) Determine the acceleration of the particle.
3) Find the magnitudes of the velocity and acceleration at t = 0.

Solution: Eqs.(1) are parametric equations of the curve. The particle moves along this curve. The position vector $\bar{r}$ of the particle is

$$\bar{r} = \bar{r}(t) = (\sin 2t, \cos 2t, e^{-t}) \qquad (2)$$

1) The velocity vector is

$$\bar{v} = \frac{d\bar{r}}{dt} \qquad (3)$$

Substituting $\bar{r}(t)$ from Eq.(2) into Eq.(3) obtain

$$\bar{v} = \frac{d\bar{r}}{dt} = (2 \cos 2t, -2 \sin 2t, -e^{-t}) \qquad (4)$$

2) The acceleration of the particle is given by

$$\bar{a} = \frac{d^2\bar{r}}{dt^2} = \frac{d\bar{v}}{dt} \qquad (5)$$

Again, substituting Eq.(4) into Eq.(5) obtain

$$\bar{a} = \frac{d\bar{v}}{dt} = (-4 \sin 2t, -4 \cos 2t, e^{-t}) \qquad (6)$$

Eq.(6) gives the acceleration of the particle as a function of time.

3) From Eq.(4), at t = 0 the velocity is

$$\bar{v}\Big|_{t=0} = (2,0,-1) \qquad (7)$$

and its magnitude

$$\left| \bar{v} \Big|_{t=0} \right| = \sqrt{4+1} = \sqrt{5} \tag{8}$$

Note, that Eq.(8) gives the speed of the particle at t = 0.

Compute the acceleration of the particle at t = 0 from Eq.(6)

$$\bar{a} \Big|_{t=0} = (0,-4,1) \tag{9}$$

The magnitude of $\bar{a}$ at t = 0 is

$$\left| \bar{a} \Big|_{t=0} \right| = \sqrt{16+1} = \sqrt{17} \tag{10}$$

• PROBLEM 1-8

A particle P of mass m moves in space subject to a force

$$\bar{F} = X\bar{i} + Y\bar{j} + Z\bar{k} \tag{1}$$

which is constant.

1) Express the position vector of the particle in terms of time t.
2) Assume that the axes in space are chosen so that $\bar{F}$ is parallel to the Z axis. Write the position vector $\bar{r} = \bar{r}(t)$ for this case.

Show that the trajectory of the particle is a parabola.

Solution: 1) Newton's Second Law states that

$$\bar{F} = m\bar{a} \tag{2}$$

where

$$\bar{a} = \frac{d^2x}{dt^2}\,\bar{i} + \frac{d^2y}{dt^2}\,\bar{j} + \frac{d^2z}{dt^2}\,\bar{k} \tag{3}$$

Combining Eqs.(1), (2) and (3)

$$X\bar{i} + Y\bar{j} + Z\bar{k} = m\frac{d^2x}{dt^2}\,\bar{i} + m\frac{d^2y}{dt^2}\,\bar{j} + m\frac{d^2z}{dt^2}\,\bar{k} \tag{4}$$

or

$$X = m\frac{d^2x}{dt^2}$$

$$Y = m \frac{d^2y}{dt^2} \tag{5}$$

$$Z = m \frac{d^2z}{dt^2}$$

Integrating Eq.(5) twice, obtain

$$\frac{1}{m} X = \frac{d^2x}{dt^2}$$

$$\frac{1}{m} Xt = \frac{dx}{dt} - a_1$$

$$\frac{1}{2m} Xt^2 = x - a_1t - a_2$$

thus

$$x = \frac{1}{2m} Xt^2 + a_1t + a_2 \tag{6}$$

In the same way

$$y = \frac{1}{2m} Yt^2 + b_1t + b_2$$

$$Z = \frac{1}{2m} Zt^2 + c_1t + c_2$$

where $a_1, a_2, b_1, b_2, c_1, c_2$ are constants, which can be determined from the additional conditions – for example the initial position and initial velocity of the particle or the position of the particle at $t = t_1$ and $t = t_2$.

2) Since $\bar{F}$ is parallel to the z-axis the x and y components of this vector are equal to zero, thus

$$X = 0$$
$$Y = 0 \tag{7}$$

Substituting Eq.(7) into Eq.(6)

$$x = a_1t + a_2$$

$$y = b_1t + b_2 \tag{8}$$

$$z = \frac{1}{2m} Zt^2 + c_1t + c_2$$

It shall be shown that the trajectory of a particle whose motion is described by Eq.(8) is a parabola.

It can be assumed that at $t = 0$ the particle is located at the origin of the system, thus Eq.(8) for $t = 0$ becomes

$$x = a_2$$
$$y = b_2 \tag{9}$$
$$z = c_2$$

At the origin $x = y = z = 0$, therefore $a_2 = b_2 = c_2 = 0$. Eq.(8) becomes

$$x = a_1 t$$
$$y = b_1 t \tag{10}$$
$$z = \frac{1}{2m} Zt^2 + c_1 t$$

From the first equation $t = x/a_1$. Substituting $t = x/a_1$ into the second and third equations

$$y = \frac{b_1}{a_1} x$$
$$z = \frac{1}{2m} Z \frac{x^2}{a_1^2} + \frac{c_1}{a_1} x \tag{11}$$

or

$$z = \frac{Z}{2ma_1^2} x^2 + \frac{c_1}{a_1} x \tag{12}$$

Denote

$$\frac{Z}{2ma_1^2} = p \quad , \qquad \frac{c_1}{a_1} = q \tag{13}$$

Eq.(12) can be written

$$z = px^2 + qx = px^2 + qx + \frac{q^2}{4p} - \frac{q^2}{4p}$$
$$= p\left(x + \frac{q}{2p}\right)^2 - \frac{q^2}{4p} \tag{14}$$

or

$$z + \frac{q^2}{4p} = p\left(x + \frac{q}{2p}\right)^2 \tag{15}$$

Transforming the coordinates

$$z' = z + \frac{q^2}{4p}$$
$$x' = x + \frac{q}{2p} \tag{16}$$

Eq.(15) becomes

$$z' = px'^2 \qquad (17)$$

which is the equation of a parabola.

• PROBLEM 1-9

Consider the motion of a system of particles $P_1, P_2, \ldots, P_n$.

Show that if there are no external forces the center of mass of the system moves with constant velocity along a straight line.

Solution: Let the position vectors of the particles $P_1, P_2, \ldots, P_n$ be $\bar{r}_1 = \overline{OP}_1$, $\bar{r}_2 = \overline{OP}_2$, ... $\bar{r}_n = \overline{OP}_n$.

The equations of motion for the i-th particle are

$$\bar{F}_i = m_i \bar{a}_i \quad (i = 1, 2, \ldots, n) \qquad (1)$$

Add equations (1)

$$\bar{F}_2 + \bar{F}_2 + \ldots + \bar{F}_n = m_1\bar{a}_1 + m_2\bar{a}_2 + \ldots + m_n\bar{a}_n \qquad (2)$$

Define the center of mass P_c of the system. Its position vector is

$$\bar{r}_c = \overline{OP}_c \qquad (3)$$

where

$$M\bar{r}_c = m_1\bar{r}_1 + m_2\bar{r}_2 + \ldots + m_n\bar{r}_n$$

and

$$M = m_1 + m_2 + \ldots + m_n \qquad (4)$$

Thus

$$\bar{r}_c = \frac{m_1\bar{r}_1 + m_2\bar{r}_2 + \ldots + m_n\bar{r}_n}{m_1 + m_2 + \ldots + m_n} \qquad (5)$$

M is the total mass of the system. The center of mass has velocity

$$\bar{v}_c = \frac{d\bar{r}_c}{dt} \qquad (6)$$

and acceleration

$$\bar{a}_c = \frac{d\bar{v}_c}{dt} = \frac{d^2\bar{r}_c}{dt^2} \qquad (7)$$

Let

$$\bar{F} = \bar{F}_1 + \bar{F}_2 + \ldots + \bar{F}_n \qquad (8)$$

Eq.(2) can be written

$$\bar{F} = M\bar{a}_c \tag{9}$$

where $\bar{F}$ is the resultant of the forces applied. Each force $\bar{F}_i$ can be expressed as a sum of the internal and external forces

$$\bar{F}_i = \bar{F}_i^{int} + \bar{F}_i^{ext} \tag{10}$$

The internal force $\bar{F}_i^{int}$ is a sum of all forces due to the interactions between the particle P_i and the other particles, thus

$$\bar{F}_i^{int} = \sum_{\substack{j=1 \\ j \neq i}}^{j=n} \bar{F}_{i,j}^{int} \tag{11}$$

where $\bar{F}_{i,j}^{int}$ is the interaction of the particle P_j on the particle P_i.

Newton's Third Law states that P_i and P_j exert equal and opposite forces on each other, both along the line P_iP_j. Thus

$$\bar{F}_{i,j}^{int} = - \bar{F}_{j,i}^{int} \tag{12}$$

From Eq.(12) it can be concluded that the sum of all the internal forces is equal to $\bar{0}$.

$$\sum_{i=1}^{i=n} \bar{F}_i^{int} = \sum_{i=1}^{i=n} \sum_{\substack{j=1 \\ j \neq i}}^{j=n} \bar{F}_{i,j}^{int} = \bar{0} \tag{13}$$

If there are no external forces Eq.(8) becomes

$$\bar{F} = \sum_{i=1}^{i=n} \bar{F}_i = \sum_{i=1}^{i=n} \bar{F}_i^{int} + \sum_{i=1}^{i=n} \bar{F}_i^{ext} = \bar{0} \tag{14}$$

For $\bar{F} = \bar{0}$ Eq.(9) is

$$M\bar{a}_c = \bar{0} \tag{15}$$

and describes the motion with constant velocity on a straight line, since

$$M\bar{a}_c = M \frac{d^2\bar{r}_c}{dt^2} = \bar{0}$$

and

$$\bar{r}_c = \bar{e}t + \bar{g} \tag{16}$$

where $\bar{e}$ and $\bar{g}$ are constant vectors.

• PROBLEM 1-10

Let XYZ be a coordinate system fixed in space and having the origin at 0. Let xyz be a coordinate system rotating with respect to XYZ, as shown in the figure.

We calculate the time derivative of a vector $\bar{A}$ in both systems. Let us denote

$\left.\frac{d\bar{A}}{dt}\right|_{XYZ}$ as the time derivative of $\bar{A}$ with respect to the fixed system XYZ

and $\left.\frac{d\bar{A}}{dt}\right|_{xyz}$ as the time derivative of $\bar{A}$ with respect to the moving system xyz.

1) Show that there exists a vector $\bar{\omega}$ such that

$$\left.\frac{d\bar{A}}{dt}\right|_{XYZ} = \left.\frac{d\bar{A}}{dt}\right|_{xyz} + \bar{\omega} \times \bar{A} \tag{1}$$

2) Let D_{XYZ} denote the time derivative operator in the fixed system and D_{xyz} denote the time derivative operator in the moving xyz system. Show that

$$D_{XYZ} = D_{xyz} + \bar{\omega} \times \tag{2}$$

Eq. (2) indicates that both operators are equivalent. That is for any vector $\bar{u}(t)$ we have

$$D_{XYZ}\bar{u}(t) = D_{xyz}\bar{u}(t) + \bar{\omega} \times \bar{u}(t) \tag{3}$$

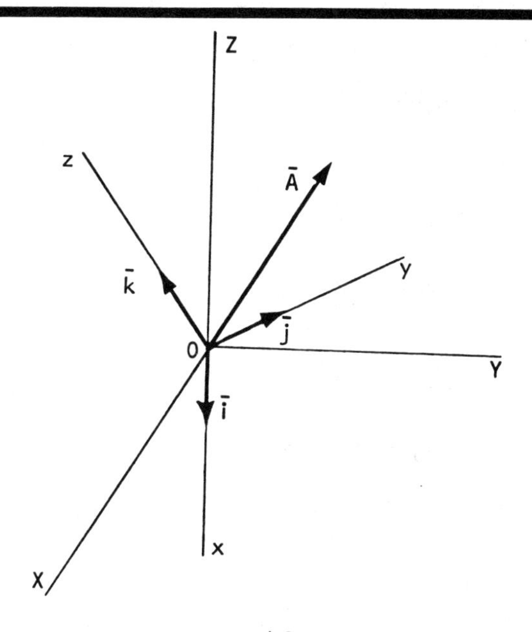

Solution: 1) Let $\bar{i}, \bar{j}, \bar{k}$ be the unit base vectors in the xyz system. As observed from the XYZ fixed system the vectors $\bar{i}, \bar{j}, \bar{k}$ change in time. The vector $\bar{A}$ can be expressed in terms of the $\bar{i}, \bar{j}, \bar{k}$ vectors

$$\bar{A} = A_1\bar{i} + A_2\bar{j} + A_3\bar{k} \tag{4}$$

Differentiating in the XYZ system,

$$\left.\frac{d\bar{A}}{dt}\right|_{XYZ} = \frac{dA_1}{dt}\,\bar{i} + \frac{dA_2}{dt}\,\bar{j} + \frac{dA_3}{dt}\,\bar{k} + A_1\frac{d\bar{i}}{dt} + A_2\frac{d\bar{j}}{dt} + A_3\frac{d\bar{k}}{dt} \tag{5}$$

Since
$$\left.\frac{d\bar{A}}{dt}\right|_{xyz} = \frac{dA_1}{dt}\,\bar{i} + \frac{dA_2}{dt}\,\bar{j} + \frac{dA_3}{dt}\,\bar{k} \tag{6}$$

we can write Eq.(5) in the form

$$\left.\frac{d\bar{A}}{dt}\right|_{XYZ} = \left.\frac{d\bar{A}}{dt}\right|_{xyz} + A_1\frac{d\bar{i}}{dt} + A_2\frac{d\bar{j}}{dt} + A_3\frac{d\bar{k}}{dt} \tag{7}$$

Remembering that $\bar{i}$ is a unit vector differentiate the equation

$$\bar{i} \cdot \bar{i} = 1$$

Thus
$$\frac{d\bar{i}}{dt} \cdot \bar{i} = 0$$

and vectors $\frac{d\bar{i}}{dt}$ and $\bar{i}$ are perpendicular. Vector $\frac{d\bar{i}}{dt}$ perpendicular to $\bar{i}$ lies in the plane of $\bar{j}$ and $\bar{k}$. Similarly $\frac{d\bar{j}}{dt}$ lies in the plane of $\bar{i}$ and $\bar{k}$ and $\frac{d\bar{k}}{dt}$ lies in the plane of $\bar{i}$ and $\bar{j}$.

One can write

$$\begin{aligned} \frac{d\bar{i}}{dt} &= p_1\bar{j} + p_2\bar{k} \\ \frac{d\bar{j}}{dt} &= p_3\bar{i} + p_4\bar{k} \\ \frac{d\bar{k}}{dt} &= p_5\bar{i} + p_6\bar{j}. \end{aligned} \tag{8}$$

Differentiate $\bar{i} \cdot \bar{j} = 0$, to obtain

$$\bar{i} \cdot \frac{d\bar{j}}{dt} + \bar{j} \cdot \frac{d\bar{i}}{dt} = 0$$

From Eq.(8)

$$\bar{i} \cdot \frac{d\bar{j}}{dt} = p_3 \qquad \text{and } \bar{j} \cdot \frac{d\bar{i}}{dt} = p_1$$

Thus $$p_1 = -p_3 \tag{9}$$

From $\bar{i} \cdot \bar{k} = 0$ we get $\bar{k} \cdot \frac{d\bar{i}}{dt} + \bar{i} \cdot \frac{d\bar{k}}{dt} = 0$

and $$p_5 = -p_2 \tag{10}$$

From $\bar{j} \cdot \bar{k} = 0$ get $\bar{j} \cdot \frac{d\bar{k}}{dt} + \bar{k} \cdot \frac{d\bar{j}}{dt} = 0$

and $$p_6 = -p_4. \tag{11}$$

Substituting Eqs.(9)(10)(11) into Eq.(8),

$$\begin{aligned} \frac{d\bar{i}}{dt} &= p_1\bar{j} + p_2\bar{k} \\ \frac{d\bar{j}}{dt} &= -p_1\bar{i} + p_4\bar{k} \\ \frac{d\bar{k}}{dt} &= -p_2\bar{i} - p_4\bar{j} \end{aligned} \tag{12}$$

Denote

$$p_1 = \omega_3, \quad p_2 = -\omega_2, \quad p_4 = \omega_1$$

System (12) can be written in the form

$$\begin{aligned} \frac{d\bar{i}}{dt} &= \omega_3\bar{j} - \omega_2\bar{k} \\ \frac{d\bar{j}}{dt} &= -\omega_3\bar{i} + \omega_1\bar{k} \\ \frac{d\bar{k}}{dt} &= \omega_2\bar{i} - \omega_1\bar{j} \end{aligned} \tag{13}$$

so that

$$A_1\frac{d\bar{i}}{dt} + A_2\frac{d\bar{j}}{dt} + A_3\frac{d\bar{k}}{dt} = (A_3\omega_2 - A_2\omega_3)\bar{i}$$

$$+ (A_1\omega_3 - A_3\omega_1)\bar{j} + (A_2\omega_1 - A_1\omega_2)\bar{k}$$

$$= \begin{vmatrix} \bar{i} & \bar{j} & \bar{k} \\ \omega_1 & \omega_2 & \omega_3 \\ A_1 & A_2 & A_3 \end{vmatrix} = \bar{\omega} \times \bar{A} \tag{14}$$

That proves Eq.(1).

Note that $\bar{\omega} = \omega_1\bar{i} + \omega_2\bar{j} + \omega_3\bar{k}$ is the angular velocity vector of the moving system with respect to the fixed system.

2) We have

$$D_{XYZ}\bar{A} = \left.\frac{d\bar{A}}{dt}\right|_{XYZ}$$

$$D_{xyz}\bar{A} = \left.\frac{d\bar{A}}{dt}\right|_{xyz} \tag{15}$$

where the first equation is the derivative in fixed system and the second in moving system.

From Eq.(1)

$$D_{XYZ}\bar{A} = D_{xyz}\bar{A} + \bar{\omega} \times \bar{A} \tag{16}$$

Eq.(16) in the operator form is

$$D_{XYZ} \equiv D_{xyz} + \bar{\omega} \times \tag{17}$$

• PROBLEM 1-11

Determine if the function

$$f(x,y) = \frac{x^2 - y^2}{x^2 + y^2} \tag{1}$$

is continuous at the origin.

Solution: Start with the definition of continuity. For this purpose repeat some basic concepts of the theory of sets. The set of points in the xy plane means any collection of points, finite or infinite in number.

A neighborhood of a point (x_0,y_0) is a set of points inside a circle having center (x_0,y_0) and radius ε. Each point of the neighborhood of (x_0,y_0) satisfies the inequality

$$\sqrt{(x - x_0)^2 + (y - y_0)^2} < \varepsilon \tag{2}$$

A set is called open if every point (x_1,y_1) of the set has a neighborhood which lies within the set. Thus, the interior of a circle is open, while the interval [0,1] is not open since point 0 does not have any neighborhood lying wholly within the interval [0,1]. A set is called a connected open set or a domain if it is open and any two points A and B of the set can be joined by a broken line lying wholly within the set. From the last definition it can be concluded that the interior of a circle is a domain. The region is a set consisting of a domain and some or all of its boundary points. A boundary point of a set is a point every neighborhood of which contains at least one point not in the set and one point in the set. For an interval [2,3] the boundary points are 2 and 3. The boundary points of the circular domain

$$x^2 + y^2 < 1$$

are the points for which

$$x^2 + y^2 = 1$$

Let $f(x,y)$ be a function over the domain D and let (x_0,y_0) be a point of D or a boundary point of D. The equation

$$\lim_{\substack{x \to x_0 \\ y \to y_0}} f(x,y) = a \tag{3}$$

means that for any $\varepsilon>0$ there exists $\delta>0$ such that for every (x,y) of the neighborhood of (x_0,y_0) of radius δ, we have

$$|f(x,y) - a| < \varepsilon \tag{4}$$

If the point (x_0,y_0) is in D and

$$\lim_{\substack{x \to x_0 \\ y \to y_0}} f(x,y) = f(x_0,y_0) \tag{5}$$

then $f(x,y)$ is said to be continuous at (x_0,y_0).

We can show that the function defined in Eq.(1)

$$f(x,y) = \frac{x^2 - y^2}{x^2 + y^2}$$

is not continuous at (0,0). Find the limit

$$\lim_{\substack{x \to 0 \\ y \to 0}} \frac{x^2 - y^2}{x^2 + y^2} \tag{6}$$

If the points (x,y) approach (0,0) on the line $x = y$ then

$$\lim_{\substack{x \to 0 \\ y \to 0 \\ x=y}} \frac{x^2 - y^2}{x^2 + y^2} = 0 \tag{7}$$

If the points (x,y) approach (0,0) on the y axis then

$$\lim_{\substack{x = 0 \\ y \to 0}} \frac{x^2 - y^2}{x^2 + y^2} = -1 \tag{8}$$

Thus, the function $f(x,y) = \frac{x^2 - y^2}{x^2 + y^2}$ has no limiting value at (0,0) and is not continuous at (0,0).

All basic theorems on limits and continuity hold without change for the functions of two or more variables.

• PROBLEM 1-12

Using the definition of the partial derivative find

$$\frac{\partial f}{\partial x}, \frac{\partial f}{\partial y} \quad \text{where}$$

1. $$f(x,y) = x^2 - y^2 \tag{1}$$

2. $$x^2 + y^2 = f^2(x,y) + 1 \tag{2}$$

3. Give the geometrical interpretation of the partial derivatives.

Solution: 1) Let f(x,y) be a function defined over a domain D of the xy plane and let (x_0,y_0) be a point in D. The function $f(x,y_0)$ depends on x alone and is defined over an interval containing x_0. We define the partial derivative of f(x,y) with respect to x at (x_0,y_0) as

$$\frac{\partial f}{\partial x}(x_0,y_0) = \lim_{\Delta x \to 0} \frac{f(x_0 + \Delta x, y_0) - f(x_0,y_0)}{\Delta x} \tag{3}$$

In a similar way the partial derivative of f(x,y) with respect to y at (x_0,y_0) can be defined as

$$\frac{\partial f}{\partial y}(x_0,y_0) = \lim_{\Delta y \to 0} \frac{f(x_0, y_0 + \Delta y) - f(x_0,y_0)}{\Delta y} \tag{4}$$

Sometimes the partial derivatives are denoted by

$$\left.\frac{\partial f}{\partial x}\right|_{(x_0,y_0)}, \qquad \left.\frac{\partial f}{\partial y}\right|_{(x_0,y_0)}.$$

Thus for the function

$$f(x,y) = x^2 - y^2,$$

$$\frac{\partial f}{\partial x} = \lim_{\Delta x \to 0} \frac{f(x + \Delta x, y) - f(x,y)}{\Delta x}$$

$$= \lim_{\Delta x \to 0} \frac{(x + \Delta x)^2 - y^2 - x^2 + y^2}{\Delta x}$$

$$= \lim_{\Delta x \to 0} \frac{2x\Delta x + (\Delta x)^2}{\Delta x} = 2x \tag{5}$$

and

$$\frac{\partial f}{\partial y} = \lim_{\Delta y \to 0} \frac{f(x,y + \Delta y) - f(x,y)}{\Delta y}$$

$$= \lim_{\Delta y \to 0} \frac{x^2 - (y + \Delta y)^2 - x^2 + y^2}{\Delta y}$$

$$= \lim_{\Delta y \to 0} \frac{-2y\Delta y - (\Delta y)^2}{\Delta y} = -2y \tag{6}$$

Note that in the same way we can define the partial derivatives of the functions of more than two variables.

2) Differentiate Eq.(2) with respect to x

$$2x = 2f \frac{\partial f}{\partial x} \tag{7}$$

and with respect to y

$$2y = 2f \frac{\partial f}{\partial y} \tag{8}$$

Thus

$$\frac{\partial f}{\partial x} = \frac{x}{f}$$

and

$$\frac{\partial f}{\partial y} = \frac{y}{f} \qquad (f \neq 0) \tag{9}$$

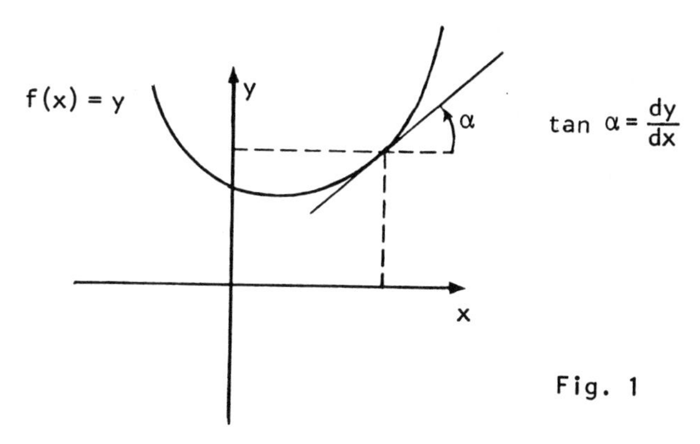

Fig. 1

3) Let us first repeat the simple property of a real-valued function $f(x) = y$ of a single real variable. The value of the derivative $\frac{dy}{dx}$ at a point is the slope of the tangent line to the graph of the function at that point, as shown in Fig. 1.

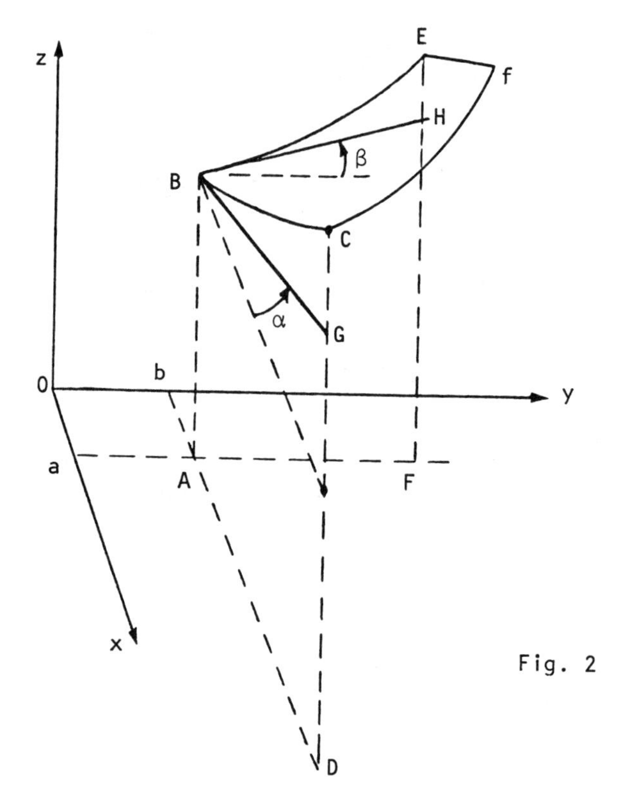

Fig. 2

Without losing generality, consider a function $z = f(x,y)$. Its graph is the set of points $(x,y,f(x,y))$ in R^3, as shown in Fig. 2. Assume that x and y are in the domain of f.

Intersect the surface f with the vertical plane $y = b$. Points A,B,C,D are located on this plane. The intersection of the surface with this plane is a curve

$$z = f(x,y), \quad y = b \tag{10}$$

The slope of the curve $f(x,b)$ at $x = a$ is

$$\frac{\partial f}{\partial x}(a,b) \tag{11}$$

At $y = b$ the slope of the curve $f(a,y)$ is equal to

$$\frac{\partial f}{\partial y}(a,b) \tag{12}$$

Then

$$\begin{aligned} \tan\alpha &= \frac{\partial f}{\partial x}(a,b) \\ \tan\beta &= \frac{\partial f}{\partial y}(a,b) \end{aligned} \tag{13}$$

The angles α, β are shown in Fig. 2. The values of $\tan\alpha$ and $\tan\beta$ are slopes of tangent lines to two curves located on the graph of the function f. These two tangent lines BG and BH determine a tangent plane to the graph of f. The plane containing both tangent lines is given by the equation

$$z = f(a,b) + (x - a)\frac{\partial f}{\partial x}(a,b) + (y - b)\frac{\partial f}{\partial y}(a,b) \tag{14}$$

It is the equation of a plane tangent to the surface f at the point

$$(a,b,f(a,b)).$$

• PROBLEM 1-13

Using the fundamental lemma find the total differential of the functions

1) $f(x_1,x_2,x_3) = x_1^2 - x_2^2 + x_3^3$

2) $f(x_1,x_2,x_3) = 2x_1x_2 + \sin x_2 + x_3^2$

Solution: The fundamental lemma states that: If $f = f(x_1,x_2,x_3)$ has continuous first partial derivatives in the region V, then f has a differential

$$df = \frac{\partial f}{\partial x_1}dx_1 + \frac{\partial f}{\partial x_2}dx_2 + \frac{\partial f}{\partial x_3}dx_3 \tag{1}$$

at every point (x_1,x_2,x_3) of D.

1) Find $\frac{\partial f}{\partial x_1}$, $\frac{\partial f}{\partial x_2}$ and $\frac{\partial f}{\partial x_3}$, then

$$\frac{\partial f}{\partial x_1} = 2x_1$$

$$\frac{\partial f}{\partial x_2} = -2x_2 \tag{2}$$

$$\frac{\partial f}{\partial x_3} = 3x_3^2$$

It can be seen that all the partial derivatives are continuous, therefore to find df apply Eq.(1).

$$df = \frac{\partial f}{\partial x_1}dx_1 + \frac{\partial f}{\partial x_2}dx_2 + \frac{\partial f}{\partial x_3}dx_3$$

$$= 2x_1dx_1 - 2x_2dx_2 + 3x_3^2dx_3 \tag{3}$$

2) The partial derivatives are

$$\frac{\partial f}{\partial x_1} = 2x_2$$

$$\frac{\partial f}{\partial x_2} = 2x_1 + \cos x_2$$

$$\frac{\partial f}{\partial x_3} = 3x_3^2$$

All of them are continuous, thus

$$df = \frac{\partial f}{\partial x_1} dx_1 + \frac{\partial f}{\partial x_2} dx_2 + \frac{\partial f}{\partial x_3} dx_3$$

$$= 2x_2 dx_1 + (2x_1 + \cos x_2) dx_2 + 3x_3^2 dx_3 \qquad (4)$$

GRADIENT, DIVERGENCE AND CURL

• PROBLEM 1-14

Evaluate ∇f at the point (2,3,5). The scalar field f is given by

$$f(x_1,x_2,x_3) = 2 \sin x_1 - x_1^2 x_2 x_3 + x_1 e^{x_2} \qquad (1)$$

Solution: By definition ∇f is equal to

$$\nabla f = \frac{\partial f}{\partial x_1}, \frac{\partial f}{\partial x_2}, \frac{\partial f}{\partial x_3} \qquad (2)$$

We shall compute the partial derivatives of f,

$$\frac{\partial f}{\partial x_1} = 2 \cos x_1 - 2x_1 x_2 x_3 + e^{x_2}$$

$$\frac{\partial f}{\partial x_2} = -x_1^2 x_3 + x_1 e^{x_2} \qquad (3)$$

$$\frac{\partial f}{\partial x_3} = -x_1^2 x_2$$

Then

$$\nabla f = (2 \cos x_1 - 2x_1 x_2 x_3 + e^{x_2}, -x_1^2 x_3 + x_1 e^{x_2}, -x_1^2 x_2) \qquad (4)$$

The value of ∇f at the point (2,3,5) is

$$\nabla f = (2 \cos 2 - 2\times3\times2\times5 + e^3, -4\times5 + 2e^3, - 4\times3)$$

$$= (-40.74, 20.17, -12)$$

Express the Laplacian in polar and cylindrical coordinates.

Solution: Consider the two-dimensional Laplacian

$$\nabla^2 f = \frac{\partial^2 f}{\partial x^2} + \frac{\partial^2 f}{\partial y^2} \tag{1}$$

Express it in terms of polar coordinates r, θ such that

$$x = r\cos\theta, \qquad y = r\sin\theta \tag{2}$$

By the chain rule

$$\frac{\partial f}{\partial x} = \frac{\partial f}{\partial r}\frac{\partial r}{\partial x} + \frac{\partial f}{\partial \theta}\frac{\partial \theta}{\partial x}$$
$$\frac{\partial f}{\partial y} = \frac{\partial f}{\partial r}\frac{\partial r}{\partial y} + \frac{\partial f}{\partial \theta}\frac{\partial \theta}{\partial y} \tag{3}$$

From Eq. (2)

$$dx = \cos\theta\, dr - r\sin\theta\, d\theta$$
$$dy = \sin\theta\, dr + r\cos\theta\, d\theta \tag{4}$$

Solving Eq. (4) for dr, dθ

$$dr = \cos\theta\, dx + \sin\theta\, dy$$
$$d\theta = -\frac{\sin\theta}{r}dx + \frac{\cos\theta}{r}dy \tag{5}$$

Hence

$$\frac{\partial r}{\partial x} = \cos\theta \qquad \frac{\partial r}{\partial y} = \sin\theta$$
$$\frac{\partial \theta}{\partial x} = -\frac{\sin\theta}{r} \qquad \frac{\partial \theta}{\partial y} = \frac{\cos\theta}{r} \tag{6}$$

Eq. (3) can be written in the form

$$\frac{\partial f}{\partial x} = \cos\theta\frac{\partial f}{\partial r} - \frac{\sin\theta}{r}\frac{\partial f}{\partial \theta}$$
$$\frac{\partial f}{\partial y} = \sin\theta\frac{\partial f}{\partial r} + \frac{\cos\theta}{r}\frac{\partial f}{\partial \theta} \tag{7}$$

Applying Eq. (7) to the function $\frac{\partial f}{\partial x}$

$$\frac{\partial^2 f}{\partial x^2} = \frac{\partial}{\partial x}\left(\frac{\partial f}{\partial x}\right) = \cos\theta\frac{\partial}{\partial r}\left(\frac{\partial f}{\partial x}\right) - \frac{\sin\theta}{r}\frac{\partial}{\partial \theta}\left(\frac{\partial f}{\partial x}\right)$$
$$= \cos\theta\frac{\partial}{\partial r}\left(\cos\theta\frac{\partial f}{\partial r} - \frac{\sin\theta}{r}\frac{\partial f}{\partial \theta}\right)$$

$$- \frac{\sin\theta}{r} \frac{\partial}{\partial\theta} \left(\cos\theta \frac{\partial f}{\partial r} - \frac{\sin\theta}{r} \frac{\partial f}{\partial\theta} \right) \tag{8}$$

$$= \cos^2\theta \frac{\partial^2 f}{\partial r^2} - \frac{2 \sin\theta \cos\theta}{r} \frac{\partial^2 f}{\partial r \partial\theta} + \frac{\sin^2\theta}{r^2} \frac{\partial^2 f}{\partial\theta^2}$$

$$+ \frac{\sin^2\theta}{r} \frac{\partial f}{\partial r} + \frac{2 \sin\theta \cos\theta}{r^2} \frac{\partial f}{\partial\theta}$$

In the same way

$$\frac{\partial^2 f}{\partial y^2} = \frac{\partial}{\partial y} \left(\frac{\partial f}{\partial y} \right) = \sin\theta \frac{\partial}{\partial r} \left(\sin\theta \frac{\partial f}{\partial r} + \frac{\cos\theta}{r} \frac{\partial f}{\partial\theta} \right)$$

$$+ \frac{\cos\theta}{r} \frac{\partial}{\partial\theta} \left(\sin\theta \frac{\partial f}{\partial r} + \frac{\cos\theta}{r} \frac{\partial f}{\partial\theta} \right) \tag{9}$$

$$= \sin^2\theta \frac{\partial^2 f}{\partial r^2} + \frac{2 \sin\theta \cos\theta}{r} \frac{\partial^2 f}{\partial r \partial\theta} + \frac{\cos^2\theta}{r^2} \frac{\partial^2 f}{\partial\theta^2}$$

$$+ \frac{\cos^2\theta}{r} \frac{\partial f}{\partial r} - \frac{2 \sin\theta \cos\theta}{r^2} \frac{\partial f}{\partial\theta}$$

Adding Eqs. (8) and (9),

$$\nabla^2 f = \frac{\partial^2 f}{\partial x^2} + \frac{\partial^2 f}{\partial y^2} = \frac{\partial^2 f}{\partial r^2} + \frac{1}{r^2} \frac{\partial^2 f}{\partial\theta^2} + \frac{1}{r} \frac{\partial f}{\partial r} \tag{10}$$

For the three-dimensional system

$$x = r \cos\theta, \quad y = r \sin\theta, \quad z = z \tag{11}$$

To find the Laplacian of $f(x,y,z)$ in terms of r,θ,z add to Eq. (10) the term

$$\frac{\partial^2 f}{\partial z^2} .$$

Hence

$$\nabla^2 f = \frac{\partial^2 f}{\partial x^2} + \frac{\partial^2 f}{\partial y^2} + \frac{\partial^2 f}{\partial z^2} = \frac{\partial^2 f}{\partial r^2} + \frac{1}{r^2} \frac{\partial^2 f}{\partial\theta^2} + \frac{1}{r} \frac{\partial f}{\partial r} + \frac{\partial^2 f}{\partial z^2} \tag{12}$$

The function f is given by

$$f(x,y,z) = \frac{z}{x^2 + y^2} \tag{1}$$

Compute ∇f in cylindrical coordinates ρ, θ, z.

Solution: Consider two systems of coordinates (x,y,z) with the base vectors $\bar{i}, \bar{j}, \bar{k}$ and (ρ, θ, z) with the base vectors $\bar{u}_\rho, \bar{u}_\theta, \bar{u}_z$. A scalar field f is expressed in cylindrical coordinates $f = f(\rho, \theta, z)$. Express ∇f in terms of cylindrical coordinates. For any vector $\bar{a}$ we can write

$$\bar{a} = a_1\bar{i} + a_2\bar{j} + a_3\bar{k} \tag{2}$$

$$\bar{a} = b_1\bar{u}_\rho + b_2\bar{u}_\theta + b_3\bar{u}_z \tag{3}$$

Scalar multiply Eq.(3) by $\bar{u}_\rho$, to obtain

$$\bar{a} \cdot \bar{u}_\rho = b_1\bar{u}_\rho \cdot \bar{u}_\rho = b_1 \tag{4}$$

since the vectors $\bar{u}_\rho$, $\bar{u}_\theta$, $\bar{u}_z$ are mutually orthogonal unit vectors.

In the same way,

$$\bar{a} \cdot \bar{u}_\theta = b_2 \quad \text{and} \quad \bar{a} \cdot \bar{u}_z = b_3 \tag{5}$$

It can be written that

$$\bar{a} = (\bar{u}_\rho \cdot \bar{a})\bar{u}_\rho + (\bar{u}_\theta \cdot \bar{a})\bar{u}_\theta + (\bar{u}_z \cdot \bar{a})\bar{u}_z \tag{6}$$

Setting $\bar{a} = \nabla f$

$$\nabla f = (\bar{u}_\rho \cdot \nabla f)\bar{u}_\rho + (\bar{u}_\theta \cdot \nabla f)\bar{u}_\theta + (\bar{u}_z \cdot \nabla f)\bar{u}_z \tag{7}$$

The coefficients of ∇f in Eq.(7) are equal to

$$\bar{u}_\rho \cdot \nabla f = \left.\frac{\partial f}{\partial s}\right|_{\theta, z=\text{const}} = \frac{\partial f}{\partial \rho} \tag{8}$$

$$\bar{u}_\theta \cdot \nabla f = \left.\frac{\partial f}{\partial s}\right|_{\rho, z=\text{const}} = \frac{1}{\rho}\frac{\partial f}{\partial \theta} \tag{9}$$

$$\bar{u}_z \cdot \nabla f = \left.\frac{\partial f}{\partial s}\right|_{\rho,\theta=\text{const}} = \frac{\partial f}{\partial z} \tag{10}$$

Thus, ∇f in cylindrical coordinates is

$$\nabla f = \frac{\partial f}{\partial \rho}\bar{u}_\rho + \frac{1}{\rho}\frac{\partial f}{\partial \theta}\bar{u}_\theta + \frac{\partial f}{\partial z}\bar{u}_z \tag{11}$$

Expressing the scalar field

$$f(x,y,z) = \frac{z}{x^2 + y^2}$$

in cylindrical coordinates we find

$$f(\rho,\theta,z) = \frac{z}{\rho^2} \tag{12}$$

Substituting Eq.(12) into Eq.(11)

$$\nabla f = -\frac{2z}{\rho^3}\bar{u}_\rho + \frac{1}{\rho^2}\bar{u}_z \tag{13}$$

• PROBLEM 1-17

Using the definition of the partial derivative of the vector, find:

1) $\frac{\partial \bar{a}}{\partial u}$

2) $\frac{\partial \bar{a}}{\partial v}$

3) $\frac{\partial^2 \bar{a}}{\partial u \partial v}$

where

$$\bar{a} = (u^2 v,\ u + e^v,\ u + v^2) \tag{1}$$

Solution: Define the partial derivatives of a vector in a similar way as the partial derivatives of the functions of several independent variables. Assume that $\bar{b}$ is a vector whose components are functions of three independent variables x,y,z. We have

$$b_1 = b_1(x,y,z)$$
$$b_2 = b_2(x,y,z) \tag{2}$$
$$b_3 = b_3(x,y,z)$$

where $\bar{b} = (b_1, b_2, b_3)$

or, in compact form

$$\bar{b} = \bar{b}(x,y,z) \tag{3}$$

The partial derivatives of $\bar{b}$ are defined as

$$\frac{\partial \bar{b}}{\partial x} = \lim_{\Delta x \to 0} \frac{\bar{b}(x + \Delta x, y, z) - \bar{b}(x,y,z)}{\Delta x} \tag{4}$$

$$\frac{\partial \bar{b}}{\partial y} = \lim_{\Delta y \to 0} \frac{\bar{b}(x, y + \Delta y, z) - \bar{b}(x,y,z)}{\Delta y} \tag{5}$$

$$\frac{\partial \bar{b}}{\partial z} = \lim_{\Delta z \to 0} \frac{\bar{b}(x, y, z + \Delta z) - \bar{b}(x,y,z)}{\Delta z} \tag{6}$$

if the limits exist.

1) Using the definition of the partial derivative

$$\frac{\partial \bar{a}}{\partial u} = \lim_{\Delta u \to 0} \frac{\bar{a}(u + \Delta u, v) - \bar{a}(u,v)}{\Delta u}$$

$$= \lim_{\Delta u \to 0} \frac{((u+\Delta u)^2 v,\ u+\Delta u+e^v,\ u+\Delta u+v^2) - (u^2 v,\ u+e^v,\ u+v^2)}{\Delta u}$$

$$= \lim_{\Delta u \to 0} \frac{(2u\Delta uv + (\Delta u)^2 v,\ \Delta u,\ \Delta u)}{\Delta u}$$

$$= \lim_{\Delta u \to 0} (2uv + \Delta u \cdot v, 1, 1) = (2uv, 1, 1) \tag{7}$$

2) Here again, from the definition

$$\frac{\partial \bar{a}}{\partial v} = \lim_{\Delta v \to 0} \frac{(u^2(v+\Delta v),\ u+e^{v+\Delta v},\ u+(v+\Delta v)^2) - (u^2 v,\ u+e^v,\ u+v^2)}{\Delta v}$$

$$= \lim_{\Delta v \to 0} \frac{(u^2 \Delta v,\ e^v \cdot e^{\Delta v} - e^v,\ 2v\Delta v + (\Delta v)^2)}{\Delta v} \tag{8}$$

$$= \lim_{\Delta v \to 0} \left(u^2,\ e^v \cdot \frac{e^{\Delta v}-1}{\Delta v},\ 2v+\Delta v\right) = (u^2,\ e^v,\ 2v)$$

Here, to compute the limit

$$\lim_{\Delta v \to 0} \frac{e^{\Delta v}-1}{\Delta v} \tag{9}$$

L'Hôpital's rule was applied which states that, if $\lim_{x\to a} f(x) = 0$ and $\lim_{x\to a} g(x) = 0$, then

$$\lim_{x\to a} \frac{f(x)}{g(x)} = \lim_{x\to a} \frac{f'(x)}{g'(x)}$$

provided the limit on the right side exists.

Now

$$\lim_{\Delta v\to 0} \frac{e^{\Delta v}-1}{\Delta v} = \lim_{\Delta v\to 0} e^{\Delta v} = 0 \tag{10}$$

3) Note that

$$\frac{\partial^2 \bar{a}}{\partial u \partial v} = \frac{\partial}{\partial u}\left(\frac{\partial \bar{a}}{\partial v}\right) \tag{11}$$

Substituting Eq. (8) into Eq. (11)

$$\frac{\partial}{\partial u}\left(\frac{\partial \bar{a}}{\partial v}\right) = \frac{\partial}{\partial u}(u^2, e^v, 2v)$$

$$= \lim_{\Delta u\to 0} \frac{((u+\Delta u)^2,\ e^v,\ 2v) - (u^2,\ e^v,\ 2v)}{\Delta u} \tag{12}$$

$$= \lim_{\Delta u\to 0} \frac{(2u\cdot\Delta u + (\Delta u)^2,\ 0,\ 0)}{\Delta u} = (2u,0,0)$$

If $\bar{a}$ has continuous partial derivatives of the second order, then

$$\frac{\partial^2 \bar{a}}{\partial u \partial v} = \frac{\partial^2 \bar{a}}{\partial v \partial u} \tag{13}$$

That is, the order of differentiation does not matter.

• PROBLEM 1-18

Evaluate div $\bar{u}$ at the point (-1,1,2)

for $\bar{u} = x^2\bar{i} + e^{xy}\bar{j} + xyz\,\bar{k}$ (1)

Solution: From the definition of the divergence

$$\text{div}\,\bar{u} = \left(\frac{\partial}{\partial x}\bar{i} + \frac{\partial}{\partial y}\bar{j} + \frac{\partial}{\partial z}\bar{k}\right)\cdot(x^2\bar{i} + e^{xy}\bar{j} + xyz\,\bar{k})$$

$$= \frac{\partial}{\partial x}(x^2) + \frac{\partial}{\partial y}(e^{xy}) + \frac{\partial}{\partial z}(xyz) \tag{2}$$

$$= 2x + xe^{xy} + xy$$

At the point (-1,1,2)

$$\operatorname{div} \bar{u}\Big|_{(-1,1,2)} = 2x + xe^{xy} + xy\Big|_{(-1,1,2)}$$

$$= -2 - e^{-1} - 1 = -3 - e^{-1} \qquad (3)$$

Here div $\bar{u}$ was evaluated directly from the definition of divergence.

• PROBLEM 1-19

Find curl $\bar{a}$ at the point (1,-2,1)

for $\bar{a} = x^2y^2\bar{i} + 2xyz\bar{j} + z^2\bar{k}$ (1)

Solution: Define a new vector field from the vector field $\bar{a}$

$$\text{curl } \bar{a} = \left(\frac{\partial a_z}{\partial y} - \frac{\partial a_y}{\partial z}\right)\bar{i} + \left(\frac{\partial a_x}{\partial z} - \frac{\partial a_z}{\partial x}\right)\bar{j} + \left(\frac{\partial a_y}{\partial x} - \frac{\partial a_x}{\partial y}\right)\bar{k} \qquad (2)$$

Definition (2) can be written in another form. The nobla operator is defined as

$$\nabla \equiv \frac{\partial}{\partial x}\,\bar{i} + \frac{\partial}{\partial y}\,\bar{j} + \frac{\partial}{\partial z}\,\bar{k} \qquad (3)$$

Using Eq.(3), Eq.(2) can be written in the form

$$\text{curl } \bar{a} = \nabla\times\bar{a} = \left(\frac{\partial}{\partial x}\,\bar{i} + \frac{\partial}{\partial y}\,\bar{j} + \frac{\partial}{\partial z}\,\bar{k}\right) \times (a_x\bar{i} + a_y\bar{j} + a_z\bar{k})$$

$$= \begin{vmatrix} \bar{i} & \bar{j} & \bar{k} \\ \frac{\partial}{\partial x} & \frac{\partial}{\partial y} & \frac{\partial}{\partial z} \\ a_x & a_y & a_z \end{vmatrix} \qquad (4)$$

The curl or rotation of $\bar{a}$ plays a very important role in vector differential calculus. It is frequently used in electromagnetism, elasticity, hydrodynamics, etc.

These three notations are equivalent

$$\text{curl } \bar{a} \equiv \text{rot } \bar{a} \equiv \nabla \times \bar{a} \qquad (5)$$

Substituting Eq. (1) into Eq. (2) we find

$$\text{curl } \bar{a} = \left[\frac{\partial a_z}{\partial y} - \frac{\partial a_y}{\partial z}\right]\bar{i} + \left[\frac{\partial a_x}{\partial z} - \frac{\partial a_z}{\partial x}\right]\bar{j} + \left[\frac{\partial a_y}{\partial x} - \frac{\partial a_x}{\partial y}\right]\bar{k}$$

$$= \left[\frac{\partial}{\partial y}(z^2) - \frac{\partial}{\partial z}(2xyz)\right]\bar{i} + \left[\frac{\partial}{\partial z}(x^2y^2) - \frac{\partial}{\partial x}(z^2)\right]\bar{j} + \left[\frac{\partial}{\partial x}(2xyz) - \frac{\partial}{\partial y}(x^2y^2)\right]\bar{k} \tag{6}$$

$$= -2xy\bar{i} + (2yz - 2x^2y)\bar{k}$$

At the point (1,-2,1) curl $\bar{a}$ is equal to

$$\text{curl } \bar{a}\Big|_{(1,-2,1)} = -2xy\bar{i} + (2yz - 2x^2y)\bar{k}\Big|_{(1,-2,1)}$$

$$= 4\bar{i} \tag{7}$$

Thus

$$\text{curl } \bar{a}\Big|_{(1,-2,1)} = 4\bar{i} = (4,0,0) \tag{8}$$

• PROBLEM 1-20

Prove the following identities

$$\text{curl grad } f = \bar{0} \tag{1}$$

$$\text{div curl } \bar{a} = 0 \tag{2}$$

Remember curl $\bar{a} \equiv$ rot $\bar{a} \equiv \nabla \times \bar{a}$

Solution: To prove Eq. (1) assume that f has continuous second partial derivatives. We have

$$\text{curl grad } f = \nabla \times (\nabla f)$$

$$= \left(\frac{\partial}{\partial x}, \frac{\partial}{\partial y}, \frac{\partial}{\partial z}\right) \times \left(\frac{\partial f}{\partial x}, \frac{\partial f}{\partial y}, \frac{\partial f}{\partial z}\right)$$

$$= \left[\frac{\partial^2 f}{\partial y \partial z} - \frac{\partial^2 f}{\partial z \partial y}, \frac{\partial^2 f}{\partial z \partial x} - \frac{\partial^2 f}{\partial x \partial z}, \frac{\partial^2 f}{\partial x \partial y} - \frac{\partial^2 f}{\partial y \partial x}\right]$$

$$= (0,0,0) = \bar{0} \tag{3}$$

Here note that the second partial derivatives are continuous and therefore the order of differentiation is immaterial.

To prove Eq. (2), assume that $\bar{a}$ has continuous second partial derivatives.

$$\text{div curl } \bar{a} = \nabla \cdot (\nabla \times \bar{a})$$

$$= \left(\frac{\partial}{\partial x}, \frac{\partial}{\partial y}, \frac{\partial}{\partial z}\right) \cdot \left[\left(\frac{\partial}{\partial x}, \frac{\partial}{\partial y}, \frac{\partial}{\partial z}\right) \times (a_1, a_2, a_3)\right]$$

$$= \left(\frac{\partial}{\partial x}, \frac{\partial}{\partial y}, \frac{\partial}{\partial z}\right) \cdot \left(\frac{\partial a_3}{\partial y} - \frac{\partial a_2}{\partial z}, \frac{\partial a_1}{\partial z} - \frac{\partial a_3}{\partial x}, \frac{\partial a_2}{\partial x} - \frac{\partial a_1}{\partial y}\right) \tag{4}$$

$$= \frac{\partial}{\partial x}\left(\frac{\partial a_3}{\partial y} - \frac{\partial a_2}{\partial z}\right) + \frac{\partial}{\partial y}\left(\frac{\partial a_1}{\partial z} - \frac{\partial a_3}{\partial x}\right) + \frac{\partial}{\partial z}\left(\frac{\partial a_2}{\partial x} - \frac{\partial a_1}{\partial y}\right)$$

$$= \frac{\partial^2 a_3}{\partial x \partial y} - \frac{\partial^2 a_2}{\partial x \partial z} + \frac{\partial^2 a_1}{\partial y \partial z} - \frac{\partial^2 a_3}{\partial y \partial x} + \frac{\partial^2 a_2}{\partial z \partial x} - \frac{\partial^2 a_1}{\partial z \partial y} = 0.$$

• PROBLEM 1-21

Consider a field $D = x\bar{a}_x$ and the surface of a unit cube centered at the origin with its edges parallel to the axes. Verify the Divergence theorem,

$$\oint_S D \cdot dS = \int_{vol} \nabla \cdot D dv.$$

<u>Solution</u>: Evaluating the surface integral first, it is obvious that D is parallel to four of the six faces, and the remaining two give

$$\oint_S D \cdot dS = \int_{-1/2}^{1/2}\int_{-1/2}^{1/2} (-1/2\bar{a}_x) \cdot (-dy\ dz\ \bar{a}_x)$$

$$+ \int_{-1/2}^{1/2}\int_{-1/2}^{1/2} (1/2\bar{a}_x) \cdot (dy\ dz\ \bar{a}_x)$$

$$= 1 \tag{1}$$

Since $\nabla \cdot D = \frac{\partial}{\partial x}(x) = 1$, the volume integral becomes

$$\int_{vol} \nabla \cdot D dv = \int_{-1/2}^{1/2}\int_{-1/2}^{1/2}\int_{-1/2}^{1/2} 1\ dx\ dy\ dz = 1 \tag{2}$$

From Eqs. (1) and (2), $\oint_S D \cdot ds = \int_{vol} \nabla \cdot D dv$, which verifies the Divergence theorem.

• PROBLEM 1-22

For an electric field $\bar{E} = 2\,r^2\,\bar{a}_r$ in spherical coordinates, verify the Divergence theorem by first evaluating the volume integral of the divergence of the field for a sphere of radius r = 3 and then evaluating the surface integral $\int_s \bar{E}\cdot d\bar{s}$, where s is the surface enclosing the volume.

Solution: Divergence theorem

$$\iiint_V \nabla\cdot\bar{E}\;dv = \iint_s \bar{E}\cdot d\bar{s}.$$

Since $\bar{E} = 2\,r^2\,\bar{a}_r$, the field is a function of r only, then, in spherical coordinates, the divergence of $\bar{E}$ is

$$\nabla\cdot\bar{E} = \frac{1}{r^2}\frac{\partial}{\partial r}(r^2 E) = \frac{1}{r^2}\frac{\partial}{\partial r}(2r^4) = 8r$$

The differential volume of the sphere of radius r is

$$dV = r^2 \sin\theta\;dr\;d\theta\;d\phi$$

Therefore

$$\int_V \nabla\cdot\bar{E}\;dv = \int_0^{2\pi}\int_0^{\pi}\int_0^{3} 8r^3 \sin\theta\;dr\;d\theta\;d\phi$$

$$= 2\pi\int_0^{\pi}\int_0^{3} 8r^3 \sin\theta\;dr\;d\theta$$

$$= 2\pi\int_0^{3} 8r^3\,[(-\cos\theta)]_0^{\pi}\;dr$$

$$= 2\pi\int_0^{3} 8r^3(2)\;dr$$

$$= 4\pi\int_0^{3} 8r^3 dr = 648\pi. \qquad (1)$$

Now, the differential area of a spherical shell is

$$\overline{ds} = r^2 \sin\theta \; d\theta \; d\phi \; \bar{a}_r$$

Therefore

$$\oint_S \bar{E}\cdot\overline{ds} = \int_0^{2\pi}\int_0^{\pi} 2r^2\bar{a}_r\cdot r^2 \sin\theta \; d\theta \; d\phi \; \bar{a}_r$$

$$= 2r^4\int_0^{2\pi}\int_0^{\pi} \sin\theta \; d\theta \; d\phi$$

$$= 2r^4\int_0^{2\pi} [-\cos\theta]_0^{\pi} \; d\phi = 2r^4(2)\int_0^{2\pi} d\phi$$

$$= 2r^4(2)(2\pi) = 8\pi r^4$$

$$= 648\pi \quad \text{for } r = 3 \qquad (2)$$

Equations (1) and (2) are equal, which verifies the Divergence theorem.

• PROBLEM 1-23

Verify the divergence theorem by considering

$$\bar{A} = 3x\bar{i}_x + (y - 3)\bar{i}_y + (2 - z)\bar{i}_z$$

and the closed surface of the box bounded by the planes $x = 0$, $x = 1$, $y = 0$, $y = 2$, $z = 0$, and $z = 3$ (see the figure).

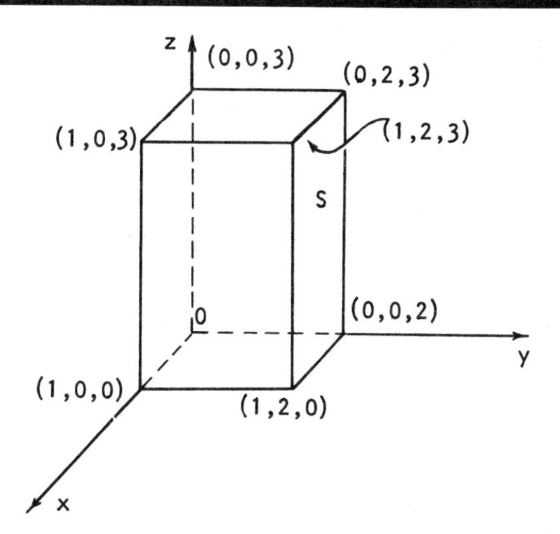

Solution: First determine $\oint_S$ A. dS by evaluating the

surface x = 0,

$$\bar{A} = (y - 3)\bar{i}_y + (2 - z)\bar{i}_z, \quad d\bar{S} = -dy\, dz\, \bar{i}_x$$

$$\bar{A} \cdot d\bar{S} = 0$$

$$\int \bar{A} \cdot d\bar{S} = 0$$

For the surface x = 1,

$$\bar{A} = 3\bar{i}_x + (y - 3)\bar{i}_y + (2 - z)\bar{i}_z, \quad dS = dy\, dz\, \bar{i}_x$$

$$\bar{A} \cdot d\bar{S} = 3\, dy\, dz$$

$$\int \bar{A} \cdot d\bar{S} = \int_{z=0}^{3} \int_{y=0}^{2} 3\, dy\, dz = 18$$

For the surface y = 0,

$$\bar{A} = 3x\bar{i}_x - 3\bar{i}_y + (2 - z)\bar{i}_z, \quad d\bar{S} = -\, dz\, dx\, \bar{i}_y$$

$$\bar{A} \cdot d\bar{S} = 3\, dz\, dx$$

$$\int \bar{A} \cdot d\bar{S} = \int_{x=0}^{1} \int_{z=0}^{3} 3\, dz\, dx = 9$$

For the surface y = 2,

$$\bar{A} = 3x\bar{i}_x - \bar{i}_y + (2 - z)\bar{i}_z, \quad d\bar{S} = dz\, dx\, \bar{i}_y$$

$$\bar{A} \cdot d\bar{S} = -dz\, dx$$

$$\int \bar{A} \cdot d\bar{S} = \int_{x=0}^{1} \int_{z=0}^{3} -\, dz\, dx = -3$$

For the surface z = 0,

$$\bar{A} = 3x\bar{i}_x + (y - 3)\bar{i}_y + 2\bar{i}_x, \quad d\bar{S} = -\, dx\, dy\, \bar{i}_z$$

$$\bar{A} \cdot d\bar{S} = -\, 2\, dx\, dy$$

$$\int \bar{A} \cdot d\bar{S} = \int_{y=0}^{2} \int_{x=0}^{1} -2\, dx\, dy = -4$$

For the surface z = 3,

$$\bar{A} = 3x\bar{i}_x + (y - 3)\bar{i}_y - \bar{i}_z, \quad d\bar{S} = dx\, dy\, \bar{i}_z$$

$$\bar{A} \cdot d\bar{S} = -\, dx\, dy$$

$$\int \bar{A} \cdot d\bar{S} = \int_{y=0}^{2} \int_{x=0}^{1} -dx\, dy = -2$$

Thus

$$\oint_S \bar{A} \cdot d\bar{S} = 0 + 18 + 9 - 3 - 4 - 2 = 18$$

Now, to evaluate $\int_{vol} (\nabla . \bar{A})\, dv$,

$$\nabla \cdot \bar{A} = \frac{\partial A_x}{\partial x} + \frac{\partial A_y}{\partial y} + \frac{\partial A_z}{\partial z}$$

$$\nabla \cdot \bar{A} = \frac{\partial (3x)}{\partial x} + \frac{\partial (y-3)}{\partial y} + \frac{\partial (2-z)}{\partial z}$$

$$\nabla \cdot \bar{A} = 3 + 1 - 1 = 3.$$

For the volume enclosed by the rectangular box,

$$\int (\nabla \cdot \bar{A})\, dv = \int_{z=0}^{3} \int_{y=0}^{2} \int_{x=0}^{1} 3dx\, dy\, dz = 18$$

which verifies the divergence theorem.

LINE, SURFACE AND VOLUME INTEGRALS

• PROBLEM 1-24

Let

$$\bar{F} = k\, r^n\, \hat{r}$$

Evaluate

$$\iint_s \bar{F} \cdot \hat{n}\, ds \text{ and } \iint_s \bar{F} \times \hat{n}\, ds$$

where s is a sphere of radius a centered at the origin.

Solution: $\hat{n} = \hat{r}$ = outward normal

$\bar{F} \cdot \hat{n} = k\, r^n$, $ds = r^2 \sin\theta\, d\theta\, d\phi$ in spherical coordinates

$$\iint_s \bar{F} \cdot \hat{n}\, ds = \int_0^{2\pi} \int_0^{1\pi} k\, a^n\, a^2 \sin\theta\, d\theta\, d\phi$$

$$= 4\pi\, k\, a^{n+2}$$

Similarly:

$$\iint_s \bar{F} \times \hat{n}\, ds = 0$$

Some commonly used differential surface elements are dx dy in rectangular coordinates, r dr dφ or r dφ dz in cylindrical coordinates, and $r^2 \sin\theta\, d\theta\, d\phi$ in spherical coordinates.

• PROBLEM 1-25

Find the line integral from P_1 to P_2 (see figure) when the force field is given by

$$\bar{F}(x, y, z) = \bar{a}_x x^2 + \bar{a}_y y^2 + \bar{a}_z z^2$$

where $\bar{a}_x$, $\bar{a}_y$ and $\bar{a}_z$ are the unit vectors in x, y, and z directions respectively.

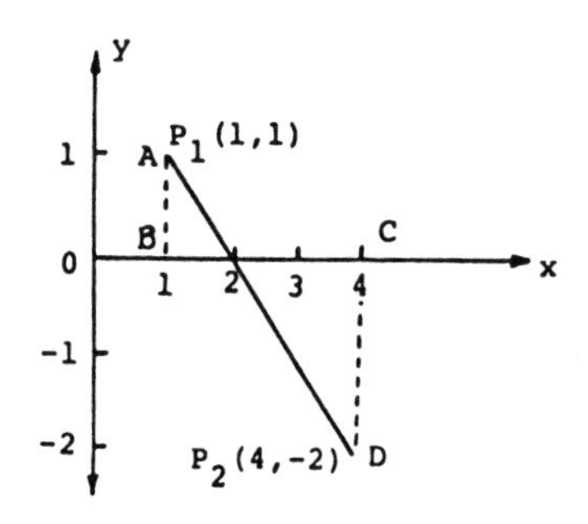

Solution: The equation of the curve from P_1 to P_2 is

$$y = -x + 2$$

If y is eliminated in the expression for $\bar{F}$, then

$$\bar{F} = x^2\bar{a}_x + (-x+2)^2\,\bar{a}_y \tag{1}$$

since z = 0 in the XY plane. The element of path length $d\bar{\ell}$ is

$$d\bar{\ell} = dx\,\bar{a}_x + dy\,\bar{a}_y + dz\,\bar{a}_z$$

But in the XY plane dz = 0, and from the equation of the curve, dy = -dx. Then

$$d\bar{\ell} = dx\,\bar{a}_x - dx\,\bar{a}_y = (\bar{a}_x - \bar{a}_y)\,dx \tag{2}$$

Using Eqs. (1) and (2)

$$I = \int_{P_1}^{P_2} \bar{F}\cdot d\bar{\ell} = \int_1^4 [x^2\bar{a}_x + (-x+2)^2\,\bar{a}_y]\cdot[\bar{a}_x - \bar{a}_y]\,dx$$

$$= \int_1^4 [x^2 - (-x+2)^2]\,dx = (2x^2 - 4x)\Big|_1^4 = 18$$

Follow the path ABCD in this integration, then

$$I = \int_A^B + \int_B^C + \int_C^D = \int_A^B \bar{F}\cdot(-\bar{a}_y dy) + \int_B^C \bar{F}\cdot(\bar{a}_x dx)$$

$$+ \int_C^D \bar{F}\cdot(-\bar{a}_y dy) = \int_0^1 (-)y^2 dy + \int_1^4 x^2 dx$$

$$+ \int_{-2}^0 (-)y^2 dy = -\frac{1}{3} + \frac{63}{3} - \frac{8}{3} = 18$$

Thus, the force field is path independent.

• PROBLEM 1-26

Let

$$\bar{F} = K\, r^n\, \hat{r}$$

in spherical coordinates. Evaluate the line integral along c, a path connecting point a($r = 1$, $\phi = 0$, $\theta = \frac{1}{2}\pi$) and point b($r = 2$, $\phi = \frac{1}{2}\pi$, $\theta = \frac{1}{4}\pi$). (See Figure.)

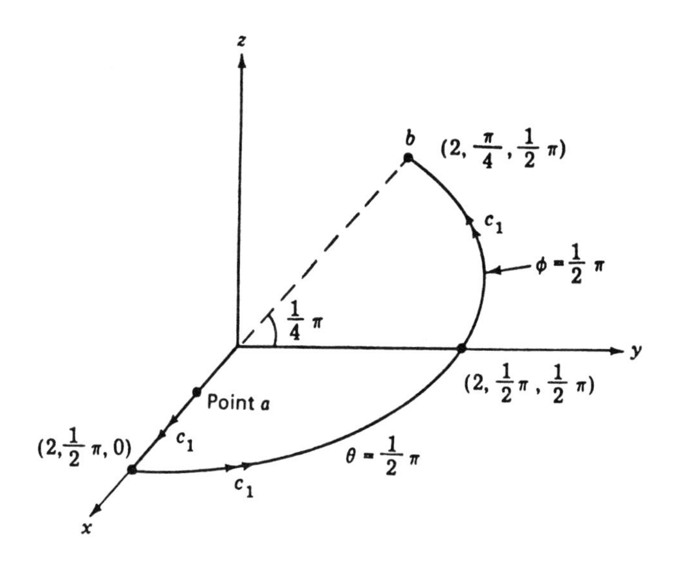

Solution: In spherical coordinates

$$\nabla \times \bar{F} = \frac{\hat{r}}{r \sin\theta}\left[\frac{\partial}{\partial\theta}(F_r \sin\theta) - \frac{\partial F_\theta}{\partial\phi}\right]$$

$$+ \frac{\hat{\theta}}{r}\left[\frac{1}{\sin\theta}\;\frac{\partial Fr}{\partial\phi} - \frac{\partial}{\partial r}(rF_\phi)\right]$$

$$+ \frac{\hat{\phi}}{r}\left[\frac{\partial}{\partial r}(rF_\theta) - \frac{\partial F_r}{\partial\theta}\right].$$

Since $\bar{F}$ here has only one component, $\bar{F}_r$, and it depends on r alone, $\nabla \times \bar{F} = 0$. If $\nabla \times \bar{F} = 0$, then the line integral is independent of path and depends only on the endpoints a and b. Therefore choose any convenient path such as path c_1 in the figure which consists of three portions. The first portion is along the x axis, over which only r varies. The second is a horizontal circular arc along which only ϕ varies. The third is a vertical circular arc along which only θ varies. In spherical coordinates,

$$\int_c \bar{F} \cdot d\bar{l} = \int_c F_r dr + F_\theta r\, d\theta + F_\phi r \sin\theta\, d\phi$$

Only the first portion of c_1 contributes, yielding:

$$\int_c \bar{F} \cdot d\bar{l} = \frac{Kr^{n+1}}{n+1}\Bigg]_1^2 = \frac{K}{n+1}(2^{n+1} - 1)$$

• PROBLEM 1-27

Let

$$\bar{F} = xy^2\hat{x} + yz^2\hat{y} + 2xz\,\hat{z}$$

Evaluate the line integral $\int_c \bar{F}.d\bar{l}$ where c is a straight line (see figure) between the points (0,0,0) and (1,2,3).

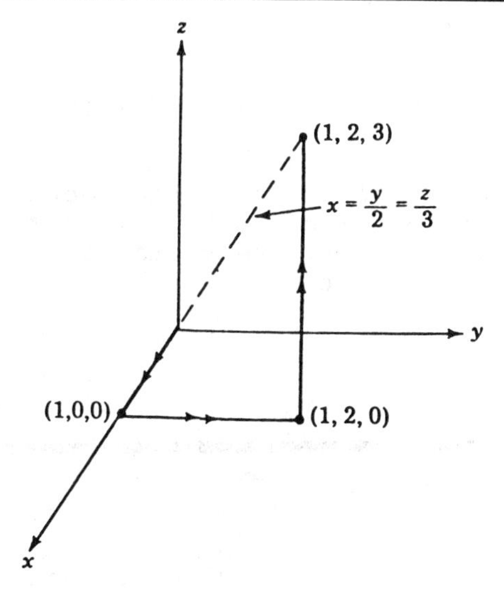

<u>Solution</u>: The equation of the straight line is:

$$x = \frac{y}{2} = \frac{z}{3} \tag{1}$$

$$\int_c \bar{F}.d\bar{l} = \int_c F_x dx + F_y dy + F_z dz = \int_c xy^2\, dx + yz^2 dy + 2xz\, dz \tag{2}$$

Using the relationships:

$$dx = \frac{dy}{2} = \frac{dz}{3} \quad \text{(From Eq. 1),}$$

Change the line integral to one in a single variable x.

$$\int_c \bar{F}.d\bar{l} = \int_0^1 (4x^3 + 36x^3 + 18x^2)dx = 16$$

There are other methods used to perform the line integration; for instance, substituting y = 2x in the first term,

$z = \frac{3}{2}y$ in the second term, and $x = \frac{1}{3}z$ in the third term

of Eq. 2 yields

$$\int_c \bar{F}.d\bar{l} = \int_0^1 4x^3dx + \int_0^2 \frac{9}{4}y^3dy + \int_0^3 \frac{2}{3}z^3dz = 16$$

Now compute the line integral over a different path between the same end points. Let path c consist of three straight segments respectively parallel to the x, y, and z axis, connecting the same end points (0,0,0) and (1,2,3), as shown in the figure.

$$\int_c \bar{F}.d\bar{l} = \int_0^1 x(0)\ dx + \int_0^2 y(0)\ dy + \int_0^3 2(1)\ z\ dz = 9$$

Note that this value differs from that obtained by integrating over a straight line path c directly connecting (0,0,0) and (1,2,3). Thus, the vector force is nonconservative, i.e., path dependent.

• **PROBLEM 1-28**

Let

$$\bar{A} = k\ r^n\ \hat{r} \qquad \text{(in spherical coordinates)}$$

Evaluate

$$\iiint_v \nabla \cdot \bar{A}\ dv$$

where v is a sphere of radius a with center at the origin.

Solution: Using the expression for divergence in spherical coordinates:

$$\nabla \cdot \bar{A} = \frac{1}{r^2}\frac{\partial}{\partial r}(r^2 A_r) + \frac{1}{r\sin\theta}\frac{\partial}{\partial\theta}(\sin\theta\, A_\theta) + \frac{1}{r\sin\theta}\frac{\partial A\phi}{\partial\phi}.$$

Here,

$$\nabla \cdot \bar{A} = \frac{1}{r^2}\frac{\partial}{\partial r}(r^2 k r^n) = \frac{k}{r^2}\frac{\partial}{\partial r}(r^{n+2}) = \frac{k}{r^2}(n+2)\, r^{n+1}.$$

In spherical coordinates,

$$dv = r^2 \sin\theta \, dr d\phi \, d\theta$$

Thus $\nabla.\bar{A}\, dv = k(n+2)r^{n+1} \sin\theta \, dr \, d\theta \, d\phi$

and

$$\iiint_v \nabla.\bar{A}\, dv = \int_0^{2\pi}\int_0^{\pi}\int_0^{a} k(n+2)r^{n+1} \sin\theta \, dr \, d\theta \, d\phi = 4\pi\, k\, a^{n+2}$$

A common mistake, made in connection with integrals of vectors, is to assume that unit vectors are constants and thereby take them outside the integral. This cannot be done for $\hat{r}$, $\hat{\phi}$, and $\hat{\theta}$. For instance,

$$\iiint_v \bar{F}\, dv = \iiint_v [F_r\hat{r} + F_\phi\hat{\phi} + F_\theta\hat{\theta}]dv$$

$$\neq \hat{r}\iiint F_r dv + \hat{\phi}\iiint F_\phi dv + \hat{\theta}\iiint F_\theta dv$$

(note the inequality sign) because $\hat{r}$, $\hat{\phi}$, and $\hat{\theta}$ vary over the volume of integration v and cannot therefore be considered as constants.

• PROBLEM 1-29

Let

$$\bar{F} = xy^2\hat{x} + y^3\hat{y} + x^2 y\hat{z}$$

Evaluate (1) $\iint_s \bar{F}.\hat{n}\, ds$

(2) $\iint_s \bar{F} \times \hat{n}\, ds$

and (3) $\iint_s \nabla \times \bar{F}.\hat{n}\, ds.$

for the surface s shown in the figure consisting of a square of length 2 lying in the xy plane.

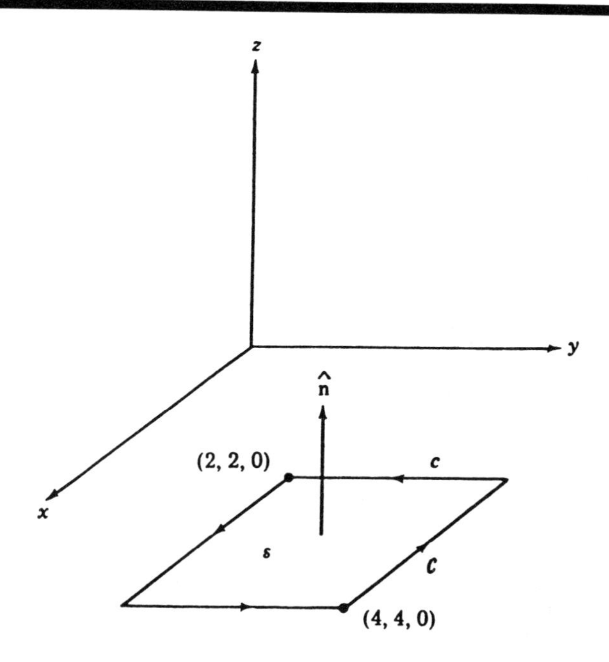

Solution: This is an open surface and n̂ is defined to be in the positive z direction.

$$\iint_s \bar{F}.\hat{n}\, ds = \int_2^4 \int_2^4 x^2 y\, dx\, dy = \left[\frac{x^3}{3}\right]_2^4 \left[\frac{y^2}{2}\right]_2^4$$

$$= \left[\frac{4^3 - 2^3}{3}\right]\left[\frac{4^2 - 2^2}{2}\right] = 112$$

Similarly,

$$\bar{F} \times \hat{n} = \begin{vmatrix} \hat{x} & \hat{y} & \hat{z} \\ xy^2 & y^3 & x^2 y \\ 0 & 0 & 1 \end{vmatrix} = \hat{x}(y^3) - \hat{y}(xy^2)$$

$$\iint_s \bar{F} \times \hat{n}\, ds = \hat{x}\left[\int_2^4 \int_2^4 y^3 dxdy\right] - \hat{y}\left[\int_2^4 \int_2^4 xy^2 dxdy\right]$$

$$= \hat{x}\ (120) - \hat{y}\ (112)$$

Lastly, to evaluate

$$\iint_s \nabla \times \bar{F}.\hat{n}\, ds$$

$$\nabla \times \bar{F} = \begin{vmatrix} \hat{x} & \hat{y} & \hat{z} \\ \partial/\partial x & \partial/\partial y & \partial/\partial z \\ xy^2 & y^3 & x^2 y \end{vmatrix}$$

$$= \hat{x}(x^2) + \hat{y}(-2xy) + \hat{z}(-2xy)$$

$$\nabla \times \bar{F}\cdot\hat{n} = -2xy$$

$$\iint_s \nabla \times \bar{F}.\hat{n}\, ds = \int_2^4\int_2^4 - 2xy\, dxdy = -72$$

The divergence theorem allows the replacing of

$$\oiint_s \bar{F}.\hat{n}\, ds$$

with

$$\iiint_v \nabla \cdot \bar{F}\, dv$$

for the special case where s is a closed surface. Stokes' Law allows the replacing of

$$\iint_s \nabla \times \bar{F}.\hat{n}\, ds$$

with

$$\oint_c \bar{F} \cdot d\bar{\ell}$$

where c is related to the direction of $\hat{n}$ by the right-hand rule.

STOKE'S THEOREM

• PROBLEM 1-30

Verify Stokes' theorem by considering

$$\bar{A} = y\bar{i}_x - x\bar{i}_y$$

and the closed path C shown in the figure.

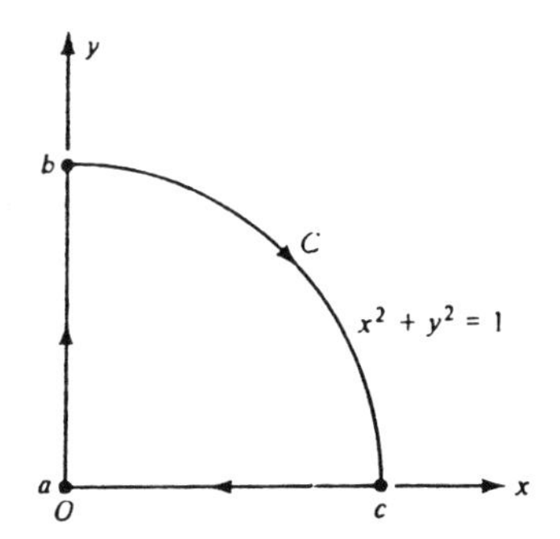

Solution: First determine $\oint_C \bar{A} \cdot d\bar{l}$ by evaluating the line integrals along the three segments of the closed path. To do this, note that $\bar{A} \cdot d\bar{l} = y\ dx - x\ dy$. Then, from a to b, $x = 0$, $dx = 0$, $A \cdot dl = 0$

$$\int_a^b \bar{A} \cdot d\bar{l} = 0$$

From b to c, $x^2 + y^2 = 1$, $y = \sqrt{1 - x^2}$

$$2x\ dx + 2y\ dy = 0, \qquad dy = -\frac{x\ dx}{y} = -\frac{x}{\sqrt{1 - x^2}}\ dx$$

$$\bar{A} \cdot d\bar{l} = \sqrt{1 - x^2}\ dx + \frac{x^2\ dx}{\sqrt{1 - x^2}} = \frac{dx}{\sqrt{1 - x^2}}$$

$$\int_b^c \bar{A} \cdot d\bar{l} = \int_0^1 \frac{dx}{\sqrt{1 - x^2}} = \left[\sin^{-1} x\right]_0^1 = \frac{\pi}{2}$$

From c to a, $y = 0$, $dy = 0$, $\bar{A} \cdot d\bar{l} = 0$

$$\int_c^a \bar{A} \cdot d\bar{l} = 0$$

Thus

$$\oint_C \bar{A} \cdot d\bar{l} = \int_a^b \bar{A} \cdot d\bar{l} + \int_b^c \bar{A} \cdot d\bar{l} + \int_c^a \bar{A} \cdot d\bar{l}$$

$$= 0 + \frac{\pi}{2} + 0 = \frac{\pi}{2}$$

Now, to evaluate $\oint_C \bar{A} \cdot d\bar{l}$ by using Stokes' theorem,

$$\nabla \times \bar{A} = \bar{i}_x\left(\frac{\partial A_z}{\partial y} - \frac{\partial A_y}{\partial z}\right) + \bar{i}_y\left(\frac{\partial A_x}{\partial z} - \frac{\partial A_z}{\partial x}\right) + \bar{i}_z\left(\frac{\partial A_y}{\partial x} - \frac{\partial A_x}{\partial y}\right)$$

Since none of the terms in $\bar{A}$ have a z component, all terms of the curl with a z drop out, therefore

$$\nabla \times \bar{A} = \nabla \times (y\bar{i}_x - x\bar{i}_y) = -2i_x$$

For the plane surface S enclosed by C,

$$d\bar{S} = -dx\, dy\, \bar{i}_z$$

Thus

$$(\nabla \times \bar{A}) \cdot d\bar{S} = -2\bar{i}_z \cdot (-dx\, dy\, \bar{i}_z) = 2\, dx\, dy$$

$$\int_S (\nabla \times \bar{A}) \cdot d\bar{S} = \int_{x=0}^{1} \int_{y=0}^{\sqrt{1-x^2}} 2\, dx\, dy$$

$$= 2 \text{ (area enclosed by C)}$$

$$= 2 \times \frac{\pi}{4} = \frac{\pi}{2}$$

thereby verifying Stokes' theorem.

• PROBLEM 1-31

Verify Stokes' theorem for the circular bounding contour in the xy plane shown in the figure with a vector field

$$\bar{A} = -y\bar{i}_x + x\bar{i}_y - z\bar{i}_z = r\bar{i}_\phi - z\bar{i}_z$$

Check the result for (a) the flat circular surface in the xy plane, (b) the hemispherical surface bounded by the contour, and (c) the cylindrical surface bounded by the contour.

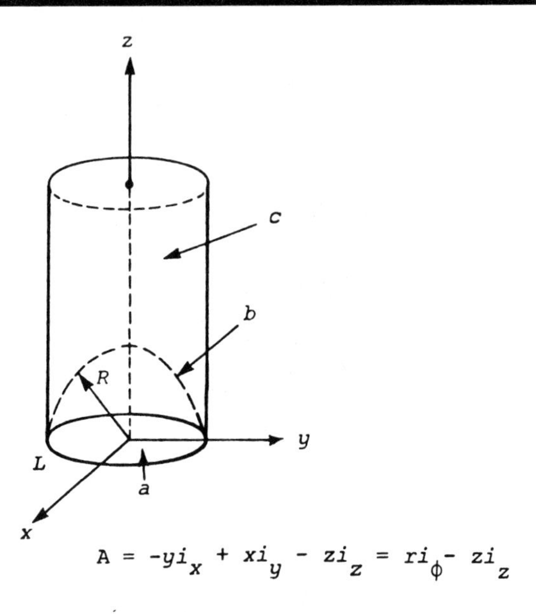

$$A = -yi_x + xi_y - zi_z = ri_\phi - zi_z$$

Solution: Stoke's theorem converts the line integral over the bounding contour L of the outer edge to a surface integral over any area S bounded by the contour:

$$\oint_L \bar{A} \cdot d\bar{\ell} = \int_S (\nabla \times \bar{A}) \cdot d\bar{S}$$

For the contour shown

$$d\bar{\ell} = R\, d\phi\, \bar{i}_\phi$$

so that

$$\bar{A} \cdot d\bar{\ell} = R^2\, d\phi$$

where on L, r = R. Then the circulation is

$$C = \oint_L \bar{A} \cdot d\bar{\ell} = \int_0^{2\pi} R^2\, d\phi = 2\pi R^2$$

The z component of $\bar{A}$ has no contribution because $d\bar{\ell}$ is entirely in the xy plane.

The curl of $\bar{A}$ is

$$\nabla \times \bar{A} = \bar{i}_z\left(\frac{\partial A_y}{\partial x} - \frac{\partial A_x}{\partial y}\right) = 2\bar{i}_z$$

(a) For the circular area in the plane of the contour,

$$\int_S (\nabla \times \bar{A}) \cdot d\bar{S} = 2\int_S dS_z = 2\pi R^2$$

which agrees with the line integral result.

(b) For the hemispherical surface

$$\int_S (\nabla \times \bar{A}) \cdot d\bar{S} = \int_{\theta=0}^{\pi/2} \int_{\phi=0}^{2\pi} 2i_z \cdot i_r R^2 \sin\theta \, d\theta \, d\phi$$

Use the dot product relation

$$\bar{i}_z \cdot \bar{i}_r = \cos\theta$$

which again gives the circulation as

$$C = \int_{\theta=0}^{\pi/2} \int_{\phi=0}^{2\pi} R^2 \sin 2\theta \, d\theta \, d\phi = -2\pi R^2 \frac{\cos 2\theta}{2}\Bigg|_{\theta=0}^{\pi/2} = 2\pi R^2$$

(c) Similarly, for the cylindrical surface, the only non-zero contributions to the surface integral are at the upper circular area that is perpendicular to $\nabla \times \bar{A}$. The integral is then the same as part (a) as $\nabla \times \bar{A}$ is independent of z.

• PROBLEM 1-32

Check Stokes' theorem for $\bar{F} = \hat{x}(x + y) - \hat{y}2x^2 + \hat{z}xy$ and the upper hemisphere of $x^2 + y^2 + z^2 = 1$, (see figure).

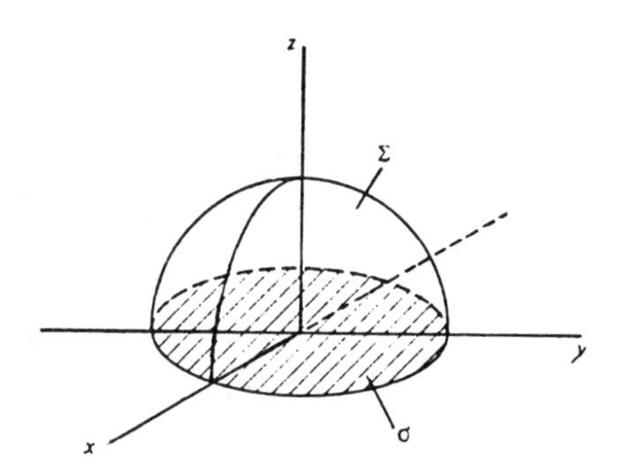

Solution:

$$\nabla \times \bar{F} = \begin{vmatrix} \hat{x} & \hat{y} & \hat{z} \\ \partial_x & \partial_y & \partial_z \\ x + y & -2x^2 & xy \end{vmatrix}$$

$$= \hat{x}(x - 0) + \hat{y}(0 - y) + \hat{z}(-4x - 1)$$

$$= \hat{x}x - \hat{y}y - \hat{z}(4x + 1)$$

$$\int_{\Sigma} (\nabla \times \bar{F}) \cdot d\bar{S} = \int_{\Sigma} [\hat{x}x - \hat{y}y - \hat{z}(4x + 1)] \cdot \hat{n}\, dS$$

$$= \int_{\sigma} [\hat{x}x - \hat{y}y - \hat{z}(4x + 1)] \cdot \hat{n} \frac{dx\, dy}{\cos(\hat{n},\hat{z})}$$

Here σ is the projection of Σ on the xy plane as shown in the figure. The gradient to the surface $f = x^2 + y^2 + z^2 - 1$ is $\nabla f = \hat{x}2x + \hat{y}2y + \hat{z}2z$. Then $|\nabla f| = \sqrt{4x^2 + 4y^2 + 4z^2} = 2\sqrt{x^2 + y^2 + z^2} = 2$, and $\hat{n} = \nabla f/|\nabla f| = \hat{x}x + \hat{y}y + \hat{z}z$. So

$$[\hat{x}x - \hat{y}y - \hat{z}(4x + 1)] \cdot \hat{n}$$

$$= [\hat{x}x - \hat{y}y - \hat{z}(4x + 1)] \cdot [\hat{x}x + \hat{y}y + \hat{z}z]$$

$$= x^2 - y^2 - (4x + 1)z$$

Also, $\hat{n} \cdot \hat{z} = (1)(1) \cos(\hat{n},\hat{z}) = (x)(0) + (y)(0) + (z)(1) = z$, so $\cos(\hat{n},\hat{z}) = z$. The surface integral is, therefore,

$$\int_{\Sigma} (\nabla \times \bar{F}) \cdot d\bar{S} = \int_{x=-1}^{1} \int_{y=-\sqrt{1-x^2}}^{\sqrt{1-x^2}} [x^2 - y^2 - (4x + 1)z] \frac{dxdy}{z}$$

$$= \int_{x=-1}^{1} dx \int_{y=-\sqrt{1-x^2}}^{\sqrt{1-x^2}} \frac{x^2 - y^2}{\sqrt{1 - x^2 - y^2}} dy$$

$$- \int_{x=-1}^{1} dx \int_{y=-\sqrt{1-x^2}}^{\sqrt{1-x^2}} (4x + 1) dy$$

$$= \int_{x=-1}^{1} x^2 dx \int_{y=-\sqrt{1-x^2}}^{\sqrt{1-x^2}} \frac{dy}{\sqrt{(1 - x^2) - y^2}}$$

$$- \int_{x=-1}^{1} dx \int_{y=-\sqrt{1-x^2}}^{\sqrt{1-x^2}} \frac{y^2\, dy}{\sqrt{(1 - x^2) - y^2}}$$

$$- \int_{x=-1}^{1} (4x + 1) dx \int_{y=-\sqrt{1-x^2}}^{\sqrt{1-x^2}} dy$$

$$= \int_{x=-1}^{1} x^2\, dx \left[\sin^{-1} \frac{y}{\sqrt{1 - x^2}} \right]_{y = -\sqrt{1 - x^2}}^{+\sqrt{1 - x^2}}$$

$$- \int_{x=-1}^{1} dx \left[- \frac{y}{2} \sqrt{(1 - x^2) - y^2} \right.$$

$$+ \frac{1 - x^2}{2} \sin^{-} \left. \frac{y}{\sqrt{1 - x^2}} \right]_{y = -\sqrt{1 - x^2}}^{+\sqrt{1 - x^2}}$$

$$- \int_{x=-1}^{1} (4x + 1)\ dx\ (2\ \sqrt{1 - x^2})$$

$$= \int_{-1}^{1} x^2\ dx \left[\frac{\pi}{2} - \left(- \frac{\pi}{2} \right) \right]$$

$$- \int_{-1}^{1} dx \left[0 + \frac{(1 - x^2)}{2} \frac{\pi}{2} - 0 \right.$$
$$\left. - \frac{(1 - x^2)}{2} \left(- \frac{\pi}{2} \right) \right]$$

$$- \int_{-1}^{1} 8x\sqrt{1 - x^2}\ dx - 2 \int_{-1}^{1} \sqrt{1 - x^2}\ dx$$

$$= \pi \int_{-1}^{1} x^2\ dx - \frac{\pi}{2} \int_{-1}^{1} (1 - x^2)dx$$

$$- 8 \int_{-1}^{1} x\sqrt{1 - x^2}\ dx - 2 \int_{-1}^{1} \sqrt{1 - x^2}\ dx$$

$$= \pi \left[\frac{x^3}{3} \right]_{-1}^{1} - \frac{\pi}{2} \left[x - \frac{x^3}{3} \right]_{-1}^{1}$$
$$- 8 - \frac{1}{3} [(1 - x^2)^{3/2}]_{-1}^{1}$$

$$- 2\left(\frac{1}{2}\right) [x\sqrt{1 - x^2} - \sin^{-1} x \Big]_{-1}^{1}$$

$$= \frac{\pi}{3}[1 - (-1)\Big] - \frac{\pi}{2}\ 1 - \frac{1}{3} - (-1) + \frac{-1}{3} \Big]$$

$$+ \frac{8}{3}(0) - \left[0 + \frac{\pi}{2} - 0 - \left(- \frac{\pi}{2}\right)\right]$$

$$= \frac{\pi}{3}(2) - \frac{\pi}{2}\left(\frac{4}{3}\right) + 0 - \pi$$

$$= -\pi$$

Now the line integral.

$$\oint \bar{F} \cdot d\bar{r} = \oint_C [\hat{x}(x + y) - \hat{y}(2x^2) + \hat{z}xy] \cdot \left[\hat{x}dx + \hat{y}dy + \hat{z}dz\right]$$

$$= \oint_C [(x + y)dx - 2x^2\, dy]$$

Let $x = \cos \phi$, $y = \sin \phi$. Then $dx = \sin \phi\, d\phi$, $dy = \cos \phi\, d\phi$

$$\oint_C \bar{F} \cdot d\bar{r} = \int_0^{2\pi} [(\cos \phi + \sin \phi)(-\sin \phi\, d\phi)$$

$$- 2 \cos^3 \phi\, d\phi]$$

$$= 0 - \pi - \frac{2}{3}[\sin \phi(\cos^2\phi) + 2 \sin \phi]_0^{2\pi}$$

$$= - \pi$$

This checks Stokes' theorem for the given case.

CHAPTER 2

ELECTRIC CHARGES

CHARGE DENSITIES AND DISTRIBUTIONS

• PROBLEM 2-1

Find the total charge contained in the cylinder shown if the volume charge density is

$\rho = 100e^{-z}(x^2 + y^2)^{-1/4} \quad C/m^3$

The total charge contained within the right circular cylinder may be obtained by evaluating $Q = \int_{vol} \rho \, dv$

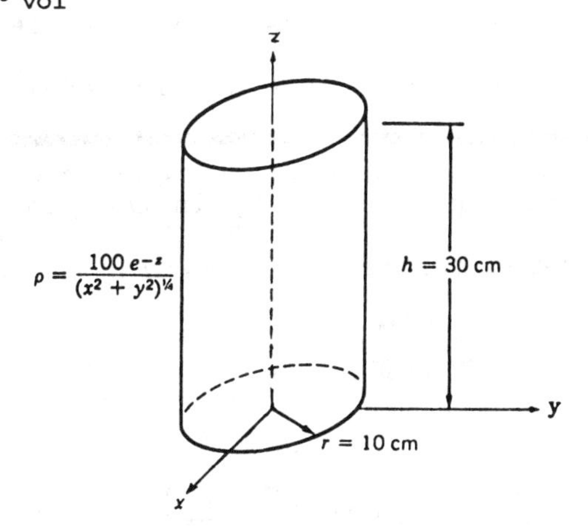

Solution: Since the volume is most easily described in cylindrical coordinates, express the charge density in cylindrical coordinates and integrate in cylindrical coordinates. Since $r = (x^2 + y^2)^{1/2}$, ρ can be rewritten as

$$\rho = 100e^{-z}r^{-1/2}$$

The volume differential in cylindrical coordinates is given by $dV = r dr d\phi dz$, therefore

$$Q = \int_{vol} \rho dV,$$

$$Q = \int_0^{0.3} \int_0^{2\pi} \int_0^{0.1} 100e^{-z} r^{-1/2} r \, dr \, d\phi \, dz$$

$$= \int_0^{0.3} \int_0^{2\pi} 100e^{-z} 2/3 r^{3/2} \Big|_0^{0.1} d\phi \, dz$$

$$= \int_0^{0.3} \int_0^{2\pi} 2.11e^{-z} d\phi \, dz$$

$$= \int_0^{0.3} 13.25e^{-z} dz$$

$$= 3.43 \text{ C}$$

• PROBLEM 2-2

> In the electrical field today, copper is the most commonly used metal. Consider an electrically neutral pin made of copper of mass m = 2.30 gm., determine the charge q in coulomb of the pin.
>
> Given: For a copper atom,
>
> nuclear charge = electron charge = $+4.60 \times 10^{-18}$ coul/atom,
>
> and atomic weight of copper M_c = 63.54 gm/mole.
>
> (Note: Avogadro's number $N_A = 6.02 \times 10^{23}$ atoms/mole.)

Solution: In order to determine the charge q in coulomb of the pin, the number of copper atoms N_c in the pin must be calculated first.

Using the relationship as follows:

$$N_A = \frac{N_c \times M_c}{\text{mass of the pin (m)}} \tag{1}$$

Now, solving eq.(1) for N_c,

$$N_c = \frac{N_A m}{M_c} = \frac{6.02 \times 10^{23} \left[\frac{\text{atoms}}{\text{mole}}\right] \times 2.30 [\text{g}]}{63.54 \left[\frac{\text{g}}{\text{mole}}\right]}$$

$$= 2.18 \times 10^{22} \text{ atoms}$$

Hence,

the charge q in coulomb $= 2.18 \times 10^{22} [\text{atoms}] \times 4.60 \times 10^{-18} \left[\frac{\text{coul}}{\text{atom}}\right]$

$$\cong 1.0 \times 10^{5} \text{ coul.}$$

• **PROBLEM 2-3**

Find the total charge within each of the following distributions illustrated in figure 1.

(a) Line charge λ_0 uniformly distributed in a circular loop of radius a.

(b) Surface charge σ_0 uniformly distributed on a circular disk of radius a.

(c) Volume charge ρ_0 uniformly distributed throughout a sphere of radius R.

(d) A line charge of infinite extent in the z direction with charge density distribution

$$\lambda = \frac{\lambda_0}{[1 + (z/a)^2]}$$

(e) The electron cloud around the positively charged nucleus Q in the hydrogen atom is simply modeled as the spherically symmetric distribution

$$\rho(r) = -\frac{Q}{\pi a^3} e^{-2r/a}$$

where a is called the Bohr radius.

Solution: (a) For a line charge in a circular loop, $dl = rd\theta$, therefore

$$q = \int_L \lambda dl = \int_0^{2\pi} \lambda_0 a d\phi = 2\pi a \lambda_0$$

(b) Using $ds = rdrd\phi$ for a surface charge distribution, the following is obtained:

$$q = \int_S \sigma dS = \int_{r=0}^{a} \int_{\phi=0}^{2\pi} \sigma_0 r dr d\phi = \pi a^2 \sigma_0$$

(c) For a spherical volume charge distribution, $dV = r^2 \sin\theta dr d\theta d\phi$, thus

$$q = \int_V \rho dV = \int_{r=0}^{R} \int_{\theta=0}^{\pi} \int_{\phi=0}^{2\pi} \rho_0 r^2 \sin\theta dr d\theta d\phi$$

$$= \frac{4}{3} \pi R^3 \rho_0$$

(d) For a line charge in the z direction, $dl = dz$ and applying the formula,

$$q = \int_L \lambda dl = \int_{-\infty}^{+\infty} \frac{\lambda_0 dz}{[1+(z/a)^2]} = \lambda_0 a \tan^{-1}\frac{z}{a}\Big|_{-\infty}^{+\infty} = \lambda_0 \pi a$$

(e) Again using $dV = r^2 \sin\theta dr d\theta d\phi$ for a spherical charge distribution, the total charge in the cloud is

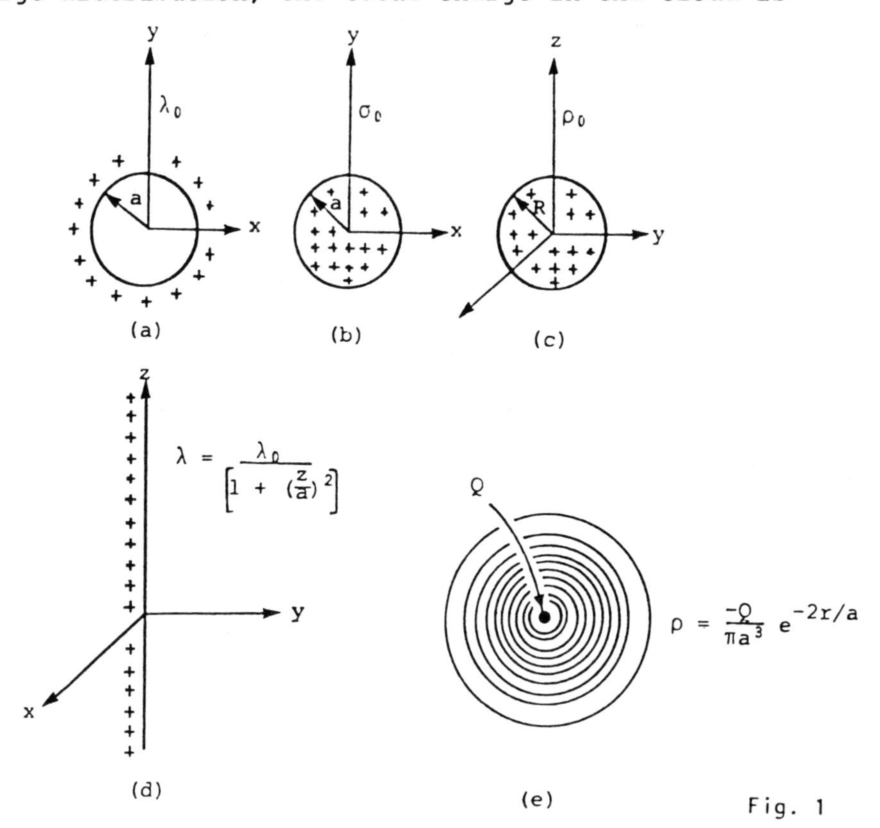

Fig. 1

Charge distributions: (a) Uniformly distributed line charge on a circular loop. (b) Uniformly distributed surface charge on a circular disk. (c) Uniformly distributed volume charge throughout a sphere. (d) Non-uniform line charge distribution. (e) Smooth radially dependent volume charge distribution throughout all space, as a simple model of the electron cloud around the positively charged nucleus of the hydrogen atom.

$$q = \int_V \rho dV$$

$$= -\int_{r=0}^{\infty} \int_{\theta=0}^{\pi} \int_{\phi=0}^{2\pi} \frac{Q}{\pi a^3} e^{-2r/a} r^2 \sin\theta dr\, d\theta\, d\phi$$

$$= -\int_{r=0}^{\infty} \frac{4Q}{a^3} e^{-2r/a} r^2 dr$$

$$= -\frac{4Q}{a^3}\left(-\frac{a}{2}\right)e^{-2r/a}\left[r^2 - \frac{a^2}{2}\left(-\frac{2r}{a} - 1\right)\right]\Bigg|_{r=0}^{\infty}$$

$$= -Q$$

COULOMB'S LAW

• **PROBLEM 2-4**

In the figure shown below, determine the resultant force R acting on the charge q_3 by the charges q_1 and q_2. Given:

$$\frac{1}{4\pi\varepsilon_0} = 9.0 \times 10^9 \text{N-m}^2/\text{coul}^2.$$

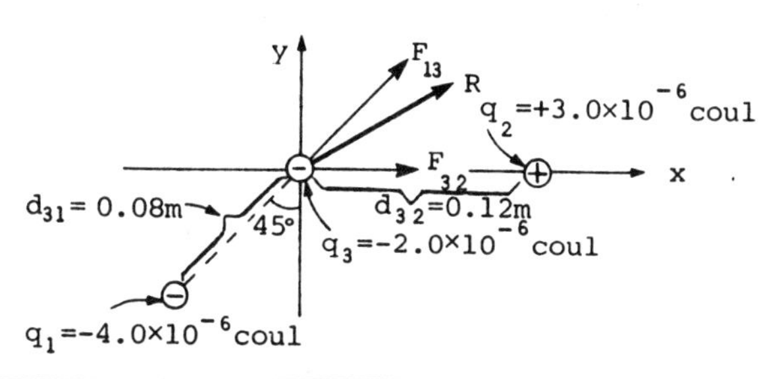

Solution: Since like charges repel and unlike charges attract, the forces acting on q_3 (i.e., F_{13} and F_{32}) are obtained as shown in the figure above.

Now, by knowing the Coulomb's Law, i.e.,

$$F = \frac{1}{4\pi\varepsilon_0}\frac{q_1 q_2}{d^2}$$

where d is the distance between q_1 and q_2, $|F_{13}|$ and $|F_{32}|$ are calculated as follows:

$$|F_{13}| = \frac{1}{4\pi\varepsilon_0} = \frac{q_1 q_3}{d_{31}{}^2}$$

$$= \frac{9.0\times10^9[\text{N-m}^2/\text{coul}^2]\times4.0\times10^{-6}[\text{coul}]\times2.0\times10^{-6}[\text{coul}]}{(0.08[\text{m}])^2}$$

$$= 11.25 \text{ N}$$

and

$$|F_{32}| = \frac{9.0\times10^9[\text{N-m}^2/\text{coul}^2]\times2.0\times10^{-6}[\text{coul}]\times3.0\times10^{-6}[\text{coul}]}{(0.12[\text{m}])^2}$$

$$= 3.75 \text{ N}$$

Now, in order to calculate the resultant force R, all the x and y components of F_{13} and F_{32} must be considered. (See Fig. 1)

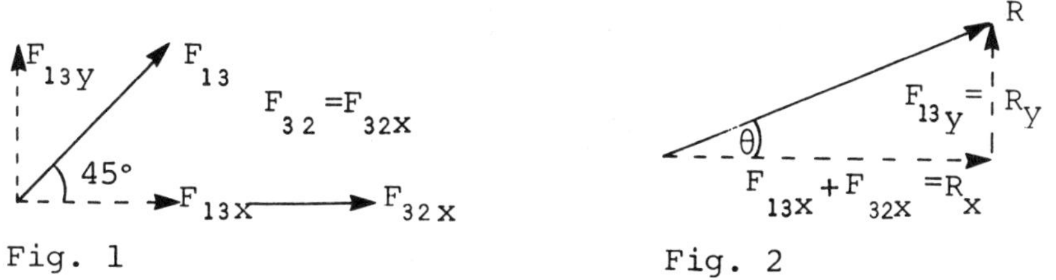

Fig. 1 Fig. 2

Hence: $R_x = F_{13x} + F_{32x}$

$$= F_{13}\cos\theta + F_{32}$$

$$= [11.25(\cos 45^0) + 3.75]N = 11.70 \text{ N}$$

$$R_y = F_{13y} + F_{32y}$$

$$= F_{13}\sin\theta + 0$$

$$= [11.25(\cos 45^0)]N = 7.95 \text{ N}$$

(See Fig. 2)

and the magnitude of the resultant force R =

$$\sqrt{(R_x)^2+(R_y)^2} = \sqrt{(11.70)^2+(7.95)^2} = 14.15 \text{ N}$$

and

$$\theta = \tan^{-1}\left(\frac{R_y}{R_x}\right) = \tan^{-1}\left(\frac{7.95}{11.70}\right) = \tan^{-1}(0.68) = 34.20^0$$

• PROBLEM 2-5

Three point-charges are located at the corners of a rectangle in free space, as shown in the figure. The scalar values of the charges are: $Q_1 = +3 \times 10^{-6}$C, $Q_2 = -2 \times 10^{-6}$C, $Q_3 = +5 \times 10^{-6}$C. The dimensions of the rectangle are 3 × 4 cm. Find $\bar{F}_3$, the force on Q_3.

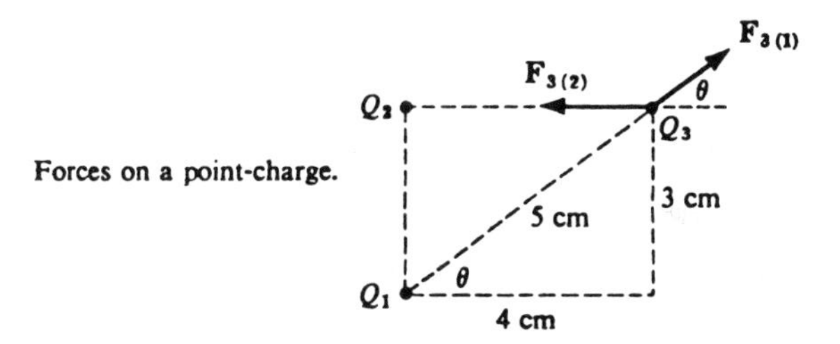

Forces on a point-charge.

Solution: Establish a Cartesian reference frame with origin at 1 and with x and y in the usual directions.

$$\bar{F}_3 = \bar{F}_{3\,(1)} + \bar{F}_{3\,(2)}$$

From Coulomb's Law

$$\bar{F}_{3\,(1)} = \frac{Q_3Q_1}{4\pi\varepsilon_0 r_{13}^2}(\bar{I}_x\cos\theta + \bar{I}_y\sin\theta)$$

$$= \frac{3 \times 5 \times 10^{-12}}{25 \times 10^{-4}(10^{-9}/9)}\left(\frac{4}{5}\,\bar{I}_x + \frac{3}{5}\,\bar{I}_y\right)$$

$$= 10.8(4\,\bar{I}_x + 3\,\bar{I}_y) = 43.2\,\bar{I}_x + 32.4\,\bar{I}_y$$

Also

$$F_{3(2)} = \frac{Q_3Q_2}{4\pi\varepsilon_0 r_{23}^2}\bar{I}_x$$

$$= \frac{-5 \times 2 \times 10^{-12}}{16 \times 10^{-4}(10^{-9}/9)}\bar{I}_x$$

$$= -\frac{900}{16}\bar{I}_x = -56.25\,\bar{I}_x$$

Therefore

$$F_3 = (43.2 - 56.3)\bar{I}_x + 32.4\,\bar{I}_y$$

$$= (-13.1\,\bar{I}_x + 32.4\,\bar{I}_y)\text{ N}$$

The resultant is 35.1 N (newtons) at an angle of 112°, assuming that the given charges and displacements are integers.

• PROBLEM 2-6

If a sphere of radius a is earthed and positive charges e, e' are placed on opposite sides of the sphere, at distances 2a, 4a respectively from the centre and in a straight line with it, show that the charge e' is repelled from the sphere if $e' < 25e/144$.

Solution: Let the line of the charges be taken as the x-axis and let the abscissae of the charges e and e' be -2a and 4a respectively. The images of these charges will be charges $-\frac{1}{2}e$ and $-\frac{1}{4}e'$ respectively situated at $-\frac{1}{2}a$ and $\frac{1}{4}a$. The force on the charge e' in the direction of the x-axis due to the electrification of the sphere and the charge e will be the same as that due to the images and the charge, namely,

$$\frac{ee'}{(6a)^2} - \frac{ee'}{2(4a + \frac{1}{2}a)^2} - \frac{e'^2}{4(4a - \frac{1}{4}a)^2};$$

this will be a repulsion if it is positive, i.e. if

$$\frac{4e'}{15^2} < e\left(\frac{1}{36} - \frac{2}{81}\right) = \frac{e}{324},$$

giving $$e' < \frac{25e}{144}.$$

• PROBLEM 2-7

A negative point charge of 10^{-6} coulomb is situated in air at the origin of a rectangular coordinate system. A second negative point charge of 10^{-4} coulomb is situated on the positive x axis at a distance of 50 cm from the origin. What is the force on the second charge?

Solution: By Coulomb's law the force

$$\bar{F} = i \frac{(-10^{-6})(-10^{-4})}{4\pi \times 0.5^2 \times 10^{-9}/36\pi}$$

$$= +i3.6 \quad \text{newtons}$$

That is, there is a force of 3.6 newtons (0.8 lb) in the positive x direction on the second charge.

• PROBLEM 2-8

The relatively heavy pith ball in the figure has a mass of 1 gram and has acquired 1 percent of the charge on a thin, ebonite rod by contact. If the charge on the rod is concentrated at one end and the distance of the pith ball from that end is 0.1 meter, what is the charge on the pith ball and the tension in the string?

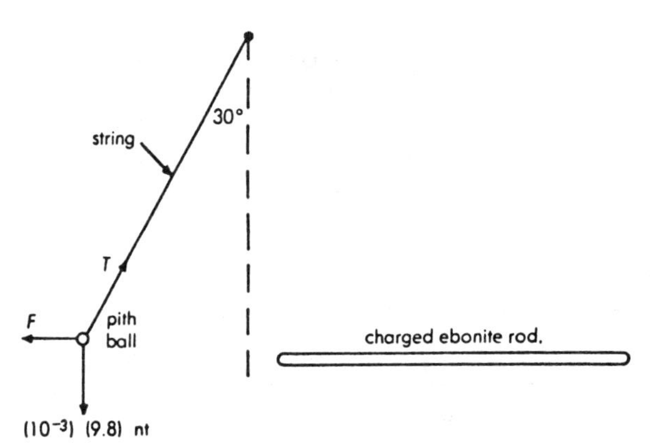

Pith ball repelled by charged rod.

Solution: Since the pith ball is in mechanical equilibrium, the sum of the forces acting on it must be zero. In the vertical direction,

$$T \cos \theta - mg = 0$$

$$T \cos \theta = mg \qquad (1)$$

and in the horizontal direction,

$$T \sin \theta - F = 0$$

$$T \sin \theta = F \qquad (2)$$

Dividing Eq. (2) by Eq. (1) yields

$$F = mg \tan \theta.$$

Thus, the force of repulsion on the pith ball is,

$$F = mg \tan \theta = (10^{-3})(9.8) \tan 30°$$

$$F = (5.65)(10^{-3})\text{nt}$$

The tension in the string is $T = F/\sin 30° = 11.3 \times 10^{-3}$nt. If the charge on the rod acts like a point charge in space, then Coulomb's law holds, and

$$F = \frac{1}{4\pi\varepsilon_0} \frac{qq'}{r^2}$$

$$5.65(10^{-3}) = 9(10^9) \frac{\left(\frac{q}{100}\right) q}{(.1)^2}$$

$$q = 7.9 \times 10^{-7} \text{ coulomb.}$$

The charge on the rod is 7.9×10^{-7} coulombs and the charge on the pith ball is 7.9×10^{-9} coulombs. These results, of course, are only approximate. Coulomb's law as stated above holds only for point charges in a vacuum.

• PROBLEM 2-9

Imagine a long, thin stick (shown in the figure) with a uniform distribution of excess charge on it. Suppose the total excess charge on the stick is Q. What will be the force of these charges on a charge q at a distance a from the stick along a line through the stick?

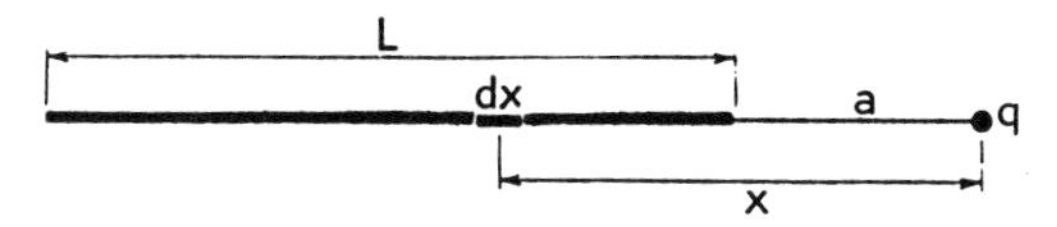

Calculation of force on a charge q due to a continuous linear distribution of charge. Fig. 1

Solution: Since the charge Q is distributed uniformly on the stick, the force F due to this distributed charge can be calculated by finding the elemental force dF due the elemental charge dQ and then integrating over the entire charge distribution. Thus what is needed is to find an expression that allows the summing up of each differential piece of Q, keeping track of its distance from q. A convenient way is to establish a representative element of charge dQ at a distance x from q. The force on q due to this element will be

$$dF = \frac{q}{4\pi\varepsilon_0} \frac{dQ}{x^2}$$

In order to integrate this, it is necessary to relate the size of dQ to the element dx. This is done by using the linear density $\mu = dQ/dx$, or

$$dQ = \mu \; dx$$

This gives the amount of charge dQ in a length dx. The equation is now

$$dF = \frac{q\mu}{4\pi\varepsilon_0} \frac{dx}{x^2}$$

which can be integrated. In the simple case of uniform density, μ is a constant, Q/L; so the integration is as follows:

$$F = + \frac{qQ}{4\pi\varepsilon_0} \frac{1}{a(L + a)} \quad \text{newtons.}$$

The positive sign indicates that the force is repulsive when q and Q have the same sign.

• PROBLEM 2-10

A semi-circular ring lying in the y-z plane has a charge density $\rho_\ell = \rho_0 \cos\theta'$ coul/m, where θ' is the angle measured from the z-axis as shown in the figure.
a) Find E for points (x,0,0) along the x-axis.
b) Show that for x>>R, the electric field is like that of a dipole, i.e., depends on the cube of the displacement from the charge distribution.

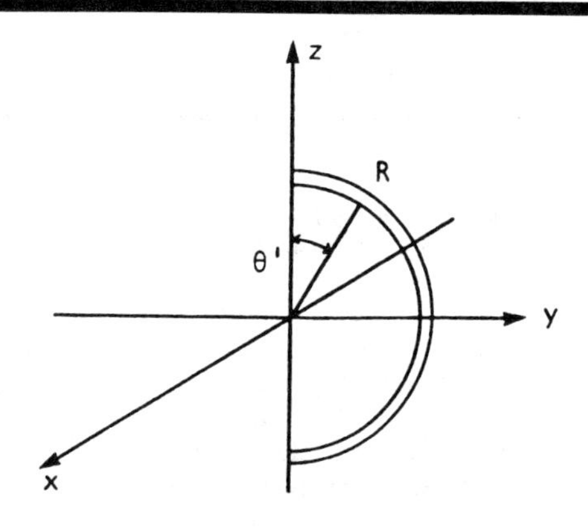

Solution: From Coulomb's law for a line charge distribution,

$$E(x,0,0) = \frac{1}{4\pi\varepsilon_0} \int \frac{\rho(r')(r - r')}{|r - r'|^3} d\ell'$$

From the figure, r', the distance from the origin to the charge distribution is given by

$$r' = \bar{a}_z R \cos\theta' + \bar{a}_y R \sin\theta';$$

r is just the distance from the origin to a point of interest or $r = \bar{a}_x x$; $d\ell'$ is just the incremental arc length given by $Rd\theta'$. Substituting these into the integral, the following is obtained:

$$E(x,0,0) = \frac{1}{4\pi\varepsilon_0} \int_0^{\pi} \frac{\rho_0 \cos\theta' [\bar{a}_x x - \bar{a}_z R\cos\theta' - \bar{a}_y R\sin\theta'] Rd\theta'}{\left(\sqrt{x^2 + R^2\cos^2\theta' + R^2\sin^2\theta'}\right)^3}$$

$$E(x,0,0) = \frac{\rho_0}{4\pi\varepsilon_0} \cdot \frac{R}{(\sqrt{x^2 + R^2})^3} \left[\int_0^{\pi} \bar{a}_x x \cos\theta' d\theta \right.$$

$$\left. - \int_0^{\pi} \bar{a}_z R \cos^2\theta' d\theta' - \int_0^{\pi} \bar{a}_y R \sin\theta' \cos\theta' d\theta' \right]$$

but

$$\int_0^{\pi} \bar{a}_x \cos\theta' d\theta' = 0$$

$$\int_0^{\pi} \bar{a}_z R \cos^2\theta' d\theta' = a_z R \frac{\pi}{2}$$

$$\int_0^{\pi} \bar{a}_y R \sin\theta' \cos\theta' d\theta' = 0$$

Therefore, the result is

$$E(x,0,0) = \frac{\rho_0}{4\pi\varepsilon_0} \cdot \frac{R}{(x^2 + R^2)^{3/2}} \cdot (- \bar{a}_z R \frac{\pi}{2})$$

For $x \gg R$

$$E(x,0,0) = \frac{\pi}{2} \cdot \frac{\rho_0 R^2}{4\pi\varepsilon_0} \cdot \frac{1}{x^3} (-\bar{a}_z)$$

proving that far away from the semi-circular ring, E is like that of a dipole.

ELECTRIC FIELD

• PROBLEM 2-11

A charge of -10^{-9}C is located at the origin in free space. What charge must be located at (2,0,0) to cause E_x to be zero at (3,1,1)?

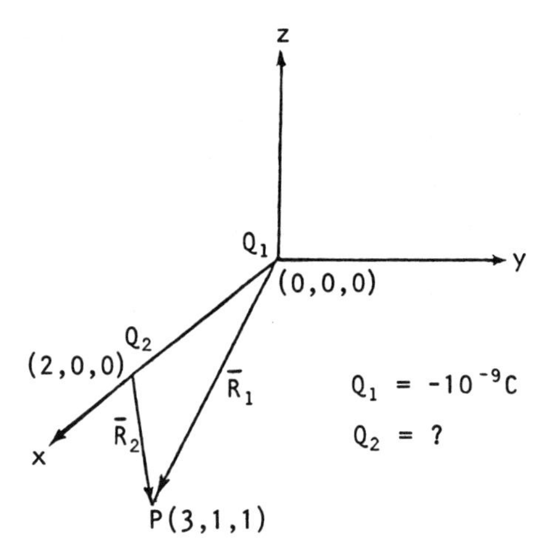

Solution: Electric field is a directed quantity. Vector method is convenient to use. Locate the charges as shown. Draw vectors $\bar{R}_1$ and $\bar{R}_2$ from Q_1 and Q_2 to the point P(3,1,1) in space, then

$$\bar{R}_1 = (3 - 0)\bar{a}_x + (1 - 0)\bar{a}_y + (1 - 0)\bar{a}_z$$

$$\bar{R}_2 = (3 - 2)\bar{a}_x + (1 - 0)\bar{a}_y + (1 - 0)\bar{a}_z$$

where $\bar{a}_x$, $\bar{a}_y$ and $\bar{a}_z$ are unit vectors in the x-, y- and z-directions respectively, and

$$|\bar{R}_1| = \sqrt{3^2 + 1 + 1} = \sqrt{11}, \quad |\bar{R}_2| = \sqrt{3}.$$

The unit vectors in the directions of $\bar{R}_1$ and $\bar{R}_2$ are respectively

$$\bar{a}_{R_1} = \frac{3\bar{a}_x + \bar{a}_y + \bar{a}_z}{\sqrt{11}}, \quad \bar{a}_{R_2} = \frac{\bar{a}_x + \bar{a}_y + \bar{a}_z}{\sqrt{3}}$$

By Coulomb's law,

$$\bar{E} = \frac{Q}{4\pi\varepsilon_0 R^2}\bar{a}_R,$$

and $\varepsilon_0 = \dfrac{1}{36\pi \cdot 10^9}\ \dfrac{F}{M}$. The electric field at P is

$$\bar{E}_p = \frac{-10^{-9}(3\bar{a}_x + \bar{a}_y + \bar{a}_z)}{10^{-9}/9\ (11)^{3/2}} + \frac{Q_2(\bar{a}_x + \bar{a}_y + \bar{a}_z)}{10^{-9}/9\ (3)^{3/2}}$$

Now, if E_x at (3,1,1) equals zero is required, then the x-component of E_p must be set to zero. Therefore

$$\frac{-3 \cdot 10^{-9}}{(11)^{3/2}} + \frac{Q_2}{(3)^{3/2}} = 0$$

Solving for Q_2,

$$Q_2 = \frac{3 \cdot 3\sqrt{3} \cdot 10^{-9}}{11\sqrt{11}} = 0.425 \cdot 10^{-9}\text{C}.$$

• PROBLEM 2-12

Consider a circular line charge in the x-y plane in which $\rho_\ell = k\sin\phi$. Calculate the electric field on the z-axis.

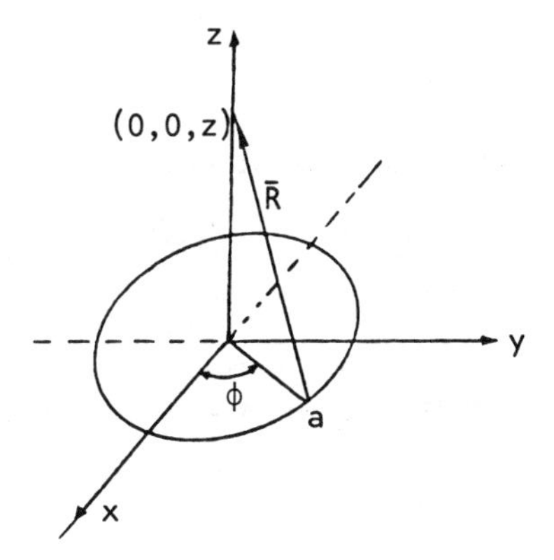

Solution: Let the point P(0,0,z) be the point in question. Draw a vector $\bar{R}$ from this point to any point on the ring. Then $\bar{R} = -a\ \bar{a}_r - z\ \bar{a}_z$, where a is the radius of the ring, and

$$\bar{a}_R = \frac{-a\ \bar{a}_r + z\ \bar{a}_z}{\sqrt{a^2 + z^2}}$$

Let $x = a\cos\phi$, $y = a\sin\phi$, so that

$$x^2 + y^2 = a^2.$$

The total line charge $dQ = \rho_\ell a\ d\phi$, $dl = a\ d\phi$. Then

$$d\bar{E} = \rho_\ell a\ d\phi \bar{a}_R / 4\pi\varepsilon_0 R^2$$

Thus

$$\bar{E} = \int_0^{2\pi} \frac{\rho_\ell a \, d\phi \bar{a}_R}{4\pi\varepsilon_0 R^2} = \int_0^{\pi} \frac{ka\sin\phi d\phi}{4\pi\varepsilon_0} \left(\frac{-a \, \bar{a}_r + z \, \bar{a}_z}{(a^2 + z^2)^{3/2}} \right)$$

Let $\bar{a}_r = \cos\phi\bar{a}_x - \sin\phi\bar{a}_y$, then

$$\bar{E} = \int_0^{2\pi} \frac{ka\sin\phi d\phi}{4\pi\varepsilon_0} \left[\frac{-a\cos\phi\bar{a}_x - a\sin\phi\bar{a}_y + z \, \bar{a}_z}{(a^2 + z^2)^{3/2}} \right]$$

$$\bar{E} = \int_0^{2\pi} \frac{kad\phi}{4\pi\varepsilon_0 (a^2 + z^2)^{3/2}} \left[-\frac{a}{2} \sin 2\phi \bar{a}_x \right.$$

$$\left. - \frac{a}{2} (1 - \cos 2\phi) \bar{a}_y + z\sin\phi\bar{a}_z \right]$$

Sin 2ϕ, cos 2ϕ, and sinϕ all integrate to 0 over an entire period, therefore only the terms without the sine and cosine functions remain in the integral or,

$$\bar{E} = - \int_0^{2\pi} \frac{k \, a^2 d\phi \bar{a}_y}{8\pi\varepsilon_0 (a^2 + z^2)^{3/2}} = - \frac{ka^2}{4\varepsilon_0 (a^2 + z^2)^{3/2}} \bar{a}_y$$

The electric field is directed in the negative y-direction.

• PROBLEM 2-13

Suppose there is a plane area on which charges are distributed uniformly with a surface density σ coul/m^2. Calculate the field at a point P a distance a from the plane as shown in Fig. 1. Assume that the dimensions of the plane are much greater than a.

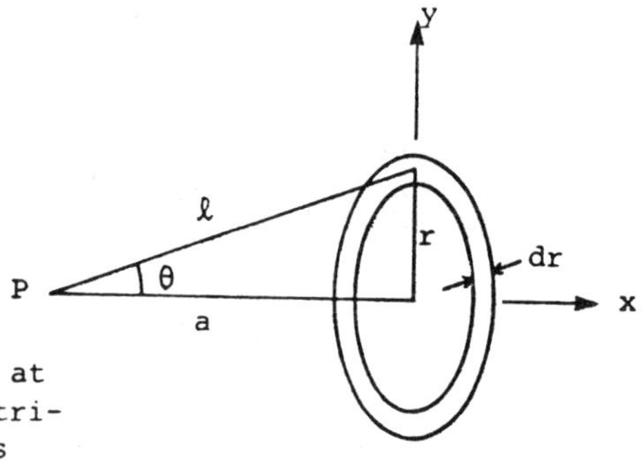

The electric field at P from a plane distribution of charge is obtained by integrating the contributions from concentric ring.

Fig. 1

Solution: Use

$$E = \frac{1}{4\pi\varepsilon_0} \int \frac{dq}{r^2} \hat{r}$$

to add up the vector contributions of all charges at the point P. Begin by calculating the contribution from the ring of charge of radius r and width dr and then integrate for all such rings that make up the total plane charge distribution. The calculation of the contribution of charges on the ring is really a two-dimensional integration problem, but because of the symmetry it can be reduced to a simple summation. The contribution of an elementary charge on the ring to the field at P makes an angle θ with the x axis as shown on the figure. Because of symmetry, however, components of $\bar{E}$ perpendicular to the x axis cancel. Thus one needs to consider only the x components of $\bar{E}$ at P. Then the field at P due to the ring of charge is

$$dE = \frac{1}{4\pi\varepsilon_0} \frac{\sigma 2\pi r dr}{\imath^2} \cos\theta$$

In order to reduce this to a single variable, substitute $a \tan\theta = r$ and $a/\cos\theta = a \sec\theta = \imath$. By differentiation $dr = a \sec^2\theta \, d\theta$. Making these substitutions yields,

$$dE = \frac{\sigma}{2\varepsilon_0} \frac{\tan\theta \sec^2\theta \cos\theta}{\sec^2\theta} d\theta = \frac{\sigma}{2\varepsilon_0} \sin\theta \, d\theta$$

This is the contribution to the field at P from the ring chosen. To obtain the total field at P from all rings making up the plane charge distribution, integrate this expression over the entire plane. The limits of integration are from $\theta = 0$ to $\theta = \pi/2$. Thus,

$$E = \frac{\sigma}{2\varepsilon_0} \int_0^{\pi/2} \sin\theta \, d\theta = -\frac{\sigma}{2\varepsilon_0} [\cos\theta]_0^{\pi/2}$$

$$= \frac{\sigma}{2\varepsilon_0} \text{ newtons/coul}$$

The resultant field is in the x direction and is independent of the distance from the plane as long as the plane is very large compared with a.

• PROBLEM 2-14

A ring-shaped conductor of radius a carries a total charge Q as shown in the figure. Find the electric field at a point a distance x from the center, along the line perpendicular to the plane of the ring, through its center.

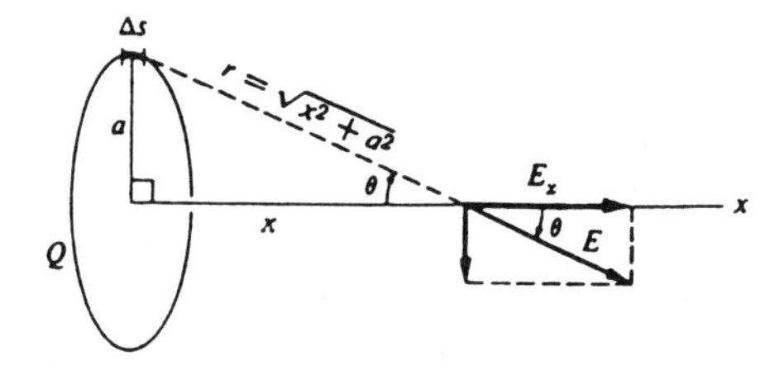

Electric field due to ring of charge.

Solution: Considering a small segment Δs of the ring, note that since the circumference is $2\pi a$, the charge Δq on this segment is

$$\Delta q = \frac{Q\Delta s}{2\pi a}$$

At the point P, this element produces an electric field of magnitude

$$\bar{E} = \left(\frac{1}{4\pi\varepsilon_0}\right)\left(\frac{\Delta q}{r^2}\right) = \left(\frac{1}{4\pi\varepsilon_0}\right)\left(\frac{Q\Delta s}{2\pi a} \cdot \frac{1}{x^2 + a^2}\right).$$

The component of the field along the x-axis is given by

$$E_x = E \cos \theta = \frac{Q\Delta s}{8\pi^2\varepsilon_0 a(x^2 + a^2)} \cdot \frac{x}{\sqrt{x^2 + a^2}}$$

$$= \frac{Qx\Delta s}{8\pi^2\varepsilon_0 a(x^2 + a^2)^{3/2}} \qquad (1)$$

To sum the contributions from all segments it is needed only to add up their lengths, since the coefficient of Δq in Eq. (1) is the same for all segments. Thus, the total field component along the axis is

$$E_x = \frac{\frac{1}{4\pi\varepsilon_0} Qx}{2\pi a(x^2 + a^2)^{3/2}} (2\pi a)$$

$$= \left(\frac{1}{4\pi\varepsilon_0}\right) \frac{Qx}{(x^2 + a^2)^{3/2}}. \qquad (2)$$

In principle, this calculation should also be performed for the components perpendicular to the axis, but it is easy to see from symmetry that these add to zero.

Equation (2) shows that at the center of the ring ($x = 0$) the total field is zero. When $x \gg a$, Eq. (2) becomes approximately equal to

$$\left(\frac{1}{4\pi\varepsilon_0}\right) Q/x^2,$$

corresponding to the fact that at distances much greater than the dimensions of the ring it appears as a point charge.

• PROBLEM 2-15

Compute the field intensity midway between two point charges: number 1 of +30 μcoulombs (microcoulombs), number 2 of +40 μcoulombs, when the charges are 10 cm apart in air.

Solution: Since the two charges exert repelling forces from opposite directions on the test charge, the field intensity will be equal to the difference of the magnitudes of the two forces. Taking the field in the direction of the 30 μcoulomb charge as positive, the result is

$$\bar{E} = \frac{-q_1}{4\pi\varepsilon_0 r_{1t}^2} + \frac{q_2}{4\pi\varepsilon_0 r_{2t}^2} = \frac{-1}{4\pi\varepsilon_0}\left[\frac{30 \times 10^{-6}}{(.05)^2} - \frac{40 \times 10^{-6}}{(.05)^2}\right],$$

$$\bar{E} = +8.99 \times 10^9 [0.012 - 0.016]$$

or $\bar{E} = 36 \times 10^6$ newtons/coulomb,

with $\bar{E}$ directed from the point of measurement toward charge 1.

• PROBLEM 2-16

The figure shows eight point charges situated at the corners of a cube. Find the electric field intensity at each point charge, due to the remaining seven point charges.

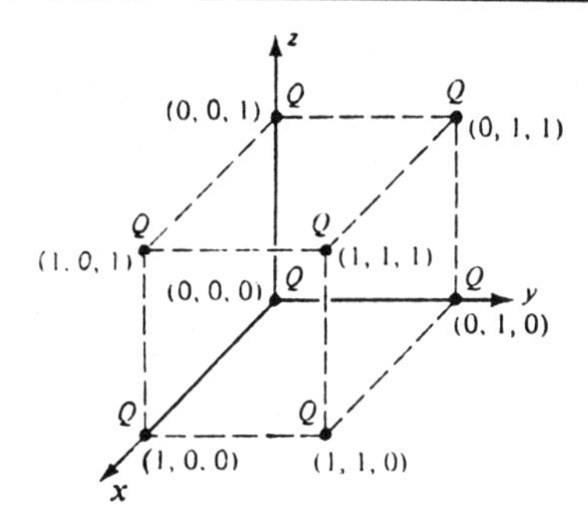

A cubical arrangement of point charges.

Solution: First note that the electric field intensity at a point $B(x_2, y_2, z_2)$ due to a point charge Q at point $A(x_1, y_1, z_1)$ is given by

$$\bar{E}_B = \frac{Q}{4\pi\varepsilon_0 (AB)^2}\,\bar{i}_{AB} = \frac{Q}{4\pi\varepsilon_0 (AB)^2}\,\frac{\overline{AB}}{(AB)} = \frac{Q(\overline{AB})}{4\pi\varepsilon_0 (AB)^3}$$

$$= \frac{Q}{4\pi\varepsilon_0}\,\frac{(x_2 - x_1)\bar{i}_x + (y_2 - y_1)\bar{i}_y + (z_2 - z_1)\bar{i}_z}{[(x_2 - x_1)^2 + (y_2 - y_1)^2 + (z_2 - z_1)^2]^{3/2}} \qquad (1)$$

Now consider the point (1,1,1). Applying (1) to each of the charges at the seven other points and using superposition the electric field intensity at the point (1,1,1) is

$$E_{(1,1,1)} = \frac{Q}{4\pi\varepsilon_0}\left[\frac{\bar{i}_x}{(1)^{3/2}} + \frac{\bar{i}_y}{(1)^{3/2}} + \frac{\bar{i}_z}{(1)^{3/2}}\right.$$

$$+ \frac{\bar{i}_y + \bar{i}_z}{(2)^{3/2}} + \frac{\bar{i}_z + \bar{i}_x}{(2)^{3/2}}$$

$$\left. + \frac{\bar{i}_x + \bar{i}_y}{(2)^{3/2}} + \frac{\bar{i}_x + \bar{i}_y + \bar{i}_z}{(3)^{3/2}}\right]$$

$$= \frac{Q}{4\pi\varepsilon_0}\left(1 + \frac{1}{\sqrt{2}} + \frac{1}{3\sqrt{3}}\right)(\bar{i}_x + \bar{i}_y + \bar{i}_z)$$

$$= \frac{3.29Q}{4\pi\varepsilon_0}\left(\frac{\bar{i}_x + \bar{i}_y + \bar{i}_z}{\sqrt{3}}\right)$$

Noting that $(\bar{i}_x + \bar{i}_y + \bar{i}_z)/\sqrt{3}$ is the unit vector directed from (0,0,0) to (1,1,1), the electric field intensity at (1,1,1) is directed diagonally away from (0,0,0) with a magnitude equal to $\frac{3.29Q}{4\pi\varepsilon_0}$ N/C. From symmetry considerations, it then follows that the electric field intensity at each point charge, due to the remaining seven point charges, has a magnitude $\frac{3.29Q}{4\pi\varepsilon_0}$ N/C, and it is directed away from the corner opposite to that charge.

● **PROBLEM 2-17**

Two point charges of equal magnitude Q and of opposite polarities are placed in the air to form an electric dipole of dipole moment $p = a \cdot Q$. Find (a) the electric field $\bar{E}$ at a point P, a distance R along the perpendicular bisector as shown in the figure, assuming $R >> a$, and (b) draw the pattern of lines of force due to this configuration.

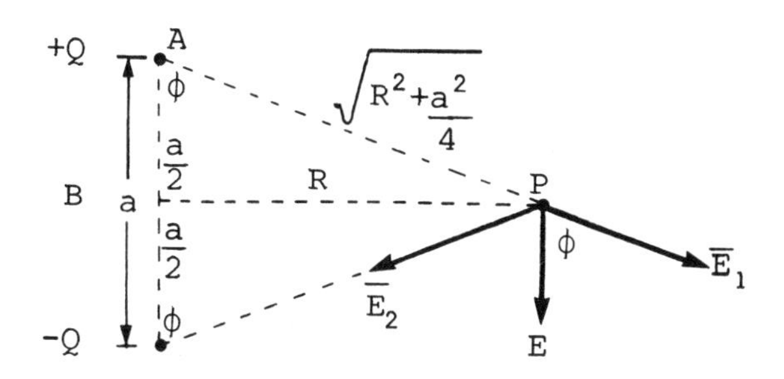

Solution: The dipole moment of an electric dipole

= The charge × the distance between these two point charges

Hence, the distance between two charges is 'a', as shown in the figure.

(a) The electric field at point $P = \bar{E} = \bar{E}_1 + \bar{E}_2$ (The vector sum) and the electric field due to a charge q at a distance r is given by:

$$E = \frac{1}{4\pi\varepsilon_0 \cdot \varepsilon_r} \cdot \frac{q}{r^2}$$

Hence, the magnitude of the electric field $\bar{E}_1$ and $\bar{E}_2$ due to the point charges +Q and -Q, respectively, is:

$$|\bar{E}_1| = |\bar{E}_2| = \frac{1}{4\pi\varepsilon_0 \cdot 1} \times \frac{Q}{R^2 + \frac{a^2}{4}} \quad \text{(Note: } \varepsilon_r = 1 \text{ for air)}$$

Now, the resultant field $|\bar{E}|$ at point P is computed using the law of parallelogram of forces as follows:

$$|\bar{E}| = 2 \cdot |\bar{E}_1| \cdot \cos\phi$$

where

$$\cos\phi = \frac{\frac{a}{2}}{\sqrt{R^2 + \frac{a^2}{4}}} \quad \text{(from } \Delta ABP\text{, see the figure)}$$

Hence, the resultant field is:

$$|\bar{E}| = 2 \times \frac{Q}{4\pi\varepsilon_0 (R + \frac{a^2}{4})} \times \frac{\frac{a}{2}}{\sqrt{R^2 + \frac{a^2}{4}}}$$

$$= \frac{2aQ}{4\pi\varepsilon_0 \times (R^2 + a^2/4)^{3/2}}$$

$$\cong \frac{2aQ}{2 \cdot 4\pi\varepsilon_0} \times \frac{1}{R^3}$$

$$\cong \frac{aQ}{4\pi\varepsilon_0 R^3} \cong \frac{p}{4\pi\varepsilon_0 \times R^3}$$

(b) The sketch:

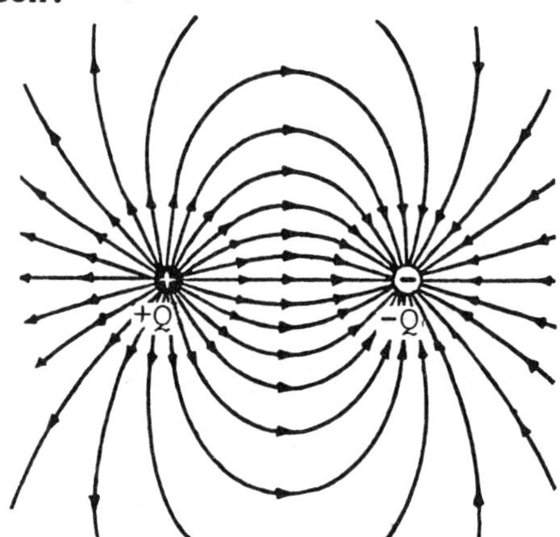

• PROBLEM 2-18

A uniform surface charge density ρ_s is distributed over a cylindrical surface r = a, extending from z = -h to z = h. Find the electrical field intensity in free space at (0,0,k). As a check on your work, if k = 2h = 2a = 2, $\bar{E} = 0.1954\rho_s/\varepsilon_0 \bar{a}_z$. Refer to the figure shown.

Solution: Draw a cylindrical surface such that its axis coincides with the z-axis. Draw a vector $\bar{R}$ from (0,0,k) to a surface strip dz on the cylinder. Use cylindrical coordinates to get

$$\bar{R} = (0 - a)\,\bar{a}_r + (k - z)\,\bar{a}_z\,,$$

and

$$\bar{a}_R = \frac{-a\,\bar{a}_r + (k - z)\bar{a}_z}{\sqrt{a^2 + (k - z)^2}}$$

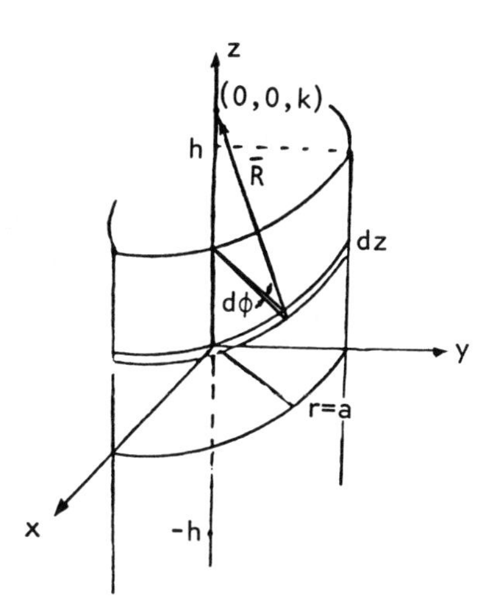

By symmetry, the electric field on the z-axis has only the z-component. To sum up the contributions due to the charge distribution, the integrations are to be carried out over the circumference and along the z-axis. Therefore

$$\bar{E} = \int_0^{2\pi} \int_{-h}^{h} \frac{a\rho_s dz d\phi [-a\,\bar{a}_r + (k - z)\bar{a}_z]}{4\pi\varepsilon_0 [a^2 + (k - z)^2]^{3/2}}$$

$$= 2\pi \int_{-h}^{h} \frac{a\rho_s\; dz(k - z)\bar{a}_z}{4\pi\varepsilon_0 [a^2 + (k - z)^2]^{3/2}}$$

$$= \frac{\rho_s a}{2\varepsilon_0} \int_{-h}^{h} \frac{(k - z)\bar{a}_z\; d_z}{[a^2 + (k - z)^2]^{3/2}} \; .$$

Let $k - z = p$, $dz = -dp$. For $z = -h$, $p = k + h$; $z = h$, $p = k - h$, therefore

$$\bar{E} = \frac{\rho_s a}{2\varepsilon_0} \int_{k+h}^{k-h} \frac{-pdp\bar{a}_z}{(a^2 + p^2)^{3/2}} = \frac{\rho_s a}{2\varepsilon_0} \left. \frac{1}{\sqrt{a^2 + p^2}} \right|_{k+h}^{k-h} \bar{a}_z$$

$$= \frac{\rho_s a}{2\varepsilon_0} \left[\frac{1}{\sqrt{a^2 + (k - h)^2}} - \frac{1}{\sqrt{a^2 + (k + h)^2}} \right] \bar{a}_z$$

If $k = 2h = 2a = 2$, then

$$\bar{E} = \frac{\rho_s a}{2\varepsilon_0} \left[\frac{1}{\sqrt{1 + 1}} - \frac{1}{\sqrt{1 + 9}} \right] = 0.1954\; \rho_s/\varepsilon_0\; \bar{a}_z$$

• **PROBLEM 2-19**

Find the electric field intensity about the finite line charge of uniform ρ_ℓ distribution along the z axis, as shown in Fig. 1. Use this result to find the electric field intensity about an infinite line charge.

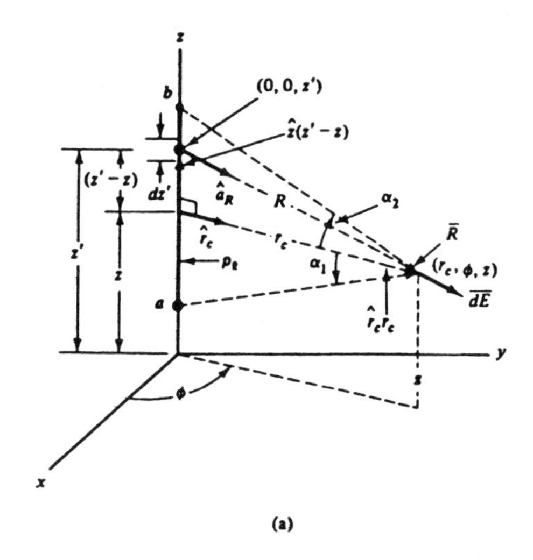

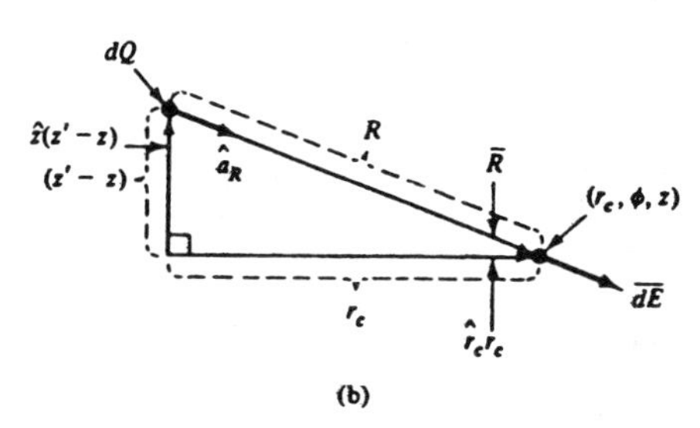

(a) Graphical construction to evaluate $\overline{dE}$ and thus $\bar{E}$ about a finite length of line charge of uniform ρ_ℓ. (b) Isolated view for evaluating $\hat{a}_R$.

Fig. 1

Solution: The differential electric field intensity $\overline{dE}$ can be found through

$$\overline{dE} = \frac{\rho_\ell d\ell}{4\pi\varepsilon_0 R^2}\,\hat{a}_R$$

Use the primed variables to locate points on the line of charge and the unprimed variables to locate the electric field point. Thus, point charge dQ is located at $P_{cyl}(r_c', \phi', z')$. With the aid of the graphical construction in Fig. 1, the following is obtained:

$$R = (r_c^2 + (z' - z)^2)^{1/2}$$

$$\bar{R} = \hat{r}_c r_c - \hat{z}(z' - z) = \hat{a}_R R$$

$$\hat{a}_R = \frac{\bar{R}}{R} = \frac{\hat{r}_c r_c - \hat{z}(z' - z)}{(r_c^2 + (z' - z)^2)^{1/2}}$$

$$d\ell = dz'$$

Substituting into the $\overline{dE}$ expression,

$$\overline{dE} = \frac{\rho_\ell dz'}{4\pi\varepsilon_0 (r_c^2 + (z' - z)^2)} \left[\frac{\hat{r}_c r_c - \hat{z}(z' - z)}{(r_c^2 + (z' - z)^2)^{1/2}} \right]$$

Now,

$$\bar{E} = \int_a^b \overline{dE} = \frac{\rho_\ell}{4\pi\varepsilon_0} \int_a^b \frac{(\hat{r}_c r_c - \hat{z}(z' - z))}{(r_c^2 + (z' - z)^2)^{3/2}} dz'$$

Over the range of the above integral, the only variable is z', while $\hat{r}_c$ is a constant, since the point of $\bar{E}$ is fixed for the integration over the line from a to b. The $\bar{E}$ integral can be rewritten as

$$\bar{E} = \frac{\rho_\ell}{4\pi\varepsilon_0} \left\{ \hat{r}_c r_c \int_a^b \frac{dz'}{(r_c^2 + (z' - z)^2)^{3/2}} \right.$$

$$\left. - \hat{z} \int_a^b \frac{(z' - z) dz'}{(r_c^2 + (z' - z)^2)^{3/2}} \right\}$$

The integrals found in the above expression are of the forms

$$\int \frac{dx}{(c^2 + x^2)^{3/2}} = \frac{x}{c^2 (c^2 + x^2)^{1/2}} \qquad (1)$$

$$\int \frac{x dx}{(c^2 + x^2)^{3/2}} = \frac{-1}{(c^2 + x^2)^{1/2}} \qquad (2)$$

Using (1) and (2), $\bar{E}$ becomes

$$\bar{E} = \frac{\rho_\ell}{4\pi\varepsilon_0} \left\{ \hat{r}_c r_c \frac{(z' - z)}{r_c^2 (r_c^2 + (z' - z)^2)^{1/2}} \Bigg|_a^b \right.$$

$$\left. + \hat{z} \frac{1}{(r_c^2 + (z' - z)^2)^{1/2}} \Bigg|_a^b \right\}$$

$$= \frac{\rho_\ell}{4\pi\varepsilon_0} \left\{ \frac{\hat{r}_c}{r_c} \left[\frac{(b - z)}{(r_c^2 + (b - z)^2)^{1/2}} \right. \right.$$

$$- \left.\frac{(a - z)}{(r_c^2 + (a - z)^2)^{1/2}}\right]$$

$$+ \hat{z}\left[\frac{1}{(r_c^2 + (b - z)^2)^{1/2}}\right.$$

$$\left.\left. - \frac{1}{(r_c^2 + (a - z)^2)^{1/2}}\right]\right\} \text{(Vm}^{-1}\text{)} \qquad (3)$$

In terms of α_1 and α_2, see Fig. 1a, eq. (3) becomes

$$\bar{E} = \frac{\rho_\ell}{4\pi\varepsilon_0}\left\{\frac{\hat{r}_c}{r_c}(\sin\alpha_2 + \sin\alpha_1) + \frac{\hat{z}}{r_c}(\cos\alpha_2\right.$$

$$\left. - \cos\alpha_1)\right\} \text{(Vm}^{-1}\text{)} \qquad (4)$$

The electric field intensity $\bar{E}$ about an infinite line of charge of uniform ρ_ℓ distributed along the z axis, can then be found.

In (3), let $b = +\infty$, $a = -\infty$ to obtain

$$\bar{E} = \frac{\rho_\ell}{2\pi\varepsilon_0 r_c}\hat{r}_c \text{ (Vm}^{-1}\text{)}$$

Also, let $\alpha_1 = \pi/2$ and $\alpha_2 = \pi/2$ in (4) to obtain

$$\bar{E} = \frac{\rho_\ell}{2\pi\varepsilon_0 r_c}\hat{r}_c \text{ (Vm}^{-1}\text{)} \qquad (5)$$

It should be noted that only the radial component exists for the infinite line of charge of uniform ρ_ℓ.

• PROBLEM 2-20

Find the electric field on the axis of a thin, uniformly charged disk of radius a and total charge q and then estimate the axial distance from the disk beyond which the disk may be regarded as a point charge if the greatest admissible error for E is 1%.

Solution: Let the axis of the disk be the z-axis with the origin at the center of the disk (Fig. 1). Using symmetry considerations, it can be concluded that $E_x = E_y = 0$. Dividing the disk into elementary rings of radius R and width dR, one can obtain the electric field due to a ring of charge and summing up all the

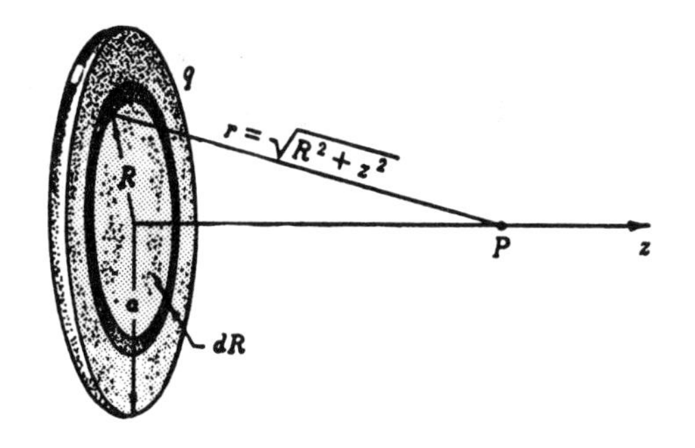

Calculation of the electric field on the axis of a charged disk. Fig. 1

rings from a radius of 0 to the maximum radius of a, the electric field for a disk of charge. For a ring of charge

$$E_z = \frac{1}{4\pi\varepsilon_0}\left[\frac{q'z}{(R^2 + z^2)^{3/2}}\right]$$

where q' = charge in the ring = $\sigma \cdot 2\pi R\, dR$

Therefore,

$$E_z = \frac{\sigma}{4\pi\varepsilon_0}\int_0^a \frac{z 2\pi R\, dR}{(R^2 + z^2)^{3/2}}$$

$$= -\frac{\sigma}{2\varepsilon_0}\frac{z}{\sqrt{R^2 + z^2}}\Bigg|_0^a = \frac{\sigma}{2\varepsilon_0}\left(1 - \frac{z}{\sqrt{a^2 + z^2}}\right),$$

and, since $\sigma\pi a^2 = q$,

$$\bar{E} = \frac{q}{2\pi\varepsilon_0 a^2}\left(1 - \frac{z}{\sqrt{a^2 + z^2}}\right)\bar{k}$$

(valid only for $z>0$ because after substituting the limits, $\sqrt{z^2} = +z$).

Expanding E in a power series for z, the following is obtained:

$$E = \frac{q}{2\pi a^2\varepsilon_0}\left[1 - \left(1 + \frac{a^2}{z^2}\right)^{-1/2}\right]$$

$$= \frac{q}{2\pi a^2\varepsilon_0}\left[1 - 1 + \frac{a^2}{2z^2} - \frac{1\cdot 3}{2\cdot 4}\frac{a^4}{z^4} + \dots\right]$$

$$= \frac{q}{4\pi\varepsilon_0 z^2} - \frac{q}{4\pi\varepsilon_0 z^2}\frac{3a^2}{4z^2} + \dots \quad .$$

The smallest z beyond which the field of the disk may be calculated from the point charge formula with an error smaller than 1% is therefore

$$\frac{3a^2}{4z_{min}^2} \approx 0.01, \quad \text{or} \quad z_{min} \approx 9a.$$

• PROBLEM 2-21

A uniform surface charge is distributed in a strip in the y = 0 plane, extending from x = -d/2 to x = d/2 and from z = -∞ to z = ∞ as shown in the figure. Find $\bar{E}$ at a general point (x,y,z).

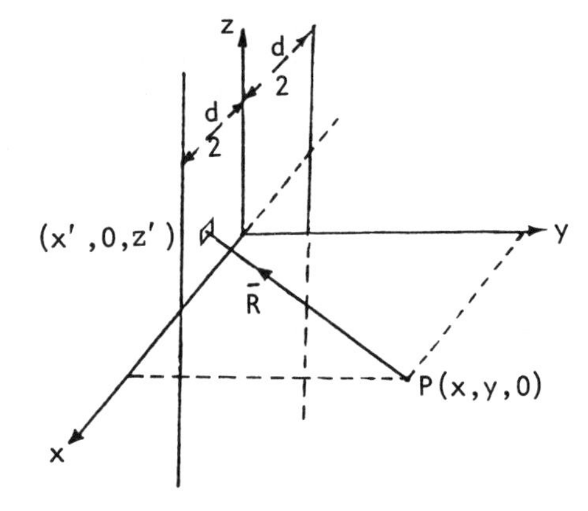

Solution: Let ρ_s be a uniform surface charge density in the y = 0 plane as shown. From symmetry, $\bar{E}$ does not vary with z. Therefore select a point on the z = 0 plane as P(x,y,0). Now, from an arbitrary point (x',0,z') on the charged sheet, draw a vector $\bar{R}$ to P(x,y,0). Then

$$\bar{R} = (x-x')\,\bar{a}_x + (y-0)\,\bar{a}_y + (0-z')\,\bar{a}_z .$$

The electric field intensity is then the sum of the charge contributions along x' and z' within the prescribed limits.

$$\bar{E} = \int_{-\infty}^{\infty}\int_{-d/2}^{d/2} \frac{\rho_s dx'dz'[(x - x')\bar{a}_x + y\bar{a}_y - z'\bar{a}_z]}{4\pi\varepsilon_0[(x - x')^2 + y^2 + z^2]^{3/2}}$$

As $\bar{E}$ is not a function of z or z', the z' $\bar{a}_z$ term must vanish. Therefore

$$\bar{E} = \int_{-\infty}^{\infty}\int_{-d/2}^{d/2} \frac{\rho_s\, dx'[(x - x')\bar{a}_x + y\,\bar{a}_y]dz'}{4\pi\varepsilon_0[(x - x')^2 + y^2 + z^2]^{3/2}}$$

$$= \int_{-d/2}^{d/2} \frac{2\rho_s \, dx'[(x - x')\bar{a}_x + y \, \bar{a}_y]}{4\pi\varepsilon_0[(x - x')^2 + y^2]}$$

Change the variables. Let

$$x - x' = u, \quad du = -dx'$$

at $x' = -d/2$, $u = x + d/2$; at $x' = d/2$, $u = x' - d/2$.

Then

$$\bar{E} = \frac{\rho_s}{2\pi\varepsilon_0} \int_{x + d/2}^{x - d/2} \frac{(u \, \bar{a}_x + y \, \bar{a}_y)du}{u^2 + y^2}$$

$$E = \frac{\rho_s}{2\pi\varepsilon_0} \int_{x + d/2}^{x - d/2} \frac{-u \, \bar{a}_x}{u^2 + y^2} + \int_{x + d/2}^{x - d/2} \frac{-y \, \bar{a}_y}{u^2 + y^2} du$$

Do each integral separately;

$$I_1 = \int_{x + d/2}^{x - d/2} \frac{-u \cdot \bar{a}_x}{u^2 + y^2} du$$

Change the variables, let $r = u^2$, $dr = 2udu$ when $u = x + \frac{d}{2}$, $r = (x + \frac{d}{2})^2$, when $u = x - \frac{d}{2}$, $r = (x - \frac{d}{2})^2$, thus

$$I_1 = \frac{1}{2} \int_{(x+d/2)^2}^{(x-d/2)^2} \frac{-dr}{r + y^2} = \frac{1}{2} \Big[-\ln(r + y^2) \Big]_{(x+d/2)^2}^{(x-d/2)^2}$$

$$I_1 = \frac{1}{2} \{-\ln[(x - \frac{d}{2})^2 + y^2] + \ln[(x + \frac{d}{2})^2 + y^2]\}$$

$$I_1 = \ln \sqrt{\frac{(x + \frac{d}{2})^2 + y^2}{(x - \frac{d}{2})^2 + y^2}} \, \bar{a}_x$$

$$I_2 = \int_{x+d/2}^{x-d/2} \frac{-y \, \bar{a}_y}{u^2 + y^2} du$$

Using $\int \frac{dx}{1 + x^2} = \tan^{-1}x$, I_2 can be solved.

Dividing the numerator and demoninator by y^2 yields

$$I_2 = \frac{1}{y} \int_{x+d/2}^{x-d/2} \frac{-du \, \bar{a}_y}{\frac{u^2}{y^2} + 1}$$

Now let $s = \frac{u}{y}$ $ds = \frac{1}{y}\,du$, thus

$$I_2 = \int_{\frac{x+d/2}{y}}^{\frac{x-d/2}{y}} \frac{-ds\,\bar{a}_y}{1+s^2} = \left[-\tan^{-1}s\right]_{\frac{x+d/2}{y}}^{\frac{x-d/2}{y}} \bar{a}_y$$

$$I_2 = \left\{\tan^{-1}\frac{x+\frac{d}{2}}{y} - \tan^{-1}\frac{x-\frac{d}{2}}{y}\right\}\bar{a}_y$$

Therefore

$$\bar{E} = \frac{\rho_s}{2\pi\varepsilon_0}\left[\ln\sqrt{\frac{(x+\frac{d}{2})^2+y^2}{(x-\frac{d}{2})^2+y^2}}\;\bar{a}_x + \left\{\tan^{-1}\frac{x+\frac{d}{2}}{y} - \tan^{-1}\frac{x-\frac{d}{2}}{y}\right\}\bar{a}_y\right]$$

• PROBLEM 2-22

A continuous volume charge defined as $\rho = (x^2 + y^2 + z^2)^{5/2}$ is distributed in the region $0 \le x \le 1$, $0 \le y \le 1$, $0 \le z \le 1$ and is zero elsewhere. Find E_x at the origin. Refer to the figure.

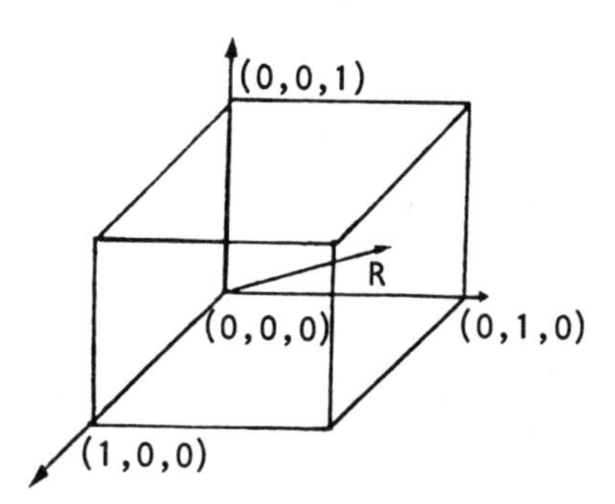

Solution: Draw a vector $\bar{R}$ from the origin to any point within the volume indicated by the given limits. Let

$$\bar{R} = x\,\bar{a}_x + y\,\bar{a}_y + z\,\bar{a}_z,$$

and

$$\bar{a}_R = \frac{x\,\bar{a}_x + y\,\bar{a}_y + z\,\bar{a}_z}{\sqrt{x^2+y^2+z^2}}$$

Then

$$\bar{E} = \int_{vol} \frac{\rho dV}{4\pi\varepsilon_0 R^2} \bar{a}_R$$

where dV = dx dy dz, and free space is assumed.

Thus

$$\bar{E} = \int_0^1 \int_0^1 \int_0^1 \frac{(x^2 + y^2 + z^2)^{5/2} dx\, dy\, dz}{4\pi\varepsilon_0 (x^2 + y^2 + z^2)}$$

$$\cdot \frac{x\, \bar{a}_x + y\, \bar{a}_y + z\, \bar{a}_z}{\sqrt{x^2 + y^2 + z^2}}$$

$$= \int_0^1 \int_0^1 \int_0^1 \frac{(x^2+y^2+z^2)(x\, \bar{a}_x + y\, \bar{a}_y + z\, \bar{a}_z)}{4\pi\varepsilon_0} dx\, dy\, dz$$

The x-component of E is

$$E_x = \frac{1}{4\pi\varepsilon_0} \int_0^1 \int_0^1 \int_0^1 (x^2 + y^2 + z^2) x\, dx\, dy\, dz$$

Integrating,

$$E_x = 9 \times 10^9 \int_0^1 \int_0^1 \left. \frac{x^4}{4} + \frac{x^2}{2} y^2 + \frac{x^2}{2} z^2 \right|_0^1 dy\, dz$$

$$= 9 \times 10^9 \int_0^1 \int_0^1 \left(\frac{1}{4} + \frac{1}{2} y^2 + \frac{1}{2} z^2\right) dy\, dz$$

$$= 9 \times 10^9 \int_0^1 \left. \frac{1}{4} y + \frac{1}{6} y^3 + \frac{1}{2} y z^2 \right|_0^1 dz$$

$$= 9 \times 10^9 \int_0^1 \left(\frac{1}{4} + \frac{1}{6} + \frac{1}{2} z^2\right) dz$$

$$= 9 \times 10^9 \left. \left(\frac{1}{4} z + \frac{1}{6} z + \frac{1}{6} z^3\right) \right|_0^1$$

$$= 9 \times 10^9 \left(\frac{1}{4} + \frac{1}{6} + \frac{1}{6}\right)$$

$$E_x = 9 \times 10^9 \cdot \frac{7}{12} = 5.25 \text{ V/m.}$$

Find the E field due to a uniformly charged sphere by direct integration.

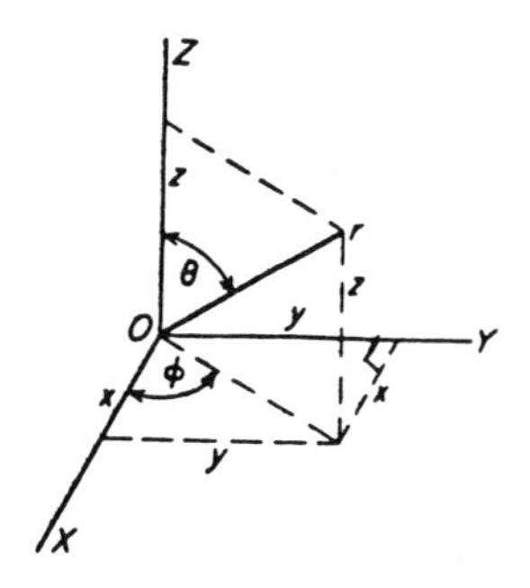

Relation of spherical and rectangular coordinates

Fig. 1

Solution: For a sphere of radius R, the charge per unit area is $q_s = q/4\pi R^2$. For convenience in integration, use spherical coordinates r, θ, ϕ. These are defined in terms of the rectangular coordinates x, y, z by

$$x = r \sin\theta \cos\phi$$

$$y = r \sin\theta \sin\phi$$

$$z = r \cos\theta$$

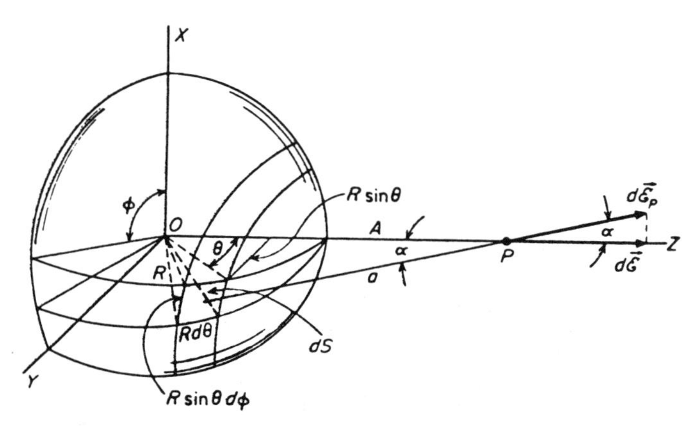

Surface element used to calculate field of charged sphere

Fig. 2

and are illustrated in Fig. 1. Then construct a thin ring on the surface of the charged sphere of Fig. 2 such that the ring is symmetrical about the z-axis (note that the coordinate system has been rotated) and subtends a half-angle θ at the center. An element of area dS is chosen with sides given by R dθ and R sin θ dϕ. The charge on this area is $q_s R^2 \sin\theta d\theta d\phi$ and the field at P due to this charge has a magnitude

$$dE_P = \frac{q_s R^2 \sin\theta d\theta d\phi}{4\pi\varepsilon_0 a^2}$$

Because of symmetry, the resultant field dE due to the ring is along OZ, so that

$$dE = dE_P \cos\alpha$$

Now integrate with respect to ϕ, using limits of 0 and 2π, to obtain

$$dE = \frac{q_s R^2 \sin\theta d\theta \cos\alpha}{2\varepsilon_0 a^2} \quad (1)$$

Now

$$\cos\alpha = \frac{A - R\cos\theta}{a}$$

and

$$a^2 = A^2 + R^2 - 2AR\cos\theta$$

from which

$$a\,da = AR\sin\theta\,d\theta \quad (2)$$

and

$$\cos\alpha = \frac{A^2 - R^2 + a^2}{2Aa} \quad (3)$$

Substituting (2) and (3) into (1) gives

$$E = \frac{q_s R}{4\varepsilon_0 A^2}\int_{A-R}^{A+R} \frac{A^2 - R^2 + a^2}{a^2}\,da = \frac{q_s R^2}{\varepsilon_0 A^2} = \frac{q}{4\pi\varepsilon_0 A^2}$$

CHAPTER 3

ELECTRIC FIELD INTENSITY

ELECTRIC FLUX

• PROBLEM 3-1

Find the electric flux Ψ_E that passes through the surface shown in the figure, when $\bar{D} = (\hat{x}y + \hat{y}x)10^{-2}\,(C/m^2)$.

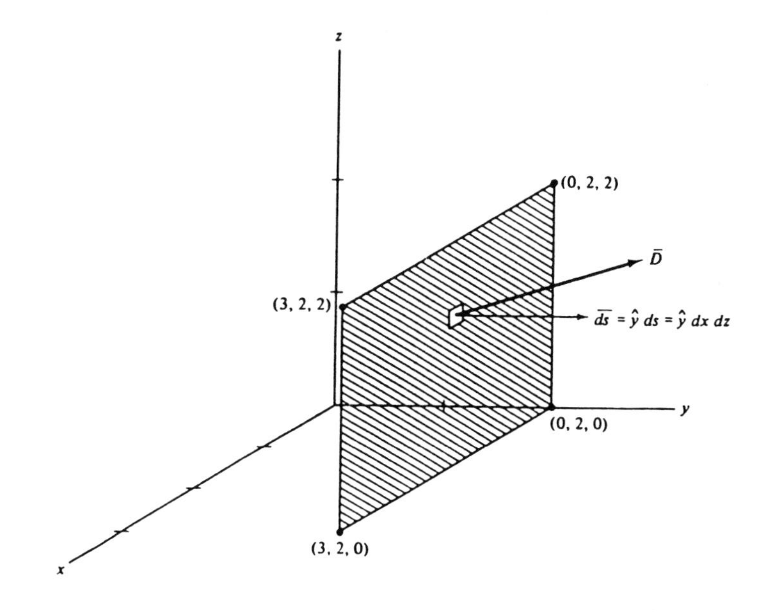

Solution: From the figure and the given flux equation,

$$d\Psi_E = \bar{D} \cdot \overline{ds} = (\hat{x}y + \hat{y}x)10^{-2} \cdot (\hat{y}\,dx\,dz)$$

$$= x10^{-2}dx\,dz$$

$$\Psi_E = \int_0^2 \int_0^3 x10^{-2}dx\,dz = 9 \times 10^{-2} \text{ (lines)}$$

• **PROBLEM 3-2**

Find the number of lines of electric flux emanating from a point charge of Q(C) at the origin by finding the electric lines of flux through an imaginary concentric sphere of radius r_s(m) as shown in the given figure.

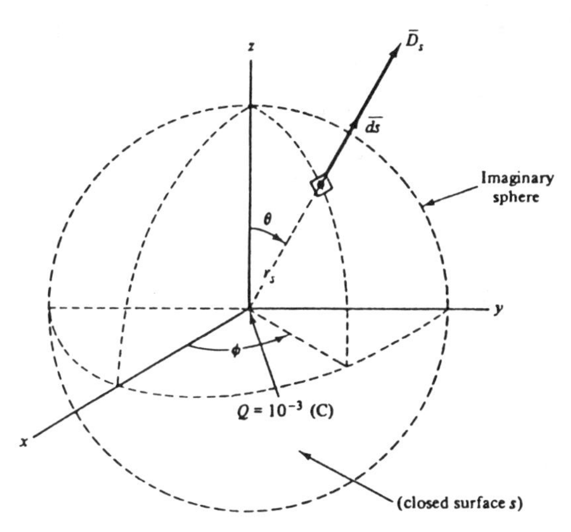

Graphical display for finding the flux density vector $\bar{D}$ on an imaginary, closed, and concentric surface of radius r_s (m).

Solution: The differential flux through ds is equal to

$$d\Psi_E = \bar{D} \cdot \overline{ds} \quad \text{(lines)}$$

Thus, $\Psi_E = \oint_s \bar{D}_s \cdot \overline{ds}$, where $\bar{D}_s$ is evaluated on the closed surface of the imaginary sphere of radius r_s. $\bar{D}_s = \varepsilon_0\bar{E}_s$, $\bar{E}_s = \hat{r}_s(Q/4\pi\varepsilon_0 r_s^2)$, and $\overline{ds} = \hat{r}_s r_s^2 \sin\theta \, d\theta \, d\phi$ in spherical coordinate system. Thus the flux is

$$\Psi_E = \oint_s \left(\varepsilon_0\hat{r}_s \frac{Q}{4\pi\varepsilon_0 r_s^2}\right) \cdot (\hat{r}_s r_s^2 \sin\theta \, d\theta \, d\phi)$$

$$= \int_0^{2\pi}\int_0^{\pi} \frac{Q}{4\pi} \sin\theta \, d\theta \, d\phi = Q \quad \text{(lines) or (C)}$$

Note that $\Psi_E = Q$(lines), where Q is the charge enclosed by the closed surface s. The solution is independent of the closed surface selected.

• PROBLEM 3-3

A cylindrical surface of length ℓ, and radius r is placed in a uniform electric field of magnitude E. Find out the total flux ϕ, coming out of this cylinder, assuming that the axis of the cylinder is parallel to the field, as shown in the figure below.

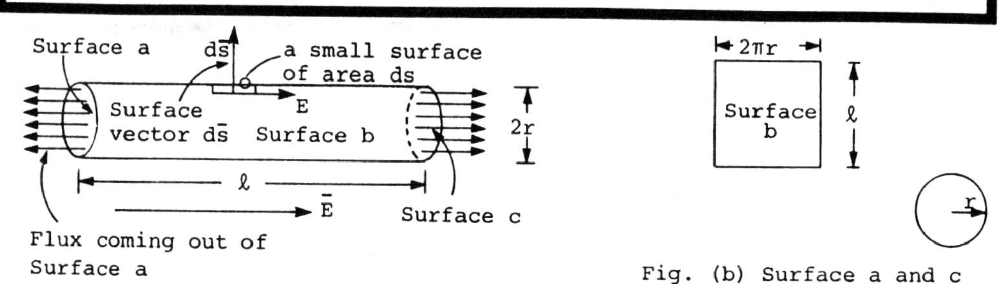

Fig. (b) Surface a and c

Fig. (a)

Solution: The total flux ϕ, coming out of this cylinder can be written as:

$$\phi = \phi_{\text{surface a}} + \phi_{\text{surface b}} + \phi_{\text{surface c}}$$

Now,

$$\phi_{\text{surface b}} = \oint_{\text{surface b}} \overline{E} \cdot \overline{ds}$$

$$= \oint_{\text{over surface b} = 2\pi \cdot r \cdot \ell} \overline{E} \cdot \overline{ds}$$

$$= \oint_{2\pi \cdot r \cdot \ell} E \cdot ds \cdot \cos\theta$$

$$= E \cdot \cos\theta \cdot 2\pi \cdot r \cdot \ell$$

$$= 0 \text{ (because } \theta = 90^0\text{, as shown in Fig. a)}$$

$$\oint_{\text{surface c}} = \oint \overline{E} \cdot \overline{ds} = \oint E \cdot ds \cdot \cos\theta \quad (\theta = 180^0)$$

$$= -\oint E \cdot ds$$

$$\oint_{\text{surface a}} = \oint_{\pi \cdot r^2} \overline{E} \cdot \overline{ds}$$

$$= \oint_{\pi \cdot r^2} E \cdot ds \cdot \cos\theta$$

$$= E \cdot \pi \cdot r^2 \cdot \ell$$

The total flux is:

$$E = 0 + (E \cdot \pi \cdot r^2) - (E \cdot \pi \cdot r^2)$$

$$= 0$$

• PROBLEM 3-4

A cube of side 1.0 mm is uniformly charged with a charge density of 10^{-6} C/m^3. If the cube is enclosed inside a spherical shell of radius 1.0 m, find the total electric flux flowing through the surface of the sphere. Can anything be said about the flux density on the surface of the sphere?

Solution: The total charge in the body of the cube is given by the product of the charge density and the volume. Therefore

$$Q = 10^{-6} \times (10^{-3})^3 = 10^{-15}\,\text{C}.$$

From Gauss's law,

FLUX OUT = CHARGE ENCLOSED.

Therefore the total flux if given by

$$\Psi = Q = 10^{-15}\,\text{C}.$$

The size of the enclosing sphere does not make any difference to the total flux flowing out through its surface so long as the charged cube is inside the surface of the enclosing sphere. As the cube is not spherically symmetrical, the flux density on the surface of a concentric sphere is not uniform. As the question does not state that the cube is at the centre of the sphere, nothing can be said about the flux density on the surface of the sphere except that it is not uniform. If the cube were at the centre of the enclosing sphere, an approximate value of flux density at the surface of the sphere may be obtained by assuming that the charge on the cube is replaced by a point charge at the centre of the cube. Then the approximate value of the flux density is given from $D = \dfrac{Q}{4\pi r^2}$ C/m^2

$$D = \frac{10^{-15}}{4\pi \times 1.0} = 7.96 \times 10^{-17}\,\text{C/m}^2.$$

As the size of the cube is only 0.1 per cent of the radius of the sphere, it may be expected that this is a good approximation.

• PROBLEM 3-5

Find the flux density due to an infinitely large thin sheet which has a uniform surface charge density of 10 $\mu C/m^2$.

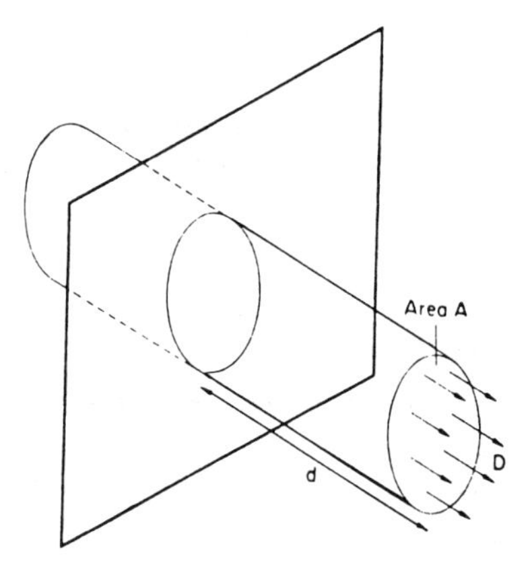

Solution: The problem is symmetrical, so that the flux density will be constant across any plane parallel to the charged plane. If a right circular cylinder perpendicular to the plane as shown in the figure is chosen as the Gaussian surface, the flux will flow through the ends of the cylinder and, because of symmetry, no flux will flow through the curved sides of the cylinder. If the cross-sectional area of the cylinder is A, the charge enclosed inside the cylinder will be the product of the charge density on the charge sheet and the cross-sectional area of the cylinder. Therefore

$$\Psi = 10^{-5}A \text{ C}.$$

The total flux will be flowing out of both ends of the cylinder and, if the charge sheet bisects the cylinder as shown in the figure, the flux density will be the same at each end of the cylinder. Therefore if the flux density is D,

$$\Psi = 2DA \text{ C}.$$

Equating the two expressions for Ψ gives the value of the flux density,

$$D = 5 \cdot 0 \times 10^{-6} \text{ C/m}^2.$$

It is seen that this expression for the flux density is independent of the distance from the charge sheet. This means that for a charge sheet of infinite extent the field is uniform on each side of the sheet with the flux density vector perpendicular to the plane of the sheet.

• PROBLEM 3-6

Parallel line charges are located +5 nC/m at (0, 0) m, +4 nC/m at (3, 0) m, -6 nC/m at (0, 4) m. Find $\bar{E}$ and $\bar{D}$ at the position (3, 4) m, by vector addition of the field quantities.

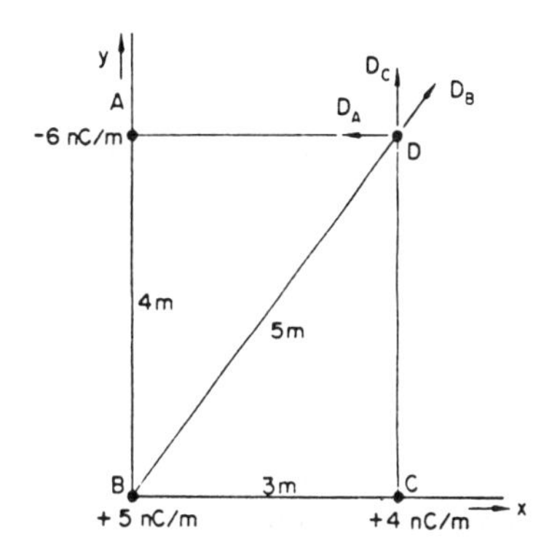

Solution: The flux density due to a line charge is given by D = Q/2πr from Gauss's law. The geometry of the problem is illustrated in the given figure. The flux densities at the point D due to the line charges at the other points are:

$$D_A = \frac{-6}{2\pi \times 3} = -\frac{1}{\pi} \text{ nC/m}^2,$$

$$D_B = \frac{5}{2\pi \times 5} = \frac{1}{2\pi} \text{ nC/m}^2,$$

$$D_C = \frac{4}{2\pi \times 4} = \frac{1}{2\pi} \text{ nC/m}^2.$$

By vector addition of these flux densities, the horizontal and vertical components of the resultant flux density are given by

$$D_x = -\frac{1}{\pi} + \frac{3}{5}\frac{1}{2\pi} = -\frac{1}{\pi}(0.7) \text{ nC/m}^2$$

and $$D_y = \frac{1}{2\pi} + \frac{4}{5}\frac{1}{2\pi} = +\frac{1}{\pi}(0.9) \text{ nC/m}^2.$$

Therefore the resultant flux density is given by

$$D = \frac{1}{\pi}\sqrt{(0.49 + 0.81)} = \frac{1.14}{\pi} = 0.36 \text{ nC/m}^2$$

and $$E = 36\pi \times 10^9 \times \frac{1.14}{\pi} \times 10^{-9} = 41 \text{ V/m}.$$

The angular direction of the flux density vector is given by

$$\text{angle} = \tan^{-1}\frac{0.7}{0.9} = 38°.$$

But the angle is obtuse if the horizontal is taken as the direction of zero angle. Therefore the direction of the flux density vector is at 128° to the horizontal.

• PROBLEM 3-7

Consider the cylindrical surface of the accompanying figure in a field

$$\bar{E} = [\bar{i}x + \bar{j}y + \bar{k}(z^2 - 1)]E_0$$

where E_0 is constant. Evaluate the flux of E across this surface in two ways. First, by direct integration, then by applying the divergence theorem.

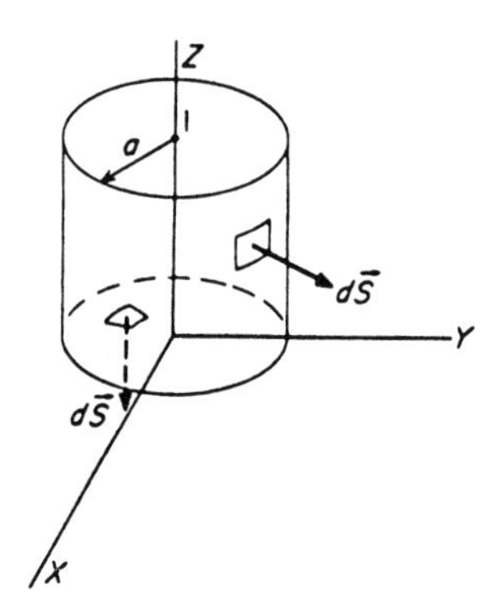

Solution: By direct integration, (a) for the bottom

$$\int \bar{E} \cdot d\bar{S} = E_0 \int (-\bar{k}) \cdot d\bar{S} = E_0 \pi a^2$$

since $d\bar{S}$ points downward, (b) for the top $E_z = 0$, and (c) for the sides

$$d\bar{S} = a\, d\theta\, dz(\bar{i} \cos\theta + \bar{j} \sin\theta)$$

so that

$$\int \bar{E} \cdot d\bar{S} = E_0 \int (\bar{i}x + \bar{j}y) \cdot (\bar{i} \cos\theta + \bar{j} \sin\theta) a\, d\theta\, dz$$

$$= E_0 \int_0^1 \int_0^{2\pi} a^2 d\theta\, dz = 2E_0 \pi a^2$$

Combining these three results

$$\oint \bar{E} \cdot d\bar{S} = 3E_0 \pi a^2$$

Next, apply the divergence theorem. Then

$$\int \text{div}\,\bar{E}\,dV = E_0 \int \left(\frac{\partial E}{\partial x} + \frac{\partial E}{\partial y} + \frac{\partial E}{\partial z}\right) dV$$

$$= E_0 \int (1 + 1 + 2z)\,dV$$

and

$$E_0 \int (2 + 2z)\,dV = E_0 \int_0^a r\,dr \int_0^{2\pi} d\theta \int_0^1 (2 + 2z)\,dz$$

$$= 3E_0 \pi a^2$$

as above.

• PROBLEM 3-8

Two hollow conducting metal spheres are concentric as shown in the given figure. The outer one has a diameter of 0.2 m and is charged initially with +0.10μC. The inner one has a diameter of 0.1 m and is charged at some time later with -0.04μC. The charge on the inner sphere must induce an equal but opposite charge on the inside surface of the outer sphere. By considering the charge distribution on the system and sketching the flux patterns, determine (a) the flux density midway between the two spheres, and (b) the electric intensity in air at 0.2 m from the centre of the system. If the inner sphere is moved to an eccentric position without touching the outer sphere, are these results changed, and how?

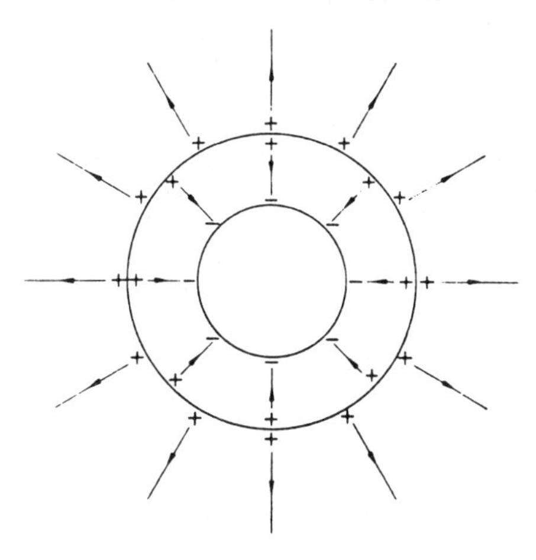

Solution: When the inner sphere is charged with -0.04 μC, an equal and opposite charge will collect on the inside surface of the outer sphere. This means that the charge on the outer sphere will be divided between its inner and outer surfaces. Therefore, charge on inner surface = + 0.04 μC,

charge on outer surface = + 0.06 μC.

By symmetry all these charges will be distributed uniformly around the surface of the spherical shells as shown in the figure.

(a) Midway between the spheres the total flux is -0.04 μC. This is uniformly distributed so that the flux density at a radius of 0.075 m is given by $D = \frac{Q}{4\pi r^2}$

$$\therefore \quad D = \frac{-0.04 \times 10^{-6}}{4\pi \times (0.075)^2} = -5.66 \times 10^{-7}\ C/m^2 = -566\ nC/m^2.$$

(b) External to the outer sphere, the total flux is +0.06 μC. This is also uniformly distributed so that the flux density at radius 0.2 is given by

$$D = \frac{0.06 \times 10^{-6}}{4\pi \times (0.2)^2}\ C/m^2.$$

Therefore $E = D/\varepsilon_0$

$$E = \frac{0.06 \times 10^{-6} \times 36\pi \times 10^9}{4\pi \times 0.04} = 1.35 \times 10^4\ V/m$$

$$= 13.5\ kV/m.$$

If the inner sphere is moved:

(a) The distribution is no longer uniform between the two spherical shells since the charges will tend to concentrate where the shells are closest. The flux density midway between the two spheres will change.

(b) Since each tube of flux inside the spherical shell terminates in a charge inside the spherical shell, the charge distribution on the outside of the spherical shell is unaffected and the flux density outside the shell remains symmetrical with the same value as before. The field outside a spherical conducting shell is unaffected by any change in distribution of the charges inside the spherical shell.

• PROBLEM 3-9

Given the vector field

$$\bar{G} = x^2\bar{a}_x + (y + z)\bar{a}_y + xy\bar{a}_z$$

(a) Find the flux of $\bar{G}$ through the rectangular surface in the xy plane bounded by the lines x = 0, x = 3, y = 1, y = 2, as shown in the figure.

(b) Calculate the flux of G through the triangular surface in the xz plane bounded by the x axis, the z axis, and the line x + z = 1, as shown in the figure.

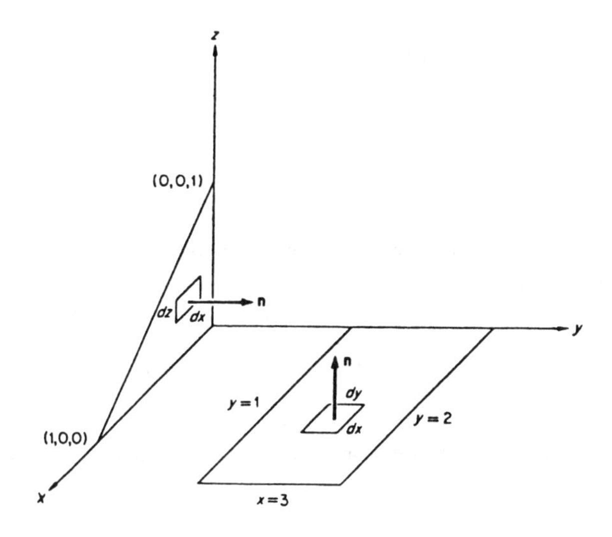

Solution: (a) From the figure it will be seen that $\bar{n} = \bar{a}_z$ and that da = dx dy. Hence

$$\bar{G}\cdot\bar{n}\,da = G_z dx\,dy = xy\,dx\,dy$$

and $$\int_S \bar{G}\cdot\bar{n}da = \int_1^2\int_0^3 xy\,dx\,dy$$

For this integral, the order of integration is immaterial.

$$\int_1^2\int_0^3 xy\,dx\,dy = \int_1^2 \left.\frac{x^2}{2}\,y\right]_{x=0}^{x=3} dy = \int_1^2 \frac{9}{2}\,y\,dy$$

$$= \left.\frac{9}{4}\,y^2\right]_1^2 = \frac{9}{4}\left[4 - 1\right] = \frac{27}{4}$$

Thus, the result is

$$\int_S \bar{G}\cdot\bar{n}\,da = \frac{27}{4}$$

(b) From the figure it will be seen that $\bar{n} = \bar{a}_y$ and that da = dx dz. Hence

$$\bar{G}\cdot\bar{n}\,da = G_y dx\,dz = (y + z)dx\,dz$$

But $y = 0$; so

$$\bar{G}\cdot\bar{n}\ da = z\ dx\ dz$$

Once again the order of integration is immaterial, but for either order the first integration has a variable limit. The result is

$$\int_0^1 z\left(\int_0^{1-z} dx\right) dz = \frac{1}{6} \quad \text{or} \quad \int_0^1 \left(\int_0^{1-x} z\ dz\right) dx = \frac{1}{6}$$

• PROBLEM 3-10

Find the flux of $\bar{F} = \hat{x}z + \hat{y}x - \hat{z}3y^2z$ out of Σ where Σ is a closed surface consisting of (1) the cylinder $x^2+y^2=4$ between $z=0$ and $z=+2$; (2) the circular area with radius $r=2$ which bounds the cylinder at $z=0$; (3) the corresponding circular area at $z=+2$. The surface is shown in the accompanying figure.

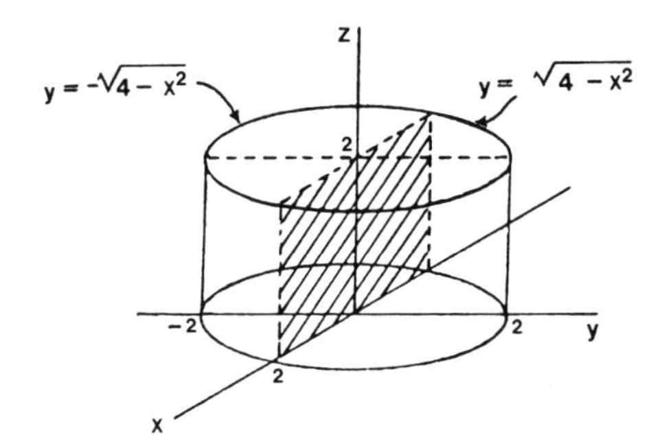

Solution: The surface consists of three parts as follows: top, bottom, and side. Over the top surface $d\bar{s} = \hat{z}ds$ so $\bar{F}\cdot d\bar{s} = -3y^2zdxdy$ and

$$\int_{top} \bar{F}\cdot d\bar{s} = \int_{-2}^{2} -3zdx \int_{-\sqrt{4-x^2}}^{\sqrt{4-x^2}} y^2dy$$

$$= -3z\int_{-2}^{2} dx \left[\frac{2}{3}(4-x^2)^{3/2}\right]$$

$$= -2(2)\int_{-2}^{2} (4-x^2)^{3/2}dx = -4(6\pi) = -24\pi$$

Over the bottom surface $d\bar{s} = -\hat{z}ds$ so $\bar{F}\cdot d\bar{s} = 3y^2zdxdy$ and

$$\int_{bottom} \bar{F}\cdot d\bar{s} = \int_{-2}^{2} 3zdx \int_{-\sqrt{4-x^2}}^{\sqrt{4-x^2}} y^2dy$$

$$= 3z\int_{-2}^{2} dx \left[\frac{2}{3}(4-x^2)^{3/2}\right]$$

But here z=0 so the value of this contribution to the flux is zero. Over the curved side, use a projection on a coordinate plane; here it is not possible to employ the xy plane. Take the xz plane, as shown in the figure.

$$\int_{side} \bar{F}\cdot d\bar{s} = \int_{side} \bar{F}\cdot\hat{n}\, ds = \int_{\sigma} \bar{F}\cdot\bar{n} \frac{dzdx}{\cos(\hat{n},\hat{y})}$$

The gradient to the cylindrical surface, $\nabla(x^2+y^2-4) = \hat{x}2x + \hat{y}2y$, is normal to this surface. Its magnitude is $\sqrt{(2x)^2 + (2y)^2} = 2\sqrt{x^2 + y^2} = 2\sqrt{4} = 4$, so the unit normal to the side surface is $\hat{x}\; x/2 + \hat{y}\; y/2$. Then $\bar{F}\cdot\hat{n} = (x/2)(y+z)$. From $\hat{n}\cdot\hat{y} = y/2$ and $\hat{n}\cdot\hat{y} = (1)(1)\cos(\hat{n},\hat{y})$, the result is $\cos(\hat{n},\hat{y}) = y/2 = \pm\frac{1}{2}\sqrt{4-x^2}$. Therefore

$$\int_{side} \bar{F}\cdot d\bar{s} = \int_{\sigma}\int \frac{x}{2}(y + z) \frac{dzdx}{y/2}$$

$$= \int_0^2 dz\int_{-2}^{2} xdx + \int_0^2 zdz\int_{-2}^{2} \frac{xdx}{\pm\sqrt{4-x^2}}$$

Instead of evaluating the flux through the side of the entire cylinder it is convenient to consider the four quadrants individually.

Out of the first quadrant the flux is

$$\int_0^2 dz\int_0^2 xdx + \int_0^2 zdz\int_0^2 \frac{xdx}{+\sqrt{4-x^2}} = 4 + 4 = 8$$

From the second quadrant the flux is

$$\int_0^2 dz\int_{-2}^{0} xdx + \int_0^2 zdz\int_{-2}^{0} \frac{xdx}{+\sqrt{4-x^2}} = -4 - 4 = -8$$

The third quadrant yields

$$\int_0^2 dz\int_{-2}^{0} xdx + \int_0^2 zdz\int_{-2}^{0} \frac{xdx}{-\sqrt{4-x^2}} = -4 + 4 = 0$$

The fourth quadrant gives, similarly,

$$\int_0^2 dz\int_0^2 xdx + \int_0^2 zdz\int_0^2 \frac{xdx}{-\sqrt{4-x^2}} = 4 - 4 = 0$$

The net flux through the entire curved side is zero. The total flux out of the entire volume is the same as that out of the top circular area: -24π. The minus sign shows the flux is actually inward.

GAUSS'S LAW

• PROBLEM 3-11

Find the field at a distance R from an infinite line charge of strength σ coulombs per meter.

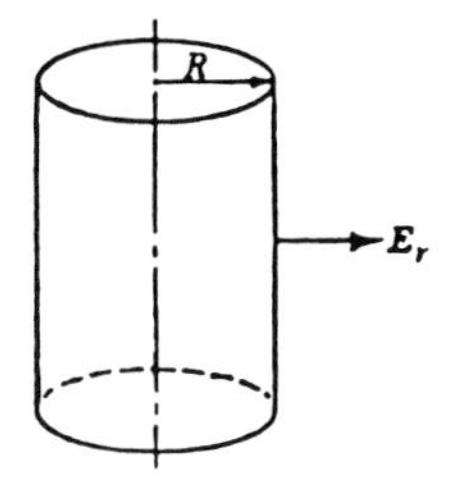

Field from a line charge.

Solution: Surround the line charge by a coaxial cylindrical surface of unit length as shown in the figure. This will be the "Gaussian surface." By symmetry, it is obvious that no flux passes through the top and bottom surfaces, and that $\bar{E}$ is uniform and radially directed outward on the curved surfaces. Applying Gauss' law:

$$\int_S \bar{E} \cdot \bar{n}\, da = 2\pi R E_r = \frac{\sigma}{\varepsilon_0}$$

or

$$E_r = \frac{\sigma}{2\pi\varepsilon_0 R}$$

• PROBLEM 3-12

Find the field from a uniformly charged infinite surface having a surface charge density of σ coulombs per square meter.

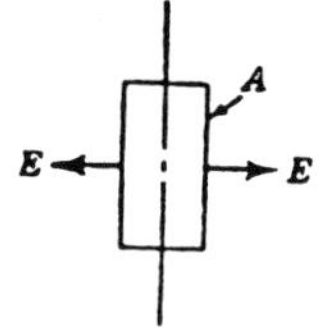

Field from a planar charge

Solution: Here, by symmetry, it is obvious that the field can be only perpendicular to the charged surface. Construct a "Gaussian pillbox," as in the figure. Clearly no flux crosses the curved surfaces, and therefore,

$$\int_S \bar{E} \cdot \bar{n}\, da = 2EA = \frac{\sigma A}{\varepsilon_0}$$

where A is the area of the end sections. This gives

$$E = \frac{\sigma}{2\varepsilon_0}$$

• PROBLEM 3-13

Find the electric field intensity due to three infinite, plane parallel sheets of charge (see the given figure) in the following regions: (1) $0 \le x \le a$, (2) $a \le x \le b$, (3) $b \le x \le \infty$.

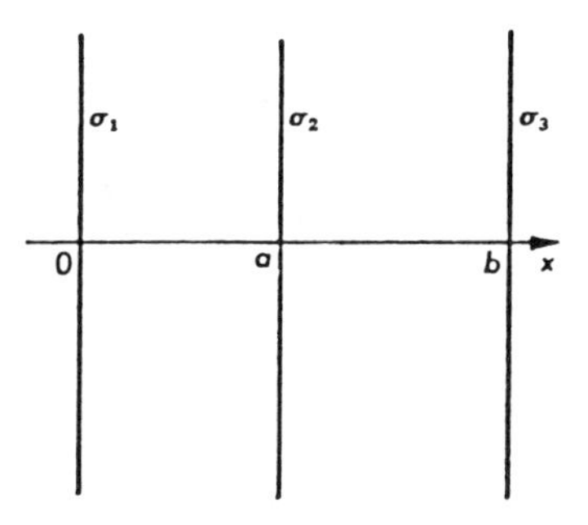

Three infinite plane parallel sheets of charge.

Solution: First, make use of the result that for a single sheet of charge

$$E = \frac{\sigma}{2\varepsilon_0}$$

This equation was derived from Gauss' law. To obtain the resultant electric field intensity in the region, $0 \le x \le a$, superpose the fields due to each of the three sheets. Thus,

$$E_x = \frac{\sigma_1}{2\varepsilon_0} - \frac{\sigma_2}{2\varepsilon_0} - \frac{\sigma_3}{2\varepsilon_0}$$

The second and third terms on the right are negative since the electric fields due to the second and third sheets are directed toward the negative x-axis. In the region, $a \le x \le b$,

$$E_x = \frac{\sigma_1}{2\varepsilon_0} + \frac{\sigma_2}{2\varepsilon_0} - \frac{\sigma_3}{2\varepsilon_0}$$

and in the region, $b \leq x \leq \infty$,

$$E_x = \frac{\sigma_1}{2\varepsilon_0} + \frac{\sigma_2}{2\varepsilon_0} + \frac{\sigma_3}{2\varepsilon_0}$$

Here, if E_x is positive, then E_x is directed toward the right.

• PROBLEM 3-14

A spherical capacitor is formed by the region between two spherical conducting shells of radii a and b, b>a. Assume a space charge density of ρ_{sa} on the outer surface of the inner sphere (see figure). Find (a) Q_a; (b) $\bar{D}$; (c) Q_b; and (d) ρ_{sb}.

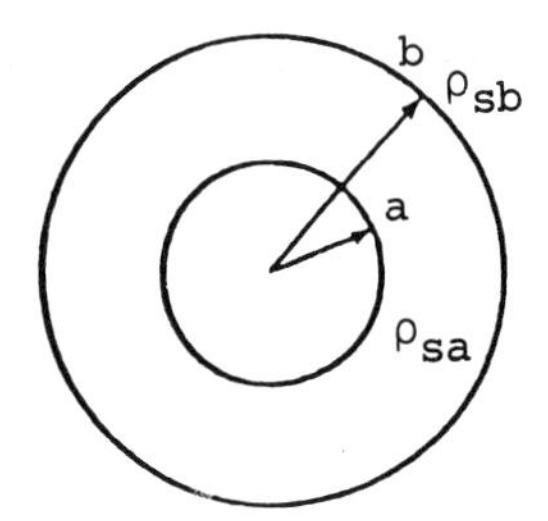

Solution: (a) On the surface of the inner sphere, r = a, and

$$Q_a = \int_s \bar{\rho}_{sa} \cdot d\bar{s}$$

where $d\bar{s} = a^2 \sin\theta d\theta d\phi \bar{a}_r$, $\bar{\rho}_{sa} = |\rho_{sa}|\bar{a}_r$.

Therefore

$$Q_a = |\rho_{sa}| \int_0^{2\pi} \int_0^{\pi} a^2 \sin\theta d\theta d\phi$$

$$= |\rho_{sa}| 2\pi a^2 \int_0^{\pi} \sin\theta d\theta = 4\pi a^2 |\rho_{sa}|$$

(b) Use Gauss's theorem. By spherical symmetry, only the r-component of $\bar{D}$ exists, therefore

$$4\pi r^2 D_r = 4\pi a^2 |\rho_{sa}|, \therefore D_r = (a^2/r^2) |\rho_{sa}|$$

(c) Q_b must be equal and opposite to Q_a,

$$Q_b = -4\pi a^2 |\rho_{sa}|$$

(d) The charge density on the spherical shell b is then

$$|\rho_{sb}| = \frac{Q_b}{4\pi b^2} = -\frac{4\pi a^2 |\rho_{sa}|}{4\pi b^2}$$

$$= -\frac{a^2 |\rho_{sa}|}{b^2}$$

• PROBLEM 3-15

Find the field produced by an infinite charged conducting plate as shown in the figure.

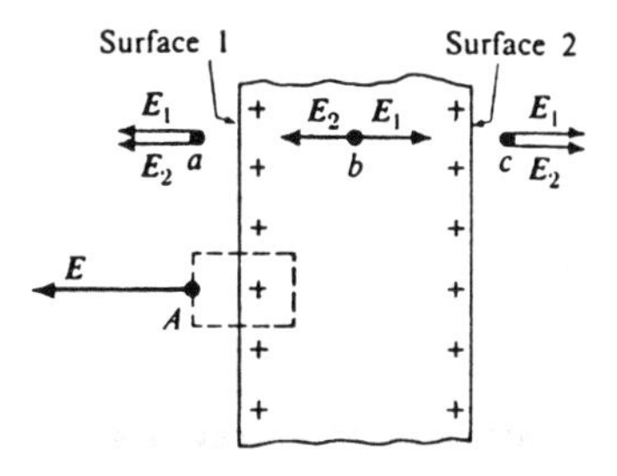

Electric field inside and outside a charged conducting plate.

Solution: When a metal plate is given a net charge, this charge distributes itself over the entire outer surface of the plate, and if the plate is of uniform thickness and is infinitely large (or not too near the edges of a finite plate), the charge per unit area is uniform and is the same on both surfaces. Hence, the field of such a charged plate arises from the superposition of the fields of two sheets of charge, one on each surface of the plate. By symmetry, the field is perpendicular to the plate, directed away from it if the plate has a positive charge, and is uniform. The magnitude of the electric intensity at any point can be found from Gauss's law, or by using the results for a sheet of charge.

The figure shows a portion of a large charged conducting plate. Let σ represent the charge per unit area in the sheet of charge on either surface. At point a outside the plate at the left, the component of electric field $\bar{E}_1$, due to the sheet of charge on the left face of the plate, is directed toward the left and its magnitude is $\sigma/2\varepsilon_0$. The component $\overline{E}_2$ due to the sheet of charge on the right face of the plate is also toward the left and its magnitude is also $\sigma/2\varepsilon_0$. The magnitude of the resultant intensity E is therefore

$$E = E_1 + E_2 = \frac{\sigma}{2\varepsilon_0} + \frac{\sigma}{2\varepsilon_0} = \frac{\sigma}{\varepsilon_0} .$$

At point b, inside the plate, the two components of electric field are in opposite directions and their resultant is zero, as it must be in any conductor in which the charges are at rest. At point c, the components again add and the magnitude of the resultant is σ/ε_0, directed toward the right.

To derive these results directly from Gauss's law, consider the cylinder shown by dotted lines. Its end faces are of area A and one lies inside and one outside the plate. The field inside the conductor is zero. The field outside, by symmetry, is perpendicular to the plate, so the normal component of E is zero over the walls of the cylinder and is equal to E over the outside end face. Hence, from Gauss's law,

$$EA = \frac{\sigma A}{\varepsilon_0}, \qquad E = \frac{\sigma}{\varepsilon_0} .$$

• PROBLEM 3-16

The figure shows a conductor with a uniform surface charge density σ_0. To the right of the conductor, there is a volume charge density ρ in free space.

$$\sigma = \sigma_0 \qquad (z = 0)$$

$$\rho = \rho_0 e^{-\alpha z} \qquad (z > 0)$$

Find the electric field due to this conductor.

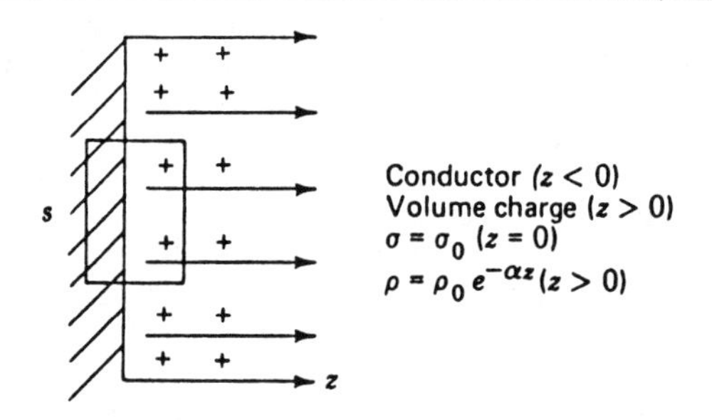

Solution: The electric field is perpendicular to the conductor and uniform in x and y. It is not uniform in z because of the volume charge. Applying Gauss' law to the surface s of the figure, note that the charge enclosed depends on the z dimension of the box.

$$\oint\!\!\oint_S \bar{E}\cdot\hat{n}\, ds = AE_z = \frac{q_{encl}}{\varepsilon_0} = \frac{\sigma_0 A}{\varepsilon_0} + \frac{A}{\varepsilon_\rho}\int_0^z \rho_0 e^{-\alpha z} dz$$

$$= \frac{\sigma_0 A}{\varepsilon_0} + \frac{A\rho_0}{\alpha\varepsilon_0}[1 - e^{-\alpha z}]$$

$$E_z = \frac{\sigma_0}{\varepsilon_0} + \frac{\rho_0}{\alpha\varepsilon_0}[1 - e^{-\alpha z}].$$

Note that the electric field is an increasing function of z.

There are some subtle points involved here. If the conductor were grounded, the total charge would be zero. This would give the following relation between σ_0 and ρ_0

$$\sigma_0 = -\frac{\rho_0}{\alpha}$$

• PROBLEM 3-17

Find the field between oppositely charged parallel plates as shown in Fig. 1.

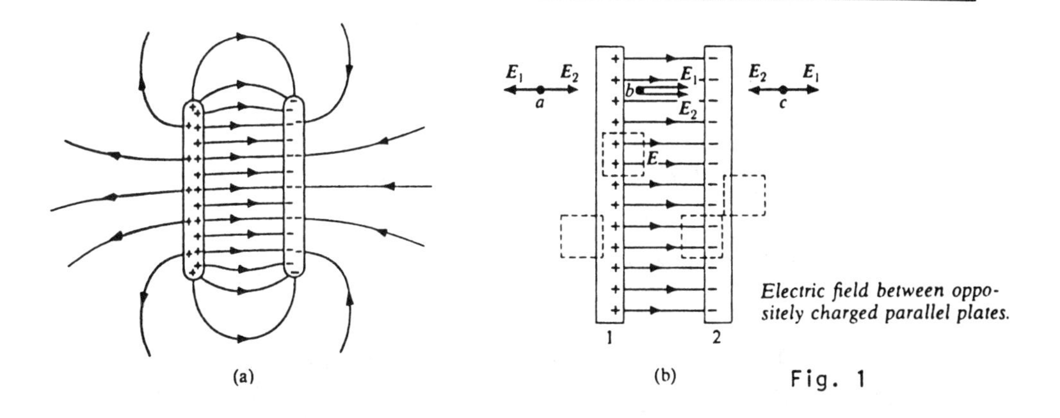

Electric field between oppositely charged parallel plates.

Fig. 1

Solution: When two plane parallel conducting plates, having the size and spacing shown, are given equal and opposite charges, the field between and around them is approximately as shown in Fig. 1(a). While most of the charge accumulates at the opposing faces of the plates and the field is essentially uniform in the space between them, there is a small quantity of charge on the outer surfaces of the plates and a certain spreading or "fringing" of the field at the edges of the plates.

As the plates are made larger and the distance between them diminished, the fringing becomes relatively less. Such an arrangement, two oppositely charged plates separated by a distance small compared with their linear dimensions, is encountered in many pieces of electrical equipment, notably in capacitors. Often the fringing is entirely negligible; even if it is not, neglecting it often provides useful approximations in cases where the work of more detailed calculations is not warranted. Therefore assume that the field between two oppositely charged plates is uniform, as in Fig. 1(b), and that the charges are distributed uniformly over the opposing surfaces.

The electric field at any point can be considered as the resultant of that due to two sheets of charge of opposite sign, or it may be found from Gauss's law. Thus at points a and c in Fig. 1(b), the components $\bar{E}_1$ and $\bar{E}_2$ are each of the magnitude of $\sigma/2\varepsilon_0$ but are oppositely directed, so their resultant is zero. At any point b between the plates the components are in the same direction and their resultant is σ/ε_0.

• PROBLEM 3-18

Assume that there is a charge density σ coul/m^2 on a thin spherical shell of radius r_0. The total charge Q on the shell is $4\pi r_0{}^2\sigma$. Use Gauss' theorem to find the field of this charge distribution both inside and outside the shell.

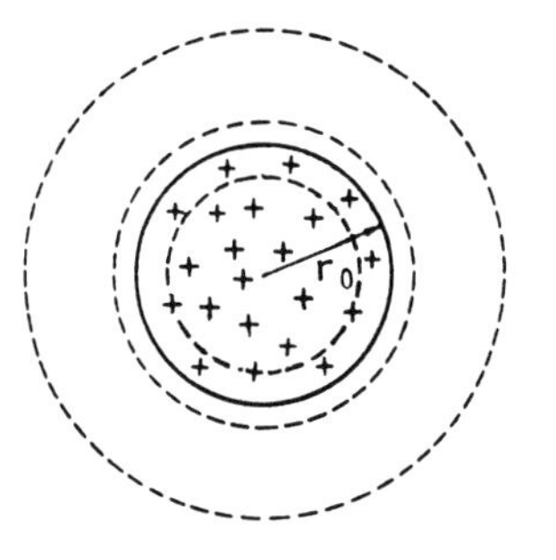

Several different gaussian surfaces (indicated by dotted lines) placed around a spherical shell of charge to investigate the field produced by the charge.

Fig. 1

Solution: In order to use Gauss' flux theorem effectively, it is necessary to determine the field from the flux. This is possible in cases for which a surface can be drawn such that the orientation of E with respect to the surface normal dS is everywhere known and the same, and for which E has a constant value over the entire surface. For example, in the present problem, Gauss' flux theorem is applied to a spherical surface just outside the spherical charge shell, as shown in Fig. 1.

For this gaussian surface symmetry considerations show that the field E has the same magnitude everywhere on the surface, and is parallel to the surface normal (or perpendicular to the surface) at each point on the surface. Using Gauss' flux theorem, the total flux through the gaussian surface is,

$$\int_{CS} \bar{E} \cdot d\bar{S} = \frac{1}{\varepsilon_0} Q \qquad (1)$$

Because of the results of the symmetry argument, the flux can be written as

$$E\int_{CS} dS = E \cdot 4\pi r_0^{\,2}.$$

By comparison with Eq. (1) this gives

$$E = \frac{1}{\varepsilon_0} \frac{Q}{4\pi r_0^{\,2}} \quad \text{newtons/coul}$$

Thus the field at the surface of the shell is exactly the same as though all the charge Q were located at the center of the sphere defined by the shell. Also, by taking as gaussian surface a spherical surface of any radius greater than r, and noting that Eq. (1) is still valid, it is seen that for any point outside a uniform shell of charge, the field behaves according to Coulomb's law for the total charge Q at the center.

The field inside the spherical shell of charge is readily shown to be zero at all points within r. Draw a spherical gaussian surface, this time, of any radius less than r of the charge shell, so that there is no charge within the volume; so by Gauss' flux theorem, the flux $\int_{CS} \bar{E} \cdot d\bar{S}$, is zero. But again by symmetry, whatever value E has, it is everywhere the same at the surface, and if it exists at all, it must (by symmetry) be perpendicular to the surface. The flux factor thus becomes $E\int_{CS} dS = E \cdot 4\pi r^2 = 0$. Since $r \neq 0$, it follows that $E = 0$ everywhere inside r_0.

• PROBLEM 3-19

Find the field in region of a charged cylindrical conductor. This problem is that of a thin cylindrical shell of charge as illustrated in Fig. 1, where the radius r is much smaller than the length l.

Solution: In order to avoid consideration of end effects, choose for a gaussian surface a cylinder of the same radius r as that of the shell of charge (so as to enclose the shell of charge) but of smaller length l_1 ($l_1 < l$). The results will then apply to the entire charged cylinder, except near the ends. From the cylindrical symmetry it is known that the electric field at the radial surface of the shell points uniformly radially outward and has the same value at all points on the radial (curved) surface (at a distance r from the axis).

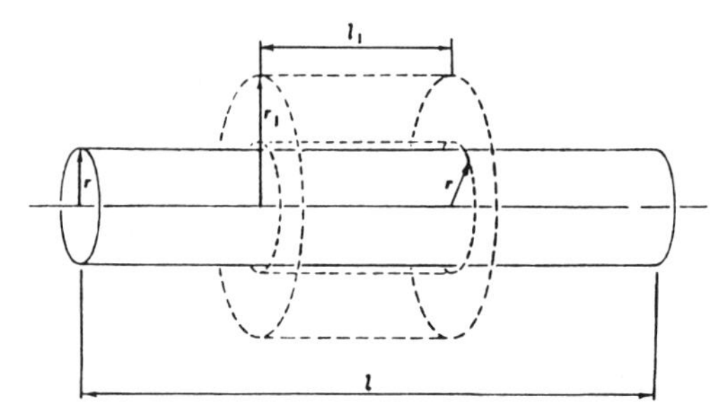

Fig. 1 A charged cylindrical conductor having a surface charge density of σ coul/m^2, surrounded by gaussian surfaces for field determinations.

Because the lines of E are parallel to the flat ends of the gaussian cylinder, the flux of E out of the ends is zero. Now evaluate the total flux over the curved surface through Gauss' theorem. Then define the surface density of the shell as σ coul/m^2. The total charge inside the gaussian surface is then $2\pi r l_1 \sigma$, and Gauss's theorem gives

$$\int_{CS} \bar{E} \cdot d\bar{S} = \frac{1}{\varepsilon_0} 2\pi r l_1 \sigma$$

or

$$E = \frac{1}{\varepsilon_0} \frac{2\pi r l_1 \sigma}{\int_{CS} dS_1} = \frac{1}{\varepsilon_0} \frac{2\pi r l_1 \sigma}{2\pi r l_1} = \frac{1}{\varepsilon_0} \sigma \quad \text{newtons/coul.}$$

This result for the field at the surface of a cylindrical shell of charge is the same as for the field at the surface of a spherical shell.

To find how the field varies out from the cylinder construct a new gaussian cylinder, this time with a radius r_1 greater than r.

Similar reasoning to that above gives for E,

$$E = \frac{1}{\varepsilon_0} \frac{2\pi r_1 l_1 \sigma}{2\pi r l_1} = \frac{1}{\varepsilon_0} \frac{\mu}{2\pi r} \quad \text{newtons/coul}$$

Here, μ is the charge on the shell per unit length of cylinder.

• PROBLEM 3-20

Consider a cylindrical electron beam as shown in the figure. The electrons have a charge density $\rho = \rho_0(1 + r^2/d^2)$ coulombs/meter^3. Find E for $d>r$ and $d<r$.

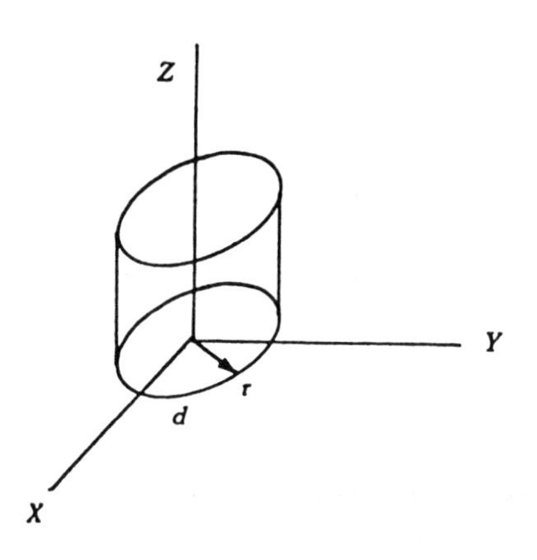

Solution: By Gauss' Flux Theorem

$$\oint_s \bar{D} \cdot d\bar{S} = \int \rho dV$$

$$= \rho_0 \int_0^{2\pi} \int_0^r \left(1 + \frac{r^2}{d^2}\right) r dr d\theta$$

now

$$\int_0^r (1 + \frac{r^2}{d^2}) r dr = \int_0^r r dr + \frac{1}{d^2} \int_0^r r^3 dr$$

$$= \frac{r^2}{2} + \frac{r^4}{4d^2}$$

when r<d

$$\oint_s \bar{D} \cdot d\bar{S} = \rho_0 \pi \left(r^2 + \frac{r^4}{2d^2}\right)$$

and

$$D_n (2\pi r) = \rho_0 \pi \left(r^2 + \frac{r^4}{2d^2}\right)$$

$$D_n = \frac{\rho_0}{2} \left(r + \frac{r^3}{2d^2}\right) ; \quad E_n = \frac{\rho_0}{2\varepsilon_v} \left(r + \frac{r^3}{2d^2}\right) \qquad (1)$$

(after dividing by $2\pi r$)

when r>d

$$\oint_s \bar{D} \cdot d\bar{S} = \rho_0 \pi \left(d^2 + \frac{d^4}{2d^2}\right) = \rho_0 \pi \cdot \frac{3}{2} d^2$$

and

$$D_n = \frac{3\rho_0 d^2}{4r} ; \qquad E_n = \frac{3\rho_0 d^3}{4\varepsilon r}$$

Thus in terms of ρ_0 ,

$$E_n = \frac{\rho_0 r}{2\varepsilon_v}\left(1 + \frac{r^2}{2d^2}\right) ; \qquad r<d$$

$$= \frac{3\rho_0 r^2}{4\varepsilon_v}\left(\frac{1}{r}\right) ; \qquad r>d$$

• PROBLEM 3-21

Consider the charge distribution given by

$$\rho = \begin{cases} -\rho_0 & \text{for } -a < x < 0 \\ \rho_0 & \text{for } 0 < x < a \end{cases}$$

as shown in Fig. 1(a), where ρ_0 is a constant. Find the electric field everywhere.

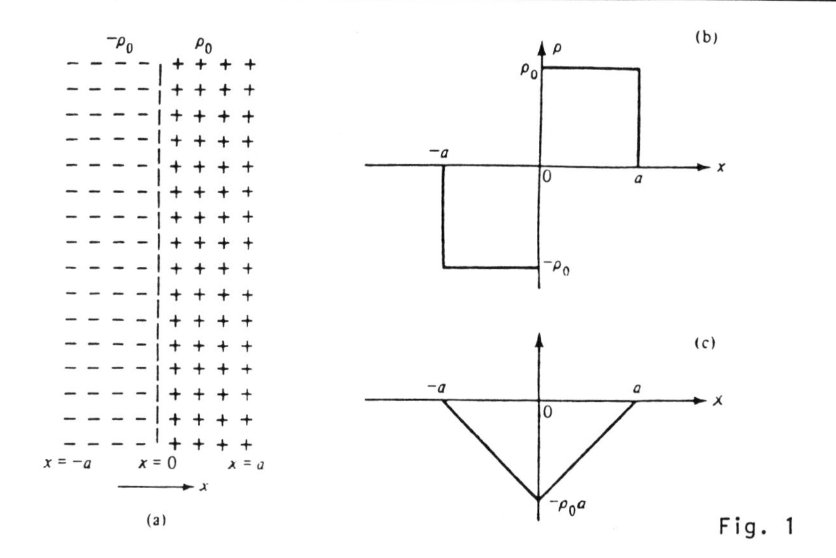

Fig. 1

Solution: Since the charge density is independent of y and z, the field is also independent of y and z, thereby giving $\partial D_y/\partial y = \partial D_z/\partial z = 0$ and reducing Gauss' law for the electric field to

$$\frac{\partial D_x}{\partial x} = \rho$$

Integrating both sides with respect to x,

$$D_x = \int_{-\infty}^{x} \rho \, dx + C$$

where C is the constant of integration.

The variation of ρ with x is shown in Fig. 1(b). Integrating ρ with respect to x, that is, finding the area under the curve of Fig. 1(b) as a function of x, the result shown in Fig. 1(c) for

$$\int_{-\infty}^{x} \rho \, dx$$

is obtained. The constant of integration C is zero since the symmetry of the field required by the symmetry of the charge distribution is already satisfied by the curve of Fig. 1(c). Thus the displacement flux density due to the charge distribution is given by

$$\bar{D} = \begin{cases} 0 & \text{for } x < -a \\ -\rho_0(x + a)\bar{i}_x & \text{for } -a < x < 0 \\ \rho_0(x - a)\bar{i}_x & \text{for } 0 < x < a \\ 0 & \text{for } x > a \end{cases}$$

The electric field intensity, $\bar{E}$, is equal to $\bar{D}/\varepsilon_0$.

• PROBLEM 3-22

A cylindrical shell of uniform volume density ρ_0 lies in the region $b \geq r \geq a$. (a) Determine $\bar{D}$ in all regions. (b) What uniform line charge should be placed at $r = 0$ to reduce the external field $(r > b)$ to zero? See the figure.

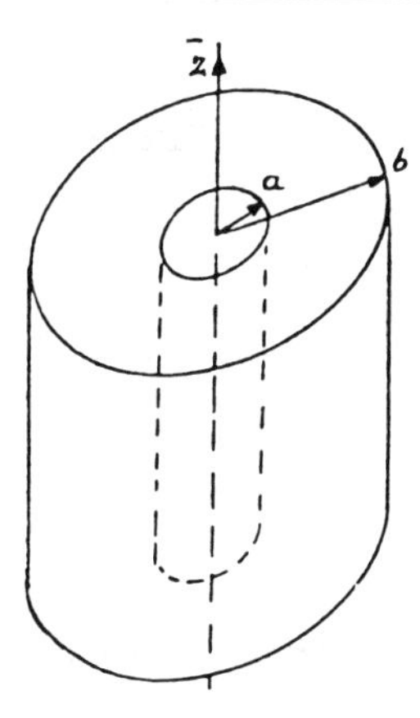

Solution: Due to circular symmetry, the problem can be solved readily by applying Gauss's theorem. Orient the z-axis of the cylinder to coincide with the coordinate z-axis. By symmetry, $\bar{D}$ has r-component only.

(a) Apply Gauss's theorem to find D_r. (i) For $r < a$,

construct a Gaussian surface, a cylindrical surface of radius r and length L. This surface $2\pi rL$ does not enclose any charge. Therefore $D_r = 0$. (ii) For $a \le r \le b$, construct a Gaussian surface with the differential surface $2\pi r dz$ for an elementary length dz. Then by Gauss's theorem,

$$2\pi r D_r dz = \int_V \rho_0 r dr d\phi dz = \int_0^{2\pi} \int_a^r \rho_0 r dr d\phi dz$$

$$= 2\pi\rho_0 \left. \frac{r^2}{2} \right|_a^r dz = 2\pi\rho_0 \frac{r^2 - a^2}{2} dz$$

Therefore

$$D_r = \frac{\rho_0 (r^2 - a^2)}{2}$$

(iii) for $r \ge b$,

$$2\pi r\ dz\ D_r = \int_V \rho_0 r dr d\phi dz = \int_0^{2\pi} \int_a^b \rho_0 r dr d\phi dz$$

$$= 2\pi\rho_0 \left[\frac{r^2}{2} \right]_a^b dz = 2\pi\rho_0 \frac{b^2 - a^2}{2} dz$$

$$D_r = \frac{\rho_0 (b^2 - a^2)}{2r} \quad \text{in } a_r \text{ direction.}$$

(b) To reduce the external field to zero, a line charge of infinite length must be placed at r = 0 such that its magnitude is equal and opposite to case (iii). Thus

$$\frac{\rho_\ell}{2\pi r} = -\frac{\rho_0 (b^2 - a^2)}{2r}$$

Therefore

$$\rho_\ell = -\rho_0 (b^2 - a^2)\pi.$$

• PROBLEM 3-23

Consider an isolated spherical conducting shell of inner radius a and outer radius b containing a charge q at its center (see Fig. 1). Find the charge density induced on the inner spherical surface and the outer surface, and the electrostatic potential of the conductor. Repeat if the conductor was not isolated but kept at a potential ϕ_0.

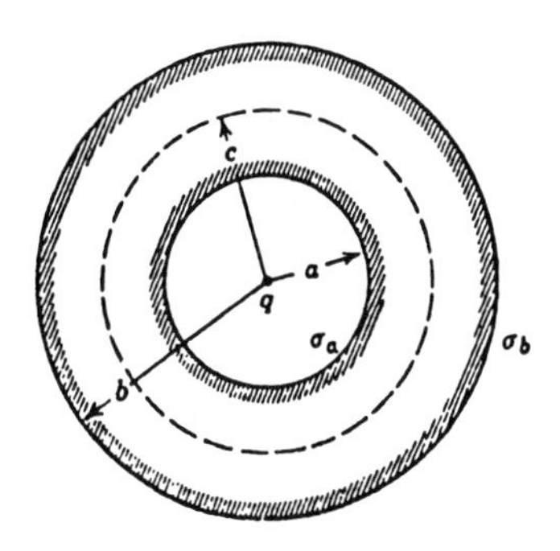

Fig. 1 A point charge q is located at the center of a spherical conducting shell of inner radius a and outer radius b. A spherical surface of radius c resides within the conducting shell (b > c > a). A surface charge density is induced on each surface of the shell as indicated.

Solution: The charge density σ_a induced on the inner spherical surface of radius a may be calculated by a direct application of Gauss's law. Imagine a spherical surface of radius c(b>c>a) as shown in the figure. On this spherical surface E vanishes identically since it resides totally within the conducting region. Since E vanishes on this surface, so does the electrostatic flux. From Gauss's law, therefore, the total charge residing within this surface must vanish. The only charges within this surface are the charge q and the charge q_a induced on Σ_a. Consequently $q_a = -q$. Since the conducting shell is isolated, the total charge induced on the outer spherical surface of radius b must be equal but opposite in sign to q_a. Therefore

$$q_b = -q_a = -(-q) = q.$$

Because of the spherical symmetry inherent in the problem, σ_b may be calculated by dividing q_b by the area of a spherical surface of radius b:

$$\sigma_b = \frac{1}{4\pi b^2}\, q.$$

The electrostatic potential ϕ of the conductor may be calculated as follows: ϕ is produced by three charges (q, q_a, and q_b). The potential at an arbitrary point within and on the conductor can be represented as

$$\phi = \phi_q + \phi_a + \phi_b$$

The electrostatic potential ϕ evaluated at a point immediately outside the conductor is identical with ϕ of the conductor. ϕ immediately outside the conductor may also be thought of as being produced by three charges

(q, q_a, and q_b). Since q_a and q_b are spherically symmetric charge distributions, for the calculation of ϕ at a point immediately outside the shell they may be considered as point charges located at the center. Since $q_a = -q_b$, the net contribution of q_a and q_b to the potential at the outer surface is identically zero. ϕ at the outer surface is therefore due solely to the charge q residing at the center and is

$$\phi = \left(\frac{1}{4\pi\varepsilon_0}\right)\frac{q}{b} .$$

Since a conductor is an equipotential region,

$$\phi = \left(\frac{1}{4\pi\varepsilon_0}\right)\frac{q}{b}$$

throughout the conducting shell.

If the conducting shell is not isolated but maintained at a potential ϕ_0, the analysis is altered somewhat. The charge residing on the inner surface is still $q_a = -q$, the argument leading to q_a being independent of whether the conductor is isolated or not.

However, the charge residing on the outer surface must be calculated by noting that the outer surface must be an equipotential with $\phi = \phi_0$. On the outer surface, $\phi (= \phi_0)$ can be thought of as due to the three charges q, $q_a = -q$, and q_b. Employing Gauss's law and the arguments of spherical symmetry,

$$\phi_0 = \left(\frac{1}{4\pi\varepsilon_0}\right)\left\{\frac{q}{b} + \frac{q_a}{b} + \frac{q_b}{b}\right\} = \left(\frac{1}{4\pi\varepsilon_0}\right)\frac{q_b}{b} .$$

Therefore

$$q_b = (4\pi\varepsilon_0)b\phi_0$$

and

$$\sigma_b = (4\pi\varepsilon_0)\frac{\phi_0}{4\pi b} .$$

If the conducting shell is grounded ($\phi_0 = 0$), then the induced charge exists solely on the inner surface of the shell since $\sigma_b = 0$.

• PROBLEM 3-24

Consider a cylindrical distribution of charge (see the figure shown) with uniform volume charge density ρ_0.

$$\rho = \rho_0 \quad (r \leq a)$$

$$= 0 \quad (r > a)$$

Find the electric field intensity everywhere due to this distribution. Also find the potentials.

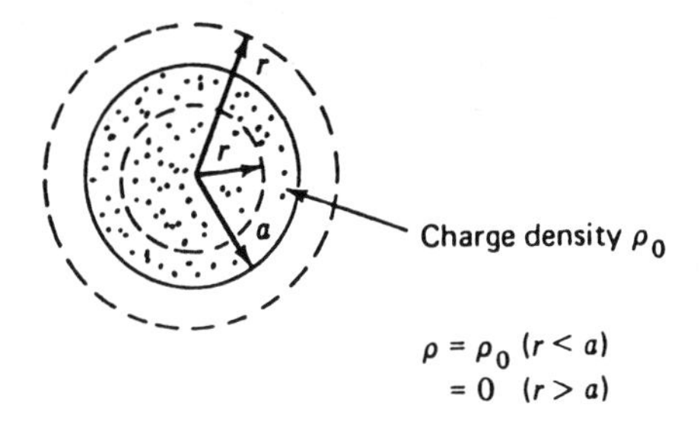

Solution: Construct a cylindrical Gaussian surface of unit length in the z direction and radius r.

Case for r < a. Applying Gauss' law:

$$\oiint_s \bar{E} \cdot \hat{n}\, ds = 2\pi r E_r = \frac{q_{encl}}{\varepsilon_0}$$

$$= \frac{1}{\varepsilon_0} \int_0^1 \int_0^{2\pi} \int_0^r \rho_0 r\, dr\, d\phi\, dz$$

$$= \frac{\pi \rho_0 r^2}{\varepsilon_0}$$

$$E_r = \frac{\rho_0 r}{2\varepsilon_0} \quad (r < a) \qquad (1)$$

Case for r > a. Applying Gauss' law:

$$\oiint_s \bar{E} \cdot \hat{n}\, ds = 2\pi r E_r = \frac{q_{encl}}{\varepsilon_0}$$

$$= \frac{1}{\varepsilon_0} \int_0^1 \int_0^{2\pi} \int_0^a \rho_0 r\, dr\, d\phi\, dz$$

$$= \frac{\pi \rho_0 a^2}{\varepsilon_0}$$

$$E_r = \frac{\rho_0 a^2}{2\varepsilon_0}\left(\frac{1}{r}\right) \quad (r > a) \qquad (2)$$

For r > a, the field is the same as that which would be produced by a line charge λ_0 where $\lambda_0 = (\pi a^2)\rho_0$.

Note that the electric field increases with r for r < a. This is due to the fact that while traveling outward from the origin along a radial line, new charges are encountered from which new electric field lines originate. In contrast, the electric field for $r > a$ has $(1/r)$ dependence corresponding to the fact that the electric field lines diverge and that no new lines are created. The $(1/r)$ dependence is related to the cylindrical geometry, and the fact that the curved surface area is proportional to r. Since the surface area increases as r, the number of lines per unit area, which represents the electric field intensity, is proportional to $(1/r)$.

Calculate the potential by integrating Eqs. (1) and (2). Make an arbitrary choice on the location of the reference point r_0. Choose r_0 to be in the region $r > a$.

Case for $(r > a)$.

$$V(r) = \int_r \bar{E} \cdot dl$$

$$V(r) = - \int_{r_0}^{r} \frac{\rho_0 a^2}{2\varepsilon_0 r} \, dr$$

$$= - \frac{\rho_0 a^2}{2\varepsilon_0} \log \left[\frac{r}{r_0} \right] \qquad (3)$$

Case for $(r < a)$.

$$V(r) = - \int_{r_0}^{a} \frac{\rho_0 a^2}{2\varepsilon_0} \frac{dr}{r} - \int_{a}^{r} \frac{\rho_0}{2\varepsilon_0} r \, dr$$

$$= - \frac{\rho_0 a^2}{2\varepsilon_0} \log \frac{a}{r_0} - \frac{\rho_0 (r^2 - a^2)}{4\varepsilon_0} \qquad (4)$$

The reference point r_0 in the region $r < a$ could have been chosen also. A particularly convenient choice would be $r_0 = a$. In any case, the same reference point must be used throughout the problem, in order to avoid discontinuities in potential. Note that Eqs. (3) and (4) agree at $r = a$.

• PROBLEM 3-25

The figure shown below gives a cross sectional view of a spherical piece of a metal. Determine the value of the electric field E at Gaussian surfaces a, b and c. Assume that the charge carried by the metal is 1.25×10^{-7} coul. The sphere has radius r_o.

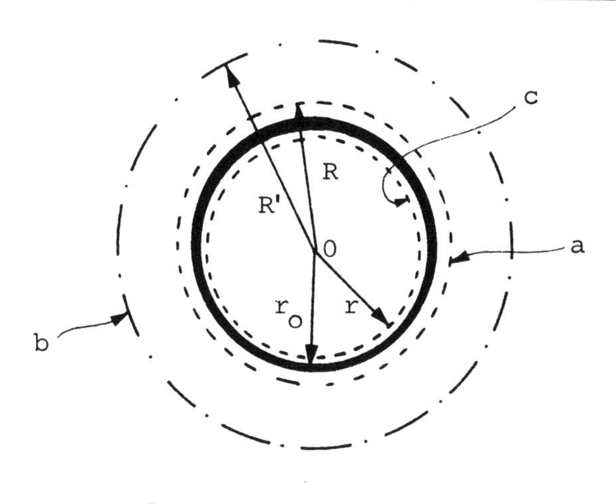

$$\frac{1}{4\pi\varepsilon_o} = 9.0\times10^9 \text{ N-m}^2/\text{coul}^2$$

$R = 0.15\text{m}$

$R' = 2.5\text{m}$

$r<r_o$

Solution: For a spherical Gaussian surface of radius R, the Gauss's Law for the electric field,

$$E = \frac{1}{4\pi\varepsilon_0}\frac{q}{R^2}$$

can be applied. Hence, at 'a' (i.e. points just outside the sphere),

$$E = \frac{1}{4\pi\varepsilon_0}\frac{q}{R^2} = \frac{9.0\times10^9[\text{N-m}^2/\text{coul}^2]\times1.25\times10^{-7}[\text{coul}]}{(0.15[\text{m}])^2}$$

$$= 5 \times 10^4 \text{ N/coul}$$

at 'b', (i.e., for R' = 2.5m),

$$E = \frac{9.0\times10^9 [N\text{-}m^2/coul^2]\ 1.25\times10^{-7}[coul]}{(2.5[m])^2}$$

$$= 180\ N\text{-}coul$$

Now, at 'c', (i.e. anywhere inside the sphere), a special case of the Gauss's Law must be considered. For any $r<r_o$, it yields zero enclosed charge.

i.e., $$\oint \bar{E} \cdot d\bar{s} = 0$$

• PROBLEM 3-26

a) Use Gauss's law to determine the electric field between two coaxial cylinders of infinitelength and of radii a and b>a, if the inner one has a total charge per unit length of $-q_\ell$ while the outer cylinder has a charge per unit length of $+ q_\ell$. See the figure shown.
b) Calculate the capacitance.

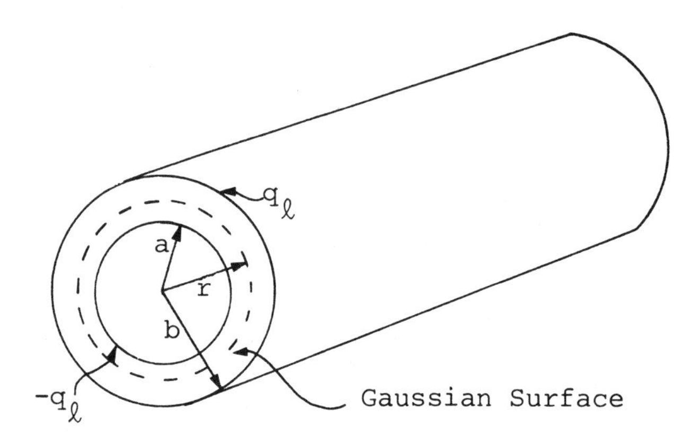

Solution: Choose a cylinder of radius r, a<r<b, and length L, as a Gaussian surface. Gauss's law is

$$\oint \bar{E} \cdot d\bar{a} = q/\varepsilon_0.$$ In cylindrical coordinates

$$\bar{E} = \hat{r}E_r + \hat{\theta}E_\theta + \hat{z}E_z$$

where $\hat{r}$, $\hat{\theta}$, and $\hat{z}$ are the unit mutually orthogonal vectors in cylindrical coordinates.

$$\oint (E_r\hat{r} + E_\theta\hat{\theta} + E_z\hat{z})\cdot(\hat{r}\ da) = \frac{q}{\varepsilon_0} = \frac{-q_\ell L}{\varepsilon_0}$$

$$\therefore \quad \oint E_r da = E_r \oint da = E_r(2\pi rL) = \frac{-q_\ell L}{\varepsilon_0}$$

$$\therefore \quad E_r = \frac{-q_\ell}{2\pi r\varepsilon_0}, \qquad \bar{E} = \frac{-\hat{r}q_\ell}{2\pi r\varepsilon_0}$$

b) The capacitance is defined as $C = \left|\frac{q}{\Delta V}\right|$ where $|\Delta V|$ is the magnitude of the potential difference.

$$\Delta V = -\int_a^b \bar{E}\cdot d\bar{r} = -\int_a^b \left(\frac{-q_\ell}{2\pi r\varepsilon_0}\right) dr$$

$$= \frac{q_\ell}{2\pi\varepsilon_0} \ell nr\Big|_a^b = \frac{q_\ell}{2\pi\varepsilon_0} \ell n \frac{b}{a}$$

$$C = \left|\frac{q_\ell L}{\frac{q_\ell}{2\pi\varepsilon_0}\ell n\frac{b}{a}}\right| = \frac{2\pi\varepsilon_0 L}{\ell n \frac{b}{a}}$$

• PROBLEM 3-27

a) Use Gauss's law to determine the electric field $\bar{E}$ between two concentric spheres of radii a and b>a, if the outer sphere carries a charge +q and the inner one a charge of -q. (See Figure.)
b) Calculate the capacitance of the two concentric spheres.

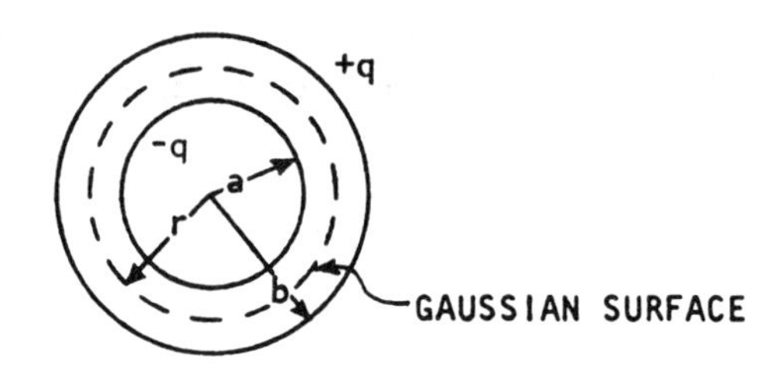

Solution: Gauss's law is

$$\oint_S \bar{E} \cdot d\bar{a} = \frac{q}{\varepsilon_0}$$

Choose as a Gaussian surface a sphere of radius r, where $a \le r \le b$. In spherical coordinates $\bar{E} = \hat{r}E_r + \hat{\theta}E_\theta + \hat{\phi}E_\phi$ where $\hat{r}$, $\hat{\theta}$, and $\hat{\phi}$ are the unit mutually orthogonal vectors, in spherical coordinates

$$\oint \bar{E} \cdot d\bar{a} = \oint (E_r\hat{r} + \hat{\theta}d\theta + \hat{\phi}d\phi) \cdot \hat{r}\, da = \frac{-q}{\varepsilon_0}$$

since the total enclosed charge is -q, and the normal to the Gaussian spherical surface is radial. Hence

$$\oint \bar{E} \cdot d\bar{a} = \oint E_r da = E_r \oint da = \frac{-q}{\varepsilon_0}$$

$$E_r(4\pi r^2) = \frac{-q}{\varepsilon_0}$$

$$E_r = \frac{-q}{4\pi r^2 \varepsilon_0}$$

$$\bar{E} = \frac{-\hat{r}\, q}{4\pi r^2 \varepsilon_0}$$

b) The capacitance is defined as $C = \left|\frac{q}{\Delta V}\right|$ where $|\Delta V|$ is the magnitude of the potential difference

$$\Delta V = -\int_a^b \bar{E} \cdot d\bar{r} = -\int_a^b \left(\frac{-q\,\hat{r}}{4\pi r^2 \varepsilon_0}\right) \cdot d\bar{r}$$

$$= \frac{q}{4\pi\varepsilon_0} \int_a^b \frac{dr}{r^2} = \frac{-q}{4\pi\varepsilon_0 r}\Bigg|_a^b = \frac{-q}{4\pi\varepsilon_0}\left(\frac{1}{b} - \frac{1}{a}\right)$$

$$\Delta V = \frac{q}{4\pi\varepsilon_0}\left(\frac{1}{a} - \frac{1}{b}\right)$$

$$C = \left|\frac{q}{\Delta V}\right| = \frac{q}{\frac{q}{4\pi\varepsilon_0}\left(\frac{1}{a} - \frac{1}{b}\right)} = \frac{4\pi\varepsilon_0 ab}{b - a}$$

CHARGES

• PROBLEM 3-28

There is an isolated positive point charge which has a value of 5×10^{-10} coulombs. What is the magnitude of E and D at a distance of 30 cm when the charge is located (a) in air; (b) in a large tank of glycerin ($\varepsilon_r = 50$)?

Solution: For air $\varepsilon = \frac{10^{-9}}{36\pi}$ and it is given that $q = 5 \times 10^{-10}$ coulomb and $r = 30$ cm. Therefore

$$E = \frac{q}{4\pi\varepsilon r^2} = \frac{5 \times 10^{-10}}{4\pi \times \left(\frac{10^{-9}}{36\pi}\right) \times 900 \times 10^{-4}}$$

$$= 50 \text{ volts/meter}$$

$$D = \varepsilon E = \frac{10^{-9}}{36\pi} \times 50 = 4.44 \times 10^{-10} \text{ coulomb/meter}^2$$

For glycerin

$$\varepsilon = \varepsilon_0\varepsilon_r = 50 \times \frac{10^{-9}}{36\pi}$$

Hence
$$E = \frac{5 \times 10^{-10}}{4\pi \times \left(\frac{50 \times 10^{-9}}{36\pi}\right) \times 900 \times 10^{-4}}$$

$$= 1 \text{ volt/meter}$$

$$D = \varepsilon_0\varepsilon_r E = \frac{50 \times 10^{-9}}{36\pi} \times 1$$

$$= 4 \cdot 42 \times 10^{-10} \text{ coulomb/meter}^2.$$

• PROBLEM 3-29

Describe the charges required to obtain the field

$$\bar{D} = \left(r + \frac{1}{r^2}\right)U(r_0 - r)1_r$$

Solution:

$$\text{div } \bar{D} = \frac{1}{r^2}\frac{\partial}{\partial r} r^2\left\{\left(r + \frac{1}{r^2}\right)U(r_0 - r)\right\}$$

$$= \frac{1}{r^2}\frac{\partial}{\partial r}(r^3 + 1)U(r_0 - r)$$

$$= \frac{1}{r^2}\{3r^2U(r_0 - r) - (r^3 + 1)\delta(r_0 - r)\}$$

$$= 3U(r_0 - r) - \left(r_0 + \frac{1}{r^2}\right)\delta(r_0 - r) \qquad (1)$$

Thus there must be a uniform charge density $\rho = 3\ C/m^3$ for $r < r_0$, and a surface charge $\rho_s = -[r_0 + (1/r_0^2)]$ on the sphere of radius $r = r_0$. However, due to the singularity of the coordinate system, it is possible that a point source at the origin might have been overlooked. If $S(r)$ is the surface of a sphere of radius r centered on the origin, then in this problem a direct calculation yields

$$\lim_{r\to 0}\oint_{S(r)} \bar{D}\cdot d\bar{S} = 4\pi$$

Thus, in addition to the two charge distributions previously determined, there is an additional point-charge $Q = 4\pi$ coulombs at the origin.

The same result could also have been obtained by realizing that for $r > r_0$, $\bar{D} \equiv 0$. Thus the charge system for $r \leq r_0$ must be neutral.

• PROBLEM 3-30

A sheet of paper charged with σ_0 coulomb per meter^2 is inserted between two parallel, oppositely charged metallic plates of area A (one side) as shown in the figure. Neglecting edge effects, find the charge densities on the surfaces of the metallic plates and the force on the left plate if the total charge on the positive metallic sheet is q.

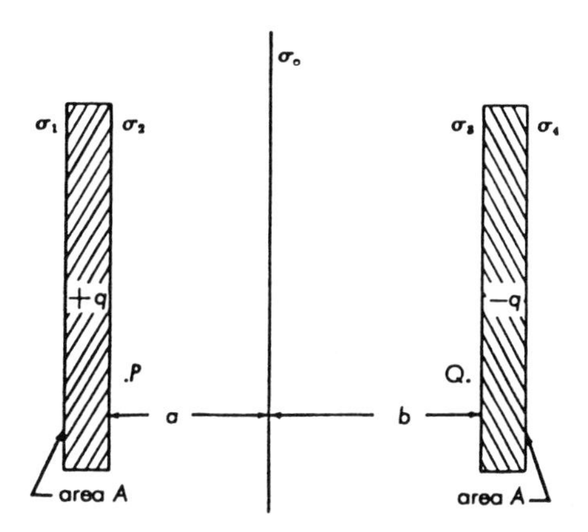

A sheet of charge between two oppositely charged metallic plates.

Solution: Use the law of conservation of charge and the law of superposition to solve this problem. Since the total charge on the metallic plates is fixed,

$$A(\sigma_1 + \sigma_2) = q \qquad A(\sigma_3 + \sigma_4) = -q$$

where the charge densities σ_1, σ_2, σ_3, σ_4 are on the surfaces indicated in the figure. The field inside each of the metallic plates is zero. Therefore, for a point inside the left plate,

$$\frac{\sigma_1}{2\varepsilon_0} - \frac{\sigma_2}{2\varepsilon_0} - \frac{\sigma_0}{2\varepsilon_0} - \frac{\sigma_3}{2\varepsilon_0} - \frac{\sigma_4}{2\varepsilon_0} = 0$$

where the positive direction for a component field is toward the right.

Similarly, for a point inside the right plate,

$$\frac{\sigma_1}{2\varepsilon_0} + \frac{\sigma_2}{2\varepsilon_0} + \frac{\sigma_0}{2\varepsilon_0} + \frac{\sigma_3}{2\varepsilon_0} - \frac{\sigma_4}{2\varepsilon_0} = 0$$

Solving the above four equations simultaneously for σ_1, σ_2, σ_3, σ_4,

$$\sigma_1 = \frac{\sigma_0}{2} \qquad \sigma_2 = \frac{q}{A} - \frac{\sigma_0}{2}$$

$$\sigma_3 = -\frac{q}{A} - \frac{\sigma_0}{2} \qquad \sigma_4 = \frac{\sigma_0}{2}$$

As a check on the correctness of these values, compute the resultant field E_P at a point P just outside the left plate.

$$E_P = \frac{\sigma_1}{2\varepsilon_0} + \frac{\sigma_2}{2\varepsilon_0} - \frac{\sigma_0}{2\varepsilon_0} - \frac{\sigma_3}{2\varepsilon_0} - \frac{\sigma_4}{2\varepsilon_0} = \frac{q}{\varepsilon_0 A} - \frac{\sigma_0}{2\varepsilon_0}$$

$$E_P = \frac{\sigma_2}{\varepsilon_0}$$

Similarly, for a point Q just outside the right plate,

$$E_Q = -\frac{\sigma_3}{\varepsilon_0}$$

The magnitude of the resultant electric field intensity at a point just outside a conductor is equal to the surface charge density divided by ε_0.

The force on the left plate can be obtained from

$$F = \frac{1}{2}\frac{\sigma_2^2}{\varepsilon_0}A - \frac{1}{2}\frac{\sigma_1^2}{\varepsilon_0}A$$

The second term on the right is negative since the outward drawn normal to the surface on which σ_1 resides points to the left.

$$F = \frac{1}{2\varepsilon_0}\left(\frac{q^2}{A} - \sigma_0 q\right)$$

If $\sigma_0 = 0$, then $F = 1/2\varepsilon_0(q^2/A)$. Thus, by measuring F and A, it is possible to determine q. A device that operates on this principle is called an absolute electrometer.

• PROBLEM 3-31

The electric displacement $\bar{D}$ in a given spherical region is given as

$$\bar{D} = \frac{5r^2}{4}\hat{r}\ \frac{C}{m^2}$$

Determine the total charge enclosed by the volume r = 4m, and $\theta = \pi/4$.

Solution: The differential form of Gauss's law is $\nabla\cdot\bar{D} = \rho$ and the total charge q is given by

$$q = \int_{\phi=0}^{2\pi}\int_{\theta=0}^{\theta=\pi/4}\int_{r=0}^{r=4}\rho dv = \int_{\phi=0}^{2\pi}\int_{\theta=0}^{\pi/4}\int_{r=0}^{4}\nabla\cdot\bar{D}\ dV$$

$$\nabla\cdot\bar{D} = \frac{1}{r^2}\frac{\partial}{\partial r}(r^2 D_r) = \frac{1}{r^2}\frac{d}{dr}\left(\frac{5r^4}{4}\right) = 5r$$

$$\therefore \quad q = \int_0^{2\pi} \int_0^{\pi/4} \int_0^4 (5r) r^2 \sin\theta dr\, d\theta\, d\phi$$

$$= \left.\frac{5r^4}{4}\right]_0^4 \left[-\cos\theta\right]_0^{\pi/4} \left[\phi\right]_0^{2\pi}$$

$$= \left(5 \cdot \frac{4^4}{4}\right)\left(1 - \frac{1}{\sqrt{2}}\right)(2\pi) = 589.1 \text{ c}$$

• PROBLEM 3-32

(a) In a certain region the components of electric field intensity are given as follows: $E_x = ax$, $E_y = ay$, and $E_z = 0$. Determine the shape of the lines of force and how much charge is contained in a cylinder of radius b and length L (with the z-axis as the symmetry axis).

(b) In a certain region, the radial and transverse components of electric field intensity are given as follows:

$$E_r = \frac{2p \cos\theta}{r^3} \qquad E_\theta = \frac{p \sin\theta}{r^3}$$

Again, find the shape of the lines of force and, in addition, the charge contained in a sphere of radius b described about the origin.

Solution: (a) The equation for the lines of force may be found by

$$\frac{E_y}{E_x} = \frac{dy}{dx}$$

$$\frac{y}{x} = \frac{dy}{dx}$$

$$\frac{dx}{x} = \frac{dy}{y}$$

$$\ln cx = \ln y$$

$$y = cx$$

The lines of force are straight lines passing through the origin.

To find the charge contained in a cylinder of radius r and length L extending along the z-axis, make use of Gauss' law.

$$\oiint \bar{E} \cdot d\bar{S} = \frac{q}{\varepsilon_0}$$

The resultant electric field intensity is radially directed away from the origin and hence coincides with the vector $d\bar{S}$ at any point on the cylindrical surface.

$$E = a\sqrt{x^2 + y^2} = ar$$

$$\iint_{(\text{cylinder})} \bar{E} \cdot d\bar{S} = (ar)(2\pi rL) = 2\pi ar^2 L = \frac{q}{\varepsilon_0}$$

There is no need to integrate over the two surfaces at the ends of the cylinder since $E_z = 0$.

(b) The equation for the lines of force may be found from

$$\frac{E_r}{E_\theta} = 2\,\frac{\cos\theta}{\sin\theta} = \frac{1}{r}\,\frac{dr}{d\theta}$$

$$2\,\frac{\cos\theta}{\sin\theta}\,d\theta = \frac{dr}{r}$$

$$2\ln(c\sin\theta) = \ln r$$

$$r = c^2 \sin^2\theta$$

This is the equation for the lines of force in polar coordinates. Note that c is a constant of integration.

To find the charge contained in a sphere of radius b described about the origin, make use of Gauss' law,

$$\oiint \bar{E} \cdot d\bar{S} = \frac{q}{\varepsilon_0}$$

In this problem, E_r and dS have the same direction. For any point on the spherical surface,

$$E_r = \frac{2p\cos\theta}{b^3} \qquad dS = b^2 \sin\theta\, d\theta\, d\phi$$

$$\oiint E_r\, dS = \int_0^{2\pi}\int_0^{\pi} \frac{2p}{b} \sin\theta\cos\theta\, d\theta\, d\phi$$

$$= \int_0^{2\pi} \frac{p}{b}\,[\sin^2\theta]_0^{\pi}\, d\phi = 0$$

The result is independent of the radius b of the spherical surface.

CHAPTER 4

POTENTIAL

WORK

• **PROBLEM 4-1**

Three point charges of values 1, 2, and 3 C are situated at the corners of an equilateral triangle of sides 1 m. It is desired to find the work required to move these charges to the corners of an equilateral triangle of shorter sides $\frac{1}{2}$ m as shown in the given figure.

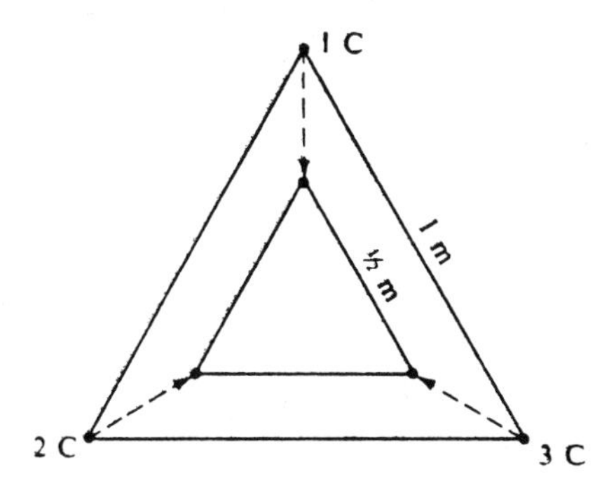

Bringing three point charges from the corners of a larger equilateral triangle to the corners of a smaller equilateral triangle.

Solution: The potential energy stored in the system of three charges at the corners of the larger equilateral triangle is given by

$$\frac{1}{2}\sum_{i=1}^{3} Q_i V_i = \frac{1}{2}\left[1\left(\frac{2}{4\pi\varepsilon_0} + \frac{3}{4\pi\varepsilon_0}\right) + 2\left(\frac{1}{4\pi\varepsilon_0} + \frac{3}{4\pi\varepsilon_0}\right)\right.$$

$$\left. + 3\left(\frac{1}{4\pi\varepsilon_0} + \frac{2}{4\pi\varepsilon_0}\right)\right]$$

$$= \frac{1}{2}\left[\frac{5 + 8 + 9}{4\pi\varepsilon_0}\right] = \frac{11}{4\pi\varepsilon_0} \text{ N-m}$$

The potential energy stored in the system of three charges at the corners of the smaller equilateral triangle is equal to twice the above value since all distances are halved. The increase in potential energy of the system in going from the larger to the smaller equilateral triangle is equal to $11/4\pi\varepsilon_0$ N-m. Obviously, this increase in energy must be supplied by an external agent and hence the work required to move the charges to the corners of the equilateral triangle of sides $\frac{1}{2}$ m from the corners of the equilateral triangle of sides 1 m is equal to $11/4\pi\varepsilon_0$ N-m.

• PROBLEM 4-2

Prescribe, by an appropriate unit vector, the direction in which a positive test charge should be moved from (1,3,-1) in the field $\bar{E} = -x\,\bar{a}_x + y\,\bar{a}_y + 3\,\bar{a}_z$ to experience (a) a maximum opposing force; (b) a maximum aiding force.

Solution: A test charge is to be moved from (1,3,-1) to another point in such a direction so as to achieve a maximum opposing or aiding force. Since the work done in moving a charge through a distance dl in field $\bar{E}$ is

$$W = -Q\int \bar{E} \cdot d\bar{l}$$

and F = W/Q, and the field at (1,3,-1) is

$$\bar{E} = -\bar{a}_x + 3\,\bar{a}_y + 3\,\bar{a}_z\,,$$

(a) the test charge must be moving in opposite direction as the field, or the new dl must be

$$dl = \bar{a}_x - 3\,\bar{a}_y - 3\,\bar{a}_z\,,$$

and the unit vector is

$$\bar{a}_\ell = \frac{\bar{a}_x - 3\,\bar{a}_y - 3\,\bar{a}_z}{\sqrt{1 + 9 + 9}} =$$

$$= 0.229\,\bar{a}_x - 0.688\,\bar{a}_y - 0.688\,\bar{a}_z$$

(b) For maximum aiding force the test charge should be moving in a direction such that the unit vector is

$$\bar{a}_\ell = \frac{-\bar{a}_x + 3\,\bar{a}_y + 3\,\bar{a}_z}{\sqrt{1 + 9 + 9}}$$

$$= -0.229\,\bar{a}_x + 0.688\,\bar{a}_y + 0.688\,\bar{a}_z$$

• PROBLEM 4-3

The force field $\bar{F} = y\,\bar{a}_x - x\,\bar{a}_y$ is nonconservative, and the work done in opposing the field,

$$-\int_B^A \bar{F} \cdot d\bar{l},$$

depends on the path followed from B to A. Let B be (0,1,0) and let A be (0,-1,0). Determine the work done in following these paths consisting of straight line segments:

(a) (0,1,0) to (1,1,0) to (1,-1,0) to (0,-1,0);

(b) (0,1,0) to (0,-1,0);

(c) (0,1,0) to (-1,1,0) to (-1,-1,0) to (0,-1,0).

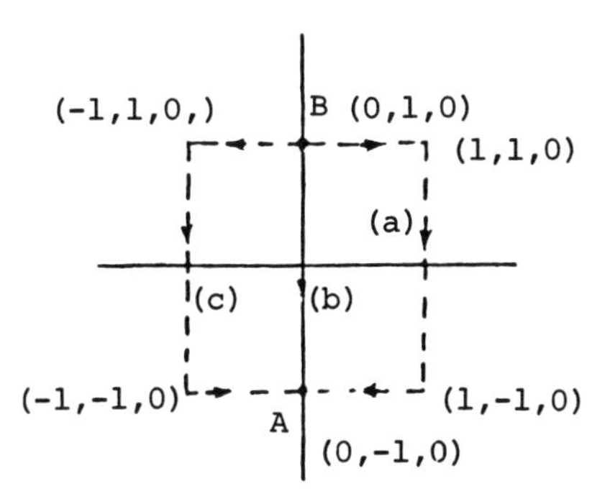

Solution: The paths are indicated as shown. Since

$$\bar{F} = y\,\bar{a}_x - x\,\bar{a}_y\,,$$

and

$$d\bar{l} = dx\,\bar{a}_x + dy\,\bar{a}_y\,,$$

and

$$W = - \int_B^A \bar{F} \cdot d\bar{l} ,$$

(a) from (0,1,0) to (1,1,0)

$$W_1 = - \int_{0,1,0}^{1,1,0} (y\ \bar{a}_x - x\ \bar{a}_y) \cdot (dx\ \bar{a}_x + dy\ \bar{a}_y)$$

$$= - \int_{0,1,0}^{1,1,0} y\ dx - x\ dy = - \int_0^1 dx = -1$$

from (1,1,0) to (1,-1,0)

$$W_2 = - \int_{1,1,0}^{1,-1,0} y\ dx - x\ dy = - \int_1^{-1} -dy = -2$$

from (1,-1,0) to (0,-1,0).

$$W_3 = - \int_{1,-1,0}^{0,-1,0} y\ dx - x\ dy = - \int_1^0 (-1)dx = -1$$

The total work done in path a is $W_1 + W_2 + W_3 = W_a$

$$= -4 \text{ Joules.}$$

(b) From (0,1,0) to (0,-1,0), $dx = 0, x = 0$

$$\therefore \quad W_b = - \int_1^{-1} -0\ dy = 0$$

(c) From (0,1,0) to(-1,1,0) to (-1,-1,0) to (0,-1,0),

$$W_c = - \left[\int_0^{-1} dx - \int_1^{-1} -dy + \int_{-1}^{0} dx \right] = - \left[-1-2-1\right] =$$

4 Joules.

• PROBLEM 4-4

A spherical cloud of charge is shown in the figure with uniform charge density ρ_0. Find the work required to assemble the charge cloud.

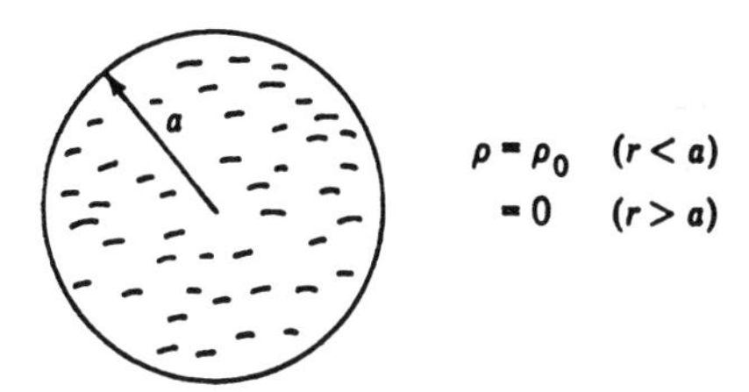

A spherical volume charge distribution.

Solution: Find the potential by applying Gauss' Law:

$$\oint\!\!\oint D \cdot \hat{n}\, ds = q_{encl}$$

$$D_r 4\pi r^2 = \rho_0 \frac{4}{3}\pi r^3 \qquad (r < a)$$

$$= \rho_0 \frac{4}{3}\pi a^3 \qquad (r > a)$$

$$E_r = \frac{\rho_0}{3\varepsilon_0} r \qquad (r < a)$$

$$= \frac{\rho_0 a^3}{3\varepsilon_0}\left(\frac{1}{r^2}\right) \qquad (r > a)$$

$$V = -\int_{\infty}^{r} \bar{E} \cdot d\bar{\ell}$$

$$V = \frac{\rho_0 a^3}{3\varepsilon_0}\,\frac{1}{r} \qquad (r > a)$$

$$= \frac{\rho_0 a^2}{2\varepsilon_0} - \frac{\rho_0 r^2}{6\varepsilon_0} \quad (r < a)$$

Substituting the potential expression for r < a into

$$W = \frac{1}{2} \iiint \rho V dv$$

yields:

$$W = \frac{1}{2} \int_0^{2\pi} \int_0^{\pi} \int_0^{a} \rho_0^2 \left[\frac{a^2}{2\varepsilon_0} - \frac{r^2}{6\varepsilon_0}\right] r^2 \sin\theta \, dr \, d\theta \, d\phi$$

$$= \frac{4\pi\rho_0^2 a^5}{15\varepsilon_0} = \frac{3q^2}{20\pi\varepsilon_0}\left(\frac{1}{a}\right)$$

where

$$q = \rho_0\left(\frac{4}{3}\pi a^3\right)$$

If the radius approaches zero while the charge q remains constant, W becomes infinite, indicating that an infinite amount of work must be done to assemble a point charge. This energy is sometimes called the self energy.

• PROBLEM 4-5

Consider the electric field given by

$$\bar{E} = y\,\bar{i}_y \, .$$

Determine the work done by the field in carrying 3μC of charge from the point A(0,0,0) to the point B(1,1,0) along the parabolic path $y = x^2$, $z = 0$ shown in Fig. 1. First use approximate methods, then use the line integral to get an exact result.

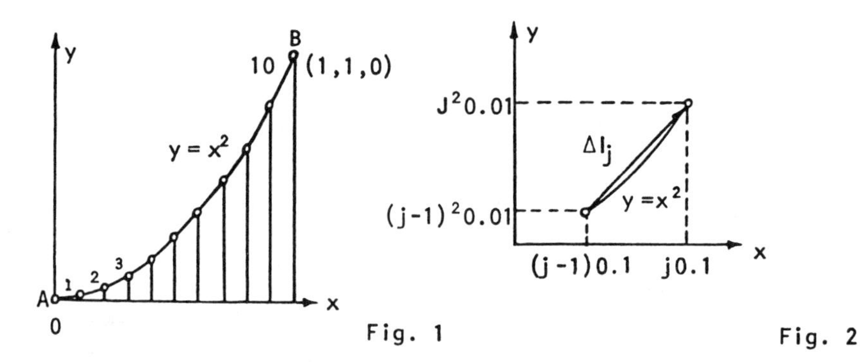

Fig. 1 Fig. 2

Solution: For the approximate method, divide the path into ten segments for convenience as shown in fig. 1. The segments will be numbered 1,2,3,...,10. The coordinates of the starting and ending points of the jth segment are as shown in fig. 2. The electric field at the start of jth segment is given by

$$\bar{E}_j = (j-1)^2 0.01\bar{i}_y$$

The length vector corresponding to the jth segment, approximated as a straight line connecting its starting and ending points is

$$\Delta\bar{l}_j = 0.1\bar{i}_x + [j^2-(j-1)^2]\,0.01\bar{i}_y$$

$$= 0.1\bar{i}_x + (2j-1)0.01\bar{i}_y$$

The required work is then given by

$$W_A^B = q \sum_{j=1}^{n} E_j \cos\alpha_j \,\Delta l_j$$

or

$$W_A^B = q \sum_{j=1}^{n} \bar{E}_j \cdot \Delta\bar{l}_j$$

$$= 3 \times 10^{-6} \sum_{j=1}^{10} \bar{E}_j \cdot \Delta\bar{l}_j$$

$$= 3 \times 10^{-6} \sum_{j=1}^{10} [(j-1)^2 0.01\bar{i}_y] \cdot [0.1\bar{i}_x + (2j-1)0.01\bar{i}_y]$$

$$= 3 \times 10^{-10} \sum_{j=1}^{10} (j-1)^2(2j-1)$$

$$= 3 \times 10^{-10}\ [0+3+20+63+144+275+468+735+1088+1539]$$

$$= 3 \times 10^{-10} \times 4335 = 1.3005\ \mu J$$

The result obtained for W_A^B, is approximate since the path from A to B was divided into a finite number of segments. By dividing it into larger and larger numbers of segments, more and more accurate results can be obtained. The value to which the result is exact is that for which $n = \infty$. The summation then becomes an integral, which represents exactly the work done by the field and is given by

$$W_A^B = q \int_A^B \bar{E} \cdot d\bar{l}$$

To compute the exact value of the work done by the electric field, note that at any arbitrary point (x,y,0) on the curve $y = x^2$, $z = 0$, the infinitesimal length vector tangential to the curve is given by

$$d\bar{l} = dx\,\bar{i}_x + dy\,\bar{i}_y$$

$$= dx\,\bar{i}_x + d(x^2)\bar{i}_y$$

$$= dx\,\bar{i}_x + 2x\,dx\,\bar{i}_y$$

The value of $\bar{E} \cdot d\bar{l}$ at the point (x,y,0) is

$$\bar{E} \cdot d\bar{l} = y\bar{i}_y \cdot (dx\,\bar{i}_x + dy\,\bar{i}_y)$$

$$= x^2\bar{i}_y \cdot (dx\,\bar{i}_x + 2x\,dx\,\bar{i}_y)$$

$$= 2x^3dx$$

Thus the required work is given by

$$W_A^B = q\int_A^B \bar{E} \cdot d\bar{l} = 3 \times 10^{-6} \int_{(0,0,0)}^{(1,1,0)} 2x^3dx$$

$$= 3 \times 10^{-6} \left[\frac{2x^4}{4}\right]_{x=0}^{x=1} = 1.5\ \mu J$$

• PROBLEM 4-6

Consider the nonuniform field

$$\bar{E} = y\bar{a}_x + x\bar{a}_y + 2\bar{a}_z$$

Determine the work expended in carrying 2 C from B (1,0,1) to A (0.8, 0.6, 1) along the shorter arc of the circle

$$x^2 + y^2 = 1 \qquad z = 1$$

Solution: Working in cartesian coordinates, the differential path dL is $dx\,\bar{a}_x + dy\,\bar{a}_y + dz\,\bar{a}_z$, and the integral becomes

$$W = -Q\int_B^A \bar{E} \cdot d\bar{L}$$

$$= -2 \int_B^A (y\bar{a}_x + x\bar{a}_y + 2\bar{a}_z) \cdot (dx\ \bar{a}_x + dy\ \bar{a}_y + dz\ \bar{a}_z)$$

$$= -2 \int_1^{0.8} y\ dx - 2 \int_0^{0.6} x\ dy - 4 \int_1^1 dz$$

where the limits on the integrals have been chosen to agree with the initial and final values of the appropriate variable of integration. Using the equation of the circular path (and selecting the sign of the radical which is appropriate for the quadrant involved), the work is:

$$W = -2 \int_1^{0.8} \sqrt{1 - x^2}\ dx - 2 \int_0^{0.6} \sqrt{1 - y^2}\ dy - 0$$

$$= -[x\sqrt{1 - x^2} + \sin^{-1} x]_1^{0.8} - [y\ \sqrt{1-y^2} + \sin^{-1} y]_0^{0.6}$$

$$= -(0.48 + 0.927 - 0 - 1.571) - (0.48 + 0.644 - 0 - 0)$$

$$= 0.96\ \text{J}$$

Now select the straight-line path from B to A, then determine the equations of the straight line. Any two of the following three equations for planes passing through the line are sufficient to define the line:

$$y - y_B = \frac{y_A - y_B}{x_A - x_B} (x - x_B)$$

$$z - z_B = \frac{z_A - z_B}{y_A - y_B} (y - y_B)$$

$$x - x_B = \frac{x_A - x_B}{z_A - z_B} (z - z_B)$$

From the first equation above,

$$y = -3(x-1)$$

and from the second,

$$z = 1$$

Thus

$$W = -2 \int_1^{0.8} y\, dx - 2 \int_0^{0.6} x\, dy - 4 \int_1^{1} dz$$

$$= 6 \int_1^{0.8} (x - 1)\, dx - 2 \int_0^{0.6} \left(1 - \frac{y}{3}\right) dy$$

$$= -0.96 \text{ J}$$

This is the same answer found using the circular path between the same two points, and it demonstrates the statement that the work done is independent of the path taken in any electrostatic field.

• PROBLEM 4-7

An electric field is represented by

$$\bar{E} = \hat{i}ay + \hat{j}ax, \text{ where}$$

$$a = 100 \text{ volts/m}^2.$$

Find (a) the potential function V, taking v = 0 at the origin.

(b) the work done by the field when a charge $q = 10^{-8}$ coul. is taken from x = -1m, y = -2m to x = 2m, y = 3m. (See the accompanying figure).

(c) the charge density ρ at any point.

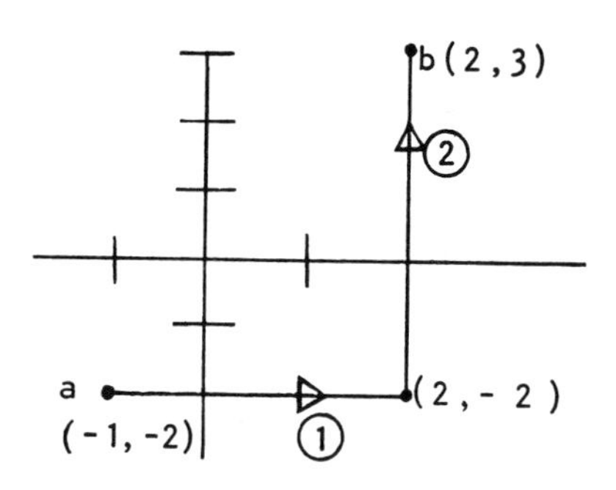

Solution: The electric field $\bar{E}$ is related to the scalar potential V by $\bar{E} = -\nabla V$

$$\therefore \quad E_x = \frac{-\partial v}{\partial x} = ay.$$

$$v = -axy + f(y)$$

where f(y) is a possible function of y.

$$E_y = -\frac{\partial v}{\partial y} = -\ axy + g(x).$$

where g(x) is a possible function of x.

$$\therefore \quad v = -\ axy + f(y) + g(x)$$

But v = 0 at x = y = 0.

therefore f(y) + g(x) = 0.

and v = - axy.

(b) $q = 10^{-8}$ coul.

$$x_a = -\ 1m,\ y_a = -\ 2m$$

$$x_b = 2m, \quad y_b = 3m$$

w, field work done, is given by the line integral

$$w = q\int_a^b \bar{E}.\ d\bar{l} = q\int_a^b (E_x dx + E_y dy)$$

$$= 10^{-8}\left\{\underset{\text{path (1)}}{\int_{-1}^{2} E_x dx} + \underset{\text{path (2)}}{\int_{-2}^{3} E_y dy}\right\}$$

$E_x = ay$, $E_y = ax$

$$w = 10^{-8}\left\{\int_{-1}^{2} aydx + \int_{-2}^{3} axdy\right\}$$

$$= 10^{-8}\left\{ayx\Big|_{x=-1}^{x=2} + axy\Big|_{y=-2}^{y=3}\right\}$$

$$= 10^{-8}\left\{(100)(-2)[2 - (-1)] + 100(2)[3 - (-2)]\right\}$$

$$= (10^{-8})(100)\left\{-2(3) + 2(5)\right\}.$$

$$w = 400 \times 10^{-8} \text{ joules}$$

(c) Use the differential form of Gauss's law in vacuum to find the charge density ρ

$$\nabla.\bar{E} = \rho/\varepsilon_0$$

where ε_0 is the permittivity of free space.

$$\rho = \varepsilon_0\, \nabla.\bar{E} = \varepsilon_0 \left(\frac{\partial E_x}{\partial x} + \frac{\partial E_y}{\partial y}\right)$$

$$\rho = \varepsilon_0 \left\{\frac{\partial(ay)}{\partial x} + \frac{\partial(ax)}{\partial y}\right\} = 0$$

$$\rho = 0.$$

POTENTIAL

• PROBLEM 4-8

Calculate the total electric potential energy of a thin spherical conducting shell of radius R and carrying charge q.

Solution: The electric energy density u is

$$u = \frac{1}{2}\varepsilon_0 E^2.$$

The magnitude E of the electric field for the charged spherical shell is

$$E = \frac{1}{4\pi\varepsilon_0}\frac{q}{r^2} \qquad r \geq R$$

$$= 0 \qquad r < R.$$

The electric potential energy dU contained within a spherical shell of radius r, thickness dr and concentric with the spherical conducting shell is

$$dU = u4\pi r^2\, dr = 2\pi\varepsilon_0 E^2 r^2\, dr.$$

Therefore, the total energy in the electric field surrounding the conducting sphere is

$$U = \int dU = \int_R^\infty 2\pi\varepsilon_0 \left(\frac{q}{4\pi\varepsilon_0 r^2}\right)^2 r^2\, dr$$

$$= \frac{q^2}{8\pi\varepsilon_0}\int_R^\infty \frac{dr}{r^2}$$

$$= \frac{q^2}{8\pi\varepsilon_0 R} \text{ joules.}$$

• PROBLEM 4-9

Four point charges Q_1, Q_2, Q_3 and Q_4 are suspended in the air such that Q_1, Q_2 and Q_3 form the vertices of an equilateral triangle of side a = 2m, while charge Q_4 is located at a distance of R = 6m, along the perpendicular bisector of the line joining Q_1 and Q_3. Assuming

$$Q_1 = Q_2 = Q_3 = +3.0 \times 10^{-8}c$$

and $$Q_4 = -1.0 \times 10^{-8}c$$

find the potential at point P, as shown in the figure below.

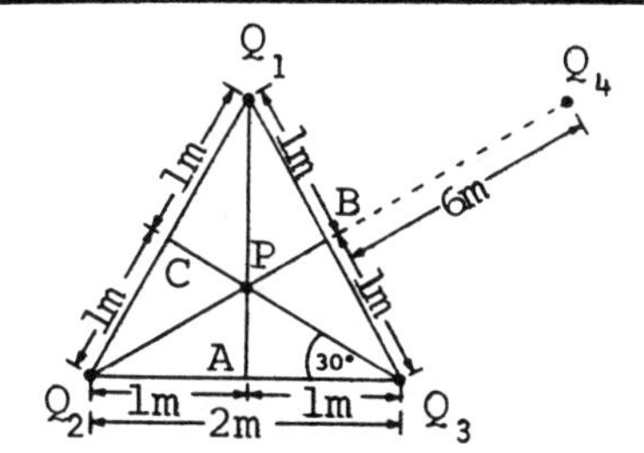

Solution: In order to find the potential due to one point charge at point P, first of all, Q_1P, Q_2P and Q_3P must be computed.

Now from ΔAPQ_3,

$$\measuredangle PQ_3A = 30^0 \quad \text{and} \quad AQ_3 = 1\text{m}$$

Hence $$PA = (\tan 30^0) \times AQ_3$$

$$= 0.577 \times 1 = 0.577\text{m}$$

From ΔQ_1AQ_3,

$$Q_1A = AQ_3 \times (\tan 60^0)$$

$$= 1.732\text{m}$$

So by substituting values of Q_1A and PA, we get

$$Q_1P = Q_2P = Q_3P = 1.73205 - 0.577$$

$$= 1.155\text{m}.$$

Similarly,

$$Q_4P = 6\text{m} + PB$$

$$= 6\text{m} + PA$$

$$= 6.577\text{m}.$$

So, the potential due to each point charge can be computed as follows:

V_{Q_1} = The potential at P due to Q = V_{Q_2}

$$= \frac{1}{4\pi\varepsilon_0} \cdot \frac{3.0 \times 10^{-8}}{Q_1P}$$

$$= (9.0 \times 10^9 \text{ N-m}^2/\text{c}^2)(3.0 \times 10^{-8})/1.155$$

$$= 233.766 \text{ v}$$

$$V_{Q_3} = +V_{Q_1} = +V_{Q_2}$$

$$= +233.766 \text{ v}$$

$$V_{Q_4} = \frac{1}{4\pi\varepsilon_0 \cdot 1} \cdot \frac{-1.0 \times 10^{-8}}{Q_4P}$$

$$= \frac{-(9.0 \times 10^9)(1.0 \times 10^{-8})}{6.577}$$

$$= -13.684 \text{ v.}$$

The net potential at P = V_P is the sum of individual voltages.

$$V_P = V_{Q_1} + V_{Q_2} + V_{Q_3} + V_{Q_4}$$

$$= 233.766 + 233.766 + 233.766 - 13.684$$

$$= 687.614\text{V}$$

• **PROBLEM 4-10**

An isolated spherical conductor of radius R carries a surface charge density σ. Find the potential energy in terms of R.

Solution:

$$W = \frac{1}{2}\int_S \sigma V \, da = \frac{1}{2} QV,$$

where

$$V = \frac{Q}{4\pi\varepsilon_0 R} .$$

Thus

$$W = \frac{Q^2}{8\pi\varepsilon_0 R}$$

We can also write that

$$W = \frac{\varepsilon_0}{2}\int_\tau E^2 \, d\tau ,$$

$$= \frac{\varepsilon_0}{2}\int_R^\infty \left(\frac{Q}{4\pi\varepsilon_0 r^2}\right)^2 4\pi r^2 \, dr = \frac{Q^2}{8\pi\varepsilon_0 R} ,$$

which is the same result as above. We have integrated from R to infinity, since there is zero E inside the sphere.

• PROBLEM 4-11

Find the potential function produced by a point charge q at the origin if the point (1,0,0) is chosen to be at ground potential.

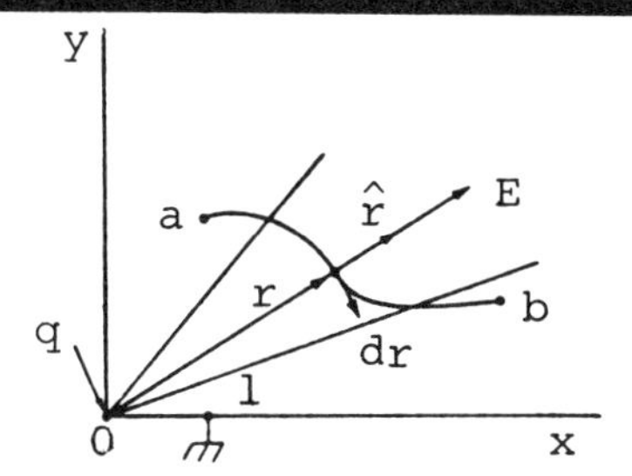

Solution: Here $\bar{E} = (1/4\pi\varepsilon_0)(q/r^2)\hat{r}$. In the figure shown let ϕ_{ab} be the potential of a relative to b. Then

$$\phi_{ab} = \phi_a - \phi_b = \int_a^b \bar{E} \cdot d\bar{r} = \frac{q}{4\pi\varepsilon_0}\int_a^b \frac{\hat{r}}{r^2} \cdot dr = \frac{q}{4\pi\varepsilon_0}\int_a^b \frac{dr}{r^2}$$

$$= \frac{q}{4\pi\varepsilon_0}\left(\frac{1}{r_a} - \frac{1}{r_b}\right)$$

Normally the ground would be taken at infinity; i.e., $\phi_b = 0$ when $r_b = \infty$. So the potential at a would then be

$$\phi_a = \frac{q}{4\pi\varepsilon_0}\left(\frac{1}{r_a}\right) \quad \text{or} \quad \phi = \frac{q}{4\pi\varepsilon_0}\left(\frac{1}{r}\right)$$

The equipotential surfaces, the loci of all points such that ϕ has some constant value, would be spheres about the origin. If another sphere, $r_b = 1$ is taken as that equipotential surface for which $\phi_b = 0$, then

$$\phi_a - 0 = \frac{q}{4\pi\varepsilon_0}\left(\frac{1}{r_a} - \frac{1}{1}\right) \quad \text{or} \quad \phi = \frac{q}{4\pi\varepsilon_0}\left(\frac{1}{r} - 1\right)$$

• PROBLEM 4-12

A sphere of radius 'a' with center at the origin contains a space charge of density $\rho = a^2 - r^2$. There are no charges exterior to the sphere so that the resulting sphere possesses spherical symmetry. Using Gauss's theorem, find the fields for the regions $r < a$, and $a < r < \infty$. Assume that $V = V_r$ at $r = a$, find the expressions for the potentials in the same regions.

Solution: Since ρ is a function of r only, evaluate the surface and volume integrals in terms of r as follows:

$$\int_s ds = \int_0^{2\pi} \int_0^{\pi} r^2 \sin\theta d\theta d\phi = 4\pi r^2$$

$$\int_v dv = \int_0^{2\pi} \int_0^{\pi} \int_0^{r} r^2 \sin\theta dr d\theta d\phi$$

$$= 4\pi \int_0^r r^2 dr.$$

(a) For $r < a$, by Gauss's theorem

$$4\pi r^2 D_r = 4\pi \int_0^r (a^2 - r^2)\, r^2 dr$$

$$= 4\pi \left[\frac{a^2 r^3}{3} - \frac{r^5}{5}\right]$$

$$\therefore D_r = \frac{a^2 r}{3} - \frac{r^3}{5} \text{ and}$$

$$E_r = \frac{a^2 r}{3\varepsilon_0} - \frac{r^3}{5\varepsilon_0}$$

$$V = -\int \bar{E}.d\bar{l} + C_1$$

$$= -\int_0^r \left(\frac{a^2 r}{3\varepsilon_0} - \frac{r^3}{5\varepsilon_0}\right) dr + C_1$$

$$= -\left[\frac{a^2 r^2}{6\varepsilon_0} - \frac{r^4}{20\varepsilon_0}\right] + C_1$$

At $r = a$, $V = V_r$, thus

$$C_1 = V_r + \frac{7a^4}{60\varepsilon_0}$$

Therefore

$$V = -\left[\frac{a^2 r^2}{6\varepsilon_0} - \frac{r^4}{20\varepsilon_0}\right] + \frac{7a^4}{60\varepsilon_0} + V_r$$

(b) For $a < r < \infty$

$$4\pi r^2 D_r = 4\pi \int_0^a (a^2 - r^2) r^2 dr$$

$$= 4\pi \left[\frac{a^5}{3} - \frac{a^5}{5}\right]$$

$$= \frac{8\pi a^5}{15}$$

$$D_r = \frac{2}{15} \frac{a^5}{r^2} \quad \text{and} \quad E_r = \frac{2}{15\varepsilon_0} \frac{a^5}{r^2}$$

$$V = -\int \frac{2}{15\varepsilon_0} \frac{a^5}{r^2} dr + C_2 = -\frac{2}{15\varepsilon_0} \frac{a^5}{r} + C_2$$

At $r = a$, $V = V_r$, solve for C_2,

$$C_2 = V_r + \frac{2}{15\varepsilon_0} a^4$$

Therefore

$$V = -\frac{2}{15\varepsilon_0} \frac{a^5}{r} + V_r + \frac{2}{15\varepsilon_0} a^4$$

• PROBLEM 4-13

Three equal spheres are placed at the corners of an equilateral triangle. When their potentials are ϕ, 0, 0 their charges are E, E', E' respectively. Show that, when each sphere is at a potential ϕ', each has a charge $(2E' + E)\phi'/\phi$.

Find the potentials when the charges are E", 0, 0.

Solution: This is an example on Green's reciprocal theorem; we have two equilibrium states in which the potentials and charges of the spheres are

(i) $\phi, 0, 0;\ E, E', E'$, (ii) $\phi', \phi', \phi';\ e', e', e'$,

where e' is the charge to be found. Using the relation $\Sigma e'\phi = \Sigma e\phi'$

$$e'\phi = (E + 2E')\phi', \tag{a}$$

giving the answer.

There is a third equilibrium state in which the potentials and charges are, say,

(iii) $\phi'', \phi_0, \phi_0; E'', 0, 0$.

Applying Green's reciprocal theorem, using the equilibrium states (i) and (iii); (ii) and (iii), respectively,

$$E''\phi = E\phi'' + 2E'\phi_0, \quad E''\phi' = e'(\phi'' + 2\phi_0).$$

Eliminating the ratio ϕ'/e' from (a) and the second equation obtain two simultaneous equations for the unknown potentials ϕ_0, ϕ'', namely,

$$\left.\begin{aligned} E\phi'' + 2E'\phi_0 &= E''\phi, \\ (E + 2E')(\phi'' + 2\phi_0) &= E''\phi. \end{aligned}\right\} \qquad \text{(b)}$$

Subtracting these equations

$$E'\phi'' + (E + E')\phi_0 = 0,$$

so that

$$\frac{\phi''}{E + E'} = -\frac{\phi_0}{E'} = \lambda \text{ say.}$$

Substitution in (b) gives

$$\lambda(E^2 + EE' - 2E'^2) = E''\phi,$$

and hence, finally,

$$\phi'' = \frac{E''(E + E')\phi}{E^2 + EE' - 2E'^2}, \quad \phi_0 = -\frac{E'E''\phi}{E^2 + EE' - 2E'^2}.$$

• PROBLEM 4-14

Consider two point electrodes that are embedded in a homogeneous and isotropic medium of conductivity σ and which are separated by a distance L. Suppose also that a current I is conducted to one electrode, designated as Q_1, and the current is conducted away from the second electrode Q_2. Find the equipotential distribution in the medium.

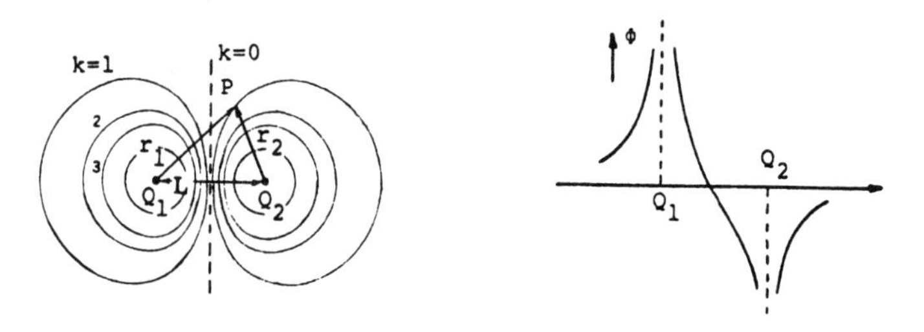

Equipotentials of two point charges of opposite sense.

Solution: The geometry of the system is illustrated in the Figure. The potential at the point P is obtained by combining the potentials at P due to each electrode separately. This simple procedure for combining potentials is possible because of the linear relationship that exists between the current and the potential. This allows simple linear superposition of the effects of each electrode. This is a broad principle that applies for linear systems and is known as the superposition principle. The potential at point P is given by:

$$\Phi_p = \frac{I}{4\pi\sigma}\left(\frac{1}{r_1} - \frac{1}{r_2}\right)$$

To find the equipotential surfaces, it is recalled that these must satisfy the condition that Φ = constant over the equipotential surface. This permits

$$\frac{1}{r_1} - \frac{1}{r_2} = \frac{4\pi\sigma\Phi}{I} = \frac{k}{L}$$

where k is a constant. It follows from this that

$$r_1 = \frac{r_2}{1 + kr_2/L}$$

This equation relates the distance from each source to the equipotential surfaces and allows these surfaces to be drawn. If k is given the values k = 0,1,2 . . . equipotential surfaces having equal potential differences will result. The figure gives a sketch of these potential surfaces.

• PROBLEM 4-15

Given $V = 100 \sin(\pi x/2) \exp(-\pi y/2)$. Describe the surfaces for V = 0. Also find the expression for the surface of constant potential V = 100 in the region $0 < x < 2$.

Solution: The expression for a constant potential can be obtained by letting V = constant and solving for x in terms of y. Thus for V = 0,

$$\text{Sin}(\pi x/2)\, e^{-\pi y/2} = 0$$

This equation can be satisfied by x = 0, 2,....2m and for any value of y. Thus the surfaces represented by:

$x = 0$, any values of y and z

and

$x = 2$,

for any values of y and z represent zero potential surfaces.

At $y = -\infty$, the potential is also zero.

To find the expression for V = 100, set

$$\sin(\pi x/2)e^{-\pi y/2} = 1$$

and solve x in terms of y,

$$x = \frac{2}{\pi}\sin^{-1}(e^{\pi y/2})$$

for any z.

This represents a surface for V = 100.

To plot this surface, check these points:

$x = 0$,	$y = -\infty$,
$x = 2$,	$y = -\infty$,
$x = 1$,	$y = 0$.
$y = -1$,	$x = .13$
	$= 1.87$

These points are sketched as shown.

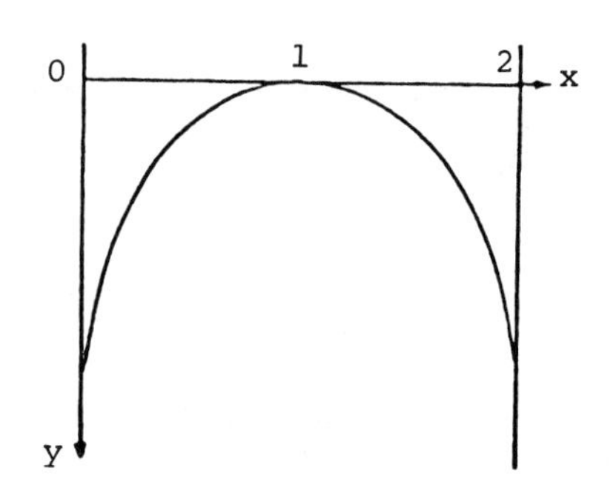

• PROBLEM 4-16

A sinusoidal voltage having a frequency of 1 MHz and a peak value of 10 volts is applied to the plates of a parallel plate capacitor which are 2 cm. apart. If an electron is released from one plate at an instant when the applied voltage is zero, find the position of the electron at any subsequent time t. Assume that the initial velocity of the electron is 10^6 m/sec in the X-direction, which is perpendicular to the plates. No magnetic field is present.

Solution: Neglecting fringing of the electric field, the electric field intensity for a parallel plate capacitor is

$$E = \frac{v}{d} = \frac{V\rho \sin(\omega t)}{d}$$

$$E = \frac{10}{0.02} \sin(2\pi \times 10^6 t) = 500 \sin(2\pi \times 10^6 t)$$

Volt/m

and by

$$m\frac{dv_x}{dt} = F = q[E_x + (\bar{V} \times \bar{B})_x]$$

for one dimension and for B = 0.

$$\frac{dv_x}{dt} = \frac{eE}{m} = 1.76 \times 10^{11} \times 500 \sin(2\pi \times 10^6 t)$$

$$= 8.80 \times 10^{13} \sin(2\pi \times 10^6 t) \quad \text{m/sec}^2$$

Integrate this differential equation to get

$$v_x = -1.40 \times 10^7 \cos(2\pi \times 10^6 t) + A$$

where A = constant of integration. A is determined from the initial condition that

$$v_x(0) = 10^6 \text{ m/sec} \qquad \text{when} \quad t = 0$$

thus

$$v_x = 1.50 \times 10^7 - 1.40 \times 10^7 \cos(2\pi \times 10^6 t) \quad \text{m/sec}$$

Integrating this equation with respect to t, subject to the initial condition:

$$t = 0, \; x = 0$$

$$x = 1.50 \times 10^7 t - 2.23 \sin (2\pi \times 10^6 t) \quad \text{m.}$$

• PROBLEM 4-17

A uniformly charged ring with total charge q and radius c is concentric and coplanar to one with charge -q and radius b, where b > a. See the figure shown. Find the potential at a distance r (far away) from the center of the ring.

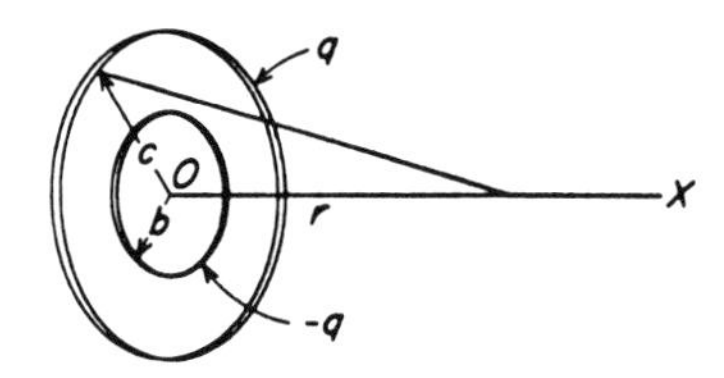

Solution: The potential at a distance r from the ring-charge +q is

$$V_+ = \frac{q}{4\pi\varepsilon_0 (r^2 + c^2)^{\frac{1}{2}}}$$

since the potential due to each element of the ring is the same. If $c \ll r$, expand the denominator as follows

$$V_+ = \frac{q}{4\pi\varepsilon_0 r} \frac{1}{(1 + c^2/r^2)^{1/2}} = \frac{q}{4\pi\varepsilon_0 r}\left(1 - \frac{c^2}{2r^2} + \cdots\right) .$$

Similarly, for the negative ring (neglecting higher order terms)

$$V_- = \frac{-q}{4\pi\varepsilon_0 r}\left(1 - \frac{b^2}{2r^2}\right)$$

and adding

$$V = \frac{q}{4\pi\varepsilon_0 r^3}\left(\frac{b^2 - c^2}{2}\right)$$

• PROBLEM 4-18

Find the potential at a distance S from the center of a small conducting sphere of radius a on which is a charge +Q uniformly distributed over the surface of the sphere. (See figure)

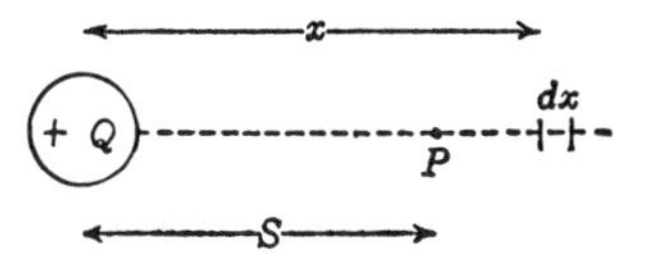

Fig. 1

Solution: For S greater than a: Since in bringing up a unit charge from a long way off ("∞") the repelling force is not constant, the work done must be found by calculus methods. At any general point distant x from Q, the repelling force will be

$$\frac{(Q)(1)}{kx^2}$$

The work done in the small distance, dx, will be given by

$$\frac{Q}{kx^2}(-dx).$$

(The minus sign appears because x is decreasing.) The expression for the whole work done from "∞" to P will therefore be

$$\text{Work} = \int_{\infty}^{S} -\frac{Q}{kx^2}\,dx = \frac{Q}{kS}$$

Hence the potential at P is equal to $\frac{Q}{kS}$.

Electric potential is not a vector quantity, so that by the principle of superposition the potential at any point due to any number of charges or to any distribution of charge is the algebraic sum of the potentials at that point produced by each charge or by each element of the distribution separately.

The case where S is less than a is most simply solved by an application of Gauss' law. By Gauss' law there can be no electric intensity within the charged conductor due to surface charges. Consequently if E within is everywhere zero,

$$\int E\,dx$$

must have a constant value throughout the volume of the conductor, i.e., the potential at all inside points is the same as the potential at the surface of the conductor, or

$$V = \frac{Q}{ka}$$

• PROBLEM 4-19

A charged uniform wire is wound into a circular ring of radius a. Find the potential on the axis if the total charge on the wire is q and if the diameter of the wire d is much less than a (d ≪ a).

Solution: Because the wire is thin compared to the dimension of interest (a), one may idealize the wire as infinitely thin. Then the potential ϕ at the point P is given by

$$\phi = \frac{1}{4\pi\varepsilon_0}\int_L \frac{\lambda}{r_{QP}}\, ds \tag{1}$$

where λ is the charge per unit length, r_{QP} is the distance between the point P and a point Q on the wire, and the integration is extended over the length L of the wire. By symmetry, it can be assumed that the charge q is uniformly distributed along the circumference.

$$\lambda = \frac{q}{2\pi a} \tag{2}$$

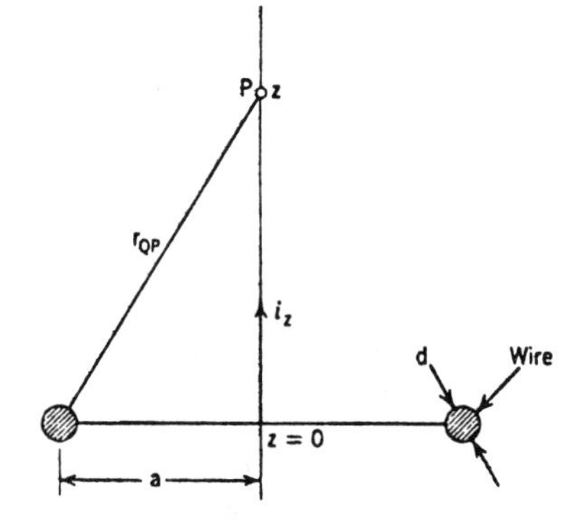

Choose a cylindrical coordinate system with its origin at the center of the ring (see the given figure). Then for any point z along the axis

$$r_{QP} = \sqrt{z^2 + a^2} \tag{3}$$

The distance r_{QP} is independent of the variable of integration:

$$ds = ad\phi$$

The integration gives

$$\Phi = \frac{\lambda}{4\pi\varepsilon_0} \int_0^{2\pi} \frac{ad\phi}{\sqrt{z^2 + a^2}} = \frac{\lambda}{4\pi\varepsilon_0} \cdot \frac{2\pi a}{\sqrt{z^2 + a^2}} \qquad (4)$$

The field on the axis is entirely z-directed, as can easily be inferred by symmetry. Thus

$$\bar{E} = -\bar{\nabla}\bar{\Phi} = -\frac{d\bar{\Phi}}{dz} = \frac{\lambda}{4\pi\varepsilon_0} \frac{2\pi\, az}{(\sqrt{z^2 + a^2})^3} \qquad (5)$$

• PROBLEM 4-20

A P-N junction device has junction located in the plane x=0 as shown in figure 1. The potential in the device is given by

$$\phi(x) = \left[1 + \frac{x}{\sqrt{x^2 + a^2}}\right] \frac{V_0}{2}$$

where a is the width of the junction region and V_0 is the overall potential difference.

a) Find the field E and the volume charge density ρ_v.

b) Plot the variation of ϕ, E, and ρ_v as a function of x.

c) What is the net charge Q in the device?

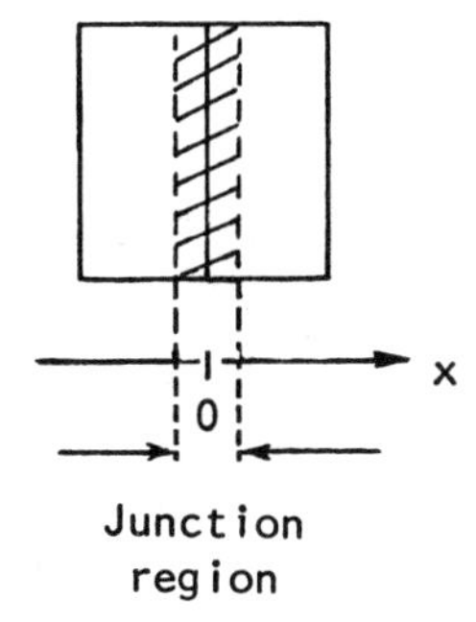

Fig. 1

Solution: a) $\bar{E} = -\bar{\nabla}\phi$

$$E_x = -\frac{\partial\phi}{\partial x} = \frac{-V_0}{2}\left[\frac{1}{\sqrt{x^2 + a^2}} - \frac{x^2}{(x^2 + a^2)^{3/2}}\right]$$

$$E_x = \frac{-V_0}{2}\left[\frac{x^2 + a^2 - x^2}{(x^2 + a^2)^{3/2}}\right]$$

$$E_x = \frac{-V_0}{2} \frac{a^2}{(x^2 + a^2)^{3/2}}$$

$$\rho_v = \varepsilon_0 \bar{\nabla} \cdot \bar{E}$$

$$\rho_v = \varepsilon_0 \frac{\partial}{\partial x} Ex$$

$$\rho_v = \frac{-V_0}{2} \varepsilon_0 a^2 \frac{\partial}{\partial x}\left[\frac{1}{(x^2 + a^2)^{3/2}}\right]$$

$$\rho_v = \frac{-V_0}{2} \varepsilon_0 a^2 \left(\frac{-3}{2}\right) \frac{2x}{(x^2 + a^2)^{5/2}}$$

$$\rho_v = \frac{3V_0\varepsilon_0}{2} \frac{a^2 x}{(x^2 + a^2)^{5/2}}$$

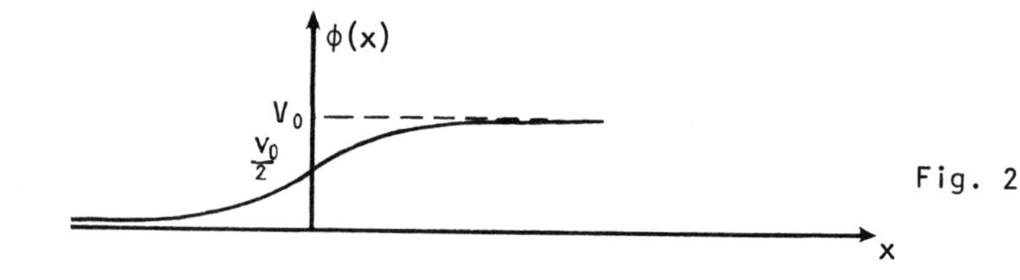

Fig. 2

b) The plots of ϕ, E, and ρ_v are shown in Figures 2, 3, and 4 respectively.

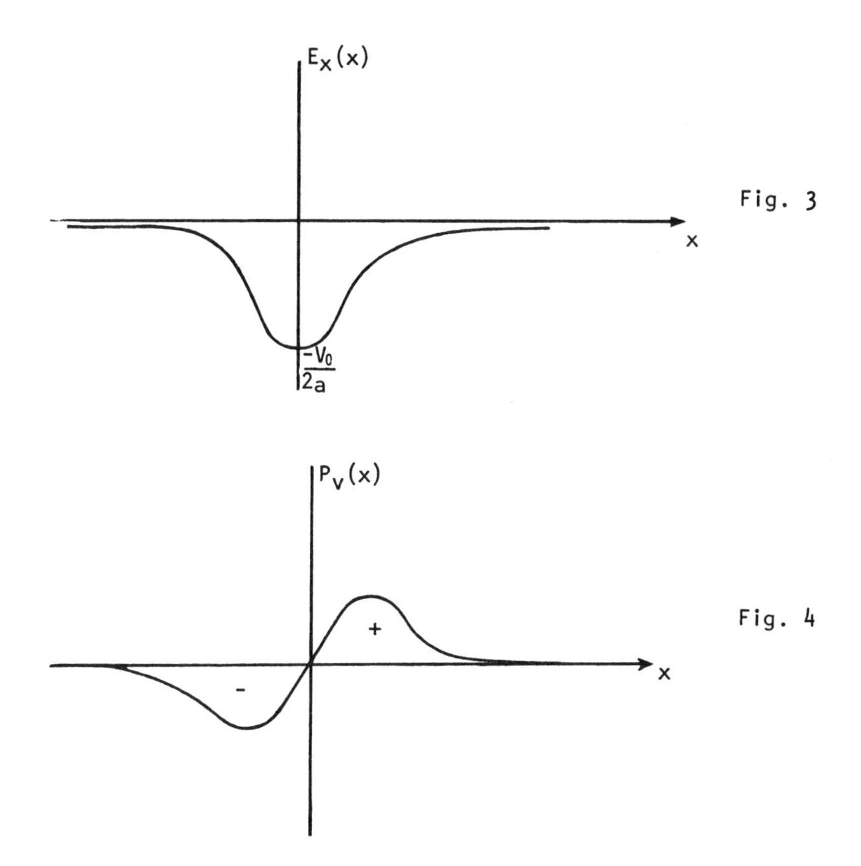

c) From the plot of ρ_v which is charge distribution in the device, it is found that the charge distribution is symmetric with respect to the origin, i.e., negative charges for $x < 0$ and positive charges in equal amounts as the negative charges for $x > 0$; therefore, the net charge in the device is 0.

• PROBLEM 4-21

Find the energy of a sphere of radius a containing charge with constant volume density ρ. Compare it to the energy of two point charges Q that are separated by a distance a.

Solution: To determine the energy the potential inside the sphere is needed. Inside the sphere the electric field is found from Gauss's law.

$$\oint \bar{E}.d\bar{l} = \frac{Q}{\varepsilon_0}$$

$$4\pi r^2 E = \frac{4\pi r^3 \rho}{3\varepsilon_0}$$

$$E = \frac{\gamma\rho}{3\varepsilon_0}$$

$$\phi = -\oint \bar{E}.d\bar{l}$$

$$= -\int_a^r \frac{r'\rho}{3\varepsilon_0}\, dr'$$

$$= \frac{-\rho}{3\varepsilon_0} \left.\frac{r^2}{2}\right]_a^r$$

$$= \frac{\rho}{6\varepsilon_0}(a^2 - r^2).$$

Outside the sphere the electric field is given by

$$E = \frac{\rho a^3}{3\varepsilon_0 r^2}$$

and

$$\phi = \frac{\rho a^3}{3\varepsilon_0 r} .$$

At the radius a, the two expressions for ϕ must be equal. Thus

$$\left.\frac{\rho}{6\varepsilon_0}(a^2 - r^2)\right]_{r=a} = \frac{\rho a^3}{3\varepsilon_0 a}$$

$$0 \neq \frac{\rho a^2}{3\varepsilon_0} .$$

Therefore add a constant term $\frac{\rho a^2}{3\varepsilon_0}$ to the first expression to satisfy the condition of continuity. Thus the potential inside is given by

$$\phi(r) = \frac{\rho}{6\varepsilon_0}(3a^2 - r^2).$$

Substitute this into

$$U_e = \frac{1}{2}\int_V \rho(r)\phi(r)dv,$$

use

$$dv = r^2 \sin\theta \; dr d\theta d\phi$$

and take all of the constants (including ρ) out of the integral.

$$U_e = \frac{\rho^2}{12\varepsilon_o}\int_0^{2\pi}\int_0^{\pi}\int_0^{a}(3a^2 - r^2)r^2 \sin\theta dr d\theta d\phi.$$

The integral over the angles gives 4π, so that

$$U_e = \frac{\pi\rho^2}{3\varepsilon_o}\int_0^a (3a^2 r^2 - r^4)dr$$

$$= \frac{4\pi\rho^2 a^5}{15\varepsilon_o}.$$

Express this in terms of the total charge of the sphere by using

$$Q = (4\pi a^3/3)\rho,$$

and we have

$$U_e = \frac{3}{5}\left(\frac{Q^2}{4\pi\varepsilon_o a}\right).$$

The energy of two point charges Q separated by a distance a is given by

$$U = \frac{Q^2}{4\pi\varepsilon_o a}.$$

Therefore the energy of the sphere is smaller than the energy of two point charges Q that are separated by a distance equal to the radius of the sphere.

Consider the line source illustrated in Fig. 1 to be embedded in a homogeneous, infinite, conducting medium. Find an expression for the potential at any point P and the equipotential surfaces.

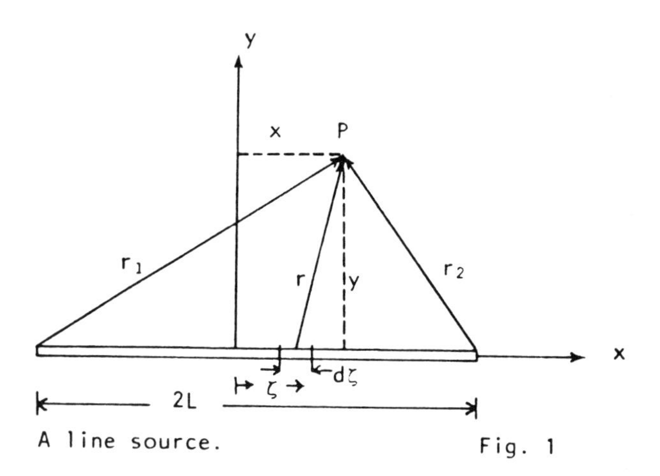

A line source. Fig. 1

Solution: To solve the problem it is supposed that the line source is a continuous and uniform current leak away from the conductor, and that this can be resolved into an array of point sources which are in a line and spaced at infinitely small distances. In particular, it is supposed that the line source can be resolved into a large number of very small line elements $d\zeta$, each of which is considered as constituting a point source. If the total current supplied by the line is I, then the current supplied by the element $d\zeta$ is $I\,d\zeta/2L$, where 2L is the total length of the line, as illustrated.

The potential at point P due to the current element $I\,d\zeta/2L$ is given by

$$d\phi = \frac{Id\zeta}{2L}\,\frac{1}{4\pi\sigma r}$$

which, by making use of the geometry of the figure, can be expressed as

$$d\phi = \frac{I}{8\pi\sigma L}\,\frac{d\zeta}{\sqrt{(x-\zeta)^2 + y^2}} \tag{1}$$

where ζ is the distance from the origin to the line element $d\zeta$. The total potential of the line source at point P is obtained directly as

$$\phi = \frac{I}{8\pi\sigma L}\int_{-L}^{L} \frac{d\zeta}{\sqrt{(x-\zeta)^2 + y^2}}$$

which integrates to the expression

$$\phi = \frac{I}{8\pi\sigma L} \ln \frac{(x+L) + \sqrt{(x+L)^2 + y^2}}{(x-L) + \sqrt{(x-L)^2 + y^2}} \qquad (2)$$

The equipotential lines in the XY plane are ellipses if $r_1 + r_2 = 2a$, where a is the semimajor axis and where the foci are at the end points of the line 2L (refer to any book on analytic geometry). In this case

$$r_1 = a + \frac{xL}{a} = \sqrt{(x+L)^2 + y^2}$$

$$r_2 = a - \frac{xL}{a} = \sqrt{(x-L)^2 + y^2}$$

If these are introduced into Eq. (2) the expression for the potential becomes

$$\phi = \frac{I}{8\pi\sigma L} \ln \frac{a+L}{a-L}$$

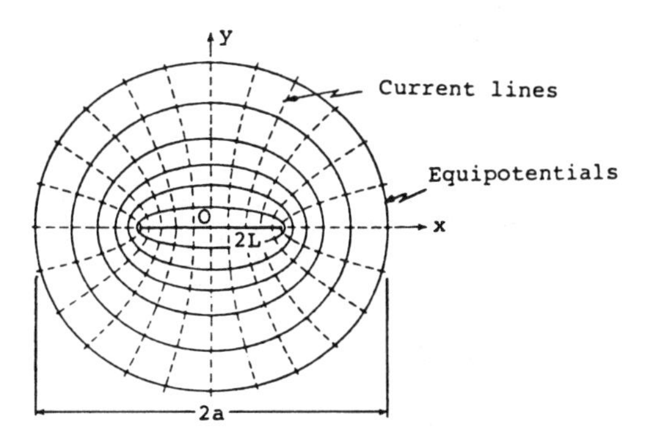

Equipotential and current flow lines for the line source. Fig. 2

Clearly, for each arbitrary value of a, the potential will be a constant. Owing to symmetry, it is clear that the equipotential surfaces are ellipsoids of revolution which are obtained by rotating the ellipses about the X-axis. Also, the current lines, which must be normal to the ellipsoids, are hyperboloids of two sheets having the same foci as the equipotential ellipsoids. The equipotential ellipses and the current flow hyperbolas are shown in Fig. 2.

• **PROBLEM 4-23**

Find the potential function produced by a uniformly charged plane circular disk at a point on the axis of symmetry perpendicular to the disk at its center.

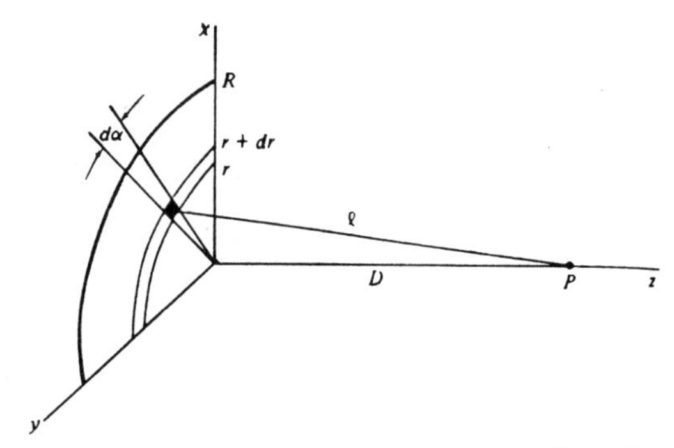

Fig. 1

Solution: Figure 1 illustrates this case. The potential at P may be found from

$$\phi = \frac{1}{4\pi\varepsilon_0} \int_\Sigma \frac{\sigma}{r}\, dS$$

This integral is a scalar integral. An element of charge in the annular ring shown produces the potential

$$d^2\phi = \frac{1}{4\pi\varepsilon_0} \frac{\sigma r\, d\alpha\, dr}{\ell}$$

Since all the elements in the ring are at the same distance ℓ from P, the entire annular ring produces the potential, obtained by integrating $d\alpha$ from $\alpha = 0$ to $\alpha = 2\pi$,

$$d\phi = \frac{\sigma r\, dr}{2\varepsilon_0 \ell}$$

Integrating next over r:

$$\phi = \frac{\sigma}{2\varepsilon_0} \int_0^R \frac{r\, dr}{\sqrt{r^2 + D^2}} = \frac{\sigma}{2\varepsilon_0} [\sqrt{D^2 + R^2} - \sqrt{D^2}]$$

For P on the +x axis this becomes

$$\phi = \frac{\sigma}{2\varepsilon_0} [\sqrt{D^2 + R^2} - D] \quad D \geq 0$$

Here the positive root, $+\sqrt{D^2}$, is taken so that $\phi \to 0$ when $R \to 0$ (i.e., when there is no disk.) For P on the -x axis the negative root must be taken if $\sqrt{D^2 + R^2}$ is always taken as positive:

$$\phi = \frac{\sigma}{2\varepsilon_0} \left[\sqrt{D^2 + R^2} + D\right] \quad D \leq 0$$

A formula applicable to both regions is

$$\phi = \frac{\sigma}{2\varepsilon_0} \left[\sqrt{D^2 + R^2} - |D|\right]$$

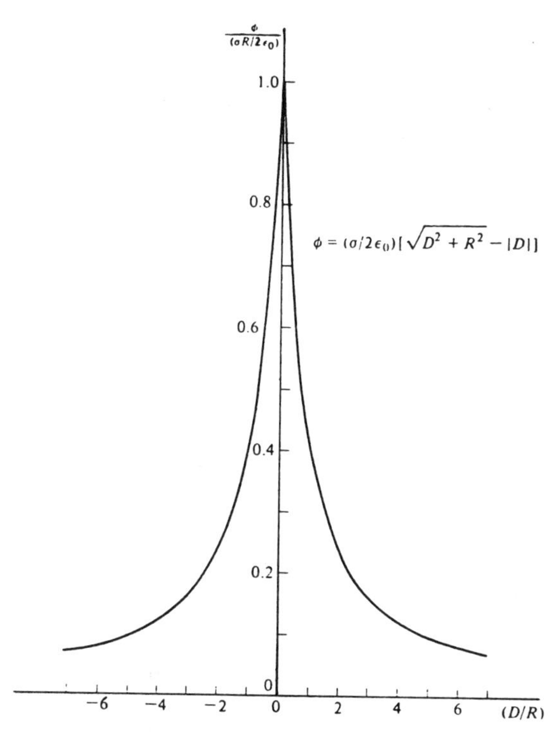

Fig. 2

Figure 2 shows how ϕ varies with distance.

• PROBLEM 4-24

A large, charged, flat metal plate with a hemispherical protusion and an equally and oppositely charged plane conductor are shown in the figure where b >>a. Find the electric field in the space between two plates and the distribution of charge on the plate with the projection.

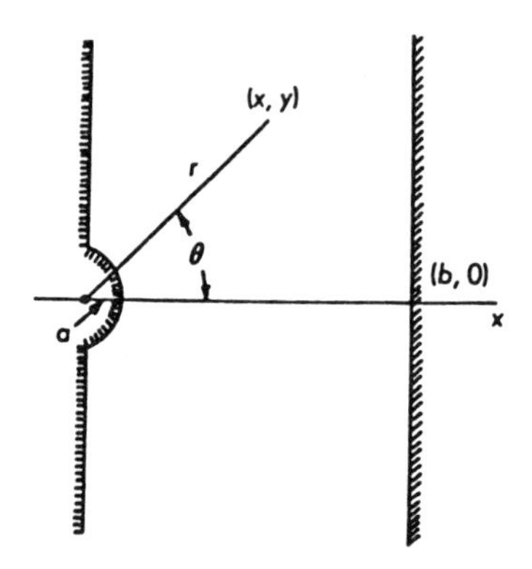

A charged flat metal plate with a hemispherical protrusion and an oppositely charged plane conductor.

Solution: The first step in the solution of this problem is to construct a satisfactory potential function. The potential function for the field between two oppositely charged, perfectly flat metal plates is $C - \sigma_0 x/\varepsilon_0$, where C is a constant and σ_0 is the charge density on the surface of the plates. Assume that the effect of the hemispherical protrusion is to superpose a dipole at the origin parallel to the x-axis. The dipole potential constitutes a correction to the unperturbed potential function. Also, the introduction of the dipole does not increase the net charge of the system. The potential function is then given by

$$V = C - \frac{\sigma_0 x}{\varepsilon_0} + \frac{ql \cos \theta}{4\pi\varepsilon_0 r^2}$$

where ql is the dipole moment and θ is the angle between the x-axis and the radius vector to the point (x,y). In polar coordinates,

$$V(r,\theta) = C - \frac{\sigma_0 r \cos \theta}{\varepsilon_0} + \frac{ql \cos \theta}{4\pi\varepsilon_0 r^2}$$

The charge distribution represented by $V(r,\theta)$ is reasonable in that it is in accord with the fact that there is no charge in the space between the conductors. Of course, one of the problems is to determine the actual charge distribution on the surface of the conductors.

A test of the correctness of this potential function V is made by determining whether it reduces to the same constant for all points on the plane (x = 0) and on the hemispherical protrusion (r = a). At r = a,

$$V(a,\theta) = C - \frac{\sigma_0 a \cos \theta}{\varepsilon_0} + \frac{ql \cos \theta}{4\pi\varepsilon_0 a^2}$$

Set $ql = 4\pi a^3 \sigma_0$, then

$$V(a,\theta) = C$$

At all points for which $x = 0$, $\theta = \pm \pi/2$,

$$V(r, \pm \frac{\pi}{2}) = C$$

Thus, the potential function V reduces to a constant everywhere on the metal surface as it should. Also, for large values of r, V is practically the same as it would be without the dipole potential. The dipole is important only in the vicinity of the curved surface. This is certainly reasonable. Consequently, the function V must represent the field.

The components of electric field intensity may be easily obtained by differentiating V.

$$E_r = - \frac{\partial V}{\partial r} = \frac{\sigma_0 \cos\theta}{\varepsilon_0} + \frac{ql \cos\theta}{2\pi\varepsilon_0 r^3}$$

$$E_\theta = - \frac{1}{r} \frac{\partial V}{\partial \theta} = - \frac{\sigma_0 \sin\theta}{\varepsilon_0} + \frac{ql \sin\theta}{4\pi\varepsilon_0 r^3}$$

At $r = a$, $E_\theta = 0$, since $ql = 4\pi a^3 \sigma_0$. This result, of course, is correct since the tangential component of electric field intensity is zero at the surface of a conductor in electrostatic equilibrium.

The charge density on the hemispherical surface is given by

$$\sigma = \varepsilon_0 E_r = - \varepsilon_0 \left(\frac{\partial V}{\partial r}\right)_{r=a}$$

$$\sigma = 3\sigma_0 \cos\theta$$

Clearly, σ has a maximum value of $3\sigma_0$ at $\theta = 0$ and a minimum value of zero at $\theta = \pm \pi/2$.

The charge density on the plane surface of the left plate is given by

$$\sigma = - \varepsilon_0 E_\theta = + \frac{\varepsilon_0}{r} \left(\frac{\partial V}{\partial \theta}\right)_{\theta=\pi/2}$$

The minus sign is used in the first equality since at

$\theta = \pi/2$, E_θ is directed toward the left, the direction of increasing θ. Use $-E_\theta$, the component of E in the direction of the outward drawn normal. Thus, on the plane surface,

$$\sigma = \sigma_0 \left(1 - \frac{a^3}{r^3}\right)$$

At $r = a$, $\sigma = 0$. This is in agreement with the result previously obtained for the hemispherical surface. At $r = \infty$, $\sigma = \sigma_0$. This indicates that, for points far from the hemispherical surface, the charge density is the same as it would be without the protrusion.

It is important that the charge density and the electric field intensity just outside the protrusion at the point (a,0) is much greater than at points just outside the plane surface. In other words, the charge density and the electric field intensity are great where the convexity of the surface is large.

This last fact helps to explain the efficacy of a lightning rod, which, in this case, is crudely represented by the protrusion. The flat plate on the right in the given figure may represent a charged cloud. Since the electric field intensity is high in the neighborhood of the projection, the air in this region becomes ionized. The negatively charged ions are attracted to the positively charged protrusion, thereby neutralizing some of the charge and consequently decreasing the electric field intensity. Of course, if there is a huge rush of ions, then a spark discharge occurs.

• PROBLEM 4-25

A small electric dipole is located as shown in the figure below. Calculate the electric field $\bar{E}$ at a point P in the field of the dipole.

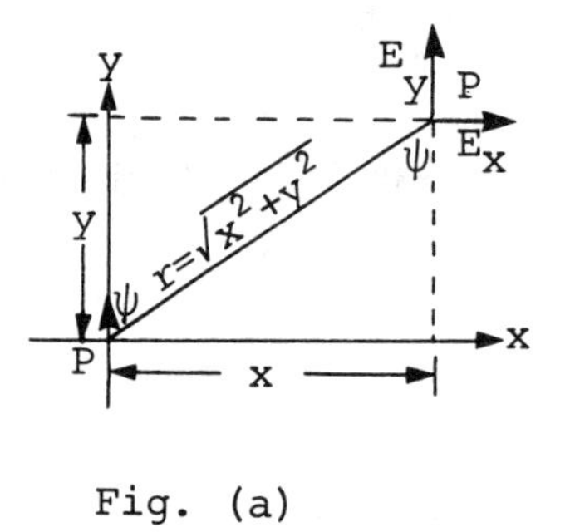

Fig. (a)

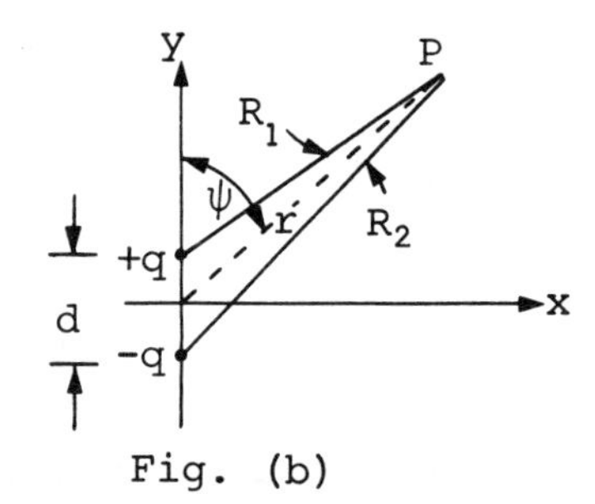

Fig. (b)

Solution: This small electric dipole p can be considered to be made up of two dipole charges (q, -q), separated by a distance d as shown in Figure b.

Now, the absolute potential of an aggregate of n charges in the free space is the sum of the potential contributions of each charge i.e.,

$$\Phi(P) = \sum_{k=1}^{n} \frac{q_k}{4\pi\varepsilon_0 \cdot R_k} \text{ Volts}$$

Now for our case, n = 2 = number of charges, substituting n=2 we get

$$\Phi(P) = \frac{q}{4\pi\varepsilon_0 \cdot R_1} + \frac{-q}{4\pi\varepsilon_0 \cdot R_2} = \frac{a}{4\pi\varepsilon_0} \cdot \frac{R_2 - R_1}{R_1 \cdot R_2}$$

In Figure (a), since r >> d, following approximations are valid,

$$R_1 R_2 \approx r^2 \quad \text{and} \quad R_2 - R_1 \approx d \cdot \cos\psi$$

By substitution

$$\Phi(P) \approx \frac{q \cdot d \cdot \cos\psi}{4\pi\varepsilon_0 \cdot r^2}$$

Now converting the polar coordinates (r, ψ) to rectangular coordinates, we get

$$\Phi(P) \approx \frac{q \cdot d}{4\pi\varepsilon_0} \cdot \frac{y}{(x^2+y^2)^{3/2}} \quad \left(\text{Because } r = \sqrt{x^2+y^2} \text{ and } \psi = \tan^{-1}\frac{x}{y};\ \text{so } \cos\psi = \frac{y}{\sqrt{x^2+y^2}}\right)$$

The electric field $\bar{E}$ existing at P will have two components E_x and E_y respectively, as shown in Figure a.

$$E_x = \left.\frac{-\partial\Phi(P)}{\partial x}\right| \text{keeping y constant} = \frac{-q \cdot d \cdot y}{4\pi\varepsilon_0}\left(-\frac{3}{2}\right)(x^2+y^2)^{-\frac{5}{2}}(2x)$$

$$= \frac{3q \cdot d}{4\pi\varepsilon_0} \cdot \frac{x \cdot y}{(x^2+y^2)^{5/2}}$$

and,

$$E_y = \frac{-\partial\Phi(P)}{\partial y}\bigg|\text{keeping x constant} =$$

$$= \frac{-q \cdot d}{4\pi\varepsilon_0}\left[\frac{(x^2+y^2)^{3/2} - y \cdot \frac{3}{2} \cdot (x^2+y^2)^{\frac{1}{2}}(2y)}{(x^2+y^2)^3}\right]$$

$$= \frac{-q \cdot d}{4\pi\varepsilon_0} \cdot \frac{x^2 - 2y^2}{(x^2+y^2)^{5/2}}$$

Thus, E_x vanishes both on the dipole axis ($x = 0$) and in the median plane ($y = 0$), i.e. $E_x = 0$ when either $x = 0$ or $y = 0$.

• PROBLEM 4-26

(a) Find the field due to the linear quadrupole shown in Figure 1. A quadrupole experiences no torque in a uniform electric field; however, a torque may be exerted on a quadrupole by a nonuniform field. A linear quadrupole is placed in an external electric field having a constant electric field gradient. The quadrupole makes an angle θ with the field gradient. (b) Deduce an expression for the torque experienced by the quadrupole. (c) Derive an expression for the potential energy of the quadrupole.

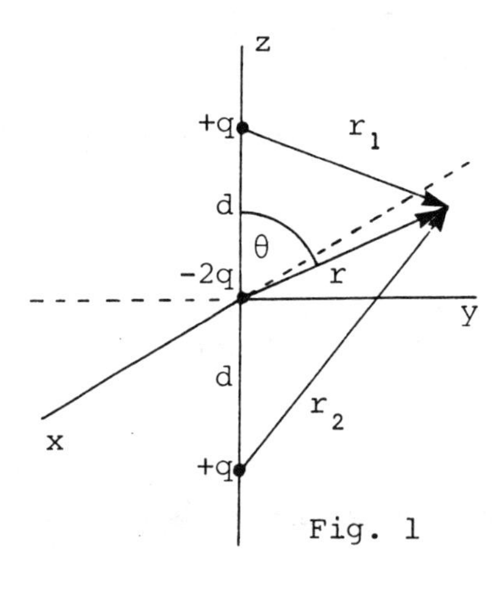

Fig. 1

Solution: (a) The electric potential at point P due to the linear quadrupole is

$$V = \frac{1}{4\pi\varepsilon_0}\left(\frac{q}{r_1} - \frac{2q}{r} + \frac{q}{r_2}\right).$$

Now, using the law of cosines,

$$r_1^2 = r^2 + d^2 - 2dr\cos\theta$$

$$= r^2\left(1 - \frac{2d\cos\theta}{r} + \frac{d^2}{r^2}\right)$$

$$r_1 = r\left(1 - \frac{2d\cos\theta}{r} + \frac{d^2}{r^2}\right)^{\frac{1}{2}}$$

$$\frac{1}{r_1} = \frac{1}{r}\left[1 - \left(\frac{2d\cos\theta}{r} - \frac{d^2}{r^2}\right)\right]^{\frac{1}{2}}$$

When $d \ll r$, the term in parentheses is $\ll 1$ and it is possible to expand the quantity in brackets using the binomial series

$$(1 - x)^{-\frac{1}{2}} = 1 + \frac{1}{2}x + \frac{3}{8}x^2 + \ldots .$$

Therefore,

$$\frac{1}{r_1} = \frac{1}{r}\left[1 + \frac{1}{2}\left(\frac{2d\cos\theta}{r} - \frac{d^2}{r^2}\right) + \frac{3}{8}\left(\frac{2d\cos\theta}{r} - \frac{d^2}{r^2}\right)^2 + \ldots\right]$$

$$= \frac{1}{r}\left[1 + \frac{d\cos\theta}{r} - \frac{d^2}{2r^2} + \frac{3}{2}\frac{d^2\cos^2\theta}{r^2} - \frac{3}{2}\frac{d^3\cos\theta}{r^3} + \frac{3}{8}\frac{d^4}{r^4} + \ldots\right]$$

$$= \frac{1}{r} + \frac{d\cos\theta}{r^2} - \frac{d^2}{2r^3} + \frac{3d^2\cos^2\theta}{2r^3} - \ldots .$$

Neglecting terms in r^{-4} and higher,

$$\frac{1}{r_1} = \frac{1}{r} + \frac{d\cos\theta}{r^2} + \frac{d^2}{2r^3}(3\cos^2\theta - 1).$$

Similarly,

$$\frac{1}{r_2} = \frac{1}{r} - \frac{d \cos \theta}{r^2} + \frac{d^2}{2r^3} (3 \cos^2 \theta - 1).$$

Upon substitution into the expression for the electric potential

$$V = \frac{qd^2 (3 \cos^2 \theta - 1)}{4\pi\varepsilon_0 r^3}.$$

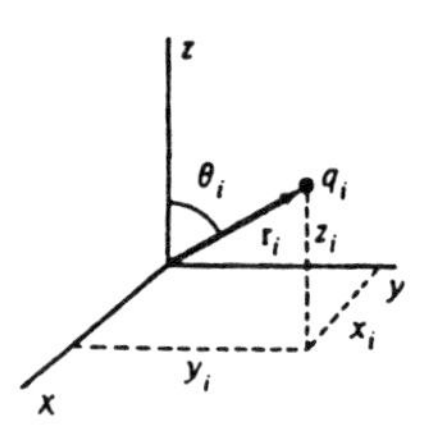

Coordinates of a charge q_i.   Fig. 2

Defining the scalar quadrupole moment eQ of a set of point charges q_i relative to the z axis (where the z axis is a symmetry axis) as (see Fig. 2)

$$eQ = \sum_i q_i (3z_i^2 - r_i^2).$$

Therefore, for the linear quadrupole (see Fig. 1)

$$eQ = q(3d^2 - d^2) - 2q(0 - 0) + q(3d^2 - d^2) = 2q(3d^2 - d^2)$$

$$= 4qd^2$$

so that

$$V = \frac{eQ}{4} \cdot \frac{(3 \cos^2 \theta - 1)}{4\pi\varepsilon_0 r^3}.$$

It follows that in polar coordinates,

$$E_r = -\frac{\partial V}{\partial r} \text{ and } E_\theta = -\frac{1}{r}\frac{\partial V}{\partial \theta}, \text{ thus}$$

$$E_r = \frac{3eQ}{4} \cdot \frac{(3 \cos^2 \theta - 1)}{4\pi\varepsilon_0 r^4}$$

and

$$E_\theta = \frac{3eQ}{4} \frac{\sin 2\theta}{4\pi\varepsilon_0 r^4}.$$

Note that the electric field produced by an electric quadrupole is proportional to its moment and inversely proportional to the fourth power of the distance at large distances.

(b) The electric field $\bar{E}$ at a point with coordinate dz is given by the expression

$$E(dz) = \bar{E}(z=0) + \left(\frac{\partial E}{\partial z}\right)_{z=0} d\bar{z}.$$

However,

$$\bar{\nabla}E = \bar{k}\left(\frac{\partial E}{\partial z}\right) = \bar{k}\left(\frac{\partial E}{\partial z}\right)_{z=0}$$

since ∇E is a constant vector in the z direction. Therefore,

$$\bar{E}(dz) = \bar{E}(z=0) + |\bar{\nabla}E|\, d\bar{z}.$$

Since our concern is only with the gradient of E and not its absolute value, put $E(z=0) = 0$ for convenience. Integrating gives the field E at a point with coordinate z as

$$\bar{E} = \int_0^z |\bar{\nabla}E|\, d\bar{z}$$

$$= |\bar{\nabla}(E)| \int_0^z d\bar{z}$$

$$= |\bar{\nabla}(E)|\, \bar{z}.$$

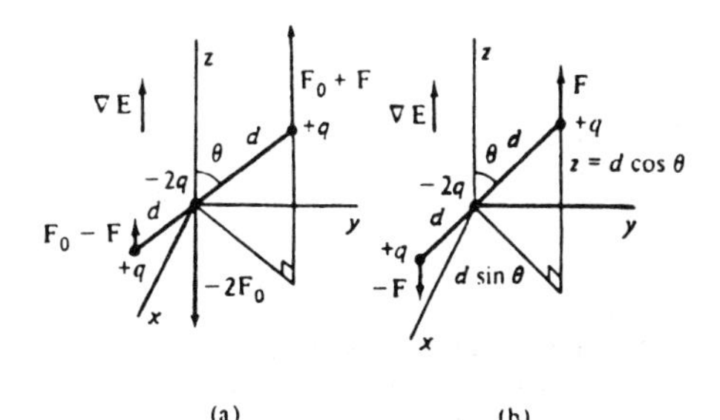

(a) An electric quadrupole in an external electric field with a constant electric field gradient in the z direction.
(b) The resultant torque experienced by the quadrupole. Fig. 3

The resulting forces on the charges constituting the quadrupole are shown in Fig. 3(a). Note that the net force experienced by the quadrupole is zero; however, it experiences a net torque. This torque τ is equivalent to that produced by two equal and opposite forces F and -F as shown in Fig. 3(b). The torque tends to align the quadrupole in the direction of the field gradient. The magnitude of the torque is

$$\tau = 2Fd \sin\theta = 2q|\nabla E|zd \sin\theta$$

$$= 2qd^2|\bar{\nabla}E|\sin\theta\cos\theta = \frac{eQ}{2}|\bar{\nabla}E| \sin\theta\cos\theta$$

$$= \frac{eQ}{4}|\bar{\nabla}E| \sin 2\theta.$$

(c) The change in potential energy of the quadrupole is equal to the work that must be done on the quadrupole to reorient it in the presence of the field gradient. The work required to change the orientation of the quadrupole from θ_1 to θ_2 is

$$W = \int dW = \int_{\theta_1}^{\theta_2} \tau\, d\theta$$

$$= \int_{\theta_1}^{\theta_2} \frac{eQ}{2}|\bar{\nabla}E| \sin\theta\cos\theta\, d\theta$$

$$= \frac{eQ}{4}|\bar{\nabla}E| \sin^2\theta_2 - \frac{eQ}{4}|\bar{\nabla}E| \sin^2\theta_1.$$

The potential energy U of the quadrupole may therefore be taken as

$$U = \frac{eQ}{4}|\bar{\nabla}E| \sin^2\theta,$$

Since work is the change in potential energy.

• PROBLEM 4-27

Consider a system comprised of one long uniform line chargeof density λ and the other of $-\lambda$. Assume that the values of L_2 and L_1, as shown in Fig. 1, are the same and very large. Find the potential and the electric field at any point P. Determine and sketch the equipotential surfaces.

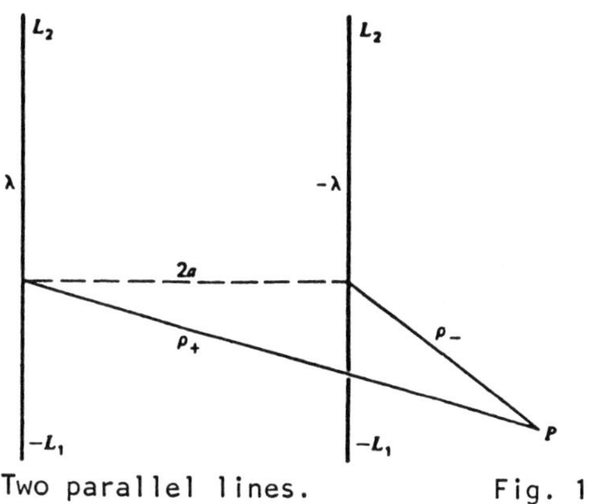

Two parallel lines. Fig. 1

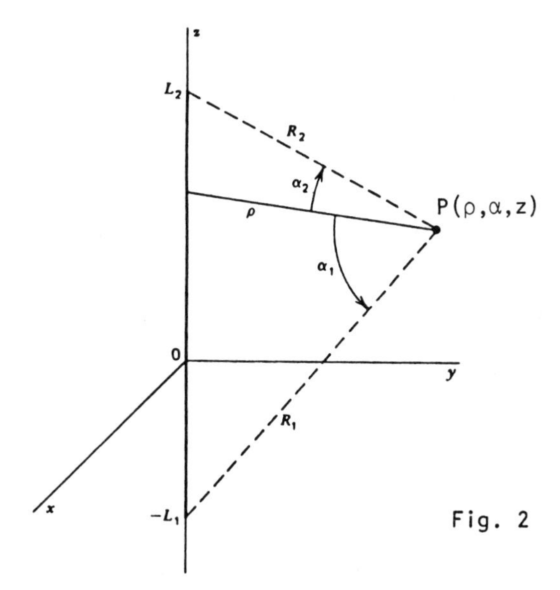

Fig. 2

Solution: The perpendicular distance from the line charge to the field point P are labeled ρ_+ and ρ_- in the figure. The potential due to a straight wire along the z-axis and endpoints L_2 and L_1 can be found from

$$\phi(r) = \frac{1}{4\pi\varepsilon_0}\int_{L'}\frac{\lambda(r')}{R}\,ds' . \tag{1}$$

With λ constant and $r' = z'\,\hat{z}$ and $r = \rho\hat{\rho} + z\,\hat{z}$, then $R = \rho\hat{\rho} + (z-z')\hat{z}$ in fig. 2, $ds' = dz'$ thus (1) becomes

$$\phi = \frac{\lambda}{4\pi\varepsilon_0}\int_{-L_1}^{L_2}\frac{dz'}{[\rho^2 + (z-z')^2]^{\frac{1}{2}}}$$

From a table of integrals, the result is

$$\phi = \frac{\lambda}{4\pi\varepsilon_0} \ln \left\{ \frac{z + L_1 + [\rho^2 + (z + L_1)^2)]^{\frac{1}{2}}}{z - L_2 + [\rho^2 + (z - L_2)^2]^{\frac{1}{2}}} \right\}$$

Now L_1 and L_2 are very large so ρ & z can be neglected in the numerator leaving $2L_1$. In the denominator, neglecting z and ρ will make it vanish so instead expand it in a power series using

$$(1 \pm x)^{\frac{1}{2}} = 1 \pm \frac{1}{2} x - \frac{1}{8} x^2 \pm \ldots$$

Taking out $L_2 - z$, the denominator becomes

$$(L_2 - z)\left\{ -1 + \left[1 + \frac{\rho^2}{(L_2 - z)^2}\right]^{\frac{1}{2}} \right\} \simeq L_2 \left[-1 + \left(1 + \frac{\rho^2}{L_2{}^2}\right)^{\frac{1}{2}} \right]$$

$$\simeq L_2 \left(-1 + 1 + \frac{\rho^2}{2L_2{}^2} \right) = \frac{\rho^2}{2L_2}$$

Only the first two terms of the power series were used.

The potential is then

$$\phi = \frac{\lambda}{4\pi\varepsilon_0} \ln \left(\frac{4L_2L_1}{\rho^2} \right) = \frac{\lambda}{2\pi\varepsilon_0} \ln \left[\frac{(4L_2L_1)^{\frac{1}{2}}}{\rho} \right]$$

Thus the individual potentials due to these two line charges are

$$\phi_+ = \frac{\lambda}{2\pi\varepsilon_0} \ln \left[\frac{(4L_2L_1)^{\frac{1}{2}}}{\rho_+} \right]$$

$$\phi_- = \frac{\lambda}{-2\pi\varepsilon_0} \ln \left[\frac{(4L_2L_1)^{\frac{1}{2}}}{\rho_-} \right]$$

so that the total potential at P is

$$\phi = \phi_+ + \phi_- = \frac{\lambda}{2\pi\varepsilon_0} \ln \left(\frac{\rho_-}{\rho_+} \right) \qquad (2)$$

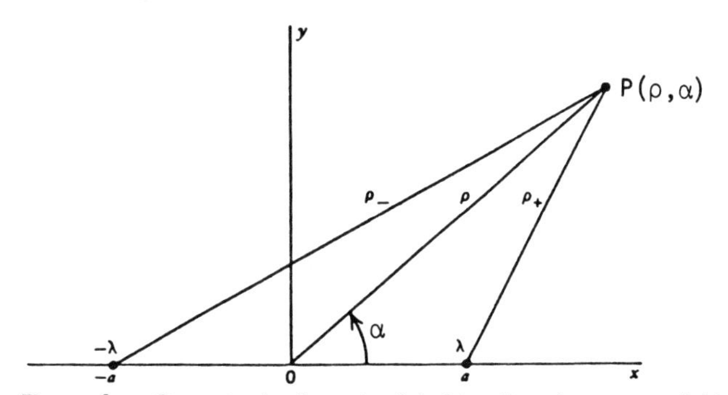

Figure 3 **Geometry for the potential of two line charges parallel to the *z* axis.**

and the factor depending on L_2 and L_1 has canceled. In a specific case, ρ_+ and ρ_- have to be evaluated in terms of the particular coordinate system that is being used. To illustrate this for the case in which the line charges are taken to be parallel to the z axis, the line of length 2a between them lying along the x axis, the origin is chosen midway between them, and P lies in the xy plane as shown in Figure 3. Use polar coordinates ρ and α to specify the location of P. Applying the law of cosines, $\rho_+{}^2 = a^2 + \rho^2 - 2a\rho\cos\alpha$ and $\rho_-{}^2 = a^2 + \rho^2 + 2a\rho\cos\alpha$, so that (2) can be written

$$\phi(\rho,\alpha) = \frac{\lambda}{4\pi\varepsilon_0}\ln\left(\frac{\rho_-{}^2}{\rho_+{}^2}\right) = \frac{\lambda}{4\pi\varepsilon_0}\ln\left(\frac{a^2+\rho^2+2a\rho\cos\alpha}{a^2+\rho^2-2a\rho\cos\alpha}\right) \tag{3}$$

and the components of E are:

$$E_\rho = \frac{\partial\phi}{\partial\rho} = \frac{\lambda}{2\pi\varepsilon_0}\left[\frac{(\rho-a\cos\alpha)}{\rho_+{}^2} - \frac{(\rho+a\cos\alpha)}{\rho_-{}^2}\right] = \frac{\lambda a(\rho^2-a^2)\cos\alpha}{\pi\varepsilon_0\rho_+{}^2\rho_-{}^2}$$

$$E_\alpha = -\frac{1}{\rho}\frac{\partial\phi}{\partial\alpha} = \frac{\lambda a(\rho^2+a^2)\sin\alpha}{\pi\varepsilon_0\rho_+{}^2\rho_-{}^2}$$

while $E_z = -\partial\phi/\partial z = 0$.

The equipotential surfaces ϕ = const. are given by (3) as

$$\frac{\rho_-{}^2}{\rho_+{}^2} = e^{4\pi\varepsilon_0\phi/\lambda} = \text{const.}$$

This equation is more easily interpreted in rectangular

coordinates; by inspecting Figure 3, it becomes

$$\frac{(x + a)^2 + y^2}{(x - a)^2 + y^2} = e^{4\pi\varepsilon_0\phi/\lambda}$$

which, with a little algebra, can be written in the form

$$(x - a \coth \eta)^2 + y^2 = \left(\frac{a}{\sinh\eta}\right)^2 \qquad (4)$$

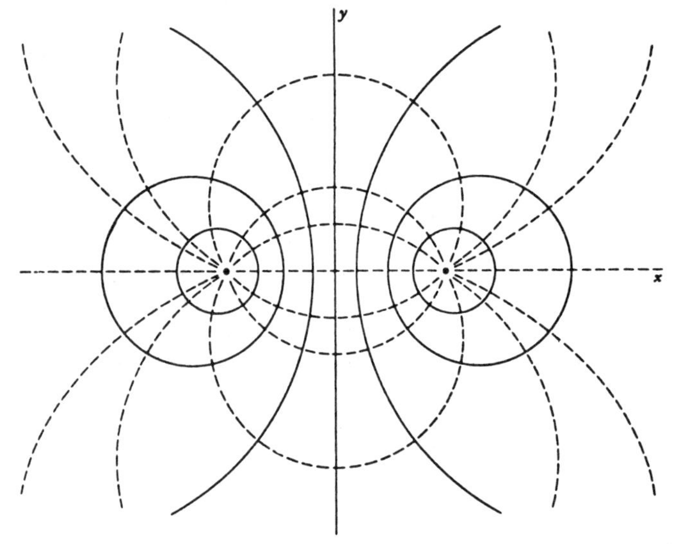

Figure 4 Equipotentials (solid) and lines of electric field (dashed) for two oppositely charged lines parallel to the *z* axis.

where $\eta = 2\pi\varepsilon_0\phi/\lambda$. This is the equation of a circle with radius $a/\sinh \eta$ and with center displaced a distance $a \coth \eta$ along the x axis. In other words, the equipotential surfaces are cylinders with axes parallel to the z axis whose intersections with the xy plane are the circles given by (4). These circles are shown as the solid curves in Figure 4. Note that the circles whose centers lie on the positive x axis correspond to $\phi > 0$, while those with the centers on the negative x axis correspond to $\phi < 0$. Also note that the yz plane $(x = 0)$ is the equipotential surface for $\phi = 0$; this is easily seen to be the case since from Figure 3 each point on the y axis has $\rho_+ = \rho_-$, making $\phi = 0$ according to (2).

POTENTIAL AND GRADIENT

• PROBLEM 4-28

Referring to the given figure the intensity of the uniform electric field is 10 volts/meter. If $x_2 - x_1$ is equal to 10 cm, find the potential difference of the two points.

Linear path in uniform electric field.

Solution:

From $E(x_2 - x_1)$ = work per unit charge (Joules/coul)

the electric poential is given by

$$V = 10 \times 0.1 = 1 \text{ volt}$$

That is, the potential of x_1 is 1 volt higher than the potential of x_2.

• PROBLEM 4-29

The potential distribution

$$V = \frac{10}{x^2 + y^2}$$

refers to a non-uniform field. Here V is in volts and x and y in centimeters. With respect to z there is no variation, therefore the distribution is two-dimensional. The equipotential contours in the figure illustrate the potential variation. Evaluate (a) expression for the gradient of the potential; (b) the value of the gradient at the point (2,1) cm; (c) the strength of electric field at this point.

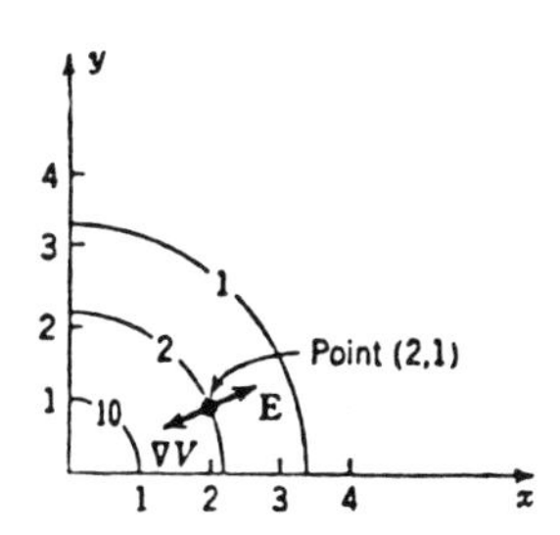

Potential distribution showing gradient of V and electric field E at a point P.

Solution: (a) Since the potential distribution is independent of z, $\partial V/\partial z = 0$ and

$$\text{grad } V = \bar{\nabla} V = \bar{i}\,\frac{\partial V}{\partial x} + \bar{j}\,\frac{\partial V}{\partial y}$$

$$= - \frac{20}{(x^2 + y^2)^2} (\bar{i}x + \bar{j}y)$$

(b) At the point (2,1)

$$\bar{\nabla}V = - \frac{20}{25} (\bar{i}2 + \bar{j}1) = 1.79/\underline{206°} \text{ volts/cm}$$

(c) The electric field has the opposite direction. Thus

$$\bar{E} = -\bar{\nabla}V = -1.79/\underline{206°} = 1.79/\underline{26°} \text{ volts/cm}$$

• PROBLEM 4-30

(a) What is the incremental potential difference between points A and B in the field $\bar{E} = 2e^{2x-2y-4z}(-2\bar{a}_x + 3\bar{a}_y - 4\bar{a}_z)$, given B at (0,0,0) and A at $(10^{-3}, \Phi, 10^{-4})$ in cylindrical coordinates? (b) What value of Φ provides the maximum values of ΔV?

Solution: (a)

$$\Delta V = - \bar{E} \cdot \Delta\bar{\ell}$$

$$\Delta\bar{\ell} = \Delta r \cos\Phi\, \bar{a}_x + \Delta r \sin\Phi\, \bar{a}_y + z\, \bar{a}_z$$

$$= 10^{-3} \cos\Phi\, \bar{a}_x + 10^{-3} \sin\Phi\, \bar{a}_y + 10^{-4}\, \bar{a}_z$$

At point B(0,0,0),

$$x = r\cos\Phi = 0, \; y = r\sin\Phi = 0, \; z = 0,$$

then

$$\bar{E} = 2(-2\,\bar{a}_x + 3\,\bar{a}_y - 4\,\bar{a}_z)$$

$$-\Delta V = \bar{E} \cdot \Delta\bar{\ell} = 2(-2\,\bar{a}_x + 3\,\bar{a}_y - 4\,\bar{a}_z) \cdot$$

$$(10^{-3}\cos\Phi\,\bar{a}_x + 10^{-3}\sin\Phi\,\bar{a}_y + 10^{-4}\,\bar{a}_z)$$

$$\Delta V = 2 \cdot 10^{-3}(2\cos\Phi - 3\sin\Phi + 0.4)$$

(b) To find maximum ΔV for varying Φ, take $dV/d\Phi$ and equate it to zero,

$$\frac{\Delta V}{\Delta \Phi} = 2 \cdot 10^{-3} (-2 \sin \Phi - 3 \cos \Phi) = 0$$

or

$$2 \sin \Phi = -3 \cos \Phi$$

$$\tan \Phi = -\frac{3}{2}$$

Therefore

$$\Phi = \tan^{-1}\left(-\frac{3}{2}\right) = -56.3°.$$

• PROBLEM 4-31

Three charges are located in the xy plane: a charge $+q_1$ at the point $(\frac{1}{2}, 0)$, a charge $+q_1$ at the point $(-1,0)$, and a charge $-3q_1$ at the point $(0,1)$.

(a) Calculate directly the electrostatic force on a charge $+q$ at the origin.

(b) By taking the gradient of the electric potential, determine the electrostatic force on a charge $+q$ at the origin.

Solution: (a) The electrostatic force F experienced by a charge $+q$ at the origin is found from Coulomb's law:

$$\bar{F} = \frac{1}{4\pi\varepsilon_0}(qq_1 - 4qq_1)\bar{i} + \frac{1}{4\pi\varepsilon_0} 3qq_1 \bar{j}$$

$$= \frac{3qq_1}{4\pi\varepsilon_0}(-\bar{i} + \bar{j})$$

(b) The electric potential at a point (x,y) near the origin is

$$V(x,y) = \frac{q_1}{4\pi\varepsilon_0}\left\{\frac{1}{[(\frac{1}{2} - x)^2 + y^2]^{\frac{1}{2}}} + \frac{1}{[(1 + x)^2 + y^2]^{\frac{1}{2}}} - \frac{3}{[x^2 + (1-y)^2]^{\frac{1}{2}}}\right\} \simeq \frac{q_1}{4\pi\varepsilon_0}\left[\frac{1}{(\frac{1}{2} - x)} + \frac{1}{(1 + x)} - \frac{3}{(1 - y)}\right]$$

The electric field at the point (x,y) is

$$\bar{E}(x,y) = -\frac{\partial V(x,y)}{\partial x}\bar{i} - \frac{\partial V(x,y)}{\partial y}\bar{j}$$

$$\simeq \frac{q_1}{4\pi\varepsilon_0}\left\{\left[\frac{-1}{(\frac{1}{2} - x)^2} + \frac{1}{(1 + x)^2}\right]\bar{i} + \frac{3}{(1 - y)^2}\bar{j}\right\}$$

The electric field E at the origin (x=y=0) is

$$\bar{E} = \frac{3q_1}{4\pi\varepsilon_0}[-\bar{i} + \bar{j}]$$

Therefore the force F experienced by a charge +q at the origin is

$$\bar{F} = \frac{3qq_1}{4\pi\varepsilon_0}[-\bar{i} + \bar{j}]$$

in agreement with the result obtained in (a).

• PROBLEM 4-32

Find the electric intensity at all points in the x-y plane which contains the two charges q_1 and q_2 having displacements of 2×10^{-9} and 5×10^{-9} coulombs respectively. The charges are located as shown in the figure. Find the potential due to the charges and then find the gradient of the potential.

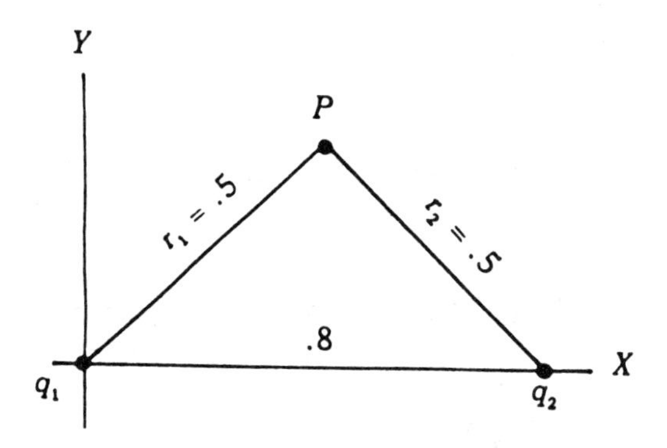

Solution: Since

$$\bar{E} = \frac{q}{4\pi\varepsilon_v r^2} \bar{1}_r$$

and

$$V(r) - V(\infty) = -\int_{\infty}^{r} \bar{E} \cdot d\bar{r}$$

then

$$V = \frac{q}{4\pi\varepsilon_v r}$$

thus

$$V_p = \frac{1}{4\pi\varepsilon_v}\left(\frac{q_1}{r_1} + \frac{q_2}{r_2}\right)$$

$$= \frac{1}{4\pi \times 8.85 \times 10^{-12}}\left(\frac{2 \times 10^{-9}}{0.5} + \frac{5 \times 10^{-9}}{0.5}\right)$$

$$= 135 \text{ volts}$$

The potential at any point on the x,y plane due to charges q_1 and q_2 located at (0,0,0) and (0.8,0,0) is

$$V_p = \frac{1}{4\pi\varepsilon_v}\left(\frac{q_1}{\sqrt{x^2 + y^2}} + \frac{q_2}{\sqrt{(x - 0.8)^2 + y^2)}}\right)$$

Differentiating partially with respect to x and y,

$$\frac{\partial V}{\partial x} = \frac{1}{4\pi\varepsilon_v}\left(\frac{q_1 x}{(x^2 + y^2)^{3/2}} + \frac{q_2(x - 0.8)}{[(x - 0.8)^2 + y^2]^{3/2}}\right)$$

$$\frac{\partial V}{\partial y} = -\frac{1}{4\pi\varepsilon_v}\left(\frac{q_1 y}{(x^2 + y^2)^{3/2}} + \frac{q_2 y}{[(x - 0.8)^2 + y^2]^{3/2}}\right)$$

and $\bar{E}$ at any point (x,y)

$$\bar{E} = -\left(\frac{\partial V}{\partial x}\bar{i} + \frac{\partial V}{\partial y}\bar{j}\right)$$

for the location given for P

$x = 0.4,$

$y = 0.3$

$$\frac{\partial V}{\partial x} = \frac{1}{4\pi\varepsilon_v}\left(\frac{2 \times 10^{-9} \times 0.4}{(0.4^2 + 0.3^2)^{3/2}} + \frac{5 \times 10^{-9} \times (0.4 - 0.8)}{[(0.4 - 0.8)^2 + 0.3^2]^{3/2}}\right)$$

$$= 9 \times 10^9 \left(\frac{0.8 \times 10^{-9}}{0.0625} - \frac{2 \times 10^{-9}}{0.0625}\right)$$

$$= - 9 \times 1.2 \times 16i = - 173\bar{i}$$

$$\frac{\partial V}{\partial y} = 9 \times 10^9 \left(\frac{2 \times 10^{-9} \times 0.3}{0.0625} + \frac{5 \times 10^{-9} \times 0.3}{0.0625}\right)$$

$$= +9 \times 4.8 \times 7\bar{j}$$
$$= 302\bar{j}$$

Therefore

$$\bar{E} = 173\bar{i} - 302\bar{j}$$

• PROBLEM 4-33

Find the electric field intensity at any point P (far away) due to a dipole located at the origin on the z-axis as shown in the figure. Use the gradient method.

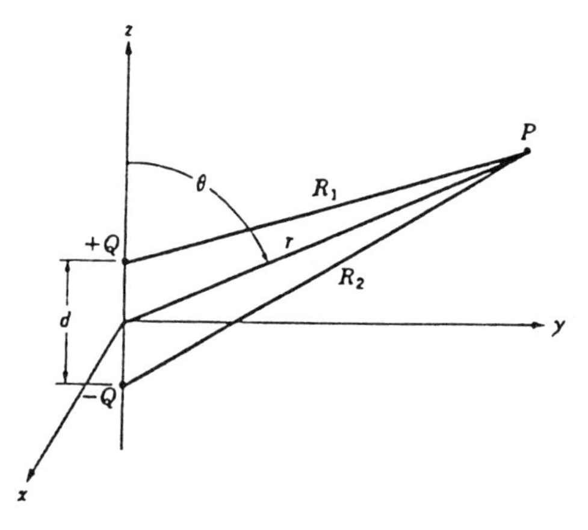

The geometry of the problem of an electric dipole. The dipole moment p = Qd is in the a_z direction.

Solution: To use the gradient method, it is necessary to get the potential at point P due to the two charges. Let the distances from Q and -Q to P be R_1 and R_2, respectively. Using superposition, write the total potential as

$$V = \frac{Q}{4\pi\varepsilon_0}\left(\frac{1}{R_1} - \frac{1}{R_2}\right) = \frac{Q}{4\pi\varepsilon_0}\,\frac{R_2 - R_1}{R_1 R_2}$$

For a distant point, $R_1 \doteq R_2$, and the R_1R_2 product in the denominator may be replaced by r^2. The approximation may not be made in the numerator, however, without obtaining the trivial answer that the potential field approaches zero very far away from the dipole. Coming back a little closer to the dipole, it can be seen from the figure that $R_2 - R_1$ may be approximated very easily if R_1 and R_2 are assumed to be parallel,

$$R_2 - R_1 \doteq d \cos\theta$$

The final result is then

$$V = \frac{Qd\cos\theta}{4\pi\varepsilon_0 r^2}$$

Using the gradient relationship in spherical coordinates,

$$\bar{E} = -\bar{\nabla}V = -\left(\frac{\partial V}{\partial r}\bar{a}_r + \frac{1}{r}\frac{\partial V}{\partial\theta}\bar{a}_\theta + \frac{1}{r\sin\theta}\frac{\partial V}{\partial\phi}\bar{a}_\phi\right)$$

The result is obtained as

$$\bar{E} = -\left(-\frac{Qd\cos\theta}{2\pi\varepsilon_0 r^3}\bar{a}_r - \frac{Qd\sin\theta}{4\pi\varepsilon_0 r^3}\bar{a}_\theta\right)$$

or

$$\bar{E} = \frac{Qd}{4\pi\varepsilon_0 r^3}(2\cos\theta\,\bar{a}_r + \sin\theta\,\bar{a}_\theta)$$

• PROBLEM 4-34

The potential function of a given field is

$$V = 8r\sin\theta$$

Find an expression for the electric field intensity in the field. Express it in rectangular coordinates.

Solution: As the potential function is specified in spherical coordinates, the field intensity will be obtained using

$$\bar{E} = -\bar{\nabla}V = -\left(\hat{r}_s \frac{\partial V}{\partial r_s} + \hat{\theta}\frac{1}{r_s}\frac{\partial V}{\partial \theta} + \hat{\phi}\frac{1}{r_s \sin\theta}\frac{\partial V}{\partial \phi}\right)$$

Since this problem does not involve $\hat{\phi}$, the last term drops out and

$$E = -\hat{r}_s \frac{\partial V}{\partial r_s} - \hat{\theta}\frac{1}{r}\frac{\partial V}{\partial \theta} = -\hat{r}_s\, 8 \sin\theta - \hat{\theta}\, 8 \cos\theta.$$

To resolve this into rectangular coordinates, the following is done:

Step 1. Write the general vector expression, using unit vectors of the new coordinate system [see (1), (2), or (3)].

$$\bar{A}_{rec} = \hat{x}A_x(x,y,z) + \hat{y}A_y(x,y,z) + \hat{z}A_z(x,y,z) \quad (1)$$

$$\bar{A}_{cyl} = \hat{r}_c A_{rc}(r_c,\phi,z) + \hat{\phi}A_\phi(r_c,\phi,z) + \hat{z}A_z(r_c,\phi,z) \quad (2)$$

$$\bar{A}_{sph} = \hat{r}_s A_{r_s}(r_s,\theta,\phi) + \hat{\theta}A_\theta(r_s,\theta,\phi) + \hat{\phi}A_\phi(r_s\theta,\phi) \quad (3)$$

Step 2. Evaluate the scalar projections onto unit vector directions of the new coordinate system. Refer to the table given.

Step 3. Change the variables from old to new coordinate system.

For this problem, it is needed to transform spherical coordinates to rectangular coordinates. First write both in their general forms.

Step 1.

$$\bar{A}_{rec} = \hat{x}A_x(x,y,z) + \hat{y}A_y(x,y,z) + \hat{z}A_z(x,y,z)$$

$$\bar{A}_{sph} = \hat{r}_s\, A_{rs}(r_s,\theta,\phi) + \hat{\theta}A_\theta(r_s,\theta,\phi) + \hat{\phi}A_\phi(r_s,\theta,\phi),$$

Step 2. (neglecting ϕ term)

$$\bar{A}_x = \bar{A}_{sph} \cdot \hat{x} = (\hat{r}_s A_{rs} + \hat{\theta} A_\theta) \cdot$$

$$(\hat{r}_s \sin\theta \cos\phi + \hat{\theta} \cos\theta \cos\phi)$$

$$= A_{rs} \sin\theta \cos\phi + A_\theta \cos\theta \cos\phi$$

$$\bar{A}_y = \bar{A}_{sph} \cdot \hat{y} = (\hat{r}_s A_{rs} + \hat{\theta} A_\theta) \cdot$$

$$(\hat{r}_s \sin\theta \sin\phi + \hat{\theta} \cos\theta \sin\phi)$$

$$= A_{rs} \sin\theta \sin\phi + A_\theta \cos\theta \sin\phi$$

$$\bar{A}_z = 0$$

Now substitute $\quad A_{rs} = -8 \sin\theta$

and

$$A_\theta = -8 \cos\theta,$$

$$\bar{A}_x = -8 \sin^2\theta \cos\phi - 8 \cos^2\theta \cos\phi$$

$$\bar{A}_y = -8 \sin^2\theta \sin\phi - 8 \cos^2\theta \sin\phi$$

$$\bar{A}_z = 0$$

or

$$\bar{A}_x - 8 \cos\phi(\sin^2\theta + \cos^2\theta) = -8 \cos\phi$$

$$\bar{A}_y - 8 \sin\phi(\sin^2\theta + \cos^2\theta) = -8 \sin\phi$$

$$\bar{A}_z = 0$$

Since ϕ is not considered in this problem (only two-dimensional), take $\phi = 0$. Thus

$$\bar{A}_x = -8$$

$$\bar{A}_y = 0$$

$$\bar{A}_z = 0 .$$

Therefore

$$E_x = -8\text{V/m}$$

and

$$E_y = 0 .$$

This is a uniform field with the field intensity vector directed parallel to the negative y-axis.

Table showing the relationships between Scalar Projections of Vectors in the Rectangular, Cylindrical, and Spherical Coordinate Systems

=	Cylindrical	Spherical	Rectangular
A_x A_y A_z	$A_{r_c} \cos\phi - A_\phi \sin\phi$ $A_{r_c} \sin\phi + A_\phi \cos\phi$ A_z	$A_{r_s} \sin\theta \cos\phi + A_\theta \cos\theta \cos\phi - A_\phi \sin\phi$ $A_{r_s} \sin\theta \sin\phi + A_\theta \cos\theta \sin\phi + A_\phi \cos\phi$ $A_{r_s} \cos\theta - A_\theta \sin\theta$	A_x A_y A_z
A_{r_c} A_ϕ A_z	A_{r_c} A_ϕ A_z	$A_{r_s} \sin\theta + A_\theta \cos\theta$ A_ϕ $A_{r_s} \cos\theta - A_\theta \sin\theta$	$A_x \cos\phi + A_y \sin\phi$ $-A_x \sin\phi + A_y \cos\phi$ A_z
A_{r_s} A_θ A_ϕ	$A_{r_c} \sin\theta + A_z \cos\theta$ $A_{r_c} \cos\theta - A_z \sin\theta$ A_ϕ	A_{r_s} A_θ A_ϕ	$A_x \sin\theta \cos\phi + A_y \sin\theta \sin\phi + A_z \cos\theta$ $A_x \cos\theta \cos\phi + A_y \cos\theta \sin\phi - A_z \sin\theta$ $-A_x \sin\phi + A_y \cos\phi$

MOTION IN ELECTRIC FIELD

• PROBLEM 4-35

An electron, starting from rest, moves unimpeded in an electric field of intensity E volts per meter. Find

(a) the force it experiences,

(b) its acceleration,

(c) the kinetic energy it attains in moving through a potential difference of V volts,

(d) the velocity it attains in moving through a p.d. of V volts.

(Take $q = 1.602 \times 10^{-19}$ coulomb, mass $= 9.11 \times 10^{-31}$ kilogram.)

Solution: (a) The force acting on the electron is

$$F = Eq = (1.602 \times 10^{-19})\ E \text{ newtons}$$

$$= (1.602 \times 10^{-14})\ E \text{ dynes.}$$

(b) The acceleration,

$$\alpha = \frac{\text{force}}{\text{mass}}$$

$$= \frac{1.602 \times 10^{-19}}{9.21 \times 10^{-31}} E = (1.758 \times 10^{11})\ E \text{ meters per sec.}^2,$$

or, in c.g.s. units,

$$= \frac{1.602 \times 10^{-14}}{9.12 \times 10^{-28}} E = (1.758 \times 10^{13})\ E \text{ cm. per sec.}^2.$$

(c) By the definition of p.d., the work done when an electron moves through V volts is Vq joules. This is converted into the kinetic energy of the electron. Neglecting the relativity change of mass at high velocities

$$\frac{1}{2} mv^2 = (1.602 \times 10^{-19})\ V \text{ joules}$$

$$= (1.602 \times 10^{-12})\ V \text{ ergs.}$$

or

(d) From (c), $\frac{1}{2}mv^2 = qV$, so that

$$v = \sqrt{\frac{2qV}{m}}$$

$$= \sqrt{\frac{2 \times 1.602 \times 10^{-19} V}{9.12 \times 10^{-31}}} = 5.93 \times 10^5 \sqrt{V} \text{ meters per sec.}$$

or, in c.g.s. units,

$$v = \sqrt{\frac{2 \times 1.602 \times 10^{-12} V}{9.12 \times 10^{-28}}} = 5.93 \times 10^7 \sqrt{V} \text{ cm. per sec.}$$

Note. The mass m of the electron can be assumed constant only if $v^2 \ll c^2$. If the usual equations of Newtonian dynamics are retained, the theory of relativity shows that the mass that must be used in the equation

$$\text{mass} = \frac{\text{force}}{\text{acceleration}}$$

when the force is in the direction of the velocity v, has the value

$$m_L = \frac{m}{\left(1 - \frac{v^2}{c^2}\right)^{3/2}}$$

(where $m = 9.11 \times 10^{-31}$ kg.), and is called the "longitudinal mass".

If the force is transverse to the velocity v (as in the case of a cathode-ray oscillograph whose beam is deflected by transverse forces) the equivalent mass has the value

$$m_t = \frac{m}{\sqrt{1 - \frac{v^2}{c^2}}}$$

and is called the "transverse mass".

• **PROBLEM 4-36**

In the figure shown, a charge q of mass m enters the space between two plane, parallel oppositely charged plates with speed v_0. Obtain an expression for the path of the particle.

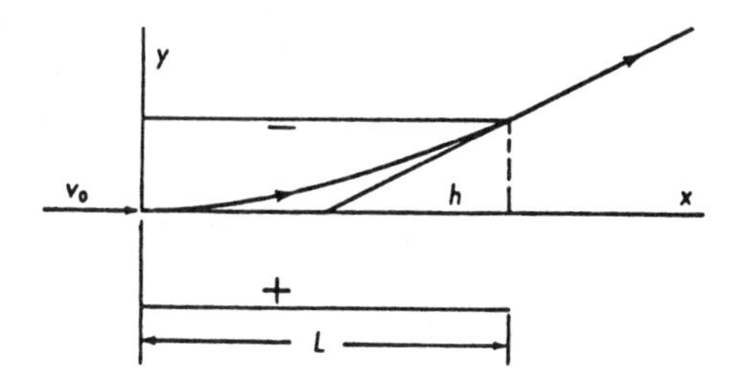

A charge q of mass m entering the space between two plane parallel oppositely charged plates with speed v_0.

Solution: In this case, if the distortion of the field at the edges of the plates may be neglected, then

$$E_x = 0$$

and

$$E_y = E = \text{constant}$$

Therefore,

$$\frac{d^2x}{dt^2} = 0 \qquad \frac{dx}{dt} = v_0 \qquad x = v_0 t$$

$$\frac{d^2y}{dt^2} = \frac{qE}{m} \quad \frac{dy}{dt} = \frac{qEt}{m} \quad y = \frac{1}{2}\left(\frac{qE}{m}\right)t^2$$

since at t = 0

$$\frac{dx}{dt} = v_0, \qquad \frac{dy}{dt} = 0 \qquad x = 0, \qquad y = 0.$$

The path of the particle may be obtained by eliminating t. Since

$$t = x/v_0,$$

$$y = \frac{1}{2}\left(\frac{qE}{mv_0^2}\right)x^2.$$

The trajectory of the particle is evidently a parabola.

The particle, when it emerges from the plates at $x = L$, appears to come from the point $x = L/2$, $y = 0$. The slope of the tangent at $x = L$ is

$$\left(\frac{dy}{dx}\right)_{x=L} = \frac{qEL}{mv_0^2}$$

At $x = L$, the height of the tangent line is

$$y = \frac{1}{2}\left(\frac{qE}{mv_0^2}\right)L^2$$

Therefore,

$$\frac{qEL}{mv_0^2} = \frac{(1/2)(qEL^2/mv_0^2)}{h}$$

where h is the base of the triangle in the figure. When the last equation is solved for h,

$$h = \frac{L}{2}$$

The particle appears to come from the point (L/2,0) when it emerges into the field-free space.

• PROBLEM 4-37

In a limited region near the origin and in the XY-plane, a certain electric field can be represented by the vector equation

$$\bar{E} = k(\bar{i}y + \bar{j}x) \qquad (1)$$

where k = 1 newton/coulomb meter, the units of k being those required to make (1) dimensionally correct. Find the increase in the potential energy of a positive test charge of 3.20×10^{-19} coulomb when the test charge is carried from B by a straight line path directly to A. Point B has coordinates $x = 3.00$ m, $y = 4.00$ m, and A is at the origin.

Solution: In the figure shown below, draw the coordinate axes, locate the points B and A, draw the straight line path from B to A, and draw a representative vector for E at any point on the path.

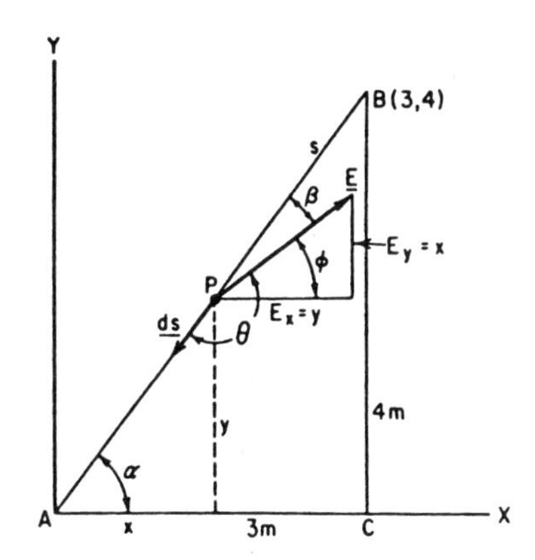

Equation (1) tells us that the x and y components of $\bar{E}$ are $E_x = ky$ and $E_y = kx$. Thus, the x component of $\bar{E}$ at a point P is directly proportional to the y coordinate of P, and the y component of $\bar{E}$ is directly proportional to the x coordinate of P. The figure shows these two components of $\bar{E}$ at the point P on the path BPA. The coordinates of P are (x,y).

At P draw an element of path length $\overline{ds}$, which must be directed toward $\bar{A}$, since the motion of q' is from B to A. θ is the angle between $\bar{E}$ and $\overline{ds}$.

To evaluate the integral on the right of

$$PE_A - PE_B = -q' \int_B^A \bar{E} \cdot d\bar{s} ,$$

express $\bar{E}$ and cos θ as functions of s. The distance s must be measured from the starting point of the motion at B so, on the diagram, s is the distance from B to P along the straight path BPA. First find the functional relationship between E and s.

From the diagram

$$s = 5 \text{ meters} - \sqrt{x^2 + y^2},$$

or

$$\sqrt{x^2 + y^2} = 5 \text{ meters} - s \qquad (2)$$

where x and y are the coordinates of any point P on the line BPA. From (1) the magnitude of E is

$$k\sqrt{x^2 + y^2}$$

$$\sqrt{x^2 + y^2} = E/k \tag{3}$$

Substitute (3) into (2) and solve for E

$$E = 5k \text{ meters} - ks \tag{4}$$

This gives the functional relationship between E and s that is needed for substitution into the integral.

Next, find the functional relationship between $\cos\theta$ and s. From the diagram, $\beta = \alpha - \phi$, so $\cos\beta = \cos\alpha \cos\phi + \sin\alpha \sin\phi$. Evaluate, from the diagram, the sines and cosines on the right, and make use of (2) to obtain

$$\cos\beta = (3y + 4x)/5(5 \text{ meters} - s) \tag{5}$$

The straight line path APB may be represented by the equation $y = 4x/3$ which, with (5), yields

$$\cos\beta = 8x/5(5 \text{ meters} - s) \tag{6}$$

Put $y = 4x/3$ into (2) and substitute the value of x obtained into (6). The result is $\cos\beta = 24/25$. But $\cos\theta = -\cos\beta$, so

$$\cos\theta = -24/25 \tag{7}$$

Equation (7) gives the functional relationship between $\cos\theta$ and s, and it shows that $\cos\theta$ is constant along this path and thus independent of s.

$$PE_A - PE_B = -q'\int_{(B)}^{(A)} (5k \text{ meters} - ks)(-24/25)ds$$

Replace the symbolic limits (A) and (B) by numerical limits expressed in terms of the variable of integration. At B, s = 0 and at A, s = 5 m. Hence

$$PE_A - PE_B = (24/25)q'\int_{0}^{5m} (5k \text{ meters} - ks)ds$$

Integrate and substitute the limits of integration

$$PE_A - PE_B = 12q' \text{ joules/coulomb} \tag{8}$$

Note that $PE_A - PE_B$ is positive, since q' is positive, as it must be, for the PE of q' at A is greater than the PE of q' at B, i.e., work has been done against the field in moving q' from B to A.

Now, to obtain the answer, substitute the value of q' into (8) and obtain

$$PE_A - PE_B = 38.4 \times 10^{-19} \text{ joule}$$

ENERGY

• PROBLEM 4-38

Four point charges are placed as shown in the figure below, forming a square of side a = 6m. Assuming that these are suspended in the air, find the total energy of the configuration, for $q = 2.0 \times 10^{-8}$C.

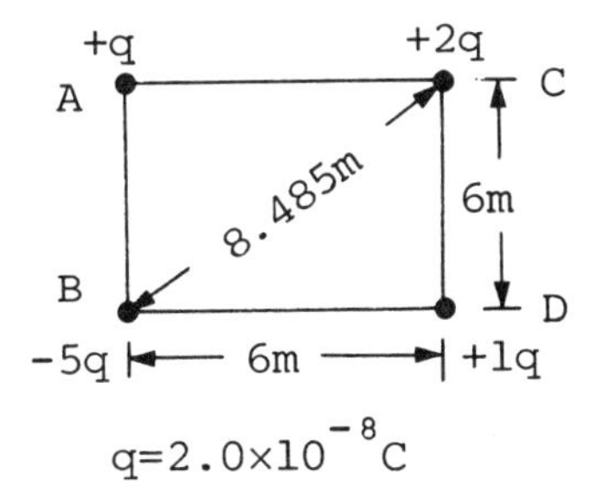

$q = 2.0 \times 10^{-8}$C

Solution: The total energy is:

$$U = U_{AB} + U_{AD} + U_{AC} + U_{BC} + U_{BD} + U_{CD}$$

Now the energy due to a pair of charges is:

$$U = \frac{1}{4\pi\varepsilon_0 \cdot 1}\left[\frac{q_i q_j}{r_{ij}}\right]$$

where i,j = A|B|C|D (either A, B, C or D) and

r_{ij} = The distance between the two charges.

By substitution,

$$U_{AB} = \frac{1}{4\pi\varepsilon_0}\left(\frac{-5q^2}{6}\right), \quad U_{AD} = \frac{1}{4\pi\varepsilon_0}\left(\frac{q^2}{8.485}\right),$$

$$U_{AC} = \frac{1}{4\pi\varepsilon_0}\left(\frac{2q^2}{6}\right), \quad U_{BC} = \frac{1}{4\pi\varepsilon_0}\left(\frac{-10q^2}{8.485}\right)$$

$$U_{BD} = \frac{1}{4\pi\varepsilon_0}\left(\frac{-5q^2}{6}\right), \quad U_{CD} = \frac{1}{4\pi\varepsilon_0}\left(\frac{2q^2}{6}\right)$$

So, taking the sum of all above terms, we get

$$U = (9.0\times10^{9}\ \text{N-m}^2/\text{c}^2)(2.0\times10^{-8})^2\left[\frac{-5}{6} + \frac{1}{8.485} + \frac{2}{6} - \frac{10}{8.485} - \frac{5}{6} + \frac{2}{6}\right]$$

$$U = 3.6 \times 10^{-6}\left[-2.0606\right]\text{J}$$

$$= -7.4185 \times 10^{-6}\text{J}$$

• PROBLEM 4-39

Calculate the electrostatic energy of a system consisting of a point charge q located at the center of a spherical conducting shell of inner radius r_1 and outer radius r_2.

Solution: Using Gauss's law, the electrostatic field has the following values:

$$\bar{E}(\bar{r}) = \left(\frac{1}{4\pi\varepsilon_0}\right)\frac{q}{r^2}\,\hat{r}$$

$$r < r_1$$

and

$$r > r_2$$

$$E(r) = 0$$

$$r_1 \leq r \leq r_2 \, .$$

The electrostatic energy of this system is calculated using

$$U = \int_V \frac{1}{2}\varepsilon_0 E^2\, dV$$

and subtracting out the self-energy of the charge q. For a sphere,

$$dv = 4\pi r^2\, dr$$

therefore

$$U = (4\pi\varepsilon_0)\frac{1}{8\pi}\left\{\int_0^{r_1}\left(\frac{1}{4\pi\varepsilon_0}\frac{q}{r^2}\right)^2 4\pi r^2\, dr + \int_{r_2}^{\infty}\left(\frac{1}{4\pi\varepsilon_0}\frac{q}{r^2}\right)^2 4\pi r^2\, dr\right.$$

$$- \int_0^\infty \left(\frac{1}{4\pi\varepsilon_0} \frac{q}{r^2}\right)^2 4\pi r^2 \, dr \Bigg\} .$$

$$U = \frac{1}{4\pi\varepsilon_0} \frac{1}{2} q^2 \left(\frac{1}{r_2} - \frac{1}{r_1}\right).$$

• PROBLEM 4-40

In a diode vacuum tube consisting of a heated, plane metal cathode and a parallel plate, electrons are emitted from the cathode and are accelerated toward the plate. In the region between the two tube elements there is a cloud of electrons called a space charge. The potential function is given by $V = V_0 (x/a)^{4/3}$, where V_0 is the potential of the plate, x is the coordinate distance from the cathode, and a is the distance of the plate from the cathode. Assuming that the electrons are emitted with negligible initial velocity, find the velocity of the electrons at any point x and the time of flight from cathode to plate.

Solution: The potential at the cathode is zero and the velocity at the cathode is also zero. Therefore

$$0 + 0 = \frac{1}{2} mv^2 + (-e)\, V_0 \left(\frac{x}{a}\right)^{4/3}$$

where -e is the charge of an electron.

$$v^2 = \left(\frac{2eV_0}{m}\right)\left(\frac{x}{a}\right)^{4/3}$$

$$v = \frac{dx}{dt} = \left(\frac{2eV_0}{m}\right)^{1/2}\left(\frac{x}{a}\right)^{2/3}$$

$$\left(\frac{2eV_0}{m}\right)^{-\frac{1}{2}} \left(\frac{x}{a}\right)^{-2/3} dx = dt$$

$$t = 3a \left(\frac{2eV_0}{m}\right)^{-\frac{1}{2}} \left(\frac{x}{a}\right)^{1/3}$$

The time of flight is the value of t at x = a. Thus

$$\text{time of flight} = 3a\left(\frac{2eV_0}{m}\right)^{-\frac{1}{2}}$$

If V_0 = 100 volts, a =0.01 meter, e = $1.6(10^{-19})$ coulomb, and m = $9.11\ (10^{-31})$ kilogram, then

$$\text{time of flight} = 5.5(10^{-9})\ \text{sec}$$

$$\text{velocity at plate} = 5.93(10^{6})\ \text{m per sec}$$

• PROBLEM 4-41

Find the energy stored in the electrostatic field between two concentric conducting spheres of radii R and 2R, respectively. The charges on the two spheres are both of magnitude Q and are opposite in sign.

Solution: The problem can be done in two ways:

(1) Use the integral formula

$$W = \frac{1}{2}\int \Phi\,\rho\,dv \tag{1}$$

The potential Φ to the outer sphere is set at $\Phi(r = 2R) = 0$. The potential of the inner sphere is then

$$\Phi = \frac{Q}{4\pi\varepsilon_0}\left(\frac{1}{R} - \frac{1}{2R}\right) = \frac{Q}{4\pi\varepsilon_0}\,\frac{1}{2R} \tag{2}$$

The charge on the inner sphere is a surface charge totaling in amount the charge Q. Since all the charge on the inner sphere is at the potential (2), integrate Eq. 1 very simply, with the result

$$W = \frac{Q^2}{16\pi\varepsilon_0 R} \tag{3}$$

(2) Alternatively use the formula involving energy density of the electrostatic field.

$$W = \frac{1}{2}\int \varepsilon_0 |E|^2\,dv \tag{4}$$

In a spherical coordinate system with its origin at the

center of the spheres the electric field is given by

$$\bar{E} = \bar{i}_r \frac{Q}{4\pi\varepsilon_0 r^2} \tag{5}$$

From Eqs. 4 and 5

$$W = \frac{\varepsilon_0}{2} \frac{Q^2}{(4\pi\varepsilon_0)^2} \int_{r=R}^{r=2R} \frac{4\pi r^2 dr}{r^4} = \frac{Q^2}{8\pi\varepsilon_0}\left(\frac{1}{R} - \frac{1}{2R}\right) = \frac{Q^2}{16\pi\varepsilon_0 R}$$

This answer checks with Eq. 3.

• PROBLEM 4-42

A spherical charge distribution of radius r_a and uniform ρ_V is centered at the origin. If the total charge is Q: (a) find the energy expended to build up the charge system through the use of

$$W = \int_V \frac{1}{2} V\rho_V \, dv \tag{1}$$

(b) repeat part (a) through the use of

$$W = \int_V \frac{1}{2} \varepsilon_0 \, E^2 dV \tag{2}$$

(c) let $r_a = 0$ and thus find the energy to assemble a point charge Q; (d) find the energy expended to bring from infinity the first point charge Q in building up a system of point charges. Assume that the point charge Q was assembled at infinity.

Solution: (a) Equation (1) must be integrated only over the volume of radius r_a since ρ_V is zero for $r_s > r_a$. Thus, the absolute potential $V = V_{r_s(\infty)}$ must also be evaluated over the same range from the knowledge of $\bar{E}_{r_s > r_a}$ and $\bar{E}_{r_s < r_a}$
From

$$V_{a(\infty)} = \int_{\infty}^{a} (-\bar{E} \cdot \bar{d\ell}) \; ,$$

$$V_{r_{s(\infty)}} = \int_{\infty}^{r_a} (-\bar{E}_{r_s > r_a} \cdot \overline{d\ell}) + \int_{r_a}^{r_s} (-\bar{E}_{r_s < r_a} \cdot \overline{d\ell})$$

Substituting for $\bar{E}_{r_s > r_a}$ and $\bar{E}_{r_s < r_a}$,

$$V_{r_{s(\infty)}} = \int_{\infty}^{r_a} - \frac{Q\hat{r}_s}{4\pi\varepsilon_0 r_s^2} \cdot \hat{r}_s \, dr_s + \int_{r_a}^{r_s} - \frac{Qr_s\hat{r}_s}{4\pi\varepsilon_0 r_a} \cdot \hat{r}_s \, dr_s$$

$$= \frac{Q}{4\pi\varepsilon_0 r_a} - \frac{Q}{8\pi\varepsilon_0 r_a^3} (r_s^2 - r_a^2) \quad (\text{for } r_s \leqq r_a) \qquad (3)$$

Substituting (3) into (1),

$$W = \int_V \rho_V \frac{VdV}{2} = \int_V \frac{1}{2} \left(\frac{Q}{\frac{4}{3}\pi r_a^3} \right) \left[\frac{Q}{4\pi\varepsilon_0 r_a} - \frac{Q}{8\pi\varepsilon_0 r_a^3} (r_s^2 - r_a^2) \right] r_s^2 \sin\theta \, dr_s \, d\theta \, d\phi$$

$$= \frac{3Q^2}{20\pi\varepsilon_0 r_a} \text{ (J)} \qquad (4)$$

(b) From (2)

$$W = \int_V \frac{1}{2} \varepsilon_0 E^2 \, dV = \int_{V(r_s > r_a)} \frac{1}{2}\varepsilon_0 E^2_{(r_s > r_a)} \, dV +$$

$$\int_{v(r_s < r_a)} \frac{1}{2}\varepsilon_0 E^2_{(r_s < r_a)}\, dV$$

$$= \int_{v(r_s > r_a)} \frac{1}{2}\varepsilon_0 \left(\frac{Q}{4\pi\varepsilon_0 r_s^2}\right)^2 (r_s^2 \sin\theta\, dr_s\, d\theta\, d\phi)$$

$$+ \int_{v(r_s < r_a)} \frac{1}{2}\varepsilon_0 \left(\frac{Qr_s}{4\pi\varepsilon_0 r_a^3}\right)^2 (r_s^2 \sin\theta\, dr_s\, d\theta\, d\phi)$$

$$= \frac{Q^2}{8\pi\varepsilon_0 r_a} + \frac{Q^2}{40\pi\varepsilon_0 r_a} = \frac{3Q^2}{20\pi\varepsilon_0 r_a} \text{ (J)} \qquad (5)$$

Note that (4) and (5) are the same.

(c) Now, let $r_a = 0$ in (4) or (5) to obtain $W = \infty$ (J). This is the energy required to build the sphere of charge as $r_a \to 0$. Thus, it is the energy to assemble a point charge of Q(C). Note that $\rho_v = \infty$ for a point charge.

(d) The energy to bring the first point charge of Q(C) from infinity to the origin is equal to zero. Here it was assummed that the point charge Q was assembled at infinity with infinite amount of energy as found in part (c).

• PROBLEM 4-43

A concentrated charge of 10 microcoulombs exists in an extensive mass of Bakelite. Determine the total energy contained in the electric field outside a radial distance of (a) 10 meters, (b) 1 meter, (c) 10 centimenters, and (d) 1 centimeter.

Solution: From Coulomb's law, the electric field intensity due to a point charge is given by

$$E = \frac{Q}{4\pi\varepsilon_0 KX^2}\,.$$

Let a differential volume of space be enclosed between a spherical shell of radius X, where X serves as a variable radial distance and another spherical shell of radius X + dX. The differential volume is

$$dv = 4\pi X^2 dX$$

and the differential energy contained therein is

$$dW = \frac{1}{2} \varepsilon_0 K E^2 \, dv$$

$$= \frac{Q^2}{8\pi\varepsilon_0 K} \frac{dX}{X^2} .$$

The total energy from X = R to x = infinity is

$$W = \frac{Q^2}{8\pi\varepsilon_0 K} \int_R^\infty X^{-2} \, dX$$

$$= - \left. \frac{Q^2}{8\pi\varepsilon_0 KX} \right]_R^\infty = \frac{Q^2}{8\pi\varepsilon_0 KR}$$

(a) The energy contained outside a radius of 10 meters is

$$W = \frac{10^{-10}}{8\pi \times 8.85 \times 10^{-12} \times 4.5 \times 10} = 0.010 \text{ joule.}$$

(b) For that outside a radius of 1 meter

$$W = 0.10 \text{ joule.}$$

(c) For that outside a radius of 10 centimeters

$$W = 1.0 \text{ joule.}$$

(d) For that outside a radius of 1 centimeter

$$W = 10 \text{ joules.}$$

• PROBLEM 4-44

A parallel plate air-capacitor is charged to a charge Q. The distance between the plates is d, the area of the plates is a.

(a) Find the energy stored in the capacitor.

(b) Find the electrostatic force per unit area upon the plates. Neglect the fringing fields.

Solution: (a) The solution to part (a) will be done in two ways.

(1) Using the expression

$$W = \frac{1}{2}\int \Phi\rho \, dv : \tag{1}$$

Assign the potential $\Phi = 0$ to one plate carrying the negative charge. The potential of the plate with the positive charge is then

$$\Phi = \frac{Q}{\varepsilon_0 a} d \tag{2}$$

where Q is the charge on this plate. Thus

$$W = \frac{1}{2}\int \Phi\rho \, dv = \frac{1}{2} \frac{Q^2}{\varepsilon_0 a} d \tag{3}$$

since the voltage is constant over the entire surface of the plate over which the charge is distributed.

(2) Using the expression

$$W = \frac{1}{2} \varepsilon_0 \int |\bar{E}|^2 \, dv \tag{4}$$

note that

$$|\bar{E}| = \frac{Q}{\varepsilon_0 a} \tag{5}$$

and is constant throughout the volume ad of the capacitor. Thus

$$W = \frac{1}{2} \varepsilon_0 \int |E|^2 \, dv = \frac{1}{2} \frac{Q^2}{\varepsilon_0 a} d \qquad (6)$$

which checks with the result of Eq. 3.

(b) The electrostatic force upon a body can be computed once the rate of change of the electrostatic energy caused by a general motion of the body is known.

If the stored electrostatic energy has changed by the amount dW, then, by conservation of energy, we must have

$$dW = -\bar{F} \cdot ds \qquad (7)$$

The minus sign is to account for a decrease (negative dW) if $F \cdot ds$ is positive and the field does work on the body.

Applying the foregoing equation to the problem at hand, the energy of a capacitor with one plate at $x = 0$ and the other plate at $x = x$ is, from Eq. 3,

$$W = \frac{1}{2} \frac{Q^2}{\varepsilon_0 a} x$$

If the plate at x is moved by the amount dx, we have

$$dW = \frac{1}{2} \frac{Q^2}{\varepsilon_0 a} dx \qquad (8)$$

The mechanical work done in the motion is

$$F_x \, dx$$

where F_x is the component of the force in the x-direction. (By symmetry we may conclude that this is the only force component.) From Eqs. (7) and (8)

$$F_x = -\frac{1}{2} \frac{Q^2}{\varepsilon_0 a} = -\frac{1}{2} \varepsilon_0 |E|^2 a$$

where E is the field in the capacitor. The force is negative, thus attracting the plate at $x = x$ to the plate $x = 0$.

CHAPTER 5

DIELECTRICS

CURRENT DENSITY

• PROBLEM 5-1

For a ρ_v distribution, in a long cylinder, moving with a velocity $\bar{U} = \hat{z}5z$ (ms^{-1}), as shown in Fig. 1, find:

(a) $\rho_v|_{z_1=2}$ when $I|_{z_1=2} = \pi(\mu A)$;

(b) $\rho_v|_{z_2=4}$ when $I|_{z_2=4} = \pi(\mu A)$;

(c) $\rho_v(z)$ when $I(z) = \pi(\mu A)$;

(d) $\bar{J}(Am^{-2})$.

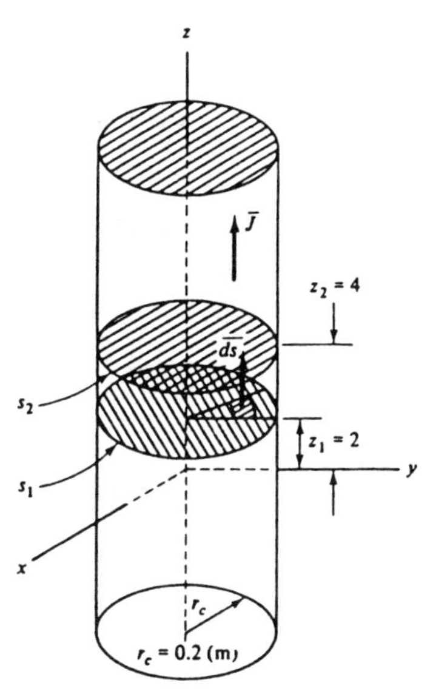

Figure 1 Graphical display for finding ρ_v in a long cylinder of charge distribution

Solution: (a) Assume that I and $\bar{J}$ are uniform over the cross section of the cylinder. From

$$J = \rho_\nu \bar{U}$$

and

$$I = \int_s \bar{J} \cdot \overline{ds} ,$$

in functional form, we have

$$I(z) = \int_s \bar{J}(z) \cdot \overline{ds} = \int_s \rho_\nu(z)\bar{U}(z) \cdot \overline{ds} \qquad (1)$$

Since $\rho_\nu(z)$ and $\bar{U}(z)$ do not vary with x and y, and $\bar{U}(z)$ is parallel to $\overline{ds}$, (1) reduces to

$$I(z) = \rho_\nu(z)U_z(z) \int_s ds = \rho_\nu(z)U_z(z)\pi r_c^2 \qquad (2)$$

Solving for $\rho_\nu(z)$ and substituting values, yields

$$\rho_\nu(z)\Big|_{z_1=2} = \frac{I(z)}{U_z(z)\pi r_c^2}\Bigg|_{z_1=2} = \frac{\pi\ 10^{-6}}{10\pi(0.2)^2} = 2.5$$

$$(\mu Cm^{-3}) \qquad (3)$$

(b) Evaluate (3) at $z_2=4$ to obtain

$$\rho_\nu(z)\Big|_{z_2=4} = \frac{\pi 10^{-6}}{20\pi(0.2)^2} = 1.25 \quad (\mu Cm^{-3}) \qquad (4)$$

(c) From (3),

$$\rho_\nu(z) = \frac{\pi 10^{-6}}{5z\pi(0.2)^2} = 5z^{-1} \quad (\mu Cm^{-3})$$

(d) Then,

$$\bar{J} = \rho_\nu\bar{U} = \frac{5}{z}\ \hat{z}5z = \hat{z}25 \quad (\mu Am^{-2})$$

• **PROBLEM 5-2**

A brass rod with a circular cross section of radius 2.1 cm carries a sinusoidal current at a frequency of 1590 cycles per second. The axial drift current density J is approximately

$$J = 0.001e^{\pi r} \sin (\omega t + \pi r) \text{ ampere/cm}^2$$

with $\omega = 10^4$ and r representing the distance in centimeters from the center of the wire. Find the dissipated power per centimeter of length. The conductivity of brass is 1.57×10^5 mhos/cm.

Solution: The dissipated power p in watts per centimeter is equal to

$$p = \int_S (J^2/\sigma)dS = \int_0^{2.1} (J^2/\sigma) 2\pi r \, dr$$

with S denoting the cross-sectional area and with all lengths expressed in centimeters. The current density squared can be written as

$$J^2 = 10^{-6}e^{2\pi r} \left[\frac{1}{2} - \frac{1}{2} \cos (2\omega t + 2\pi r)\right]$$

Substitution for J^2 and σ in the expression for p gives

$$p = 31.8 \times 10^{-13} \int_0^{2\cdot 1} 2\pi r[1 - \cos(2\omega t + 2\pi r)] 2\pi r \, dr$$

Changing the variable of integration from r to x for convenience, with $x = 2\pi r$, the dissipated power per centimeter becomes

$$p = 5.06 \times 10^{-13} \int_0^{4.2\pi} xe^x [1 - \cos(2\omega t + x)]dx$$

The integral of xe^x is readily evaluated by integration by parts.

$$\int xe^x \sin(\omega t + x) \, dx = \left[\frac{1}{2} e^x\right] [(x \cos x + x \sin x - \sin x) \sin \omega t + (x \sin x - x \cos x + \cos x) \cos \omega t]. \qquad (1)$$

Equation (1), with ωt replaced with $2\omega t + 90°$, can be utilized to evaluate the integral of

$$xe^{x} \cos (2\omega t + x).$$

The resulting power is

$$p = [3.31 - 2.44 \sin (2\omega t + 83.3°)] \times 10^{-6} \text{ watt/cm.}$$

The time-average dissipated power is obviously 3.31 μw/cm, for the time average of the sinusoidal term is zero.

• PROBLEM 5-3

A brass washer has a 2-cm inside diameter and a 4-cm outside diameter and is 0.6 cm thick; $\sigma = 1.5 \times 10^7$ mho/m. The washer is cut in half along a diameter, and good electrical connections are made to the two rectangular faces of one part. The resultant density is

$$\bar{J} = (10^4 /r)\bar{a}_\phi ,$$

where the z-axis is the axis of the washer, see the figure.

(a) What total current is flowing?

(b) What is the maximum $\bar{E}$ at any point within the half washer?

(c) What is the potential difference between the two faces?

(d) What is the total resistance between the two faces?

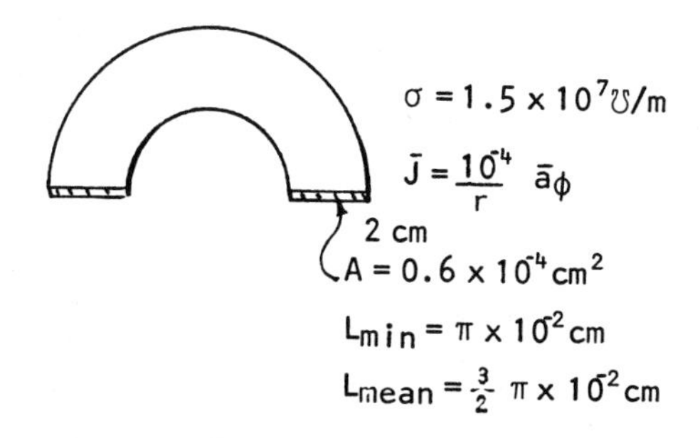

Solution:

(a) $$I = \int_S \bar{J} \cdot d\bar{S} , \qquad d\bar{S} = dr\, dz\, \bar{a}_\phi$$

$$I = 10^4 \int_0^{0.006} \int_1^2 \frac{dr}{r}\, dz = 60 \ln 2 = 41.6\text{A}$$

(b) Since $\bar{J} = \sigma\bar{E}$, and $\bar{E}$ is strongest at $r = 1$ cm.

and

$$\bar{J}_{max} = (10^4/10^{-2})\ \bar{a}_\phi$$

Therefore

$$\bar{E}_{max} = \frac{10^6}{1.5 \times 10^7}\ \bar{a}_\phi = 0.0667\ \bar{a}_\phi \text{ V/m.}$$

(c) The potential difference between the two faces is

$$V = E_{max}L_{min} = 0.0667 \times \pi \times 10^{-2} = 2.1 \times 10^{-3} \text{ V}$$

(d) The resistance is $R = L/\sigma A$, where L is the mean path length midway between the edges and is

$$\frac{3\pi}{2} \times 10^{-2} \text{ cm.,}$$

therefore

$$R = \frac{\frac{3\pi}{2} 10^{-2}}{1.5 \times 10^7 \times 0.6 \times 10^{-4}} = 5.23 \times 10^{-5}\ \Omega$$

• **PROBLEM 5-4**

(a) A PN junction rectifier made of germanium has a rectangular cross section of area 10^{-6} m^2. Let x represent the distance from the junction to an arbitrary point in the N-type region. The hole density p in the N region is

$$[1.8 + 400 \exp(-x/0.0005)]10^{18}$$

holes per cubic meter, and the free electron density n is

$$[320 + 400 \exp(-x/0.0005]\ 10^{18} \text{electrons/m}^3$$

If the hole mobility μ_p is 0.19 m^2/volt-sec and the electron mobility μ_n is 0.39 m^2/volt-sec, determine the conductivity as a function of x.

(b) In the rectifier the hole diffusion current density J_p is $-0.005e\ dp/dx\ i$ amperes/m^2, and the electron diffusion current density J_n is $0.01e\ dn/dx\ i$. If the total current density, consisting of both drift and diffusion current densities, is everywhere equal to 1000i amperes/m^2, determine the voltage drop v from x = 0 to x = 0.0015 m.

Solution: (a) The conductivity σ is the sum of $pe\mu_p$ and $ne\mu_n$, with e representing the electronic charge

$$1.6 \times 10^{-19} \text{ coulomb.}$$

Adding these terms together, after substituting the values given for p, n, and the mobilities, gives

$$\sigma = 20 + 37.1 \exp(-x/0.0005) \text{mhos/m}$$

(b) p and n are given as functions of x. Differentiation of these expressions shows that

$$dp/dx = dn/dx = -8 \times 10^{23} \exp(-x/0.0005).$$

From this result J_p is found to be 640 exp (-x/0.0005)i and J_n to be -1280 × exp (-x/0.0005)i. Therefore, the total difusion current density is -640 × exp (-x/0.0005)i. As the sum of the drift and diffusion current densities is everywhere equal to 1000i, it follows that the drift current density is [1000 + 640 × exp (-x/0.0005)]i.

The drift current density equals σE. Therefore, E_x is $1/\sigma$ times the magnitude of the drift current density. Using the conductivity calculated in part (a),

$$E_x = \frac{1000 + 640 \exp(-x/0.0005)}{20 + 37.1 \exp(-x/0.0005)}$$

The voltage drop v is found by integrating E_x with respect to x from 0 to 0.0015. The integration is readily performed by changing variables, with $y = -x/0.0005$, and referring to integral tables. The result shows that the voltage drop is 59 millivolts.

The current over a cross section of the rectifier equals the current density of 1000 amperes/m^2 multiplied by the area of $10^{-6} m^2$. This gives a current of 0.001 ampere, or

1 ma. The ratio of the voltage drop v to the current is 59 ohms.

It should be noted that the diffusion current densities are not related to the electric field and, consequently, do not contribute to the voltage drop. The voltage drop is due entirely to the drift current densities.

Find the current density $\bar{J}$ between two concentric spherical electrodes, then find the resistance from the previous result.

The figure illustrates two concentric spheres of respective radii a and b. The inner sphere is kept at the potential V_o relative to the outer sphere. The medium between the spheres is assumed to be homogeneous of conductivity σ.

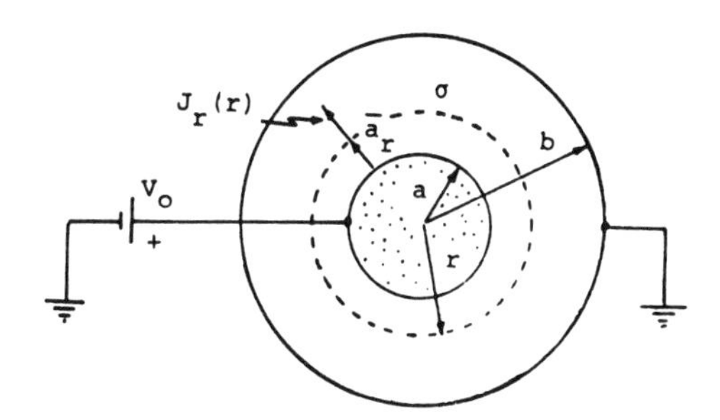

Solution: The configuration has perfect spherical symmetry, and it may be assumed that the current density $\bar{J}$ is in the radial direction and is a function of r only. Let the total current supplied by the center electrode be I; thus I is the total source strength. Now apply Gauss' law to the spherical surface of radius r, to get

$$\oint_S \bar{J} \cdot d\bar{S} = I$$

Because of the symmetry, the integration can be performed to get

$$\oint_S \bar{J} \cdot d\bar{S} = J_r(r) \oint \bar{a}_r \cdot d\bar{S} = 4\pi r^2 J_r(r) = I$$

The existence of symmetry is a crucial requirement in order to assume that the radial current density $J_r(r)$ is constant everywhere on the spherical surface of radius r and to permit the performing of integration.

From the above result,

$$J_r(r) = \frac{1}{4\pi r^2}$$

Now writing in spherical coordinates

$$\bar{J}_r(r) = \sigma\bar{E} = -\sigma \frac{\partial\phi}{\partial r}\bar{a}_r$$

and so

$$\frac{\partial\phi}{\partial r} = \frac{-1}{4\pi\sigma r^2}$$

from which, by integration,

$$\phi = \frac{1}{4\pi\sigma r} + C$$

From the figure the boundary condition is: $r = b$, $\phi = 0$; from this $C = -I/4\pi\sigma b$. Further, for ϕ to equal V_o at $r = a$, the current I must then have the value

$$I = \frac{4\pi\sigma ab}{b-a} V_o$$

and the current density is

$$J_r(r) = \frac{\sigma abV_o}{(b-a)r^2}$$

The ratio of V_o to the total current I is the total or effective lumped resistance to current flow.

$$R = \frac{V_o}{I} = \frac{b-a}{4\pi\sigma ab}$$

• PROBLEM 5-6

Consider a conducting sphere of radius R surrounded by free space. The parameters of the sphere are scalar constants ε, μ, and σ. The parameters of free space are, of course, ε_0, μ_0, and $\sigma_0 = 0$.

Assume that at $t = 0$ charge is distributed uniformly throughout the conducting sphere with a charge density ρ_0. Find the current distribution inside the sphere and the electric field intensity vector both inside and outside the sphere as functions of position and time.

Solution: First, find the initial fields $\bar{D}_0$, $\bar{E}_0$, and $\bar{J}_0$, defined as the fields at $t = 0$. Initially, the charge distribution is uniform and spherically symmetric. From Gauss's law,

$$\oint \bar{D} \cdot d\bar{\ell} = Q$$

$$D(4\pi r^2) = \frac{4}{3} \pi r^3 \rho_0$$

$$\bar{D} = \frac{\rho_0 r}{3} \hat{a}_r$$

Thus it is clear that, for $r < R$,

$$\bar{D}_0 = \frac{\rho_0 r}{3} \hat{a}_r$$

$$\bar{E}_0 = \frac{\rho_0 r}{3\varepsilon} \hat{a}_r$$

Consequently, inside the sphere,

$$\bar{J}_0 = \sigma E_0 = \frac{\rho_0 \sigma r}{3\varepsilon} \hat{a}_r$$

and outside the sphere,

$$\bar{D}_0 = \frac{\rho_0 R^3}{3r^2} \hat{a}_r$$

$$\bar{E}_0 = \frac{\rho_0 R^3}{3\varepsilon_0 r^2} \hat{a}_r$$

$$\bar{J}_0 = 0$$

Applying the equation of continuity for $r < R$,

$$\left.\frac{\partial \rho}{\partial t}\right|_{t=0} = -\nabla \cdot \bar{J}_0 = -\frac{\sigma}{3\varepsilon} \rho_0 \qquad r < R$$

The rate of change of charge density is independent of r, and hence the charge density remains uniform. With this result write down the fields. For $r < R$,

$$\bar{D} = \frac{\rho_0 r e^{-(\sigma/\varepsilon)t}}{3} \hat{a}_r$$

$$\bar{E} = \frac{\rho_0 r e^{-(\sigma/\varepsilon)t}}{3\varepsilon} \hat{a}_r$$

$$\bar{J} = \frac{\rho_0 \sigma r e^{-(\sigma/\varepsilon)t}}{3\varepsilon_0} \hat{a}_r$$

and for $r > R$,

$$\bar{D} = \frac{\rho_0 R^3}{3r^2}\,\hat{a}_r$$

$$\bar{E} = \frac{\rho_0 R^3}{3\varepsilon_0 r^2}\,\hat{a}_r$$

$$\bar{J} = 0$$

It is also interesting to calculate the surface charge density on the sphere. Using the boundary condition

$$\vec{n} \cdot (\bar{D}_2 - \bar{D}_1) = \rho_S$$

and the previous results,

$$\rho_S = \frac{\rho_0 R}{3}\left(1 - e^{-(\sigma/\varepsilon)t}\right)$$

• PROBLEM 5-7

A uniform thick metal plate in the shape of a quarter circle, as shown in the figure, has a constant d-c voltage applied at the planes marked $V = 0$ and $V = V_1$. Compute the current density distribution in the plate.

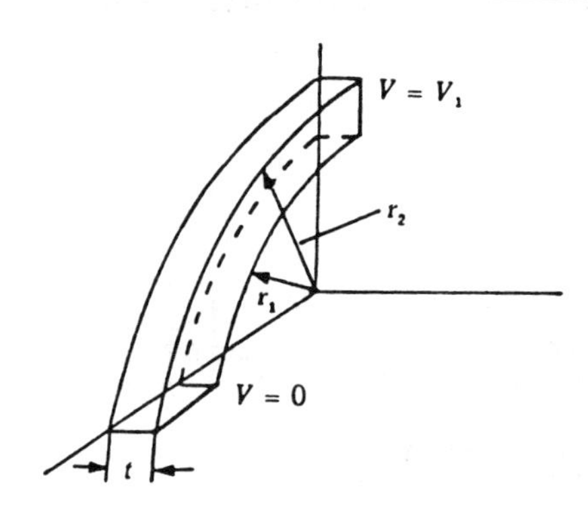

Solution:

$$\bar{J} = \gamma\bar{E}$$

Now, since there is no charge inside the conductor

$$\nabla^2 V = 0$$

$$\frac{\partial^2 V}{\partial r^2} + \frac{1}{r}\frac{\partial V}{\partial r} + \frac{1}{r^2}\frac{\partial^2 V}{\partial \phi^2} = 0$$

in cylindrical coordinates

Since V cannot change as a function of r because the plate is a conductor

$$\frac{1}{r^2}\frac{\partial^2 V}{\partial \phi^2} = 0$$

And since

$$r \neq 0,$$

$$\frac{\partial^2 V}{\partial \phi^2} = 0$$

Integrating twice

$$V = C_1\phi + C_2$$

But

$$V = 0 \text{ at } \phi = 0.$$

Therefore

$$C_2 = 0$$

$$V = V_1 \text{ at } \phi = 90° = \frac{\pi}{2}$$

Therefore

$$V_1 = C_1 \frac{\pi}{2}$$

or

$$C_1 = \frac{2V_1}{\pi}$$

Therefore

$$V = \frac{2V_1}{\pi}\phi$$

Now

$$E = -\nabla V = -\bar{1}_\phi \frac{1}{r}\frac{\partial V}{\partial \theta} = -\bar{1}_\phi \frac{2V_1}{\pi r}$$

But

$$\bar{J} = \gamma\bar{E} = -\bar{1}_\phi \frac{\gamma 2V_1}{\pi r}$$

A second approach for the same problem is as follows:

By definition

$$V_1 = -\int_r \bar{E} \cdot d\bar{r}$$

Since $\bar{E}$ is constant

$$V_1 = -E \int_r dr = -\frac{2\pi r}{4} E$$

for the quarter circle,

or

$$\bar{E} = -\frac{2V_1}{\pi r} \bar{1}_\phi$$

But

$$\bar{J} = \gamma\bar{E}$$

so that

$$\bar{J} = -\frac{2V_1\gamma}{\pi r} \bar{1}_\phi$$

RESISTANCE

• PROBLEM 5-8

Two perfectly conducting coaxial cylinders of length ℓ, inner radius a, and outer radius b are maintained at a potential difference v and enclose a material with Ohmic conductivity σ, as in the figure. Find the resistance of the coaxial cylinder configuration.

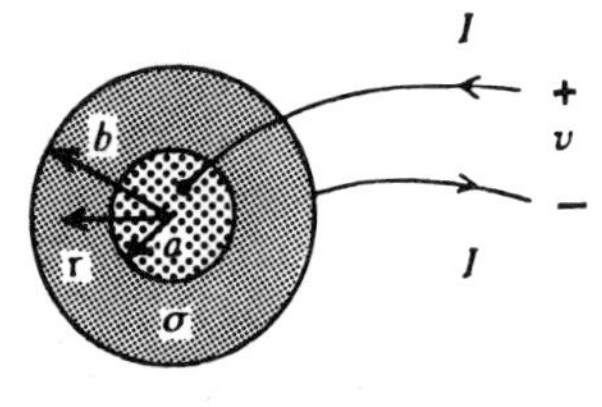

Depth l

Solution: The electric field must be perpendicular to the electrodes so that with no free charge Gauss's law requires (in cylindrical coordinates)

$$\nabla \cdot (\varepsilon \bar{E}) = 0 \Rightarrow \frac{1}{r}\frac{\partial}{\partial r}(rE_r) = 0 \Rightarrow E_r = \frac{c}{r}$$

where c is an integration constant found from the voltage condition

$$\int_a^b E_r dr = c \ln r \Big|_a^b = v \Rightarrow c = \frac{v}{\ln(b/a)}$$

The current density is then

$$J_r = \sigma E_r = \frac{\sigma v}{r \ln(b/a)}$$

with the total current at any radius r being a constant

$$I = \int_{z=0}^{\ell} \int_{\phi=0}^{2\pi} J_r r \, d\phi \, dz = \frac{\sigma v 2\pi \ell}{\ln (b/a)}$$

so that the resistance is

$$R = \frac{v}{I} = \frac{\ln (b/a)}{2\pi\sigma\ell} .$$

• PROBLEM 5-9

The figure shows a perfectly conducting hemispherical electrode sunk into an extended region of space, the conductivity of which is nonzero but finite. Determine the resistance between this electrode and a concentric, hemispherical, perfectly conducting surface of infinite radius.

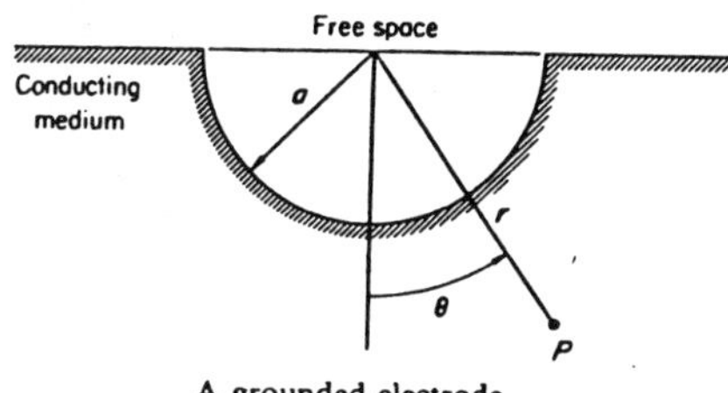

A grounded electrode.

Solution: Establish a spherical system of coordinates (upside down) where ϕ (not shown) measures angles on the horizontal plane separating the conductor from the free space above. Since the spherical electrode is a perfect conductor, the potential will be constant at every point within and on the surface of this electrode. The potential of the concentric infinite hemisphere will also be constant, and may be arbitrarily set equal to zero. Because of symmetry about a vertical axis, all field quantities will be independent of the variable ϕ. Accordingly, $\Phi = A/r$, where A is a constant, is an admissible general solution of Laplace's equation that satisfies all the requirements.

The analysis is facilitated by assuming that the electrode is held at some arbitrary potential V. The actual value of V is immaterial since in this case the interest is only in the ratio of V to the total current crossing the outer surface of the electrode. Thus, at $r = a$, $\Phi = V$, and from this it follows that $A = aV$. Then $\Phi = V\, a/r$.

One other condition remains to be fulfilled. At every point on the interface between the conducting region and the free space above, the normal component of $\bar{J}$ must vanish. The vector

$$\bar{J} = \sigma\bar{E} = \sigma(-\nabla\Phi) = -\sigma\left(\frac{\partial\Phi}{\partial r}\,\bar{a}_r\right) = \frac{\sigma V a}{r^2}\,\bar{a}_r$$

satisfies this condition. Moreover, the total current which crosses any concentric hemisphere of radius r is

$$I = (\text{current density})(\text{area}) = \frac{\sigma V a}{r^2}\,\frac{4\pi r^2}{2} = 2\pi\sigma a V$$

Therefore the required expression for resistance is

$$R = \frac{V}{I} = \frac{V}{2\pi\sigma a V} = \frac{1}{2\pi\sigma a}\,.$$

• PROBLEM 5-10

Find the resistance of a conical section shown in the figure.

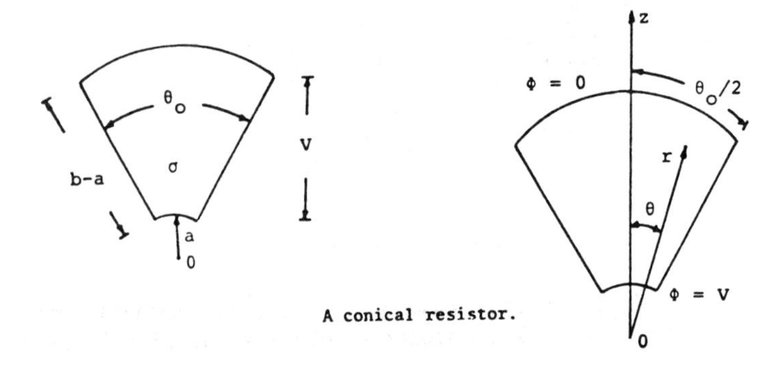

A conical resistor.

Solution: The figure illustrates a cone-shaped resistor with end surfaces that are portions of spherical surfaces. The resistance between the spherical end surfaces is to be found. The conical section can be viewed as a portion of a solid sphere, and so spherical coordinates are appropriate. A suitable solution for Φ is

$$\Phi = C_1 + C_2/r$$

(since V in a solid sphere varies with the reciprocal of the radius), where C_1 and C_2 are constants. At $r = a$, $\Phi = V$, while at $r = b$, $\Phi = 0$. Hence

$$V = C_1 + C_2/a$$

and

$$0 = C_1 + C_2/b.$$

These yield

$$C_1 = -\frac{C_2}{b}; \; C^2 = \frac{abV}{b-a}$$

Consequently the potential distribution is given by

$$\Phi = C_1 + C_2/r = -\frac{C_2}{b} + \frac{C_2}{r}$$

$$\Phi = \frac{C_2(b-r)}{br} = \frac{abV}{(b-a)} \times \frac{(b-r)}{br}$$

$$\Phi = \frac{abV}{(b-a)r} - \frac{aV}{b-a}$$

The current field is given by

$$\bar{J} = -\sigma\nabla\Phi = -\sigma\frac{\partial\Phi}{\partial r}\,\overline{a_r} \qquad = \frac{\sigma abV}{(b-a)r^2}\,\bar{a}_r$$

The total current leaving the surface at $r = a$ is

$$I = \int_0^{2\pi}\int_0^{\theta_0/2} \frac{\sigma abV\ a^2}{(b-a)a^2} \sin\theta \; d\theta \; d\phi = \frac{2\pi\sigma abV}{b-a}\left(1 - \cos\frac{\theta_0}{2}\right)$$

Thus the resistance R is given by

$$R = \frac{V}{I} = \frac{b-a}{2\pi\sigma ab(1-\cos\theta_0/2)}$$

Note that if $\theta_0 = 2\pi$, the resistance between two concentric spheres is obtained, which is $(b-a)/(4\pi ab\sigma)$.

POLARIZATION

• PROBLEM 5-11

A cylindrical slab as shown in the figure (a & b) possesses a polarization which varies linearly with the r-coordinate

$$\bar{P}(r) = \bar{a}_r p_o r$$

Find the polarization charge inside the slab and its surface charge.

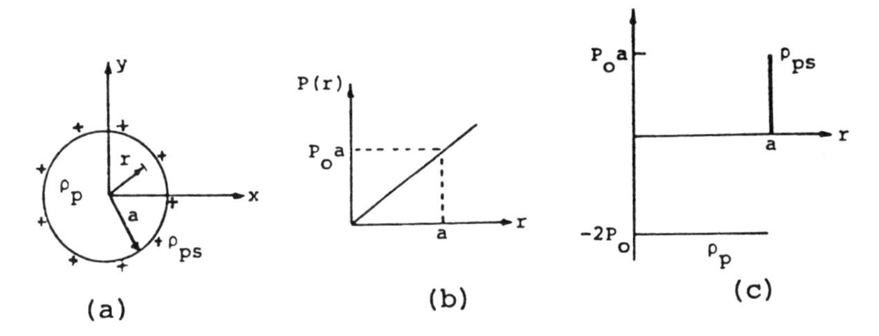

Polarized cylindrical slab.

Solution: Here the point variables are x,y,z, and the charge is given by

$$\rho_p = -\nabla \cdot \bar{p} = \frac{1}{r}\frac{\partial}{\partial r}(rp_o r) = -2\,p_o$$

which shows that the polarization charge is constant. The polarization surface charge at r = a is found to be,

$$\rho_{ps} = \bar{a}_r \cdot \bar{p} = p_o a$$

The figure (c) shows a sketch of the results obtained.

• PROBLEM 5-12

Find the polarization $\bar{P}$, the polarization charge density ρ_p and the polarization surface charge density ρ_{ps} when a charged cylinder of radius a having charge Q (Coulomb/unit length) is embedded in a dielectric medium.

Solution: Choose the center of the cylinder as the origin of a cylindrical coordinate system. By Gauss' law, the flux density inside the dielectric is readily found to be

$$\bar{D} = \frac{Q}{2\pi r}\,\bar{a}_r \tag{1}$$

The electric field intensity in the dielectric is then

$$\bar{E} = \frac{Q}{2\pi\varepsilon r}\,\bar{a}_r \tag{2}$$

and the polarization is,

$$\bar{P} = \bar{D} - \varepsilon_0\bar{E} = (\varepsilon-\varepsilon_0)\,\bar{E} = \left(1 - \frac{1}{\varepsilon_r}\right)\bar{D} \tag{3}$$

At distances larger than the radius of the cylinder $\rho = 0$ and $\nabla \cdot \bar{D} = 0$. Then by Eq. (3) $\nabla \cdot \bar{P} = 0$, which indicates that $\rho_p = 0$. However, there does exist a polarization charge density given by

$$\rho_{ps} = -\bar{a}_r \cdot \bar{P} = -\left(1 - \frac{1}{\varepsilon_r}\right)\rho_s$$

where

$$\rho_s = \frac{Q}{2\pi a}$$

is the surface charge density on the cylinder due to the free charges.

• PROBLEM 5-13

An insulating medium is situated between two parallel planes of charge. By considering the polarization of a representative number of atoms in the insulator, obtain an expression for the relative permittivity of the insulator in terms of the induced charges on the surface of the insulator.

Solution: The insulator with a large number of polarized atoms is shown in Fig. 1 between two parallel lines of charge. In so far as the field outside the insulator is concerned, the polarized atoms may be replaced by an induced charge on the surface of the insulator. The system of induced charges is shown in Fig. 2 together with the flux densities due to both the applied charges and the induced charges. The total flux density in the insulating medium is given by

$$D_t = D - D_i = \sigma - \sigma' .$$

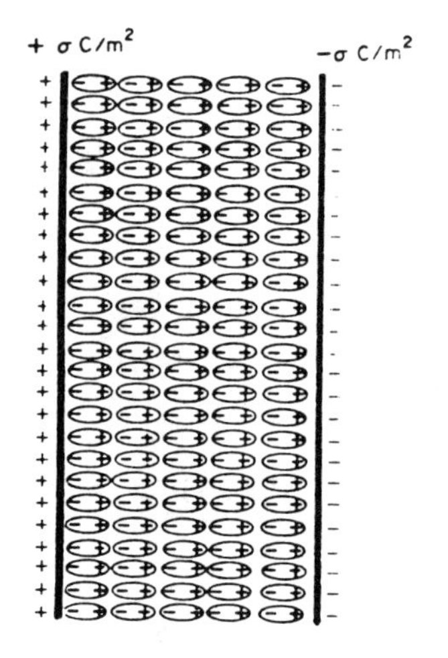

Polarization of the atoms of a dielectric when between two charge sheets.

Fig. 1

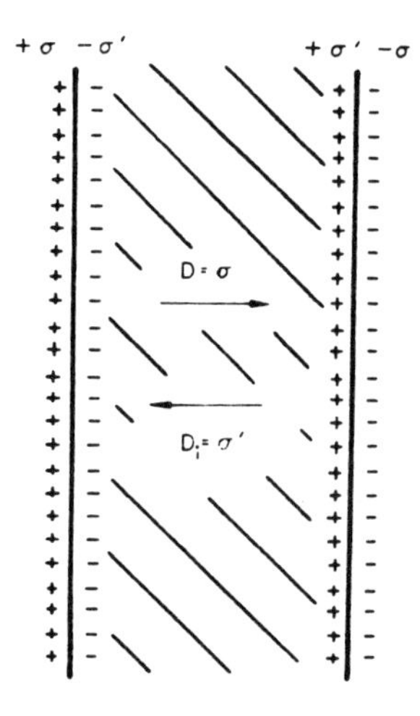

Resultant effect of the polarization of the atoms of the dielectric between two charge sheets.

Fig. 2

Then inside the insulator the electric field intensity is given by

$$E = \frac{D_t}{\varepsilon_0} = \frac{\sigma - \sigma'}{\varepsilon_0} = \frac{D}{\varepsilon_0}\left(\frac{\sigma - \sigma'}{\sigma}\right) .$$

But in terms of the relative permittivity and the externally applied flux density, the field intensity is given by

$$E = \frac{D}{\varepsilon_0 \varepsilon_r} .$$

Therefore ε_r is given by

$$\varepsilon_r = \left(\frac{\sigma}{\sigma - \sigma'}\right) .$$

• PROBLEM 5-14

A uniform dielectric slab is contained in a uniform electric field, as shown in Fig. 1. Find the polarization charge inside the dielectric and also the surface charge. Specify the electric field distribution.

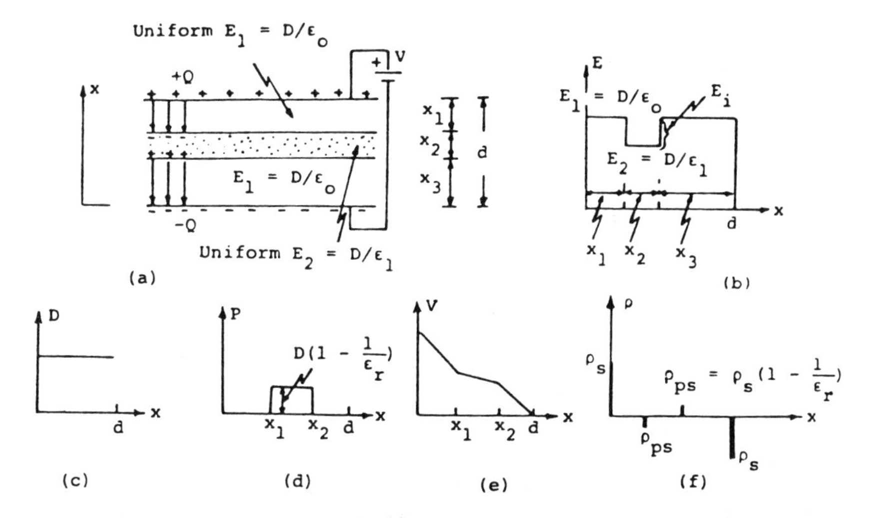

A dielectric slab in an electric field.

Fig. 1

Solution: It is assumed that the dielectric is totally polarized. Because the dielectric is homogeneous, only the dipoles near the surface are important. That is, the volume distribution $\rho_p = 0$ yields $\nabla \cdot \bar{P} = 0$. This means that for the uniform dielectric $\bar{P}$ is everywhere constant in magnitude and normal to the slab surface.

In the free space region outside of the dielectric the field is $\bar{E}_1$, as shown. Because of the induced surface charges which arise from the alignment of the dipoles of the dielectric, an electric field is produced which is in a direction opposite to the external field. The surface charge density is

$$\rho_{ps} = \bar{P} \cdot \bar{n} = p_n$$

The effect of the surface layer is a reduction in the electric field within the dielectric, with

$$\bar{E}_2 = \bar{E}_1 - \frac{\rho_{ps}}{\varepsilon_0}\bar{n}$$

or

$$\varepsilon_0 E_1 = \varepsilon_0 E_2 + P = \varepsilon E_2$$

which is simply that

$$\bar{D}_1 = \bar{D}_2$$

A sketch of the electric field distribution is shown in Fig. 1.

Fig. 1c, d, e, f shows the sketch of D,P,V, and ρ respectively.

• PROBLEM 5-15

An electret has a permanent electric dipole moment in the absence of free charges. Given an electret sphere of radius R, with electric polarisation vector $\bar{P} = \hat{r}\, P_o r$.

Determine the bound charge density ρ', electric displacement vector $\bar{D}$, and the electric field $\bar{E}$ as functions of r.

Solution: The bound charge density is determined from $\rho' = -\nabla.\bar{P}$.

Using spherical coordinates

$$\rho' = -\nabla\cdot\bar{P} = -\frac{1}{r^2}\frac{d}{dr}\left[r^2 P_r(r)\right] \;;$$

$$P_r = P_0 r$$

since $P \neq f(\theta, \phi)$

$$\rho' = -\frac{1}{r^2}\frac{d}{dr}\left[r^3 P_0\right]$$

$$= -\frac{1}{r^2} 3r^2 P_0 = -3P_0 = \rho'$$

To find $\bar{D}$, use Gauss's law

$$\oint \bar{D}.d\bar{a} = Q$$

where Q is the free charge enclosed by the surface. In this problem, no free charge is enclosed. Hence $\bar{D} = 0$.

To find $\bar{E}$, use

$$\oint \bar{E}.d\bar{a} = \frac{q_{TOTAL}}{\varepsilon_0} = \frac{1}{\varepsilon_0}\left(q_{free} + q_{bound}\right)$$

Use a spherical Gaussian surface of radius r.

$$\oint \bar{E}.d\bar{a} = \int \left[E_r.\hat{r} + E_\theta\hat{\theta} + E_\phi\hat{\phi}\right].\hat{r}\, da$$

$$= \oint E_r da = E_r \oint da = E_r(4\pi r^2)$$

$$= \frac{q_{bound}}{\varepsilon_0}$$

$$q_{bound} = \int \rho\, dv$$

for r < R

$$E_r(4\pi r^2) = \frac{1}{\varepsilon_0}\int \rho' dv = \frac{-3P_0}{\varepsilon_0}\int dv$$

$$E_r(4\pi r^2) = \frac{-3P_0}{\varepsilon_0}\int_0^r 4\pi r^2 dr$$

$$= \frac{-4\pi P_0 r^3}{\varepsilon_0}$$

$$r < R, \quad \bar{E} = -\hat{r}\,\frac{P_0 r}{\varepsilon_0}$$

for r > R, $q_{TOTAL} = 0$ since the dielectric dipoles consist of equal positive and negative charges.

$$\bar{E}\ (r > R) = 0.$$

A cylinder of radius a and height L is centered about the z axis and has a uniform polarization along its axis, $\overline{P} = \overline{P}_0 i_z$, as shown in Figure 1. Find the electric field $\overline{E}$ and displacement vector $\overline{D}$ everywhere on its axis.

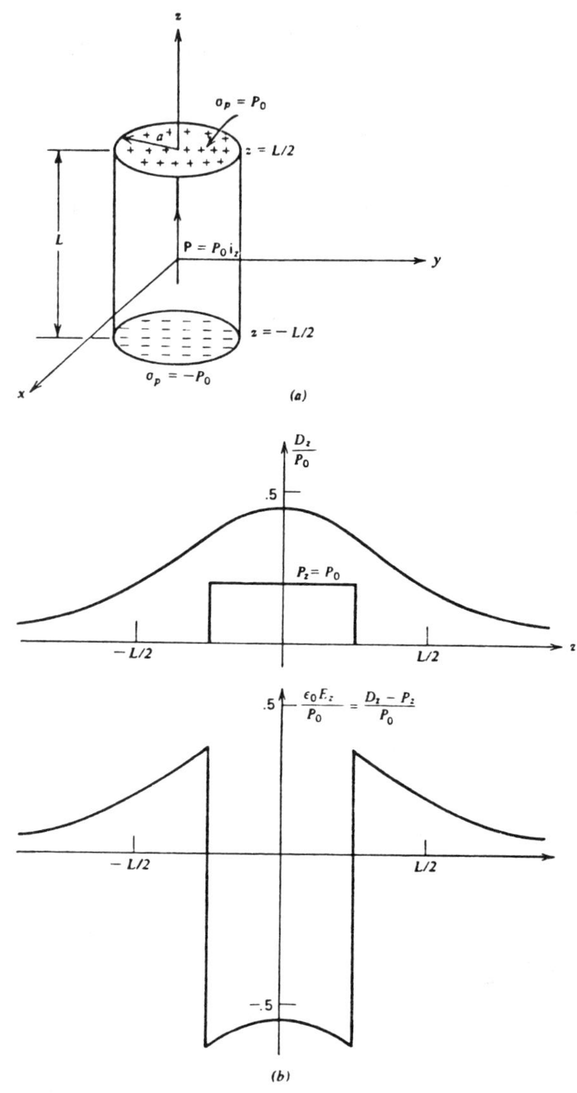

(a) The electric field due to a uniformly polarized cylinder of length L is the same as for two disks of surface charge of opposite polarity $\pm P_0$ at $z = L/2$. (b) The perpendicular displacement field D_z is continuous across the interfaces at $z = \pm L/2$ while the electric field E_z is discontinuous.

Fig. 1

Solution: With a constant polarization $\overline{P}$, the volume polarization charge density is zero:

$$\rho_p = -\nabla \cdot \overline{P} = 0$$

Since $\bar{P} = 0$ outside the cylinder, the normal component of $\bar{P}$ is discontinuous at the upper and lower surfaces yielding uniform surface polarization charges:

$$\sigma_p(z = L/2) = P_0$$

$$\sigma_p(z = -L/2) = -P_0$$

The solution for a single disk of surface charge is

$$D_z = \frac{\sigma_0}{2} - \frac{\sigma_0 z}{2(a^2 + z^2)^{1/2}} \quad \text{for} \quad z > 0$$

$$D_z = -\frac{\sigma_0}{2} + \frac{\sigma_0 z}{2(a^2 + z^2)^{1/2}} \quad \text{for} \quad z < 0$$

Superpose the results for the two disks taking care to shift the axial distance appropriately by $\pm L/2$ yielding the concise solution for the displacement field:

$$D_z = \frac{P_0}{2}\left(\frac{(z + L/2)}{[a^2 + (z + L/2)^2]^{1/2}} - \frac{(z - L/2)}{[a^2 + (z - L/2)^2]^{1/2}}\right)$$

The electric field is then

$$E_z = \begin{cases} D_z/\varepsilon_0 & |z| > L/2 \\ (D_z - P_0)/\varepsilon_0 & |z| < L/2 \end{cases}$$

These results can be examined in various limits. If the radius a becomes very large, the electric field should approach that of two parallel sheets of surface charge $\pm P_0$:

$$\lim_{a\to\infty} E_z = \begin{cases} 0, & |z| > L/2 \\ -P_0/\varepsilon_0, & |z| < L/2 \end{cases}$$

with a zero displacement field everywhere.

In the opposite limit, for large z ($z \gg a$, $z \gg L$) far from the cylinder, the axial electric field dies off as the dipole field with $\theta = 0$

$$\lim_{z\to\infty} E_z = \frac{p}{2\pi\varepsilon_0 z^3}, \quad p = P_0 \pi a^2 L$$

with effective dipole moment p given by the product of the total polarization charge at $z = L/2$, $(P_0 \pi a^2)$, and the length L.

A very long coaxial conductor pair contains a concentric, homogenous dielectric sleeve with a permittivity ε as shown. Air fills the remaining regions between the conductors. Assuming positive and negative surface charges $\pm Q$ Coulombs are on every axial lengh ℓ of the inner and outer conductors respectively, determine the following, making use of the symmetry. In each region between the conductors, find:

(a) $\bar{D}$ field

(b) $\bar{E}$ field

(c) the $\bar{P}$ field

(d) ρ_s on the conductor surfaces, and the surface polarization charge density at $r = b$ and $r = c$.

(e) if a = 1 cm, b = 2 cm, c = 3 cm, d = 4 cm, e = 4.2 cm, $Q/\ell = 10^{-2}$ μC/m, and $\varepsilon = 2.1\,\varepsilon_0$, find the values of $\bar{D}$, $\bar{E}$, and $\bar{P}$ at the inner surface $r = b$ just inside the dielectric.

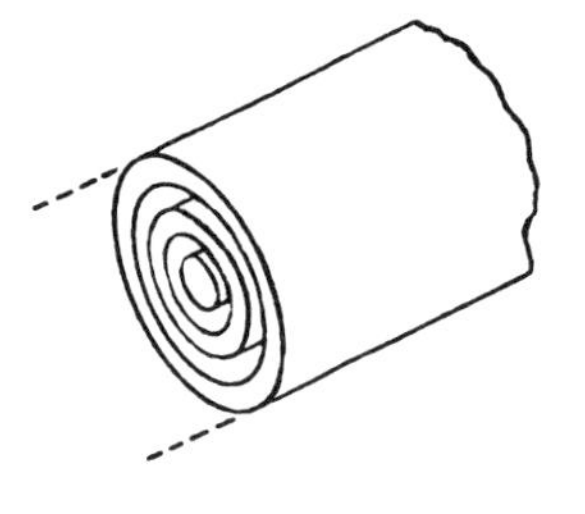

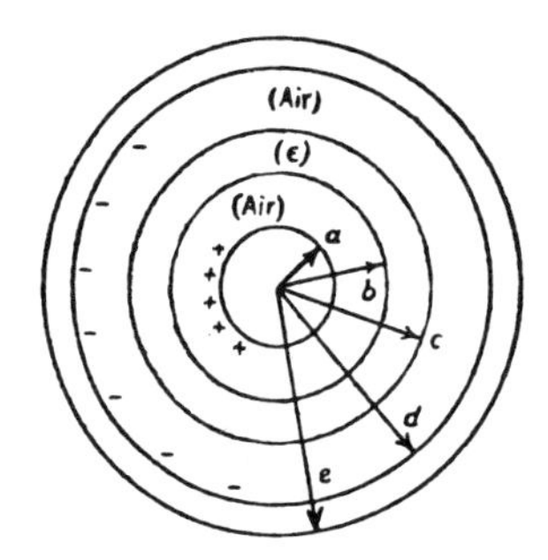

Solution: (a, b & c): For $r < a$, $Q_{enclosed} = 0$

$$\therefore \quad \bar{D} = \bar{E} = \bar{P} = 0$$

For $a < r < b$,

using Gauss's Law

$$\oint \bar{D} \cdot d\bar{A} = Q_{enclosed}$$

$$dA = 2\pi r\ell$$

$$D \times 2\pi r\ell = Q$$

$$\bar{D} = \frac{Q}{2\pi r\ell}\,\hat{r}$$

[$\hat{r}$ meaning in the radial direction]

$$E = \frac{D}{\varepsilon} = \frac{Q}{2\pi\varepsilon r\ell}$$

P = 0 in vacuum

For $b < r \leq c$

$\bar{D}$ is same as for air

$$\bar{D} = \frac{Q\hat{r}}{2\pi r\ell}$$

$$E = \frac{D}{\varepsilon} = \frac{Q}{2\pi\varepsilon r\ell}$$

$$P = D - \varepsilon_0 E$$

$$P = \frac{Q}{2\pi r\ell} = \frac{\varepsilon_0 Q}{2\pi r\ell\varepsilon}$$

$$\bar{P} = \frac{Q\hat{r}}{2\pi r\ell}\left[1 - \frac{\varepsilon_0}{\varepsilon}\right]$$

For $c \leq r \leq d$

the results are the same as for $a < r \leq b$ since it is again the air region

For $d < r \leq e$

The charge enclosed is 0,

therefore

$$\bar{D} = \bar{E} = \bar{P} = 0$$

For $r > e$

Assuming that the net charge on the outer conductor is 0,

$$Q_{enclosed} = Q$$

$$\bar{D} = \frac{Q}{2\pi r\ell}\,\hat{r}$$

$$\bar{E} = \frac{Q}{2\pi\varepsilon_0 r\ell}\,\hat{r}$$

$$\bar{P} = 0$$

d) On the conductor surfaces, that is, at r = a and r = d,

$$\rho_s \text{ (at } r = a) = Q/\text{area}$$

$$\rho_s = \frac{Q}{2\pi \ell a} \quad [c/m^2]$$

$$\rho_s \text{ (at } r = d) = \frac{-Q}{\text{area}} = \frac{-Q}{2\pi \ell d} \quad [c/m^2]$$

The surface polarization charge density is found on the dielectric

$$\rho_s = \bar{P} \cdot \hat{n}$$

at r = b

$$\rho_s = \frac{-Q}{2\pi b \ell}\left[1 - \frac{\varepsilon_0}{\varepsilon}\right] \quad c/m^2$$

at r = c

$$\rho_s = \frac{Q}{2\pi c \ell}\left[1 - \frac{\varepsilon_0}{\varepsilon}\right] \quad c/m^2$$

e) Substituting the values of a, b, c, d, e, and Q/ℓ and ε into the equations for r = b,

$$\bar{D} = \frac{Q}{2\pi r \ell}\hat{r} = \frac{10^{-2}}{2\pi r} = \frac{10^{-2}}{2\pi(0.02)} = 0.796\ \mu c/m^2\ \hat{r}$$

$$\bar{E} = \frac{Q}{2\pi \varepsilon r \ell}\hat{r} = \frac{(10^{-2})(10^{-6})}{2\pi(\varepsilon)(0.2)(2.1)} = 4.29\ k_r/m\ \hat{r}$$

$$\bar{P} = \bar{D} - \varepsilon_0 \bar{E} = 4.2 \times 10^{-2}\ \mu c/m^2\ \hat{r}$$

• PROBLEM 5-18

Given a point charge q inside a spherical dielectric shell of inner radius a, outer radius b as shown in the figure.

a) Use the Gauss law to find the electric displacement $\bar{D}$ and electric field $\bar{E}$ for r < a, a < r < b, b < r.

b) Find the electric polarisation $\bar{P}$ and the bound charge density ρ' for r < a, a < r < b, r > b.

Solution: Gauss's law in dielectrics is $\oint \bar{D}.d\bar{a} = q$ = net free charge enclosed. Using spherical coordinates

$$\oint (D_r\hat{r} + D_\theta\hat{\theta} + D_\phi\hat{\phi}).\hat{r}da = q.$$

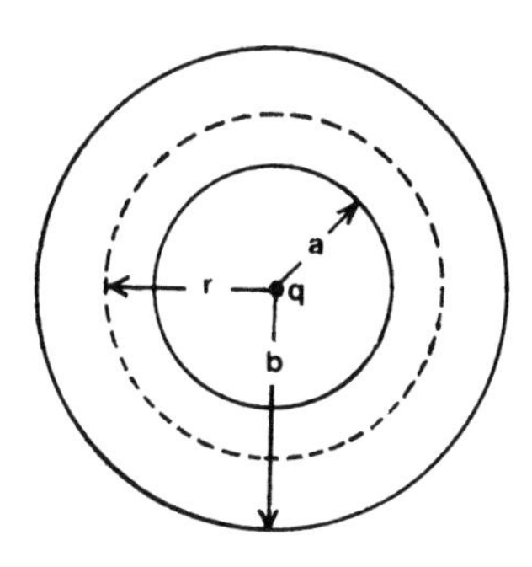

Use a spherical surface of radius r as the Gaussian surface

$$D_r \oint da = D_r(4\pi r^2) = q$$

$$D_r = \frac{q}{4\pi r^2}$$

for $r < a$, $a < r < b$, $r > b$

for $r < a$, and $r > b$

$$\bar{E} = \frac{\bar{D}}{\varepsilon_0} = \frac{\hat{r}q}{4\pi\varepsilon_0 r^2}$$

where ε_0 is permittivity of free space for $a < r < b$.

$$\bar{E} = \frac{\bar{D}}{\varepsilon} = \frac{\hat{r}q}{4\pi\varepsilon r^2}$$

b) The electric polarisation is given by

$$\bar{P} = \bar{D} - \varepsilon_0\bar{E} \text{ for } a < r < b$$

$$\bar{P} = \frac{q}{4\pi r^2} - \frac{\varepsilon_0 q}{4\pi\varepsilon r^2}$$

$$= \frac{q}{4\pi r^2}\left(1 - \frac{\varepsilon_0}{\varepsilon}\right) = \bar{P}$$

ρ', bound charge is given by

$$\rho' = -\nabla.\bar{P}$$

but $\bar{P} = f(r)$ only

hence $\rho' = -\nabla.\bar{P}$

$$= -\frac{1}{r^2}\frac{\partial}{\partial r}(r^2 P_r)$$

$$= -\frac{1}{r^2}\frac{\partial}{\partial r}\left[\frac{r^2 q}{4\pi r^2}\left(1 - \frac{\varepsilon_0}{\varepsilon}\right)\right] = 0.$$

In regions $r < a$, $r > b$

$\bar{P} = 0$ (no dielectric)

hence $\rho' = 0$.

BOUNDARY CONDITIONS

• PROBLEM 5-19

Referring to the figure and considering medium 2 to be a conductor, find α_1.

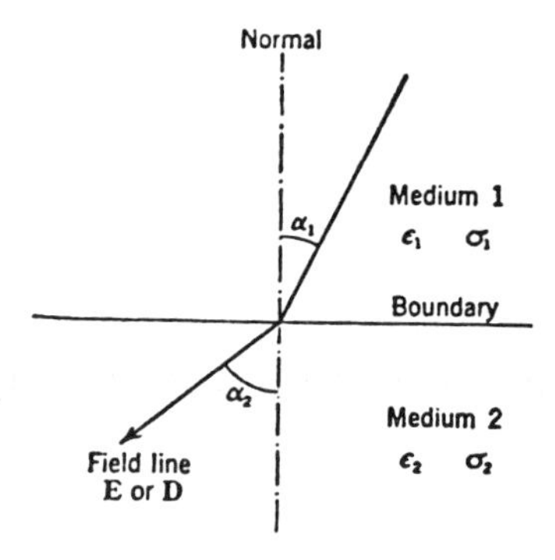

Boundary between a conductor and a dielectric.

Solution: Since medium 2 is a conductor, $D_2 = E_2 = 0$ under static conditions. According to the boundary relations,

$$D_{n1} = \rho_s$$

or

$$E_{n1} = \frac{\rho_s}{\varepsilon_1}$$

and

$$E_{t1} = 0$$

Therefore

$$\alpha_1 = \tan^{-1} \frac{E_{t1}}{E_{n1}} = \tan^{-1} 0 = 0$$

It follows that a static electric field line or flux tube at a dielectric-conductor boundary is always perpendicular to the conductor surface (when no currents are present).

• PROBLEM 5-20

(a) Derive a refractive law for $\bar{E}$ at an interface separating two nonconductive regions.

(b) Deduce from boundary conditions the direction of $\bar{E}$ just outside a perfect conductor.

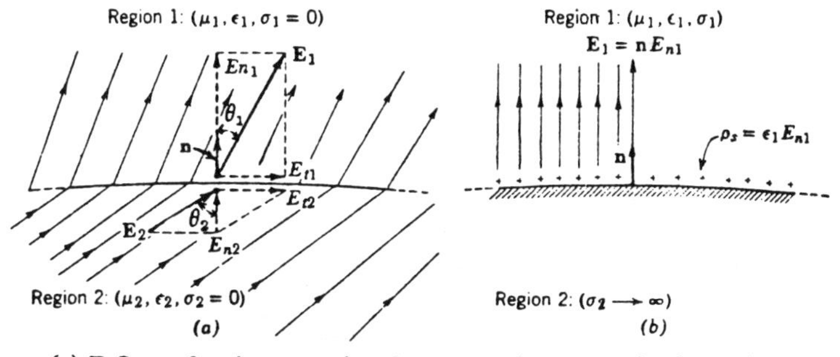

(a) E flux refraction at an interface separating nonconductive regions.
(b) E is everywhere normal to the surface of a perfect conductor. conductor.

Solution: (a) The boundary conditions for the tangential and the normal components of E at an interface separating nonconductive regions are

$$\varepsilon_1 E_{n1} = \varepsilon_2 E_{n2}$$

and

$$E_{t1} = E_{t2}.$$

From the latter and the geometry of (a),

$$E_{n2} \tan \theta_2 = E_{n1} \tan \theta_1$$

$$\therefore \qquad \tan \theta_2 = \frac{E_{n1}}{E_{n2}} \tan \theta_1 = \frac{\varepsilon_2 E_{n2}}{\varepsilon_1 E_{n2}} \tan \theta_1$$

$$\tan \theta_2 = \frac{\varepsilon_2}{\varepsilon_1} \tan \theta_1$$

(b) A perfectly conductive region 2 implies null fields inside it. Then by

$$E_{t1} = E_{t2} ,$$

E_{t1} in the adjacent region 1 must vanish also. The remaining normal component in region 1 is given by $D_{n1} = \rho_s$, yielding $\rho_s = \varepsilon_1 E_{n1}$ as shown in (b).

• PROBLEM 5-21

A thin dielectric cylinder of dielectric constant ε is placed in an initially uniform field $\bar{E}$, the axis of the cylinder being parallel to E (Fig. 1). Neglecting end effects, find the final field inside and outside the cylinder.

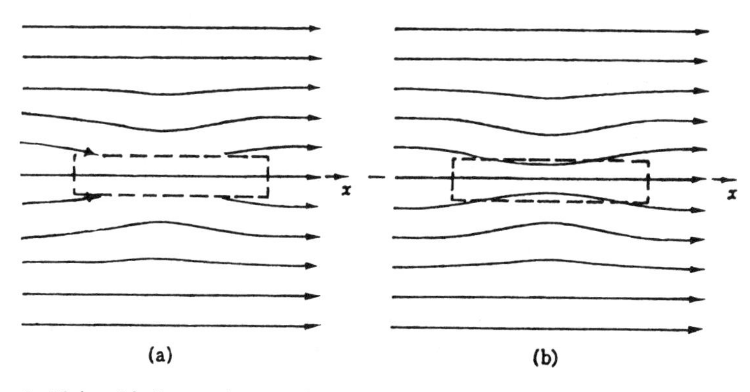

A thin dielectric cylinder parallel to an initially homogeneous electric field. (a) E map. (b) D map. If $\varepsilon \gg 1$, the field is strongly distorted, and the solution given is not valid.

Fig. 1

Solution: By inspection recognize that, except near the cylinder's ends, the field outside the cylinder is undisturbed

$$\bar{E}_{outside} = \bar{E},$$

and the field inside the cylinder is uniform and parallel to the axis. Since we neglect the end effects, the boundary conditions need be satisfied at the curved surface of the cylinder only, where

$$E_{t\ \text{outside}} = E_{t\ \text{inside}}$$

and

$$D_{n\ \text{outside}} = D_{n\ \text{inside}}$$

Since it was assumed that the field is parallel to the cylinder's axis, the normal component of the field is zero, and the boundary condition for D is satisfied automatically. In the boundary condition for E we can drop the subscript "t". We then have

$$E_{\text{inside}} = E_{\text{outside}}$$

and since

$$E_{\text{outside}} = E,$$

this condition will be satisfied if

$$\bar{E}_{\text{inside}} = \bar{E}.$$

• PROBLEM 5-22

A thin dielectric disk of dielectric constant ε is placed in an initially uniform field $\bar{E}$, normal to the field as shown in Fig. 1. Neglecting end effects, find the final field inside and outside the cylinder.

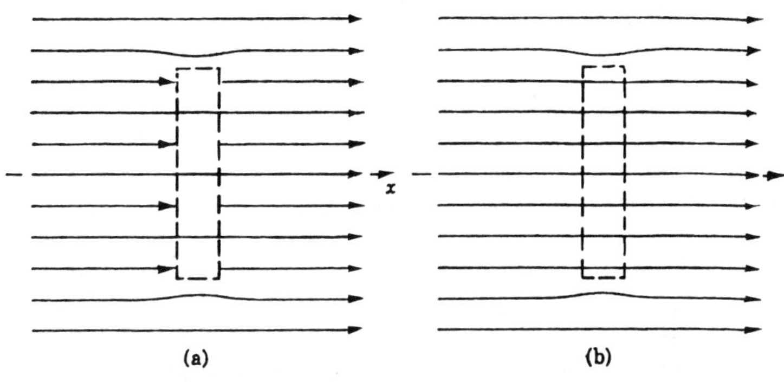

Fig. 1

A thin dielectric disk normal to an initially homogeneous electric field. (a) The map of the electric field E. (b) The map of the displacement field D.

Solution: The final field must satisfy boundary conditions

at the surface of the disk, and must be equal to the initial field at large distances from the disk. Once these requirements are satisfied, the problem is solved, since no other independent solution satisfying these requirements can exist. The geometry of the problem suggests that except at the edges of the disk the field outside the disk will remain undisturbed

$$\bar{E}_{outside} = \bar{E},$$

and the field inside the disk will be uniform and normal to the disk. In this case, the requirement that the final field be equal to the initial field at large distances from the disk will be met. The boundary condition at the bases of the disk will also be satisfied, since $E_{t\ outside}$ and $E_{t\ inside}$ will both be zero on the bases. As far as the boundary conditions at the side surface (curved surface) of the disk are concerned, they may be disregarded altogether, since in a thin disk the side surface is responsible only for edge effects, which by the statement of the problem are to be neglected. All that is needed in order to complete the solution is, then, to satisfy the boundary condition at the bases of the disk, where we must have

$$D_{n\ outside} = D_{n\ inside} ,$$

or, since the field is normal to the disk,

$$D_{outside} = D_{inside}$$

But $D_{outside} = \varepsilon_o E_{outside}$, and $D_{inside} = \varepsilon_o \varepsilon E_{inside}$.
Therefore, the boundary condition will be satisfied if

$$\bar{E}_{inside} = \frac{1}{\varepsilon}\bar{E}$$

• PROBLEM 5-23

A conducting sphere of radius a carrying a charge q is submerged halfway into a nonconducting liquid of dielectric constant ε (Fig. 1). Find the electric field outside the sphere and the charge density on the surface of the sphere.

Solution: Constructing a concentric spherical Gaussian surface S of radius r enclosing the sphere, and applying Gauss's law to this surface, yields

$$\oint \bar{D} \cdot d\bar{S} = \int_{S_1} \bar{D}_{liquid} \cdot d\bar{S} + \int_{S_2} \bar{D}_{air} \cdot d\bar{S} = q,$$

where S_1 and S_2 are the parts of the Gaussian surface passing through the liquid and through the air, respectively.

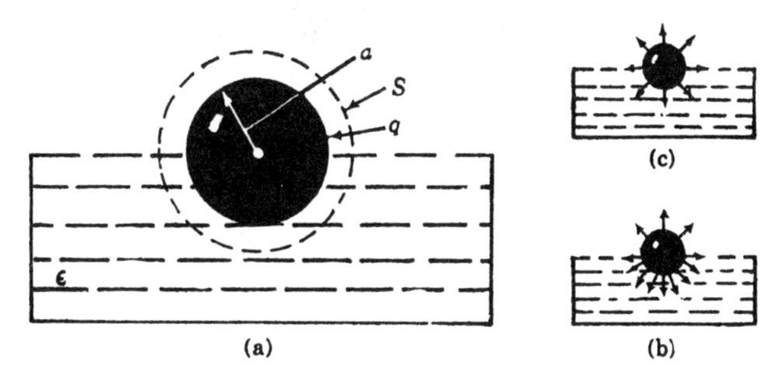

(a) Charged conducting sphere floating in a nonconducting liquid. (b) Field lines of E. (c) Field lines of D.

Fig. 1

The geometry of the problem suggests that the field is everywhere radial, so that $\bar{D} \cdot d\bar{S} = D\,dS$. It also suggests that D_{liquid} is constant at all points of S_1 and D_{air} is constant at all points of S_2, so that D can be factored out from under the integral signs. We can therefore write

$$D_{liquid} \int_{S_1} dS + D_{air} \int_{S_2} dS = q,$$

or

$$(D_{liquid} + D_{air})\, 2\pi r^2 = q, \qquad (1)$$

where $2\pi r^2$ is the area of S_1 and S_2. Now, by the displacement law, $D_{liquid} = \varepsilon_0 \varepsilon E_{liquid}$ and $D_{air} = \varepsilon_0 E_{air}$. Since the field is radial, it is tangent to the boundary between the liquid and the air, and hence, by the boundary conditions, $E_{liquid} = E_{air}$. The subscripts on E are then not needed, and we can write $D_{liquid} = \varepsilon_0 \varepsilon E$, $D_{air} = \varepsilon_0 E$. Substituting these expressions into Eq. (1), the following is obtained:

$$(\varepsilon_0 \varepsilon E + \varepsilon_0 E)\, 2\pi r^2 = \varepsilon_0 (\varepsilon + 1) E 2\pi r^2 = q,$$

or

$$E = \frac{q}{2\pi\varepsilon_0 (\varepsilon + 1) r^2},$$

which gives the electric field both in the liquid and in the air. The displacement is then

$$D_{liquid} = \frac{\varepsilon q}{2\pi(\varepsilon + 1) r^2}$$

and

$$D_{air} = \frac{q}{2\pi(\varepsilon + 1)r^2} .$$

The surface charge density σ on the sphere is equal to the displacement at the surface of the sphere, so that

$$\sigma_1 = \frac{\varepsilon q}{2\pi(\varepsilon + 1)a^2}$$

and

$$\sigma_2 = \frac{q}{2\pi(\varepsilon + 1)a^2}$$

on the submerged and the exposed half of the sphere, respectively.

• PROBLEM 5-24

An electric field in a medium whose relative permittivity is 7 passes into a medium of relative permittivity 2. If $\bar{E}$ makes an angle of 60° with the boundary normal, what angle does the field make with the normal in the second dielectric? See the figure.

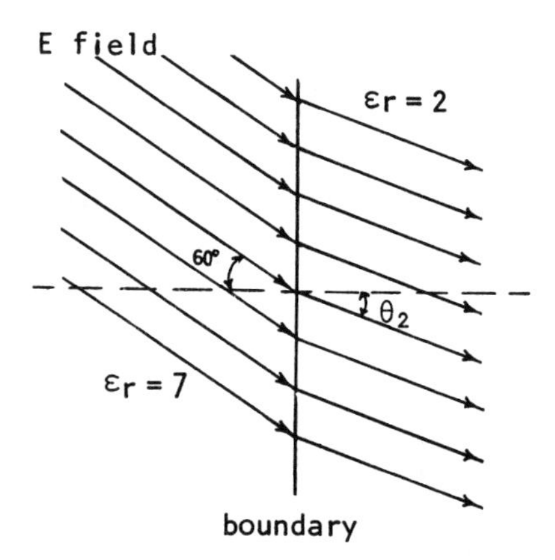

Solution:

$$\frac{E_{t_1}}{E_{n_1}} = \tan 60°$$

$$\frac{E_{t_2}}{E_{n_2}} = \tan \theta_2$$

But from boundary conditions

$$E_{t_2} = E_{t_1}$$

Also

$$D_{n_1} = \varepsilon_1 E_{n_1}$$

$$D_{n_2} = \varepsilon_2 E_{n_2}$$

and

$$D_{n_1} = D_{n_2}$$

then

$$E_{n_2} = E_{n_1} \frac{\varepsilon_1}{\varepsilon_2} = 3.5\ E_{n_1}$$

Thus

$$E_{t_1} = E_{n_1} \tan 60°$$

$$E_{t_2} = E_{n_2} \tan \theta_2 = E_{n_1} \times 3.5 \tan \theta_2$$

Setting E_{t_1} equal to E_{t_2} :

$$E_{n_1} \tan 60° = E_{n_1} \times 3.5 \tan \theta_2$$

$$\tan \theta_2 = \frac{\tan 60°}{3.5} = 0.495$$

$$\therefore \quad \theta_2 = 26.4°$$

Therefore, the electric field makes an angle of 26.4° with the normal in the second dielectric.

• PROBLEM 5-25

A sheet of dielectric $\varepsilon = 10\varepsilon_0$ is placed in a uniform electric field $E_0 = 142$ v/m. Find the components of $\bar{E}$ inside the sheet that are parallel to, and perpendicular to the sheet. Sketch the direction of $\bar{E}$ and find its magnitude. The diagram is shown in fig. 1.

Solution: Let $\bar{E}$ = resultant field inside the sheet. (1) is the area in the air and (2) is the space inside the sheet.

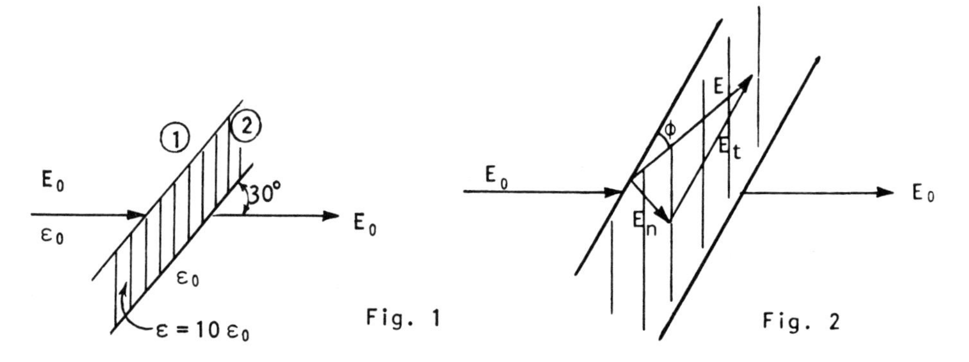

Fig. 1

Fig. 2

Using boundary conditions for the sheet separating the air and the dielectric:

$$E_{t_1} = E_{t_2}$$

$$D_{n_1} = D_{n_2}$$

Therefore,

$$E_{t_1} = E_{t_2} = E_0 \text{Cos } 30° = (142)(0.866)$$

$$E_t = 123 \text{ v/m}$$

(Parallel to sheet, E field = 123 v/m)

$$D_{n_1} = D_{n_2}$$

$$\varepsilon_0 E_{n_1} = 10\varepsilon_0 E_{n_2}$$

$$E_{n_2} = \frac{E_{n_1}}{10} = \frac{E_0 \text{Sin } 30°}{10} = \frac{\frac{1}{2}(142)}{10}$$

$$E_{n_2} = 7.1 \text{ v/m}$$

(normal to sheet, E field = 7.1 v/m)

Fig. 2 shows a sketch of the field inside the sheet.

$$E = \sqrt{E_t^2 + E_n^2} = \sqrt{(123)^2 + (7.1)^2}$$

$$E = 123.2 \text{ v/m}$$

$$\phi = \tan^{-1} \frac{E_n}{E_t}$$

$$= \tan^{-1} \frac{7.1}{123} = 3.3°$$

where ϕ is as shown in fig. 2.

• PROBLEM 5-26

> Two isotropic dielectric media 1 and 2 are separated by a charge-free plane boundary as shown in the figure. Their permittivities are ε_1 and ε_2, and the conductivities $\sigma_1 = \sigma_2 = 0$. Referring to the figure shown, find α_2 given α_1 .

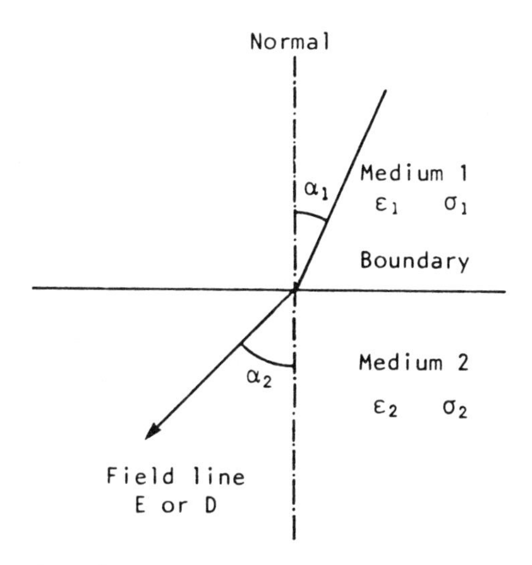

Boundary between two dielectric media showing change in direction of field line. Fig. 1

Solution: Let

D_1 = magnitude of $\bar{D}$ in medium 1

D_2 = magnitude of $\bar{D}$ in medium 2

E_1 = magnitude of $\bar{E}$ in medium 1

E_2 = magnitude of $\bar{E}$ in medium 2

In an isotropic medium, $\bar{D}$ and $\bar{E}$ have the same direction. According to the boundary relations,

$$D_{n_1} = D_{n_2}$$

and

$$E_{t_1} = E_{t_2} \quad (1)$$

Referring to the figure,

$$D_{n_1} = D_1 \cos \alpha_1$$

and

$$D_{n_2} = D_2 \cos \alpha_2 \quad (2)$$

while

$$E_{t_1} = E_1 \sin \alpha_1$$

and

$$E_{t_2} = E_2 \sin \alpha_2 \quad (3)$$

Substituting (2) and (3) into (1) and dividing the resulting equations yields

$$\frac{D_1 \cos \alpha_1}{E_1 \sin \alpha_1} = \frac{D_2 \cos \alpha_2}{E_2 \sin \alpha_2} \quad (4)$$

But $D_1 = \varepsilon_1 E_1$, and $D_2 = \varepsilon_2 E_2$, so that (4) becomes

$$\frac{\tan \alpha_1}{\tan \alpha_2} = \frac{\varepsilon_1}{\varepsilon_2} = \frac{\varepsilon_{r_1} \varepsilon_0}{\varepsilon_{r_2} \varepsilon_0} = \frac{\varepsilon_{r_1}}{\varepsilon_{r_2}} \quad (5)$$

where

ε_{r1} = relative permittivity of medium 1

ε_{r2} = relative permittivity of medium 2

ε_0 = permittivity of vacuum

Suppose, for example, that medium 1 is air ($\varepsilon_{r1} = 1$), while medium 2 is a slab of sulfur ($\varepsilon_{r2} = 4$). Then when $\alpha_1 = 30°$, the angle α_2 in medium 2 is 66.6°.

• **PROBLEM 5-27**

At a boundary between polystyrene and air, a particular field condition yields a value for the electric field intensity in the polystyrene of 2500 volts per meter making an angle with the normal to the boundary surface of 20°, as shown in Fig. 1(a). Determine the magnitudes of (a) the electric field intensity, (b) the electric flux density in the air just past the boundary, and (c) the angle these vectors make with the normal to the boundary surface. Further, show to scale both $\bar{D}$ and $\bar{E}$ vectors on both sides of the boundary.

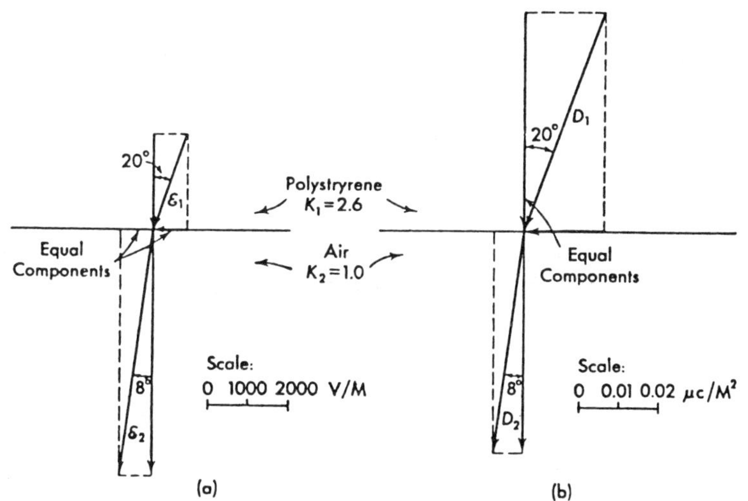

Fig. 1

$\mathcal{E}$ and D vectors at a polystyrene-air boundary for an incident angle of 20° in the Polystyrene.

Table 1 Approximate Dielectric Constants

Substance	K	Substance	K
Acetone	27	Lava	5
Alcohol (0° C)		Marble	8
Amyl	17	Mica	6
Ethyl	25	Micarta	4
Methyl	35	Mineral oil	2.5
Amber	2.9	Paper	
Ammonia	22	Dry manila	2
Asbestos paper	2.5	Impregnated	3.5
Asphalt	2.5	Paraffin	2.2
Bakelite	4.5	Petroleum	2.1
Beeswax	2	Polystyrene	2.6
Benzene	2.3	Porcelain	5
Carbon bisulphide	2.6	Pyranol	5
Carbon tetrachloride	2.2	Quartz (fused)	5
Castor oil	4.7	Resin	2.5
Ebonite	2.6	Rubber	
Ether (Ethyl)	4.5	Pure	2.5
Fiber	2.5	Hard	3
Glass		Shellac	3
Flint	10	Slate	7
Hard crown	7	Steatite	6
Lead	6.6	Sulphur	4
Pyrex	4.5	Turpentine	2.2
Glycerine	40	Water (distilled)	80
Gutta-percha	4	Wood (dry)	4

Solution: (a) Let number 1 subscripts be used to indicate polystyrene and number 2 be used for air. The dielectric constant of polystyrene is 2.6 from Table 1.

Applying the boundary condition that the tangential components of electric field are equal at the boundary between two dielectrics yields

$$\frac{\tan \alpha_1}{\tan \alpha_2} = \frac{K_1}{K_2}$$

$$\tan \alpha_2 = \frac{K_2}{K_1} \tan \alpha_1 = \frac{1}{2.6} \tan 20° = 0.140$$

$$\alpha_2 = 8° \quad \text{(Ans. c)}$$

Applying the boundary conditon that the normal components of flux density are equal at the boundary between two dielectric yields.

$$E_1 \operatorname{Sin} \alpha_1 = E_2 \operatorname{Sin} \alpha_2$$

$$E_2 = E_1 \frac{\operatorname{Sin} \alpha_1}{\operatorname{Sin} \alpha_2} = 2500 \frac{\sin 20°}{\sin 8°} = 6150 \text{ volts per}$$

meter (Ans. a)

(b) Then

$$D_2 = \varepsilon_0 K_2 E_2$$

(c) To complete the data for plotting the vectors, all of which are shown in Fig. 1,

$$D_1 = \varepsilon_0 K_1 E_1$$

$$D_1 = 8.85 \times 10^{-12} K_1 \xi_1 = 8.85 \times 10^{-12} \times 2.6 \times 2500$$

$$= 0.0575 \times 10^{-6} \text{ coulomb per square meter.}$$

• PROBLEM 5-28

The space between two coaxial conducting cylinders of length L = 25 cm is half-filled with a dielectric having relative dielectric constant $\varepsilon_r = 8$. The cylinders have radius 0.5 cm and 2 cm, as shown, and are connected to a 100 V battery.

a) Find the fields $\bar{E}$ and $\bar{D}$ in the air and in the dielectic.

b) Find the surface charge induced on the inner conductor at points adjacent to the air, and at points adjacent to the dielectric.

c) Find the total charge on the inner conductor, and the capacitance.

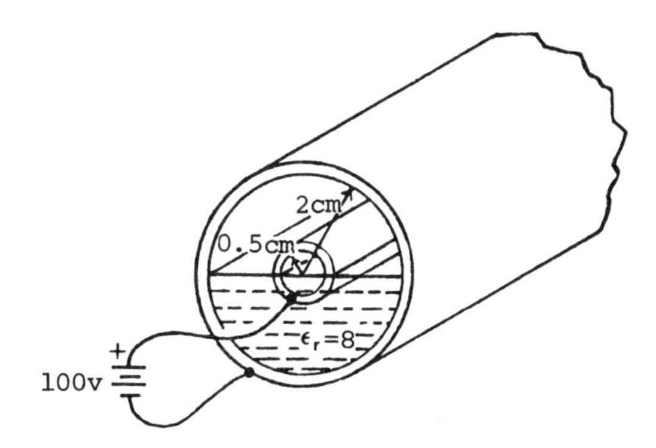

Solution:

a) From boundary conditions, the following are true:

$$E_{tan}\Big|_{①} = E_{tan}\Big|_{②}$$

and $D_n\Big|_{①} = D_n\Big|_{②}$

where ① and ② are used to represent the two regions (air and dielectric).

The boundary condition concerning the normal components are already met since there are no components normal to either regions,

$$D_n\Big|_{①} = D_n\Big|_{②} = 0.$$

The $\bar{E}$ field in both the dielectric and the air will be the same, since the tangential components must be equal. The E field of a coaxial cylinder is of the form

$$E(r) = \frac{K}{2\pi r}$$

where the constant K must be determined by

$$-\int \bar{E} \cdot d\bar{l} = V_2 - V_1$$

Putting in the limits of integration, the following is obtained:

$$-\int_{0.005}^{0.02} E(r)dr = V(0.02) - V(0.005) = -100$$

$$\int_{0.005}^{0.02} \frac{K}{2\pi r}\, dr = 100$$

$$\frac{K}{2\pi}\ln r \Big|_{0.005}^{0.02} = 100$$

$$\therefore\ K = \frac{200\ \pi}{\ln 4}$$

Therefore $E(r) = \frac{200\ \pi}{\ln 4} \cdot \frac{1}{2\pi r} = \frac{100}{\ln 4} \cdot \frac{1}{r}$

In the air:

$$D = \varepsilon_0 E = \frac{\varepsilon_0\ 100}{\ln 4} \cdot \frac{1}{r}$$

In the dielectric:

$$D = \varepsilon_r \varepsilon_0 E = \frac{8\varepsilon_0 100}{\ln 4} \cdot \frac{1}{r} = \frac{\varepsilon_0 800}{\ln 4} \cdot \frac{1}{r}$$

b) In inner conductor, the surface charge induced is:

$$\rho_s = \bar{D} \cdot \bar{n}$$

At points adjacent to the air:

$$\rho_s = \varepsilon_0 \frac{100}{\ln 4} \cdot \frac{1}{0.005}$$

At points adjacent to the dielectric:

$$\rho_s = 8\varepsilon_0 \frac{100}{\ln 4} \cdot \frac{1}{0.005}$$

c) The total charge on the inner conductor is the sum of the induced charges on points adjacent to the air and on points adjacent to the dielectric,

$$Q = \left[\rho_s(\text{air}) + \rho_s(\text{dielectric})\right] \cdot \pi r l$$

$$Q = \left[\varepsilon_0 \frac{100}{\ln 4} \cdot \frac{1}{0.005} + 8\varepsilon_0 \frac{100}{\ln 4} \cdot \frac{1}{0.005}\right]$$

$$\cdot\ \pi(0.005)\ (0.25)$$

$$Q = 9\varepsilon_0 \frac{100}{\ln 4} \pi(0.25)$$

$$C = \frac{Q}{V} = \frac{9\varepsilon_0 100\pi(0.25)}{\ln 4 \; 100}$$

$$C = \frac{9\varepsilon_0 \pi(0.25)}{\ln 4} \text{ farad.}$$

DIELECTRICS

• PROBLEM 5-29

Calculate the critical angle for an electromagnetic wave passing from the following dielectrics into air.

Material	$\varepsilon/\varepsilon_0$ (Ratio of permittivity to that of air)
Distilled water	81·1
Glass	9

Solution:

If the dielectric is medium 1 and air is medium 2, then the critical angle is given by the formula

$$\theta_c = \sin^{-1} \sqrt{\varepsilon_0/\varepsilon}$$

Then for distilled water

$$\theta_c = \sin^{-1} \sqrt{\varepsilon_0/\varepsilon} = \sin^{-1} \sqrt{1/81 \cdot 1} = 6.3°$$

For glass $\theta_c = \sin^{-1} \sqrt{1/9} = 19.4°$

• PROBLEM 5-30

Given that the dielectric constant of teflon is K=2.0 determine values for the electric susceptibility and the permittivity. Deduce the magnitude of the dipole moment and the surface charge density of a slab of teflon that just fills the space between the plates of a parallel plate capacitor for a potential difference of 1000 V between the plates of the capacitor.

Solution: The electric susceptibility of teflon is

$$\chi_e = K - 1$$

$$= 1.0$$

and the permittivity is

$$\varepsilon = K\varepsilon_0$$

$$= 2.0 \times 8.9 \times 10^{-12}$$

$$= 1.8 \times 10^{-11} C^2 \cdot N^{-1} \cdot m^{-2}.$$

The polarization P of the teflon slab in the capacitor is

$$P = \varepsilon_0 \chi_e E = \varepsilon_0 \chi \frac{V}{d}$$

$$= \frac{8.9 \times 10^{-12} \times 1.0 \times 10^3}{0.50 \times 10^{-2}}$$

$$= 1.8 \times 10^{-6} C \cdot m^{-2}.$$

The surface charge density of the teflon slab is therefore

$$\sigma_p = 1.8 \times 10^{-6} C \cdot m^{-2}.$$

The magnitude of the dipole moment of the slab is

$$PAL = 1.8 \times 10^{-6} \times 10^{-2} \times 0.50 \times 10^{-2}$$

$$= 9.0 \times 10^{-11} C \cdot m.$$

• PROBLEM 5-31

An infinite hollow circular cylinder of dielectric of radii a and b (b > a) and dielectric constant K is set with its axis perpendicular to a uniform filad of strength E. Find the field in the hollow space.

Solution: Let Φ_1, Φ_2 and Φ_3 be te potential in the regions

$$r > b, \qquad b > r > a, \qquad a > r > 0$$

respectively; since the potential due to the applied electric field, E, supposed parallel to the axis $\theta = 0$, is

$$-Er \cos \theta,$$

the only harmonics entering in the solution are of the form

$$r \cos\theta, \qquad r^{-1}\cos\theta.$$

Since $\Phi_1 \sim -Er\cos\theta$ at $r = \infty$ and Φ_3 is finite at the origin,

$$\Phi_1 = -E(r\cos\theta + Ar^{-1}\cos\theta),$$

$$\Phi_2 = E(Br\cos\theta + Cr^{-1}\cos\theta),$$

$$\Phi_3 = EDr\cos\theta.$$

The boundary conditions are

$$\text{(i)} \quad \Phi_1 = \Phi_2 \text{ and } K\frac{\partial\Phi_2}{\partial r} = \frac{\partial\Phi_1}{\partial r} \quad \text{at} \quad r = b,$$

$$\text{(ii)} \quad \Phi_2 = \Phi_3 \text{ and } K\frac{\partial\Phi_2}{\partial r} = \frac{\partial\Phi_3}{\partial r} \quad \text{at} \quad r = a,$$

Thus

$$-(b + Ab^{-1}) = Bb + Cb^{-1}, \quad K(B - Cb^{-2}) = -1 + Ab^{-2}$$

and

$$Ba + Ca^{-1} = Da, \qquad K(B - Ca^{-2}) = D.$$

Solve these equations for D

$$D = -\frac{4K}{(K+1)^2 - (K-1)^2(a/b)^2}$$

showing that the (uniform) field inside the cavity is altered in the ratio

$$4K : \{(K+1)^2 - (K-1)^2(a/b)^2\}.$$

• PROBLEM 5-32

A sphere of radius a is made of homogeneous dielectric with dielectric constant ε_R. The sphere is centêred at the origin in free space. The potential field within the sphere is $V_{in} = -(3r E_0 \cos\theta)/(\varepsilon_R + 2)$, outside, $V_{out} = -r E_0 \cos\theta + (a3E_0/r^2)[(\varepsilon_R - 1)/(\varepsilon_R + 2)]\cos\theta$. (a) Show that $\bar{E}_{in}$ is uniform. (b) Show that for $r >> a$, $\bar{E}_{out} \cong \bar{E}_0 \bar{a}_z$. and (c) Show that the fields satisfy all dielectric boundary conditions at $r = a$.

Solution: (a) In spherical coordinates,

$$\nabla V = \frac{\partial V}{\partial r}\bar{a}_r + \frac{1}{r}\frac{\partial V}{\partial \theta}\bar{a}_\theta + \frac{1}{r\sin\theta}\frac{\partial V}{\partial \Phi}\bar{a}_\Phi$$

For $V_{in} = -(3E_0 r\cos\theta)/(\varepsilon_R + 2)$,

$$\bar{E}_{in} = -\nabla V_{in} = \frac{3E_0\cos\theta}{\varepsilon_r + 2}\bar{a}_r - \frac{3E_0\sin\theta}{\varepsilon_r + 2}\bar{a}_\theta$$

But $\cos\theta\,\bar{a}_r - \sin\theta\bar{a}_\theta = \bar{a}_z$

Therefore $\bar{E}_{in} = 3E_0/(\varepsilon_r + 2)\bar{a}_z$ which is uniform

(b) If $r >> a$,

$$V_{out} \cong -rE_0\cos\theta$$

$$\bar{E}_{out} = -\nabla V_{out} = E_0(\cos\theta\,\bar{a}_r - \sin\theta\,\bar{a}_\theta) = E_0\,\bar{a}_z$$

This is the same as the original electric field in the z-direction. In other words, the dielectric sphere has no effect on the field at far distance from it.

(c) In general,

$$\bar{E}_{out} = -\nabla V_{out} = \left(E_0\cos\theta + \frac{2a^3E_0}{r^3}\frac{\varepsilon_R - 1}{\varepsilon_R + 2}\cos\theta\right)\bar{a}_r$$

$$-\left(E_0\sin\theta - \frac{a^3E_0}{r^3}\frac{\varepsilon_R - 1}{\varepsilon_R + 2}\sin\theta\right)\bar{a}_\theta$$

At the boundary, $r = a$,

$$\bar{E}_{out} = \frac{3\varepsilon_r E_0\,\text{Cos}\,\theta}{\varepsilon_r + 2}\bar{a}_r - \frac{3E_c\sin\theta}{\varepsilon_R + 2}\bar{a}_\theta$$

The tangential component of -E, i.e., the θ-component of the electric field

$$E_{in}(\tan) = E_{out}(\tan) = -3E_0\sin\theta/(\varepsilon_R + 2)$$

The normal component of D, i.e. the r-component of D,

$$\varepsilon_0\varepsilon_r E_{in}(\text{norm}) = 3\varepsilon_0\varepsilon_r E_0\cos\theta/(\varepsilon_R + 2) = \varepsilon_0 E_{out}(\text{norm})$$

Thus, all boundary conditions are satisfied.

• **PROBLEM 5-33**

Two parallel conducting plates infinite in the y, z directions are located at x = - d and x = +d. The space between them is filled with a dielectric medium and a space-dependent permittivity

$$\varepsilon = \frac{4\varepsilon_0}{\left(\frac{x}{d}\right)^2 + 1}$$

The plate at x = d is held at a time-independent potential difference V_0 with respect to the plate at x = - d.

(a) Find the electric field and the potential distribution between the plates.

(b) Find the polarization $\bar{P}$ and the density of polarization charge, ρ_p.

Solution: In a linear dielectric medium such as the one discussed here it is convenient to introduce the displacement flux density $\bar{D}$. $\bar{E}$ and $\bar{D}$ are related by

$$\varepsilon\bar{E} = \bar{D}$$

The convenience of the use of $\bar{D}$ lies in the fact that in the absence of a free surface charge the normal component of $\bar{D}$ continuous on interfaces between material media and air (whereas $\bar{E}$ experiences a discontinuity in its normal component due to the surface charge of polarization). Start by observing that $\bar{D}$ has only an x-component, D_x, due to the one-dimensionality of the problem, and that D_x is continuous throughout the entire space. Thus

$$E_x = \frac{D_x}{\varepsilon} = \frac{D_x}{4\varepsilon_0}\left[\left(\frac{x}{d}\right)^2 + 1\right]$$

inside the dielectric.

The potential at a point x is

$$V(x) = -\int_{-d}^{x} E_x dx = -\frac{D_x d}{4\varepsilon_0}\left[\frac{1}{3}\left(\frac{x}{d}\right)^3 + \frac{x}{d} + \frac{4}{3}\right]$$

where V(-d) is set arbitrarily at V(-d) = 0. Since V(d) = V_0, D_x can be found.

$$D_x = -\frac{3}{2}\frac{\varepsilon_0 V_0}{d}$$

The polarization P is found with

$$\bar{D} = \varepsilon_0 \bar{E} + \bar{P}$$

and thus

$$\bar{P} = i_x \bar{P}_x = \bar{i}_x (D_x - \varepsilon_0 E_x)$$

$$P_x = \frac{D_x}{4}\left[3 - \left(\frac{x}{d}\right)^2\right]$$

The polarization charge density is

$$\rho_p = - \operatorname{div} \bar{P} = - \frac{\partial P_x}{\partial x} = \frac{x D_x}{2d^2}$$

• PROBLEM 5-34

Medium 1 has $\kappa_1 = 1.5$ and extends to the left of the yz plane. Medium 2 has $\kappa_2 = 2.5$ and extends to the right of the yz plane.

$$\bar{E}_1 = (\hat{x}2 - \hat{y}3 + \hat{z}) \qquad \text{Vm}^{-1}$$

Find $\bar{D}_1$, $\bar{E}_2$, and $\bar{D}_2$.

Solution:

$$\bar{D} = \varepsilon \bar{E} = \kappa \varepsilon_0 \bar{E} \qquad \text{Wb m}^{-2} \quad \text{So } D_1 = \kappa_1 \varepsilon_0 E_1$$

$$\bar{D}_1 = 1.5\varepsilon_0 (\hat{x}2 - \hat{y}3 + \hat{z}) = \hat{x}3\varepsilon_0 - \hat{y}4.5\varepsilon_0 + \hat{z}1.5\varepsilon_0$$

The normal direction at this interface is the x direction, so

$$\bar{E}1_n = \hat{x}2$$

$$\bar{E}1_t = - \hat{y}3 + \hat{z}$$

$$\bar{D}1_n = \hat{x}3\varepsilon_0$$

$$\bar{D}1_t = - \hat{y}4.5\varepsilon_0 + \hat{z}1.5\varepsilon_0$$

From boundary conditions, $E_{1t} = E_{2t}$ and $D_{1n} = D_{2n}$

$$\bar{E}_{2t} = -\hat{y}3 + \hat{z} \qquad \bar{D}_{2n} = \hat{x}3\varepsilon_0$$

$\bar{E}_{2n}$ is found from $\bar{D}_{2n}$: $\bar{D}_{2n} = \kappa_2 \varepsilon_0 \bar{E}_{2n}$. So

$$E_{2n} = \frac{1}{2.5\varepsilon_0}(\hat{x}3\varepsilon_0) = \hat{x}1.2$$

Similarily $\bar{D}_{2t}$ is found from $\bar{E}_{2t}$: $\bar{D}_{2t} = \kappa_2 \varepsilon_0 \bar{E}_{2t}$,

$$\bar{D}_{2t} = 2.5\varepsilon_0(-\hat{y}3 + \hat{z}) = -\hat{y}7.5\varepsilon_0 + \hat{z}2.5\varepsilon_0$$

Summarizing,

$$\bar{E}_1 = \hat{x}2 - \hat{y}3 + \hat{z} \qquad \bar{E}_2 = \hat{x}1.2 - \hat{y}3 + \hat{z}$$

$$\bar{D}_1 = \varepsilon_0(\hat{x}3 - \hat{y}4.5 + \hat{z}1.5) \qquad \bar{D}_2 = \varepsilon_0(\hat{x}3 - \hat{y}7.5 + \hat{z}2.5)$$

The laws for the behavior of the components of $\bar{E}$ are different from those for $\bar{D}$ but the behavior of $\bar{E}$ and $\bar{D}$ are similar in this respect - $\bar{E}_1$ is parallel to $\bar{D}_1$ and $\bar{E}_2$ is parallel to $\bar{D}_2$.

• PROBLEM 5-35

The figure shows a parallel plate capacitor with two sheets of dielectric filling the space between its metal plates B and G. The dielectrics are paraffin and glass with relative dielectric constants of 2.40 and 4.00 respectively. Their thicknesses are shown in the figure, and their faces are parallel to the metal plates. Each metal plate has an area of 2.00 m^2 and, by connection to a battery, the left plate is given a positive conduction charge of 4.00×10^{-5} coulomb and the right plate is given a negative conduction charge of the same magnitude.

(a) What are the values of D, E, and P in each dielectric?

(b) What is the surface density of polarization charge on each face of each dielectric?

(c) How many more lines of force per unit area, which approach the dielectric interface I from the left, terminate on the negative polarization charges on the face of the glass than originate on positive polarization charges on the face of the paraffin?

(d) What is the susceptibility of each dielectric?

(e) What is the potential difference between the metal plates of the capacitor?

(f) What is the capacitance of the capacitor?

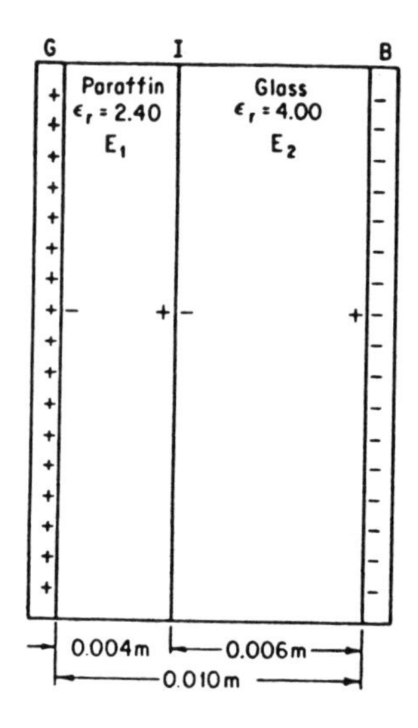

Parallel plate capacitor with two sheets of dielectric filling the space between the plates.

Fig. 1

Solution:

(a) $\sigma = q/A = 2.00 \times 10^{-5}$ coulomb/m^2 (1)

Now, σ lines of D originate on each unit area of plate G perpendicular to the plate and, since the face of the paraffin is parallel to the face of the plate, σ lines of D enter each unit area of the face of the paraffin. The number of lines of D per unit area perpendicular to D is equal to D, so D in the paraffin is equal to σ or

$$D = \sigma = 2.00 \times 10^{-5} \text{ coulomb/m}^2 \tag{2}$$

Also, $\bar{D}$ is normal to all dielectric interfaces; hence, from the boundary conditions, $\bar{D}_1 = \bar{D}_2$ in such a case. Thus $\bar{D}$ has the same magnitude and direction in both dielectrics.

$$E = D/\varepsilon_r \varepsilon_0 \tag{3}$$

In the paraffin, let E_1 represent the electric field intensity and substituting into (3) obtain

$$E_1 = 9.42 \times 10^5 \text{ newton/coulomb} \tag{4}$$

In the glass, let E_2 represent the electric field intensity, and again substitution into (3), gives

$$E_2 = 5.65 \times 10^5 \text{ newton/coulomb} \tag{5}$$

$$P = D - \varepsilon_0 E \tag{6}$$

In the paraffin, using (6),

$$P_1 = 1.16 \times 10^{-5} \text{ coulomb/m}^2 \tag{7}$$

In the glass, again using (6)

$$P_2 = 1.50 \times 10^{-5} \text{ coulomb/m}^2 \tag{8}$$

(b) Since $\bar{P}$ is parallel to $\bar{D}$ and $\bar{E}$, then $\bar{P}$ is normal to all dielectric interfaces in this problem, so $\sigma_p = P$. Also, since P has the same value at all points in a given dielectric in this problem, the surface density on the left face of each dielectric is equal in magnitude to the surface density of polarization on the right face of the same dielectric. Hence on the left face of the paraffin, from (7)

$$\sigma_{P_1} = -\ 1.16 \times 10^{-5} \text{coulomb/m}^2 \tag{9}$$

and on the right face of the paraffin

$$\sigma_{P_1} = +\ 1.16 \times 10^{-5} \text{ coulomb/m}^2 \tag{10}$$

Similarly on the left face of the glass, from (8),

$$\sigma_{P_2} = -\ 1.50 \times 10^{-5} \text{ coulomb/m}^2 \tag{11}$$

and on the right face of the glass

$$\sigma_{P_2} = +\ 1.50 \times 10^{-5} \text{ coulomb/m}^2 \tag{12}$$

(c) Evidently, $\sigma_{P_2}/\varepsilon_0$ lines of force per unit area originate or terminate on the σ_p polarization charges per unit area. Hence $\sigma_{P_2}/\varepsilon_0$ lines of force per unit area terminate on the glass and $\sigma_{p1}/\varepsilon_0$ lines of force per unit area originate on the paraffin, so the difference is

$$\frac{1}{\varepsilon_0}\sigma_{P_2} - \frac{1}{\varepsilon_0}\sigma_{P_1} = \frac{1 \text{ newton m}^2}{8.85 \times 10^{-12} \text{ coulomb}^2} (1.50 - 1.16) \times$$

$$10^{-5} \frac{\text{coulomb}}{\text{m}^2}$$

$$= 3.80 \times 10^{-5} \text{ newton/coulomb} \tag{13}$$

As a check on arithmetic, this difference must equal $E_1 - E_2$, and from (4) and (5),

$$E_1 - E_2 = 3.77 \times 10^{-5} \text{ newton/coulomb} \quad (14)$$

Thus, (13) and (14) check within slide rule error.

(d) $$\eta = P/\varepsilon_0 E \quad (15)$$

For paraffin, using (4), (7), and (15),

$$\eta_1 = \frac{1.16 \times 10^{-5} \text{ coulomb/m}^2}{8.85 \times 10^{-12} \dfrac{\text{coulomb}^2}{\text{newton m}^2} \times 9.42 \times 10^5 \dfrac{\text{newton}}{\text{coulomb}}}$$

$$= 1.40 \quad (16)$$

or more simply, $\eta = \varepsilon_r - 1$

$$\eta_1 = 2.40 - 1 = 1.40 \quad (17)$$

Similarly, $\eta_2 = 4.00 - 1 = 3.00$ (18)

(e) $$V_G - V_B = -\int_{(B)}^{(I)} E_2 \cos\theta \, ds - \int_{(I)}^{(G)} E_1 \cos\theta \, ds$$

For evaluation of these line integrals, select a path which is normal to both plates so the path will be parallel to the lines of force. Along this path, E_1 and E_2 are both constant, so the integrals reduce to

$$V_G - V_B = E_2 d_2 + E_1 d_1 = 5.65 \times 10^5 \text{v/m} \times 6 \times 10^{-3}\text{m} + 9.42$$

$$\times 10^5 \text{v/m} \times 4 \times 10^{-3}\text{m} = 7160\text{v} \quad (19)$$

(f) $C = q/V$. Hence from the value of q given in the problem and (19)

$$C = 4.00 \times 10^{-5} \text{ coulomb}/7.16 \times 10^3 \text{v} =$$

$$55.9 \times 10^{-10}\text{f} = 5590 \; \mu\mu\text{f}$$

A capacitor with parallel conducting plates at x = 0 and x = d is filled with a material whose dielectric constant varies with position as

$$\varepsilon(x) = \varepsilon_0 \left(\frac{2d}{x + d}\right)$$

A static voltage source V_0 is connected across the capacitor plates.

a) Find E(x), D(x) and ρ(x) between the plates neglecting fringing effects.

b) Find the volume and surface distributions of both free and polarization charges.

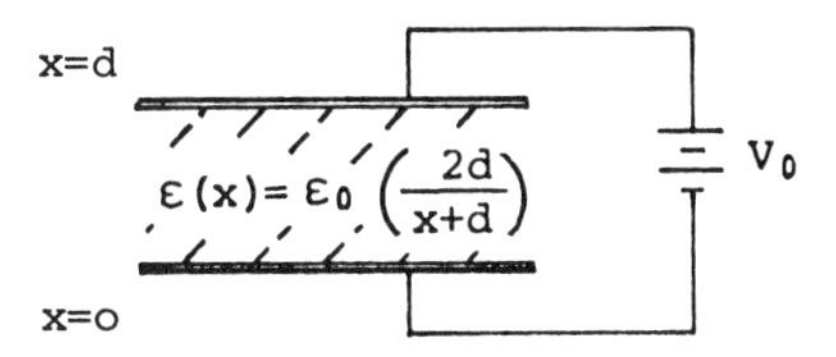

Solution: A diagram corresponding to the capacitor is shown in the figure.

a) Using $\nabla \cdot \bar{D} = \rho_{free}$ (free charge)

ρ_{free} between the plates is zero,

therefore

$$\nabla \cdot [\bar{a}_z \; D(z)] = \frac{\partial}{\partial z} D(z) = 0$$

$$\therefore \quad D(z) = \text{constant (call it } D_0)$$

$$E = \frac{1}{\varepsilon} D_0$$

Using the formula for potential difference,

$$V = \int_a^b -\bar{E} \cdot d\bar{\ell}$$

$$V_0 = \int_0^d \frac{1}{\varepsilon} D_0 \, dx$$

$$V_0 = D_0 \int_0^d \frac{x+d}{2d\varepsilon_0}\, dx = (D_0/2d\varepsilon_0) \left. \frac{(x+d)^2}{2} \right]_0^d$$

$$V_0 = \frac{D_0}{2d\varepsilon_0} \left[\frac{(2d)^2}{2} - \frac{d^2}{2} \right] = \frac{D_0}{2d\varepsilon_0} \left[\frac{4d^2}{2} - \frac{d^2}{2} \right]$$

$$V_0 = \frac{3}{4} \frac{D_0 d}{\varepsilon_0}$$

Solving for D_0 ,

$$D_0 = \frac{4\varepsilon_0}{3d} V_0$$

$$E(x) = \frac{1}{\varepsilon} D_0$$

$$E(x) = \frac{x+d}{\varepsilon_0 2d} \left[\frac{4}{3} \cdot \frac{\varepsilon_0}{d} \cdot V_0 \right]$$

$$E(x) = \frac{2}{3} \cdot \frac{x+d}{d^2} \cdot V_0$$

The polarization $P(x)$ is

$$P(x) = D(x) - \varepsilon_0 E(x) = D_0 - \varepsilon_0 E(x)$$

$$P(x) = \frac{4}{3} V_0 \frac{\varepsilon_0}{d} - \frac{2}{3} \varepsilon_0 \frac{(x+d)}{d^2} V_0$$

$$P(x) = \frac{2}{3} V_0 \frac{\varepsilon_0}{d} \left[2 - \left(\frac{x+d}{d} \right) \right]$$

b) The free volume charge is given by

$$\nabla \cdot \bar{D} = \rho_{v(free)} = 0$$

The free surface charge is

$$\rho_{s(free)} = \hat{n} \cdot \bar{D}$$

where $\hat{n}$ is the normal to the surface. In this case

$$\hat{n} = \hat{a}_x \text{ (x axis is positive upwards)}$$

$$\rho_{s(free)} = \hat{a}_x \frac{4}{3} \frac{\varepsilon_0}{d} V_0$$

The volume and surface charges due to polarization effects are given by

$$\rho_{v(eq)} = -\nabla \cdot \bar{P}$$

$$\rho_{s(eq)} = \hat{n} \cdot \bar{P}$$

$$\rho_{v(eq)} = -\frac{\partial}{\partial x} P(x) = -\frac{\partial}{\partial x}\left\{\frac{2}{3} V_0 \frac{\varepsilon_0}{d}\left[2 - \left(\frac{x+d}{d}\right)\right]\right\}$$

$$\rho_{v(eq)} = +\frac{2}{3} V_0 \frac{\varepsilon_0}{d^2}$$

$$\rho_{s(eq)} = \hat{a}_x \cdot P = \rho$$

$$\rho_{s(eq)} = \frac{2}{3} V_0 \frac{\varepsilon_0}{d}\left[1 - \left(\frac{x+d}{d}\right)\right]$$

The only surface charge is at x = d and x = 0.

$$\text{at } x = d \quad \rho_{s(eq)} = \frac{2}{3} V_0 \frac{\varepsilon_0}{d}[-1] = -\frac{2}{3} V_0 \frac{\varepsilon_0}{d}$$

$$\text{at } x = 0 \quad \rho_{s(eq)} = \frac{2}{3} V_0 \frac{\varepsilon_0}{d}[0] = 0$$

• PROBLEM 5-37

Consider an electric field $\bar{E}_1$ produced by a pair of electrodes far enough apart to produce parallel flux lines in a medium of permittivity ε_1. Now consider an uncharged dielectric sphere of permittivity ε_2 immersed in the first medium. (See figure.) Find the field inside the sphere (considered uniform) if in addition to $\bar{E}_1$ outside the sphere, there is a dipole with strength $\bar{P}$ inside the sphere (centrally located).

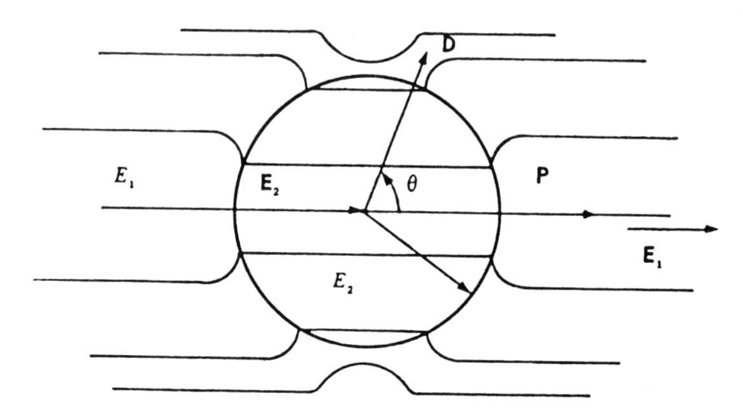

Solution: The potential inside the sphere, if r is the radius of the sphere is

$$V_2 = - E_2 r \cos\theta \tag{1}$$

where θ is the angle between the $\bar{E}$ and $\bar{D}$ vectors.

The contribution to the potential at r due to the dipole moment is

$$\frac{P}{4\pi\varepsilon_v r^2} \cos\theta$$

and the potential outside the sphere due to the applied field is

$$-E_1 r \cos\theta$$

Therefore the total potential outside the sphere is

$$V_1 = - E_1 r \cos\theta + \frac{P}{4\pi\varepsilon_v r^2} \cos\theta \tag{2}$$

Now, by the conservation of energy at the boundary the potential must be continuous, therefore since $V_1 = V_2$, equating (1) and (2)

$$- E_2 r = - E_1 r + \frac{P}{4\pi\varepsilon_v r^2} \tag{3}$$

Also the gradients must be equal at the boundary so that

$$\varepsilon_1 \frac{\partial V_1}{\partial r} = \varepsilon_2 \frac{\partial V_2}{\partial r}$$

and taking the partial derivatives of (1) and (2) with respect to r,

$$- \varepsilon_1 E_1 \cos\theta - \frac{2P}{4\pi\varepsilon_v r^3} \cos\theta = - \varepsilon_1 E_2 \cos\theta$$

or

$$- \varepsilon_2 E_2 = - \varepsilon_1 E_1 - \frac{2\varepsilon_1 P}{4\pi\varepsilon_v r^3} \qquad (4)$$

Rewriting (3) and (4) as (5) and (6) and multiplying through (5) by ε_2

$$- \varepsilon_2 E_2 r^3 = - \varepsilon_2 E_1 r^3 + \frac{\varepsilon_2 P}{4\pi\varepsilon_v} \qquad (5)$$

$$- \varepsilon_2 E_2 r^3 = - \varepsilon_1 E_1 r^3 - \frac{2\varepsilon_1 P}{4\pi\varepsilon_v} \qquad (6)$$

Subtracting (6) from (5)

$$0 = - \varepsilon_2 E_1 r^3 + \varepsilon_1 E_1 r^3 + \frac{P}{4\pi\varepsilon_v} (\varepsilon_2 + 2\varepsilon_1)$$

or

$$0 = E_1 r^3 (\varepsilon_1 - \varepsilon_2) + \frac{P}{4\pi\varepsilon_v} (\varepsilon_2 + 2\varepsilon_1)$$

Solving:

$$\frac{P}{4\pi\varepsilon_v} = -E_1 r^3 \frac{(\varepsilon_1 - \varepsilon_2)}{\varepsilon_2 + 2\varepsilon_1}$$

and substituting in (3)

$$-E_2 r = -E_1 r - E_1 r \frac{(\varepsilon_1 - \varepsilon_2)}{2\varepsilon_1 + \varepsilon_2}$$

or

$$E_2 = E_1 \left(1 + \frac{\varepsilon_1 - \varepsilon_2}{2\varepsilon_1 + \varepsilon_2}\right) = E_1 \left(\frac{2\varepsilon_1 + \varepsilon_2 + \varepsilon_1 - \varepsilon_2}{2\varepsilon_1 + \varepsilon_2}\right)$$

Finally

$$E_2 = E_1 \left(\frac{3\varepsilon_1}{2\varepsilon_1 + \varepsilon_2}\right)$$

• PROBLEM 5-38

Consider a Class A dielectric sphere of radius R with a point charge Q embedded at the center, as in Figure 1. Find D, E, P, and the bound charge densities at the surfaces of the sphere if the relative dielectric constant of the sphere is ε_r.

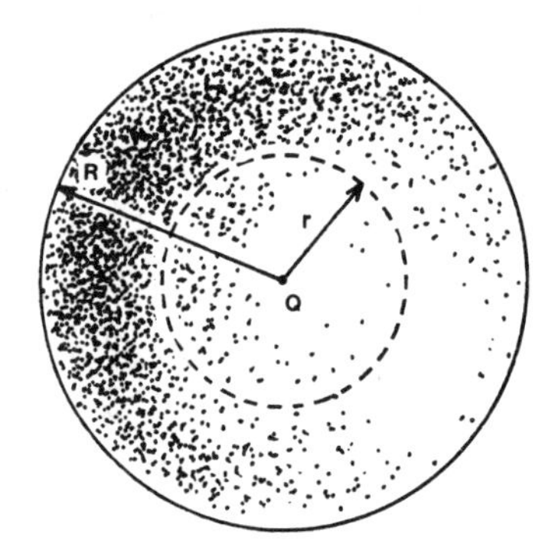

Dielectric sphere with a point charge Q at the center.

Fig. 1

Solution: $\bar{D}$ can be found from Gauss's law. Draw an imaginary sphere of radius r < R and use it as a Gaussian surface. Then

$$\bar{D} = \frac{Q}{4\pi r^2}\bar{r}_1, \tag{1}$$

$$\bar{E} = \frac{Q}{4\pi\varepsilon_0\varepsilon_r r^2}\bar{r}_1. \tag{2}$$

$$\bar{P} = \frac{\varepsilon_r - 1}{\varepsilon_r}\bar{D} = \frac{\varepsilon_r - 1}{\varepsilon_r}\frac{Q}{4\pi r^2}\bar{r}_1 \tag{3}$$

At the outer surface of the sphere,

$$\sigma_b = \bar{P}\cdot\bar{n} = \bar{P} = \frac{\varepsilon_r - 1}{\varepsilon_r}\frac{Q}{4\pi R^2}$$

The total amount of bound charge on the outer surface is thus

$$Q_{bo} = \frac{\varepsilon_r - 1}{\varepsilon_r} Q.$$

Since there is no volume density of free charge, there should be no volume density of bound charge. This is correct:

$$\rho_b = - \nabla \cdot \bar{P} = - \frac{1}{r^2} \frac{\partial}{\partial r} (r^2 P) = 0.$$

There is also a bound charge on the surface of the cavity containing the free charge Q at the center. This can be calculated as follows. Let the radius of the cavity be δ. Then, if $-\sigma_{bc}$ is the density of bound charge on the cavity surface, the total bound charge on the cavity is

$$Q_{bc} = -\sigma_{bc} 4\pi\delta^2 = - P_\delta 4\pi\delta^2 = - \left(\frac{\varepsilon_r - 1}{\varepsilon_r}\right) D_\delta 4\pi\delta^2,$$

$$= - \left(\frac{\varepsilon_r - 1}{\varepsilon_r}\right) Q.$$

The dielectric as a whole therefore remains neutral, as must be expected. The net charge at the center is then

$$Q_{net} = Q - \left(\frac{\varepsilon_r - 1}{\varepsilon_r}\right) Q = \frac{Q}{\varepsilon_r}$$

and is smaller than the free charge by the factor ε_r. This accounts for the reduction of the electric field intensity within the dielectric by the factor ε_r.

It is instructive to compare the electric field intensity outside the sphere and that inside. Outside, from Gauss's law,

$$\bar{D} = \frac{Q}{4\pi r^2} \bar{r}_1,$$

$$\bar{E} = \frac{Q}{4\pi\varepsilon_0 r^2} \bar{r}_1,$$

which is to be compared to Eqs. 1 and 2 for the field inside the sphere. At the surface of the sphere the electric field intensity is discontinuous, the magnitude just outside the surface being ε_r times as large as the magnitude just inside the surface. The difference is due to the bound charges, which produce an opposin; field within the dielectric.

• PROBLEM 5-39

An ungrounded spherical configuration of concentric spherical dielectric shells, enclosed by a conductor shell, is shown in Fig. 1. Regions # 1 and # 3 are free space, region # 2 is a dielectric whose relative premittivity equals ε_r, and region # 4 is a conductor. If a charge Q is placed at the center, find: (a) $\overline{D}$ in all regions through the use of Gauss's law; (b) $\overline{E}$ in all regions through the use of $\overline{E} = \overline{D}/\varepsilon$; (c) $\overline{P}$ in all regions through the use of $\overline{P} = \overline{D} - \varepsilon_0\overline{E}$; (d) ρ_s on the conductor surfaces; (e) ρ_{sb} on the dielectric surfaces; (f) ρ_{vb} within the dielectric.

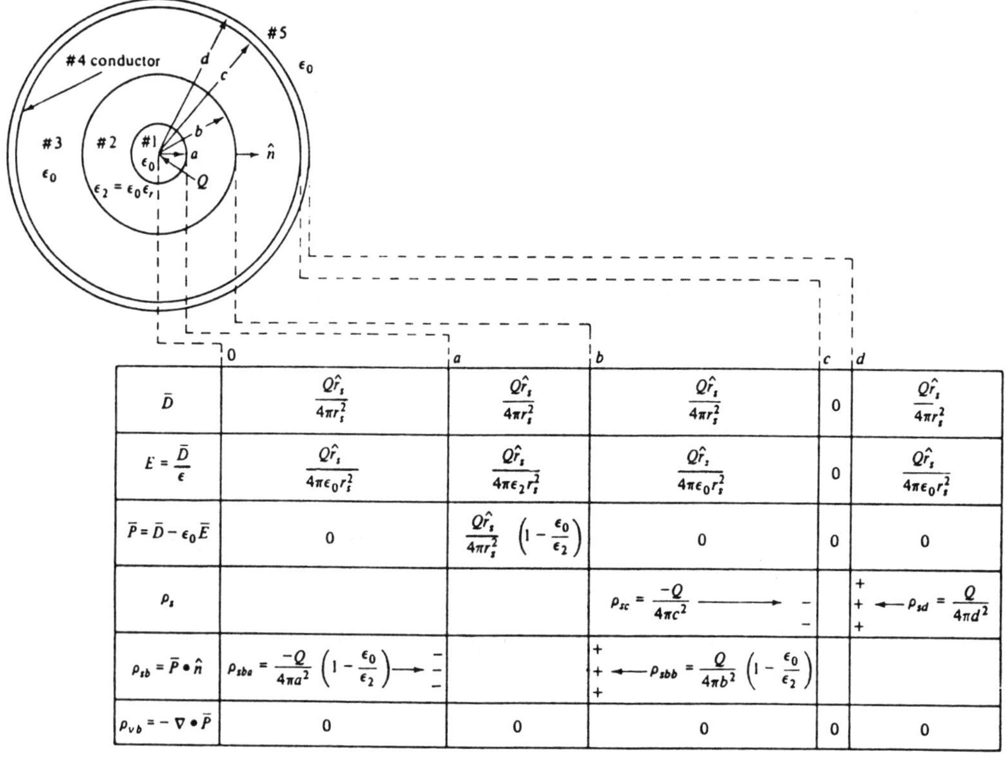

	0 – a	a – b	b – c	c – d	d –
$\overline{D}$	$\frac{Q\hat{r}_s}{4\pi r_s^2}$	$\frac{Q\hat{r}_s}{4\pi r_s^2}$	$\frac{Q\hat{r}_s}{4\pi r_s^2}$	0	$\frac{Q\hat{r}_s}{4\pi r_s^2}$
$E = \frac{\overline{D}}{\epsilon}$	$\frac{Q\hat{r}_s}{4\pi\epsilon_0 r_s^2}$	$\frac{Q\hat{r}_s}{4\pi\epsilon_2 r_s^2}$	$\frac{Q\hat{r}_s}{4\pi\epsilon_0 r_s^2}$	0	$\frac{Q\hat{r}_s}{4\pi\epsilon_0 r_s^2}$
$\overline{P} = \overline{D} - \epsilon_0\overline{E}$	0	$\frac{Q\hat{r}_s}{4\pi r_s^2}\left(1 - \frac{\epsilon_0}{\epsilon_2}\right)$	0	0	0
ρ_s			$\rho_{sc} = \frac{-Q}{4\pi c^2}$ → − −		+ + + ← $\rho_{sd} = \frac{Q}{4\pi d^2}$
$\rho_{sb} = \overline{P} \bullet \hat{n}$	$\rho_{sba} = \frac{-Q}{4\pi a^2}\left(1 - \frac{\epsilon_0}{\epsilon_2}\right)$ → − − −		+ + + ← $\rho_{sbb} = \frac{Q}{4\pi b^2}\left(1 - \frac{\epsilon_0}{\epsilon_2}\right)$		
$\rho_{vb} = -\nabla \bullet \overline{P}$	0	0	0	0	0

Graphical display of an ungrounded spherical configuration containing a free charge Q at its center and concentric spherical shells of dielectrics all enclosed by a conductor shell.

Fig. 1

Solution: (a) From Gauss's law $\oint_s \overline{D} \cdot \overline{ds} = Q_{en}$, where Q_{en} is the free charge enclosed by the Gaussian surface. It should be noted that any bound charges, due to polarization, that are enclosed by the Gaussian surface are not included in Q_{en}. Applying Gauss's law to region # 1,

$$\oint_s \overline{D}_1 \cdot \overline{ds} = D_{1r} \oint_s ds = D_{1r} (4\pi r_s^2) = Q$$

Thus,

$$\overline{D}_1 = \frac{Q\hat{r}_s}{4\pi r_s^2} \qquad (0 < r_s < a) \tag{1}$$

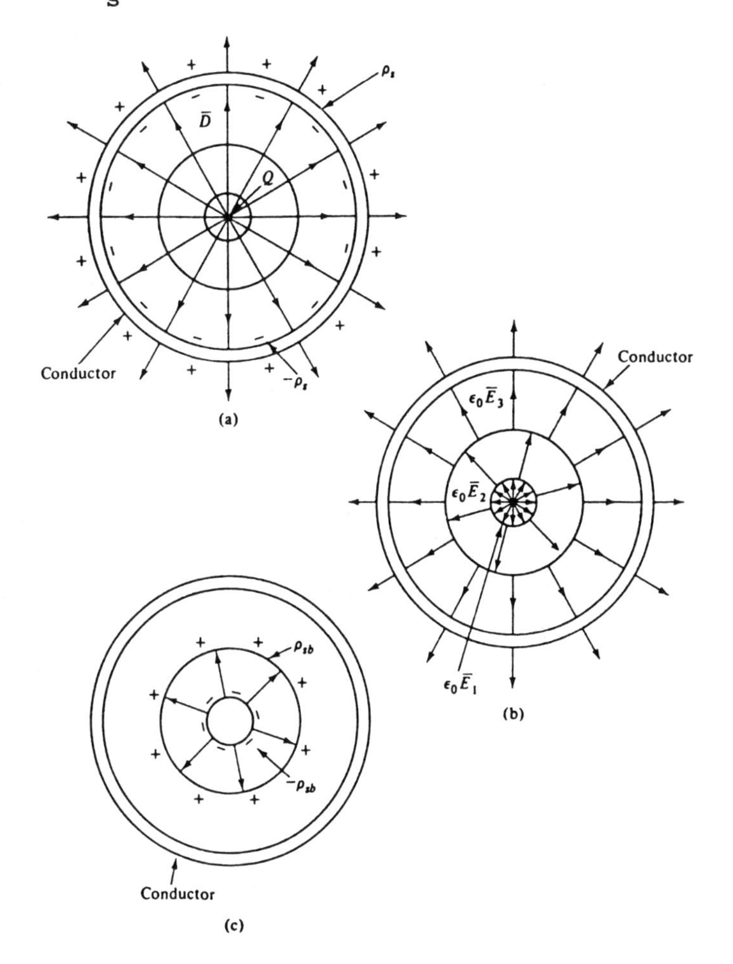

Flux plots of: (a) $\overline{D}$; (b) $\varepsilon_0\overline{E}$; (c) $\overline{P}$.

Fig. 2

Repeating the same procedure in regions # 2 and # 3 yields

$$\overline{D}_2 = \frac{Q\hat{r}_s}{4\pi r_s^2} \qquad (0 \leq r_s \leq b) \tag{2}$$

$$\overline{D}_3 = \frac{Q\hat{r}_s}{4\pi r_s^2} \qquad (b \leq r_s \leq c) \tag{3}$$

In region # 4, $\overline{D}_4$ = 0 which is true for a conductor under static conditions. Applying Gauss's law in region # 4

$$\oint_s \overline{D}_4 \cdot \overline{ds} = D_{4r} \oint_s ds = Q + Q' = Q_{en}$$

where Q' is the free charge induced on the conductor surface. Since $\overline{D}_4$ = 0, Q' = - Q. Now, if -Q is induced at r_s = c, then Q must be induced at r_s = d. Now apply Gauss's law to region # 5, Q_{en} = Q and

$$\overline{D}_s = \frac{Q\hat{r}_s}{4\pi r_s^2} \qquad (d \leqq r_s) \tag{4}$$

From (1), (2), (3), and (4) it should be noted that $\overline{D}$ has the same functional form and is independent of any bound charges. The values of $\overline{D}$ are tabulated in Fig. 1

(b) The expressions for $\overline{E}$ are found through $\overline{E} = \overline{D}/\varepsilon$. The values of $\overline{E}$ in all regions are found tabulated in Fig. 1.

(c) The expressions for $\overline{P}$ are found through $\overline{P} = \overline{D} - \varepsilon_0\overline{E}$. From the values for $\overline{D}$ and $\overline{E}$ found previously, the values for $\overline{P}$ are found and tabulated in Fig. 1.

(d) The free surface charge density ρ_s is found only on the conductor surface. Since the charge at r_s = c is - Q, then

$$\rho_{sc} = \frac{-Q}{4\pi c^2}\,(Cm^{-2}) \tag{5}$$

at r_s = d,

$$\rho_{sd} = \frac{Q}{4\pi d^2}\,(Cm^{-2}) \tag{6}$$

(e) The bound surface charge density is found only on the surfaces of # 2 region. From $\overline{P} \cdot \hat{n} = \rho_{sb}$, at r_s = a,

$$\rho_{sba} = \overline{P} \cdot \hat{n} = \frac{\hat{r}_s Q}{4\pi a^2}\left(1 - \frac{\varepsilon_0}{\varepsilon_2}\right) \cdot (-r_s)$$

$$= \frac{-Q}{4\pi a^2}\left(1 - \frac{\varepsilon_0}{\varepsilon_2}\right)\,(Cm^{-2}) \tag{7}$$

and at r_s = b,

$$\rho_{sbb} = \frac{Q}{4\pi b^2}\left(1 - \frac{\varepsilon_0}{\varepsilon_2}\right) \quad (Cm^{-2}) \tag{8}$$

(f) The bound volume charge density $\rho_{vb} = - \nabla \cdot \overline{P}$. The $\nabla \cdot \overline{P}$ in region # 2 equals zero; thus $\rho_{vb} = 0$. Flux plots of $\overline{D}$, $\varepsilon_0\overline{E}$, and $\overline{P}$ are shown in Fig. 2.

CHAPTER 6

CAPACITANCE

CAPACITANCE

• PROBLEM 6-1

Find the capacitance of a single, isolated conducting sphere of radius a.

Solution: Assuming that the sphere carries a charge q, by using Gauss's law

$$\bar{D} = \frac{q}{4\pi r^2}\bar{r}_u \,.$$

Using the displacement law,

$$\bar{E} = \frac{q}{4\pi\varepsilon_0 r^2}\bar{r}_u .$$

The potential of the sphere with respect to $\phi_\infty = 0$ is then

$$\phi_a = \int_a^\infty \bar{E}\cdot d\bar{l} = \int_a^\infty \frac{q}{4\pi\varepsilon_0 r^2}dr = -\left.\frac{q}{4\pi\varepsilon_0 r}\right|_a^\infty = \frac{q}{4\pi\varepsilon_0 a} \,.$$

Using now the capacitance equation,

$$C = \frac{q}{\phi_a} = \frac{q 4\pi\varepsilon_0 a}{q},$$

or

$$C = 4\pi\varepsilon_0 a.\text{F}$$

• **PROBLEM 6-2**

1. If a spherical conductor, of radius a, with no other conductor in the neighborhood, is coated with a uniform thickness d of shellac of which k is the specific inductive capacity, show that the capacity of the conductor is increased in the ratio k(a + d): ka + d.

Solution: Let e be the charge on the conductor; by symmetry the displacement vector inside the dielectric and the electric field outside it will be radial. If D and E are their magnitudes,

$$D = \frac{e}{r^2} \text{ (inside)}, \qquad E = \frac{e}{r^2} \text{ (outside)}.$$

Also D = kE inside the dielectric, so that there the electric field is given by

$$E = \frac{e}{kr^2} \quad (a < r < a + d),$$

and

$$E = \frac{e}{r^2} \quad (r > a + d).$$

If ϕ is the potential, we have $E = -\partial\phi/\partial r$. Hence, the potential of the sphere V is given by (taking $\phi = 0$ at infinity)

$$0 - V = \int_a^{a+d} \left(-\frac{e}{kr^2}\right) dr + \int_{a+d}^{\infty} \left(-\frac{e}{r^2}\right) dr =$$

$$= -\frac{e}{a} \frac{d + ka}{k(a + d)}.$$

The capacity of the sphere is $\frac{e}{V} = \frac{k(a + d)}{d + ka} a$, and since the capacity of the sphere in the absence of the dielectric is a, the required result is obtained.

PARALLEL PLATE CAPACITORS

• **PROBLEM 6-3**

Two square conducting plates are parallel, one meter on a side, and 1 mm apart in air. Find the capacitance C. One plate is now rotated slightly about an axis, parallel to an edge and passing through the center of the plate, until the separation is 0.5 mm at one side and 1.5 mm at the other. By considering the plates as parts of radial planes, find the new capacitance C' and the percentage of change.

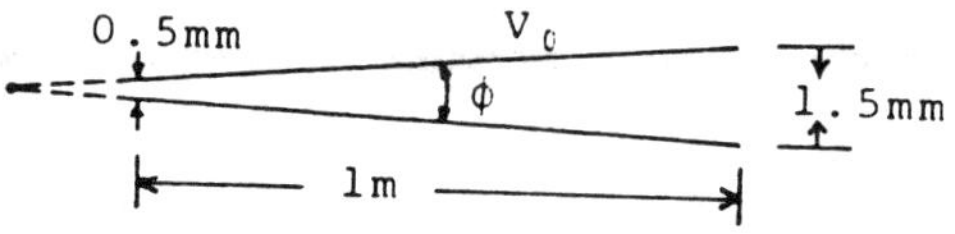

Solution: The capacitance of the original parallel plate capacitor is

$$C = \frac{\varepsilon_0 A}{d} = 8.854 \times 10^{-12} \frac{1}{10^{-3}} = 8.854 \times 10^{-9} F$$

Now, rotate one of the plates as shown. Consider the plates as parts of radial planes, then

$$V = V_0 \phi/\alpha$$

and

$$\bar{E} = -\frac{1}{r}\frac{\partial V}{\partial \phi}\bar{a}_\phi$$

where

$$\alpha = \frac{(1.5 - 0.5) \times 10^{-3}}{1} = 10^{-3}$$

Therefore

$$\bar{E} = -\frac{V_0}{\alpha r}\bar{a}_\phi$$

and

$$\bar{D} = \varepsilon_0 \bar{E} = -\frac{\varepsilon_0 V_0}{\alpha r}\bar{a}_\phi = -|\rho_s|\bar{a}_\phi$$

where

$$\rho_s = \frac{\varepsilon_0 V_0}{\alpha r}$$

Now, the total charge Q is

$$Q = \int_s \rho_s ds = \int_{0.5}^{1.5} \frac{\varepsilon_0 V_0}{\alpha r} dr = \frac{\varepsilon_0 V_0}{\alpha} \ln 3$$

Therefore

$$C' = \frac{Q}{V_0} = \frac{\varepsilon_0 \ln 3}{\alpha} = \frac{8.854 \times 10^{-12}}{10^{-3}} \times 2.3 \times 0.447$$

$$= 9.714 \times 10^{-9} F$$

The percentage of change is

$$\frac{9.714 - 8.854}{8.854} \times 100 = 9.7\%$$

• PROBLEM 6-4

A parallel-plate condenser has plates 20 cm. square and 0.5 cm. apart. Find

(a) its capacitance,

and, when charged to a p.d. of 1000 volts:

(b) its charge,

(c) the energy stored.

(d) the force of attraction between the plates.

Solution: (K = 1 for air.)

$$C = \frac{\varepsilon_0 A}{d} \qquad (A = 0.04 \text{ sq. metre, } d = 0.005 \text{ metre})$$

$$= \frac{8.854 \times 10^{-12} \times 0.04}{0.005}$$

$$= 70{\cdot}8 \times 10^{-12} \text{ farad}$$

$$= 70{\cdot}8 \text{ picofarads (pF).}$$

(b) $Q = CV = 70{\cdot}8 \times 10^{-12} \times 10^3 = 7.08 \times 10^{-8}$ coulomb.

(c) $W = \frac{1}{2}CV^2 = 3.54 \times 10^{-5}$ joule.

(d) $F = \frac{CV^2}{2d} = \frac{70{\cdot}8 \times 10^{-12} \times 10^6}{0.01}$

$$= 7{\cdot}08 \times 10^{-3} \text{ newton (708 dynes).}$$

• PROBLEM 6-5

Find the capacitance of the parallel-plate capacitor shown in Fig. 1, when uniform ρ_s but of opposite polarity is found on the conductor plates. Assume zero $\bar{E}$ field fringing at the edges.

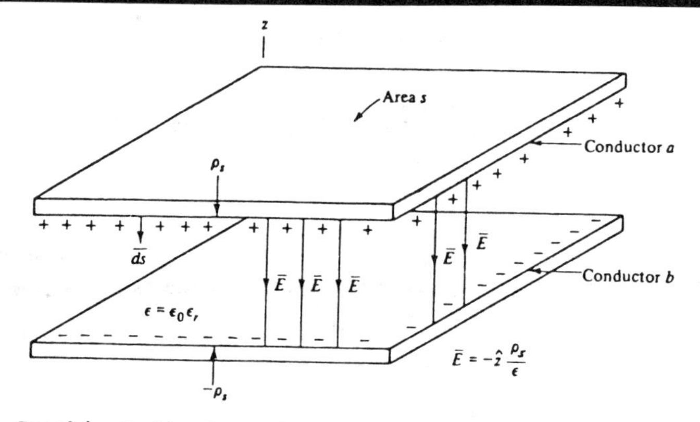

Graphical display of two-conductor parallel-plate capacitor of uniform ρ_s.

Fig. 1

Solution: From

$$C = \frac{|Q|}{|V|} = \frac{\left|\int_{sa} \rho_s ds\right|}{\left|-\int_b^a \bar{E}\cdot d\bar{l}\right|} = \frac{\left|\int_{sa} \varepsilon\bar{E}\cdot ds\right|}{\left|-\int_b^a \bar{E}\cdot d\bar{l}\right|}$$

and $\bar{E}|_{cond.} = \bar{E}$ (between conductors) $= -\hat{z}\rho_s/\varepsilon$,

$$C = \frac{\left|\int_{sa} \varepsilon\bar{E}\cdot\overline{ds}\right|}{\left|-\int_b^a \bar{E}\cdot d\ell\right|} = \frac{\left|\int_{top} \varepsilon\left(\frac{-\hat{z}\rho_s}{\varepsilon}\right)\cdot(-\hat{z}ds)\right|}{\left|-\int_0^\ell \left(-\hat{z}\frac{\rho_s}{\varepsilon}\right)\cdot(\hat{z}\ dz)\right|}$$

Thus,

$$C = \frac{\rho_s\ s}{\rho_s\ \ell/\varepsilon} = \frac{\varepsilon s}{\ell} \quad \text{(F)}$$

• PROBLEM 6-6

A parallel-plate capacitor of plate separation d has the region between its plates filled by a block of solid dielectric of permittivity ε. The dimensions of each plate are length b, width w. The plates are maintained at the constant potential difference ΔU. If the dielectric block is withdrawn along the b dimension until only the length x remains between the plates (see Figure), calculate the force tending to pull the block back into place.

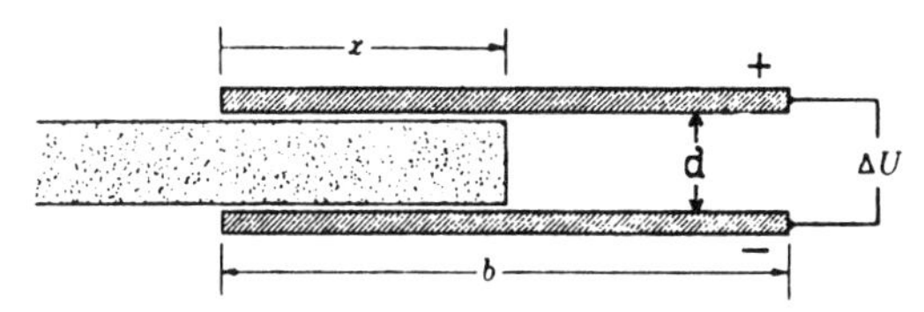

Dielectric slab partially withdrawn from between two charged plates.

Solution: The energy of the system may be calculated by

$$W = \frac{1}{2}\int_V \varepsilon E^2 dv$$

where the region of integration need include only those parts of space where $E \neq 0$. Neglecting fringing effects at the edge of the capacitor,

$$W = \frac{1}{2}\varepsilon\left(\frac{\Delta U}{d}\right)^2 dwx + \frac{1}{2}\varepsilon_0\left(\frac{\Delta U}{d}\right)^2 dw(b - x).$$

The force may be calculated from

$$F_x = \left(\frac{\partial W}{\partial x}\right)_u$$

Thus

$$F_x = \frac{\partial}{\partial x}\left[\frac{1}{2}\varepsilon\left(\frac{\Delta U}{d}\right) dwx + \frac{1}{2}\varepsilon_0\left(\frac{\Delta U}{d}\right)^2 dw(b - x)\right]$$

$$F_x = \frac{1}{2}\varepsilon\frac{(\Delta U)^2 w}{d} - \frac{1}{2}\varepsilon_0\frac{(\Delta U)^2}{d} w$$

or

$$F_x = \frac{1}{2}(\varepsilon - \varepsilon_0) w \frac{(\Delta U)^2}{d}$$

in the direction of increasing x.

• PROBLEM 6-7

The plates of a parallel-plate capacitor are separated by a dielectric whose relative permittivity varies continuously with thickness as follows

$$\varepsilon_r = \frac{1}{1 - \frac{x^2}{3d^2}}$$

Compute the capacitance.

Solution: The capacitance for a parallel plate capacitor is given by

$$C = \frac{a\varepsilon}{d}$$

But here

$$\varepsilon = \varepsilon_r \varepsilon_v$$

so that

$$C = \varepsilon_r \varepsilon_v \cdot \frac{a}{d} = \frac{1}{1 - \frac{x^2}{3d^2}} \varepsilon_v \frac{a}{d}$$

Now since this is a continuously varying dielectric, integrate over the interval of variation. Thus

$$\frac{1}{C} = \frac{d}{\varepsilon_r \varepsilon_v} = \frac{1}{\varepsilon_v a} \int_0^d \left(1 - \frac{x^2}{3d^2}\right) dx$$

$$= \frac{1}{\varepsilon_v a}\left[x - \frac{x^3}{3 \cdot 3d^2}\right]_0^d = \frac{1}{\varepsilon_v a}\left(d - \frac{d^3}{9d^2}\right)$$

$$= \frac{1}{\varepsilon_v a}\left(d - \frac{1}{9}d\right)$$

$$C = \frac{9}{8} \frac{\varepsilon_v a}{d} \text{ Farad}$$

• PROBLEM 6-8

(a) Find the capacity of a capacitor consisting of two parallel metal plates each having an area of 1500 sq. cm. and separated by a layer of air 0 5 cm. thick. Neglect the capacity between the external surface of the plates.

(b) When a sheet of ebonite 0·18 cm. thick is introduced between the plates and their distance apart increased to 0·61 cm. it is found that the capacity remains unaltered. Determine the capacity of the condenser with the ebonite in position and a spacing of 0·5 cm. between the plates.

Solution: (a)

$$C = \frac{\varepsilon_0 A}{d} = \frac{8.854 \times 10^{-12} \times 0{\cdot}15}{5 \times 10^{-3}}$$

$$= 2{\cdot}656 \times 10^{-10} \text{ farad}$$

$$= 265{\cdot}6 \text{ pF.}$$

(b) First find the capacitance of a parallel-plate capacitor, whose plates are distant d apart, and between which there is a sheet of insulating material of thickness d_1 and dielectric constant K.

The arrangement is equivalent to an air-capacitor C_1, of spacing $(d - d_1)$, in series with a capacitor C_2, of spacing d_1, whose dielectric constant is K, the two capacitors having the same area.

If C is the total capacitance,

$$\frac{1}{C} = \frac{1}{C_1} + \frac{1}{C_2} = \frac{(d - d_1)}{\varepsilon_0 A} + \frac{d_1}{K\varepsilon_0 A}$$

$$= \frac{1}{\varepsilon_0 A}\left\{d - d_1 + \frac{d_1}{K}\right\}.$$

The capacitance without the insulating sheet is

$$C_a = \frac{\varepsilon_0 A}{d},$$

so that the ratio

$$\frac{C_a}{C} = \frac{Kd - (K - 1)d_1}{Kd}.$$

In this problem first find K for the sheet of ebonite. When the sheet is introduced and the spacing of the plates increased, the charge is unaltered, and since C is unaltered so is the potential difference between the plates. The field intensity E in the air space is unaltered since the charge is constant.

Let d be the original spacing of the plates = 0·5cm.,

d_1 be the thickness of the ebonite sheet = 0·18 cm.,

d_2 be the thickness of the air space to keep the capacitance unchanged = 0·43 cm.

Then the p.d., V, $= dE = d_2E + \frac{d_1E}{K}$,

whence

$$K = \frac{d_1}{d - d_2}$$

$$= 2.57.$$

The ratio

$$\frac{C_a}{C} = \frac{2.57 \times 0.5 - 1.57 \times 0.18}{2.57 \times 0.5} = \frac{1}{1.283},$$

so that

$$C = 1.283 \times C_a = 1.283 \times 265.6 \simeq 340.8 \text{ pF}.$$

• PROBLEM 6-9

A parallel plate capacitor is constructed using two very thin and rectangular plates of aluminum, of area A, as shown in the figure below. A bettery is used to charge this capacitor to a dc voltage of v_0, using air as the dielectric media initially. When the capacitor was fully charged, the battery was removed and a dielectric slab of dielectric constant $\varepsilon_r = k'$ was inserted. Assuming $A = 500 \text{ cm}^2$, $d = 6.0$ cm, $b = 0.25$ cm, $k' = 6.0$ and $v = 80$ volts, calculate

(a) The capacitance c_0 before the slab is introduced and the free charge q.

(b) $\bar{E}$ in the gap and in the dielectric.

(c) The potential difference between the plates and the capacitance with the slab in place.

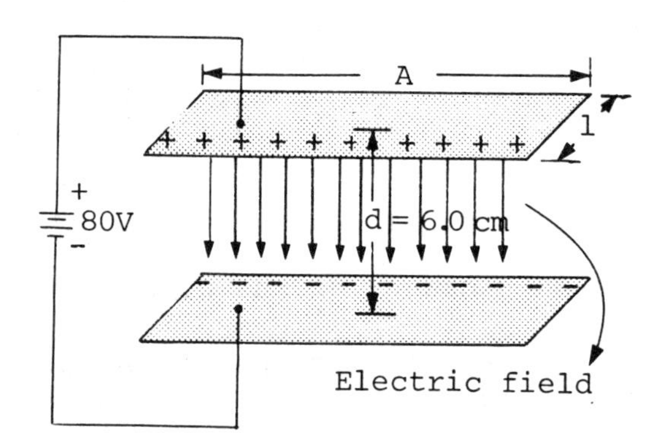

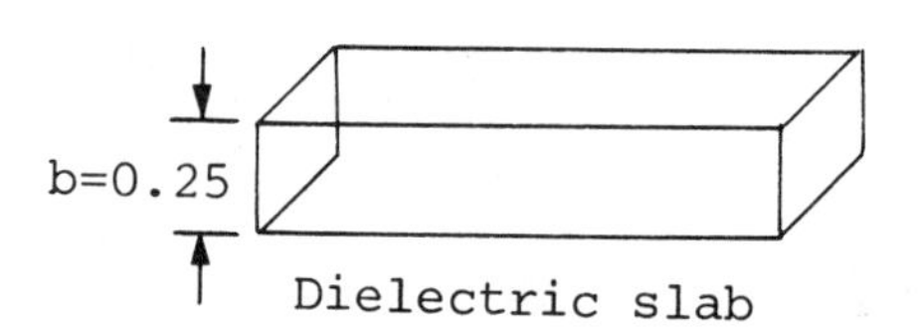

Solution: The capacitance c_0, with air as the dielectric media, is given by

$$c_0 = \frac{\varepsilon_0 A}{d} = \frac{(8.9\times 10^{-12}\,\text{coul}^2/\text{nt-m}^2)(5\times 10^{-2})\text{m}^2}{6\times 10^{-2}\text{m}}$$

$$= 7.417\times 10^{-12}\,\text{F}.$$

The free charge $q = c_0 \cdot v$ (using standard relation for $c = \frac{q}{v}$)

$$= 7.416\times 10^{-12}\times 80$$

$$= 5.933\times 10^{-10}\ \text{coulomb}$$

(b) Application of the Gauss's law to the Gaussian surface of the gap results.

$$\varepsilon_0 \cdot \oint_{gap} k \cdot \bar{E}_{gap} \cdot d\bar{s} = \varepsilon_o \cdot 1 \cdot \bar{E}_{gap} \oint_{gap} d\bar{s} = \varepsilon_o \cdot E_{gap} \cdot A = q$$

$$\therefore E_{gap} = \frac{q}{\varepsilon_0 A}$$

$$= \frac{5.933\times 10^{-10}}{(8.9\times 10^{-12})(5\times 10^{-2})}$$

$$= 1333.26\ \text{volts/m}$$

Similarly, applying the Gauss's law to the dielectric slab results

$$E_{dielectric} = \frac{q}{k\cdot\varepsilon_0\cdot A} = \frac{E_{gap}}{k}$$

$$= \frac{1332.13}{6.0} = 222.2\ \text{volts/m}.$$

(c) The potential difference between the plates is given by

$$V = -\int \bar{E}\cdot d\bar{\ell}$$

Now, in order to calculate the voltage difference between the plates, the line integral must be evaluated over a straight path between the plates because in this case lines of force of E are pointing straight downward.

So, taking the dot product of $\bar{E}$ and $d\bar{\ell}$,

$$v = -\int \bar{E} \cdot \cos 180^0 \cdot d\ell \quad \text{(since } \bar{E} \text{ and } d\bar{\ell} \text{ are } \parallel \text{ and in opposite direction, } \theta = 180^0\text{)}$$

$$= E_{gap} \cdot (d-b) + E_{dielectric} \cdot b$$

So, by substituting respective quantities, we get

$$v = 1333.26(5.75 \times 10^{-2}) + 222.2(0.25 \times 10^{-2})$$

$$= 77.22 \text{ volts}$$

$$c_{\text{with the slab}} = \frac{q}{v} = \frac{5.933 \times 10^{-10}}{77.22}$$

$$= 7.68 \text{ pf}$$

• PROBLEM 6-10

It is desired to increase the capacitance of a parallel-plate capacitor by inserting a material of relative permittivity 5·0 into the space between the plates. There is only sufficient dielectric material to fill half the space between the plates. What is the most efficient shape of dielectric to use?

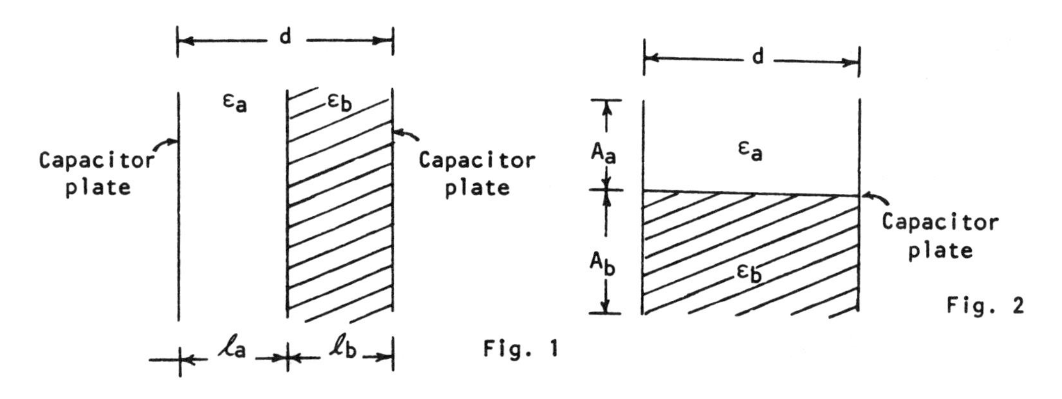

Fig. 1

Fig. 2

Solution: Since the parallel-plate capacitor has a uniform field between the plates, the most efficient dielectric shape will be with the boundary either parallel to or perpendicular to the plates. The dielectric will either be in the shape shown in Fig. 1 or that shown in Fig. 2. For the shape shown in Fig. 1, the capacitance for this configuration is given by

$$C = \frac{A}{\frac{\ell_a}{\varepsilon_o \varepsilon_a} + \frac{\ell_b}{\varepsilon_o \varepsilon_b}}$$

or

$$C = \frac{\varepsilon_o \varepsilon_a \varepsilon_b A}{\varepsilon_b \ell_a + \varepsilon_a \ell_b}$$

But $\varepsilon_a = 1$, $\varepsilon_b = 5{\cdot}0$, $\ell_a = \ell_b = \frac{1}{2}d$, therefore

$$C = \frac{5\varepsilon_0 A}{3d} = 1{\cdot}67 \frac{\varepsilon_0 A}{d}.$$

For the shape shown in Fig. 2, which is the other alternative, the capacitance is given by

$$C = \frac{\varepsilon_o \varepsilon_a A_a}{d} + \frac{\varepsilon_o \varepsilon_b A_b}{d}$$

or

$$C = \frac{\varepsilon_o \varepsilon_a A_a + \varepsilon_o \varepsilon_b A_b}{d}$$

But $\varepsilon_a = 1$, $\varepsilon_b = 5.0$, $A_a = A_b = \frac{1}{2}A$, therefore

$$C = \frac{(\frac{1}{2}\varepsilon_0 + \frac{5}{2}\varepsilon_0)A}{d}$$

or

$$C = \frac{3\varepsilon_0 A}{d}$$

Therefore the system shown in Fig. 2 is the best way of using the dielectric material which increases the capacitance of the capacitor by three compared with there being no dielectric material.

• PROBLEM 6-11

The parallel plates in the figure have an area of 2000 cm^2 or $2 \times 10^{-1} m^2$, and are 1 cm or 10^{-2} m apart. The original potential difference between them, V_0, is 3000 V, and it decreases to 1000 V when a sheet of dielectric is inserted between the plates. Compute (a) the original capacitance C_0, (b) the charge Q on each plate, (c) the capacitance C after insertion of the dielectric, (d) the dielectric constant K of the dielectric, (e) the permittivity ε of the dielectric, (f) the induced charge Q_i on each face of the dielectric, (g) the original electric field E_0 between the plates, and (h) the electric field E after insertion of the dielectric.

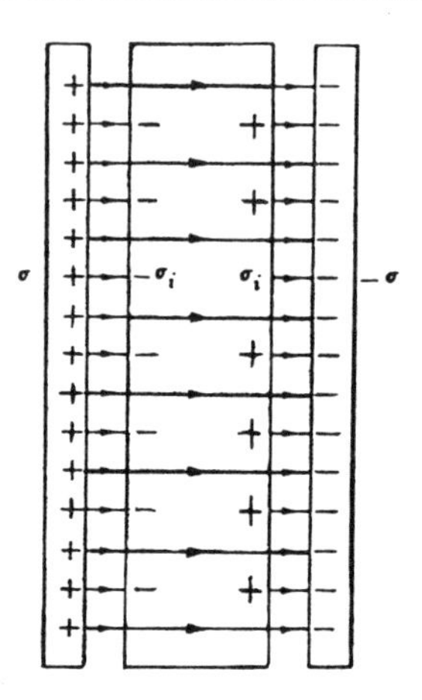

Induced charges on the faces of a dielectric in an external field.

Solution: For a parallel plate capacitor, the capacitance, C, is given by

a) $C_0 = \varepsilon_0 \frac{A}{d}$

$$= (8.85 \times 10^{-12} C^2 N^{-1} m^{-2}) \frac{2 \times 10^{-1} m^2}{10^{-2} m}$$

$$= 17.7 \times 10^{-11} F = 177 pF.$$

b) The charge on each plate can be calculated by

$$Q = C_0 V_0$$

$$= (17.7 \times 10^{-11} F)\ (3 \times 10^3 V)$$

$$= 53.1 \times 10^{-8} C.$$

c) Then calculate the capacitance after the dielectric was inserted:

$$C = \frac{Q}{V} = \frac{53.1 \times 10^{-8} c}{10^3 V}$$

$$= 53.1 \times 10^{-11} F.$$

d) The dielectric constant, K, can be found from

the ratio of the capacitance after and before inserting the dielectric,

$$K = \frac{C}{C_0}$$

$$= \frac{53.1 \times 10^{-11}F}{17.7 \times 10^{-11}F} = 3.$$

The dielectric constant could also be found from the ratio of the voltages:

$$K = \frac{V_0}{V} = \frac{3000V}{1000V} = 3.$$

e) The permittivity is

$$\varepsilon = K\varepsilon_0 = (3)(8.85 \times 10^{-12}C^2N^{-1}m^{-2})$$

$$= 26.6 \times 10^{-12}C^2N^{-1}m^{-2}.$$

f) Before inserting the dielectric, $E_0 = \sigma/\varepsilon_0$; after inserting the dielectric, $E = (\sigma - \sigma_i)/\varepsilon_0$ therefore $K = E_0/E = \sigma/(\sigma - \sigma_i)$ and since

$$Q_i = A\sigma_i, \qquad Q = A\sigma$$

and $$\sigma - \sigma_i = \frac{\sigma}{K}, \qquad \sigma_i = \sigma\left(1 - \frac{1}{K}\right),$$

then, $$Q_i = \left(1 - \frac{1}{K}\right)Q$$

$$= (53.1 \times 10^{-8}C)\left(1 - \frac{1}{3}\right)$$

$$= 35.4 \times 10^{-8}C.$$

g) The electric field before inserting the dielectric is

$$E_0 = \frac{V_0}{d} = \frac{3000\ V}{10^{-2}m} = 3 \times 10^5 V\ m^{-1}.$$

h) The electric field after inserting the dielectric is

$$E = \frac{V}{d} = \frac{1000\ V}{10^{-2}m} = 1 \times 10^5 V\ m^{-1};$$

or

$$E = \frac{\sigma}{\varepsilon} = \frac{Q}{A\varepsilon}$$

$$= \frac{53.1 \times 10^{-8} C}{(2 \times 10^{-1} m^2)(26.6 \times 10^{-12} C^2 N^{-1} m^{-2})}$$

$$= 1 \times 10^5 V\ m^{-1}$$

or

$$E = \frac{\sigma - \sigma_i}{\varepsilon_0} = \frac{Q - Q_i}{A\varepsilon_0}$$

$$= \frac{(53.1 - 35.4) \times 10^{-8} C}{(2 \times 10^{-1} m^2)(8.85 \times 10^{-12} C^2 N^{-1} m^{-2})}$$

$$= 1 \times 10^5 V\ m^{-1};$$

or, from

$$K = \frac{E_0}{E} ,$$

$$E = \frac{E_0}{K} = \frac{3 \times 10^5 V\ m^{-1}}{3}$$

$$= 1 \times 10^5 V\ m^{-1}.$$

• PROBLEM 6-12

A barium titante capacitor is made by cutting a hollow $BaTiO_3$ (high dielectric, $\varepsilon_R = 5000$) cylinder in half along a diameter and depositing silver electrodes on the two rectangular faces of one half. Find the capacitance between these faces. It is given that the inner and outer radii of the cylinder are 1 and 2 centimeters respectively with a length of 3 centimeters.

Solution: For a dielectric material with high ε_R, an assumption can be made that the electric field is practically confined inside the material. The problem is analogous to current flow. One can find the current flow, the potential difference and the resistivity as if it were a true current problem. Then by using analogue, convert R into C.

Let

$$\bar{J}_\phi = (k/r)\ \bar{a}_\phi = \sigma E_\phi \bar{a}_\phi$$

or

$$E_\phi = (k/\sigma r)$$

Then

$$V_0 = \int_0^{\pi} \frac{k}{\sigma r} r d\phi = k\pi/\sigma$$

$$\bar{I} = \int_s \bar{J}_\phi \cdot d\bar{S} = \int_0^{.03} \int_{.01}^{.02} \frac{K}{r}\, dr\, dz = 0.3\ k \ln 2$$

and

$$R = V_0/I = \frac{k\pi/\sigma}{0.03\ K \ln 2} = \frac{\pi/\sigma}{0.03 \ln 2}$$

To convert R into C, consider the analogue expressions

$$R = \frac{V_0}{I} = \frac{-\int \bar{E}_\phi \cdot d\bar{l}}{\sigma \int_s \bar{E}_\phi \cdot d\bar{S}}$$

$$C = \frac{Q}{V_0} = \frac{\varepsilon \int_s \bar{E}_\phi \cdot d\bar{S}}{-\int \bar{E}_\phi \cdot d\bar{l}}$$

Multiply these expressions together,

$$R\ C = \varepsilon/\sigma, \qquad \text{or} \qquad C = \varepsilon/R\sigma$$

Therefore

$$C = \frac{10^{-9}}{36\ \pi} \times \frac{5000}{\pi} \times 0.03 \ln 2 = 393 \times 10^{-12} F$$

• PROBLEM 6-13

Consider two conductors of some shape and fill the conductor with a dielectric of permittivity ε. Put equal and opposite charges on them as shown in Fig. 1. Then instead of the dielectric, fill the region between the plates with a conductor of conductivity σ as shown in Fig. 2. Find the relation between the capacitance in case A to the resistance in case B if the potential between the plates is kept constant.

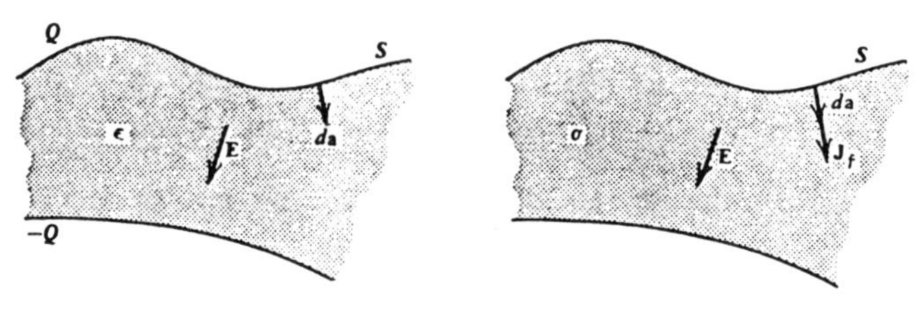

Fig. 1 Fig. 2

Solution: The potential difference can be found by evaluating the integral

$$\Delta\phi = \int_{+}^{-} \bar{E}.d\bar{s} \tag{1}$$

over any convenient path between the plates. Write the free charge Q on the positive plate as an integral over its surface S.

$$Q = \int_S \sigma_f \, da = \int_S \varepsilon \bar{E}.d\bar{a} = \varepsilon \int_S \bar{E}.d\bar{a} \tag{2}$$

The capacitance can be written as

$$C = \frac{Q}{V}$$

$$C = \frac{\varepsilon \int_S \bar{E}.d\bar{a}}{\int_{+}^{-} \bar{E}.d\bar{s}} \tag{3}$$

For the second case, ϕ satisfies Laplace's equation under these circumstances, and since the boundary conditions are exactly the same in the two cases, then from the uniqueness theorem $\phi(r)$ will be identical for each. In other words, the potential difference will again be given by (1) with exactly the same values of E at each point on the path of integration. The total current I passing between the plates can be expressed as a surface integral over the same upper plate; the result is that

$$I = \int_S \bar{J}_f.d\bar{a} = \int_S \sigma \bar{E}.d\bar{a} = \sigma \int_S \bar{E}.d\bar{a} \tag{4}$$

Put (1) and (4) into $I = \frac{\phi}{R}$. Then the resistance of this system is given by

$$\frac{1}{R} = \frac{\sigma \int_S \bar{E}.d\bar{a}}{\int_{+}^{-} \bar{E}.d\bar{s}} \tag{5}$$

Comparing (3) and (5), $1/R\sigma = C/\varepsilon$ or

$$RC = \frac{\varepsilon}{\sigma}$$

showing that these two properties of the system are not independent but are related in this simple way.

COAXIAL AND CONCENTRIC CAPACITORS

• PROBLEM 6-14

Find the capacitance of a coaxial capacitor of length ℓ shown in Figure 1. Assume zero $\bar{E}$ field fringing at the ends and uniform ρ_ℓ along the conductors.

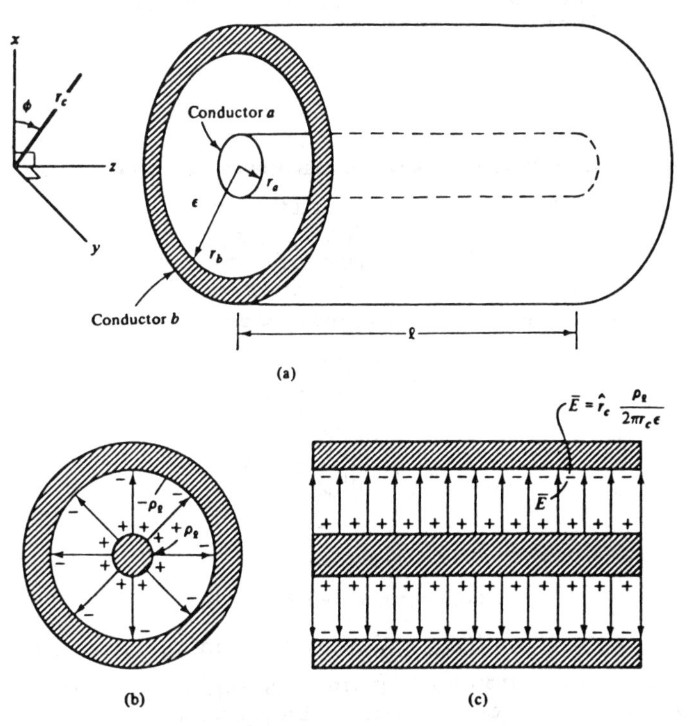

(a) Graphical display of a two-conductor cylindrical capacitor. (b) End view of $\bar{E}$ flux lines. (c) $\bar{E}$ flux lines along the length of the capacitor.

Fig. 1

Solution: Through the application of Gauss's law for $r_a \leq r_c \leq r_b$, the $\bar{E}$ field is,

$$\bar{E} = \hat{r}_c \frac{\rho_\ell}{2\pi r_c \varepsilon} \tag{1}$$

where the form is found to be the same as about an infinite

line charge. The potential difference V_{ab} becomes

$$V_{ab} = -\int_b^a \bar{E} \cdot \overline{d\ell} = -\int_b^a \left(\hat{r}_c \frac{\rho_\ell}{2\pi r_c \varepsilon}\right) \cdot (\hat{r}_c dr_c)$$

$$= \frac{\rho_\ell}{2\pi\varepsilon} \ln\left(\frac{r_b}{r_a}\right) \tag{2}$$

Through the use of (2), the capacitance of length ℓ is,

$$C = \left|\frac{Q}{V_{ab}}\right| = \frac{|\rho_\ell \ell|}{\left|\frac{\rho_\ell}{2\pi\varepsilon}\ln\left(\frac{r_b}{r_a}\right)\right|} = \frac{2\pi\varepsilon\ell}{\ln\left(\frac{r_b}{r_a}\right)} \quad \text{(F)}$$

The capacitance is found to be independent of ρ_ℓ or V_{ab} since ε was assumed to be constant (independent of $\bar{E}$ as required in a linear dielectric).

• PROBLEM 6-15

Determine the capacitance per unit length between two infinitely long concentric conducting cylinders as shown in the figure. The outside radius of the inner conductor is a and the inside radius of the outer conducter is b.

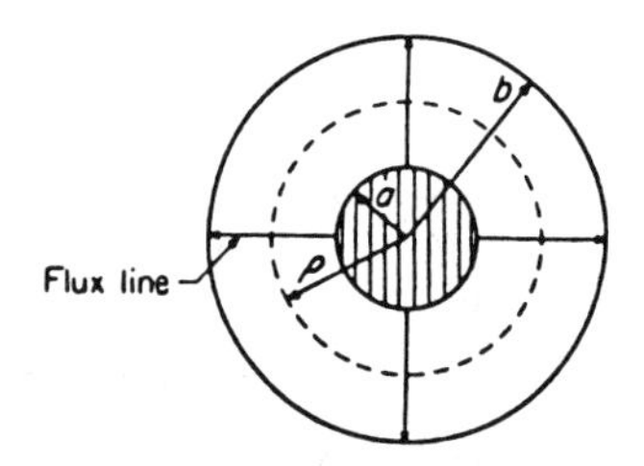

Concentric conductors.

Solution: Assume a charge distribution ρ_L coulombs per meter on the inner conductor and an equal and opposite charge on the outer conductor. Because of symmetry the lines of electric flux will be radial and the displacement through any cylindrical shell will be ρ_L coulombs per unit length. The magnitude of the displacement density will be

$$D = \frac{\rho_L}{2\pi\rho}$$

and the magnitude of the electric field strength will be

$$E = \frac{\rho_L}{2\pi\rho\varepsilon}$$

The voltage between the conductors is

$$V = \int_a^b E\ d\rho = \int_a^b \frac{\rho_L}{2\pi\rho\varepsilon} d\rho = \frac{\rho_L}{2\pi\varepsilon} \ln \rho \Big]_a^b = \frac{\rho_L}{2\pi\varepsilon} \ln\frac{b}{a}$$

The capacitance per meter will be

$$C = \frac{\rho_L}{V} = \frac{2\pi\varepsilon}{\ln b/a} \qquad \text{F/m}$$

For the air dielectric for which $\varepsilon = 1/(36\pi \times 10^9)$

$$C = \frac{10^{-9}}{18 \ln b/a} \qquad \text{F/m}$$

• PROBLEM 6-16

Find the potential difference between the coaxial cylinders of the given figure. Also find the capacitance per unit length for the coaxial cylinder.

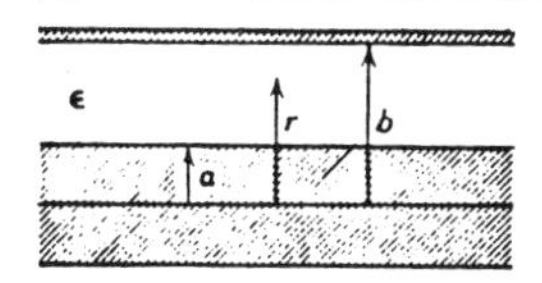

Solution: The potential difference in this case is

$$\phi a - \phi b = -\int_b^a \frac{q\ell dr}{2\pi\varepsilon r} = \frac{q\ell}{2\pi\varepsilon} \ell n \left(\frac{b}{a}\right) \text{ Volts}$$

Now the capacitance per unit length is given by the charge on one conductor divided by the potential difference. Therefore,

$$C = \frac{q\ell}{\phi a - \phi b} = \frac{2\pi\varepsilon}{\ell n (b/a)} \quad \text{farads/meter.}$$

• PROBLEM 6-17

Compute the capacitance per unit length of a coaxial cable of inner and outer radius a and b, respectively, and in which the permittivity is a linear function of the radial coordinate r (see the figure).

$$\varepsilon = \varepsilon_1 + (\varepsilon_2 - \varepsilon_1)\frac{r - a}{b - a}$$

where ε_1 and ε_2 are constant. Let V be the potential difference between the conductors and let q be the charge per unit length on the inner conductor. Also find the polarization charge density within the dielectric.

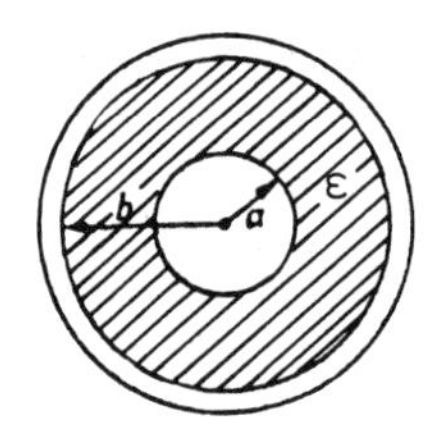

Solution: Gauss' law applied to a cylinder concentric with the inner conductor yields

$$\bar{D} = \frac{q}{2\pi r} i_r .$$

Thus

$$\bar{E} = \frac{q}{2\pi \varepsilon r} i_r .$$

Determine the capacitance per unit length from the equation

$$V = \int_a^b \frac{q}{2\pi \varepsilon r} \, dr = \frac{q(b - a)}{2\pi(\varepsilon_1 b - \varepsilon_2 a)} \ln \frac{\varepsilon_1 b}{\varepsilon_2 a},$$

which yields

$$C = \frac{2\pi(\varepsilon_1 b - \varepsilon_2 a)}{(b - a) \ln (\varepsilon_1 b/\varepsilon_2 a)}$$

To find the polarization charge density within the dielectric, proceed as follows. First find the polarization from the relation

$$\bar{P} = (\varepsilon - \varepsilon_0)\bar{E} = \left(1 - \frac{\varepsilon_0}{\varepsilon}\right) \frac{q}{2\pi r} \bar{i}_r .$$

Then find the polarization charge density by using ρ_p = -div P and in cylindrical coordinates

$$\text{div } P = \frac{1}{r}\left[\frac{\partial}{\partial r}(rP_r) + \frac{\partial P_\phi}{\partial \phi} + \frac{\partial}{\partial z}(rP_z)\right] ,$$

$$\rho_p = -\frac{1}{r}\frac{\partial}{\partial r}(rP_r) = -\frac{q\varepsilon_0(\varepsilon_2 - \varepsilon_1)}{2\pi\varepsilon^2 r(b - a)} .$$

• PROBLEM 6-18

Find the capacitance of a conducting sphere surrounded by an isolated thick spherical conducting shell (Fig 1). The thick outer shell is isolated and considered to be initially uncharged. A charge +Q is to be placed on the inner sphere.

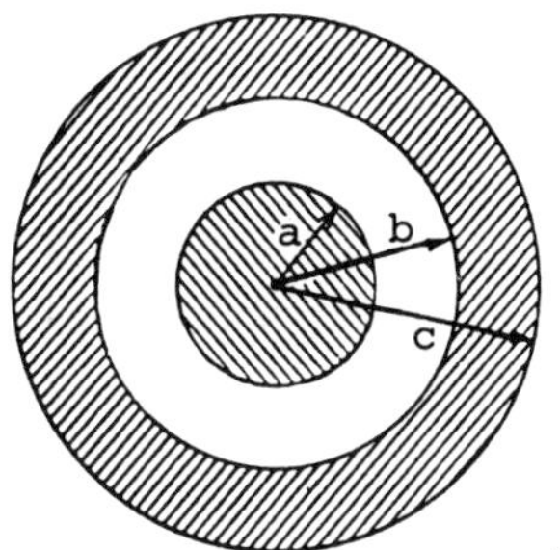

Thick conducting shell surrounding a spherical conductor.

Fig. 1

Solution: By Gauss' flux theorem, it can be deduced that a negative charge -Q is induced on the inner surface of the shell. This leaves a charge +Q on the outer surface of the shell. Thus outside the metal shell the field is identical with the field due to a point charge +Q at the center. Therefore calculate the potential V_c at the surface by integrating the field from ∞ in the usual manner. Within the metal shell (between c and b) the field must be zero; therefore V_b is the same as V_c (since $E_r = 0 = -\partial V/\partial r$). The field between b and a is also that of a point charge +Q; so the increase in potential at a over the value at b is obtained by integration from b to a. Then the potential at a, V_a, is given by

$$V_a = -\int_{\infty}^{c} \bar{E} \cdot dr - \int_{c}^{b} \bar{E} \cdot d\bar{r} - \int_{b}^{a} \bar{E} \cdot d\bar{r}$$

$$= -\frac{Q}{4\pi\varepsilon_0}\int_{\infty}^{c}\frac{dr}{r^2} - 0 - \frac{Q}{4\pi\varepsilon_0}\int_{b}^{a}\frac{dr}{r^2}$$

$$= \frac{Q}{4\pi\varepsilon_0 c} + \frac{Q}{4\pi\varepsilon_0}\left(\frac{1}{a} - \frac{1}{b}\right)$$

The capacitance can be obtained in the usual fashion.

$$C = \frac{Q}{V} = \frac{1}{\frac{1}{4\pi\varepsilon_0 C} + \frac{1}{4\pi\varepsilon_0}\left(\frac{1}{a} - \frac{1}{b}\right)}$$

• PROBLEM 6-19

Find the capacitance between concentric spherical conductors. Consider the outer sphere of radius b grounded as shown in Fig. 1. Place a charge Q on the inner sphere (through a small hole in the outer sphere).

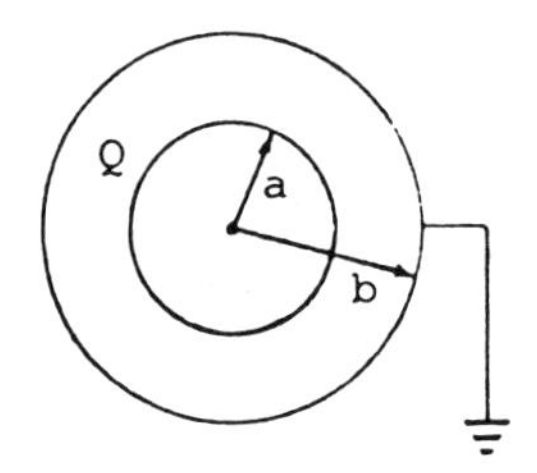

Two concentric metal spheres forming a capacitor.

Fig. 1

Solution: An equal charge $-Q$ will be induced on the inner surface of the outer sphere (this follows from Gauss' flux theorem). A field exists only between the two spheres. The potential difference V_{ab} is given by integrating the field:

$$V_{ab} = -\int_b^a E \cdot dr = \frac{-Q}{4\pi\varepsilon_0}\int_b^a \frac{dr}{r^2} = \frac{Q}{4\pi\varepsilon_0}\, \frac{1}{a} - \frac{1}{b}$$

Then

$$C = \frac{Q}{V} = \frac{4\pi\varepsilon_0}{1/a - 1/b} = 4\pi\varepsilon_0 \frac{ab}{b - a} \quad \text{farads.}$$

Since the potential of a conducting sphere of radius r is known to be $V = \frac{1}{4\pi\varepsilon_0}\frac{Q}{r}$, this capacitance can be calculated alternatively by taking the difference between the potentials of the two spheres directly, instead of integrating the field over the distance between spheres. Then,

$$C = \frac{Q}{V_{ab}} = \frac{Q}{(1/4\pi\varepsilon_0)(Q/a - Q/b)} = 4\pi\varepsilon_0 \frac{ab}{b - a} \quad \text{farads}$$

The same result is obtained.

• PROBLEM 6-20

Find the capacitance of a spherical capacitor of inner radius $\underline{a}$ and outer radius $\underline{b}$ charged with Q Coulombs, and containing a nonhomogeneous dielectric which is assumed to vary according to

$$\varepsilon = \varepsilon_1 + \varepsilon_2 \frac{1}{r^2}$$

where ε_1 and ε_2 are constants and $\varepsilon_1 > \varepsilon_2$.

Solution: By an application of Gauss' law, the electric flux density $\bar{D}$ is given by

$$\bar{D} = \frac{Q}{4\pi r^2}\bar{a}_r$$

and the electric field intensity in the dielectric is

$$\bar{E} = \frac{Q}{4\pi\varepsilon r^2}\bar{a}_r$$

The potential difference is

$$V = \int_a^b \frac{Q}{4\pi\varepsilon r^2}dr = \frac{Q}{4\pi}\int_a^b \frac{dr}{\varepsilon_1 r^2 + \varepsilon_2}$$

$$V = \frac{Q}{4\pi\varepsilon_2}\int_a^b \frac{dr}{\frac{\varepsilon_1 r^2}{\varepsilon_2} + 1}$$

Changing the variables, let

$$x = \sqrt{\frac{\varepsilon_1}{\varepsilon_2}}\,xr \quad dx = \sqrt{\frac{\varepsilon_1}{\varepsilon_2}}dr,$$

then

$$V = \frac{Q}{4\pi\varepsilon_2}\sqrt{\frac{\varepsilon_2}{\varepsilon_1}}\int_{a'}^{b'} \frac{dx}{1 + x^2}$$

when

$$r = a,\ x = \sqrt{\frac{\varepsilon_1}{\varepsilon_2}}a$$

and when

$$r = b,\ x = \sqrt{\frac{\varepsilon_1}{\varepsilon_2}}b,$$

therefore,

$$V = \frac{Q}{4\pi}\frac{1}{\sqrt{\varepsilon_1\varepsilon_2}}\int_{\sqrt{\frac{\varepsilon_1}{\varepsilon_2}}a}^{\sqrt{\frac{\varepsilon_1}{\varepsilon_2}}b} \frac{dx}{1 + x^2}$$

Now

$$\int_a^b \frac{dx}{1 + x^2} = \tan^{-1} x\Big]_a^b ,$$

thus

$$V = \frac{Q}{4\pi}\frac{1}{\sqrt{\varepsilon_1\varepsilon_2}}\left[\tan^{-1}\left(b\sqrt{\frac{\varepsilon_1}{\varepsilon_2}}\right) - \tan^{-1}\left(a\sqrt{\frac{\varepsilon_1}{\varepsilon_2}}\right)\right]$$

The capacitance of the system is then given by the ratio Q/V.

• PROBLEM 6-21

A spherical shell of paraffin is contained between two spherical conductors. The inner sphere has a radius of 0.60 meter, and the inner radius of the outer sphere is 0.70 meter.

(a) What is the capacitance of the space between the conductors?

(b) What is the largest electric potential difference which may be applied to the system if dielectric breakdown is to be avoided?

(c) If the electric potential difference of (b) could be established between the two spheres, how much energy would be stored in the electric field?

(d) If the heat loss from a particular building under certain conditions is 40,000 Btu per hour, for how long would the energy in the electric field of (c) maintain the temperature in this building if this energy were released?

Solution: (a) If the radius of the inner sphere is R_1 and the inner radius of the outer sphere is R_2, then the potential difference between the conductors is found from

$$\phi_{21} = -\int_{R_2}^{R_1} \bar{E} \cdot d\bar{\ell}$$

where for a spherical configuration $E = \frac{Q}{4\pi\varepsilon_0 K} \cdot \frac{1}{r^2}$ and $d\ell = dr$. Therefore

$$\phi_{21} = \frac{Q}{4\pi\varepsilon_0 K} \int_{R_2}^{R_1} \frac{dr}{r^2} = \frac{Q}{4\pi\varepsilon_0 K} \left(\frac{1}{R_1} - \frac{1}{R_2}\right) \qquad (1)$$

The capacitance is defined as

$$C = \frac{Q}{\phi}$$

Thus

$$C = \frac{4\pi\varepsilon_0 K}{\left[\frac{1}{R_1} - \frac{1}{R_2}\right]} = \frac{1.112 \times 10^{-10} \times 2.2}{\left[\frac{1}{0.60} - \frac{1}{0.70}\right]}$$

$$= 0.00103 \text{ microfarad.}$$

(b) The voltage gradient which will be critical for breakdown in paraffin is 29×10^6 volts per meter from Table 1. If this voltage gradient is to be permitted at the inner edge of the dielectric where R_1 is 0.6 meter, then

$$\frac{Q}{4\pi\varepsilon_0 K} = ER_1^2 = 29 \times 10^6 \times (0.6)^2 = 10.44 \times 10^6$$

and from eq. (1)

$$\Phi_{21} = \frac{Q}{4\pi\varepsilon_0 K}\left(\frac{1}{R_1} - \frac{1}{R_2}\right) = 10.44 \times 10^6 \left(\frac{1}{0.60} - \frac{1}{0.70}\right)$$

$$= 44 \times 10^6 \text{ volts.}$$

Table 1 Approximate Dielectric Strengths

Substance	Kv/M	Substance	Kv/M
Air	3,000	Mineral oil	20,000
Bakelite	21,000	Paraffin	29,000
Ebonite	60,000	Polystyrene	30,000
Glass		Porcelain	11,000
Pyrex	90,000	Pyranol	20,000
Other	30,000	Quartz (fused)	30,000
Gutta-percha	14,000	Rubber	25,000
Marble	6,000	Slate	3,000
Mica	60,000	Steatite	8,000

(c) The magnitude of the charge on one sphere is

$$Q = 10.44 \times 10^6 \times 4\pi\varepsilon_0 K$$

$$= 10.44 \times 10^6 \times 1.112 \times 10^{-10} \times 2.2$$

$$= 0.00255 \text{ coulomb}$$

and the energy is

$$W = \tfrac{1}{2}Q\Phi_{21} = 0.5 \times 0.00255 \times 44 \times 10^6$$

$$= 56 \times 10^3 \text{ joules.}$$

(d) The 56×10^3 joules are equivalent to approximately 53 Btu, and hence if 40,000 Btu are required for maintaining the building temperature for one hour, the energy in the electric field would maintain the temperature of the building for approximately 4.8 seconds.

MULTIPLE DIELECTRIC CAPACITORS, SERIES AND PARALLEL COMBINATIONS

• PROBLEM 6-22

A thin parallel-plate capacitor contains two dielectrics of dielectric constant ε_1 and ε_2, as shown in Fig. 1. Neglecting edge effects find the capacitance.

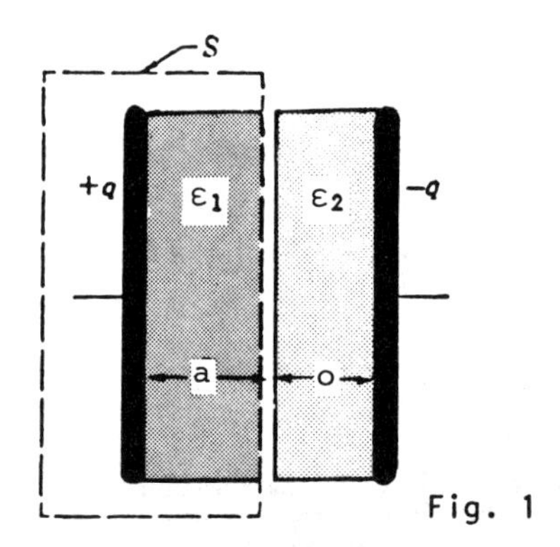

Fig. 1

Solution: Let the charge of the capacitor be q. By symmetry, the field in the capacitor is homogeneous (except near the edges). Constructing a Gaussian surface S in the shape of a box enclosing the positive plate, and observing that if the edge effects are neglected the only contribution to the integral $\oint D \cdot dS$ comes from the portion of the Gaussian surface lying directly between the plates, from Gauss's law

$$\oint \bar{D} \cdot d\bar{S} = \int D \, dS = DA = q,$$

where A is the area of the enclosed plate. Hence, between the plates,

$$D = \frac{q}{A} .$$

The electric field is then, by the displacement law,

$$E_1 = \frac{q}{\varepsilon_0 \varepsilon_1 A} \quad \text{and} \quad E_2 = \frac{q}{\varepsilon_0 \varepsilon_2 A}$$

in dielectrics 1 and 2, respectively. The voltage between the plates is

$$V = \int \bar{E} \cdot d\bar{l} = \int_{\text{Dielectric 1}} \bar{E}_1 \cdot d\bar{l}$$

$$+ \int_{\text{Dielectric 2}} \bar{E}_2 \cdot d\bar{l}$$

$$= \frac{q}{\varepsilon_0 \varepsilon_1 A} \int_0^a dl + \frac{q}{\varepsilon_0 \varepsilon_2 A} \int_a^{a+b} dl,$$

or

$$V = \frac{q}{\varepsilon_0 A} \left(\frac{a}{\varepsilon_1} + \frac{b}{\varepsilon_2} \right) .$$

The capacitance is therefore

$$C = \frac{q}{V} = \frac{\varepsilon_0 A}{(a/\varepsilon_1 + b/\varepsilon_2)} .$$

As a check, note that this formula reduces to the expression for the capacitance of an empty capacitor if $\varepsilon_1 = \varepsilon_2 = 1$.

• PROBLEM 6-23

A thin parallel-plate capacitor of plate separation d contains two dielectrics of dielectric constant ε_1 and ε_2, as shown in Fig. 1. Neglecting edge effects, find the capacitance.

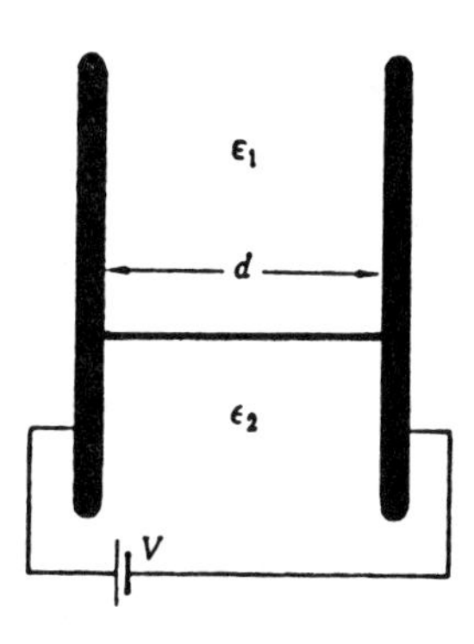

Fig. 1

Solution: Let the voltage between the plates be V. By symmetry, the electric field between the plates is then $E = V/d$. The displacement is

$$D_1 = \varepsilon_0 \varepsilon_1 \frac{V}{d} \quad \text{and} \quad D_2 = \varepsilon_0 \varepsilon_2 \frac{V}{d}$$

in dielectrics 1 and 2, respectively. The charge on that part of the positive plate which is in contact with dielectric 1 (area A_1) is then

$$q_1 = D_1 A_1 = \varepsilon_0 \varepsilon_1 \frac{V}{d} A_1 .$$

The charge on that part of the positive plate which is in contact with dielectric 2 (area A_2) is

$$q_2 = D_2 A_2 = \varepsilon_0 \varepsilon_2 \frac{V}{d} A_2$$

The total charge is $q = q_1 + q_2$, or

$$q = \varepsilon_0 \frac{V}{d} (\varepsilon_1 A_1 + \varepsilon_2 A_2) .$$

The capacitance is therefore

$$C = \frac{q}{V} = \frac{\varepsilon_0 (\varepsilon_1 A_1 + \varepsilon_2 A_2)}{d} \text{ F} .$$

Again, if $\varepsilon_1 = \varepsilon_2 = 1$, the capacitance reduces to that of the empty capacitor.

• PROBLEM 6-24

A capacitor has a sandwich of five dielectric layers between metal plates. Each layer is 1 mm thick and has area $0.5m^2$. The plates carry charge $\pm 4\mu C$, as shown.

a) Find E and D in each layer.

b) Find the voltage between the plates.

c) Find the capacitance C.

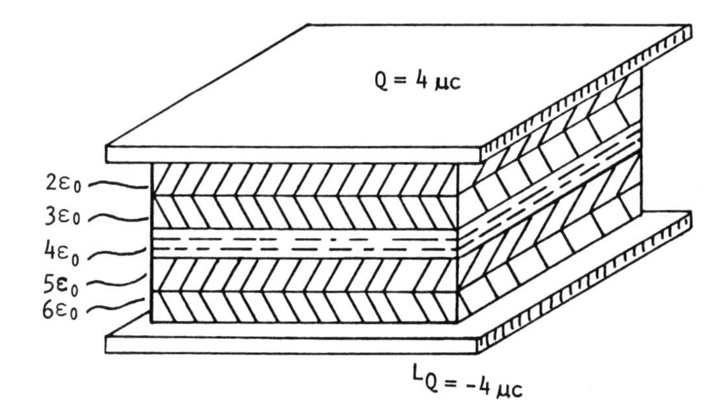

Solution: a) The E and D field will be normal to all the plates and since from boundary conditions,

$$\bar{D}_n\Big|_1 = \bar{D}_n\Big|_2 ,$$

the D fields will be the same on all the plates.

$$D = \frac{Q}{A} = \frac{4 \times 10^{-6}}{0.5} = 8 \times 10^{-6} \text{ coul./m}^2$$

$$E = \frac{D}{\varepsilon}$$

In layer 1 (from the top):

$$E_1 = \frac{D}{2\varepsilon_0} = \frac{4 \times 10^{-6}}{\varepsilon_0}$$

In layer 2:

$$E_2 = \frac{D}{3\varepsilon_0} = \frac{8}{3\varepsilon_0} \times 10^{-6}$$

In layer 3:

$$E_3 = \frac{D}{4\varepsilon_0} = \frac{2}{\varepsilon_0} \times 10^{-6}$$

In layer 4:

$$E_4 = \frac{D}{5\varepsilon_0} = \frac{8}{5\varepsilon_0} \times 10^{-6}$$

In layer 5:

$$E_5 = \frac{D}{6\varepsilon_0} = \frac{4}{3\varepsilon_0} \times 10^{-6}$$

b) The voltage V is the sum of the voltages between each plate:

$$V = E_1 d_1 + E_2 d_2 + E_3 d_3 + E_4 d_4 + E_5 d_5$$

where d_1, d_2, d_3, d_4, and d_5 are the thicknesses of each plate; in this case d_1 through d_5 are given as 0.001 meter.

Thus,

$$V = 0.001\ (E_1 + E_2 + E_3 + E_4 + E_5).$$

On substituting the values of E_1, E_2, E_3, E_4, and E_5 into this equation and using

$$\varepsilon_0 \approx \frac{36\pi}{10^{-9}} ,$$

the voltage between the plates is

$$V = 1310 \text{ volts.}$$

c) The capacitance C is found from

$$C = \frac{Q}{V} = \frac{4 \times 10^{-6}}{1310}$$

$$C = 3.05 \times 10^{-9} F$$

An alternate method for determining the capacitance is done by noting that the layers represent five capacitors in series, of capacitance, C_1, C_2, C_3, C_4, and C_5 where for a parallel plate capacitor

$$C = \frac{\varepsilon A}{d} .$$

Therefore,

$$C_1 = \frac{2\varepsilon_0 A}{d}$$

$$C_2 = \frac{3\varepsilon_0 A}{d}$$

$$C_3 = \frac{4\varepsilon_0 A}{d}$$

$$C_4 = \frac{5\varepsilon_0 A}{d}$$

$$C_5 = \frac{6\varepsilon_0 A}{d}$$

The capacitors in series then add up according to the rule:

$$C = \frac{1}{\frac{1}{C_1} + \frac{1}{C_2} + \frac{1}{C_3} + \frac{1}{C_4} + \frac{1}{C_5}}$$

On substituting the values of C_1 through C_5, the same result as the previous result for C will be obtained.

• PROBLEM 6-25

Two capacitor plates of area 240π cm^2 are separated 10 mm. The space between them is filled with several sheets of different dielectric, each of the same thickness, with the dielectric interfaces parallel to the plates. Find the capacitance if

a) there are two dielectrics, $\varepsilon_{R_1} = 3.5$ and $\varepsilon_{R_2} = 8.5$;

b) there are 10 dielectrics,

ε_R: 1.5, 2.5, ... 10.5;

c) there is one nonhomogeneous dielectric, with ε_R varying linearly from 1 to 11.

Solution: (a) If there are two dielectrics with

$$\varepsilon_{R_1} = 3.5 \qquad \text{and} \qquad \varepsilon_{R_2} = 8.5$$

of same thickness, i.e., each with a thickness of 5 mm., they form a capacitor consisting of two capacitors in series,

with

$$C_1 = \frac{\varepsilon_0 \varepsilon_{R_1} A}{d} = \frac{10^{-9}}{36\pi} \frac{240\pi \times 10^{-4}}{5 \cdot 10^{-3}} 3.5 ,$$

$$C_2 = \frac{10^{-9} \times 240\pi \times 10^{-4}(8.5)}{36\pi \quad 5 \cdot 10^{-3}}$$

$$\frac{1}{C} = \frac{1}{C_1} + \frac{1}{C_2} = \frac{3 \times 10^{10}}{14} + \frac{3 \times 10^{10}}{34}$$

$$C = 330 \times 10^{-12} F$$

b) If there are 10 dielectrics, 1 mm thick each, there are ten capacitors in series with

$$C_1 = \frac{6}{3} 10^{-10}$$

$$C_2 = \frac{10}{3} 10^{-10}$$

$$C_3 = \frac{14}{3} 10^{-10}$$

$$C_4 = \frac{18}{3} 10^{-10}$$

$$C_5 = \frac{22}{3} 10^{-10}$$

$$C_6 = \frac{26}{3} 10^{-10}$$

$$C_7 = \frac{30}{3} 10^{-10}$$

$$C_8 = \frac{34}{3} 10^{-10}$$

$$C_9 = \frac{38}{3} 10^{-10}$$

$$C_{10} = \frac{42}{3} 10^{-10}$$

and

$$C = \frac{1}{\frac{1}{C_1} + \frac{1}{C_2} + \dots + \frac{1}{C_{10}}} \qquad = 282 \times 10^{-12}\ \text{F.}$$

c) If ε_R varies linearly with distance from 1 to 10, then

$$\varepsilon = \varepsilon_0 \left(1 + \frac{10x}{10 \times 10^{-3}}\right) = \varepsilon_0 (1 + 1000x)$$

Then

$$V = -\int \bar{E} \cdot d\bar{l} = -\int \frac{\rho_s}{\varepsilon}\, dx = -\int_0^{0.01} \frac{\rho_s A}{A\varepsilon_0 \varepsilon_r}\, dx$$

$$= -\int_0^{0.01} \frac{Q}{A\varepsilon_0} \frac{dx}{1 + 1000x} = \frac{Q}{\varepsilon_0 A} \ln 11$$

$$C = \frac{Q}{V} = \frac{\varepsilon_0 A}{\ln 11} = \frac{10^{-9}}{36\pi} \times \frac{240\pi \times 10^{-4}}{\ln 11}$$

$$= 278 \times 10^{-12}\text{F} = 278\ \text{PF}$$

• PROBLEM 6-26

Three initially charged capacitors C_1, C_2, C_3 are connected in series to a battery of potential difference V as shown in the figure. Find the charge on the positive plate of C_1 after C_2 is placed in the circuit.

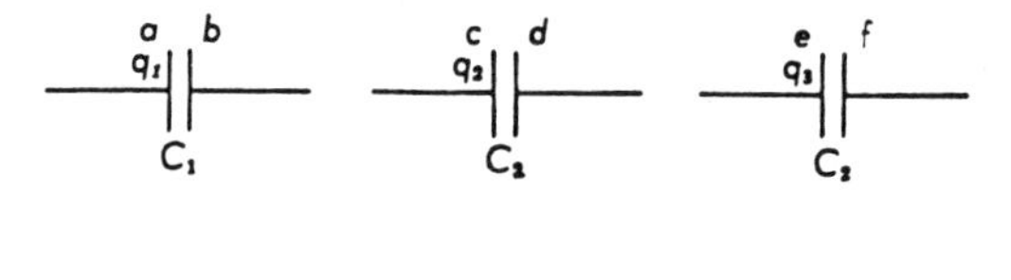

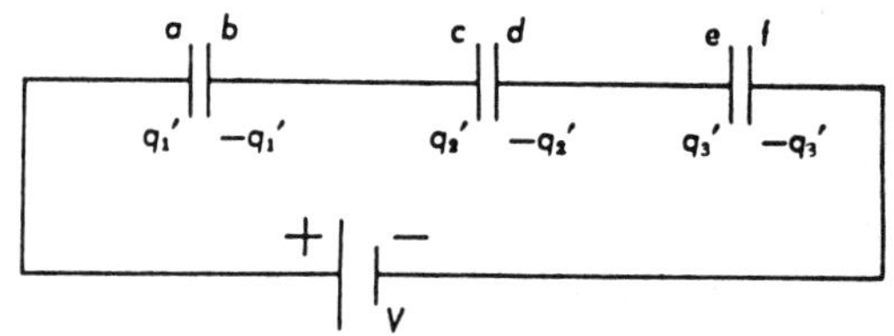

Three initially charged capacitors connected in series to a battery.

Solution: The two basic principles of Physics that are applicable here are the law of conservation of charge and the conservative nature of an electrostatic field.

Firstly, using the law of conservation of charge, since no charge can escape from the plates b and c because of the connection,

$$q_2' - q_1' = q_2 - q_1.$$

Similarly, the total charge on the plates d and e must remain constant. Consequently,

$$q_3' - q_2' = q_3 - q_2$$

Now make use of the conservative nature of an electrostatic field. The total change in potential after traversing a closed loop must be zero. Thus, starting at the positive terminal of the battery and considering potential rises as positive and potential drops as negative,

$$-\frac{q_1'}{C_1} - \frac{q_2'}{C_2} - \frac{q_3'}{C_3} + V = 0$$

From these three equations find q_1', q_2', q_3'. The equation from which q_1' may be found is

$$V = q_1'\left(\frac{1}{C_1} + \frac{1}{C_2} + \frac{1}{C_3}\right) + (q_2 - q_1)\left(\frac{1}{C_2} + \frac{1}{C_3}\right) + (q_3 - q_2)\left(\frac{1}{C_3}\right)$$

In this case, if the capacitors were initially equally charged, then

$$V = \frac{q_1'}{C_e}$$

which is the same result as for initially uncharged capacitors.

• PROBLEM 6-27

There are two sheet conductors as shown in the figure. The conductor separation at gg is 4 cm and at ff it is 1 cm. Assume that the conductors have a depth (into the page) of 20 cm. Assuming that the conductors end at ff and gg, and considering the fringing of the field to be negligible, find the capacitance C of the resulting capacitor. The medium in the capacitor is air.

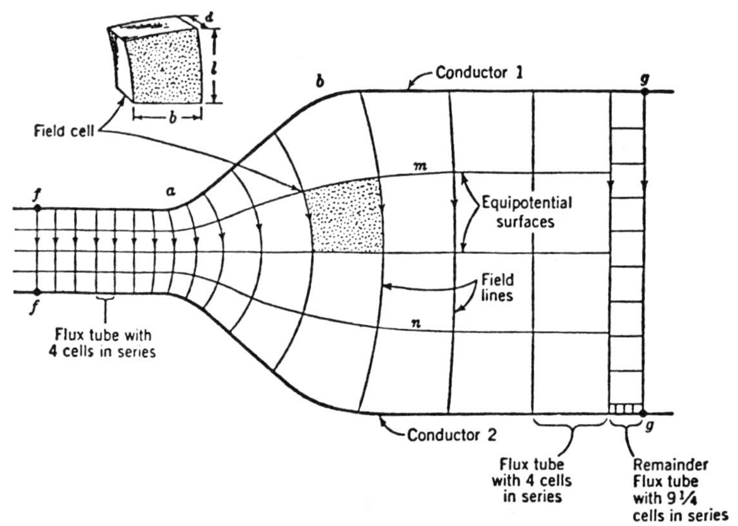

Cross section of two sheet conductors with completed field map. A three-dimensional view of a field cell is also given.

Solution: The method of solution will be to evaluate the series-parallel combination of capacitors formed by the individual cells.

Each cell has a capacitance

$$C_0 = \varepsilon_0 d = 8.85 \times 0.2 = 1.77 \ \mu\mu f$$

The capacitance between the ends of each flux tube with 4 cells in series is then

$$\frac{1.77}{4} = 0.442 \mu\mu f$$

The capacitance between the ends of the remainder flux tube with 9.25 cells in series is

$$\frac{1.77}{9.25} = 0.191 \mu\mu f$$

There are fifteen 4-cell tubes and one remainder (9.25-cell) tube. Hence the total capacitance C between ff and gg is the sum of the capacitances of all the flux tubes, or

$$C = 15 \times 0.442 + 0.191 = 6.82 \ \mu\mu f$$

The above calculation is somewhat simplified if each cell is arbitrarily assigned a capacitance of unity. On this basis the total capacitance in arbitrary units is given by

$$\frac{15}{4} + \frac{1}{9.25} = 3.86 \text{ units}$$

and the total actual capacitance C is the product of this result and the actual capacitance of a cell, or

$$C = 3.86 \times (8.85 \times 0.2) = 6.82\mu\mu f$$

where

N = number of cells (or flux tubes) in parallel

n = number of cells in series

C_0 = capacitance of one cell

and where all cells are of the same kind. Thus in the above example, counting in terms of the 10-volt cells, we have

$$C = \frac{15.43}{4} \times 8.85 \times 0.2 = 6.82\mu\mu f$$

• PROBLEM 6-28

Two condensers A and B have capacities C_1 and C_2 . Condenser A is charged by a battery and, after the battery is removed the condenser is discharged by allowing a spark to pass between the plates. Condenser A is then charged as before, the battery removed and the condenser B connected in parallel with A, a spark appearing as the connection is made. A and B are then separated and each is discharged by a spark. Show that the energies of the four sparks are in the ratio

$$(C_1 + C_2)^2 : C_2(C_1 + C_2) : C_1^2 : C_1C_2 \ .$$

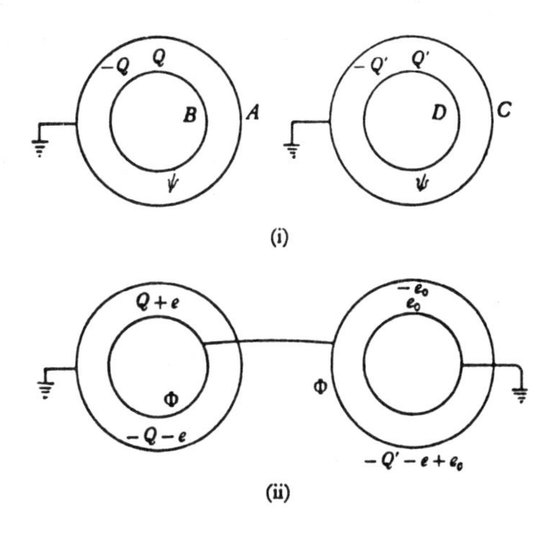

Solution: (i) Suppose that the condenser A is raised to a potential V; on discharging, the energy of the condenser reappears as energy E_1 of the spark which is thus equal to $\frac{1}{2} C_1V^2$.

(ii) When condenser A is charged again to a potential V, and then connected to B, the charge Q on the positive plate of the condenser A will divide itself into a charge Q_1 on A, say, and a charge Q_2 on B, so that

$$Q = Q_1 + Q_2 .$$

Suppose that after connection the common potential of the high-potential plates of the two condensers is V_1, the low-potential plates being assumed to be at zero potential throughout, this equation gives at once

$$VC_1 = V_1C_1 + V_1C_2,$$

and hence

$$V_1 = \frac{C_1}{C_1 + C_2} V . \tag{a}$$

The energy of the condenser A before connection to B is

$$\frac{1}{2} C_1V^2 .$$

After connection to B in parallel, the energy of the system is

$$\frac{1}{2} (C_1 + C_2)V_1^2 ,$$

since the capacity of the system is $C_1 + C_2$. The energy of the spark is thus the difference of these two expressions, viz.

$$E_2 = \frac{1}{2} C_1V^2 - \frac{1}{2} (C_1 + C_2) V_1^2 = \frac{1}{2} \frac{C_1C_2}{C_1 + C_2} V^2 ,$$

by (a)

(iii) When A and B are separated, their energies E_3, E_4 are respectively $\frac{1}{2}C_1V_1^2$ and $\frac{1}{2}C_2V_1^2$. Hence

$$E_3 = \frac{1}{2}C_1\left(\frac{C_1}{C_1+ C_2}\right)^2 V^2, \quad E_4 = \frac{1}{2}C_2 \frac{C_1}{C_1+ C_2}^2 V^2.$$

Thus $$E_1 : E_2 : E_3 : E_4 = C_1 : \frac{C_1C_2}{C_1 + C_2} : \frac{C_1^3}{(C_1 + C_2)^2} : \frac{C_2C_1^2}{(C_1 + C_2)^2}$$

or $$(C_1 + C_2)^2 : C_2(C_1 + C_2) : C_1^2 : C_1C_2 .$$

POTENTIAL

• PROBLEM 6-29

A conducting slab with an applied potential difference is composed of two different materials, as shown in the figure. Find the electrostatic potential and surface charge at the interface.

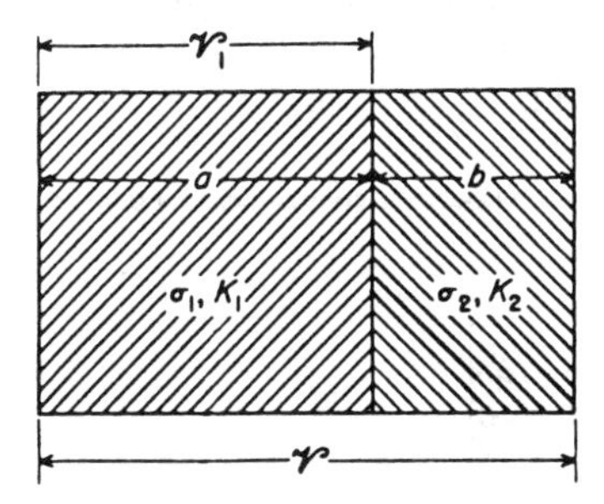

Solution: The potential across the entire conductor can be expressed as

$$V = aE_1 + bE_2$$

Also relate E_1 and E_2 via the boundary condition

$$D_{n1} - D_{n2} = q_s$$

or

$$q_s = K_1\varepsilon_0 E_1 - K_2\varepsilon_0 E_2$$

Another relation comes from the fact that

$$J_1 = J_2$$

or

$$\sigma_1 E_1 = \sigma_2 E_2$$

Eliminating E_1 and E_2 from these three equations gives

$$q_s = \frac{V(K_2\sigma_1 - K_1\sigma_2)\varepsilon_0}{a\sigma_2 + b\sigma_1}$$

Also

$$V_1 = aE_1 = \frac{Va\sigma_2}{a\sigma_2 + b\sigma_1}$$

• PROBLEM 6-30

Two very long, coaxial conducting cylinders have radii of 5 and 15 millimeters where the charges reside. The dielectric between these two radii is rubber with a dielectric constant of 2.5. The electric potential difference between the two conductors is 2500 volts. Determine and plot (a) the electric field intensity and (b) the electric potential with respect to the outer conductor, versus radial distance across the dielectric.

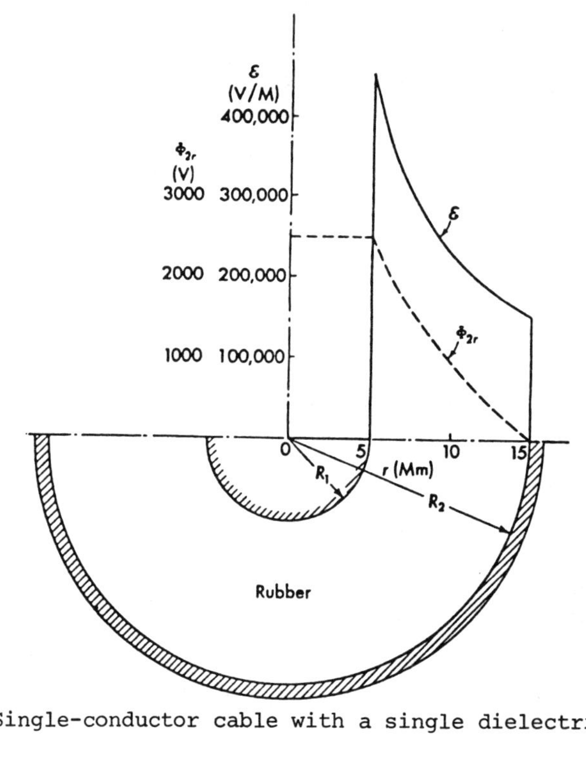

Single-conductor cable with a single dielectric.

Solution: The electric potential difference across the dielectric, calling the inner conductor 1 and the outer one 2, is found from

$$\Phi_{21} = \frac{\Lambda}{2\pi\varepsilon_0 K} \ln \frac{R_2}{R_1} \tag{1}$$

$$\Phi_{21} = 2500 = \frac{\Lambda}{2\pi\varepsilon_0 K} \ln \frac{15}{5} = 1.099 \frac{\Lambda}{2\pi\varepsilon_0 K}$$

or

$$\frac{\Lambda}{2\pi\varepsilon_0 K} = \frac{2500}{1.099} = 2275.$$

From the electric field intensity at any radial distance r for a coaxial cylinder

$$E = \frac{\Lambda}{2\pi\varepsilon_0 Kr} = \frac{2275}{r}$$

and again from eq. 1

$$\Phi_{2r} = \frac{\Lambda}{2\pi\varepsilon_0 K} \ln \frac{R_2}{r} = 2275 \ln \frac{0.015}{r}$$

where r is the radial distance in meters to the point in the dielectric under consideration and Φ_{2r} is the electric potential difference from the outside shell to that point.

A plot of these two functions is shown in the figure. The electric field intensity is zero inside a radius of 5 millimeters and also outside of 15 millimeters.

The largest value of E is 455,000 volts per meter, and this exists in the dielectric at its inner radius. The electric potential difference from the outer conductor to the radial distance r gradually increases as r decreases from 15 to 5 millimeters and then holds constant at 2500 volts from 5 millimeter to the center of the cylinder.

• PROBLEM 6-31

Consider two concentric, spherical, brass shells insulated from each other and isolated in a large space with an arrangement for introducing a charge Q to the inner sphere only. The sphere radii are r_1, r_2, r_3, and r_4 in increasing order. What is the potential of the inner sphere?

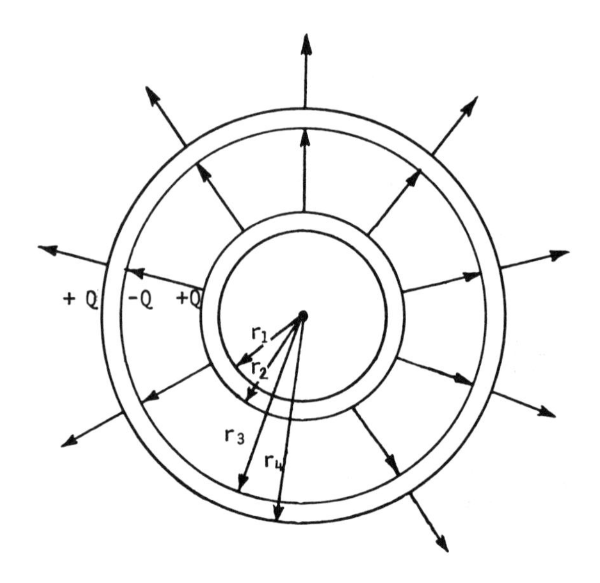

Solution: The accompanying figure shows a sketch of this arrangement. The excess charge +Q on the inner sphere will find its way to the outer surface (radius r_2); hence the inner radius r_1 is not significant. This follows from the fact that there is no net charge(that is, $\rho = 0$) for $r < r_2$. Therefore, for $r < r_2$, $E = 0$ and Ψ is constant at $\Psi(r_2)$. Consequently the problem reduces to a uniform, spherical surface charge of total value Q at radius r_2 surrounded by a conducting shell of radii r_3 and r_4.

It is tempting to assume that

$$\Psi(r_2) = Q/4\pi\varepsilon r_2 ,$$

since this value holds for an isolated, spherical surface charge. However, in this problem the charge is not isolated and so this answer should be modified by the induced charge on the outer sphere. Besides, two different dielectrics may be used, one between the conductors and the other outside. The simpler answer does not allow for this complication.

Recall that the outer sphere must be neutral and have zero internal field. These conditions are satisfied if the outer sphere carries a uniform surface charge -Q at r_3 and another uniform surface charge +Q at r_4. These are the induced charges, which necessarily will appear on the actual conductor and which must be added to the source charges. Now the problem to be solved has three concentric, uniform, spherical charges:

+Q at r_2, -Q at r_3 , +Q at r_4 .

From Gauss's law

$$E = 0 \qquad r < r_2$$

$$E = \frac{Q}{4\pi\varepsilon_1 r^2} \qquad r_2 < r < r_3$$

$$E = 0 \qquad r_3 < r < r_4$$

$$E = \frac{Q}{4\pi\varepsilon_2 r^2} \qquad r_4 < r$$

Also

$$\Psi(a) = -\int_{\infty}^{a} E \cdot dl$$

Inside the conductors, the field is zero and the potential will be the same as the potential on the outer surface of the conductor; thus

$$\Psi(r) = \frac{Q}{4\pi\varepsilon_2 r} \qquad r \geqq r_4$$

$$\Psi(r) = \frac{Q}{4\pi\varepsilon_2 r_4} \qquad r_4 \geqq r \geqq r_3$$

Between the conductors, the potential will be the sum of the potentials due to the electric field outside the conductors and between the conductors, or

$$\Psi(r) = \frac{Q}{4\pi\varepsilon_2 r_4} + \frac{Q}{4\pi\varepsilon_1}\left[\frac{1}{r} - \frac{1}{r_3}\right] \qquad r_3 \geqq r \geqq r_2 \qquad (1)$$

and finally to get the potential at the inner sphere, substitute r_2 for r in equation (1):

$$\psi(r_2) = \frac{Q}{4\pi}\left[\frac{1}{\varepsilon_2 r_4} + \frac{1}{\varepsilon_1}\left(\frac{1}{r_2} - \frac{1}{r_3}\right)\right]$$

• PROBLEM 6-32

A spherically symmetric charge system consists of a solid conducting sphere of radius 0.2 mm at the center of a sphere of dielectric material of relative permittivity 6.0

and outer radius 2.0 mm which is surrounded by a layer of dielectric material of relative permittivity 2.0 to an outer radius 4.0 mm. If a charge of 1.0 nC is placed on the solid conducting sphere, find the absolute potential of the surface of the conducting sphere and of the outer surface of the outer dielectric material. Refer to the given figure.

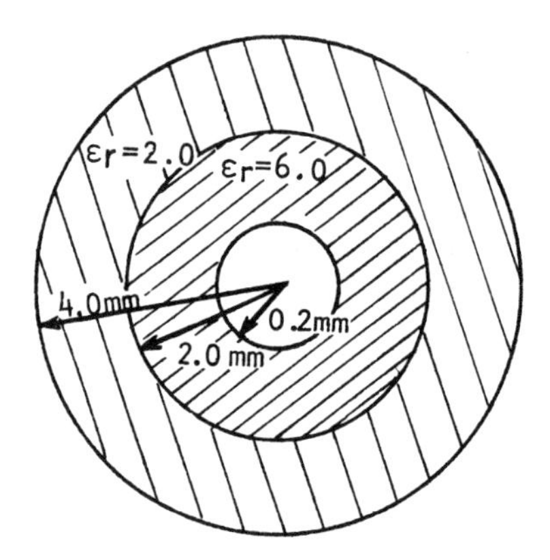

Solution: The charge uniformly distributed on the surface of the solid conducting sphere may be replaced by a point charge of the same value. In order to find the absolute potential, it is necessary to consider that the zero of potential is at an infinite distance away from the point charge so that the potential difference is obtained by integrating from an outer radius of infinity to the point under consideration. In this problem there are three regions of field having different relative permittivities. The boundaries between the different materials lie along equipotential lines so that the flux density is the same on each side of the boundary. The flux density due to a point charge is given by

$$D = \frac{Q}{4\pi r^2}$$

and

$$E = \frac{Q}{4\pi\varepsilon_0\varepsilon_r r^2}\ .$$

The potential difference between any two radii r_1 and r_2 is given by

$$\phi = \phi_1 - \phi_2 = \int_2^1 - E \cdot dl = \int_{r_2}^{r_1} - \frac{Q}{4\pi\varepsilon_0\varepsilon_r r^2}\, dr$$

$$= \frac{Q}{4\pi\varepsilon_0\varepsilon_r}\left(\frac{1}{r_1} - \frac{1}{r_2}\right).$$

Inserting numbers from the problem, the absolute potential of the outer surface of the outer sphere is given by

$$\phi_1 = \frac{10^{-9}}{4\pi\varepsilon_0}\left(\frac{1}{4.0 \times 10^{-3}} - \frac{1}{\infty}\right) = \frac{10^{-9} \times 36\pi \times 10^9}{4\pi \times 4.0 \times 10^{-3}} =$$

$$2.25 \times 10^3 \text{ V} = 2.25 \text{ kV}.$$

The absolute potential of the surface of the inner sphere is given by

$$\phi_2 = (\text{p.d. across the inner dielectric}) + (\text{p.d. across the outer dielectric}) + \phi_1 .$$

$$\phi_2 = \frac{Q}{4\pi\varepsilon_0 \times 6.0}\left(\frac{1}{0.2} - \frac{1}{2.0}\right) \times 10^3 + \frac{Q}{4\pi\varepsilon_0 \times 2.0}\left(\frac{1}{2.0} - \frac{1}{4.0}\right) \times 10^3 + \phi_1$$

$$= \frac{10^{-9} \times 36\pi \times 10^9 \times 10^3}{4\pi}(0.75 + 0.125) + 2.25 \times 10^3$$

$$= (7.88 + 2.25) \times 10^3 \text{ V} = 10.13 \text{ kV}.$$

STORED ENERGY AND FORCE IN CAPACITORS

• PROBLEM 6-33

A conducting sphere of radius a is embedded in a homogeneous linear dielectric medium with permittivity ε, as shown in the figure. The sphere has a total charge Q. Find the stored energy. Repeat if the permittivity ε varies according to

$$\varepsilon = \varepsilon_0\left(1 + \frac{a}{r}\right)^2$$

Solution: By Gauss's law, write

$$4\pi r^2 D_r = Q$$

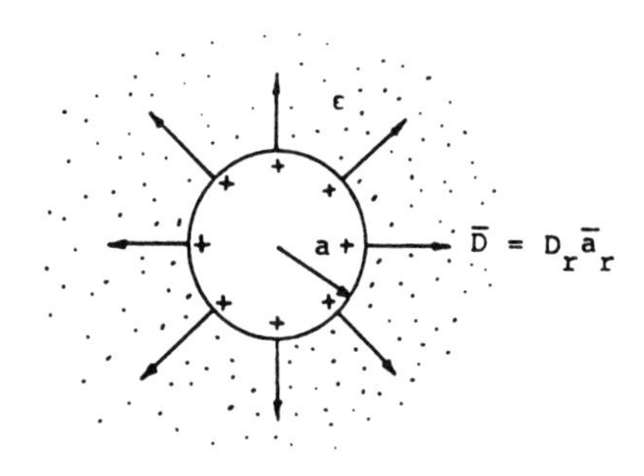

A charged sphere inside a dielectric.

and so

$$E_r = \frac{Q}{4\pi\varepsilon r^2}$$

The stored energy is given by

$$W_e = \frac{1}{2} \int_{space} \varepsilon |\bar{E}|^2 \, dV$$

In spherical coordinates, $dV = r^2 \sin\theta \, dr d\theta d\phi$

Thus

$$W_e = \frac{\varepsilon}{2} \int_0^{2\pi} \int_0^{\pi} \int_a^{\infty} \frac{Q^2 r^2 \sin\theta \, d\theta \, d\phi \, dr}{(4\pi\varepsilon r^2)^2} = \frac{Q^2}{8\pi\varepsilon a}$$

If the permittivity ε varies according to the form,

$$\varepsilon = \varepsilon_0 \left(1 + \frac{a}{r}\right)^2$$

$Q = 4\pi r^2 D_r$ is still true from Gauss's law. However, in this inhomogeneous case the stored energy is given by

$$W_e = \frac{1}{2} \int_0^{2\pi} \int_0^{\pi} \int_a^{\infty} \frac{Q^2 r^2 \sin\theta}{(4\pi r^2)^2 \, \varepsilon_0} \frac{d\theta \, d\phi \, dr}{\left(1 + \frac{a}{r}\right)^2}$$

where the integrand is $D_r^2/\varepsilon(r)$. After integrating over θ and ϕ,

$$W_e = \frac{Q^2}{8\pi} \int_a^\infty \frac{dr}{\varepsilon_0 (r+a)^2} = \frac{Q^2}{8\pi\varepsilon_0} \left(\frac{-1}{r+a}\right) \Bigg|_a^\infty \frac{Q^2}{16\pi\varepsilon_0 a} .$$

• PROBLEM 6-34

Calculate the change in the stored energy of a parallel plate capacitor as a result of inserting a dielectric slab.

Solution: The initial energy U_0 is

$$U_0 = \frac{1}{2} D_0 E_0 (Ad)$$

$$= \frac{1}{2} \varepsilon_0 \left(\frac{V_0}{d}\right)^2 (Ad)$$

The stored energy U after the dielectric slab is inserted is

$$U = \frac{1}{2} DE (Ad)$$

$$= \frac{1}{2} \varepsilon \left(\frac{V}{d}\right)^2 (Ad)$$

$$= \frac{1}{2} K\varepsilon_0 \left(\frac{V_0}{dK}\right)^2 (Ad)$$

$$= \frac{U_0}{K} .$$

The energy is reduced by a factor K when the dielectric slab is inserted. The change in energy ΔU is

$$\Delta U = U_0 - U = U_0 \left(1 - \frac{1}{K}\right) .$$

For K = 5.6 (pyrex glass),

$$\frac{\Delta U}{U_0} = 1 - \frac{1}{5.6} = 0.82$$

• **PROBLEM 6-35**

Verify that for a charged spherical conductor the energy

$$\frac{1}{2} Q^2/C = \iiint (E^2/8\pi)\,d\Omega ,$$

where E is the electric field due to the charge and the integration is throughout all space.

Solution: The electric field at an external point distant r from the center is $E = Q/r^2$; since this is the same at all points equidistant from the center, divide the space into elementary spherical shells, concentric with the sphere. For a typical shell of radii r and r + dr

$$d\Omega = 4\pi r^2 dr$$

and

$$\frac{E^2}{8\pi}\,d\Omega = \frac{Q^2}{2r^4}\,r^2 dr = \frac{Q^2}{2r^2}\,dr.$$

Since E = 0 within the sphere, we have

$$\iiint \frac{E^2}{8\pi}\,d\Omega = \int_a^\infty \frac{Q^2}{2r^2}\,dr = \frac{1}{2}\,\frac{Q^2}{a}$$

where a is the radius of the sphere. But for a spherical conductor C = a, and the expression is thus equal to the energy of the conductor.

• **PROBLEM 6-36**

Calculate the electrostatic energy in the system of the figure shown using

$$W = \frac{1}{2} \iiint_{\substack{\text{all}\\ \text{space}}} \mathbf{E}\cdot\mathbf{D}\,dv$$

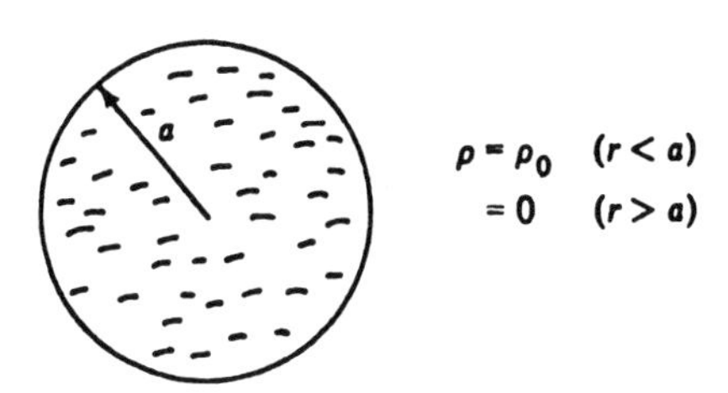

Solution:

$$W = \frac{1}{2} \iiint E \cdot D \, dv = \frac{1}{2} \iiint \varepsilon |E|^2 \, dv$$

From Gauss's law

$$\oiint D \cdot \hat{n} ds = q_{encl.}$$

$$D_r 4 \pi r^2 = \rho_0 \frac{4}{3} \pi r^3 \qquad (r < a)$$

$$= \rho_0 \frac{4}{3} \pi a^3 \qquad (r > a)$$

$$D_r = \frac{r \rho_0}{3} \qquad (r < a)$$

$$D_r = \frac{\rho_a a^3}{r^2} \qquad (r > a)$$

$$E_r = \frac{\rho_0}{3\varepsilon_0} r \qquad (r < a)$$

$$E_r = \frac{\rho_0 a^3}{3\varepsilon_0} \left(\frac{1}{r^2}\right) \qquad (r > a)$$

$dv = r^2 \sin\theta \, dr \, d\theta \, d\phi$

Thus,

$$W = \frac{1}{2} \int_0^{2\pi} \int_0^{\pi} \int_0^{a} \varepsilon_0 \left(\frac{\rho_0 r}{3\varepsilon_0}\right)^2 r^2 \sin\theta \, dr \, d\theta \, d\phi$$

$$+ \frac{1}{2} \int_0^{2\pi} \int_0^{\pi} \int_a^{\infty} \varepsilon_0 \left(\frac{\rho_0 a^3}{3\varepsilon_0 r^2}\right)^2 r^2 \sin\theta \, dr \, d\theta \, d\phi$$

$$= \frac{4\pi\rho_0^2 a^5}{15\varepsilon_0}$$

• **PROBLEM 6-37**

If the energy density is taken to be $u_1 = \frac{1}{2}\varepsilon_0 E_1^2$ for one charge distribution and

$$u_2 = \frac{1}{2}\varepsilon_0 E_2^2$$

for a second, the energy density when both charge distributions are present would be

$$u = \frac{1}{2}\varepsilon_0 (E_1 + E_2)^2.$$

So

$$u \neq u_1 + u_2 .$$

Does the superposition principle apply to energy?

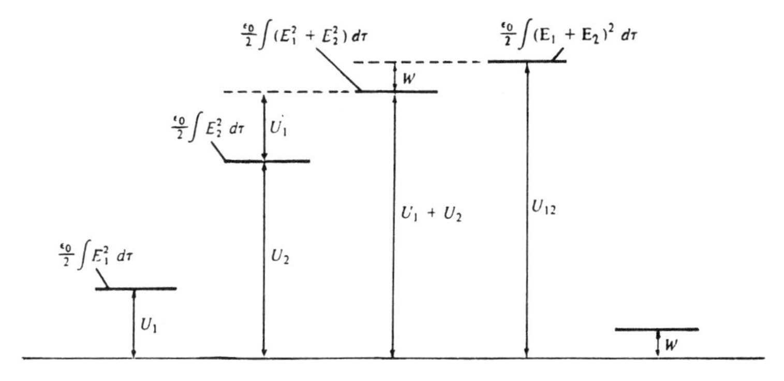

Solution:

$$U_1 + U_2 = (\varepsilon_0/2)\int (E_1^2 + E_2^2)d\tau$$

is the total energy stored in the separate, individual fields; here E_1 is produced by $\rho_1(r)$, while E_2 is produced by $\rho_2(r)$. In the combined field the energy stored is

$$U_{12} = (\varepsilon_0/2)\int (E_1 + E_2)^2 d\tau,$$

since $E_1 + E_2$ is the combined field produced by $\rho_1(r)$ and $\rho_2(r)$ acting together. The difference between these two potential energies is

$$\Delta U = U_{12} - (U_1 + U_2) = \varepsilon_0 \int E_1 \cdot E_2 d\tau = \varepsilon_0 \int \nabla\phi_1 \cdot \nabla\phi_2 d\tau$$

In the vector identity $\nabla \cdot (aA) = a\nabla \cdot A + \nabla a \cdot A$ set $a = \phi_1$ and $A = \nabla\phi_2$; then

$$\nabla \cdot (\phi_1 \nabla \phi_2) = \phi_1 \nabla^2 \phi_2 + \nabla\phi_1 \cdot \nabla\phi_2$$

Substituting into ΔU gives

$$\Delta U = \varepsilon_0 \int \nabla \cdot (\phi_1 \nabla \phi_2) d\tau - \varepsilon_0 \int \phi_1 \nabla \cdot \nabla\phi_2 d\tau$$

$$= \varepsilon_0 \oint_{\Sigma} (\phi_1 \nabla \phi_2) \cdot dS + \varepsilon_0 \int \phi_1 \nabla \cdot E_2 \, d\tau$$

If Σ is taken to be of very large radius, the first integral may be made arbitrarily small. Then

$$\Delta U = \int \phi_1(r) \rho_2(r) d\tau$$

But this is just the work required to bring the second charge distribution $\rho_2(r)$ into the potential field ϕ (r) created by the first charge distribution.

That the energy densities of the two fields are not additive is due to the work required to bring the charge density which produces E_2 into the field E_1. This situation is illustrated in the given figure.

• PROBLEM 6-38

Two parallel, perfectly conducting electrodes of area A and a distance x apart are shown in Figure 1. For each of the following two configurations, find the force on the upper electrode in the x direction when the system is constrained to constant voltage V_0 or constant charge Q_0.

(a) The electrodes are immersed within a liquid dielectric with permittivity ε, as shown in Figure 1a.

(b) A solid dielectric with permittivity ε of thickness s is inserted between the electrodes with the remainder of space having permittivity ε_0, as shown in Figure 1b.

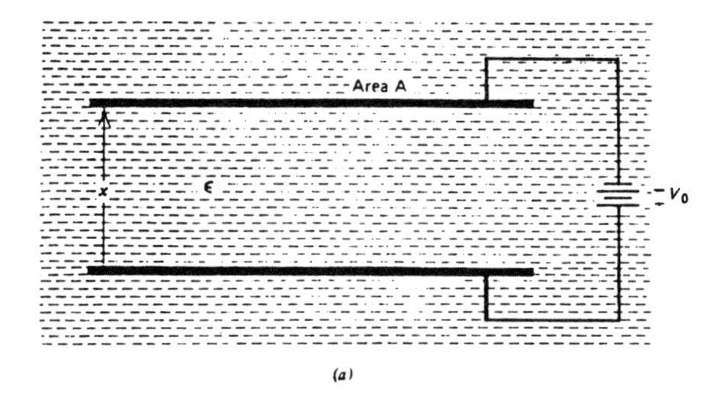

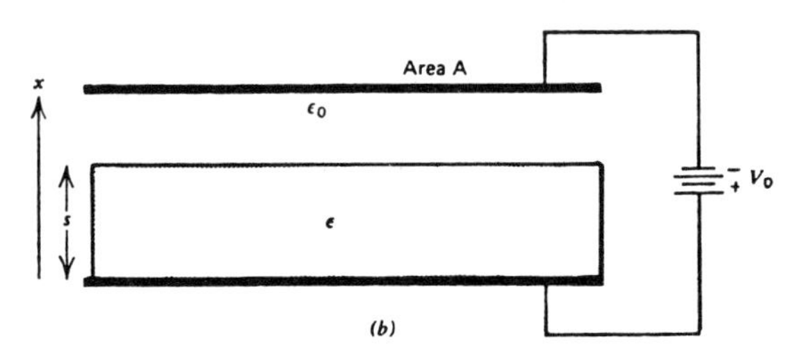

A parallel plate capacitor (a) immersed within a dielectric fluid or with (b) a free space region in series with a solid dielectric.

Fig. 1

Solution: (a) The capacitance of the system is

$$C(x) = \varepsilon A/x$$

so that the force for constant voltage is

$$f_x = \frac{1}{2} V_0^2 \frac{dC(x)}{dx} = -\frac{1}{2} \frac{\varepsilon A V_0^2}{x^2}$$

The force being negative means that it is in the direction opposite to increasing x, in this case downward. The capacitor plates attract each other because they are oppositely charged and opposite charges attract. The force is independent of voltage polarity and gets infinitely large as the plate spacing approaches zero. The result is also valid for free space with $\varepsilon = \varepsilon_0$. The presence of the dielectric increases the attractive force.

If the electrodes are constrained to a constant charge Q_0 the force is then attractive but independent of x:

$$f_x = -\frac{1}{2} Q_0^2 \frac{d}{dx} \frac{1}{C(x)} = \frac{1}{2} \frac{Q_0^2}{\varepsilon A}$$

For both these cases, the numerical value of the force is the same because Q_0 and V_0 are related by the capacitance, but the functional dependence on x is different. The presence of a dielectric now decreases the force over that of free space.

(b) The total capacitance for this configuration is given by the series combination of capacitance due to the dielectric block and the free space region:

$$C(x) = \frac{\varepsilon\varepsilon_0 A}{\varepsilon_0 s + \varepsilon(x - s)}$$

The force on the upper electrode for constant voltage is

$$f_x = \frac{1}{2} V_0^2 \frac{d}{dx} C(x) = -\frac{\varepsilon^2\varepsilon_0 A V_0^2}{2[\varepsilon_0 s + \varepsilon(x - s)]^2}$$

If the electrode just rests on the dielectric so that x = s, the force is

$$f_x = -\frac{\varepsilon^2 A V_0^2}{2\varepsilon_0 s^2} .$$

This result differs from that of part (a) when x = s by the factor $\varepsilon_r = \varepsilon/\varepsilon_0$ because in this case moving the electrode even slightly off the dielectric leaves a free space region in between. In part (a) no free space gap develops as the liquid dielectric fills in the region, so that the dielectric is always in contact with the electrode. The total force on the electrode-dielectric interface is due to both free and polarization charge.

With the electrodes constrained to constant charge, the force on the upper electrode is independent of position and also independent of the permittivity of the dielectric block:

$$f_x = \frac{-1}{2} Q_0^2 \frac{d}{dx} \frac{1}{C(x)} = -\frac{1}{2} \frac{Q_0^2}{\varepsilon_0 A}$$

• PROBLEM 6-39

Calculate the force as a function of x in Fig 1 where the charges on the plates are constant at +q and -q and Fig. 2 where the voltage between the plate is a constant V.

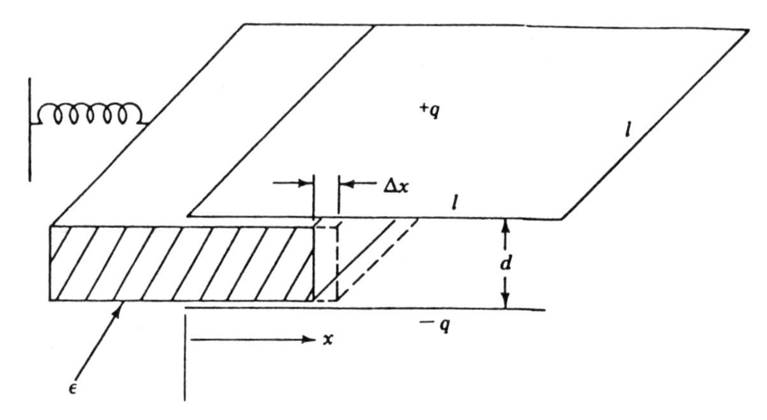

A dielectric block between parallel plates(constant charge case).

Fig. 1

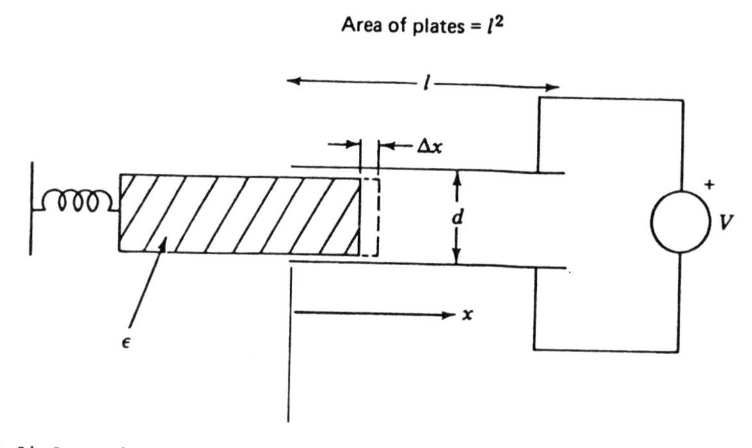

A dielectric block between parallel plates (constant voltage case).

Fig. 2

Solution:

(a) As the dielectric block in Fig. 1 is moved, the charge distribution on the parallel plates may change but the total charge q remains constant. The force exerted on the block may be calculated by considering a small ΔX to be small enough so that the force is essentially constant as the block is moved a distance ΔX. The mechanical work done is $F_x \Delta X$ and this is equal to $-\Delta W$, the change in electrostatic energy of the system. By conservation of energy,

$$F_x \Delta X + \Delta W = 0$$

$$F_x = -\frac{\Delta W}{\Delta X}$$

ecomes

$$F_x = -\frac{dW}{dX}$$

as ΔX approaches 0.

The tangential E field is continuous across the dielectric boundary and is thus uniform throughout the capacitor. The D field is equal to εE in the dielectric and equal to $\varepsilon_0 E$ in the free space region. The surface charge density is equal to εE in the dielectric region and $\varepsilon_0 E$ in the air region. Since the net charge on the upper capacitor plate is q,

$$\varepsilon E(\ell x) + \varepsilon_0 E(\ell - x)\ell = q$$

$$E = \frac{q}{\varepsilon \ell x + \varepsilon_0(\ell - x)\ell}$$

Now calculate the energy stored in the capacitor.

$$W = \frac{1}{2} \iiint E \cdot D \, dv$$

$$= \frac{1}{2}\left[\frac{q}{(\varepsilon - \varepsilon_0)\ell x + \varepsilon_0 \ell^2}\right]^2 [\varepsilon(\ell x)d + \varepsilon_0(\ell - x)\ell d]$$

$$= \frac{1}{2}\left[\frac{q^2 d}{(\varepsilon - \varepsilon_0)\ell x + \varepsilon_0 \ell^2}\right]$$

Now calculate the force.

$$F_x = -\frac{dW}{dx}$$

$$F_x = +\frac{q^2 d(\varepsilon - \varepsilon_0)\ell}{[(\varepsilon - \varepsilon_0)\ell x + \varepsilon_0 \ell^2]^2}$$

The positive sign indicates that the force is in the positive x direction.

(b) In this case there are three possible sources of energy

storage, namely in the electrostatic system, in a mechanical system such as a spring, and in the battery. If fields are compared before and after the insertion of the dielectric, it can be noted that the electric field is constant (since the voltage is unchanged) and that the electrostatic energy is increased by the factor $\varepsilon/\varepsilon_0$ after the insertion. The change in energy ΔW is equal to

$$\frac{1}{2}\,(V/d)^2 dl^2(\varepsilon - \varepsilon_0).$$

Energy could have come either from the battery or from mechanical work (perhaps the dielectric block had to be pushed in).

Since the electric field is not changed by the insertion of the dielectric, the displacement field is increased by the factor $\varepsilon/\varepsilon_0$. Therefore the charge on the plates has been increased by the same factor $\varepsilon/\varepsilon_0$. To move this amount of additional charge through a potential V requires an amount of work equal to

$$\frac{\varepsilon - \varepsilon_0}{\varepsilon_0}\,qV,$$

where q is the original charge on the plates. Since the original charge is

$$q = CV = \frac{\varepsilon_0 \ell^2}{d}\,V$$

the amount of work supplied by the battery is

$$(\varepsilon - \varepsilon_0)qV^2\,\frac{\ell^2}{d}$$

which is just twice that supplied to the electrostatic field. Thus the battery supplies positive energy, half of which goes into stored energy in the capacitor and half of which is available to do mechanical work.

This turns out to be a general principle for the constant voltage case. When the dielectric is inserted, the battery supplies power, half going into the mechanical energy and half going into the electrostatic energy of the capacitor. When the dielectric is pulled out, the battery absorbs energy, half of it being supplied by mechanical means and half of it coming from the capacitor, whose electrostatic energy is

calculate the forces involved by considering a small Δx of the dielectric block. The energy balance is now:

$$F_x \Delta x + \Delta W - (\Delta q)V = 0$$

The last term is the work supplied by the battery and thus has a minus sign. The term $(\Delta q)V$ is twice ΔW, as pointed out before, and so

$$F_x \Delta x + \Delta W - 2\Delta W = 0 \qquad F_x = \frac{\Delta W}{\Delta x} \rightarrow \frac{dW}{dx}$$

for the constant voltage case.

$$E = \frac{V}{d}, \quad D = \varepsilon_0 E = \frac{\varepsilon_0 V}{d}$$

in the air region and

$$D = \frac{\varepsilon V}{d}$$

in dielectric.

$$W = \frac{1}{2} \iiint E \cdot D \, dv = \frac{1}{2} \frac{V^2}{d} [\varepsilon \ell x + \varepsilon_0 (\ell - x)\ell]$$

$$F_x = \frac{dW}{dx} = \frac{1}{2} \frac{V^2}{d} \ell(\varepsilon - \varepsilon_0)$$

CHAPTER 7

POISSON'S AND LAPLACE'S EQUATIONS

LAPLACE'S EQUATION

• PROBLEM 7-1

In what manner must the permittivity vary in an inhomogeneous charge free space so that Laplace's equation continues to hold?

Solution: In a charge-free space,

$$\nabla \cdot \bar{D} = 0$$

But

$$\bar{D} = \varepsilon \bar{E},$$

therefore

$$\nabla \cdot (\varepsilon \bar{E}) = 0$$

Since

$$\bar{E} = - \nabla V,$$

then

$$\nabla \cdot (\varepsilon \nabla V) = 0 .$$

Now, let ε be varying spacially, then

$$\nabla \cdot (\varepsilon \nabla V) = (\nabla V) \cdot \nabla \varepsilon + \varepsilon \nabla^2 V = 0 .$$

For

$$\nabla^2 V = 0$$

to hold, it is required that

$$(\nabla V) \cdot (\nabla \varepsilon) = 0$$

Now, the product of two vectors can be zero if they are perpendicular to each other, then the permittivity must be varied in such a manner that the gradient of ε be perpendicular to the field.

• PROBLEM 7-2

If $x = a \operatorname{Sinh}\eta \ \sin\psi \ \cos\phi$, $y = a\operatorname{Sinh}\eta \ \sin\psi \ \sin\phi$, and $z = a \operatorname{Cosh}\eta \ \cos\psi$, then the surfaces η = constant are prolate spheroids (football shape), the surfaces ψ = constant are hyperboloids of two sheets (somewhat like cones with round vertices), and ϕ is the same angle as used previously. In prolate spheroidal coordinates, Laplace's equation is

$$\nabla^2 V = \frac{1}{a^2(\sinh^2\eta + \sin^2\psi)} \left(\frac{\partial^2 V}{\partial \eta^2} + \operatorname{Coth}\eta \ \frac{\partial V}{\partial \eta} + \frac{\partial^2 V}{\partial \psi^2} + \cot\psi \ \frac{\partial V}{\partial \psi} \right)$$

$$+ \frac{1}{a^2 \sinh^2\eta \ \sin^2\psi} \ \frac{\partial^2 V}{\partial \phi^2} = 0$$

(a) Assume V varies only with η, find the general form of V.

(b) Find V if it is a function of ψ alone.

(c) Find V if $V = f(\phi)$.

Solution: (a) If V varies only with η, then

$$\nabla^2 V = \frac{1}{a^2 (\sinh^2\eta + \sin^2\psi)} \left(\frac{\partial^2 V}{\partial \eta^2} + \coth\eta \ \frac{\partial V}{\partial \eta} \right) = 0$$

Let

$$u = \frac{\partial V}{\partial \eta},$$

then

$$\nabla^2 V = \frac{1}{a^2 (\sinh^2\eta + \sin^2\psi)} \left(\frac{\partial u}{\partial \eta} + u \coth\eta \right) = 0$$

or

$$\frac{du}{d\eta} = - u \coth\eta,$$

or

$$\frac{du}{u} = - \coth\eta \, d\eta$$

Solving for u,

$$\ln u = - \ln \text{Sinh}\eta + A,$$

or

$$u = \frac{A}{\sinh\eta} = \frac{dV}{d\eta}$$

then

$$dV = A d\eta / \text{Sinh}\ \eta,$$

and

$$V = A \ln \left(\tanh \frac{\eta}{2} + B\right)$$

(b) If V varies with ψ only, then

$$\nabla^2 V = \frac{1}{a^2 (\sinh^2\eta + \sin^2\psi)} \left(\frac{\partial^2 V}{\partial \psi^2} + \cot\psi \frac{\partial V}{\partial \psi} \right) = 0$$

Let

$$W = \frac{\partial V}{\partial \psi},$$

then,

$$\frac{dw}{w} = - \cot \psi \, d\psi$$

Solving for w,

$$w = c/\sin\psi \ .$$

Solving for V,

$$V = C \ln \left(\tan \frac{\psi}{2} \right) + D.$$

(c) If

$$V = f(\phi),$$

then

$$\frac{\partial^2 V}{\partial \phi^2} = 0$$

or

$$\frac{d^2 V}{d\phi^2} = 0$$

and

$$V = E\phi + F.$$

where

A, B, C, D, E and F are arbitrary constants.

• PROBLEM 7-3

A square-wave potential in x is in the xz plane and the plane at y = b is at zero potential as shown in the figure. The problem is independent of z. Find the potential at all points between y = 0 and y = b.

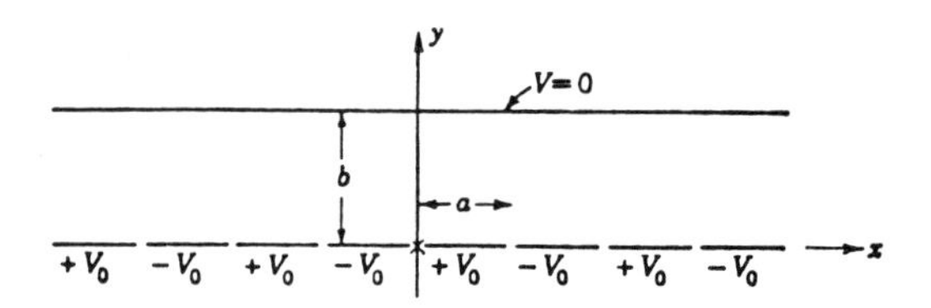

Solution: The boundary condition for a square-wave potential at $y = 0$ is

$$V = \sum_{\text{odd } n} \frac{4V_0}{\pi n} \sin \frac{\pi n x}{a} \tag{1}$$

A suitable potential to choose is

$$V = \sum_{\text{odd } n} A_n \sinh \frac{\pi n}{a} (b - y) \sin \frac{\pi n x}{a} \tag{2}$$

Choose the term $\sinh (\pi n/a)(b - y)$ since this is one form of the exponential solution that must be used for Y; in particular it meets the boundary condition at $y = b$. To match the boundary condition at $y = 0$, let $y = 0$ in (2) and compare with (1). This shows that

$$A_n \sinh \frac{\pi n b}{a} = \frac{4V_0}{\pi n}$$

or

$$A_n = \frac{4V_0}{\pi n} \frac{1}{\sinh \pi n b/a} \tag{3}$$

The solution is thus

$$V = \sum_{\text{odd } n} \frac{4V_0}{\pi n} \frac{\sinh (\pi n/a)(b - y)}{\sinh \pi n b/a} \sin \frac{\pi n x}{a} \tag{4}$$

A solution for y in the form $y = Ae^{kx} + Be^{-kx}$ could also have been used but it would have been more difficult to obtain the constants A and B.

• PROBLEM 7-4

Use Laplace's equation in cylindrical coordinates to find the capacitance of a coaxial cable.

Solution: Laplace's equation in cylindrical co-ordinates is

$$\nabla^2 V = \frac{1}{r}\frac{\partial}{\partial r}\left(r\,\frac{\partial V}{\partial r}\right) + \frac{1}{r^2}\,\frac{\partial^2 V}{\partial \phi^2} + \frac{\partial^2 V}{\partial z^2} = 0$$

Assume variation with respect to r only. Laplace's equation becomes

$$\frac{1}{r}\frac{\partial}{\partial r}\left(r\,\frac{\partial V}{\partial r}\right) = 0$$

or

$$\frac{1}{r}\frac{d}{dr}\left(r\,\frac{dV}{dr}\right) = 0$$

Multiply by r and then integrate,

$$r\,\frac{dV}{dr} = A$$

rearrange, and integrate again,

$$V = A \ln r + B \qquad (1)$$

The equipotential surfaces are given by r = constant and are cylinders, and the problem is that of the coaxial capacitor or coaxial transmission line. Choose a potential difference of V_0 by letting $V = V_0$ at $r = a$, $V = 0$ at $r = b$, $b > a$, and obtain

$$V = V_0\,\frac{\ln\,(b/r)}{\ln\,(b/a)}$$

from which

$$\bar{E} = \frac{V_0}{r} \frac{1}{\ln (b/a)} \bar{a}_r$$

$$D_{n(r=a)} = \frac{\varepsilon V_0}{a \ln (b/a)}$$

$$Q = \frac{\varepsilon V_0 2\pi a L}{a \ln (b/a)}$$

$$C = \frac{2\pi\varepsilon L}{\ln (b/a)}$$

• PROBLEM 7-5

> Assume that V is a function only of ϕ in cylindrical coordinates. Look at the physical problem first for a change and see that equipotential surfaces are given by ϕ = constant. These are radial planes. Boundary conditions might be $V = 0$ at $\phi = 0$ and $V = V_0$ at $\phi = \alpha$, leading to the physical problem detailed in the figure. Find the electric field intensity between these two planes.

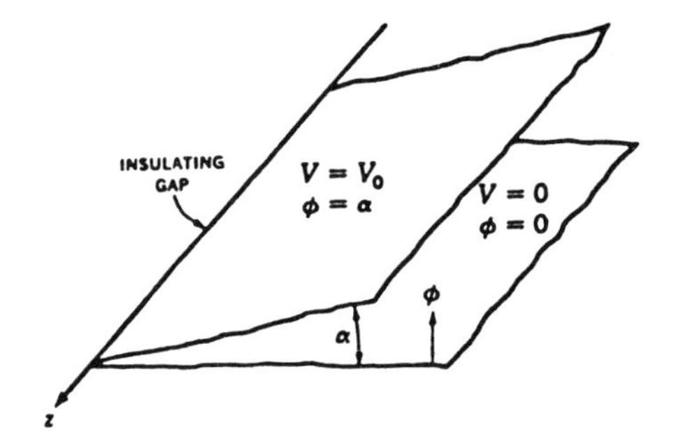

Two infinite radial planes with an interior angle α. An infinitesimal insulating gap exists at $r = 0$. The potential field may be found by applying Laplace's equation in cylindrical coordinates.

Solution: Laplace's equation is now

$$\frac{1}{r^2} \frac{\partial^2 V}{\partial \phi^2} = 0$$

or

$$\frac{d^2V}{d\phi^2} = 0$$

from which

$$V = A\phi + B$$

The boundary conditions determine A and B, and

$$V = V_0 \frac{\phi}{\alpha} \qquad (1)$$

Taking the gradient of (1) produces the electric field intensity,

$$\bar{E} = -\frac{V_0 \bar{a}_\phi}{\alpha r}$$

• PROBLEM 7-6

Use Laplace's equation to find the electric field between two concentric spherical shells of radius 0.5 m and 2.0 m. The inner sphere is at a potential of 0 volts while the outer sphere is at a potential of 100 volts.

Solution: In this problem the potential is a function of r only.

∴ Laplace's equation is

$$\frac{1}{r^2}\frac{d}{dr}\left(r^2 \frac{dV}{dr}\right) = 0$$

∴

$$r^2 \frac{dV}{dr} = C_1 = \text{constant}$$

∴

$$\frac{dV}{dr} = \frac{C_1}{r^2}$$

Integrating

$$V = \frac{-C_1}{r} + C_2 \qquad (1)$$

at

$$r = 0.5,\ V = 0,$$

$\therefore$

$$0 = -\frac{C_1}{\frac{1}{2}} + C_2 \qquad (2)$$

at

$$r = 2,\ V = 100$$

$\therefore$

$$100 = \frac{-C_1}{2} + C_2 \qquad (3)$$

equation (2) - equation (3) yields

$$-200 = -C_1 - 2C_1 = -3C_1$$

$\therefore$

$$C_1 = \frac{200}{3}$$

adding equation (2) and (3) yields

$$C_2 = \frac{400}{3}$$

Plugging the numerical values of C_1 and C_2 into equation (1)

$$V = \frac{-200}{3r} + \frac{400}{3}$$

$$\bar{E} = -\nabla V = -\hat{r}\frac{dV}{dr} = \frac{-200}{3r^2}\hat{r}$$

• PROBLEM 7-7

Find the capacitance between a conducting cone with its vertex separated from a conducting plane by an infinitesimal insulating gap and its axis normal to the plane by evaluating Laplace's equation in spherical coordinates and restricting the potential to a function of θ only.

Solution: In spherical coordinates, Laplace's equation is

$$\nabla^2 V = \frac{1}{r}\frac{\partial}{\partial r}\left(r^2 \frac{\partial V}{\partial r}\right) + \frac{1}{r^2 \sin\theta}\frac{\partial}{\partial \theta}\left(\sin\theta \frac{\partial V}{\partial \theta}\right) + \frac{1}{r^2 \sin^2\theta}\frac{\partial^2 V}{\partial \phi^2} = 0$$

if the potential is restricted to a function of θ only, then Laplace's equation reduces to

$$\frac{1}{r^2 \sin\theta}\frac{d}{d\theta}\left(\sin\theta \frac{dV}{d\theta}\right) = 0$$

from which

$$\sin\theta \frac{dV}{d\theta} = A$$

The second integral is then

$$V = \int \frac{A\, d\theta}{\sin\theta} + B$$

From integral tables,

$$V = A \ln(\tan \theta/2) + B$$

The equipotential surfaces are cones, and if V = 0 at

$$\theta = \frac{\pi}{2}$$

and

$$V = V_0 \text{ at } \theta = \alpha,\ \alpha < \pi/2,$$

then

$$V = V_0 \frac{\ln (\tan \theta/2)}{\ln (\tan \alpha/2)}$$

In order to find the capacitance first find the field strength:

$$\bar{E} = - \bar{\nabla} V = - \frac{1}{r} \frac{\partial V}{\partial \theta} \bar{a}_\theta =$$

$$- \frac{V_0}{r \sin \theta \ln (\tan \alpha/2)} \bar{a}_\theta$$

The surface charge density on the cone is then

$$\rho_s = \frac{-\varepsilon V_0}{r \sin \alpha \ln (\tan \alpha/2)}$$

producing a total charge Q,

$$Q = \frac{-\varepsilon V_0}{\sin \alpha \ln (\tan \alpha/2)} \int_0^\infty \int_0^{2\pi} \frac{r \sin \alpha \, d\phi \, dr}{r}$$

This leads to an infinite value of charge and capacitance, and it becomes necessary to consider a cone of finite size. The answer will now be only an approximation, because the theoretical equipotential surface is $\theta = \alpha$, a conical surface extending from $r = 0$ to $r = \infty$, whereas the physical conical surface extends only from $r = 0$, to, say, $r = r_1$. The approximate capacitance can then be found.

$$Q = \frac{-\varepsilon V_0}{\sin \alpha \ln (\tan \alpha/2)} \int_0^{r_1} \int_0^{2\pi} \frac{r \sin \alpha}{r} d\phi \, dr$$

$$Q = \frac{- \varepsilon V_0 \; 2\pi r_1}{\ln (\tan \alpha/2)}$$

$$C = \frac{|Q|}{|V|} = \frac{+\ \varepsilon V_0\ 2\pi r_1/[\ln\ (\tan\ \alpha/2)]}{V_0\ \ln\ (\tan\ \theta/2)/[\ln\ (\tan\ \alpha/2)]}$$

$\therefore$

$$C = \frac{2\pi\ \varepsilon r_1}{\ln\ (\tan\ \theta/2)}$$

• PROBLEM 7-8

Consider a coaxial cable for which the conductor radii are $r_1 = 0.5$ cm, $r_3 = 1.0$ cm. For r in [0.5, 0.75] cm, $\varepsilon_r = 2$; for r in [0.75, 1.0] cm, $\varepsilon_r = 3$. Describe the potential distribution in the dielectric when a source of 100 V is applied between the conductors, with the inner conductor positive. In particular, evaluate the potential at the interface between the two dielectrics.

Solution: Because of the discrete inhomogeneity in the dielectric, the arbitrary constants in the solution will differ in each medium. The solutions must satisfy

$$\psi(r_3) = 0,$$

$$\psi(r_1) = 100,$$

and in addition the boundary conditions must be satisfied at the dielectric interface. Thus, from Laplace's equation in cylindrical coordinates and since ψ is a function of r only,

$$\frac{1}{r}\frac{d}{dr}\left(r\frac{d\psi}{dr}\right) = 0$$

from which

$$r\frac{d\psi}{dr} = k_1$$

and

$$\psi(r) = k_1 \ln r + k_2$$

$$\Psi(r) = k_1 \ln r + k_2, \qquad 0.5 \le r \le 0.75 \text{ cm}$$

$$\Psi(r) = k_3 \ln r + k_4, \qquad 0.75 \le r \le 1.0 \text{ cm}$$

The calculation is somewhat simplified if lengths are left in centimeters.

At the inner conductor

$$100 = k_1 \ln 0.5 + k_2$$

or

$$-0.710k_1 + k_2 = 100$$

At the outer conductor

$$0 = k_3 \ln 1.0 + k_4$$

or

$$k_4 = 0$$

At the dielectric interface, equating potentials,

$$k_1 \ln 0.75 + k_2 = k_3 \ln 0.75 + k_4$$

or

$$-0.312k_1 + k_2 + 0.312k_3 = 0$$

Conservation of the normal component of D at the interface merely means that the coefficients of ln r must have a ratio that is the inverse of that of the corresponding permittivities. Thus

$$\frac{k_1}{k_2} = \frac{3}{2}$$

Solving for the constants,

$$k_1 = -165$$

$$k_2 = -17.2$$

$$k_3 = -110$$

Finally,

$$\Psi(r) = -165 \ln r - 17.2,$$

$$0.5 \leq r \leq 0.75 \text{ cm}$$

and

$$\Psi(r) = -110 \ln r,$$

$$0.75 \leq r \leq 1.0 \text{ cm.}$$

In particular, at the dielectric interface, $r = 0.75$, and from either equation,

$$\Psi(0.75) = 34.3 \text{ V}$$

• PROBLEM 7-9

Find the potential at all points above an infinite checkerboard in the xy plane such that the black squares are at $+V_0$ and the white squares are at $-V_0$ as shown in the figure. Let the size of the squares be $a \times b$ and let the plane at $z = c$ be at zero potential.

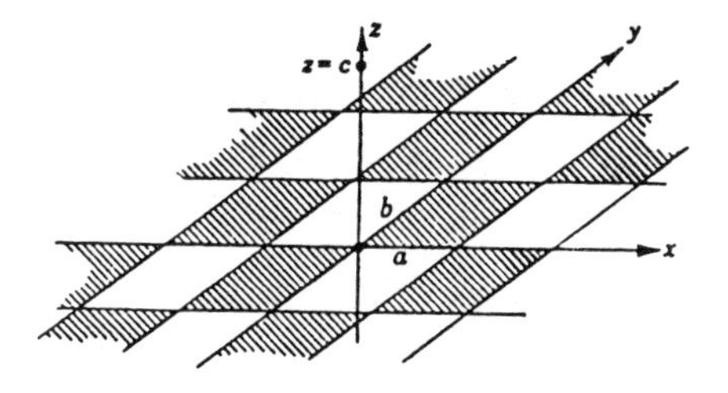

Solution: Here the boundary condition is expressed by a double Fourier series:

$$V(z = 0) = \sum_{\ell m} A_{\ell m} \sin \frac{\pi \ell x}{a} \sin \frac{\pi m y}{b} \tag{1}$$

where the origin is chosen at the corner of one of the black squares. Multiply through by

$$\sin \frac{\pi p x}{a} \sin \frac{\pi q y}{b} \, dx \, dy$$

and integrate between the limits 0 to a and 0 to b along the x- and y- axes respectively.

All the terms on the right side of (1) drop out except that one for which $p = \ell$ and $q = m$, and

$$\int_0^b \int_0^a V_0 \sin \frac{\pi \ell x}{a} \sin \frac{\pi m y}{b} \, dx \, dy$$

$$= A_{\ell m} \int_0^b \int_0^a \sin^2 \frac{\pi \ell x}{a} \sin^2 \frac{\pi m y}{b} \, dx \, dy \qquad (2)$$

or

$$\frac{2a}{\pi \ell} \frac{2b}{\pi m} V_0 = A_{\ell m} \frac{ab}{4} \qquad (3)$$

leading to

$$A_{\ell m} = \frac{16 V_0}{\pi^2 \ell m} \qquad \text{for } \ell \text{ and } m \text{ odd}$$

$$A_{\ell m} = 0 \qquad \text{for } \ell \text{ or } m \text{ even} \qquad (4)$$

Equation (1) then becomes

$$V(z = 0) = \sum_{\text{odd } \ell m} \frac{16 V_0}{\pi^2 \ell m} \sin \frac{\pi \ell x}{a} \sin \frac{\pi m y}{b} \qquad (5)$$

For the solution in z choose

$$Z = \sinh k_z(c - z) \tag{6}$$

which will meet the boundary condition at z = c.

Consideration of (5) leads us to choose trigonometric solutions for X and Y, which forces us to choose an exponential solution such as (6) for Z. k_z is determined from the auxiliary condition

$$-k_x^2 - k_y^2 + k_z^2 = 0 \tag{7}$$

Comparison with (5) shows that to match the boundary condition at z = 0 choose

$$k_x = \frac{\pi \ell}{a}$$

$$k_y = \frac{\pi m}{b}$$

and hence

$$k_z = \sqrt{\left(\frac{\pi \ell}{a}\right)^2 + \left(\frac{\pi m}{b}\right)^2} \tag{8}$$

The solution is then of the form

$$V = \sum_{\ell m} B_{\ell m} \sinh k_z(c - z) \sin \frac{\pi \ell x}{a} \sin \frac{\pi m y}{b} \tag{9}$$

Comparison with (5) with z = 0 shows that

$$B_{\ell m} \sinh k_z c = \frac{16 V_0}{\pi^2 \ell m}$$

and the answer is

$$V = \sum_{\text{odd } \ell m} \frac{16V_0}{\pi^2 \ell m} \frac{\sinh k_z(c - z)}{\sinh k_z c}$$

$$\sin \frac{\pi \ell x}{a} \sin \frac{\pi m y}{b} \tag{10}$$

where

k_z is given by (8).

• PROBLEM 7-10

A hollow infinitely long circular dielectric cylinder is placed in a uniform electric field, as shown in the figure. Determine the effectiveness of the dielectric in screening region 3 from the primary field in region 1. The primary potential is

$$U_p = - E_0 x.$$

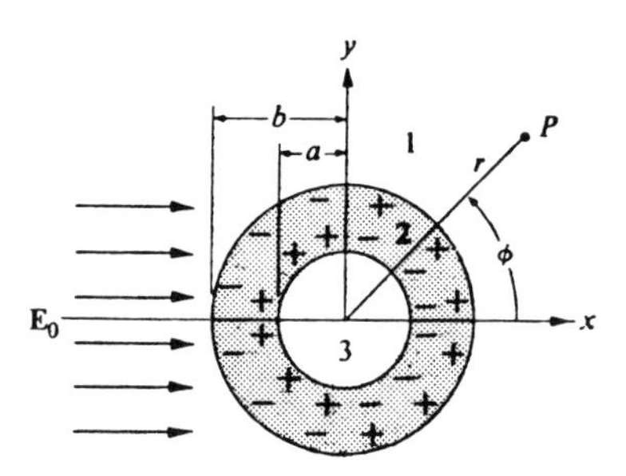

A hollow dielectric cylinder placed in a uniform electric field.

Solution: Since the electric field is independent of z, Laplace's equation reduces to the two-dimensional form

$$\nabla^2_T U = \frac{\partial^2 U}{\partial x^2} + \frac{\partial^2 U}{\partial y^2} = 0. \tag{1}$$

Now $r^2 = x^2 + y^2$ and it can be shown that $U = \ln r$ is a solution of Eq. (1). Substitute this function into Eq. (1), and successively differentiate both sides of the equation with respect to x and y to obtain further solutions. For example,

$$\frac{\partial}{\partial x}(\ell n\ r) = \frac{x}{r^2}$$

is a solution of Eq. (1). A potential of the form $\ell n\ r$ results from a line source which is a uniform distribution of charge along an infinitely long straight line. A potential of the form x/r^2 results from two line sources infinitely close together and oppositely charged to the same magnitude.

The primary field $\bar{E}_0$ results in a polarization of the dielectric. As seen from a field point P in region 1, the polarization charges appear equivalent to a two-dimensional dipole located at the origin. For the secondary potential therefore set

$$U_s = \frac{A\cos\phi}{r}.$$

The resultant potential in region 1 is

$$U_1 = -E_0 r\cos\phi + \frac{A\cos\phi}{r}.$$

When P is in region 3 the polarization charges appear uniformly distributed, resulting in a uniform field. Thus set

$$U_3 = -E_3 r\cos\phi$$

where $\bar{E}_3$ represents the assumed uniform electric field intensity. In region 2 the potential is taken as a linear combination of the potential of a uniform and a two-dimensional dipole field. Thus

$$U_2 = Br\cos\phi + \frac{C\cos\phi}{r}.$$

The assumed potentials are solutions of Laplace's equation in two dimensions. Note that the solutions chosen are based on physical considerations concerning the type of field that is expected. Determine the unknown constants A, B, and C so that the boundary conditions are satisfied, then the unique solution is determined.

The boundary conditions at r = a are $U_2 = U_3$ and

$$-\varepsilon_0 \frac{\partial U_3}{\partial r} = -\varepsilon \frac{\partial U_2}{\partial r},$$

which yield

$$B = -(\varepsilon_r + 1)E_3/2\varepsilon_r\,,$$

$$C = -a^2(\varepsilon_r - 1)E_3/2\varepsilon_r\,. \tag{2}$$

The boundary conditions at $r = b$ are $U_1 = U_2$

and

$$-\varepsilon_0 \frac{\partial U_1}{\partial r} = -\varepsilon \frac{\partial U_2}{\partial r}\,,$$

which yield

$$E_0 + A/b^2 = -\varepsilon_r B + \varepsilon_r C/b^2,$$

$$-E_0 + A/b^2 = B + C/b^2. \tag{3}$$

Eliminating A between Eqs. (3) and substituting in Eqs. (2) yield the ratio

$$\frac{E_3}{E_0} = \frac{4\varepsilon_r}{(\varepsilon_r + 1)^2}\left[1 - \left(\frac{\varepsilon r - 1a}{\varepsilon_r + 1b}\right)^2\right]^{-1},$$

which is a measure of the effectiveness of the dielectric cylinder in screening region 3 from the applied field $\overline{E}_0$. Now consider the case of a solid cylinder where $a = 0$. Region 3 no longer exists and now $C = 0$. Solving Eqs.

(3) for B yields $B = -2E_0/(\varepsilon_r + 1)$. The potential within the dielectric is

$$U_2 = -\frac{2E_0 x}{\varepsilon_r + 1}$$

which results in a uniform electric field intensity of magnitude

$$E_2 = \frac{2E_0}{\varepsilon_r + 1} < E_0\,.$$

The reduction in the electric field intensity inside the cylinder is due to the polarization charges on the surface of the cylinder which set up an electric field intensity

$$\bar{E}_p = -\frac{\varepsilon_r - 1}{\varepsilon_r + 1}\bar{E}_0$$

Since the polarization is proportional to the field

$$\bar{E}_2 = \bar{E}_p + \bar{E}_0 ,$$

the surface charges tend to depolarize the dielectric. The reduction in the polarizing field strength can be expressed in the form

$$E_2 = E_0 - N_x \frac{P_x}{\varepsilon_0} ,$$

where N_x is defined as the depolarization factor for the x-axis.

• PROBLEM 7-11

Use Laplace's equation to find the capacitance of a parallel plate capacitor.

Solution: Assume that V is a function only of x. Laplace's equation reduces to

$$\frac{\partial^2 V}{\partial x^2} = 0$$

and the partial derivative may be replaced by an ordinary derivative, since y and z are not involved,

$$\frac{d^2 V}{dx^2} = 0$$

Integrate twice, obtaining

$$\frac{dV}{dx} = A$$

and

$$V = Ax + B \tag{1}$$

where A and B are constants of integration. Equation (1) contains two such constants, as we should expect for a second-order differential equation. These constants can be determined only from the boundary conditions.

Since the field varies only with x and is not a function of y and z, if x is a constant, then V is a constant, or in other words, the equipotential surfaces are described by setting x constant. These surfaces are parallel planes normal to the x axis. The field is thus that of a parallel-plate capacitor, and as soon as the potential on any two planes is specified, the constants of integration may be evaluated.

To be very general, let $V = V_1$ at $x = x_1$ and $V = V_2$ at $x = x_2$. These values are then substituted into (1), giving

$$V_1 = Ax_1 + B \qquad V_2 = Ax_2 + B$$

$$A = \frac{V_1 - V_2}{x_1 - x_2} \qquad B = \frac{V_2 x_1 - V_1 x_2}{x_1 - x_2}$$

and

$$V = \frac{V_1(x - x_2) - V_2(x - x_1)}{x_1 - x_2} \tag{2}$$

A simpler answer would have been obtained by choosing simpler boundary conditions. If $V = 0$ at $x = 0$ and $V = V_0$ at $x = d$, then

$$A = \frac{V_0}{d} \qquad B = 0$$

and

$$V = \frac{V_0 x}{d} \tag{3}$$

Capacitance is given by the ratio of charge to potential difference, so choose now the potential difference as V_0, which is equivalent to one boundary condition, and then choose whatever second boundary condition seems to help the form of the equation the most. This is the essence of the second set of boundary conditions which produced (3). The potential difference was fixed as V_0 by choosing the potential of one plate zero and the other V_0; the location of these plates was made as simple as possible by letting $V = 0$ at $x = 0$.

Using (3), then, the total charge on either plate is still needed before the capacitance can be found.

The necessary steps are these after the choice of boundary conditions has been made:

1. Given V, use $\bar{E} = -\bar{\nabla}V$ to find $\bar{E}$.
2. Use $\bar{D} = \varepsilon\bar{E}$ to find $\bar{D}$.
3. Evaluate $\bar{D}$ at either capacitor plate, $\bar{D} = \bar{D}_s = D_n\bar{a}_n$.
4. Recognize that $\rho_s = D_n$.
5. Find Q by a surface integration over the capacitor plate,

$$Q = \int_s \rho_s \, dS.$$

Here

$$V = V_0 \frac{x}{d}$$

$$\bar{E} = -\frac{V_0}{d}\bar{a}_x$$

$$\bar{D} = -\varepsilon\frac{V_0}{d}\bar{a}_x$$

$$\bar{D}_s = \bar{D}\Big|_{x=0} = -\varepsilon\frac{V_0}{d}\bar{a}_x$$

$$\bar{a}_n = \bar{a}_x$$

$$D_n = -\varepsilon\frac{V_0}{d} = \rho_s$$

$$Q = \int_S \frac{-\varepsilon V_0}{d} dS = -\varepsilon \frac{V_0 S}{d}$$

and the capacitance is

$$C = \frac{|Q|}{V_0} = \frac{\varepsilon S}{d} .$$

• PROBLEM 7-12

The x,z-plane is composed of four separate charged planes, having potentials as follows:

first quadrant $(x > 0, z > 0)$:

$$V = V_0$$

second quadrant $(x < 0, z > 0)$:

$$V = 0$$

third quadrant $(x < 0, z < 0)$:

$$V = -V_0$$

fourth quadrant $(x > 0, z < 0)$:

$$V = 0$$

Find the corresponding electric field $\bar{E}$ as a function of x,y, and z.

Solution: Handle this problem with the superposition principle, which is based on the idea that the field due to a group of charges is the sum of the fields due to the individual charges, with a similar statement holding for potentials. Hence, regarding the arrangement of Fig. 1a as equivalent to the sum of two sets of half-planes, as shown in Fig. 1b and Fig. 1c, and specified by

$$x > 0, \qquad V = V_0/2$$

$$x < 0, \qquad V = -V_0/2$$

$$z > 0, \qquad V = V_0/2$$

$$z < 0, \qquad V = -V_0/2 .$$

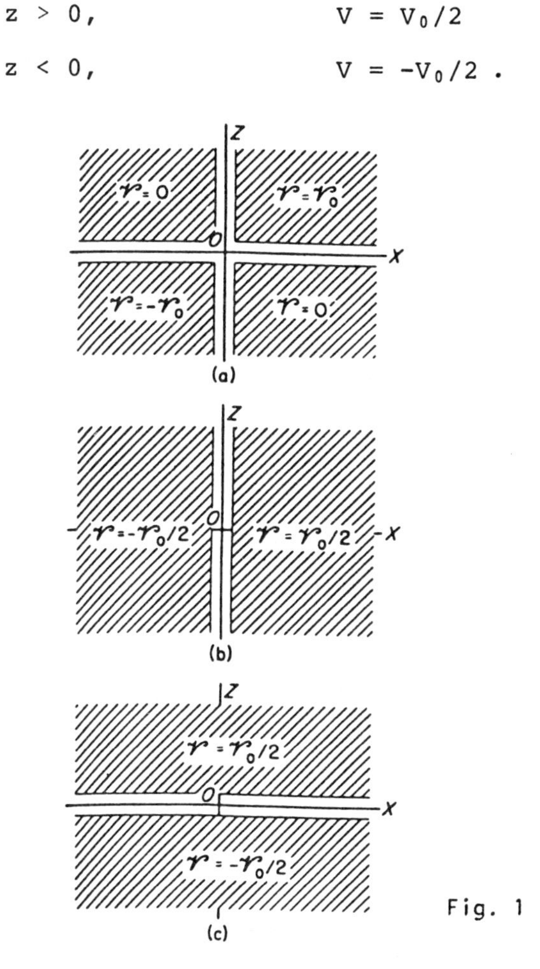

Fig. 1

Four quarter-planes whose potentials are the sum of those for two sets of half-planes.

For the set of half-planes $x > 0$ and $x < 0$, the z-axis is an axis of symmetry in cylindrical coordinates. Laplace's equation,

$$\nabla^2 V = \frac{1}{\rho}\frac{\partial}{\partial \rho}\left(\rho \frac{\partial V}{\partial \rho}\right) + \frac{1}{\rho^2}\frac{\partial^2 V}{\partial \theta^2} + \frac{\partial^2 V}{\partial z^2} ,$$

becomes

$$\frac{\partial^2 V}{\partial \theta^2} = 0$$

since V depends only on θ and ρ is arbitrary. The solution of this equation is

$$V = A\theta + B$$

with boundary conditions

$$V = \frac{V_0}{2} \text{ at } \theta = 0$$

$$V = \frac{-V_0}{2} \text{ at } \theta = \pi$$

Hence,

$$A = \frac{-V_0}{\pi}$$

$$B = \frac{V_0}{2}$$

and

$$V = \frac{-V_0\theta}{\pi} + \frac{V_0}{2}$$

The corresponding field is

$$\bar{E}_\theta = -\frac{1}{\rho}\frac{\partial V}{\partial \theta}\bar{i}_\theta = \frac{V_0}{\pi\rho}\bar{i}_\theta =$$

$$\frac{V_0}{\pi\sqrt{x^2 + y^2}}(\cos\theta\bar{j} - \sin\theta\bar{i})$$

$$= \frac{V_0}{\pi(x^2 + y^2)}(x\bar{j} - y\bar{i})$$

In the same way, the solution using the x-axis as the polar axis in terms of the corresponding angle ϕ is

$$\bar{E}_\phi = \frac{V_0}{\pi(y^2 + z^2)}(z\bar{i} - y\bar{k})$$

and the total field is

$$\bar{E} = \bar{E}_\theta + \bar{E}_\phi$$

• PROBLEM 7-13

Consider an infinitely long rectangular slot cut in a semi-infinite plane conducting slab held at zero potential, as shown by the cross-sectional view, transverse to the slot, in the figure. With reference to the coordinate system shown in the figure, assume that a potential distribution

$$V = V_0 \sin(\pi y/b),$$

where V_0 is a constant, is created at the mouth $x = a$ of the slot by the application of a potential to an appropriately shaped conductor away from the mouth of the slot not shown in the figure. Find the potential distribution in the slot.

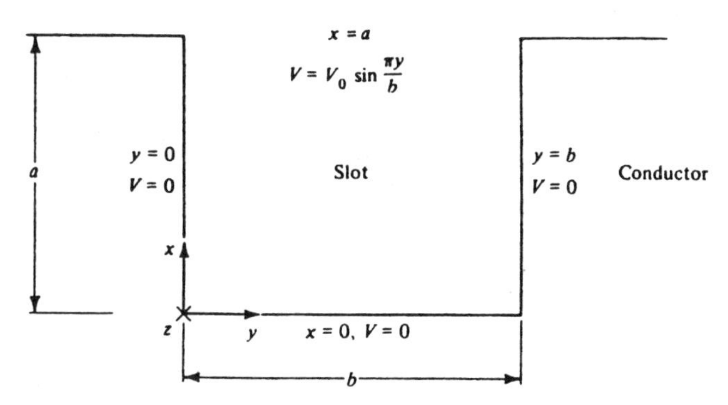

Cross-sectional view of a rectangular slot cut in a semi-infinite plane conducting slab at zero potential. The potential at the mouth of the slot is $V_0 \sin(\pi y/b)$ volts.

Solution: Since z is not involved in this problem, Laplace's equation reduces to

$$\frac{\partial^2 V}{\partial x^2} + \frac{\partial^2 V}{\partial y^2} = 0.$$

Using separation of variables, one can assume a solution of the form

$$V(x,y) = X(x)\, Y(y).$$

Substituting this into the Laplace's equation,

$$Y \frac{\partial^2 X}{\partial x^2} + X \frac{\partial^2 Y}{\partial y^2} = 0.$$

Dividing the above equation throughout by XY,

$$\frac{1}{X} \frac{\partial^2 X}{\partial x^2} + \frac{1}{Y} \frac{\partial^2 Y}{\partial y^2} = 0. \qquad (1)$$

Note that the first term is a function of x alone and the second term is a function of y alone. Now, the above equation must hold for all values of x and y, which is possible only if each of the terms in the equation is a constant, i.e.,

$$\frac{1}{X} \frac{\partial^2 X}{\partial x^2} = \alpha^2$$

$$\frac{1}{Y} \frac{\partial^2 Y}{\partial y^2} = \beta^2 \qquad (2)$$

Substituting Eqs. (2) into Eq. (1),

$$\alpha^2 + \beta^2 = 0$$

or

$$\beta^2 = -\alpha^2$$

Eqs. (2) then become

$$\frac{1}{X} \frac{\partial^2 X}{\partial x^2} = \alpha^2$$

$$\frac{1}{Y} \frac{\partial^2 Y}{\partial y^2} = -\alpha^2$$

Since only one variable is involved in each of the above equations, the partial derivatives can be replaced by total derivatives to get

$$\frac{d^2 X}{dx^2} = \alpha^2 X \qquad (3)$$

and

$$\frac{d^2Y}{dy^2} = -\alpha^2 Y \tag{4}$$

Thus there are now two ordinary differential equations involving separately the variables x and y, starting with the partial differential equation involving both of the variables x and y.

The solutions for (3) and (4) are given by

$$X(x) = \begin{cases} Ae^{\alpha x} + Be^{-\alpha x} & \text{for } \alpha \neq 0 \\ A_0 x + B_0 & \text{for } \alpha = 0 \end{cases} \tag{5}$$

where A, B, A_0, and B_0 are arbitrary constants, and

$$Y(y) = \begin{cases} C \cos \alpha y + D \sin \alpha y & \text{for } \alpha \neq 0 \\ C_0 y + D_0 & \text{for } \alpha = 0 \end{cases} \tag{6}$$

where C, D, C_0, and D_0 are arbitrary constants. Substituting (5) and (6) into

$$V(x,y) = X(x)Y(y) ,$$

the solution is

$$V(x,y) = \begin{cases} (Ae^{\alpha x} + Be^{-\alpha x})(C \cos \alpha y + D \sin \alpha Y) & \text{for } \alpha \neq 0 \\ (A_0 x + B_0)(C_0 y + D_0) & \text{for } \alpha = 0 \end{cases} \tag{7}$$

Equation (7) is the general solution for Laplace's equation in the two dimensions x and y.

Since the slot is infinitely long in the z direction with uniform cross section, the problem is two dimensional in x and y and the general solution for V given by (7) is applicable. The boundary conditions are

$$V = 0 \qquad \text{for } y = 0,\ 0 < x < a \tag{8}$$

$$V = 0 \qquad \text{for } y = b,\ 0 < x < a \tag{9}$$

$$V = 0 \qquad \text{for } x = 0,\ 0 < y < b \tag{10}$$

$$V = V_0 \sin \frac{\pi y}{b} \qquad \text{for } x = a,\ 0 < y < b \tag{11}$$

The solution corresponding to $\alpha = 0$ does not fit the boundary conditions since V is required to be zero for two values of y and in the range $0 < x < a$. Hence ignore this solution and consider only the solution for $\alpha \neq 0$.

Applying the boundary condition (8)

$$0 = (Ae^{\alpha x} + Be^{-\alpha x})(C) \qquad \text{for } 0 < x < a$$

The only way of satisfying this equation for a range of values of x is by setting $C = 0$. Next, applying the boundary condition (10),

$$0 = (A + B)D \sin \alpha y \qquad \text{for } 0 < y < b$$

This requires that $(A + B)D = 0$, which can be satisfied by either $A + B = 0$ or $D = 0$. However, rule out $D = 0$ since it results in a trivial solution of zero for the potential. Hence set

$$A + B = 0$$

or

$$B = -A$$

Thus the solution for V reduces to

$$V(x,y) = (Ae^{\alpha x} - Ae^{-\alpha x})D \sin \alpha y$$

$$= A' \sinh \alpha x \sin \alpha y \tag{12}$$

where

$$A' = 2AD$$

Next, applying boundary condition (9) to (12),

$$0 = A' \sinh \alpha x \sin \alpha b$$

for

$$0 < x < a$$

To satisfy this equation without obtaining a trivial solution of zero for the potential, set

$$\sin \alpha b = 0$$

or

$$\alpha b = n\pi \qquad n = 1,2,3,\ldots$$

$$\alpha = \frac{n\pi}{b} \qquad n = 1,2,3,\ldots$$

Since several values of α satisfy the boundary condition, several solutions are possible for the potential. To take this into account, write the solution as the superposition of all these solutions multiplied by different arbitrary constants. In this manner,

$$V(x,y) = \sum_{n=1,2,3,\ldots}^{\infty} A_n' \sinh \frac{n\pi x}{b} \sin \frac{n\pi y}{b}$$

$$\text{for } 0 < y < b \tag{13}$$

Finally, applying the boundary condition (11) to (13)

$$V_0 \sin \frac{\pi y}{b} = \sum_{n=1,2,3,\ldots}^{\infty} A_n' \sinh \frac{n\pi a}{b} \sin \frac{n\pi y}{b}$$

$$\text{for } 0 < y < b \tag{14}$$

On the right side of (14) is an infinite series of sine terms in y, but on the left side there is only one sine term in y. Equating the coefficients of the sine terms having the same arguments,

$$A'_n \sinh \frac{n\pi a}{b} = \begin{cases} V_0 & \text{for } n = 1 \\ 0 & \text{for } n \neq 1 \end{cases}$$

or

$$A'_1 = \frac{V_0}{\sinh (\pi a/b)}$$

$$A'_n = 0 \qquad \text{for } n \neq 1$$

Substituting this result in (13), the required solution for V is

$$V(x,y) = V_0 \frac{\sinh (\pi x/b)}{\sinh (\pi a/b)} \sin \frac{\pi y}{b} \tag{15}$$

Now compute the potential at any point inside the slot given the values of a, b, and V_0. For example, for a = b, that is, for a square slot, (15) gives the potential at the center of the slot to be $0.1993V_0$.

• PROBLEM 7-14

Let the potential in the xz plane be given as a square wave in the x direction, each "wavelength" being of width 2a and amplitude V_0 as shown in the figure. The potential is independent of z. Find the potential at all points.

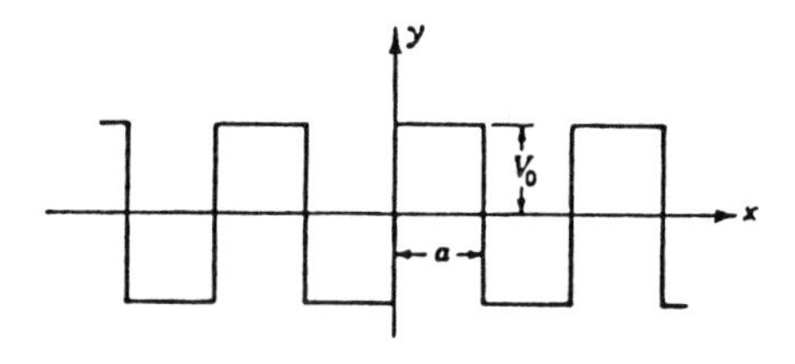

Solution: Before proceeding with the solution proper, we must obtain the boundary condition in a suitable mathematical form. This is done by Fourier series. At y = 0 the potential is a square wave as in the figure. In general this can be expressed as

$$V = \sum_{n=0}^{\infty} \left(A_n \sin \frac{\pi n x}{a} + B_n \cos \frac{\pi n x}{a} \right) \tag{1}$$

Here the function is clearly odd, so only odd functions will appear in the expansion, making $B_n = 0$. It can be easily seen that only odd n will appear. The method of Fourier series is to multiply both sides of (1) by

$$\sin \frac{\pi p x}{a} ,$$

where p is any integer, and integrate over the whole period, namely from x = -a to x = +a.

The following formulas apply:

$$\int_{-\pi}^{\pi} \sin nx \sin px \, dx = 0 \qquad \text{if } n \neq p$$

$$\int_{-\pi}^{\pi} \sin nx \cos px \, dx = 0 \qquad \text{in any case} \tag{2}$$

The result of this operation in view of (2) is that all the terms of the infinite sum on the right side of (1) vanish, except that one term for which n = p. Equation (1) then becomes, letting p = n,

$$\int_{-a}^{+a} V(x) \sin \frac{\pi n x}{a} \, dx = A_n \int_{-a}^{+a} \sin^2 \frac{\pi n x}{a} \, dx \tag{3}$$

Here it is possible to integrate from 0 to a, rather than from -a to +a and will get exactly the same value of A_n.

The reason for this is that V(x) is odd, as is $\sin \pi nx/a$ so that both integrands in (3) are even. The contribution to the integrals from -a to 0 is equal to that from 0 to a and hence only the integral from 0 to a need be computed.

A similar situation holds for an even function, where again only the integral over one of the symmetric half-periods need be computed. If the function is neither even nor odd, the integration must be done over the full period. For this problem replace (3) by

$$V_0 \int_0^a \sin \frac{\pi n x}{a}\, dx = A_n \int_0^a \sin^2 \frac{\pi n x}{a}\, dx \tag{4}$$

These integrals are easily evaluated, the one on the right becoming

$$\int_0^a \sin^2 \frac{\pi n x}{a}\, dx = \frac{a}{\pi n} \int_0^{\pi n} \sin^2 \theta \, d\theta$$

$$= \frac{a}{\pi n} \left(\frac{\theta}{2} - \frac{1}{4} \sin 2\theta \right) \Bigg|_0^{\pi n}$$

$$= \frac{a}{\pi n} \; \frac{\pi n}{2} = \frac{a}{2} \tag{5}$$

The left-hand integral of (4) becomes

$$\int_0^a \sin \frac{\pi n x}{a}\, dx = \frac{a}{\pi n} \int_0^{\pi n} \sin \theta \, d\theta =$$

$$\frac{a}{\pi n} (- \cos \pi n + 1)$$

$$= \frac{2a}{\pi n} \qquad \text{for n odd} \tag{6}$$

$$= 0 \qquad \text{for n even}$$

Equation (4) thus becomes, for n odd,

$$V_0 \frac{2a}{\pi n} = A_n \frac{a}{2}$$

or (7)

$$A_n = \frac{4V_0}{\pi n}$$

Clearly $A_n = 0$ for n even. The boundary condition at

$x = 0$ then is

$$V(x) = \sum_{\text{odd } n} \frac{4V_0}{\pi n} \sin \frac{\pi n x}{a} \quad (8)$$

Now proceed to choose the solutions of Laplaces' equation that will fit this problem.

From separation of variables, the solutions are in the form

$$X = Ae^{kx} + Be^{-kx} \quad \text{if } k \text{ is real}$$

$$X = A \sin kx + B \cos kx \quad \text{if } k \text{ is imaginary}$$

$$X = A_0 x + B_0 \quad \text{if } k \text{ is zero}$$

where the X's are the solutions to the equation

$$\frac{1}{X} \frac{d^2x}{dx^2} = k^2$$

The general solution to the problem is in the form

$$\phi(x,y,z) = X(x)\, Y(y) Z(z) \; .$$

In this problem $Z = 1$ since the problem is independent of z. Equation (8) suggests that trigonometric solutions be used for X and this forces us to use exponential solutions for Y. The solution is of the form

$$V = (Ae^{-ky} + Be^{+ky})(C \cos kx + D \sin kx) \quad (9)$$

The same k is used in the exponential solution as in the trigonometric solution because of the auxiliary condition

$$-k_x{}^2 + k_y{}^2 = 0$$

or (10)

$$k_x = \pm k_y = \text{say } \pm k$$

Since the potential must obviously vanish at $|y| = \infty$, then $B = 0$ for positive y and $A = 0$ for negative y. To meet the boundary condition at $x = 0$, choose an infinite set of solutions such as (9) with $C = 0$ and also choose $k = \pi n/a$ where n will be odd only. The solution for positive y thus becomes

$$V = \sum_{\text{odd } n} A_n e^{-\pi n y/a} \sin \frac{\pi n x}{a} \tag{11}$$

Letting y become 0 and comparing with (8) it is found

$$A_n = 4V_0/\pi n$$

and the answer becomes

$$V = \sum_{\text{odd } n} \frac{4V_0}{\pi n} e^{-\pi n y/a} \sin \frac{\pi n x}{a} \tag{12}$$

For negative y the exponent must have a positive sign. This solution seems rather formal, but some useful information can be obtained from it. Consider values of y such that $\pi y/a$ is somewhat larger than 1. Then values of n greater than 1 will not contribute very much to V and the potential will be approximately

$$V = \frac{4V_0}{\pi} e^{-\pi y/a} \sin \frac{\pi x}{a} \tag{13}$$

Thus the potential drops off as $e^{-\pi y/a}$, showing that the potential variation caused by alternating strips falls off very rapidly with distance beyond a distance about equal to a.

• PROBLEM 7-15

The potential in the xz plane is given by $V = V_0 \sin px$ and is independent of z as shown in the figure. Find the potential at all points.

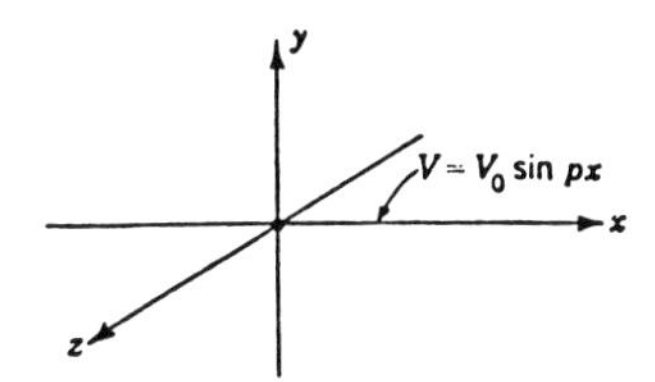

Solution: Laplace's equation is a partial differential equation and as such is generally quite difficult to solve. There are an infinite number of solutions to such equations. The method of solving a problem is to select from this infinite number of solutions a suitable set that applies to the problem at hand. Such a set of solutions will often have an infinite number of terms in it.

Laplace's equation restricts us to solving problems in regions where there is no space charge. There will of course be charges in the problem such as a few point or line charges and various conductors which will have fixed potentials. These, however, will be boundary conditions at the boundaries of the charge-free region where we are seeking solutions to Laplace's equation.

Before proceeding with a general solution to Laplace's equation it is well to mention that one special group of solutions can be written down by inspection. Clearly,

$$\phi = ax^2 + by^2 + cz^2 \tag{1}$$

is a solution, for on substitution in $\nabla^2\phi = 0$,

$$a + b + c = 0 \tag{2}$$

Not all of these constants can be positive, but any choice of a, b, and c which satisfies (2) will lead to a solution of Laplace's equation. Such solutions are hyperboloids of one or two sheets. A similar solution is easily obtained in cylindrical coordinates for problems which have rotational symmetry.

A general solution of Laplace's equation in rectangular coordinates will now be obtained. This equation is

$$\frac{\partial^2\phi}{\partial x^2} + \frac{\partial^2\phi}{\partial y^2} + \frac{\partial^2\phi}{\partial z^2} = 0 \tag{3}$$

A method of solution is called "separation of variables." A solution of the form

$$\phi(x,y,z) = X(x)Y(y)Z(z) \tag{4}$$

is assumed and an attempt is made to find the functions X, Y, and Z.

Substituting (4) into (3),

$$YZ \frac{\partial^2 X}{\partial x^2} + XZ \frac{\partial^2 Y}{\partial y^2} + XY \frac{\partial^2 Z}{\partial z^2} = 0 \tag{5}$$

Dividing by XYZ, the following is obtained

$$\frac{1}{X} \frac{\partial^2 X}{\partial x^2} + \frac{1}{Y} \frac{\partial^2 Y}{\partial y^2} + \frac{1}{Z} \frac{\partial^2 Z}{\partial z^2} = 0 \tag{6}$$

Thus the variables have been separated, for the first term is a function of x alone, the second a function of y alone, and the third a function of z alone. Now Equation (6) must hold no matter what value the variables x, y, and z have. This can only happen if each of the three terms is a constant, or

$$\begin{aligned} \frac{1}{X} \frac{\partial^2 X}{\partial x^2} &= a^2 \\ \frac{1}{Y} \frac{\partial^2 Y}{\partial y^2} &= b^2 \\ \frac{1}{Z} \frac{\partial^2 Z}{\partial z^2} &= c^2 \end{aligned} \tag{7}$$

with the auxiliary condition

$$a^2 + b^2 + c^2 = 0 \tag{8}$$

Before proceeding with the easy solution of Equation (7) note that the restriction (8) means that the arbitrary constants a, b, and c cannot all be real or imaginary and still satisfy (8). Solve one of the equations of (7) and see just what this implies. Since only one variable is involved, replace the partial derivatives by total derivatives and get

$$\frac{1}{X} \frac{d^2 X}{dx^2} = a^2$$

or (9)

$$\frac{d^2X}{dx^2} = a^2X$$

If a is real and not zero, the solutions are

$$X = Ae^{ax} + Be^{-ax}$$

or (10)

$$X = A \sinh ax + B \cosh ax$$

If a is imaginary, let $a = j\alpha$ and the equation becomes

$$\frac{d^2X}{dx^2} = -\alpha^2X \tag{11}$$

The potential itself must be real so a real solution to (11) is

$$X = A \sin \alpha x + B \cos \alpha x \tag{12}$$

The special case $a = 0$ results in the zero order or linear solution

$$X = A_0x + B_0 \tag{13}$$

All the above solutions involve two arbitrary constants because the equation is a second-order differential equation. Similar solutions exist for Y and Z. Always choose the constants so as to satisfy (8). This means that if none of a, b, or c is zero, at least one of the solutions must be trigonometric and at least one must be exponential, the third solution being of either kind. The values of the constants chosen must satisfy (8). If one of the constants is zero, one solution will be exponential, the other trigonometric, and the third a linear solution, as in (13).

Which solutions to choose depends on the problem to be solved. In general certain boundary conditions can be expressed in terms of trigonometric functions, as by a Fourier series, and this shall guide our choice of the type of solution to use. It is not convenient to use the constants a, b, and c as above in solving an actual problem, but rather certain other constants k_x, k_y, and k_z.

Thus if the desired solutions are of the form

$$\phi = \begin{Bmatrix} \sin \\ \cos \end{Bmatrix} k_x x \begin{Bmatrix} \sin \\ \cos \end{Bmatrix} k_y y \quad \left\{ e^{\pm k_z z} \right\} \tag{14}$$

the auxiliary condition (15) is expressed by

$$-k_x{}^2 - k_y{}^2 + k_z{}^2 = 0 \tag{15}$$

In (14) a special nomenclature of indicating a linear combination of two functions was used. Thus

$$\begin{Bmatrix} \sin \\ \cos \end{Bmatrix} k_x x$$

means A sin $k_x x$ + B cos $k_x x$,

and

$$\left\{ e^{\pm k_z z} \right\}$$

means

$$Ae^{k_z z} + Be^{-k_z z}$$

where A and B are arbitrary constants.

Choose the type of solution wanted and then make up an auxiliary condition (15) from these chosen solutions. Trigonometric solutions will lead to negative contributions to the auxiliary equation, and exponentials to positive contributions. It is seen from (15) that only two of the arbitrary constants can be chosen at will, the third being determined by (15).

Since this problem is independent of z, $k_z = 0$ and $Z = A_0 z + B_0$; choose $A_0 = 0$, and set $B_0 = 1$.

The boundary condition at y = 0 suggests a trigonometric solution for X with $k_x = p$. The auxiliary condition then becomes

$$-k_x{}^2 + k_y{}^2 = -p^2 + k_y{}^2 = 0$$

or

$$k_y = \pm p$$

The solution then is of the form

$$V = (A \sin px + B \cos px)(Ce^{py} + De^{-py})$$

It is clear that in the region of positive y the potential must become zero at large y. Therefore C = 0 for positive y. The boundary condition at y = 0 is then satisfied by taking B = 0 A = V_0, giving the solution

$$V = V_0 e^{-py} \sin px \qquad y \geq 0$$

The solution for negative y is clearly

$$V = V_0 e^{+py} \sin px \qquad y \leq 0$$

To check this note that the solution meets the boundary conditions at y = 0 and at y = $\pm\infty$ and that it is also a solution of Laplace's equation. Such a solution is unique and these are only conditions that the solution must satisfy.

POISSON'S EQUATION

• PROBLEM 7-16

The region between two concentric conducting cylinders with radii 2 and 5 cm contains a uniform volume charge distribution of -10^{-8} C/m^3. If E_r and V are both zero at the inner cylinder, find V at the outer cylinder. Refer to the diagram.

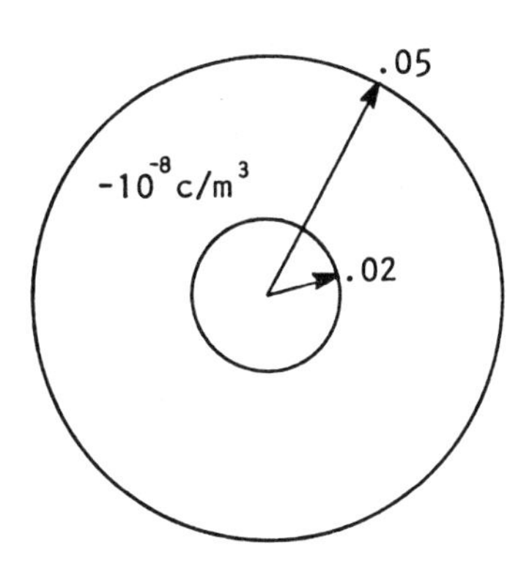

Solution: Assume air dielectric, Poisson's equation is

$$\nabla^2 V = -\rho/\varepsilon_0 = -360\pi$$

In cylindrical coordinates and by symmetry.

$$\frac{d}{dr}\left(r\,\frac{dV}{dr}\right) = 360\pi$$

Integrating once,

$$\frac{dV}{dr} = 180\pi r + \frac{A}{r}$$

Apply the boundary condition that E_r at r = 2 cm is zero and solve for A

$$A = -180\pi(4 \times 10^{-4})$$

Substituting back into the equation for dV/dr and integrating again,

$$V = 180\pi\left[\frac{r^2}{2} - (4 \times 10^{-10})\ \ln r + B\right]$$

Again, apply the boundary condition that $V_r = 0$ at r = 2 cm, and solve for B,

$$B = 4 \times 10^{-4}\ \ln 0.02 - 2 \times 10^{-4}$$

Therefore

$$V = 180\pi\left[\frac{r^2}{2} - 4 \times 10^{-4}\ (\ln r - \ln 0.02) - 2 \times 10^{-4}\right]$$

On the outer cylinder, r = 5 cm,

$$V = 180\pi\left[\frac{1}{2}\ \{(0.05)^2 - 4 \times 10^{-4}\} - 4 \times 10^{-4}\ \ln 50(0.05)\right]$$

$$= 0.387 \text{ volt.}$$

• **PROBLEM 7-17**

Suppose that the region between plates of the parallel-plate capacitor shown in the figure is filled with charge continuously distributed with a density

$$\rho(x) = \rho_0 x,$$

where ρ_0 is a constant. Find the resulting electric field, neglecting fringing effects, and the capacitance.

Solution: Since the problem is one-dimensional, Poisson's equation reduces to

$$\frac{d^2\phi}{dx^2} = -\frac{\rho_0}{\varepsilon_0} x$$

The general solution is of the form

$$\phi = -\frac{\rho_0}{6\varepsilon_0} x^3 + Bx + C$$

Applying the boundary conditions

$$\phi = \begin{cases} 0 & \text{at} \quad x = 0 \\ V & \text{at} \quad x = d \end{cases}$$

gives

$$\phi = \frac{V}{d} x + \frac{\rho_0}{6\varepsilon_0} (xd^2 - x^3)$$

The electric field intensity calculated from $\bar{E} = -\bar{\nabla}\phi$ is

$$\bar{E} = \left[-\frac{V}{d} + \frac{\rho_0}{6\varepsilon_0} (3x^3 - d^2)\right] \bar{a}_x$$

The distribution of charge on the bottom plate $(x = 0)$

$$\rho_s = -\varepsilon_0 \left(\frac{V}{d} + \frac{\rho_0}{6\varepsilon_0} d^2\right)$$

is seen to differ from the distribution on the top plate $(x = d)$:

$$\rho_s = \varepsilon_0 \left(\frac{V}{d} + \frac{\rho_0}{3\varepsilon_0} d^2 \right)$$

Therefore the total charges on the top and bottom plates are obviously different from each other. This suggests that the term capacitance is meaningless here because it can no longer be defined uniquely.

• PROBLEM 7-18

Consider a sphere of radius a containing charge of constant density so that ρ = const. inside, while $\rho = 0$ outside. Find ϕ everywhere by solving Poisson's equation.

Solution: Poisson's equation is $\nabla^2\phi = \rho/\varepsilon_0$. If ϕ is a function of r only, then the equation reduces to

$$\frac{1}{r^2} \frac{d}{dr} \left(r^2 \frac{d\phi}{dr} \right) = \frac{-\rho(r)}{\varepsilon_0} .$$

Outside the sphere, this equation becomes

$$d(r^2 d\phi_0/dr)/dr = 0,$$

which can be integrated twice to give

$$\phi_0(r) = A_0 + \frac{B_0}{r} \tag{1}$$

where A_0 and B_0 are constants of integration. Inside the sphere

$$\frac{d}{dr} \left(r^2 \frac{d\phi_i}{dr} \right) = - \frac{\rho r^2}{\varepsilon_0}$$

which can be easily integrated twice to give

$$\phi_i(r) = - \frac{\rho r^2}{6\varepsilon_0} + A_i + \frac{B_i}{r} \tag{2}$$

where A_i and B_i are constants. Now all that remains to be done is to evaluate the constants of integration from the boundary conditions.

Since all of the charges are contained within a finite volume, ϕ is to vanish at infinity. Thus, as

$$r \to \infty, \quad \phi_0 \to 0,$$

and we see from (1) that $A_0 = 0$ and therefore

$$\phi_0 = B_0/r.$$

Since there are no point charges at the origin (ϕ is finite at the origin) equation (2) still is applicable and shows us, that, in (2) $B_i = 0$. ϕ is continuous at $r = a$, so that so that $\phi_0(a) = \phi_i(a)$ which gives

$$A_i = (B_0/a) + (\rho a^2/6\varepsilon_0),$$

making ϕ_i now have the form

$$\phi_i(r) = \frac{\rho}{6\varepsilon_0}(a^2 - r^2) + \frac{B_0}{a} \tag{3}$$

Finally, since there is no surface charge, $E_n = E_r$ is continuous at $r = a$, so that

$$-(\partial\phi_0/\partial r)_{r=a} = -(\partial\phi_i/\partial r)_{r=a},$$

which leads to

$$(B_0/a^2) = (\rho a/3\varepsilon_0)$$

or

$$B_0 = \rho a^3/3\varepsilon_0 .$$

Putting this into (1) (with $A_0 = 0$) and (3),

$$\phi_0(r) = \frac{\rho a^3}{3\varepsilon_0 r},$$

$$\phi_i(r) = \frac{\rho}{6\varepsilon_0}(3a^2 - r^2)$$

• PROBLEM 7-19

Two infinite parallel plates separated by a distance d are at potentials 0 and V_0 as shown in the figure. The space charge density between the plates is given as

$$\rho = \rho_0 \frac{\Psi}{d},$$

where the distance Ψ is measured from the plate at zero potential. Find the potential V between the plates using Poisson's equation.

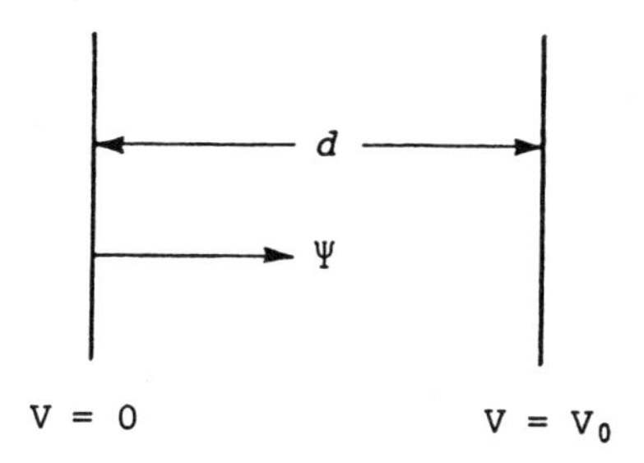

Solution:

Poisson's equation is

$$\nabla^2 V = -\rho/\varepsilon_0$$

and since

$$V = f(x) \text{ only}$$

Poisson's equation becomes

$$\frac{d^2V}{d\Psi^2} = \frac{d}{d\Psi}\left(\frac{dV}{d\Psi}\right) = \frac{-\rho}{\varepsilon_0} = -\frac{\rho_0}{\varepsilon_0} \cdot \frac{\Psi}{d}$$

or

$$d\left(\frac{dV}{d\Psi}\right) = \frac{-\rho_0}{\varepsilon_0 d} \Psi \, d\Psi$$

Integrating

$$\int d\left(\frac{dV}{d\Psi}\right) = \frac{-\rho_0}{\varepsilon_0 d} \int \Psi d\Psi$$

$$\frac{dV}{d\Psi} = \frac{-\rho_0}{\varepsilon_0 d} \frac{\Psi^2}{2} + C_1$$

Integrating again

$$V = \frac{-\rho_0}{\varepsilon_0 d} \frac{\Psi^3}{6} + C_1 \Psi + C_2 \qquad (1)$$

where C_1 and C_2 are integration constants to be determined from the boundary conditions.

At

$$\Psi = 0, \qquad V = 0,$$

equation (1) becomes

$$0 = 0 + 0 + C_2$$

$$C_2 = 0$$

$\therefore$

equation (1) becomes

$$V = \frac{-\rho_0}{\varepsilon_0 d} \frac{\Psi^3}{6} + C_1\Psi \qquad (2)$$

at $\Psi = d$, $V = V_0$, and equation (2) becomes

$$V_0 = \frac{-\rho_0}{\varepsilon_0 d}\frac{d^3}{6} + C_1 d$$

hence

$$C_1 = \frac{V_0}{d} + \frac{\rho_0 d}{6\varepsilon_0} \tag{3}$$

Substituting equation (3) for C_1 into equation (2)

$$V = \frac{-\rho_0 \Psi^3}{6\varepsilon_0 d} + \left(\frac{V_0}{d} + \frac{\rho_0 d}{6\varepsilon_0}\right)\Psi$$

• PROBLEM 7-20

Consider a large spherical region of radius R with a uniform charge q_V per unit volume. Find the electric field at the interior and exterior of the sphere by solving Poisson's and Laplace's equation respectively.

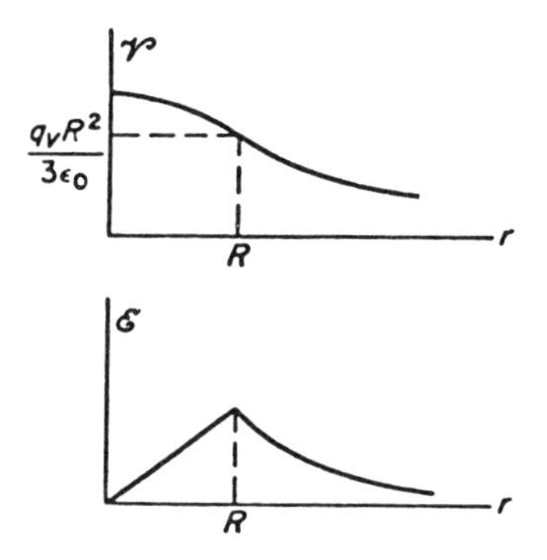

Potential and field for a uniformly charged sphere of radius R

Solution: In spherical coordinates and since the potential is a function of r only,

$$\frac{d^2V}{dr^2} + \frac{2}{r}\frac{dV}{dr} = \begin{cases} \dfrac{-q_V}{\varepsilon_0} & \text{for } r \leqq R \\ 0 & \text{for } r \geqq R \end{cases} \tag{1}$$

For the exterior of the sphere $(r > R)$,

$$\frac{d^2V}{dr^2} + \frac{2}{r}\frac{dV}{dr} = 0$$

Introduce the notation

$$\frac{dV}{dr} = V'$$

and now

$$\frac{dV'}{V'} = \frac{-2dr}{r}$$

Integrating,

$$\ln V' = -2 \ln r + C$$

let $C = \ln A$, where both C & A are constants.

Then

$$\ln V' = \ln \frac{A}{r^2}$$

and $dV = \dfrac{Adr}{r^2}$

or

$$V_E = -\frac{A}{r} + B \qquad (2)$$

For the interior of the sphere $(r < R)$, use a technique often found useful in solving awkward differential equations. The form of (2) leads to a solution

$$V_I = \frac{f(r)}{r} \qquad (3)$$

where f(r) is a new function to be determined. Differentiating (3) and substituting, eq. (1) is converted into

$$\frac{d^2f}{dr^2} = \frac{-q_V r}{\varepsilon_0}$$

Integrating twice

$$f = \frac{-q_V r^3}{6\varepsilon_0} + Cr + D$$

and

$$V_I = \frac{-q_V r^2}{6\varepsilon_0} + C + \frac{D}{r}$$

As boundary conditions, first use the fact that $V_E = 0$ at $r = \infty$, making $B = 0$. Next, unless $D = 0$, $V_I = \infty$ at $r = 0$. Finally, impose the condition that V_E and V_I must have the same value $q/4\pi\varepsilon_0 R$ at $r = R$. Thus

$$\frac{q_V(\frac{4}{3}\pi R^3)}{4\pi\varepsilon_0 R} = \frac{-q_V R^2}{6\varepsilon_0} + C = \frac{-A}{R}$$

or

$$C = \frac{q_V R^2}{2\varepsilon_0}, \quad A = \frac{-q_V R^3}{3\varepsilon_0}$$

and

$$V_I = \frac{q_V}{2\varepsilon_0}\left(R^2 - \frac{r^2}{3}\right)$$

$$V_E = \frac{q_V R^3}{3\varepsilon_0 r}$$

Using

$$E = E_r = -\ \mathrm{grad}_r V = -\ \frac{\partial V}{\partial r}$$

the interior and exterior fields are given by

$$E_I = \frac{q_V r}{3\varepsilon_0}, \quad E_E = \frac{q_V R^3}{3\varepsilon_0 r^2}$$

Plots of both E and V are shown in the accompanying figure and note that the field increases linearly from the center to the surface of the sphere and then drops off again as r increases further.

• PROBLEM 7-21

Two similar, parallel, conducting plates are spaced 3 cm apart in air, a distance that is small compared with their surface dimensions. A 30-V source is connected between the plates. At 1 cm from the negative plate there is a thin sheet of charge for which $\rho_s = -a\varepsilon_0$ coulombs, where a is a positive constant. Describe the potential distribution between the plates. Use Poisson's equation.

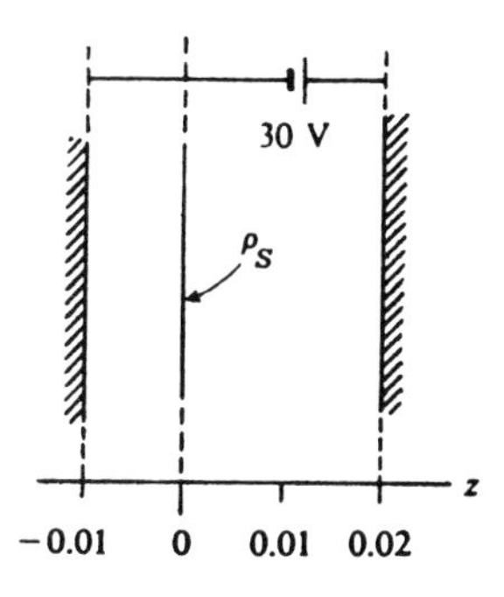

Fig. 1

Solution: If the x,y-plane is taken in the plane of the surface charge and the problem is idealized to the case of infinite planes for the conductors and the charge as illustrated in Fig. 1, the potential is obviously a function of z only. Hence the equation to be satisfied in V is

$$\bar{\nabla}^2\Psi = -\frac{\rho}{\varepsilon_0} \tag{1}$$

or

$$\frac{d^2\Psi}{dz^2} = a\delta(z) \tag{2}$$

The particular integral Ψ_p is the potential due to the surface charge alone. The electric field of such charge has a magnitude $|\rho_s/2\varepsilon_0|$. In this case the field magnitude is $|a/2|$, and since the charge is negative, the direction is $-\bar{1}_z$ for positive z and $+\bar{1}_z$ for negative z. The potential therefore increases in either direction in direct proportion to the distance from the surface charge. The particular solution to Poisson's equation may therefore be written

$$\Psi_p = \frac{a|z|}{2} \tag{3}$$

It is desirable to check this solution by substitution back into Eq. 2. A first differentiation of Eq. 3 yields

$$\frac{d\Psi_p}{dz} = \begin{cases} \frac{a}{2}, & \text{for } z > 0 \\ \frac{-a}{2}, & \text{for } z < 0 \end{cases} \tag{4}$$

Graphs of Ψ_p and $d\Psi_p/dz$ are illustrated in Fig. 2.

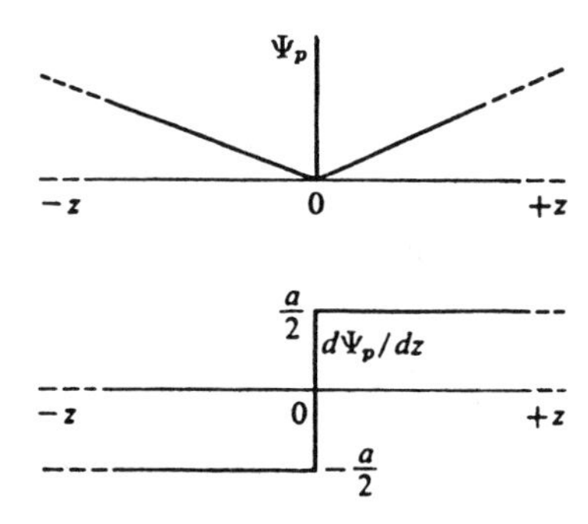

Fig. 2

From the figure it is apparent that

$$\frac{d^2\Psi_p}{dz^2} = 0, \qquad z \neq 0 \tag{5}$$

Furthermore

$$\int_{-\infty}^{+\infty} \frac{d^2\Psi_p}{dz^2}\, dz = \frac{d\Psi_p}{dz}\Bigg|_{z\to-\infty}^{z\to\infty} = \frac{a}{2} - \left(-\frac{a}{2}\right) = a \tag{6}$$

Equations 5 and 6 define the function $a\delta(z)$. Hence Eq. 1 is satisfied by the Ψ_p given in Eq. 3. Notice that Eq. 3 causes the surfaces of the conductors to be equipotentials, but does not in general duplicate the specified difference in potential. Basically this is the reason that the complementary solution Ψ_c is needed.

Since Laplace's equation reduces to

$$\frac{d^2\Psi_c}{dz^2} = 0 \tag{7}$$

the complementary solution is

$$\Psi_c = k_1 z + k_2 \tag{8}$$

The complete solution to the Poisson problem is therefore

$$\Psi = \frac{a|z|}{2} + k_1 z + k_2 \tag{9}$$

Letting $z = 0$, it is seen that k_2 is the absolute potential at the surface charge. Since this value is unspecified, ignore it for the moment and proceed to impress the boundary conditions. In this case, there is only one condition, which is

$$\Psi(0.02) - \Psi(-0.01) = 30 \tag{10}$$

Since

$$\Psi(0.02) = 0.01a + 0.02k_1 + k_2 \tag{11}$$

and

$$\Psi(-0.01) = 0.005a - 0.01k_1 + k_2 \tag{12}$$

substitution of Eqs. 11 and 12 into Eq. 10 reveals that the boundary condition is satisfied if

$$k_1 = 1000 - \frac{a}{b} \tag{13}$$

Thus a satisfactory solution is

$$\Psi = \frac{a|z|}{2} + \left(1000 - \frac{a}{6}\right) z + k_2$$

Obviously the value of k_2 has no bearing on the electric field between the plates and in fact serves no purpose other than to relate the potential between the plates to potentials in the outside world. Since the concern is not with points outside the plates in this problem, it is convenient to let $k_2 = 0$. With this arbitrary condition, the potential function between the plates (not the absolute potential necessarily) takes its simplest form:

$$\Psi = \frac{a|z|}{2} + \left(1000 - \frac{a}{6}\right) z$$

Obviously a must be at least of the order of 1000 in order to be significant. For a = 2000, for instance,

$$\Psi(z) = 1000 \; |z| + 667 \; z$$

Then

$$\Psi(0.02) = 20 + 13.3 = 33.3 \text{ V}$$

and

$$\Psi(-0.01) = 10 - 6.67 = 3.3 \text{ V}$$

We may, in fact, find a value for a to determine a special condition such as $\Psi(-0.01) = 0$. For this case,

$$\Psi(-0.01) = 0 = \frac{0.01a}{2} - 10 + \frac{0.01a}{6}$$

and

$$a = 1500$$

Notice that when a has this value, there will be no field to the left of the origin.

Consider a p-n junction between two halves of a semiconductor bar extending in the x-direction. Assume that the region for $x < 0$ is doped p type and that the region for $x > 0$ is n type. The degree of doping is identical on each side of the junction. The type of charge distribution is shown in fig. 1a and can be approximated by the equation

$$\rho = 2\rho_0 \operatorname{sech} \frac{x}{a} \tanh \frac{x}{a} \tag{1}$$

which has a maximum charge density $\rho_{max} = \rho_0$ at $\frac{x}{a} = 0.881$.

(a) Solve Poisson's equation to obtain the E field and plot this with respect to x/a.

(b) Solve for the potential V if the potential at the center of the junction is taken to be zero. Plot this potential function.

(c) Find the capacitance of this bar.

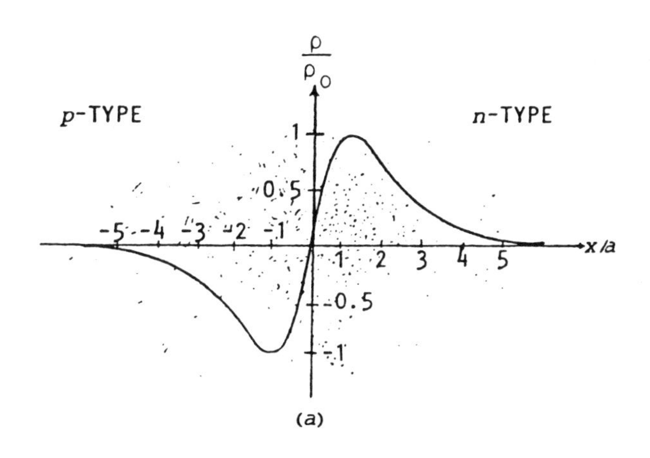

(a)

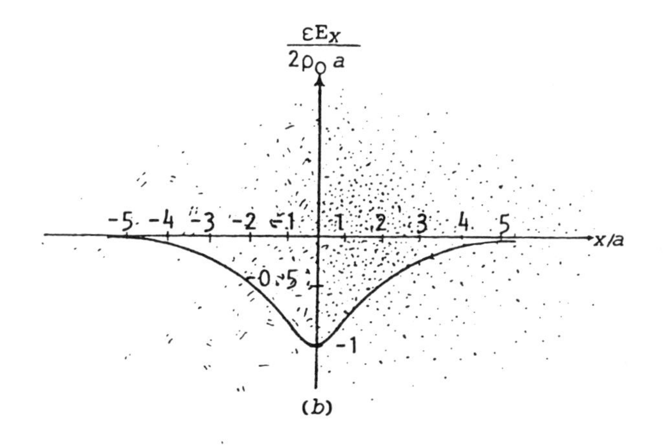

(b)

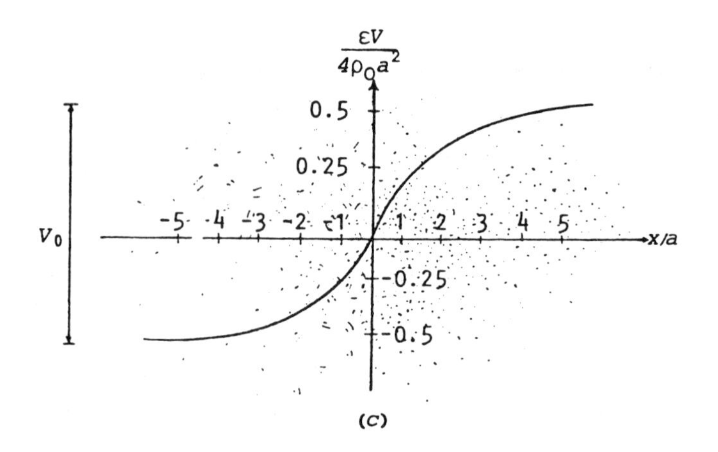

(a)The charge density,(b)the electric field intensity, and (c)the potential are plotted for a *pn* junctions as functions of distance from the center of the junction. The *p-type* material is on the left, and the *n-type* is on the right.

Solution: (a) Solving Poisson's equation,

$$\nabla^2 V = -\frac{\rho}{\varepsilon}$$

Substituting for ρ,

$$\frac{d^2V}{dx^2} = -\frac{2\rho_0}{\varepsilon} \operatorname{sech} \frac{x}{a} \tanh \frac{x}{a}$$

in this one-dimensional problem in which variations with y and z are not present. Integrate once,

$$\frac{dV}{dx} = \frac{2\rho_0 a}{\varepsilon} \operatorname{sech} \frac{x}{a} + C_1$$

and obtain the electric field intensity,

$$E_x = -\frac{2\rho_0 a}{\varepsilon} \operatorname{sech} \frac{x}{a} - C_1$$

To evaluate the constant of integration C_1, note that no net charge density and no fields can exist far from the junction. Thus, as $x \to \pm\infty$, E_x must approach zero. Therefore $C_1 = 0$, and

$$E_x = -\frac{2\rho_0 a}{\varepsilon} \operatorname{sech} \frac{x}{a} \qquad (2)$$

(b) Integrating again,

$$V = \frac{4\rho_0 a^2}{\varepsilon} \tan^{-1} e^{x/a} + C_2$$

At x = 0, V = 0; therefore solving for C_2,

$$0 = \frac{4\rho_0 a^2}{\varepsilon} \; \frac{\pi}{4} + C_2$$

and finally,

$$V = \frac{4\rho_0 a^2}{\varepsilon} \left(\tan^{-1} e^{x/a} - \frac{\pi}{4} \right) \tag{3}$$

Fig. 1 shows the charge distribution, electric field intensity, and the potential, as given by (1), (2), and (3), respectively.

The potential is constant at a distance of about 4a or 5a from the junction. The total potential difference V_0 across the junction is obtained from (3),

$$V_0 = \frac{2\pi\rho_0 a^2}{\varepsilon} \tag{4}$$

(c) The total positive charge is

$$Q = S \int_0^\infty 2\rho_0 \operatorname{sech} \frac{x}{a} \tanh \frac{x}{a} \, dx = 2\rho_0 a S$$

where S is the area of the junction cross section. Make use of (4) to eliminate the distance parameter a, the charge becomes

$$Q = S \sqrt{\frac{2\rho_0 \varepsilon V_0}{\pi}} \tag{5}$$

$$I = \frac{dQ}{dt} = C \frac{dV_0}{dt}$$

and thus

$$C = \frac{dQ}{dV_0}$$

By differentiating (5) the capacitance is

$$C = \sqrt{\frac{\rho_0 \varepsilon}{2\pi V_0}}\, S = \frac{\varepsilon S}{2\pi a} \quad (6)$$

ITERATION METHOD

• PROBLEM 7-23

Consider a square region with conducting boundaries (Fig. 1). The potential of the top is 100 V and that of the sides and bottom is zero. Using the iteration method, calculate the potential at the grid points dividing the square region into 16 squares.

Cross section of a square trough with sides and bottom at zero potential and top at 100 V. The cross section has been divided into 16 squares, with the potential estimated at every corner.

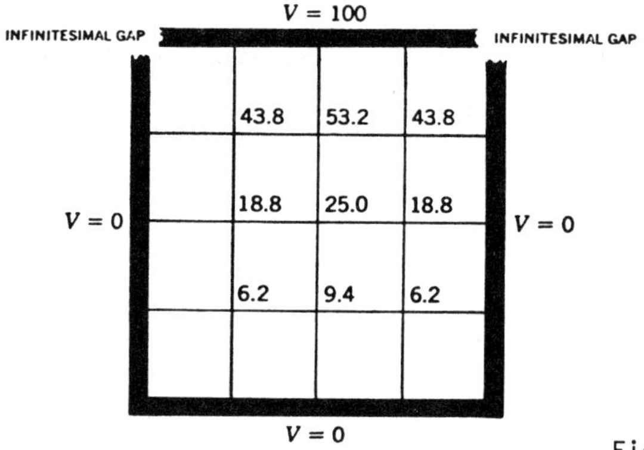

Fig. 1

Solution: Before using the iteration method, some estimate of the potential must be made at every corner. At the center of the figure the potential estimate is then 1/4(100 + 0 + 0 + 0) = 25.0.

The potential may now be estimated at the centers of the four double-sized squares by taking the average of the potentials at the four corners, or applying

$$V = 1/4(V_1 + V_2 + V_3 + V_4) \quad (1)$$

along a diagonal set of axes. Use of this "diagonal average" is made only in preparing initial estimates. For the two upper double squares, select a potential of 50 V for the gap (the average of 0 and 100), and then V = 1/4(50 + 100 + 25 + 0) = 43.8 (to the nearest tenth of a volt), and for the lower ones,

$V = 1/4(0 + 25 + 0 + 0) = 6.2$

The potential at the remaining four points may now be obtained by applying eq. (1) directly. The complete set of estimated values is shown in Fig. 1.

The initial traverse is now made to obtain a corrected set of potentials, beginning in the upper left corner (with the 43.8 value, not with the boundary where the potentials are known and fixed), working across the row to the right, and then dropping down to the second row and proceeding from left to right again. Thus the 43.8 value changes to $1/4(100 + 53.2 + 18.8 + 0) = 43.0$. The best or newest potentials are always used when applying (1), so both points marked 43.8 are changed to 43.0, because of the evident symmetry, and the 53.2 value becomes $1/4(100 + 43.0 + 25.0 + 43.0) = 52.8$.

Because of the symmetry, little would be gained by continuing across the top line. Each point of this line has now been improved once. Dropping down to the next line, the 18.8 value becomes

$$1/4(43.0 + 25.0 + 6.2 + 0) = 18.6$$

and the traverse is completed in this manner. The values at the end of this traverse are shown as the top numbers in each column of Fig. 2. Additional traverses must now be made until the value at each corner shows no change. The values for the successive traverses are usually entered below each other in column form, as shown in Fig. 2, and the final value is shown at the bottom of each column. Only four traverses are required in this example.

The results of each of the four necessary traverses are shown in order in the columns. The final values, unchanged in the last traverse, are at the bottom of each column.

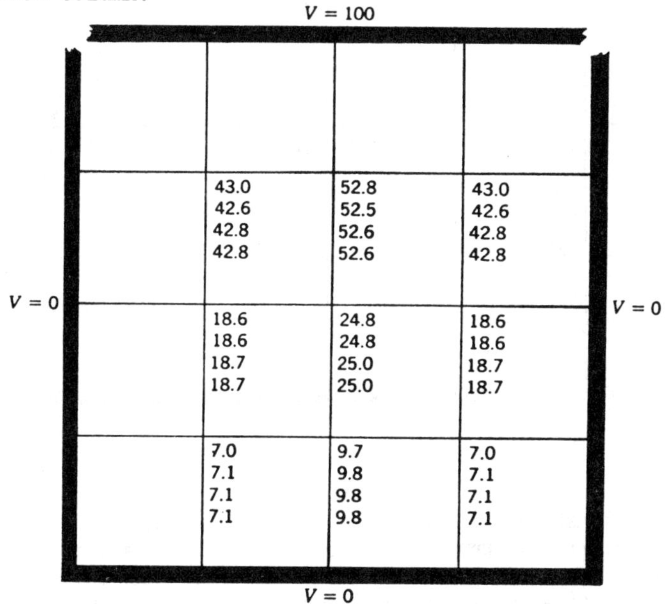

Fig. 2

Since there is a large difference in potential from square to square, the results should not be expected to be accurate to the tenth of a volt shown (and perhaps not to the nearest volt). Increased accuracy comes from dividing each square into four smaller squares, and not from finding the potential to a larger number of significant figures at each corner.

The problem divided into smaller squ..s. Values obtained on the nine successive traverses are listed in order in the columns.

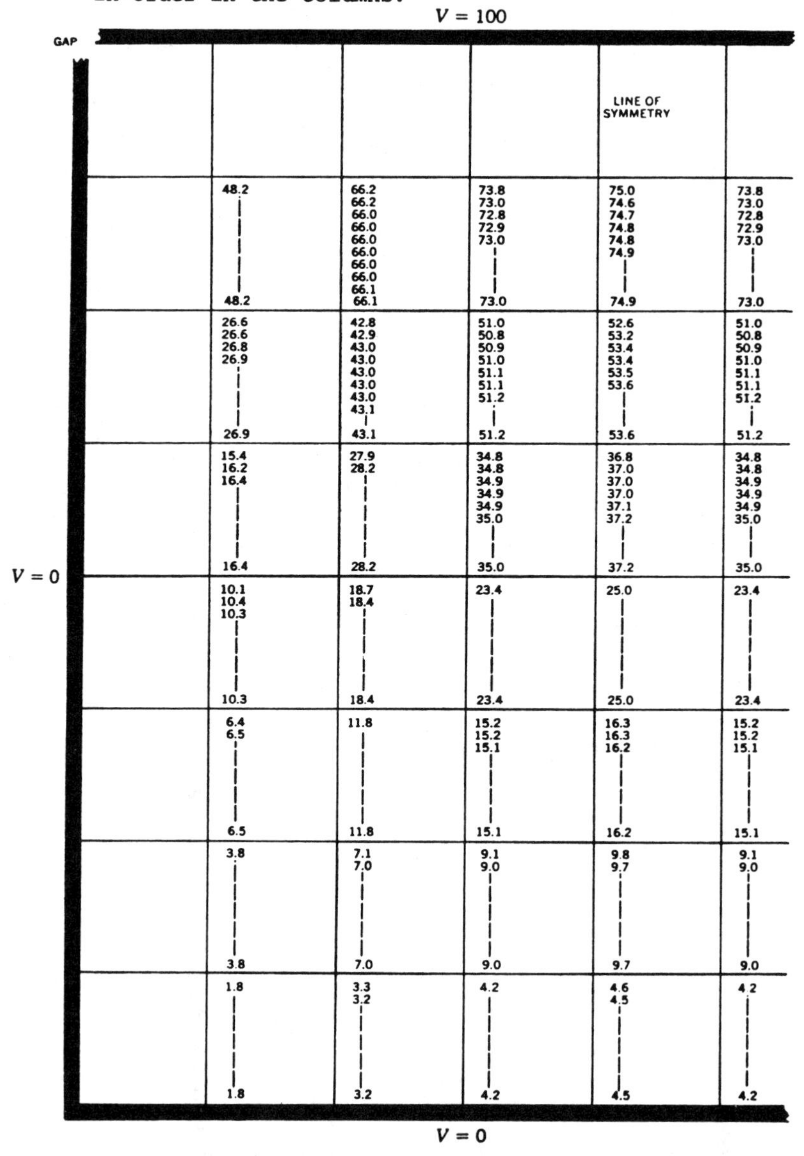

Fig. 3

In Fig. 3, which shows only one of the symmetrical halves plus an additional column, this subdivision is accomplished, and the potential at the newly created corners is estimated by applying (1) directly where possible and diagonally when necessary. The set of estimated values appears at the top of each column, and the values produced by the successive traverses appear in order below. Here nine sets of values are required, and it might be noted that no values change on the last traverse (a necessary condition for the last traverse) and only one value changes on each of the preceding three traverses. No value in the bottom four rows changes after the second traverse; this results in a great saving in time, for if none of the four potentials in (1) changes, the answer is of course unchanged. When more accurate results are required, the square region will need to be further subdivided. For a very large number of subdivisions, use of a digital computer will be most applicable.

IMAGES

• PROBLEM 7-24

Place a positive point-charge q at a distance d from the grounded, conducting plane of Fig. 1a. The presence of the charge in front of the plane causes a redistribution of the electrons in the metal, giving it a negative surface charge which is a maximum at the point 0. The creation of this surface charge is called charging by induction, and it results in an attractive force between the point-charge and the metal. Compute this force, as well as the induced charge-density and the field at the plane.

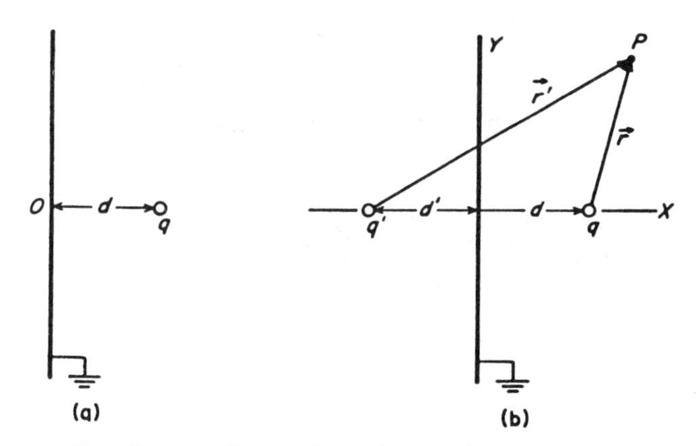

The image of a point-charge in a grounded, conducting plane.

Fig.1

Solution: To solve this problem, place a fictitious charge q' at a distance d' behind the plane (Fig. 1b) and temporarily ignore the existence of the plane. Then the potential V at some arbitrary point P is

$$V = \frac{1}{4\pi\varepsilon_0}\left[\frac{q}{r} + \frac{q'}{r'}\right] \quad (1)$$

In order to have V = 0 along the surface x = 0 (corresponding to a grounded plane), Eq. (1) shows that the following conditions should be imposed,

$$q' = -q, \quad d' = d \quad (2)$$

Geometrically, this result corresponds to the position of the image in a plane mirror. Replacing the plane, the lines of force from the charge to the plane and the equipotentials can now be drawn (Fig. 2).

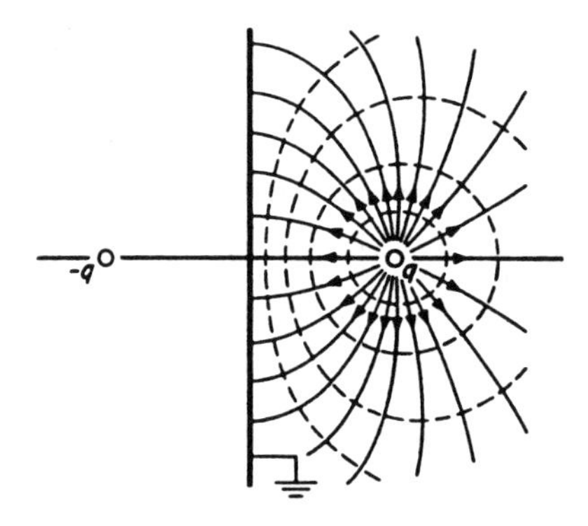

The lines of force and equipotentials for a point-charge and a grounded, conducting plane.

Fig.2

At the plane, E must be normal and this normal field is then (Fig. 3)

$$\bar{E}_n = -2\left(\frac{q \cos \theta}{4\pi\varepsilon_0 r^2}\right) = -\left(\frac{qd}{2\pi\varepsilon_0 r^3}\right) \quad (3)$$

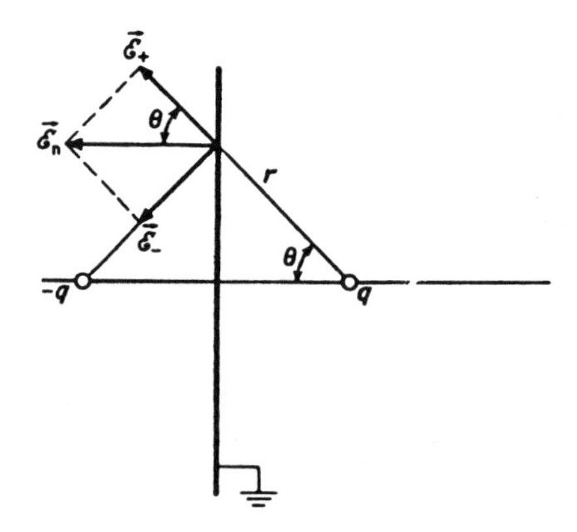

Computation of surface charge induced on a plane by a point charge.

Fig.3

The surface charge-density induced on the plane may be obtained from

$$\bar{E}_n = \frac{q_s}{\varepsilon_0}$$

and by (3)

$$q_s = \frac{-qd}{2\pi r^3} \tag{4}$$

This shows that the induced charge-density falls off as the cube of the distance from the plane to the position of q.

To verify that the total induced charge is -q, consider an element of area dS on the plane (Fig. 4). The corresponding charge dq_p on the plane is

$$dq_p = q_s dS$$

Then

$$q_p = \frac{-qd}{2\pi} \int_0^{2\pi} d\phi \int_0^{\infty} \frac{\rho d\rho}{(\rho^2 + d^2)^{3/2}} = -q$$

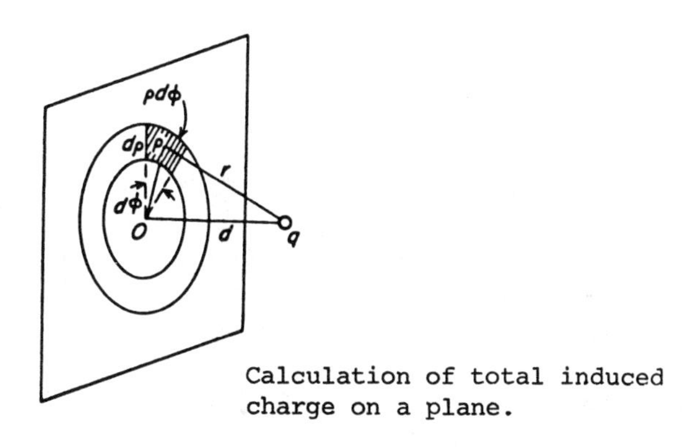

Calculation of total induced charge on a plane.

Fig.4

The force between the charge and the plane is the same as the force between the charge and its image, or

$$F = \frac{-q^2}{16\pi\varepsilon_0 d^2} \tag{5}$$

• **PROBLEM 7-25**

Find an exact expression for the capacitance of a right circular cylinder of radius p whose axis is parallel to and distant s from a plane conducting surface.

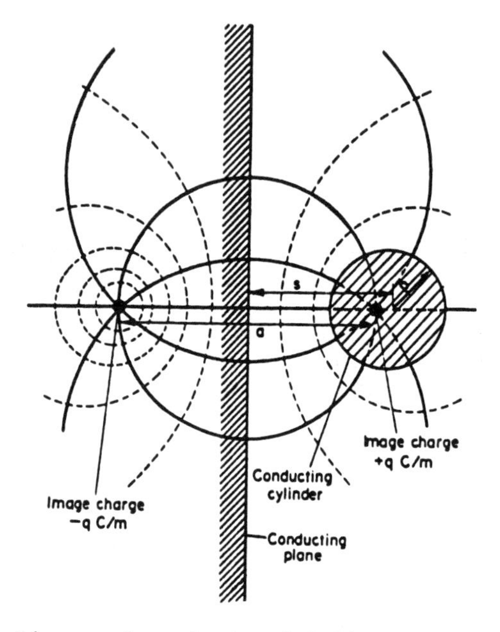

Flux lines and equipotential lines for a line source parallel to the axis of a perfectly conducting cylinder, which is the same field pattern as for a perfectly conducting cylinder parallel to a plane conducting surface.

Fig.1

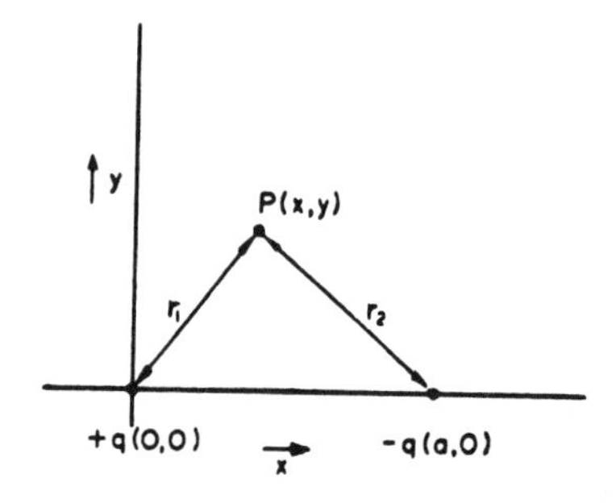

Dimensions of a point relative to a line source and a line sink.

Fig.2

Solution: The equipotential lines due to a line source parallel to an equal line sink are circular. Therefore the circular cylinder parallel to a plane surface may be represented by a line source and its image line sink as shown in fig. 1. The surface of the cylinder coincides with one of the circular equipotential lines shown in fig. 1 and the plane surface coincides with the equipotential line half-way between the line source and its image; both source and sink are images. The difficulty is that the center of the equipotential circles does not coincide with the line of the line source. It is necessary to find an expression for the center of the equipotential circle. Consider a line source distant a from an equal line sink as shown in Fig. 2. The equation for the equipotential line is

$$\phi = \frac{q}{2\pi\varepsilon_0} \log_e \frac{r_2}{r_1} ,$$

where ϕ is a constant. Rearranged, this becomes

$$\frac{r_2^2}{r_1^2} = \exp \frac{4\pi\varepsilon_0\phi}{q} = c \text{ (say)}.$$

From the geometry of Fig. 2.,

$$r_1^2 = x^2 + y^2,$$

$$r_2^2 = (a - x)^2 + y^2.$$

Therefore

$$(a - x)^2 + y^2 = c(x^2 + y^2),$$

which is the equation of a circle. Rewriting this equation gives

$$\left(x + \frac{a}{c - 1}\right)^2 + y^2 = \frac{ca^2}{(c - 1)^2} ,$$

which shows that the circle has a radius of $ac^{\frac{1}{2}}/(c - 1)$ and centre coordinates

$$\{-a/(c - 1), 0\}.$$

From the geometry of the problem, the radius of the cylinder is p and the position of its centre relative to the coordinate system in Fig. 2 is $\frac{1}{2}a - s$. Therefore

$$p = \frac{ac^{\frac{1}{2}}}{c - 1}$$

and

$$\tfrac{1}{2}a - s = \frac{-a}{c - 1} \ .$$

From the second of these equations,

$$c - 1 = \frac{2a}{2s - a} \ .$$

Therefore

$$c = \frac{2s + a}{2s - a} \ .$$

Substituting into the other of these equations gives

$$p^2\left(\frac{2a}{2s - a}\right)^2 = a^2\left(\frac{2s + a}{2s - a}\right) .$$

Therefore

$$4a^2p^2 = a^2(4s^2 - a^2)$$

and

$$a^2 = 4(s^2 - p^2).$$

On the surface of the cylinder,

$$r_1 = p - \frac{a}{c - 1} = \tfrac{1}{2}a - s + p,$$

$$r_2 = a - r_1 = \tfrac{1}{2}a + s - p.$$

Substituting for the value of a gives

$$\frac{r_2}{r_1} = \frac{\sqrt{(s + p)} + \sqrt{(s - p)}}{\sqrt{(s + p)} - \sqrt{(s - p)}} \ .$$

On the surface of the plane conducting surface, $r_1 = r_2$, so that this has a potential function value of zero and the potential difference between the cylinder and the plane is given by

$$\phi = \frac{q}{2\pi\varepsilon_0} \log_e \frac{\sqrt{(s+p)} + \sqrt{(s-p)}}{\sqrt{(s+p)} - \sqrt{(s-p)}} .$$

Therefore the capacitance is given by

$$C = 2\pi\varepsilon_0 \Big/ \log_e \frac{\sqrt{(s+p)} + \sqrt{(s-p)}}{\sqrt{(s+p)} + \sqrt{(s-p)}} .$$

• PROBLEM 7-26

A point charge Q is a distance d from the center of a grounded (U = 0) spherical conductor of radius a (Fig. 1). Consider a conductor as grounded when it is conductively connected to an unlimited source of charge at infinity. The point charge Q can therefore attract a charge -kQ from infinity and the sphere acquires a surface charge density σ. Use the method of images to evaluate σ.

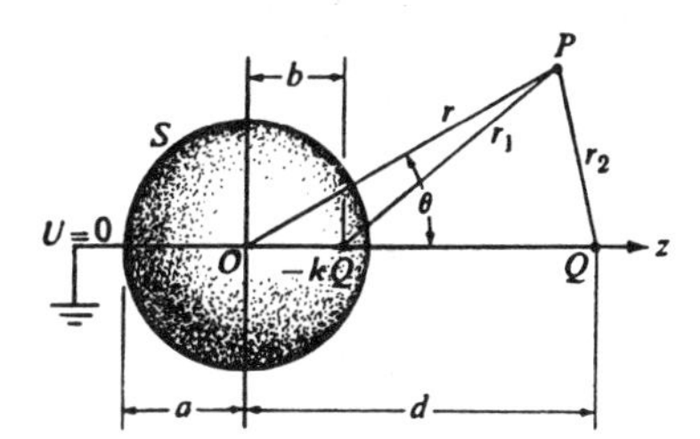

A point charge situated near a grounded spherical conductor.

Fig.1

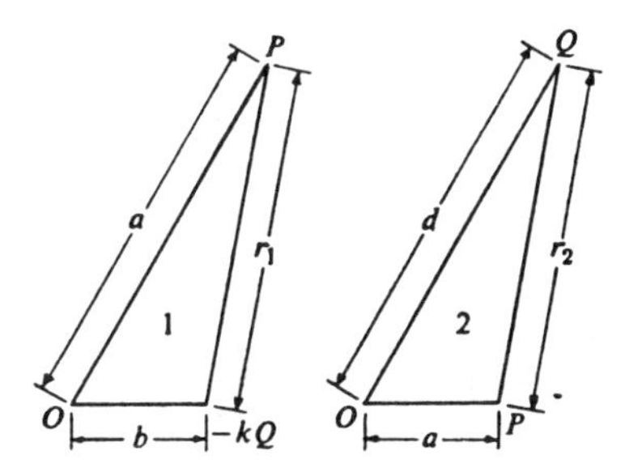

Similar triangles for satisfying boundary conditions.

Fig.2

Solution: If the field point P is outside the sphere, the distribution of charge on the surface of the sphere can be replaced by a single image charge -kQ at a distance b from the center. Thus

$$U = \frac{Q}{4\pi\varepsilon_0}\left(\frac{1}{r_2} - \frac{k}{r_1}\right).$$

The unknown quantities k and b are determined by the boundary condition which requires that U = 0 on S. Then

$$k = \frac{r_1}{r_2} = \text{const.}$$

The ratio r_1/r_2 must be constant for any position of P on the sphere. This condition can be satisfied by choosing the triangles 1 and 2 illustrated in Fig. 2 to be similar. This yields

$$\frac{a}{d} = \frac{r_1}{r_2} = \frac{b}{a} = k$$

Thus $b = a^2/d$ and $k = a/d$. To determine σ introduce polar coordinates so that

$$r_2^2 = r^2 + d^2 - 2rd\cos\theta,$$

$$r_1^2 = r^2 + b^2 - 2rb\cos\theta,$$

and use

$$\sigma = -\varepsilon_0 \frac{\partial u_1}{\partial r} \quad :$$

$$\sigma = -\varepsilon_0 \left.\frac{\partial u}{\partial r}\right]_{r=a} =$$

$$-\frac{Q}{4\pi}\left[a\left(\frac{k}{r_1^3} - \frac{1}{r_2^3}\right) + \left(\frac{d}{r_2^3} - \frac{kb}{r_1^3}\right)\cos\cdot\theta\right].$$

Since

$$\frac{d}{r_2^3} - \frac{kb}{r_1^3} = \frac{1}{r_2^3}\left(d - \frac{b}{k^2}\right) = 0$$

and

$$\frac{k}{r_1^3} - \frac{1}{r_2^3} = \frac{k(1 - k^2)}{r_1^3},$$

$$\sigma = -\frac{Qak(1 - k^2)}{4\pi r_1^3}.$$

• PROBLEM 7-27

A spherical conductor of radius a is placed in a uniform electric field

$$\vec{E}_0 = \bar{i}_z E_0$$

(see the figure). Use the method of images to determine the surface charge density and the maximum electric field intensity.

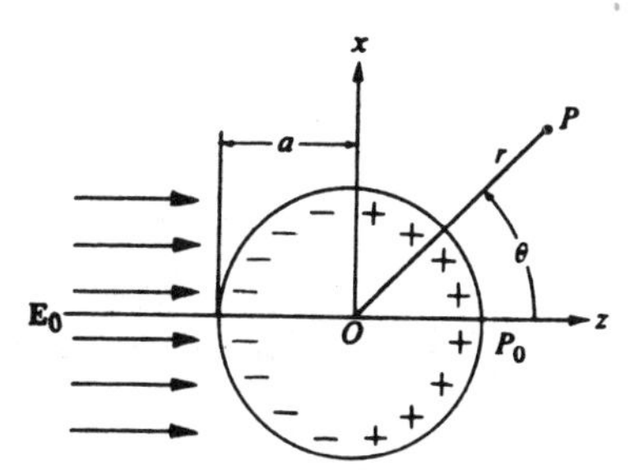

A spherical conductor placed in a uniform electric field.

Solution: The uniform field E_0 can be considered as a primary field; it can be determined from the potential

$$U_p = -E_0 z = -E_0 r \cos\theta \quad .$$

The primary field polarizes the sphere, resulting in a secondary field due to the surface charges. As seen from an external field point P, the surface charges appear equivalent to a dipole located at the origin. The dipole has a potential of the form

$$U_s = \frac{A \cos \theta}{r^2},$$

where A is a constant. The resultant potential is

$$U = -E_0 r \cos \theta + \frac{A \cos \theta}{r^2}.$$

The boundary condition requires that U = 0 when r = a. This yields

$$A = E_0 a^3 .$$

The surface charge density can be determined from

$$\sigma = -\varepsilon_0 \left.\frac{\partial U}{\partial r}\right]_{r=a} = 3\varepsilon_0 E_0 \cos \theta .$$

The maximum field strength occurs at the point P_0 and has a magnitude

$$E_x = -\left.\frac{\partial U}{\partial x}\right]_{x=a} = 3E_0 .$$

Thus the effect of the secondary field is to increase the primary field strength at P_0 by a factor of 3.

• PROBLEM 7-28

The figure shown illustrates an open-wire transmission line composed of parallel conducting circular cylinders of radii a_1 and a_2, respectively, which are charged to a potential difference of V volts. Use the method of images to determine the potential and the capacitance per unit length of the system.

Solution: Let q be the charge per unit length on cylinder 2. Replace the surface charge distribution on the conductors by line sources a distance b apart, as shown in the figure. The potential at a field point P due to the line sources is

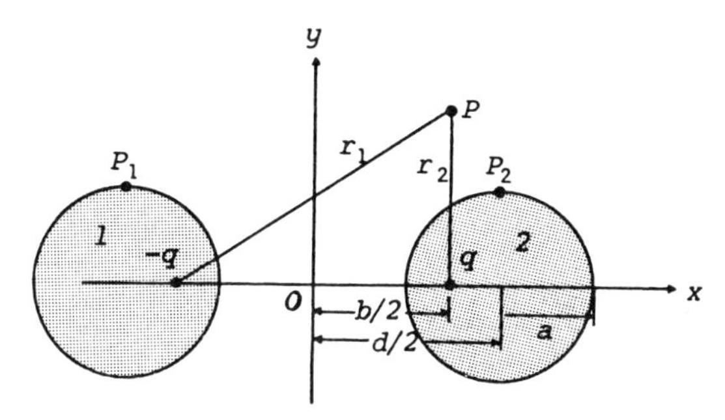

A two wire transmission line.

$$U = \frac{q}{2\pi\varepsilon_0} \ln \frac{r_1}{r_2}$$

Due to symmetry, the surface potentials of cylinders 1 and 2 can be taken as

$$U_1 = - V/2$$

and

$$U_2 = V/2,$$

respectively. When the field point P is taken to coincide with any point P_2 on the surface of conductor 2,

$$\frac{V}{2} = \frac{q}{2\pi\varepsilon_0} \ln k, \qquad (1)$$

where k is a constant given by

$$k = \frac{r_1}{r_2} > 1 . \qquad (2)$$

Now let

$$r_1^2 = (x + b/2)^2 + y^2 ,$$

$$r_2^2 = (x - b/2)^2 + y^2 ,$$

and square Eq. (2). This yields the equation of the circle defining the surface of conductor 2. Thus

$$(x - d/2)^2 + y^2 = a^2 ,$$

where

$$d = b \frac{k^2 + 1}{k^2 - 1} , \tag{3}$$

$$a = \frac{bk}{k^2 - 1} \tag{4}$$

Equations (3) and (4) relate the known radius a and the distance between centers d to the unknown quantities k and b of the image line. Eliminating b from these equations yields

$$k^2 - \frac{d}{a} k + 1 = 0 . \tag{5}$$

Since $k = r_1 / r_2 > 1$, the desired root of Eq. (5) is

$$k = \frac{d}{2a} + \sqrt{(d/2a)^2 - 1} . \tag{6}$$

Equation (3) yields

$$b = \sqrt{d^2 - 4a^2} .$$

The potential function for this problem is now completely determined. To determine the capacitance per unit length,

$$C = q/V,$$

use Eqs. (1) and (6) :

$$C = \frac{\pi \varepsilon_0}{\ln[(d + \sqrt{d^2 - 4a^2})/2a]}$$

A long straight conductor of radius a having a charge density q runs parallel to a plane conducting surface (the ground) with its center a distance d above the surface. Assume that the radius of the conductor is negligibly small compared with its height above the ground. Find (a) the flux density at the surface of the ground immediately below the conductor, (b) the potential difference between the surface of the conductor and the surface of the ground, and (c) the capacitance per unit length of the conductor-ground system.

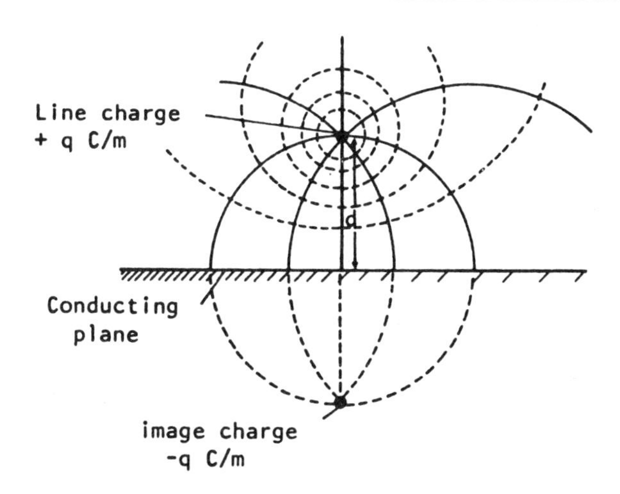

Solution: The line charge parallel to a plane conducting surface can be analyzed using the method of images. Take an image charge of equal and opposite strength on the opposite side of the plane conducting surface and an equal length away and remove the plane conducting surface. The line charge with its image is shown in the given figure.

(a) The flux density immediately below the conductor at the surface of the conducting plane will now be given by the vector sum of the flux densities due to the line charge and its image. These flux densities are prallel, so that

$$D = \frac{q}{2\pi d} - \frac{-q}{2\pi d} = \frac{q}{\pi d} \quad C/m^2 .$$

(b) The potential function value at any point in the field due to a line source is

$$\phi = \frac{q}{2\pi\varepsilon_0} \log_e \frac{r_2}{r_1} ,$$

At the surface of the conducting wire, $r_2 = 2d - a$ and $r_1 = a$. On the conducting ground, $r_1 = r_2$ and the potential value is zero. Therefore the potential difference between the surface of the wire and the surface of the ground is given by

$$\phi = \frac{q}{2\pi\varepsilon_0} \log_e \frac{2d-a}{a} \approx \frac{q}{2\pi\varepsilon_0} \log_e \frac{2d}{a} .$$

(c) The capacitance is given by

$$C = \frac{q}{\phi}$$

Therefore

$$C = \frac{2\pi\varepsilon_0}{\log_e (2d/a)}$$

CHAPTER 8

STEADY MAGNETIC FIELDS

BIOT-SAVART'S LAW

• **PROBLEM** 8-1

A thin wire carrying current I_0 has a right-angle bend, as shown in Fig. 1. Using the Biot-Savart Law, find B along the positive x axis

Note that the B field due to an infinitely long wire from $-\infty$ to $+\infty$ is

$$B = \frac{\mu_0 I}{2\pi\rho}\,\hat{a}_\phi$$

where ρ is the distance away from the wire and $\hat{a}_\phi$ is the unit vector in the ϕ-direction.

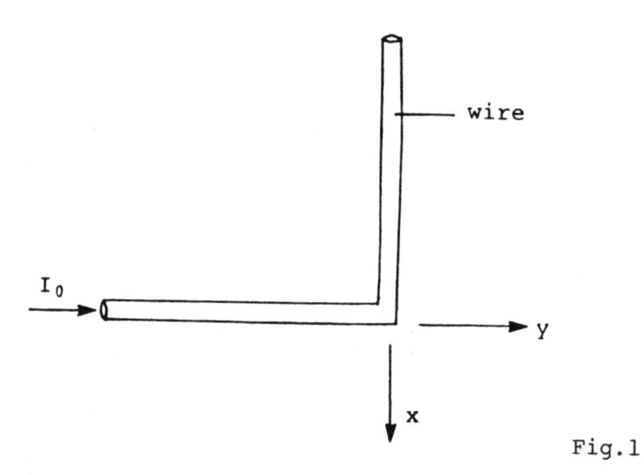

Fig.1

Solution:

Using Biot-Savart Law,

$$B = \frac{\mu_0 I_0}{4\pi} \oint_{C_1} \frac{d\ell \times (r_2 - r_1)}{|r_2 - r_1|^3}$$

where $r_2 - r_1$, $d\ell$ are all shown in Fig. 2.

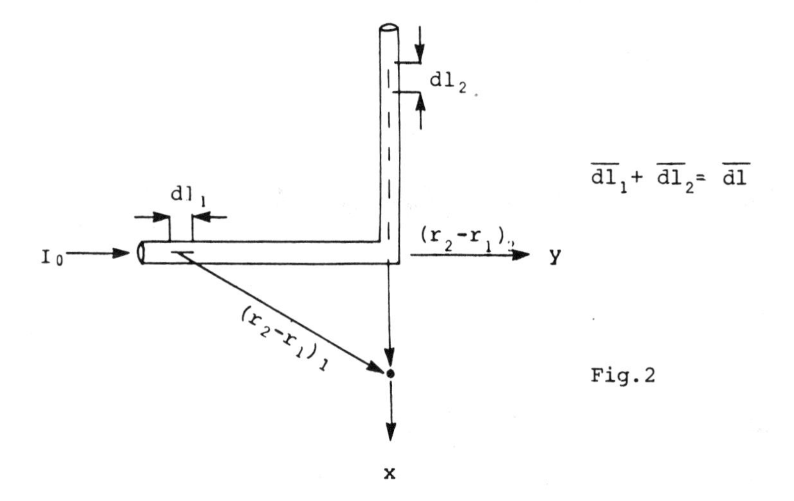

Fig.2

Note that $d\ell_2$ does not contribute to the B field since it is parallel to $(r_2 - r_1)_2$ and the cross product is zero. Therefore only the horizontal wire will contribute to the B field.

$$B = \frac{\mu_0 I}{4\pi} \int_C \frac{d\ell_1 \times (r_2 - r_1)_1}{|r_2 - r_1|_1^3}$$

$$= \frac{\mu_0 I}{4\pi} \int_{-\infty}^{0} \frac{(\hat{a}_y\, dy) \times (r_2 - r_1)_1}{|r_2 - r_1|_1^3}$$

$$= \frac{\mu_0 I}{4\pi} \int_{-\infty}^{0} \frac{(\hat{a}_y\, dy) \times (r_2 - r_1)}{|r_2 - r_1|^3}$$

$$= \frac{\mu_0 I}{2\pi} \int_{-\infty}^{\infty} \frac{(\hat{a}_y\, dy) \times (r_2 - r_1)}{|r_2 - r_1|^3}$$

As given

$$\frac{\mu_0 I}{4\pi} \int_{-\infty}^{\infty} \frac{(\hat{a}_y\, dy) \times (r_2 - r_1)}{|r_2 - r_1|^3} = \frac{\mu_0 I}{2\pi\rho} \hat{a}_\phi$$

for an infinite wire,

$$\therefore \qquad B = \frac{-\mu_0 I}{4\pi\rho} \hat{a}_z$$

where ρ = (distance along x axis) = x

$$\therefore \qquad B = \frac{-\mu_0 I}{4\pi x} \hat{a}_z$$

Note that at the x axis, the B field is going into the paper which accounts for the $-\hat{a}_z$. This can be obtained by using the right hand rule.

• PROBLEM 8-2

Use the Biot-Savart law to calculate the magnetic induction $\bar{B}$ at a distance R from an infinitely long straight wire carrying a current I. Refer to the accompanying figure.

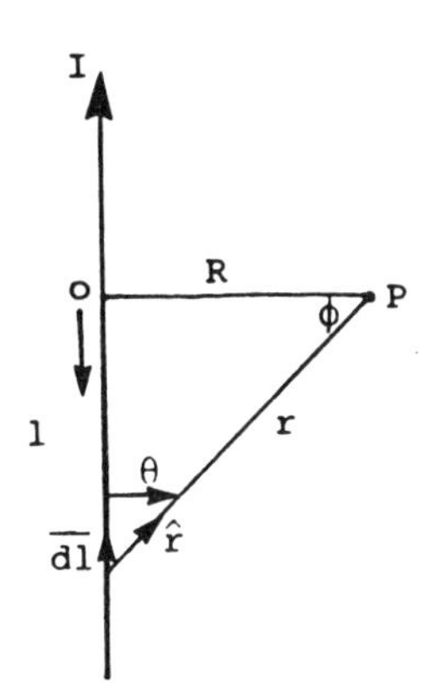

Solution: The Biot Savart law is

$$\bar{B} = \frac{\mu_0}{4\pi} \oint \frac{I\bar{d\ell} \times \hat{r}}{r^2}$$

where $\hat{r}$ is a unit vector pointing from the current element $Id\ell$ to point P. $\bar{B}$ is in the direction of $d\bar{\ell} \times \hat{r}$, which is perpendicular to the plane of the paper and points into the paper as indicated by the tail of the arrow in the figure.

$$B_p = \frac{\mu_0 I}{4\pi} \oint \frac{d\ell \sin\theta}{r^2} \quad \text{into the paper} \tag{1}$$

In order to integrate, express all variables in terms of a single variable. Here ϕ is taken as the variable.

$$\sin\theta = \cos\phi$$

$$\frac{\ell}{R} = \tan\phi$$

$$\ell = R\tan\phi$$

$$d\ell = R\sec^2\phi d\phi \tag{2}$$

$$\cos\phi = \frac{R}{r}$$

$$r = \frac{R}{\cos\phi} \tag{3}$$

Limits of ϕ are from $-\pi/2$ to $+\pi/2$

$$\therefore\ B = \frac{\mu_0 I}{4\pi} \int_{-\frac{\pi}{2}}^{\frac{\pi}{2}} \frac{(R \sec^2 \phi\ d\phi)\ \cos^2 \phi \cos \phi}{R^2}$$

$$= \frac{\mu_0 I}{4\pi R} \int_{-\frac{\pi}{2}}^{\frac{\pi}{2}} \cos \phi\ d\phi = \frac{\mu_0 I}{4\pi R} [\sin \phi]_{-\frac{\pi}{2}}^{\frac{\pi}{2}}$$

$$= \frac{\mu_0 I}{4\pi R} \cdot 2$$

$$= \frac{\mu_0 I}{2\pi R} = B$$

into the paper.

AMPERE'S LAW

• PROBLEM 8-3

Calculate the magnetic induction $\overline{B}$ at a distance R from an infinitely long straight wire, using Ampere's circuital law. See the figure.

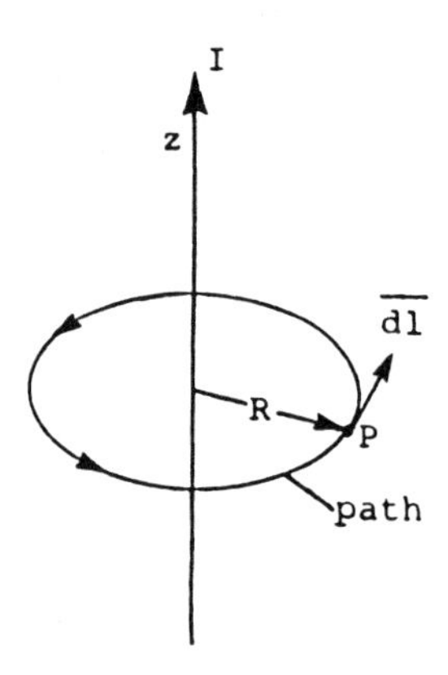

Solution: Ampere's circuital law is

$$\oint \overline{B} \cdot \overline{dl} = \mu_0 I$$

where μ_0 is the permeability of free space and I is the total current crossing any surface bounded by the line integral path. Using cylindrical coordinates

$$\overline{B} = \hat{r} B_r + \hat{\theta} B_\theta + \hat{z} B_z .$$

Take for the path a circle of radius R passing through P and with center at the wire, and the plane of the circle perpendicular to the axis of wire.

$$\therefore \overline{dl} = \hat{\theta} R d\theta$$

$$\therefore \qquad \oint \overline{B} \cdot \overline{dl} = \oint [\hat{r} B_r + \hat{\theta} B_\theta + \hat{z} B_z] \cdot \hat{r} R d\theta$$

$$= R B_\theta \oint d\theta = B_\theta R 2\pi = \mu_0 I.$$

Hence

$$B_\theta = \frac{\mu_0 I}{2\pi R}.$$

One can also argue that the $\overline{B}$ field are circles whose centers lie on the axis of the wire due to symmetry.

Also, according to the Biot-Savart law

$$d\overline{B} = \frac{\mu_0 I d\overline{z} \times \hat{r}}{4\pi r^2}$$

$\therefore \overline{B}$ must be in the θ direction only, since if $\overline{C} = \overline{A} \times \overline{B}$, vector $\overline{C}$ is perpendicular to the plane containing $\overline{A}$ and $\overline{B}$.

• PROBLEM 8-4

A cylindrical conductor of radius a and of unit length is given, with a total current I flowing axially in the cylinder as shown in fig. 1. Find (a) the magnetic flux density (magnetic induction) inside r < a, and (b) total magnetic flux inside.

Solution: Ampere's circuital law for magnetic media is

$$\oint \overline{H} \cdot d\overline{\ell} = \int_s J \cdot d\overline{a} = I$$

where H is the magnetic field intensity, C is the curve bounding the surface s. J is the current density. I is the current of free charges.

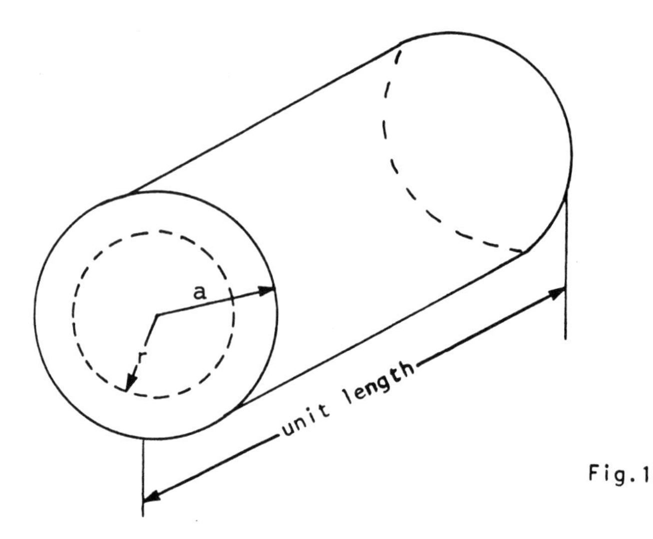

Fig.1

Ampere's circuital law says that the line integral of $\overline{H}$ about any closed path is equal to the direct current enclosed by that path. In this problem, choose a circular path of radius r, along which H is tangential and constant. Hence Ampere's law becomes

$$\oint \overline{H}\cdot d\overline{l} = \int_0^{2\pi} H_\phi r d\phi = H_\phi r \int_0^{2\pi} d\phi$$

$$= H_\phi \, r(2\pi)$$

$$H_\phi(2\pi r) = \int_s J \cdot d\overline{a} = J \cdot \int_s d\overline{a}$$

$$= J \; (\pi r^2)$$

but

$$J = I/\pi a^2$$

$$\therefore \; H_\phi(2\pi r) = \frac{I}{\pi a^2} \; (\pi r^2)$$

$$= \frac{Ir^2}{a^2}$$

$$H_\phi = \frac{Ir}{2\pi a^2}$$

or

$$\overline{H} = \frac{\hat{\phi} \; Ir}{2\pi a^2}$$

The magnetic flux density $\overline{B}$ is related to the magnetic intensity $\overline{H}$ by $\overline{B} = \mu\overline{H}$ where μ is the permeability.

Hence $\bar{B} = \frac{\mu \ Ir \ \hat{\phi}}{2\pi a^2}$

The total magnetic flux Φ inside the cylinder is determined from

$$\Phi = \int_s \bar{B} \cdot d\bar{a}$$

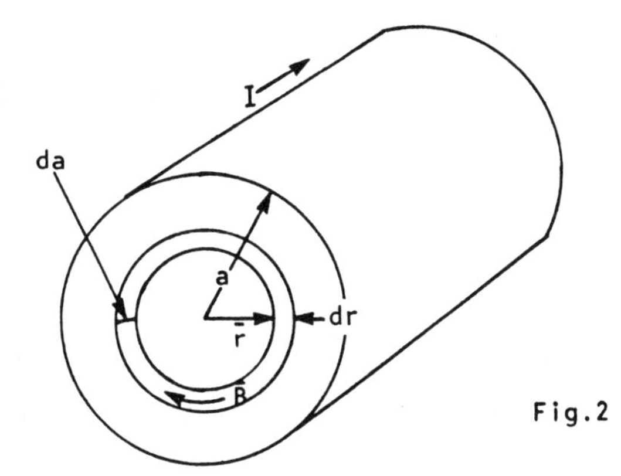

Fig.2

The surface area in the integral is dr x l(See fig. 2)

$$\Phi = \frac{\mu I}{2\pi a^2} \int_0^a r \ dr = \frac{\mu I}{2\pi a^2} \frac{a^2}{2}$$

$$= \frac{\mu I}{4\pi}$$

$$= \Phi$$

• PROBLEM 8-5

> Figure 1 is a sectional view of two long parallel plates of width w. The plate on the left carries a current I toward the reader, and that on the right carries an equal current <u>away</u> from the reader. Find the field between the two parallel plates.

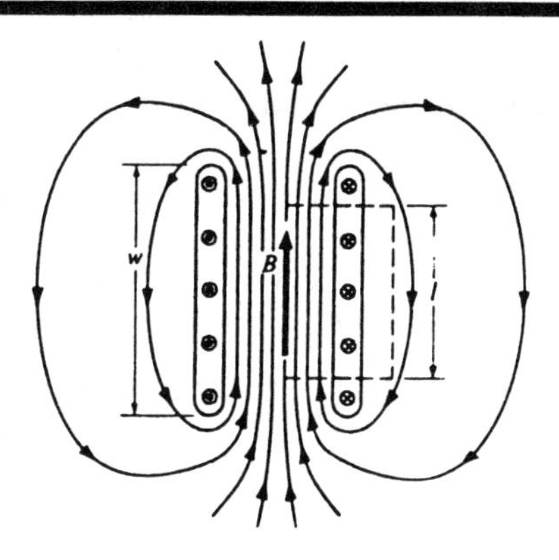

Fig. 1

Solution: In the region between the plates and not too near their edges, the B-field is uniform. The field outside the plates is small, and becomes smaller as the width w is increased.

Apply Ampere's law to the dotted rectangle, and assume the field outside the plates to be zero. Then

$$B \cdot dl = Bl.$$

If I is the total current in either plate, the current per unit width is I/w and the current through the rectangle is $I\ell/w$. Hence

$$Bl = \frac{\mu_0 I\ell}{w}, \qquad B = \frac{\mu_0 I}{w}.$$

It is interesting to compare the nature of the magnetic field between two long, current-carrying plates, with the electric field between two large charged plates. Both fields are approximately uniform, the magnetic lines being parallel to the plates and the electric lines perpendicular. In the magnetic case,

$$B = \mu_0 (I/w),$$

whereas in the electric case

$$E = (1/\varepsilon_0)(Q/A).$$

• PROBLEM 8-6

Through the use of Ampere's circuital law, find the $\overline{H}$ field in all regions of an infinte length coaxial cable carrying a uniform and equal current I in opposite directions in the inner and outer conductors. Assume the inner conductor to have a radius of a(m) and the outer conductor to have an inner radius of b(m) and an outer radius of c(m). Assume that the cable's axis is along the z-axis.

Solution: Through the use of symmetrical pairs of filamentary currents as shown in Fig. (b), it can be argued that the $\overline{H}$ field in all regions will be in the $\hat{\phi}$ direction and thus $\overline{H} = \hat{\phi} H_\phi$.

(a) For a concentric amperian closed loop drawn in the region ($r_c < a$) of the inner conductor, Ampere's circuital law becomes

$$\oint_{\ell} \overline{H} \cdot \overline{dl} = \oint_{\ell} (\hat{\phi} H_{\phi}) \cdot (\hat{\phi} r_c d\phi)$$

$$= H_{\phi} r_c \int_0^{2\pi} d\phi = H_{\phi} 2\pi r_c = I_{en}. \quad (1)$$

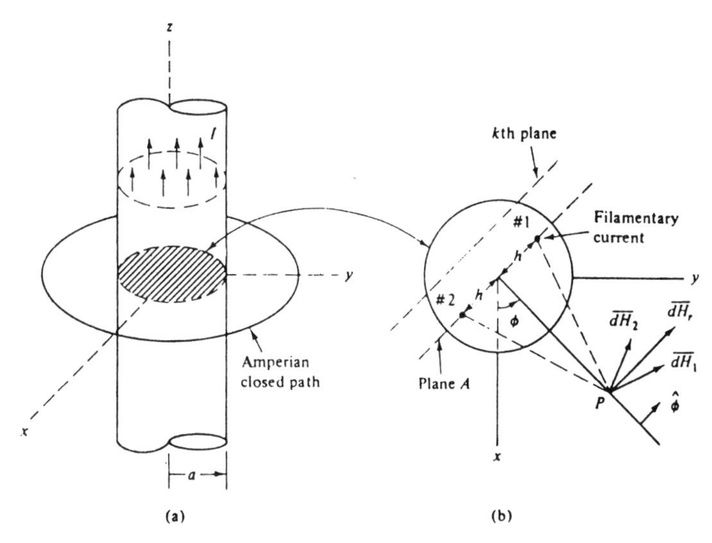

(a) Graphical display for finding the $\vec{H}$ field inside and outside a conductor of finite cross section. (b) Symmetrical pairs of filamentary currents produce a resultant field in the $\hat{\phi}$ direction.

Fig. 1

Now, $I_{en} = Ir_c^2/a^2$ for the amperian closed loop inside the inner conductor. Thus, the last two terms in (1) become

$$H_{\phi} 2\pi r_c = I_{en} = \frac{Ir_c^2}{a^2}$$

yielding

$$H_{\phi} = \frac{Ir_c}{2\pi a^2}$$

and

$$\overline{H} = \hat{\phi} \frac{Ir_c}{2\pi a^2} (Am^{-1}) \quad (r_c < a). \quad (2)$$

(b) For a concentric amperian closed loop drawn in the region $(a < r_c < b)$, Ampere's circuital law becomes

$$\oint_{\ell} \bar{H} \cdot \overline{dl} = \oint_{\ell} (\hat{\phi} H_{\phi}) \cdot (\hat{\phi} r_c d\phi)$$

$$= H_{\phi} r_c \cdot \int_0^{2\pi} d\phi = H_{\phi} 2\pi r_c = I \qquad (3)$$

where $I_{en} = I(A)$. Solving for H_{ϕ} from the last two terms in (3),

$$\bar{H} = \frac{\hat{\phi} I}{2\pi r_c} \ (Am^{-1}) \qquad (a < r_c < b). \qquad (4)$$

(c) For a concentric amperian closed loop in the region $(b < r_c < c)$, Ampere's circuital law becomes

$$\oint_{\ell} \bar{H} \cdot \overline{dl} = \oint_{\ell} (\hat{\phi} H_{\phi}) \cdot (\hat{\phi} r_c d\phi)$$

$$= H_{\phi} r_c \int_0^{2\pi} d\phi = H_{\phi} 2\pi r_c = I_{en}. \qquad (5)$$

Now, $I_{en} = I - I[(r_c^2 - b^2)/(c^2 - b^2)]$ for the amperian closed loop inside the outer conductor. It should be noted that some of the enclosed current flows in the reverse direction. Substituting for I_{en} into (5) and solving for $H\phi$ from the last two terms,

$$\bar{H} = \hat{\phi} \frac{I}{2\pi r_c} \left(\frac{c^2 - r_c^2}{c^2 - b^2} \right) (Am^{-1}) \qquad (b < r_c < c). \qquad (6)$$

(d) For a concentric amperian closed loop drawn in the region $(c < r_c)$, the enclosed current is found to be zero, and thus $\bar{H}$ is zero.

• PROBLEM 8-7

A solenoid is constructed by winding wire in a helix around the surface of a cylindrical form, usually of circular cross section. The turns of the winding are ordinarily closely spaced and may consist of one or more layers. For simplicity, the figure represents a solenoid by a relatively small number of circular turns, each carrying a current I. Find the field at any point due to the solenoid.

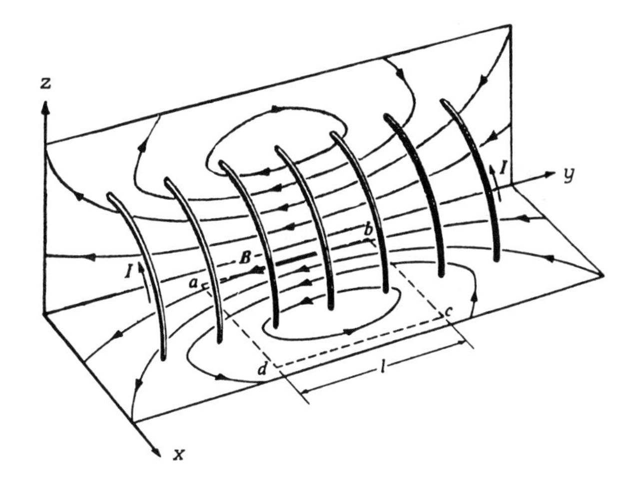

Lines of induction surrounding a solenoid. The dotted rectangle abcd is used to compute the flux density B in the solenoid from Ampère's law.

Solution: The resultant field at any point is the vector sum of the B-vectors due to the individual turns. The diagram shows the lines of induction in the xy- and yz - planes. Exact calculations show that for a long, closely-wound solenoid, half of the lines passing through a cross section at the center emerge from the ends and half "leak out" through the windings between center and end.

If the length of the solenoid is large compared with its cross-sectional diameter, the internal field near its center is very nearly uniform and parallel to the axis, and the external field near the center is very small. The internal field at or near the center can then be found by use of Ampere's law.

Select as a closed path the dotted rectangle abcd in the figure. Side ab, of length ℓ, is parallel to the axis of the solenoid. Sides bc and da are to be taken very long so that side cd is far from the solenoid and the field at this side is negligibly small.

By symmetry, the B field along side ab is parallel to this side and is constant, so that for this side

$$B_{||} = B$$

and

$$\oint B \cdot dl = Bl.$$

Along sides bc and da, $B_{||} = 0$ since B is perpendicular to these sides; and along side cd, $B_{||} = 0$ also since $B = 0$. The sum around the entire closed path therefore reduces to Bl.

Let n be the number of turns per unit length in the windings. The number of turns in length ℓ is then $n\ell$. Each of these turns passes once through the rectangle abcd and carries a current I, where I is the current in the windings. The total current through the rectangle is then $n\ell I$, and from Amperes law,

$$B\ell = \mu_0 n\ell I,$$

$$B = \mu_0 nI \quad \text{(solenoid)}.$$

Since side ab need not lie on the axis of the solenoid, the field is uniform over the entire cross section.

• PROBLEM 8-8

Figure 1(a) represents a toroid, wound with wire carrying a current I. Find the field at all points inside the toroid.

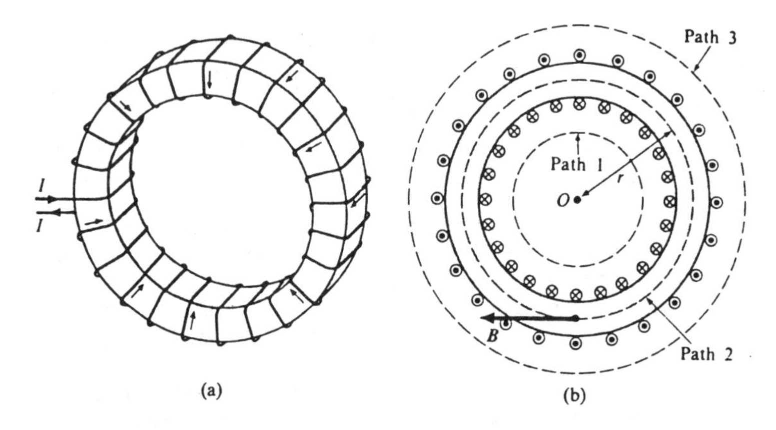

(a) A toroidal winding. (b) Closed paths (dotted circles) used to compute the flux density B set up by a current in a toroidal winding. The field is very nearly zero at all points except those within the space enclosed by the windings.

Fig. 1

Solution: The dotted lines in Fig. 1(b) are a number of paths to which Ampere's law is applied. Consider first path 1. By symmetry, if there is any field at all in this region, it will be tangent to the path at all points, and $\oint B \cdot dl$ will equal the product of B and the circumference of the path. The current through the path, however, is zero, and hence from Ampere's law (since the circumference is not zero), the field B must be zero.

Similarly, if there is any field at path 3, it will also be tangent to the path at all points. Each turn of the winding passes twice through the area bounded by this path, carrying equal currents in opposite directions. The net current through the area is therefore zero, and hence $B = 0$ at all points of the path. The field of the toroid is therefore confined wholly to the space enclosed by the windings. The toroid may be thought of as a solenoid that has been bent into a circle.

Finally, consider path 2, a circle of radius r. Again, by symmetry, the B field is tangent to the path, and $\oint B \cdot dl$ equals $2\pi rB$. Each turn of the winding passes once through the area bounded by path 2, and the total current through the area is NI, where N is the total number of turns in the winding. Then, from Ampere's law,

$$2\pi rB = \mu_0 NI,$$

and

$$B = \frac{\mu_0}{2\pi}\left(\frac{NI}{r}\right) \quad \text{(toroid)}.$$

The magnetic field is not uniform over a cross section of the core, because the path length ℓ is larger at the outer side of the section than at the inner side. However, if the radial thickness of the core is small compared with the toroid radius r, the field varies only slightly across a section. In that case, considering that $2\pi r$ is the circumferential length of the toroid and that $N/2\pi r$ is the number of turns per unit length n, the field may be written

$$B = \mu_0 nI,$$

just as at the center of a long straight solenoid.

MAGNETIC FLUX AND FLUX DENSITY

• PROBLEM 8-9

A circular coil of diameter 20 mm is mounted with the plane of the coil perpendicular to the direction of a uniform magnetic flux density of 100 mT. Find the total flux threading the coil.

Solution: The flux is given by

$$\psi = \int B \cdot ds.$$

As the flux density is uniform, the flux density is given by

$$\Psi = B \cdot A.$$

The vector area is normal to the plane of the area so that the vector flux density is parallel to the vector area and the required flux is the product of the flux density and the area. Therefore

$$\Psi = 10^{-1} \times \pi \times 10^{-4}$$

$$= 3.14 \times 10^{-5} \text{ Wb} = 31.4 \ \mu\text{Wb}.$$

• PROBLEM 8-10

Two very long, straight, thin copper wires, placed D m apart, are carrying equal currents I in opposite directions. If the wires are placed perpendicular to the plane of the page, find the expression for the magnetic field B, in terms of distance x from the wire 'a', as shown in the figure below.

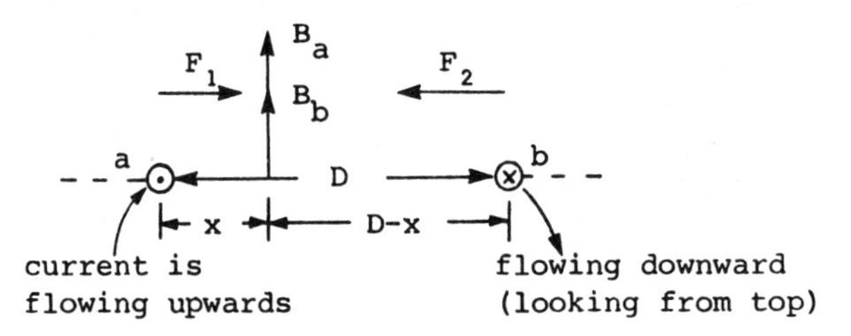

Solution: Since the wires are carrying opposite directed currents, they will try to attract each other. The force acting on wire 'a', F_1, due to field created by b will try to move wire 'a' closer to wire 'b'. Same is true for force F_2.

Now, using the "Right-hand rule", which states that if a current carrying conductor is placed in a field B, such that B is perpendicular to the direction of current flow, then the force exerted on the conductor will have direction perpendicular to both I and B.

So field B_a and B_b have same direction as shown in the figure.

Now, using the Biot-Savart's law, the field is given by,

$$B(x) = \text{Magnetic field due to wire 'a'} = \frac{\mu_0 \cdot I}{2\pi \cdot x}$$

$$B(D-x) = \text{Magnetic field due to wire 'b'} = \frac{\mu_0 \cdot I}{2\pi(D-x)}$$

The total field is the sum of individual fields B(x) and B(D-x). Hence,

$$B = \frac{\mu_0 \cdot I}{2\pi}\left[\frac{1}{x} + \frac{1}{D-x}\right]$$

$$= \frac{\mu_0 \cdot I}{2\pi}\left[\frac{D}{x(D-x)}\right]$$

• PROBLEM 8-11

> Find the flux density at a distance of 20cm from a northpole of strength 550 amp-meters in a medium with relative permeability 50. Also find the force on another northpole of equal strength at this distance. Assume that the south poles are at a large distance.

Solution: The magnitude of the flux density is given by

$$B = \frac{\mu_0 \mu_r Q_m}{4\pi r^2}$$

$$= \frac{4\pi \times 10^{-7} \times 50 \times 550}{4\pi \times (0.2)^2}$$

$$= 6.875 \times 10^{-2} \text{ Newton/amp-meter.}$$

Since the pole is positive, the direction of B is radially away from the pole. Another pole of equal strength at this point is acted on by a force of magnitude

$$F = Q_m B$$

$$= 550 \times 6.875 \times 10^{-2}$$

$$= 37.8 \text{ Newtons.}$$

The direction of the force F is the same as for the flux density B.

• PROBLEM 8-12

> A very long, straight, thin copper ribbon of width d carries a steady current of I amperes. Find an expression for the flux-density at a point in the plane of the ribbon and distant a from its nearer edge.

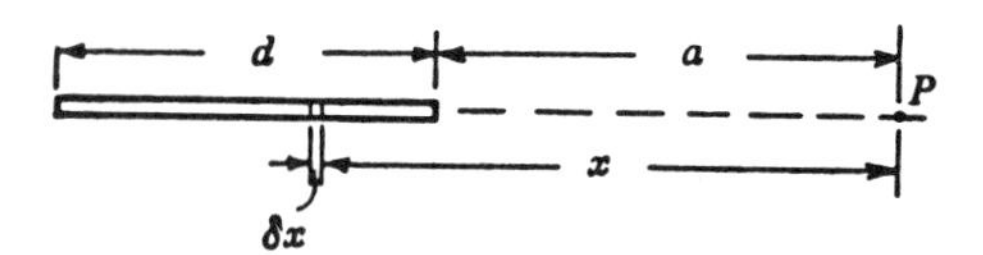

Solution: Consider the field at P (see figure) due to the current in an elementary filament of the conductor of width δx and distant x from P. The current in the filament is

$$\delta I = \frac{I}{d}\delta x,$$

so that the field at P due to the filament is

$$B = \frac{\mu_0 I}{2\pi R}$$

$$\delta B_0 = \mu_0 \frac{I}{2\pi d}\frac{\delta x}{x},$$

whence the field due to the whole current I is

$$B_0 = \mu_0 \frac{I}{2\pi d}\int_a^{a+d}\frac{dx}{x} = \mu_0 \frac{I}{2\pi d}\log_e \frac{a+d}{a}.$$

• **PROBLEM 8-13**

The figure depicts a short copper wire ℓ units long, with cross-sectional area a, situated in air coincident with the z-axis at the origin. The current density in the positive z direction is J. Assuming the hypothetical situation that J is uniform throughout the wire and constant with respect to time, find the magnetic flux density B everywhere at a large distance from the wire, using the vector potential to obtain the solution.

Solution: The vector potential A at any point P produced by the wire is given by

$$A = \frac{\mu}{4\pi}\iiint \frac{J}{r}dv,$$

where the ratio J/r is integrated throughout the volume occupied by the wire. Since B is to be found only at a large distance r from the wire, it suffices to find A at a large distance. Specifically, the distance r should be large compared with the length of the wire ($r >> \ell$). Then, at any point P the distance r to different parts of the wire can be considered constant and

$$A = \frac{\mu_0}{4\pi r} \iiint J dv. \tag{1}$$

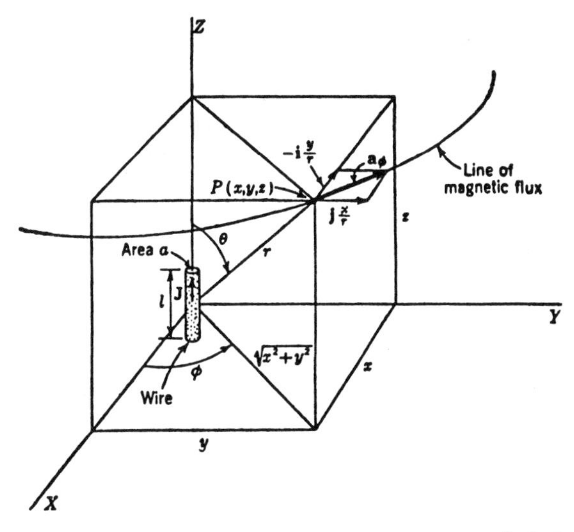

Construction for finding the vector potential A and flux density B due to a short current-carrying wire.

Now J is everywhere in the z direction and also is uniform. Thus $J = \hat{k}J_z$, and

$$\iiint J dv = \hat{k} \int_{-\ell/2}^{\ell/2} \iint_a J_z ds\, d\ell$$

$$= \hat{k} \int_{-\ell/2}^{\ell/2} I\, d\ell \tag{2}$$

where $I = J_z a$ = current in wire. Completing the integration in (2) and substituting this result in (2), obtain

$$A = \hat{k}\frac{\mu_0 I \ell}{4\pi r} = \hat{k}A_z \tag{3}$$

where A = vector potential at distance r from wire (webers/meter)

$\hat{k}$ = unit vector in positive z direction (dimensionless)

μ_0 = permeability of air ($4\pi \times 10^{-7}$ henry/meter)

I = current in wire (amp)

ℓ = length of wire (meters)

r = distance from wire (meters)

Equation (3) gives the vector potential A at a large distance from the wire. It is everywhere in the positive z direction as indicated by the unit vector $\hat{k}$ and is inversely proportional to the distance r from the wire. It is not a function of angle (ϕ or θ in Figure).

Having found the vector potential A, the flux density B is obtained by taking the curl of A. In rectangular components the curl of A is given by

$$\nabla \times A = \hat{i}\left(\frac{\partial A_z}{\partial y} - \frac{\partial A_y}{\partial z}\right) + \hat{j}\left(\frac{\partial A_x}{\partial z} - \frac{\partial A_z}{\partial x}\right)$$

$$+ \hat{k}\left(\frac{\partial A_y}{\partial x} - \frac{\partial A_x}{\partial y}\right) \tag{4}$$

Since A has only a z component, equation (4) reduces to

$$\nabla \times A = i\,\frac{\partial A_z}{\partial y} - j\,\frac{\partial A_z}{\partial x} \tag{5}$$

Now

$$r = \sqrt{x^2 + y^2 + z^2}\,.$$

Therefore

$$\frac{\partial A_z}{\partial y} = \frac{\mu_0 I \ell}{4\pi}\frac{\partial}{\partial y}\left(x^2 + y^2 + z^2\right)^{-\frac{1}{2}}$$

$$= -\frac{\mu_0 I \ell}{4\pi}\frac{y}{r^3}$$

and

$$\frac{\partial A_z}{\partial x} = \frac{\mu_0 I \ell}{4\pi} \frac{\partial}{\partial x} (x^2 + y^2 + z^2)^{-\frac{1}{2}}$$

$$= - \frac{\mu_0 I \ell}{4\pi} \frac{x}{r^3}$$

Introducing these relations in (5) and noting the geometry in the figure, we have

$$\nabla \times A = \frac{\mu_0 I \ell}{4\pi r^2} \left(- i \frac{y}{r} + j \frac{x}{r} \right)$$

$$= \hat{a}_\phi \frac{\mu_0 I \ell}{4\pi r^2} \frac{\sqrt{x^2 + y^2}}{r}$$

or

$$B = \nabla \times A = \hat{a}_\phi \frac{\mu_0 I \ell \sin \theta}{4\pi r^2}$$

where

B = magnetic flux density (webers/meter^2) at distance r and angle θ

$\hat{a}_\phi$ = unit vector in ϕ direction (see Figure) (dimensionless)

θ = angle between axis of wire and radius vector r (dimensionless)

μ_0 = permeability of air (= $4\pi \times 10^{-7}$ henry/meter)

I = current in wire (amp)

ℓ = length of wire (meters)

r = distance from wire to point where B is being evaluated (meters)

• PROBLEM 8-14

Find the flux between the conductors of the coaxial line shown in the figure. The conductor extends a length L.

Cross section of a coaxial cable carrying a uniformly distributed current I in the inner conductor and $-I$ in the outer conductor.

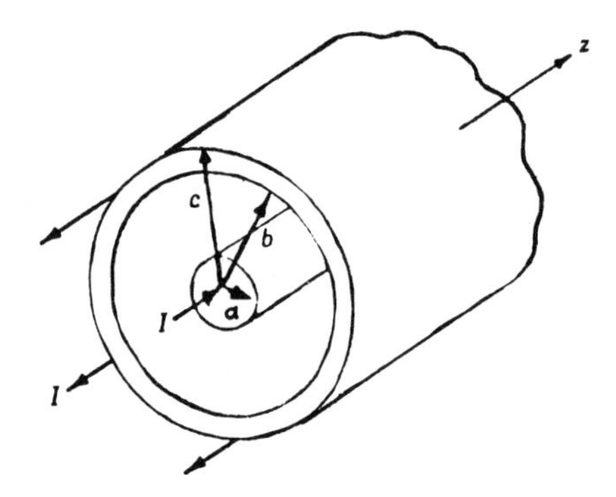

Solution: By applying Ampere's law to the region between the conductors,

$$\oint H \cdot d\ell = I_{enclosed}$$

In the region between the conductors, the current enclosed is I, therefore

$$H \cdot 2\pi r = I$$

$$H = \frac{I}{2\pi r} \qquad (a < r < b)$$

and therefore

$$B = \mu_0 H = \frac{\mu_0 I}{2\pi r} \hat{a}_\phi$$

The magnetic flux contained between the conductors in a length L is the flux crossing any radial plane extending from r = a to r = b and from, say, z = 0 to z = L

$$\Phi = \int_S B \cdot dS = \int_0^L \int_a^b \frac{\mu_0 I}{2\pi r} \hat{a}_\phi \cdot dr\, dz \hat{a}_\phi$$

or

$$\Phi = \frac{\mu_0 IL}{2\pi} \ln \frac{b}{a}$$

• **PROBLEM 8-15**

Consider an infinite coaxial cable with an insulating magnetic material occupying the space between the two conductors (see Fig. 1). The inner conductor is carrying a uniform current density J and the outer conductor a current density J' such that the current I is equal in magnitude and opposite in direction to that of the inner conductor. Determine B and H in the three regions of the cable as well as the exterior. Letting r be the distance from the axis, solve this problem for the following four regions, each considered in turn.

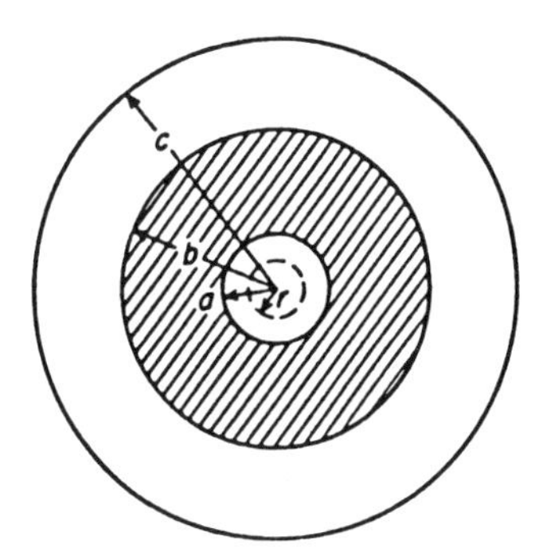

A coaxial cable filled with magnetic material.

Fig.1

Solution: (a) The inner conductor,

$$r < a.$$

From Ampere's law, the field is given by

$$\oint B \cdot d\ell = \mu_0 I$$

or

$$B = \frac{\mu_0 I}{2\pi r} \tag{1}$$

The current I(r) in the region of radius r is related to the current I in the entire inner conductor by the equation

$$\frac{I(r)}{I} = \frac{r^2}{R^2} \tag{2}$$

so that

$$B = \frac{\mu_0 Ir}{2\pi a^2} \qquad (3)$$

A plot of (3) for

$r < a$

and of (1) for

$r > a$

is given in Fig. 2.

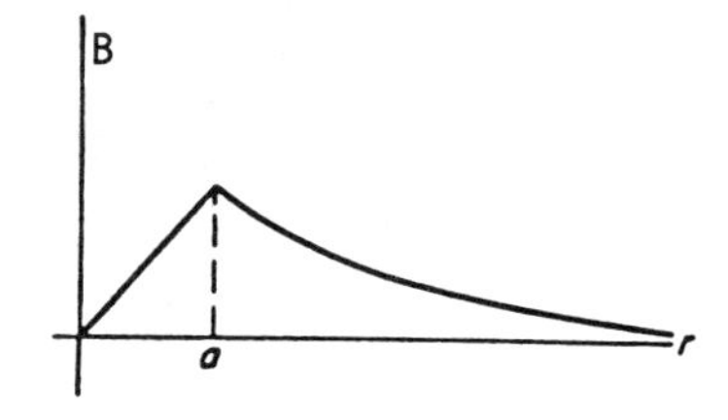

Fig.2

For the inner conductor, (3) then gives

$$B = \frac{\mu_0 Jr}{2} \qquad (4)$$

from which

$$H = \frac{Jr}{2} \qquad (5)$$

and both B and H are tangential to a circle of radius r.

(b) The insulating region,

$a < r < b$

In this region, the current I is the entire current in the inner conductor, so that (1) gives

$$B = \frac{\mu_0 J^2 a}{2r} \qquad (6)$$

and

$$H = \frac{J^2 a}{2r} \tag{7}$$

Note that the boundary condition on H becomes

$$H_{t_1} = H_{t_2} \tag{8}$$

and this relation is satisfied in this example.

(c) The outer conductor,

$$b < r < c$$

The field in this region depends on the currents J and J' and on the polarization of the magnetic medium. The contribution of J' will have the form (4), all the other contributions will depend on 1/r. Hence, assume that

$$B = \frac{\mu_0 J' r}{2} + \frac{A\mu_0}{r} \tag{9}$$

where A is to be determined. Then

$$H = \frac{J' r}{2} + \frac{A}{r} \tag{10}$$

and matching this to (7) at r = b in accordance with (8) shows that

$$A = \frac{1}{2}(Ja^2 - J'b^2) \tag{11}$$

so that

$$B = \frac{\mu_0 J' r}{2} + \frac{\mu_0}{2r}(Ja^2 - J'b^2) \tag{12}$$

Using the condition that the currents are equal but opposite gives

$$J'\pi(c^2 - b^2) = -J\pi a^2$$

and (8) becomes

$$H = \frac{B}{\mu_0} = \frac{a^2 J}{2r}\left(\frac{c^2 - r^2}{c^2 - b^2}\right) \tag{13}$$

(d) The external region,

$$r > c$$

From (13), $H = 0$ when $r = c$. Hence, H is zero everywhere outside the cable, in order to satisfy the boundary condition on H_t.

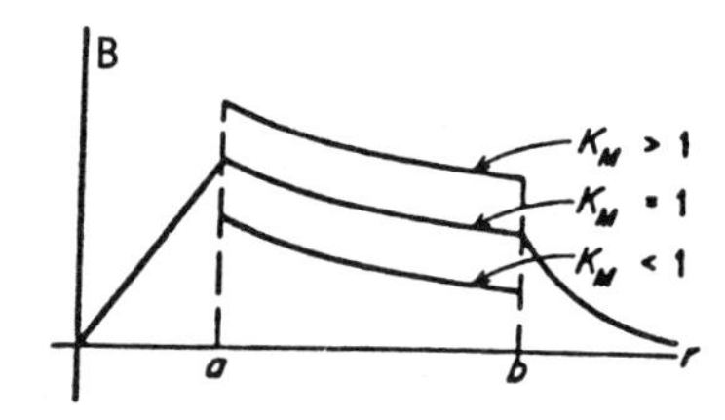

Fig.3

Plot B for a cable with the magnetic medium replaced by air, yields the continuous curve of Fig. 3. For the medium,

$$B = \mu_0 K_M H$$

so that the curve between $r = a$ and $r = b$ is displaced as shown for paramagnetic or diamagnetic materials.

• PROBLEM 8-16

A constant current $K_0 i_z$ flows in the $y = 0$ plane, as shown in the figure. Find the magnetic field everywhere due to this sheet of surface current.

Solution: Break the sheet into incremental line currents $K_0 dx$, each of which gives rise to a magnetic field as given by

$$B_\phi = \frac{\mu_0 I_2}{2\pi r}$$

The unit vector $\hat{i}_\phi$ is equivalent to the Cartesian components

$$\hat{i}_\phi = -\sin\phi \hat{i}_x + \cos\phi \hat{i}_y$$

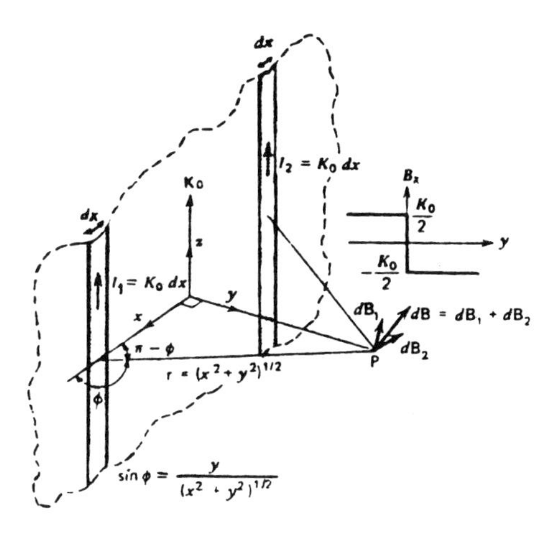

The symmetrically located line charge elements a distance x on either side of a point P have y magnetic field components that cancel but x components that add. The total magnetic field is then

$$B_x = -\int_{-\infty}^{+\infty} \frac{\mu_0 K_0 \sin\phi}{2\pi(x^2 + y^2)^{\frac{1}{2}}}\,dx$$

$$= \frac{-\mu_0 K_0 y}{2\pi} \int_{-\infty}^{+\infty} \frac{dx}{(x^2 + y^2)}$$

$$= \frac{-\mu_0 K_0}{2\pi} \tan^{-1} \frac{x}{y} \Bigg|_{-\infty}^{+\infty}$$

$$= \begin{cases} -\mu_0 K_0/2, & y > 0 \\ \mu_0 K_0/2, & y < 0 \end{cases}$$

The field is constant and oppositely directed on each side of the sheet.

• **PROBLEM 8-17**

If a second current sheet with current flowing in the opposite direction

$$-K_0 i_z$$

is placed at y = d parallel to a current sheet $K_0 i_z$

at y = 0, as in Figure 1, find the magnetic field due to these two sheets, and the force per unit area on each slab.

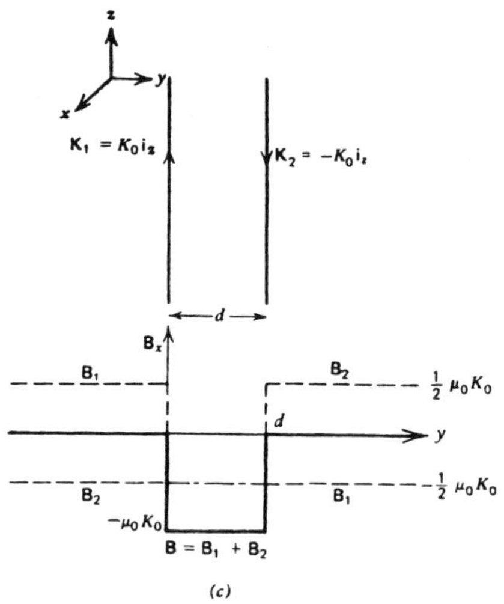

Fig. 1

Solution: The magnetic field due to each sheet alone is

$$B_1 = \begin{cases} \dfrac{-\mu_0 K_0}{2}\,\hat{i}_x\,, & y > 0 \\[2ex] \dfrac{\mu_0 K_0}{2}\,\hat{i}_x\,, & y < 0 \end{cases}$$

$$B_2 = \begin{cases} \dfrac{\mu_0 K_0}{2}\,\hat{i}_x & y > d \\[2ex] \dfrac{-\mu_0 K_0}{2}\,\hat{i}_x\,, & y < d \end{cases}$$

Thus in the region outside the sheets, the fields cancel while they add in the region between:

$$B = B_1 + B_2 = \begin{cases} -\mu_0 K_0 \hat{i}_x, & 0 < y < d \\ 0, & y < 0,\ y > d \end{cases}$$

The force on a surface current element on the second sheet is

$$df = -K_0 \hat{i}_z \, dS \times B$$

However, since the magnetic field is discontinuous at the current sheet, it is not clear which value of magnetic field to use. To take the limit properly, model the current sheet at y = d as a thin volume current with density J_0 and thickness Δ, as in Figure 2, where

$$K_0 = J_0 \Delta.$$

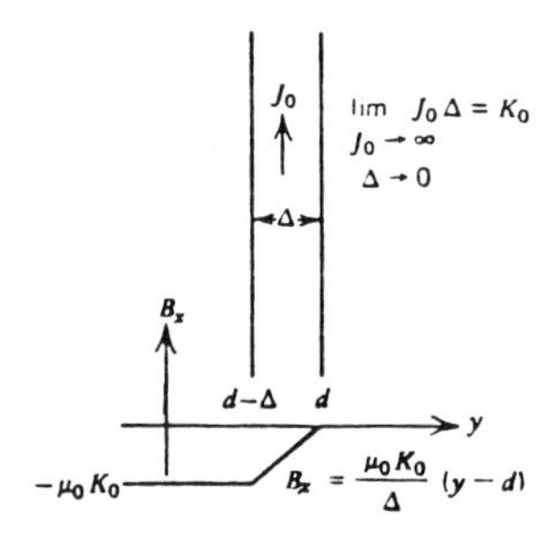

Fig. 2

The results for a single slab show that in a slab of uniform volume current, the magnetic field changes linearly to its values at the surfaces

$$B_x(y = d - \Delta) = -\mu_0 K_0$$

$$B_x(y = d) = 0$$

so that the magnetic field within the slab is

$$B_x = \frac{\mu_0 K_0}{\Delta}(y - d)$$

The force per unit area on the slab is then

$$F_S = -\int_{d-\Delta}^{d} \frac{\mu_0 K_0}{\Delta} J_0 (y - d)\hat{i}_y \, dy$$

$$= \frac{-\mu_0 K_0 J_0}{\Delta} \frac{(y - d)^2}{2} \hat{i}_y \Bigg|_{d-\Delta}^{d}$$

$$= \frac{\mu_0 K_0 J_0 \Delta}{2} \hat{i}_y = \frac{\mu_0 K_0{}^2}{2} \hat{i}_y$$

The force acts to separate the sheets because the currents are in opposite directions and thus repel one another.

• PROBLEM 8-18

Find the magnetic field vector at the center of a cylindrical hole which is displaced from the center of a cylindrical conductor, as shown in the figure.

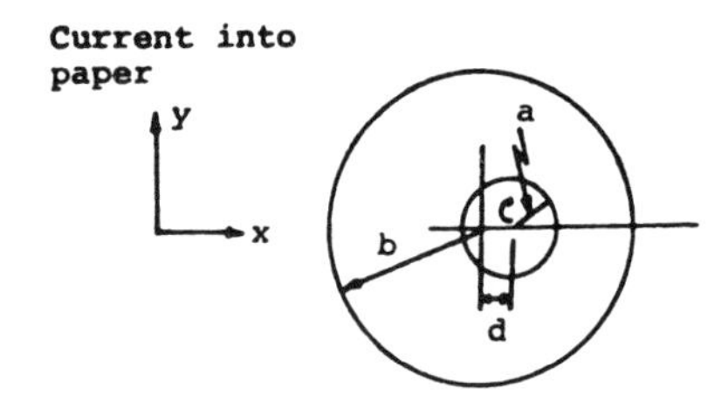

An off-centered hole in a cylindrical conductor.

Solution: The solution of this problem is readily accomplished if it is first noted that any effect of the hole in the conductor can be accounted for by considering the zero current in the hole as being made up of two equal currents in opposite directions. The problem then becomes the superposition of two problems: (a) to find the field at point C due to a homogeneous conductor carrying a current I', and (b) the field at point C due to a homogeneous conductor of radius a carrying a current of equal current density in the reverse direction to that in (a). The resultant magnetic vector is the vector sum of the effects under (a) and (b).

Let I equal the actual current in the original conductor, I' = the current in the homogeneous conductor having the same current density. Then the current density in the solid conductor is given by

$$J' = \frac{I'}{\pi b^2} \cdot$$

The current in the original conductor is then the current in the homogeneous conductor minus the current in the hole of radius a, that is,

$$I = I' = \frac{\pi a^2 I'}{\pi b^2} = \frac{(\pi b^2 - \pi a^2)I'}{\pi b^2}$$

or

$$\frac{I'}{\pi b^2} = \frac{I}{\pi(b^2 - a^2)}$$

The magnetic vector at C due to an equivalent solid conductor is from Ampere's Law

$$\oint B \cdot d\ell = \mu_0 I'_{\text{enclosed}}$$

$$\oint B \cdot d\ell = \mu_0 \left(\frac{\pi d^2}{\pi b^2}\right) I'$$

but

$$I' = \frac{b^2 I}{(b^2 - a^2)},$$

so

$$B(2\pi d) = \mu_0 \left(\frac{\pi d^2}{\pi b^2}\right) \frac{b^2 I}{(b^2 - a^2)}$$

or

$$B = \frac{\mu_0 d^2 I}{2\pi d(b^2 - a^2)}$$

and finally

$$\bar{B} = \frac{\mu_0 Id}{2\pi(b^2 - a^2)} (-\bar{a}_y)$$

The magnetic vector at C due to the equivalent solid conductor of radius a carrying a specified known current opposite to that above is zero, since no current is contained within a path of zero radius. Hence the resultant magnetic vector at point C, which is the summation of the two foregoing factors is then simply

$$\bar{B} = \frac{\mu_0 Id}{2\pi(b^2 - a^2)} (-\bar{a}_y)$$

• PROBLEM 8-19

Find the B field due to a slab of volume current if the z-directed current $J_0 i_z$ is uniform over a thickness d as shown in figure 1. Find also the force per unit area on the slab.

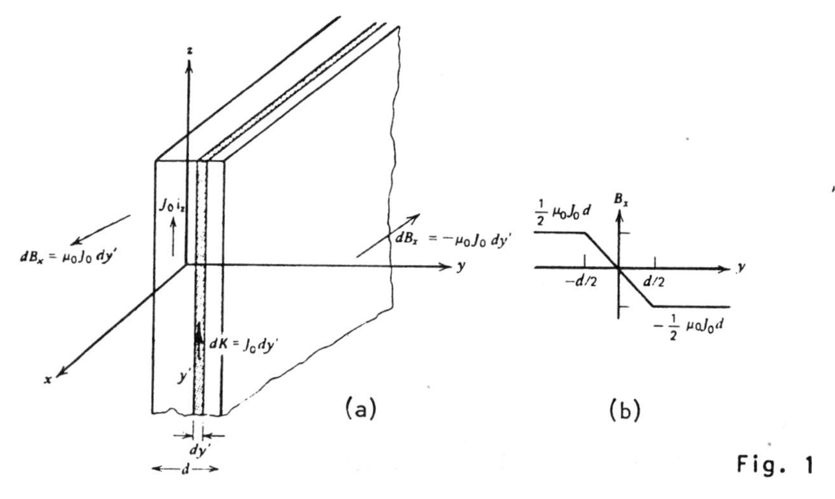

Fig. 1

Solution: Break the slab into incremental current sheets $J_0 dy'$. The magnetic field from each current sheet is given by

$$B_x = \begin{cases} -\mu_0 K_0/2, & y > 0 \\ \mu_0 K_0/2, & y < 0 \end{cases}$$

When adding the contributions of all the differential

sized sheets, those to the left of a field point give a negatively x directed magnetic field while those to the right contribute a positively x-directed field:

$$B_x = \begin{cases} \displaystyle\int_{-d/2}^{+d/2} \frac{-\mu_0 J_0 dy'}{2} = \frac{-\mu_0 J_0 d}{2}, & y > \frac{d}{2} \\ \displaystyle\int_{-d/2}^{+d/2} \frac{\mu_0 J_0 dy'}{2} = \frac{\mu_0 J_0 d}{2}, & y < -\frac{d}{2} \\ \displaystyle\int_{-d/2}^{y} \frac{-\mu_0 J_0 dy'}{2} + \int_{y}^{+d/2} \frac{\mu_0 J_0 dy'}{2} = -\mu_0 J_0 y, & -\frac{d}{2} \le y \le \frac{d}{2}. \end{cases}$$

The B field is plotted in Fig. 1b.

The total force per unit area on the slab is zero:

$$F_{S_y} = \int_{-d/2}^{+d/2} J_0 B_x dy$$

$$= -\mu_0 J_0{}^2 \int_{-d/2}^{+d/2} y dy$$

$$= -\mu_0 J_0{}^2 \frac{y^2}{2} \Bigg|_{-d/2}^{+d/2} = 0.$$

A current distribution cannot exert a net force on itself.

• PROBLEM 8-20

Show that the magnetic field at a point distant a from the centre of a single circular turn of radius R, in the plane of the turn, is given by

$$B_0 = \mu_0 \frac{IR}{\pi(R^2 - a^2)} \int_0^{\frac{1}{2}\pi} \sqrt{1 - \left(\frac{a}{R}\right)^2 \sin^2\phi}\, d\phi,$$

for points inside the turn, and plot a curve showing the variation of B_0 over a diameter of the circle.

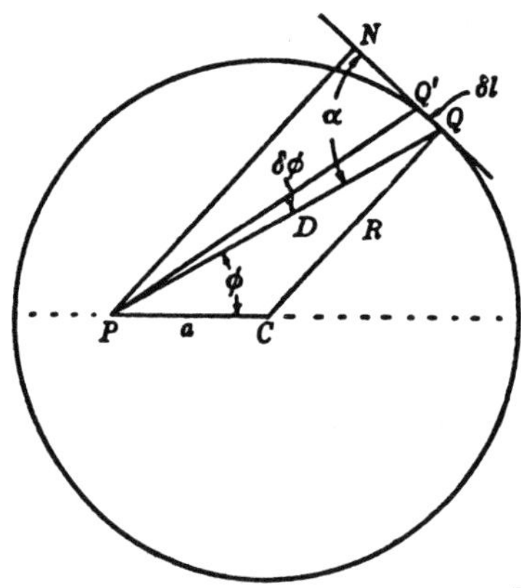

Fig. 1

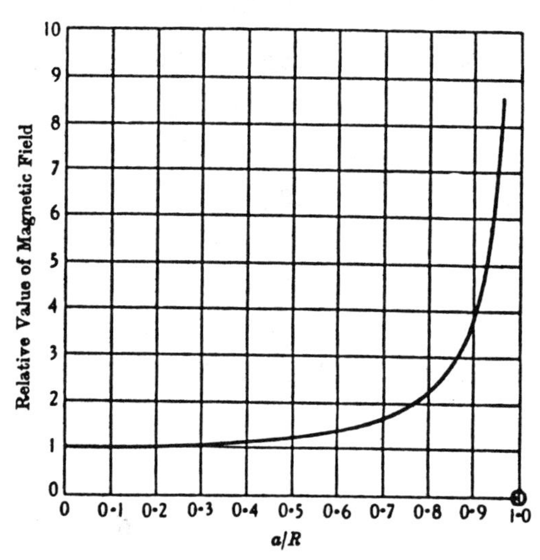

Variation of magnetic field over plane of circular loop. Fig. 2

Solution: Consider the field at P due to the current I in a length $\delta \ell$ of the wire between Q and Q'. Let PQ = D and let PN be the normal to the tangent at Q (fig. 1), then

$$\delta B_0 = \mu_0 \frac{I\delta \ell}{4\pi D^2} \sin\alpha .$$

Now the area of the triangle

$$PQQ' = \tfrac{1}{2}D^2\delta\phi = \tfrac{1}{2}PN\delta\ell = \tfrac{1}{2}D\sin\alpha\delta\ell,$$

so that

$$\sin\alpha\ \delta l = D\delta\phi$$

and

$$B_0 = \frac{\mu_0 I}{4\pi}\int_0^{2\pi}\frac{d\phi}{D}\ .$$

Now

$$D = PQ = a\cos\phi + \sqrt{R^2 - a^2\sin^2\phi}$$

and

$$\frac{1}{D} = \frac{-a\cos\phi + \sqrt{R^2 - a^2\sin^2\phi}}{R^2 - a^2}$$

whence

$$B_0 = \frac{\mu_0 I}{4\pi(R^2 - a^2)}\int_0^{2\pi}\sqrt{R^2 - a^2\sin^2\phi}\,d\phi$$

$$\left(\text{since} \int_0^{2\pi}\cos\phi\,d\phi = 0\right)$$

$$= \frac{\mu_0 IR}{\pi(R^2 - a^2)}\int_0^{1/2\pi}\sqrt{1 - k^2\sin^2\phi}\,d\phi,$$

where $k = a/R$.

The integral may be evaluated by first expanding the expression by the Binomial Theorem. It is, however, the "standard elliptic integral" denoted by "E," values of which are to be found in any good set of mathematical tables.

In order to plot a curve showing the variation of B_0 over a diameter, take the simple case where R = I = unity. Thus

$$B_0 = \frac{\mu_0}{\pi(1 - k^2)}(E),$$

where k is the distance of P from the centre C.

The value of the integral (E) is usually given in tables in terms of an angle θ, where k = sin θ, the modulus of the integral. Tabulate the results as follows, for various values of θ:

$\theta°$	E	$k = \sin\theta$	$(1-k^2)$	$B_0/4\mu_0$ $[E/(1-k^2)]$	Relative values
0	1·5708	0·0 (centre)	1·000	1·571 ($\frac{1}{2}\pi$)	1·000
10	1·5590	0·1736	0·970	1·607	1·023
20	1·5238	0·3420	0·883	1·727	1·100
30	1·4675	0·5000	0·750	1·958	1·247
40	1·3931	0·6428	0·587	2·37	1·51
50	1·3055	0·7661	0·414	3·15	2·00
60	1·2110	0·8660	0·250	4·84	3·09
70	1·1184	0·9397	0·117	9·56	6·1
80	1·0401	0·9848	0·030	34·45	22·0
90	1·0000	1·0000	0·000	Infinite	Infinite

Thus up to a radius of 17.4% of the radius of the loop the magnetic field in the plane of the loop does not vary more than 2.3%. The infinite value at k = 1 is of course never reached owing to the finite diameter of the wire. The variation over a diameter is shown graphically in Fig. 2.

• PROBLEM 8-21

Use the Biot-Savart's law. Write an expression and indicate the direction for the magnetic field $\overline{B}$ at point P in the figure. In the figure, a circular loop of wire, carrying a current i is given.

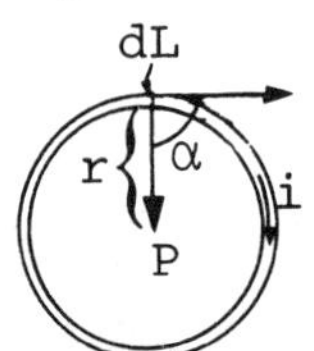

Solution: Applying the Biot-Savart's law, the magnitude and direction of the magnetic field dB at a certain point due to a current element $i d\bar{L}$ is given as:

$$d\bar{B} = \frac{\mu_0 i}{4\pi} \frac{d\bar{L} \times \bar{r}}{r^3}$$

where $\bar{r}$ = a displacement vector from a current element to P.

and $d\bar{L}$ = the length of the current element in the direction of the current flow.

Since the contribution from any current element will be in the same direction (i.e., into the plane of the page) all these contributions must be added directly.

The magnitude of dB is given by

$$dB = \frac{\mu_0 i}{4\pi} \frac{dL \sin\alpha}{r^2}$$

where α is the angle between $d\bar{L}$ and $\bar{r}$.

Now, since a circular loop is considered, $\alpha = 90^0$ and hence $\sin 90^0 = 1$.

The magnetic field at point P can be calculated by integrating dB over the entire loop.

Thus,

$$B_{(at\ P)} = \int dB = \int \frac{\mu_0 i}{4\pi} \frac{dL}{r^2}$$

$$= \frac{\mu_0 i}{4\pi r^2} \int dL$$

$$= \frac{\mu_0 i}{4\pi r^2} \times 2\pi r$$

$$= \frac{\mu_0 i}{2r} \quad \text{(directed into the page)}$$

(Note: $\frac{\mu_0 i}{4\pi r^2}$ = constant and $\int dL = 2\pi r$ = the circumference of the loop.)

• PROBLEM 8-22

> A hollow cylinder of length L and radius a has a uniform surface current $K_0 i_\phi$ as shown in the figure. Find the magnetic field along the z axis due to this hollow cylinder of surface current.

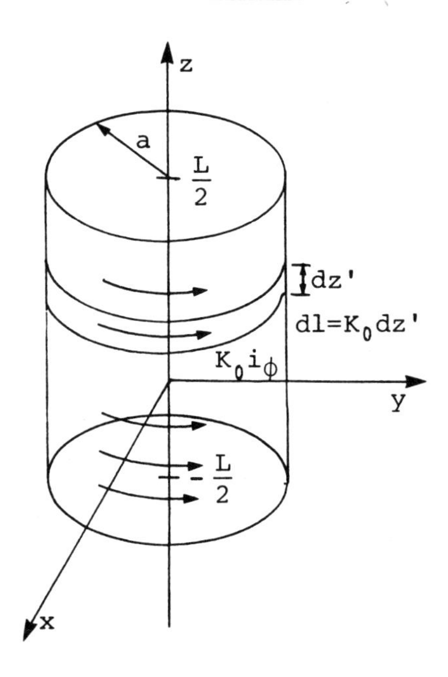

Solution: The current per unit length is

$$K_0 = NI/L.$$

The magnetic field along the z axis at the position z due to each incremental hoop at z' is found from the field due to a single hoop: .

$$B_z = \frac{\mu_0 I a^2}{2(z^2 + a^2)^{3/2}},$$

by replacing z by (z - z') and I by $K_0 dz'$:

$$dB_z = \frac{\mu_0 a^2 K_0 dz'}{2[(z - z')^2 + a^2]^{3/2}} .$$

The total axial magnetic field is then

$$B_z = \int_{z=-L/2}^{+L/2} \frac{\mu_0 a^2 K_0}{2} \frac{dz'}{[(z - z')^2 + a^2]^{3/2}}$$

$$= \frac{\mu_0 a^2 K_0}{2} \left. \frac{(z' - z)}{a^2[(z - z')^2 + a^2]^{1/2}} \right|_{z'=-L/2}^{+L/2}$$

$$= \frac{\mu_0 K_0}{2} \left(\frac{-z + L/2}{[(z - L/2)^2 + a^2]^{1/2}} + \frac{z + L/2}{[(z + L/2)^2 + a^2]^{1/2}} \right).$$

As the cylinder becomes very long, the magnetic field far from the ends becomes approximately constant

$$\lim_{L \to \infty} B_z = \mu_0 K_0 .$$

• **PROBLEM 8-23**

(a) A single turn of wire forms the perimeter of a square of side 2a and carries a current I. Show that the magnetic force due to this circuit at a point on the central normal to its plane and distant x from the plane is

$$\frac{8a^2I}{(a^2 + x^2)\sqrt{2a^2 + x^2}}.$$

(b) A non-metallic former of length ℓ has a square cross-section of side d. It is closely and uniformly wound throughout its length with T turns (total) of negligible thickness. Show that when a current I amperes flows in the winding, the magnetic force at the centre of the former is

$$\frac{1 \cdot 6(IT)}{\ell} \tan^{-1} \frac{\ell}{\sqrt{\ell^2 + 2d^2}}.$$

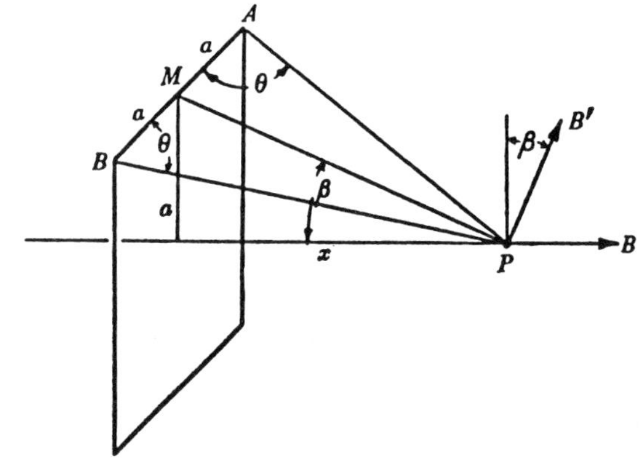

Fig. 1

Solution:(a)The flux density at P (Fig. 1) due to one side AB of the square is

$$B' = \frac{\mu_0 I}{4\pi PM} \int_{\theta}^{\pi-\theta} \sin\alpha \, d\alpha.$$

Now PM = $\sqrt{a^2 + x^2}$, and $\cos\theta = \dfrac{a}{\sqrt{2a^2 + x^2}} = -\cos(\pi - \theta)$

whence

$$B' = \frac{\mu_0 aI}{2\pi\sqrt{a^2 + x^2}\sqrt{2a^2 + x^2}}$$

in a direction making an angle β with the normal to the axis of the coil.

B' may be resolved along and normal to this axis, the normal component being cancelled by that due to the opposite side. Thus the field at P due to all four sides will be along the axis and equal to four times the axial component of B'. That is

$$B = 4B' \sin \beta,$$

where

$$\sin \beta = \frac{a}{\sqrt{a^2 + x^2}},$$

$$= \frac{2\mu_0 a^2 I}{\pi (a^2 + x^2)\sqrt{2a^2 + x^2}} .$$

(b) This part of the problem is most readily solved by putting the expression for B in terms of the angle β.

Now

$$(a^2 + x^2) = a^2 \operatorname{cosec}^2\beta$$

and

$$(2a^2 + x^2) = \left(\frac{a^2}{a^2 + x^2} + 1\right)(a^2 + x^2),$$

so that

$$\sqrt{2a^2 + x^2} = \frac{a\sqrt{\sin^2\beta + 1}}{\sin \beta}$$

and

$$B = \mu_0 \frac{2I \sin^3 \beta}{\pi a \sqrt{\sin^2\beta + 1}} .$$

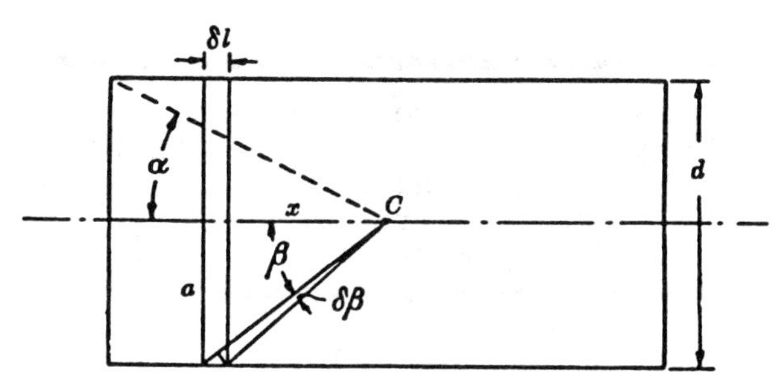

Fig. 2

Consider the field at the center of the coil, C, due to a length $\delta\ell$ of the coil (Fig. 2). The number of turns in this element is

$$\frac{T\delta\ell}{\ell}$$

so that the field at C due to the whole coil is

$$B_C = \mu_0 \int_0^{\ell} \frac{2IT \sin^3\beta}{\pi a\ell\sqrt{\sin^2\beta + 1}} d\ell;$$

but

$$\delta\ell \sin \beta = a \operatorname{cosec} \beta\delta\beta$$

or

$$\delta\ell = \frac{a\delta\beta}{\sin^2 \beta},$$

so that

$$B_C = \frac{\mu_0 2IT}{\pi\ell} \int_{\alpha}^{\pi-\alpha} \frac{\sin\beta d\beta}{\sqrt{2 - \cos^2 \beta}}$$

$$= \frac{\mu_0 2IT}{\pi\ell} \left[\int_{\alpha}^{\pi-\alpha} \sin^{-1}\left(\frac{\cos\beta}{\sqrt{2}}\right)\right]$$

$$= \frac{\mu_0 4IT}{\pi\ell} \sin^{-1} \frac{\ell}{\sqrt{2(\ell^2 + d^2)}}$$

$$= \frac{\mu_0 4IT}{\pi\ell} \tan^{-1} \frac{1}{\sqrt{1^2 + 2d^2}} .$$

(Note. If the expression for B in terms of x is retained, integrate by making the substitution:

$$z = \frac{x}{\sqrt{\tfrac{1}{2}d^2 + x^2}} .)$$

• PROBLEM 8-24

Often it is desired to have an accessible region in space with an essentially uniform magnetic field. This can be arranged by placing another coil at z = d, parallel to a coil at z = 0, as in figure 1. Find the magnetic field due to these two coils along the z axis.

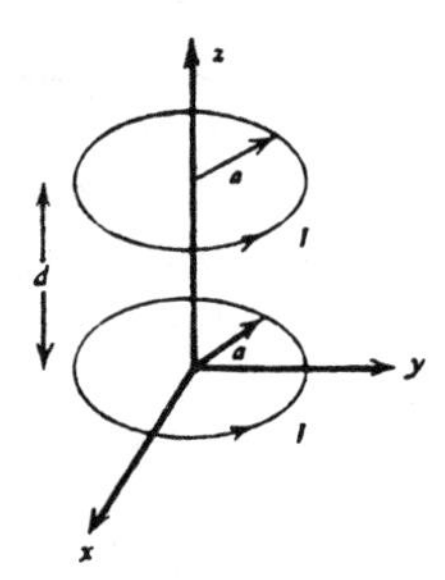

Fig. 1

Solution: The total magnetic field along the z axis is found by superposing the field for each hoop:

B for a single hoop is given by

$$B_z = \frac{\mu_0 I a^2}{2(z^2 + a^2)^{3/2}}$$

therefore for two hoops:

$$B_z = \frac{\mu_0 I a^2}{2}\left(\frac{1}{(z^2 + a^2)^{3/2}} + \frac{1}{((z - d)^2 + a^2)^{3/2}}\right)$$

Figure 2 shows a plot of this B field.

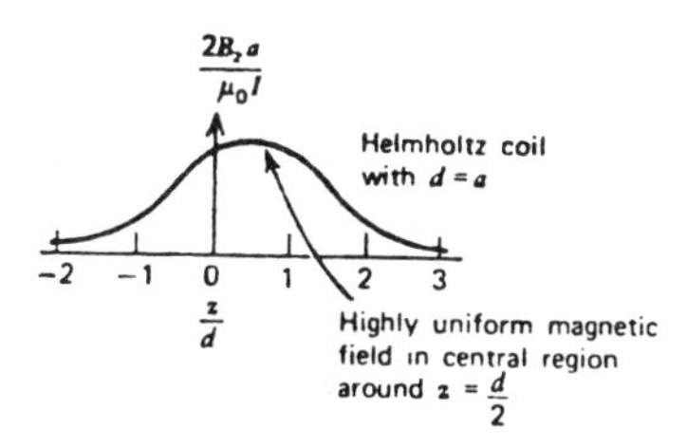

Fig. 2

Then the slope of B_z,

$$\frac{\partial B_z}{\partial z} = \frac{3\mu_0 I a^2}{2}\left(\frac{-z}{(z^2 + a^2)^{5/2}} - \frac{(z - d)}{((z - d)^2 + a^2)^{5/2}}\right)$$

is zero at $z = d/2$. The second derivative,

$$\frac{\partial^2 B_z}{\partial z^2} = \frac{3\mu_0 I a^2}{2}\left(\frac{5z^2}{(z^2 + a^2)^{7/2}} - \frac{1}{(z^2 + a^2)^{5/2}} + \frac{5(z - d)^2}{((z - d)^2 + a^2)^{7/2}} - \frac{1}{((z - d)^2 + a^2)^{5/2}}\right)$$

can also be set to zero at $z = d/2$, if $d = a$, giving a highly uniform field around the center of the system, as plotted in Figure 2. Such a configuration is called a Helmholtz coil.

• PROBLEM 8-25

Consider the magnetic field given by

$$B = 3xy^2 \hat{i}_z \text{ Wb/m}^2$$

Determine the magnetic flux crossing the portion of the xy plane lying between $x = 0$, $x = 1$, $y = 0$, and $y = 1$. First use approximation methods, then use exact methods to calculate the flux.

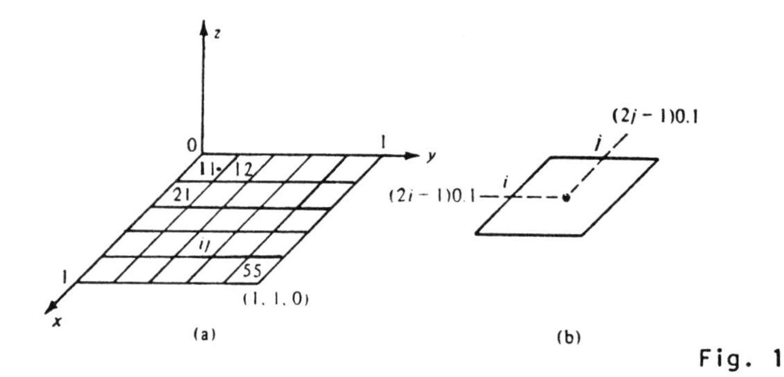

Fig. 1

Solution: For the approximate result, divide the surface into 25 equal areas as shown in Fig. 1(a). Designate the squares as 11, 12, . . ., 15, 21, 22, . . ., 55 where the first digit represents the number of the square in the x direction and the second digit represents the number of the square in the y direction. The x and y coordinates of the midpoint of the ijth square are (2i - 1)0.1 and (2j - 1)0.1, respectively, as shown in Fig. 1 (b). The magnetic field at the center of the ijth square is then given by

$$B_{ij} = 3(2i - 1)(2j - 1)^2 0.001\hat{i}_z$$

Since the surface was divided into equal areas and since all areas are in the xy plane,

$$\Delta S_{ij} = 0.04\hat{i}_z \qquad \text{for all i and j}$$

The required magnetic flux is then given by

$$[\psi]_s = \sum_{i=1}^{5} \sum_{j=1}^{5} B_{ij} \cdot \Delta S_{ij}$$

$$= \sum_{i=1}^{5} \sum_{j=1}^{5} 3(2i - 1)(2j - 1)^2 0.001\hat{i}_z \cdot 0.04\hat{i}_z$$

$$= 0.00012 \sum_{i=1}^{5} \sum_{j=1}^{5} (2i - 1)(2j - 1)^2$$

$$= 0.00012(1 + 3 + 5 + 7 + 9)(1 + 9 + 25 + 49 + 81)$$

$$= 0.495 \text{ Wb}$$

The result obtained for $[\psi]_s$ is approximate since the surface was divided into a finite number of areas. By dividing it into larger and larger numbers of squares, more and more accurate results can be obtained. The value to which the result is exact is that for which the number of squares in each direction is infinity. The summation then becomes an integral that represents exactly the magnetic flux crossing the surface and is given by

$$[\psi]_S = \int_S B \cdot dS$$

To calculate the exact result using the surface integral, note that at any arbitrary point (x, y) on the surface, the infinitesimal surface vector is given by

$$dS = dx\ dy\ \hat{i}_z$$

The value of $B \cdot dS$ at the point (x, y) is

$$B \cdot dS = 3xy^2\, \hat{i}_z \cdot dx\ dy\ \hat{i}_z$$

$$= 3xy^2\ dx\ dy$$

Thus the required magnetic flux is given by

$$[\psi]_S = \int_S B \cdot dS$$

$$= \int_{x=0}^{1} \int_{y=0}^{1} 3xy^2\ dx\ dy = 0.5\ \text{Wb}$$

• PROBLEM 8-26

A linear current circuit carries a current I. A rectangular circuit has two sides of length L parallel to the wire, one at a distance a and the other at a distance b from the wire. See the given figure. Find the magnetic flux through the rectangular circuit.

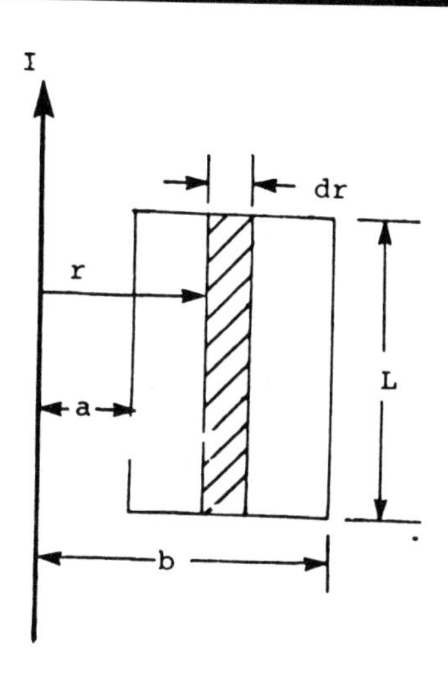

Solution: The magnetic flux is given by the surface integral

$$\Phi = \int_S \bar{B} \cdot d\bar{a}$$

$\bar{B}$ due to a long straight wire carrying current I is

$$\bar{B} = \frac{\hat{\theta}\mu_0 I}{2\pi r}$$

(from Ampere's Circuital law).

The flux through the elementary rectangle Ldr is $d\phi = \bar{B} \cdot d\bar{a}$.

$$= \frac{\mu_0 I}{2\pi r}\, Ldr$$

The total flux through the rectangular circuit is

$$\Phi = \int_{r=a}^{r=b} d\Phi = \int_{r=a}^{r=b} \frac{\mu_0 ILdr}{2\pi r}$$

$$= \frac{\mu_0 IL}{2\pi} \ln r \Big]_a^b$$

$$= \frac{\mu_0 IL}{2\pi} \ln\left(\frac{b}{a}\right)$$

• PROBLEM 8-27

A solenoid is a coil of wire wound on the surface of a cylinder as shown in cross section in Fig. 1(a). Derive a formula for the magnetic flux density on the axis of the solenoid due to the current flowing in the wires of the winding.

Solution: In the figure the wires are drawn much too large in cross section relative to the diameter of the cylinder; that is, there should be a great many turns whereas only a few are shown. On a properly wound solenoid the wires are very close together, being separated only by the thin layer of insulation on each wire. Thus, although the winding is a helix, each turn of wire may be considered to be a plane circular loop like the one shown in Fig. 2.

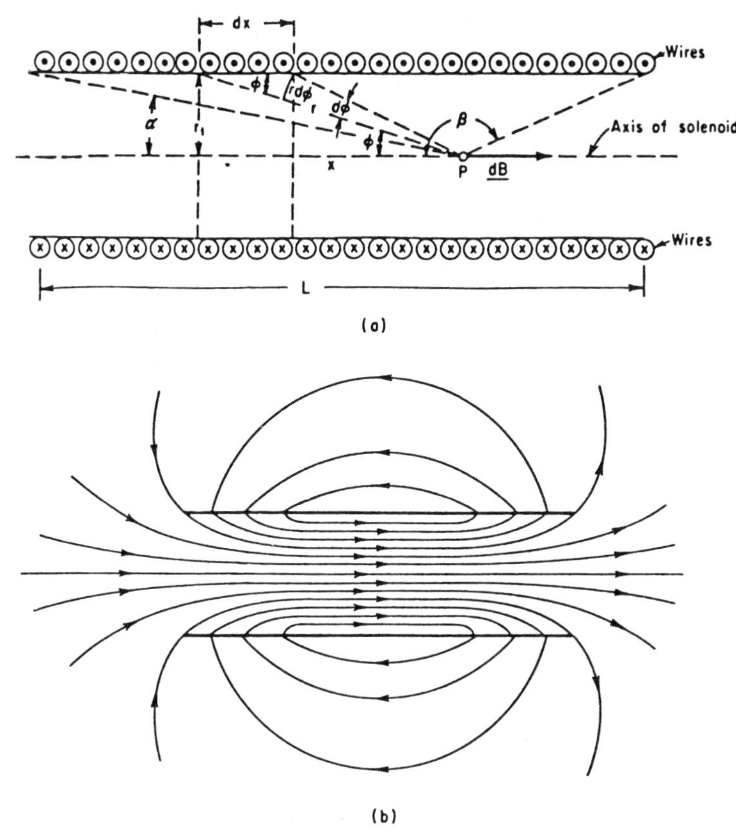

Figure 1. (a) A solenoid shown in cross section. Current is flowing out of the page in the wires on the upper surface (as shown by the dots in the circles which represent the wires in cross section) and into the page in the wires on the lower surface (as shown by the crosses in the circles which represent the wires in cross section). (b) An approximate map of the magnetic flux lines of a solenoid.

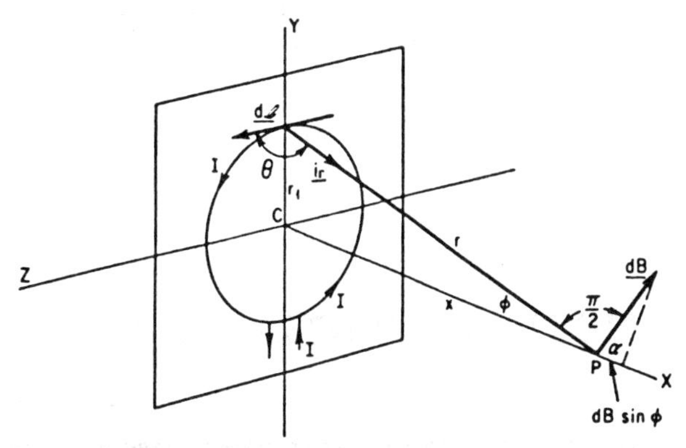

Figure 2 Magnetic field on the axis of a plane circular coil of wire carrying a current *I*.

Hence, use

$$B = \mu_0 I (\sin \phi) 2\pi r_1 / 4\pi r^2 \tag{1}$$

for the magnetic flux density on the axis due to each turn

of wire. For any point on the axis, such as P in Fig. 1(a), the total magnetic flux density is the vector sum of the contributions due to all the turns of wire on the solenoid. All the magnetic flux densities at P, due to all the turns, will be parallel and in the same direction, so use integration to find the vector sum.

Let L represent the total length of the solenoid and N represent the total number of turns of wire on the solenoid, so that N/L represents the number of turns of wire per unit length. Let r_1 be the radius of the solenoid. Consider the element of length dx of the solenoid as shown in Fig. 1 (a). There are (N/L) dx turns of wire on the length dx, and each turn contributes a flux density given by (1), so

$$dB = [\mu_0 I r_1 (\sin\phi) N/2r^2 L]dx \tag{2}$$

is the element of magnetic flux density at P due to the element of length dx of the solenoid, where ϕ represents the angle between the axis and a line drawn to dx from P. Thus, $d\phi$ represents the element of angle which dx subtends at P. In (2), of course, r is the distance from dx to P. Then from Fig. 1 (a), $r\,d\phi = dx \sin\phi$ and $\sin\phi = r_1/r$. Use these relationships to eliminate r, and (2) becomes

$$dB = (\mu_0 NI/2L) \sin\phi\, d\phi$$

which has changed the variable from x to ϕ. Now add all the dB's at P due to all the dx's from the left end of the solenoid to the right end. This means that ϕ is to go from α to β in Fig. 1 (a). Hence, integrating from α to β

$$B = \mu_0 NI(\cos\alpha - \cos\beta)/2L \tag{3}$$

and B is parallel to the axis of the solenoid. Equation (3) gives the magnetic flux density on the axis of the solenoid not only inside the solenoid but beyond the ends as well.

Fig. 1 (b) shows an approximate plot of the flux lines of a solenoid.

VECTOR MAGNETIC POTENTIAL

• PROBLEM 8-28

Calculate the vector potential arising from a uniform distribution of current in a long circular cylinder of radius a.

Solution: Two methods are given:

(a) Use cylindrical polar coordinates (r, ϕ, z) with A =

$\{0\ 0\ A\}$, where A depends only on r. Find B = curl A and equate it to the expression for B due to a long circular cylinder:

$$B = \frac{\mu_0 Ir}{2\pi a^2} \quad \text{for} \quad r < a$$

$$B = \frac{\mu_0 I}{2\pi r} \quad \text{for} \quad r > a$$

thus:

$$\text{curl } A = \left\{0 \quad -\frac{dA}{dr} \quad 0\right\}.$$

$$\therefore \frac{dA}{dr} = -\frac{\mu_0 Ir}{2\pi a^2} \quad \text{for} \quad r \leq a.$$

$$\therefore A = -\frac{\mu_0 Ir^2}{4\pi a^2} + C_1 \quad \text{for} \quad r \leq a.$$

Also

$$\frac{dA}{dr} = -\frac{\mu_0 I}{2\pi r} \quad \text{for} \quad r \geq a.$$

$$\therefore A = -\frac{\mu_0 I}{2\pi} \ln r + C_2 \quad \text{for } r \geq a.$$

The constant C_2 is either ignored or given some convenient value. Then the fact that A is continuous at r = a determines C_1.

(b) Alternatively, use the fact that $\nabla^2 A = -\mu_0 J$ and write $A = A\hat{k}$ since A must be directed parallel to the current. Hence

$$\nabla^2 A = -\frac{\mu_0 I}{4\pi a^2} \quad \text{for} \quad r \leq a.$$

In terms of cylindrical coordinates

$$\frac{1}{r}\frac{d}{dr}\left(r\frac{dA}{dr}\right) = -\frac{\mu_0 I}{4\pi a^2}, \quad \frac{dA}{dr} = -\frac{\mu_0 Ir}{2\pi a^2} - \frac{C_3}{r}.$$

$$\therefore B = \left\{0 \quad \frac{\mu_0 Ir}{2\pi a^2} + \frac{C_3}{r} \quad 0\right\} = \mu_0 H.$$

Since $\oint H \cdot ds = Ir^2/a^2$, $C_3 = 0$.

$$\therefore \frac{dA}{dr} = -\frac{\mu_0 Ir}{2\pi a^2}$$

and the remainder follows as above.

Find the magnetic field established by a current flowing through two long parallel wires using the magnetic potential. Assume currents flowing in opposite directions, as shown in Fig. 2.

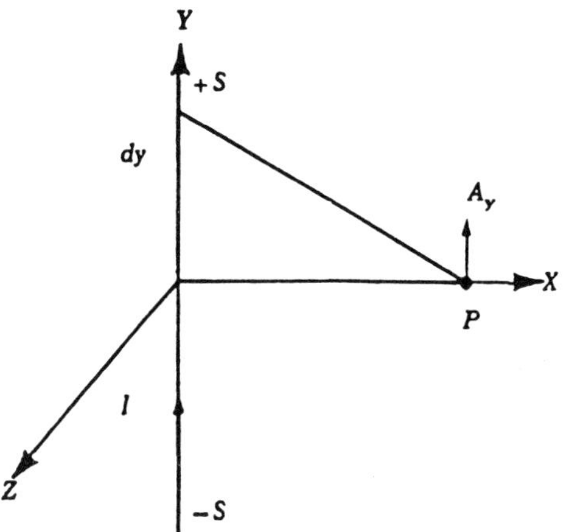

Fig. 1

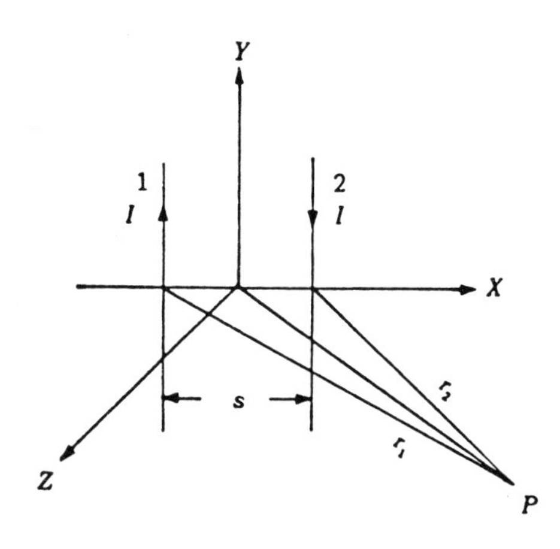

Fig. 2

Solution: First find the vector potential about a long straight wire with current flow I, as shown in Fig. 1.

The vector potential is

$$A = \frac{1\mu_0}{4\pi}\int_V \frac{J}{r}\, dV$$

The current density J integrated over the cross section is the current I all in the y direction.

Rewriting the equation

$$A_y = \frac{I\mu_0}{4\pi}\int_{-s}^{+s} \frac{1}{r}\, dy$$

At y point P on the x-axis the potential is, by the geometry of the figure

$$A_y = 2 \cdot \frac{I\mu_0}{4\pi}\int_0^s \frac{I}{\sqrt{x^2 + y^2}}\, dy$$

$$= \frac{I\mu_0}{2\pi}[\ln y + \sqrt{x^2 + y^2}]_0^s$$

$$= \frac{I\mu_0}{2\pi}[\ln (s + \sqrt{x^2 + s^2}) - \ln x]$$

If the wire is very long, s >> x so that

$$A_y \cong \frac{I}{2\pi}\mu_0(\ln 2s - \ln x)$$

From the previous result

$$A_y \cong \frac{I}{2\pi}\mu_0(\ln 2s - \ln x)$$

is the potential due to one wire, so that the total potential at P due to two wires is

$$A_y \cong \frac{I}{2\pi}\mu_0(\ln 2s - \ln r_1) - \frac{I}{2\pi}(\ln 2s - \ln r_2)$$

or

$$A_y \cong \frac{I}{2\pi}\mu_0(\ln r_2 - \ln r_1)$$

But

$$r_1 = \sqrt{(x + s/2)^2 + x^2}$$

$$r_2 = \sqrt{(x - s/2)^2 + x^2}$$

So that

$$\frac{\partial r_1}{\partial x} = \frac{x + s/2}{r_1}\,; \quad \frac{\partial r_2}{\partial x} = \frac{x - s/2}{r_2}\,.$$

and

$$\frac{\partial r_1}{\partial z} = \frac{z}{r_1}\,; \quad \frac{\partial r_2}{\partial z} = \frac{z}{r_2}$$

$$H_z = \text{curl}_z\, A_y = \frac{\partial A_y}{\partial x}$$

$$= \frac{I}{2\pi}\left[\frac{x + s/2}{r_1^2} - \frac{x - s/2}{r_2^2}\right]$$

and

$$H_x = \text{curl}_x\, A_y = -\frac{\partial A_y}{\partial z}$$

$$= -\frac{1}{2\pi}\left(\frac{z}{r_1^2} - \frac{z}{r_2^2}\right)$$

• PROBLEM 8-30

Compute the vector magnetic potential within the outer conductor for the coaxial line whose magnetic field intensity $A_z = 0$ at $r = 5a$, where a is the radius of the inner conductor. 5a is the inside radius of the outer conductor. The outside radius of the outer conductor is 6a.

Solution: The current density on the outer conductor is

$$J_z \text{ (outside cond.)} = -\frac{I}{\pi(c^2 - b^2)} = -\frac{I}{\pi(6^2 - 5^2)a^2}$$

Let $\bar{A}_z$ be the vector magnetic potential and is in the same direction as $\bar{J}_z$. Solve the Poisson's equation

$$-\frac{1}{\mu_0}\nabla^2\bar{A}_z = -\frac{I}{11\,\pi\,a^2}$$

In cylindrical coordinates,

$$\nabla^2\bar{A}_z = \frac{1}{r}\frac{d}{dr}\left(r\frac{d\bar{A}_z}{dr}\right)$$

Substituting,

$$\frac{1}{\mu_0}\frac{1}{r}\frac{d}{dr}\left(r\frac{d\bar{A}_z}{dr}\right) = -\frac{I}{11\pi a^2}$$

Integrating once,

$$\frac{d\bar{A}_z}{dr} = \frac{\mu_0 Ir}{22\pi a^2} + \frac{C_1}{r}$$

Integrating again,

$$\bar{A}_z = \frac{\mu_0 Ir^2}{44\pi a^2} + C_1 \ln r + C_2.$$

At $r = b = 5a$, $A_z = 0$. Solve for C_2.

$$C_2 = -\frac{\mu_0 I(25)}{44\pi} - C_1 \ln 5a$$

and

$$\bar{A}_z = \frac{\mu_0 I}{44\pi}\left(\frac{r^2}{a^2} - 25\right) + C_1 \ln \frac{r}{5a}\ .$$

Now evaluate C_1. Since

$$\nabla \times \bar{A} = -\left(\frac{\mu_0 Ir}{22\pi a^2} + \frac{C_1}{r}\right)\bar{a}_\phi$$

Therefore

$$H_\phi = -\frac{Ir}{22\pi a^2} + \frac{C_1}{\mu_0 r}$$

at $r = 5a$, $2\pi(5a)H_\phi = I$,

Substituting,

$$C_1 = -\frac{\mu_0 I(36)}{22\pi}$$

Thus

$$\bar{A}_z = \frac{\mu_0 I}{44\pi}\left(\frac{r^2}{a^2} - 25\right) - \frac{\mu_0 I(36)}{22\pi} \ln \frac{r}{5a}$$

• PROBLEM 8-31

Find the A and B fields of a thin wire loop of radius a and carrying a steady current I, as in Figure 1 (a). Make approximations to provide valid answers at large distances from the loop (assume $a << r$).

Solution: Without detracting from the generality, the field point P can be located directly above the y axis as shown in Figure 1(b). The A field at P is given by

$$A(u_1, u_2, u_3) = \int_\ell \frac{\mu_0 I d\ell'}{4\pi R}$$

in which $I\, d\ell' = \hat{a}_\phi Ia\, d\phi'$. The variable direction of a_ϕ in the integrand is handled by pairing the effects of the current elements $I\, d\ell'_1$ and $I\, d\ell'_2$ at the symmetrical locations about the y axis in Figure 1 (b). From the geometry,

$$I\,d\ell'_1 = \hat{a}_\phi Ia\,d\phi'$$

$$= (-\hat{a}_x \sin\phi' + \hat{a}_y \cos\phi')Ia\,d\phi'$$

$$I\,d\ell'_2 = (-\hat{a}_x \sin\phi' - \hat{a}_y \cos\phi')Ia\,d\phi' \quad (1)$$

to provide a cancellation of the y components of the potential contributions of the pair of elements at P, leaving a net dA at P that is -x directed. Thus the integral becomes

$$-A_x = A_\phi = 2\int_{\phi'=-\pi/2}^{\pi/2} \frac{\mu_0 Ia \sin\phi'\,d\phi'}{4\pi R} \quad (2)$$

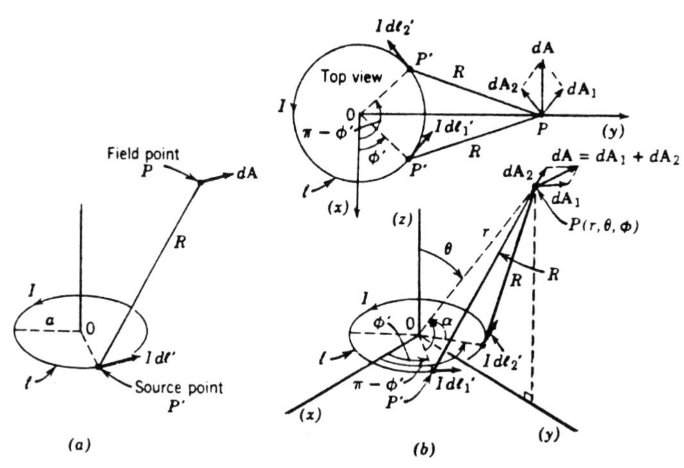

Circular loop, showing the spherical coordinate geometry adopted for finding the static magnetic field at P.
(a) Circular loop carrying a current I.
(b) Making use of symmetry to obtain fields at P. Fig. 1

From the law of cosines applied to the triangle POP' in the figure, $R^2 = a^2 + r^2 - 2ar\cos\alpha = a^2 + r^2 - 2ar\sin\theta\sin\phi'$.

If $r >> a$, one can approximate, making use of the binomial theorem,

$$(1 - x)^{\frac{1}{2}} = 1 - \frac{x}{2} + \frac{x^2}{3} + \ldots .$$

Therefore,

$$R \cong r\left[1 - 2\frac{a}{r}\sin\theta\sin\phi'\right]^{\frac{1}{2}}$$

$$\cong r\left[1 - \frac{a}{r}\sin\theta\sin\phi' + \ldots\right]$$

The reciprocal, for small a, is similarly approximated

$$\frac{1}{R} \cong \frac{1}{r} + \frac{a}{r^2}\sin\theta\sin\phi'$$

Thus (2) becomes

$$A_\phi \cong \frac{2\mu_0 Ia}{4\pi}\int_{\phi'=-\pi/2}^{\pi/2}\left[\frac{1}{r} + \frac{a}{r^2}\sin\theta\sin\phi'\right]\sin\phi'\,d\phi' \qquad (3)$$

The integral of the $(\sin\phi')/r$ term is zero, so integrating the second term yields the answer

$$A_\phi \cong \frac{\mu_0 a^2 I\sin\theta}{4r^2}$$

Taking $B = \nabla \times A$ in spherical coordinates therefore yields

$$B \cong \frac{\mu_0 a^2 I}{4r^3}[a_r 2\cos\theta + a_\theta \sin\theta]$$

• PROBLEM 8-32

Find the vector magnetic potential in the plane bisecting a straight piece of thin wire of finite length 2L in free space, assuming a direct current I as in the figure shown. Find B from A.

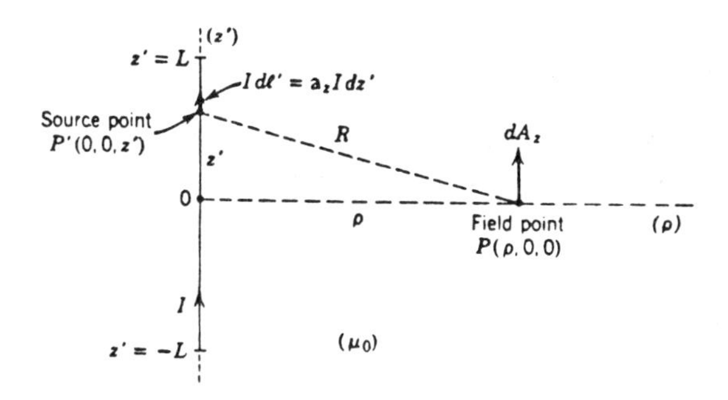

Solution: The fixed field point is on the plane $z = 0$ at $P(\rho, 0, 0)$. The typical current source element at $P'(0, 0, z')$ is $I\,d\ell' = \hat{a}_z I\,dz'$, and R from P' to P is $R = \sqrt{\rho^2 + (z')^2}$, putting the line integral $A(u_1, u_2, u_3) = \int_\ell \frac{\mu_0 I\,d\ell'}{4\pi R}$ in the form

$$A(\rho, 0, 0) = \int_{z'=-L}^{L} \frac{\mu_0 a_z I\,dz'}{4\pi\sqrt{\rho^2 + (z')^2}} \qquad (1)$$

The unit vector $\hat{a}_z$ has the same direction at all P', yielding at P

$$A = \hat{a}_z \frac{\mu_0 I}{4\pi}\int_{-L}^{L} \frac{dz'}{\sqrt{\rho^2 + (z')^2}}$$

$$= \hat{a}_z \frac{\mu_0 I}{4\pi} [\ell n(z' + \sqrt{\rho^2 + (z')^2})\Big]_{-L}^{L}$$

$$= \hat{a}_z \frac{\mu_0 I}{4\pi} \ell n \frac{\sqrt{L^2 + \rho^2} + L}{\sqrt{L^2 + \rho^2} - L} \tag{2}$$

Find B at P using

$$\nabla \times A = \begin{vmatrix} \frac{\hat{a}_\rho}{\rho} & \hat{a}_\phi & \frac{\hat{a}_z}{\rho} \\ \frac{\partial}{\partial \rho} & \frac{\partial}{\partial \phi} & \frac{\partial}{\partial z} \\ A_\rho & A_\phi & A_z \end{vmatrix}$$

Thus, in circular cylindrical coordinates

$$B = \nabla \times A = \begin{vmatrix} \frac{\hat{a}_\rho}{\rho} & \hat{a}_\phi & \frac{\hat{a}_z}{\rho} \\ \frac{\partial}{\partial \rho} & 0 & 0 \\ 0 & 0 & A_z \end{vmatrix}$$

$$= -a_\phi \frac{\partial A_z}{\partial \rho} = a_\phi \frac{\mu_0 I}{2\pi\rho} \frac{L}{\sqrt{L^2 + \rho^2}} \tag{3}$$

For $\rho << L$, (3) simplifies to

$$B = \frac{\mu_0 I}{2\pi\rho} \hat{a}_\phi$$

a result very nearly correct when near a finite-length wire, or correct at any ρ distance for an infinitely long wire.

• PROBLEM 8-33

An important result for antennas is that if a current flows along the z axis, then for distant fields

$$\mu\bar{H} = \nabla \times \bar{A} \tag{1}$$

reduces to

$$\mu H_\phi = - \sin\theta \frac{\partial A_z}{\partial r} \tag{2}$$

Use equation (1) to obtain equation (2).

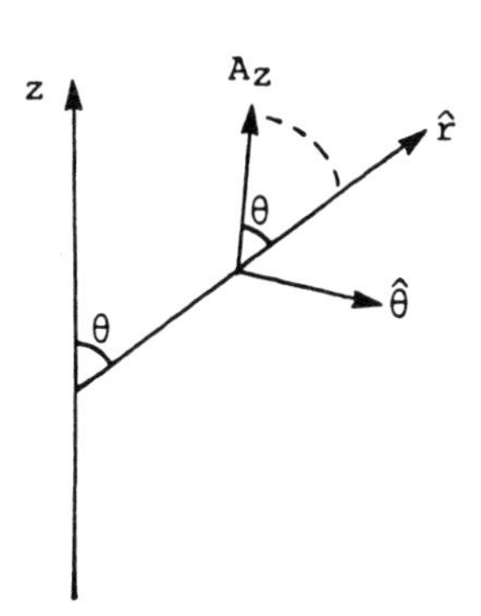

<u>Solution</u>: It is given that $J = \hat{k} J_z$. The vector potential for a current density J is

$$\bar{A} = \frac{\mu}{4\pi} \int \frac{J(\bar{r}')}{R} dv'$$

$$= \frac{\mu \hat{k}}{4\pi} \int \frac{J_z}{R} dv'$$

hence $A_x = A_y = 0$, $\bar{A} = \hat{k} A_z$.

The spherical components of $\bar{A}$ are obtained by:

$$A_\theta = \hat{\theta} \cdot \hat{k} A_z = A_z \cos\left(\frac{\pi}{2} + \theta\right) = -A_z \sin\theta \tag{3}$$

$$A_r = \hat{r} \cdot \hat{k} A_z = A_z \cos\theta \tag{4}$$

$$A_\phi = \hat{\phi} \cdot \hat{k} A_z = 0 \quad (\hat{\phi} \perp \hat{k}) \tag{5}$$

Expressing $\mu\bar{H} = \nabla \times \bar{A}$ in spherical coordinates,

$$\mu H_r = (\nabla \times \bar{A})_r = \frac{1}{r\sin\theta}\left[\frac{\partial}{\partial\theta}(\sin\theta\, A_\phi) - \frac{\partial A_\theta}{\partial\phi}\right] = 0$$

since $A_\phi = 0$ (eq. 5), and $A_\theta \neq f(\phi)$ (eq. 3)

$$\therefore H_r = 0$$

$$\mu H_\phi = \frac{1}{r}\left[\frac{\partial}{\partial r}(rA_\theta) - \frac{\partial A_r}{\partial\theta}\right]$$

$$= \frac{1}{r}\left[\frac{\partial}{\partial r}(r(-A_z \sin\theta)) - \frac{\partial}{\partial\theta}(A_z \cos\theta)\right]$$

$$= \frac{1}{r}\left[-A_z \sin\theta - r\frac{\partial}{\partial r}(A_z \sin\theta) + A_z \sin\theta - \cos\theta \frac{\partial A_z}{\partial\theta}\right]$$

$$= \frac{1}{r}\left[-r\frac{\partial}{\partial r}(A_z \sin\theta) - \cos\theta \frac{\partial A_z}{\partial\theta}\right]$$

$$= -\frac{\partial}{\partial r}(A_z \sin\theta) - \frac{\cos\theta}{r}\frac{\partial A_z}{\partial\theta}$$

$$= -\sin\theta\frac{\partial A_z}{\partial r} - \frac{\cos\theta}{r}\frac{\partial A_z}{\partial\theta}$$

for distant fields (large r), the 1st term dominates

$$\mu H_\phi \cong -\sin\theta\frac{\partial A_z}{\partial r}$$

$$\mu H_\theta = \left[\frac{1}{r\sin\theta}\frac{\partial A_r}{\partial\phi} - \frac{1}{r}\frac{\partial}{\partial r}(r A_\phi)\right] = 0$$

since $A_r = A_z \cos\theta \neq f(\phi)$, and $A_\phi = 0$.

• PROBLEM 8-34

Given a magnetic field $\bar{B}(t) = B_0 t\,\hat{k}$ where B_0 is a constant. Assume symmetry about the z axis, determine the vector potential $\bar{A}$ and find the induced electric field $\bar{E}$ by assuming the scalar potential $V = 0$.

Solution: $\bar{B} = \nabla \times \bar{A}$

$$\therefore \int_S \bar{B}\cdot d\bar{a} = \int_S (\nabla\times\bar{A})\cdot d\bar{a} = \oint \bar{A}\cdot d\bar{\ell}$$

using Stokes' theorem

$$\therefore \int \bar{A}\cdot\overline{d\ell} = \int_S \bar{B}\cdot d\bar{a}$$

Use a circle for the path integral, and let the circle be centered on the z axis. Using cylindrical coordinates

$$\int \bar{A}\cdot\overline{d\ell} = \oint[A_r\hat{r} + A_\theta\hat{\theta} + A_z\hat{z}]$$

$$\cdot\ (\hat{r}dr + r\hat{\theta}d\theta + dz\hat{z})$$

$$= A_\theta(z\pi r) = \int_S \bar{B}\cdot d\bar{a} = \int_S B_0 t\hat{z}\cdot\hat{z}\,da$$

$$= B_0 t\int da = B_0 t(\pi r^2)$$

$$A_\theta = \frac{B_0 tr}{2}$$

$$\bar{A} = \frac{\hat{\theta} B_0 tr}{2}$$

The electric field in terms of the vector potential $\bar{A}$ and scalar potential V is given by

$$\bar{E} = -\frac{\partial \bar{A}}{\partial t} - \nabla V$$

and since it was given that V = 0

$$\bar{E} = -\frac{\partial \bar{A}}{\partial t} = -\frac{\hat{\theta} B_0 r}{2}$$

H-FIELD

• PROBLEM 8-35

A spherical shell of soft iron of inner and outer radii a and b is placed in a uniform external magnetic field H. If μ is the permeability of the shell and (r, θ) are spherical polar coordinates with origin at the center of the shell and initial line parallel to the applied field, show that the magnetic potential may be given by the following forms:

$$r < a, \quad \Omega_1 = Ar \cos\theta,$$

$$a < r < b, \quad \Omega_2 = Br\cos\theta + (C\cos\theta)r^{-2},$$

$$b < r, \quad \Omega_3 = -Hr\cos\theta + (D\cos\theta)r^{-2},$$

and find the magnetic field in the cavity.

Solution: The expressions given above are all finite solutions of Laplace's equation $\nabla^2\Omega = 0$, which is one of the equations to be satisfied inside and outside the material. The other conditions to be satisfied are the boundary conditions at the interfaces r = a, r = b, namely,

$$\Omega_1 = \Omega_2 \quad \text{for } r = a, \qquad \Omega_2 = \Omega_3 \quad \text{for } r = b;$$

$$\frac{\partial \Omega_1}{\partial n} = \mu\frac{\partial \Omega_2}{\partial n} \quad \text{for } r = a, \qquad \mu\frac{\partial \Omega_2}{\partial n} = \frac{\partial \Omega_3}{\partial n} \quad \text{for } r = b.$$

The first two conditions give

$$Aa = Ba + \frac{C}{a^2}, \qquad Bb + \frac{C}{b^2} = -Hb + \frac{D}{b^2}, \tag{1}$$

and since here $\partial/\partial n = \partial/\partial r$, the last two conditions give

$$A = \mu\left(B - \frac{2C}{a^3}\right), \quad \mu\left(B - \frac{2C}{b^3}\right) = -H - \frac{2D}{b^3}. \qquad (2)$$

The four equations (1) and (2) can be solved for the unknowns A, B, C, D and by the uniqueness theorem in potential theory it is the only solution. Solve for A then

$$A = -9\mu H\{(\mu + 2)(2\mu + 1) - 2a^3b^{-3}(\mu - 1)^2\}^{-1};$$

and the field in the cavity is $-\partial\Omega/\partial x$, where $x = r \cos \theta$, and thus uniform and equal to $-A$.

• PROBLEM 8-36

(a) Derive an expression for dH for the incremental magnetic field intensity produced at P(x,y,z) by a current element $Idl\, \bar{a}_z$ located at the origin in cartesian coordinates.

(b) Also in cylindrical coordinates.

Solution: The Biot-Savart law states that

$$d\bar{H} = \frac{Id\bar{\ell} \times \bar{a}_r}{4\pi r^2}$$

(a) in cartesian coordinates, vector $\bar{r}$ from the origin to P(x,y,z) is

$$\bar{r} = x\,\bar{a}_x + y\,\bar{a}_y + z\,\bar{a}_z,$$

and

$$\bar{a}_r = \frac{x\bar{a}_x + y\bar{a}_y + z\bar{a}_z}{\sqrt{x^2 + y^2 + z^2}},$$

since $d\bar{l} = dl\,\bar{a}_z$, therefore

$$d\bar{H} = \frac{I\,d\ell\,[x\,\bar{a}_z \times \bar{a}_x + y\,\bar{a}_z \times \bar{a}_y + z\,\bar{a}_z \times \bar{a}_z}{4\pi[x^2 + y^2 + z^2]^{3/2}}.$$

Now,

$$\bar{a}_z \times \bar{a}_x = \bar{a}_y, \quad \bar{a}_z \times \bar{a}_y = -\bar{a}_x, \quad \bar{a}_z \times \bar{a}_z = 0,$$

then

$$d\bar{H} = \frac{Id\ell[x\bar{a}_y - y\bar{a}_x]}{4\pi(x^2 + y^2 + z^2)^{3/2}}$$

(b) In cylindrical coordinates,

$$\bar{r} = r\,\bar{a}_r + z\,\bar{a}_z$$

$$\bar{a}_r = \frac{r\,\bar{a}_r + z\,\bar{a}_z}{(r^2 + z^2)^{\frac{1}{2}}}$$

Therefore

$$d\bar{H} = \frac{I\,d\ell\,\bar{a}_z \times [r\,\bar{a}_r + z\,\bar{a}_z]}{4\pi(r^2 + z^2)^{3/2}}$$

$$= \frac{I\,d\ell\,[r\,\bar{a}_z \times \bar{a}_r + z\,\bar{a}_z \times \bar{a}_z]}{4\pi(r^2 + z^2)^{3/2}}$$

Again, $\bar{a}_z \times \bar{a}_r = \bar{a}_\phi$,

$$d\bar{H} = \frac{I\,d\ell\,r}{4\pi(r^2 + z^2)^{3/2}}\,\bar{a}_\phi$$

• PROBLEM 8-37

A filamentary conductor is formed into an equilateral triangle with sides of length a carrying current I. Find the magnetic field intensity at the center of the triangle, as shown in the figure.

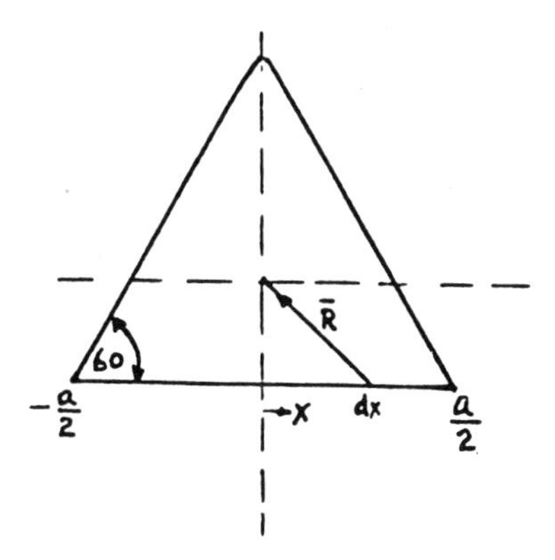

Solution: Determine the center of the equilateral triangle as shown. From the center, draw a vector $\bar{R}$ to one side of the triangle, then

$$\bar{R} = -x\,\bar{a}_x + \frac{a}{2\sqrt{3}}\,\bar{a}_y ,$$

and

$$\bar{a}_R = \frac{-x\ \bar{a}_x + \frac{a}{2\sqrt{3}}\ \bar{a}_y}{\left(x^2 + \frac{a^2}{12}\right)^{\frac{1}{2}}} .$$

Let

$$d\bar{l} = dx\ \bar{a}_x \ ,$$

then

$$d\bar{H} = \frac{I\ dx\ \bar{a}_x \times (-x\bar{a}_x + \frac{a}{2\sqrt{3}}\ \bar{a}_y)}{4\pi\left(x^2 + \frac{a^2}{12}\right)^{3/2}} .$$

Since

$$\bar{a}_x \times \bar{a}_x = 0 \ ,$$

$$\bar{a}_x \times \bar{a}_y = \bar{a}_z \ ,$$

therefore,

$$d\bar{H} = \frac{I\ dx\ \frac{a}{2\sqrt{3}}\ \bar{a}_z}{4\pi\ (x^2 + \frac{a^2}{12})^{\frac{3}{2}}}$$

The magnetic field intensity due to one side carrying a current I is

$$\bar{H} = \frac{I}{4\pi} \int_{-\frac{a}{2}}^{\frac{a}{2}} \frac{\frac{a}{2\sqrt{3}}\ dx}{(x^2 + \frac{a^2}{12})^{3/2}}\ \bar{a}_z$$

$$= \frac{I}{4\pi}\ \frac{a}{2\sqrt{3}}\ \bar{a}_z \int_{-\frac{a}{2}}^{\frac{a}{2}} \frac{dx}{(x^2 + \frac{a^2}{12})^{3/2}}$$

$$= \frac{Ia}{8\sqrt{3}\,\pi}\ \bar{a}_z \left[\frac{x}{\frac{a^2}{12}\sqrt{x^2 + \frac{a^2}{12}}} \right]_{-\frac{a}{2}}^{\frac{a}{2}} = \frac{3I}{2\pi a}\ \bar{a}_z$$

By symmetry, all sides of the equilateral triangle contribute equally, the total magnetic field intensity is

$$\bar{H} = \frac{9I}{2\pi a}\ \bar{a}_z$$

• PROBLEM 8-38

Find the contribution to the magnetic field intensity at P caused by (a) the semicircular section; (b) the two horizontal conductors; and (c) the short vertical section, in the diagram shown.

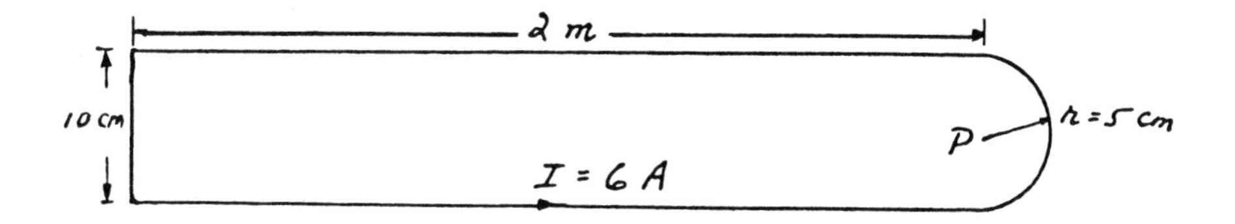

Solution: (a) H caused by the semicircular section:

$$d\bar{H} = \frac{I\ r\ d\phi(-\bar{a}_\phi \times \bar{a}_r)}{4\pi r^2} = \frac{Id\phi}{4\pi r}\ \bar{a}_z$$

$$\bar{H} = \int_0^\pi \frac{I}{4\pi r}\ d\phi\bar{a}_z = \frac{I}{4r}\ \bar{a}_z = \frac{6\bar{a}_z}{4(0.05)} = 30\ \bar{a}_z\ \text{A/m}$$

(b) H caused by two horizontal conductors:

Let

$$\bar{r} = -x\,\bar{a}_x + 0.05\,\bar{a}_y\,,$$

$$\bar{a}_r = \frac{-x\,\bar{a}_x + 0.05\,\bar{a}_y}{[x^2 + (0.05)^2]^{1/2}}\,.$$

$\therefore$

$$\bar{H} = 2\int_{-2}^{0} \frac{I\,dx\,\bar{a}_x \times [-x\,\bar{a}_x + 0.05\,\bar{a}_y]}{4\pi[x^2 + (0.05)^2]^{3/2}}$$

$$= \frac{I}{2\pi}\,\bar{a}_z \int_{-2}^{0} \frac{0.05\,dx}{[x^2 + (0.05)^2]^{3/2}}$$

$$= \frac{3}{\pi}\,\bar{a}_z\left[\frac{2}{\sqrt{4 + (.05)^2}}\right] = 0.955\,\bar{a}_z \text{ A/m}$$

(c) H caused by the vertical short bar:

Let

$$\bar{r} = -2\,\bar{a}_x + y\bar{a}_y\,,$$

$$\bar{a}_r = \frac{-2\,\bar{a}_x + y\,\bar{a}_y}{(4 + y^2)^{1/2}}\,,$$

$$\bar{H} = \int_{-0.05}^{0.05} \frac{I\,dy\,\bar{a}_y \times (-2\bar{a}_x + y\bar{a}_y)}{4\pi(4 + y^2)^{3/2}}$$

$$= \int_{-0.05}^{0.05} \frac{2I \, dy \, \bar{a}_z}{4\pi(4 + y^2)^{3/2}}$$

$$= \frac{I}{2\pi} \bar{a}_z \int_{-.05}^{.05} \frac{dy}{(4 + y^2)^{3/2}}$$

$$= \frac{3}{\pi} \bar{a}_z \left[\frac{y}{4\sqrt{4 + y^2}} \right]_{-0.05}^{0.05}$$

$$= 0.0119 \, \bar{a}_x \quad \text{A/m}$$

• PROBLEM 8-39

By first evaluating the mmf, find the value of H in a long solenoid of N turns and a length ℓ which carries a current I (Fig. a). The diameter of the solenoid is small compared with ℓ.

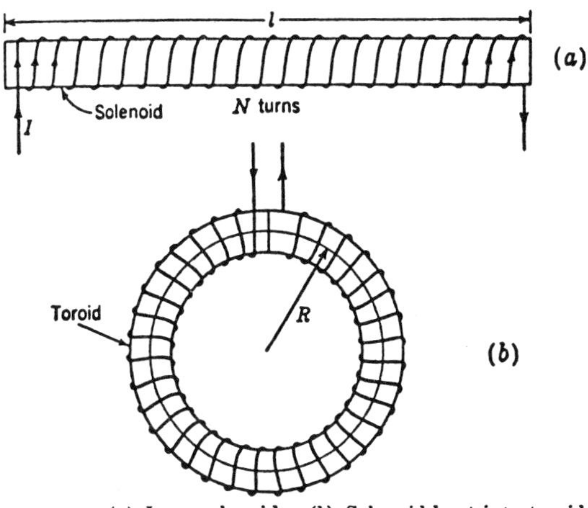

(a) Long solenoid. (b) Solenoid bent into toroid.

Solution: For a long solenoid the field inside is essentially uniform and will not be appreciably changed if the solenoid is bent into a circle and closed on itself, forming a toroid (as in Fig. b). Then, integrating H once around a path entirely inside the coil (at a radius R), we link all of the turns obtaining the mmf, or

$$\oint H \cdot dl = F = NI \tag{1}$$

Let the magnitude of H inside be H_i. Then (1) becomes

$$\oint H \cdot dl = H_i \oint d\ell = H_i 2\pi R$$

$$= H_i \ell = NI$$

or

$$H_i = \frac{NI}{\ell} = K$$

• PROBLEM 8-40

Show that

$$\nabla \times \bar{H} = \bar{J}$$

for the field in each conductor of a coaxial cable:

$$H_\phi = \frac{Ir}{2\pi a^2},$$

$r < a$ (field in the inside conductor)

$$H_\phi = \frac{I}{2\pi r} \frac{c^2 - r^2}{c^2 - b^2},$$

$c > r > b$ (field in the outside conductor)

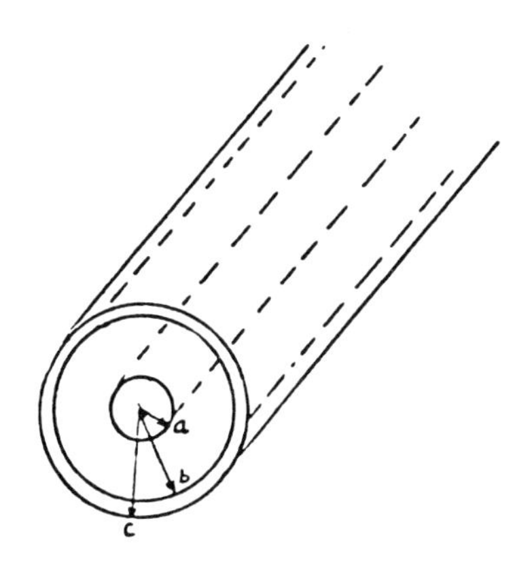

Solution: A cross-section of the coaxial cable is as shown. In cylindrical coordinates, the curl of H is

$$\nabla \times \bar{H} = \left(\frac{1}{r}\frac{\partial Hz}{\partial \phi} - \frac{\partial H\phi}{\partial z}\right)\bar{a}_r + \left(\frac{\partial Hr}{\partial z} - \frac{\partial Hz}{\partial r}\right)\bar{a}_\phi$$

$$+ \frac{1}{r}\left(\frac{\partial (rH\phi)}{\partial r} - \frac{\partial Hr}{\partial \phi}\right)\bar{a}_z$$

By symmetry,

$$H_r = H_z = 0,$$

and

H_ϕ is not a function of z.

Therefore,

$$\nabla \times \bar{H} = \frac{1}{r}\left(\frac{\partial (rH\phi)}{\partial r}\right)\bar{a}_z$$

for $r < a$ $\nabla \times \bar{H} = \frac{1}{r}\left[\frac{\partial}{\partial r}\left(\frac{Ir^2}{2\pi a^2}\right)\right]\bar{a}_z = \frac{I}{\pi a^2}\bar{a}_z = \bar{J}_z$ inside

$c > r > b,$

$$\nabla \times \bar{H} = \frac{1}{r}\left[\frac{\partial}{\partial r} \frac{I}{2\pi} \frac{c^2 - r^2}{c^2 - b^2}\right]\bar{a}_z$$

$$= \frac{I}{2\pi r} \frac{(-2r)}{c^2 - b^2} \bar{a}_z$$

$$= \frac{-I}{\pi (c^2 - b^2)} = \bar{J}_z \qquad \text{(outside)}$$

Thus the statement is proved.

• PROBLEM 8-41

Find the magnetic field due to a line source, by applying Stoke's Theorem. Refer to fig. 1.

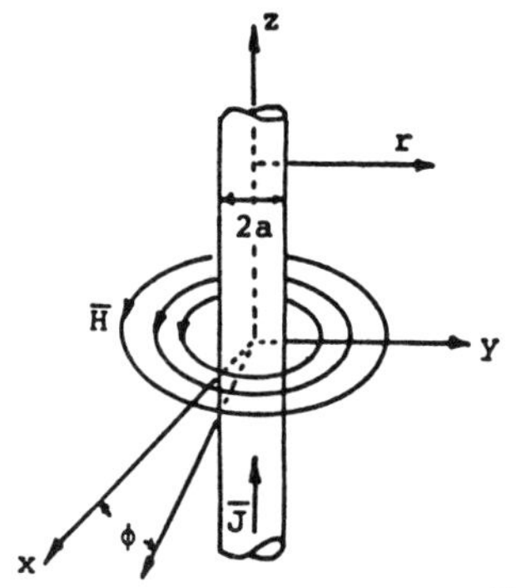

Fig. 1

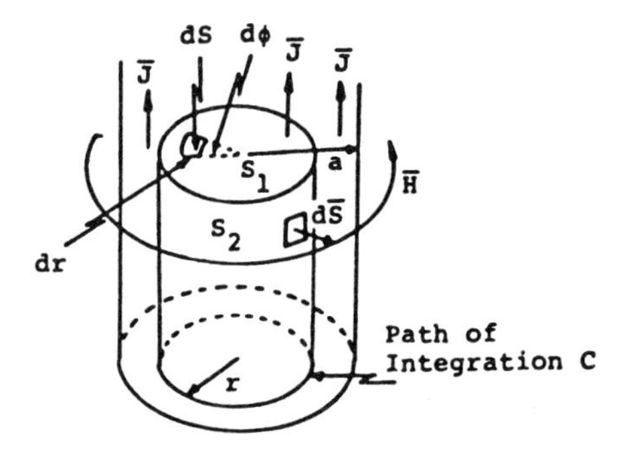

Fig. 2

Application of Stokes theorem to a current source.

Solution: Consider an enlarged view of the conductor of radius a and consider as a Stoke's surface a small concentric cylinder of radius r with S_1 denoting the end cup and S_2 denoting the cylindrical surface. Observe the symmetry of the field, and see that $\bar{H}$ has a ϕ-component which is a function of r only. Apply Stokes' theorem to the cup consisting of a cylindrical cup of side area S_2 and top area S_1 that is defined by the circular boundary C of radius r (see Fig. 2). This yields

$$\int_S (\nabla \times \bar{H}) \cdot d\bar{s} = \oint_C \bar{H} \cdot d\bar{\ell} = H_\phi(r) \int_0^{2\pi} r\, d\phi =$$

$$2\pi r\, H_\phi(r)$$

However, since

$$\nabla \times \bar{H} = \bar{J} ,$$

the integral relation becomes

$$2\pi r\, H_\phi(r) = \int_S \bar{J} \cdot d\bar{S} = \int_{S_1} \bar{J} \cdot d\bar{S} +$$

$$\int_{S_2} \bar{J} \cdot d\bar{S}$$

But $\bar{J}$ is perpendicular to $d\bar{S}$ over S_2 and the second integral vanishes. Thus

$$2\pi r\, H_\phi(r) = \int_0^{2\pi} \int_0^{r} J_z r\, dr\, d\phi$$

For $r < a$,

$$I/\pi a^2 (\pi r^2)$$

is obtained for the surface integral, while for

$$r > a$$

the result is I, because

$$J_z = 0$$

for

$$r > a.$$

The solution for H_ϕ is

$$H_\phi = \begin{cases} \dfrac{Ir}{2\pi a^2} & r \leq a \\ \\ \dfrac{I}{2\pi r} & r \geq a \end{cases}$$

CHAPTER 9

FORCES IN STEADY MAGNETIC FIELDS

FORCES ON MOVING CHARGES

• **PROBLEM 9-1**

A test charge q C, moving with a velocity $\bar{v} = (\bar{i}_x + \bar{i}_y)$ m/sec, experiences no force in a region of electric and magnetic fields. If the magnetic flux density $\bar{B} = (\bar{i}_x - 2\bar{i}_z)\mathrm{Wb/m^2}$, find $\bar{E}$.

Solution: From $\bar{F} = q(\bar{E} + \bar{v} \times \bar{B})$, the electric field intensity $\bar{E}$ must be equal to $-\bar{v} \times \bar{B}$ for the charge to experience no force. Thus

$$\bar{E} = -(\bar{i}_x + \bar{i}_y) \times (\bar{i}_x - 2\bar{i}_z)$$

$$= (2\bar{i}_x - 2\bar{i}_y + \bar{i}_z)\,\mathrm{volts/m}$$

• **PROBLEM 9-2**

Given a magnetic induction

$$\bar{B} = \hat{i}\ 10^{-2}\ \mathrm{w/m^2}$$

find the force on an electron whose velocity is 10^7 m/sec

a) in the x-direction

b) in the y-direction

c) in the z-direction

d) in the xy-plane at 45° to the x-axis.

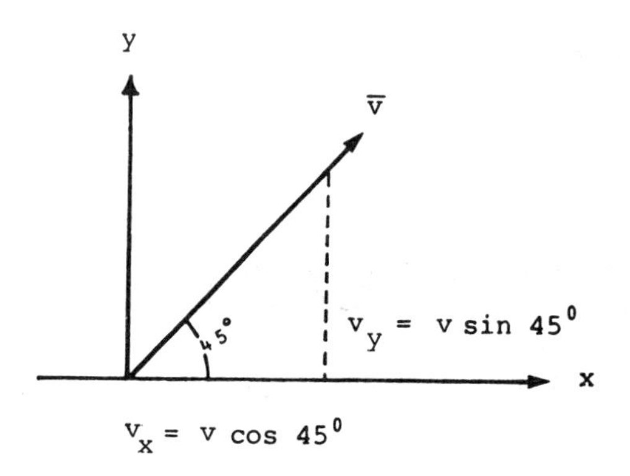

Solution: a) $\bar{v} = \hat{i}\ 10^7$m/sec. Magnetic force is $\bar{F} = q(\bar{v} \times \bar{B})$. For electron, $q = -1.6 \times 10^{-19}$ coul.

$$\bar{F} = (-1.6 \times 10^{-19}\text{ coul})(\hat{i}\ 10^7 \times \hat{i}\ 10^{-2}) = 0 = \bar{F}$$

since $\hat{i} \times \hat{i} = 0$.

b) $\bar{v} = \hat{j}\ 10^7$ m/sec

$$\bar{F} = -1.6 \times 10^{-19}\text{ coul } (\hat{j}\ 10^7 \times \hat{i}\ 10^{-2})$$

$$= -1.6 \times 10^{-19} \times 10^7 \times 10^{-2}\ (-\hat{k})$$

$$\bar{F} = 1.6 \times 10^{-14}\text{ Nt in z direction.}$$

c) $\bar{v} = 10^7\ \frac{m}{sec}\ \hat{k}$

$$\bar{F} = -1.6 \times 10^{-19}(10^7\ \frac{m}{sec}\ \hat{k} \times \hat{i}\ 10^{-2})$$

$$= -1.6 \times 10^{-14}\ \hat{j}$$

$$\bar{F} = 1.6 \times 10^{-14}\text{ Nt in - y direction}$$

d) Refer to the given figure

$$\bar{v} = \hat{i}v_x + \hat{j}\ v_y$$

$$\bar{v} = [\hat{i}\cos 45° + \hat{j}\sin 45°]\ 10^7\ \frac{m}{sec}$$

$$\bar{v} = [\hat{i}\ (.707) + \hat{j}\ (.707)]\ 10^7\ \frac{m}{sec}$$

$$\bar{F} = (-1.6 \times 10^{-19})\ 10^7[\hat{i}\ (.707) + \hat{j}\ (.707)] \times 10^{-2}\hat{i}$$

$$= -\ 1.6 \times 10^{-19} \times 10^7(.707)\ (-\hat{k})$$

$$\bar{F} = 1.1312 \times 10^{-14}\text{ Nt in -z direction}$$

• PROBLEM 9-3

The forces experienced by a test charge q for three different velocities at a point in a region characterized by electric and magnetic fields are given by

$$\bar{F}_1 = q[E_0\bar{i}_x + (E_0 - v_0B_0)\bar{i}_y] \quad \text{for } \bar{v}_1 = v_0\bar{i}_x$$

$$\bar{F}_2 = q[(E_0 + v_0B_0)\bar{i}_x + E_0\bar{i}_y] \quad \text{for } \bar{v}_2 = v_0\bar{i}_y$$

$$\bar{F}_3 = q[E_0\bar{i}_x + E_0\bar{i}_y] \quad \text{for } \bar{v}_3 = v_0\bar{i}_z$$

where v_0, E_0, and B_0 are constants. Find E and B at the point.

Solution: From Lorentz force equation,

$$q\bar{E} + qv_0\bar{i}_x \times \bar{B} = q[E_0\bar{i}_x + (E_0 - v_0B_0)\bar{i}_y] \quad (1)$$

$$q\bar{E} + qv_0\bar{i}_y \times \bar{B} = q[(E_0 + v_0B_0)\bar{i}_x + E_0\bar{i}_y] \quad (2)$$

$$q\bar{E} + qv_0\bar{i}_z \times \bar{B} = q[E_0\bar{i}_x + E_0\bar{i}_y] \quad (3)$$

Eliminating $\bar{E}$ by subtracting (1) from (2) and (3) from (2)

$$(\bar{i}_y - \bar{i}_x) \times \bar{B} = B_0(\bar{i}_x + \bar{i}_y) \quad (4)$$

$$(\bar{i}_y - \bar{i}_z) \times \bar{B} = B_0\bar{i}_x$$

It follows from these two equations that B is perpendicular to both $(\bar{i}_x + \bar{i}_y)$ and $\bar{i}_x$. Hence it is equal to $C(\bar{i}_x + \bar{i}_y) \times \bar{i}_x$ or $-C\bar{i}_z$ where C is to be determined. To do this, substitute $B = -C\bar{i}_z$ in (4) to obtain

$$(\bar{i}_y - \bar{i}_x) \times (-C\bar{i}_z) = B_0(\bar{i}_x + \bar{i}_y)$$

$$-C(\bar{i}_x + \bar{i}_y) = B_0(\bar{i}_x + \bar{i}_y)$$

or $C = -B_0$. Thus $B = B_0 i_z$. Substituting this result in (3), the $\bar{E}$ field is obtained,

$$\bar{E} = E_0(\bar{i}x + \bar{i}y)$$

$$\bar{B} = B_0\bar{i}_z$$

A region is characterized by crossed electric and magnetic fields, $\bar{E} = E_0\bar{i}_y$ and $\bar{B} = B_0\bar{i}_z$ as shown in Fig. 1, where E_0 and B_0 are constants. A small test charge q having a mass m starts from rest at the origin at t = 0. Obtain the parametric equations of motion of the test charge.

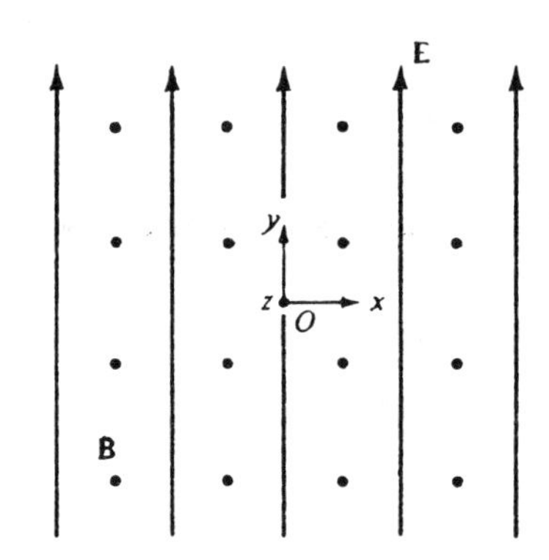

Fig. 1

A region of crossed electric and magnetic fields.

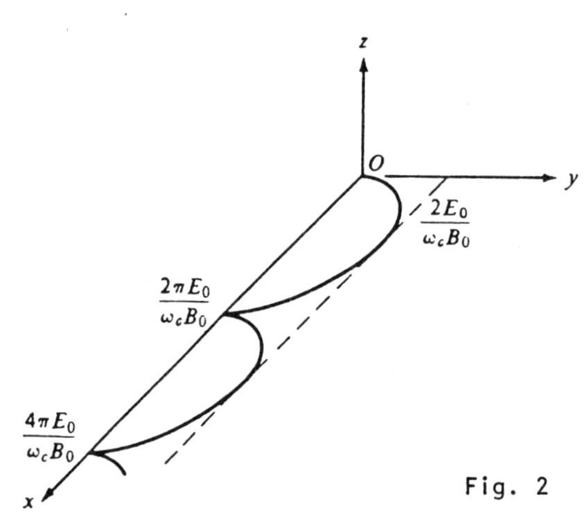

Fig. 2

Path of a test charge q in crossed electric and magnetic fields $\mathbf{E} = E_0\mathbf{i}_y$ and $\mathbf{B} = B_0\mathbf{i}_z$.

Solution: The force exerted by the crossed electric and magnetic fields on the test charge is

$$\bar{F} = q(\bar{E} + \bar{v} \times \bar{B}) = q[E_0\bar{i}_y + (v_x\bar{i}_x + v_y\bar{i}_y + v_z\bar{i}_z) \times (B_0\bar{i}_z)] \tag{1}$$

The equations of motion of the test charge can therefore be written as

$$\frac{dv_x}{dt} = \frac{qB_0}{m} v_y \ . \tag{2}$$

$$\frac{dv_y}{dt} = -\frac{qB_0}{m} v_x + \frac{q}{m} E_0 \tag{3}$$

$$\frac{dv_z}{dt} = 0 \tag{4}$$

Eliminating v_y from (2) and (3),

$$\frac{d^2 v_x}{dt^2} + \left(\frac{qB_0}{m}\right)^2 v_x = \left(\frac{q}{m}\right)^2 B_0 E_0 \tag{5}$$

The solution for (5) is

$$v_x = \frac{E_0}{B_0} + C_1 \cos \omega_c t + C_2 \sin \omega_c t \tag{6}$$

where C_1 and C_2 are arbitrary constants and $\omega_c = qB_0/m$. Substituting (6) into (2),

$$v_y = -C_1 \sin \omega_c t + C_2 \cos \omega_c t \tag{7}$$

Using initial conditions given by

$$v_x = v_y = 0 \quad \text{at } t = 0$$

to evaluate C_1 and C_2 in (6) and (7),

$$v_x = \frac{E_0}{B_0} - \frac{E_0}{B_0} \cos \omega_c t \tag{8}$$

$$v_y = \frac{E_0}{B_0} \sin \omega_c t \tag{9}$$

Integrating (8) and (9) with respect to t,

$$x = \frac{E_0}{B_0} t - \frac{E_0}{\omega_c B_0} \sin \omega_c t + C_3 \tag{10}$$

$$y = -\frac{E_0}{\omega_c B_0} \cos \omega_c t + C_4 \tag{11}$$

Using initial conditions given by

$$x = y = 0 \quad \text{at } t = 0$$

to evaluate C_3 and C_4 in (10) and (11),

$$x = \frac{E_0}{B_0} t - \frac{E_0}{\omega_c B_0} \sin \omega_c t = \frac{E_0}{\omega_c B_0} (\omega_c t - \sin \omega_c t) \qquad (12)$$

$$y = - \frac{E_0}{\omega_c B_0} \cos \omega_c t + \frac{E_0}{\omega_c B_0} = \frac{E_0}{\omega_c B_0} (1 - \cos \omega_c t) \qquad (13)$$

Equation (4), together with the initial conditions $v_z = 0$ and $z = 0$ at $t = 0$, yields a solution

$$z = 0 \qquad (14)$$

The equations of motion of the test charge in the crossed electric and magnetic field region are thus given by (12), (13), and (14). These equations represent a cycloid in the $z = 0$ plane, as shown in Fig. 2.

• PROBLEM 9-5

A long hollow cylinder consists of an insulating material of dielectric constant K, its inner and outer cylindrical surfaces being faced with thin brass coatings. It is arranged to rotate in a constant magnetic field, produced by a stationary and co-axial solenoid surrounding the cylinder, parallel to the axis. An electrometer is connected by wire to sliding contacts on the inner and outer surfaces of the cylinder as shown in the figure.

Show that the potential difference measured by the electrometer will be proportional to $1 - 1/K$.

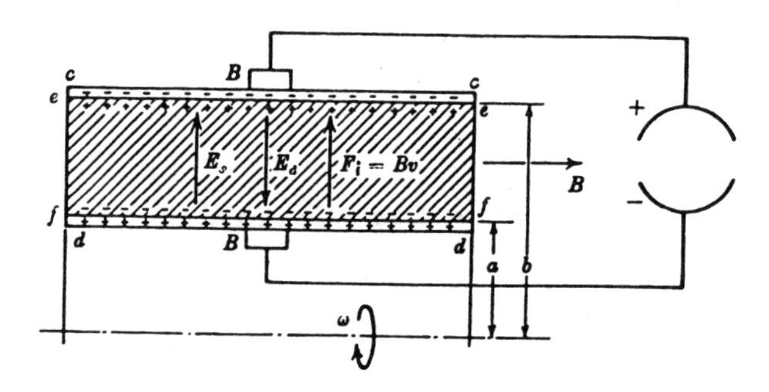

Solution: In the figure the shaded area represents a section of the cylinder wall, the inner and outer radii being a and b, while c - c and d - d represent the metal coatings, whose thickness may be assumed to be negligible. Consider a point in the insulator distant r from the axis. If the angular velocity is ω, this point will be moving with velocity $v = \omega r$ through the magnetic field B and hence the material of the insulator experiences an electromagnetic force, or motional intensity,

$$F_i = Bv.$$

The insulator is therefore polarized, the outer surface e - e becoming positively charged and the inner surface f - f negatively charged. These induced surface charges will tend to induce free charges on the inner surfaces of the metal coatings, which will become charged, provided that charges can flow to them from the outer surfaces or from connected conductors, such as the electrometer plates and leads.

Under conditions of equilibrium there will be three components of electric force acting on the material of the insulator:

(1) the electromagnetic motional force $F_i = B\omega r$,

(2) the electrostatic field E_s of the charges on the brass coatings,

(3) the electrostatic field E_d of the charges displaced in the insulator by its polarization.

The polarizing force acting on the insulator is due to the external sources, and is

$$F_0 = F_i + E_s = B\omega r + \frac{q}{2\pi\varepsilon_0 r},$$

where q is the charge per unit axial length of the coatings. The field of the displaced charges in the insulator is then, by $E_d = -(1 - 1/k)E_0$,

$$E_d = \left(B\omega r + \frac{q}{2\pi\varepsilon_0 r}\right)\left(1 - \frac{1}{K}\right), \text{ opposed to } F_0.$$

The resultant electrostatic field between the stationary brushes B - B is therefore

$$E = E_d - E_s = \left(B\omega r + \frac{q}{2\pi\varepsilon_0 r}\right)\left(1 - \frac{1}{K}\right) - \frac{q}{2\pi\varepsilon_0 r}$$

$$= B\omega r\left(1 - \frac{1}{K}\right) - \frac{q}{2\pi K\varepsilon_0 r}.$$

The potential difference between the brushes is then

$$V = \int_a^b E\,dr = B\omega\left(1 - \frac{1}{K}\right)\frac{b^2 - a^2}{2} - \frac{q}{2\pi K\varepsilon_0}\log_e\frac{b}{a},$$

or

$$V = \phi n(1 - \frac{1}{K}) - \frac{Q}{C},$$

where ϕ is the total flux passing through a cross-section of the cylinder, n is the speed in rev/sec., Q is the total charge on each metal coating, and C is the capacitance of the capacitor formed by the two coatings and the intervening dielectric.

If C' is the capacitance of the electrometer, and if the stray capacitance of the outer surface of the coatings and of the leads is negligible, Q = C'V, since V is also the potential difference across the electrometer. Whence

$$V = \phi n(1 - \frac{1}{K}) \frac{C}{C + C'} .$$

FORCES ON DIFFERENTIAL CURRENT ELEMENTS

• PROBLEM 9-6

A current of 40 amperes exists in an element of length $d\ell$, oriented in the X-Y plane at 45° from both the X and Y axes. The flux density at the element of current is 10^{-2} newton per meter-ampere and is oriented in the Y-Z plane at 45° from both the Y and Z axes. Determine the vector $(d\bar{F})/(d\bar{\ell})$. Refer to the figure.

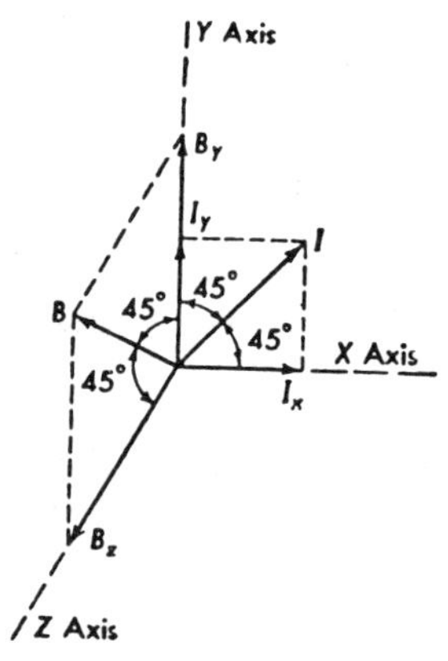

Components of flux densities and currents.

Solution: Resolving the current and flux density into components along the coordinate axes yields

$$\bar{I} = 40(\bar{i}\ 0.707 + \bar{j}\ 0.707 + \bar{k}\ 0)$$

$$= \bar{i}\ 28.28 + \bar{j}\ 28.28 + \bar{k}\ 0 \text{ amperes}$$

$$\bar{B} = 10^{-2}\ (\bar{i}\ 0 + \bar{j}\ 0.707 + \bar{k}\ 0.707)$$

$$= \bar{i}\ 0 + \bar{j}\ 0.707 \times 10^{-2} + \bar{k}\ 0.707 \times 10^{-2}$$

newton per meter-ampere.

These components are shown in the figure.

Consider now I_x acting with B_y. The component force will be oriented along the Z axis and will be +0.20 newton per meter.

I_x acting with B_z will produce a component force of -0.20 newton per meter acting along the Y axis.

I_y acting with B_y will produce no component force; and I_y acting with B_z will produce +0.20 newton per meter acting along the X axis.

Adding these component forces and writing as a vector gives

$$\frac{d\bar{F}}{d\bar{\ell}} = \bar{i}\ 0.20 - \bar{j}\ 0.20 + \bar{k}\ 0.20 = 0.3464(\bar{i}\ 0.577 - \bar{j}\ 0.577 + \bar{k}\ 0.577) \text{ newton per meter.}$$

• PROBLEM 9-7

Two elements of current in air are normal to each other and in the same plane, as is illustrated in the figure. Find the magnitude and direction of the differential force upon each element of current.

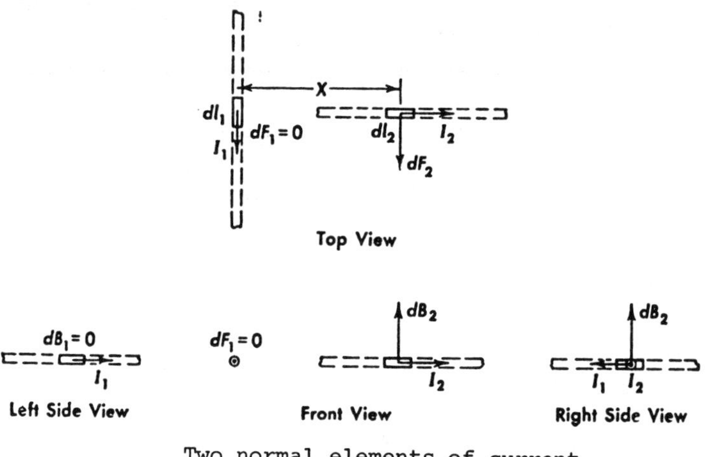

Two normal elements of current.

Solution: First consider the element of current $I_1 d\ell_1$ as a source of magnetic flux density at the location of the second current element. Applying the formula for a differential line current;

$$dB_p = \frac{10^{-7} \ I \cos \alpha}{r^2} \ d\ell \tag{1}$$

gives

$$dB_2 = \frac{10^{-7} I_1 d\ell_1}{X^2}.$$

This differential magnetic flux density is vertically upward, as shown in the front view of the figure.

An application of $d\bar{F} = I\bar{B} \times d\bar{\ell}$ gives

$$dF_2 = \frac{10^{-7} I_1 I_2 d\ell_1 \ d\ell_2}{X^2}.$$

This differential force is directed forward as shown in the top view of the figure.

Now consider the element of current $I_2 \, d\ell_2$ as a source of magnetic flux density at the location of the first current element. The angle α of Eq. 1 is now 90°, and hence the flux density at this point is zero. Therefore current element $I_1 d\ell_1$ experiences no force as caused by the differential element of current $I_2 d\ell_2$. However, any complete system does attempt to distort itself, and equal and opposite forces do exist upon two isolated free-body parts of the system when total forces are considered upon the complete current path of each part of the system.

• PROBLEM 9-8

Three differential elements of current (expressed in meter-amperes), $I_1 d\ell_1 = 1000 d\ell$, $I_2 d\ell_2 = 500 d\ell$, and $I_3 d\ell_3 = 500 d\ell$ are parallel and directly opposite one another in air at coordinates as follow: and I_3 at (0.10, -0.05) meter. See the figure. Find the magnitude and direction of the differential force upon each element of current.

Solution: First consider the element of current $I_1 d\ell_1$ as a source of magnetic-flux density at the location of the second and third current elements. Applying the formula for a differential line current;

$$dB_p = \frac{10^{-7} \ I \cos \alpha \ d\ell}{r^2},$$

gives the magnitudes of dB_{21} and dB_{31} in webers per square meter, as

$$dB_{21} = dB_{31} = \frac{10^{-7} \times 1000 d\ell}{0.01 + 0.0025} = 0.008 d\ell.$$

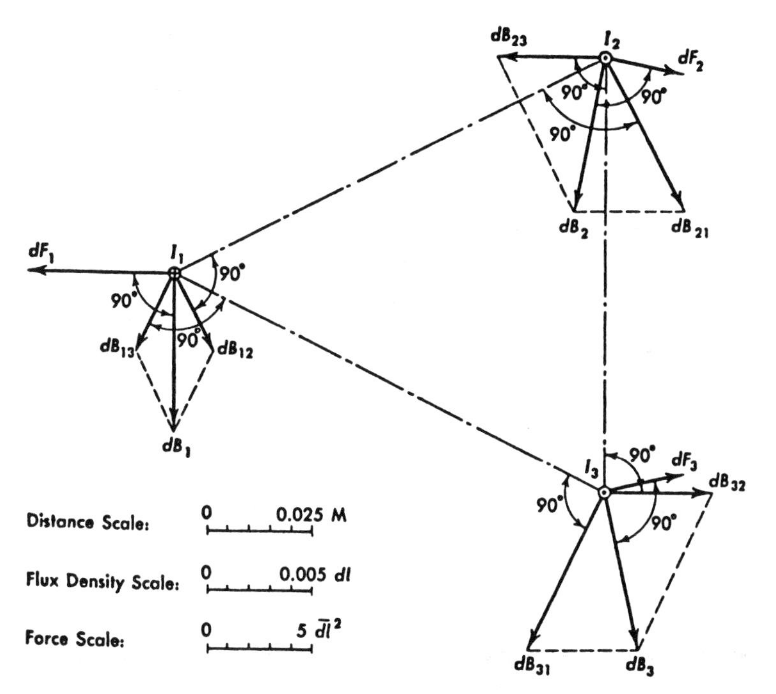

Three parallel differential elements of current directly opposite one another.

The vectors, however, do not have the same direction. For the current direction shown in the figure, and using the unit vectors i and j to indicate components,

$$d\bar{B}_{21} = 0.008d\ell\,(+\,\bar{i}\ 0.447 - \bar{j}\ 0.895)$$

$$d\bar{B}_{31} = 0.008d\ell\,(-\,\bar{i}\ 0.447 - \bar{j}\ 0.895).$$

Similarly, the effects of I_2 and I_3, each of 500 amperes, at the location of I_1 give flux densities of

$$d\bar{B}_{12} = 0.004d\ell(+\,\bar{i}\ 0.447 - \bar{j}\ 0.895)$$

$$d\bar{B}_{13} = 0.004d\ell(-\,\bar{i}\ 0.447 - \bar{j}\ 0.895).$$

The magnitudes of dB_{32} and dB_{23} are

$$dB_{32} = dB_{23} = \frac{10^{-7} \times 500d\ell}{0.01} = 0.005d\ell.$$

The vectors are

$$d\bar{B}_{32} = 0.005d\ell\,(+\bar{i} + \bar{j}0)$$

$$d\bar{B}_{23} = 0.005d\ell(-\bar{i} + \bar{j}0).$$

The resultant densities, by vector addition are

$$d\bar{B}_1 = d\bar{B}_{12} + d\bar{B}_{13} = +\bar{i} \quad 0 \quad - \bar{j}0.00716d\ell$$

$$d\bar{B}_2 = d\bar{B}_{21} + d\bar{B}_{23} = -\bar{i}0.00146d\ell - \bar{j}0.00716d\ell$$

$$d\bar{B}_3 = d\bar{B}_{31} + d\bar{B}_{32} = +\bar{i}0.00146d\ell - \bar{j}0.00716d\ell.$$

Applying $d\bar{F} = I\bar{B} \times d\bar{\ell}$ for the differential forces expressed in newtons gives

$$d\bar{F}_1 = -\bar{i}\ 7.16\ \overline{d\ell}^2 + \bar{j}\ \ 0 \quad = 7.16\ \overline{d\ell}^2\ \underline{/-180°}$$

$$d\bar{F}_2 = +\bar{i}\ 3.58\ \overline{d\ell}^2 - \bar{j}\ 0.73\ \overline{d\ell}^2 = 3.65\ \overline{d\ell}^2\ \underline{/-11.5°}$$

$$d\bar{F}_3 = +\bar{i}\ 3.58\ \overline{d\ell}^2 + \bar{j}\ 0.73\ \overline{d\ell}^2 = 3.65\ \overline{d\ell}^2\ \underline{/+11.5°}$$

FORCES ON CONDUCTORS CARRYING CURRENTS

• PROBLEM 9-9

Describe the phenomenon known as "pinch effect" exhibited in a conductor carrying current. Deduce an expression for the magnitude of this effect. If a cylindrical column of mercury 1 cm. diameter carries a current of 100 amperes, calculate the intensity of the mechanical pressure due to the pinch effect: (1) at a radius of 0.25 cm., (2) at the axis of the conductor. Find also the total axial mechanical force arising from this effect.

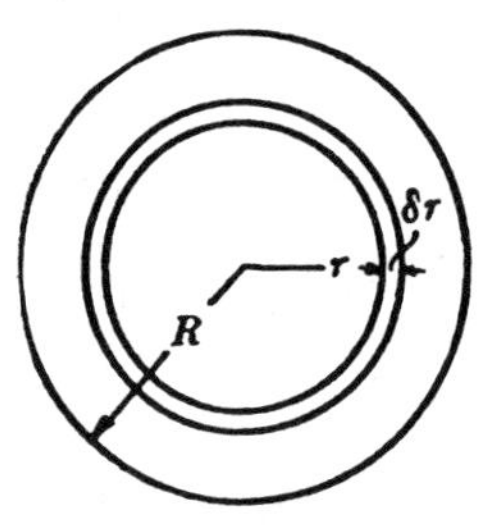

Pinch effect

Solution: Consider a conductor of circular section as shown in the figure, carrying a steady current I. The flux-density inside the conductor, at a point distant r from the axis, is

$$B_r = \frac{\mu\mu_0 Ir}{2\pi R^2},$$

where μ is the relative permeability of the wire.

The moving charges (which constitute the current) at

this point will be moving in this magnetic field, and will experience a force directed towards the axis of the wire. This force is transmitted to the material structure of the wire in such a way as to tend to decrease its section. It is therefore named the "pinch effect", and is best shown experimentally when the conductor is liquid, e.g. a column of mercury.

To calculate the effect, consider the total inward force on the current flowing in an elementary cylinder of radius r and thickness δr. The current in this element is

$$\delta I = \frac{2\pi r \delta r}{\pi R^2} I = \frac{2rI}{R^2}\delta r.$$

The force on this current (radially inwards) is equal to $B_r \delta I$ per unit of axial length, and this acts uniformly over a cylindrical surface of area $2\pi r$. Thus the inward pressure due to the element is

$$\delta p = \frac{B_r \delta I}{2\pi r} = \frac{\mu\mu_0 I^2}{2\pi^2 R^4} r \ \delta r,$$

and the total pressure at a point distant a from the axis will be

$$p_a = \int_0^R dp = \frac{I^2 \mu\mu_0}{4\pi^2 R^4} (R^2 - a^2). \tag{1}$$

In a column of mercury this acts as a hydrostatic pressure; i.e., it acts equally in all directions at any point. If the column is in contact at both ends with metal plates, these plates will experience a force due to the axial pressure.

The total axial force acting upon the cross-section of the elementary cylinder is

$$\delta F = p_r 2\pi r \delta r = \frac{\mu\mu_0 I^2}{2\pi^2 R^4}(R^2 r - r^3)\delta r.$$

The total axial-force over the complete cross-section is then

$$F = \int_{r=0}^{r=R} dF = \frac{\mu\mu_0 I^2}{8\pi}. \tag{2}$$

In the case of the mercury column in this problem:

$R = 5 \times 10^{-3}$ meter, $I = 100$ amperes, $\mu = 1$, $\mu_0 = 4\pi \times 10^{-7}$.

(1) At $a = 2.5 \times 10^{-3}$ meter, from equation (1),

$$p_a = \frac{30}{\pi} \text{ newtons per sq. meter}$$

$$= \frac{300}{\pi} \text{ dynes per sq. cm.}$$

$$= 0.0975 \text{ gm. per sq. cm.} = 0.00138 \text{ lb. per sq. in.}$$

(2) At a = 0 (at the axis),

$$p_a = \frac{40}{\pi} \text{ newtons per sq. meter}$$

(3) The total axial force {equation (2)} is

$$F = 5 \times 10^{-4} \text{ newtons}$$

$$= 50 \text{ dynes.}$$

• PROBLEM 9-10

> Find the force per meter between two infinite and parallel filamentary current carrying conductors that are separated d (m) and carry a current I (A) in opposite directions.

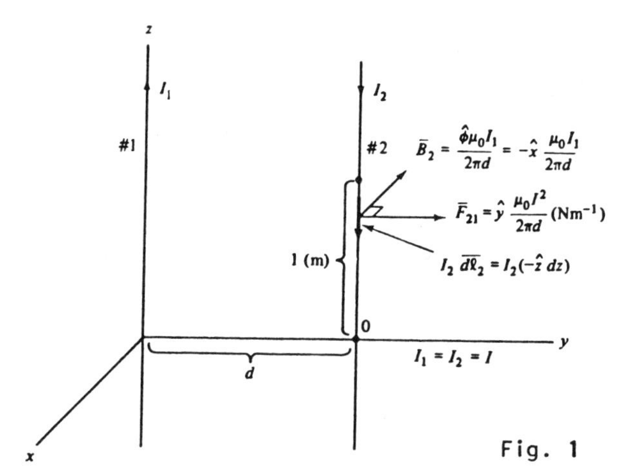

Fig. 1

Graphical display for finding the force per meter between two infinite and parallel filamentary current-carrying conductors.

Solution: Assume that conductor #1 is along the z axis and that conductor #2 is at y = d, as shown in Fig. 1. The B_2 field at the location of conductor #2, from

$$\bar{H} = \frac{\hat{\phi} I}{2\pi r_c} \tag{1}$$

is

$$\bar{B}_2 = \mu_0 \bar{H}_2 = \mu_0 \frac{\hat{\phi} I_1}{2\pi r_c}\Bigg|_{\substack{r_c = d \\ \phi = \pi/2}} = \frac{-\hat{x}\mu_0 I_1}{2\pi d} \quad (\text{Wb/m}^2) \tag{2}$$

Since $\overline{dF} = I\ \overline{d\ell} \times \overline{B}$, substituting (2) into (1) and integrating over 1 meter

$$\overline{F}_2 = \int_0^1 I_2 \overline{dl}_2 \times \left(\frac{-\hat{x}\mu_0 I_1}{2\pi d}\right) = \int_0^1 I_2(-\hat{z}\ dz_2) \times \left(\frac{-\hat{x}\mu_0 I_1}{2\pi d}\right)$$

$$= \hat{y}\mu_0 \frac{I_1 I_2}{2\pi d} = \hat{y}\ \frac{\mu_0 I^2}{2\pi d} \quad (N/m)$$

• PROBLEM 9-11

The figure shows an edge-on view of two long parallel planes carrying surface currents of constant density K but in opposite senses, one directed out of the page and one into it. The width of the sheets is w and x is the distance between them as measured in the sense shown. The sheets are very long and if w >> x, then neglect the edge effects and treat them as if they were infinite plane sheets. Find the force on one sheet due to the other.

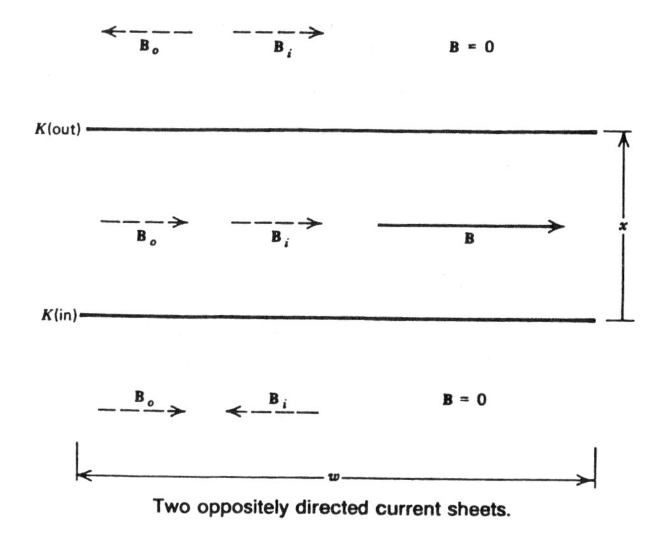

Two oppositely directed current sheets.

Solution: From Ampere's law, the B produced by each sheet has magnitude $\frac{1}{2}\mu_0 K$, is perpendicular to the direction of K and hence is in the plane of the page, and has opposite directions on the two sides of the sheet. These inductions are shown as the dashed arrows in the figure and are labeled by their source current ("out" or "in"). The resultant induction $\overline{B} = \overline{B}_o + \overline{B}_i$ is shown as the solid arrow and see that $\overline{B} \neq 0$ only in the region between the sheets where it has the constant magnitude $\mu_0 K$. Putting this into

$$u_m = \frac{B^2}{2\mu_0}$$

the energy density is found to be $u_m = \frac{1}{2}\mu_0 K^2 = \text{const.}$ between the plates and zero everywhere else. Therefore, the magnetic energy of a length ℓ of this system, and corresponding volume $w\ell x$, as obtained from

$$U_m = \int_{\text{all space}} u_m \, d\tau$$

is

$$U_m = \int \tfrac{1}{2}\mu_0 K^2 \, d\tau = \tfrac{1}{2}\mu_0 K^2 w\ell x.$$

The magnetic force on the upper sheet, which corresponds to an increase in x, as obtained from $\bar{F}_m = (\nabla U_m)_I$ is then found to be

$$\bar{F}_m = \frac{\partial U_m}{\partial x}\hat{x} = \frac{1}{2}\mu_0 K^2 (w\ell)\hat{x}$$

Note that this force is in the positive x direction and thus is repulsive, as is to be expected for these oppositely directed currents. Also note that F_m is proportional to $w\ell$, the area of the sheet, so introduce a force per unit area f_m,

$$f_m = \frac{F_m}{w\ell} = \frac{1}{2}\mu_0 K^2 = u_m$$

and see that its magnitude is exactly equal to the magnetic energy density. In this case, $\hat{x}$ is the normal to the surface directed from the $B \neq 0$ region to the $B = 0$ region which makes the force per unit area in vector form as

$$\bar{f}_m = f_m\hat{x} = u_m\hat{x}$$

The direction of the force is such as to tend to move the sheet into the $B = 0$ region and hence may be properly described as a pressure. This is an effect more typically associated with kinetic energy than with potential energy.

Combining all of the relations involved, write f_m variously as

$$f_m = u_m = \frac{B^2}{2\mu_0} = \frac{1}{2}\mu_0 K^2 = \frac{1}{2} KB$$

where K and B are evaluated at the position where f_m is wanted.

• PROBLEM 9-12

A semicircular loop of wire carrying current I is situated in a magnetic field B perpendicular to the plane defined by the loop. Deduce the force exerted on the loop.

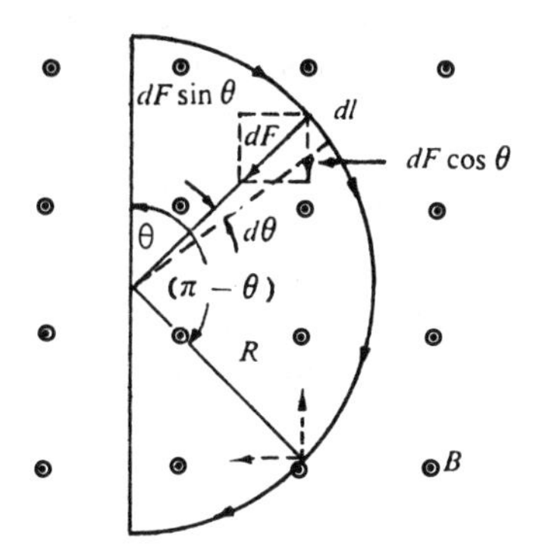

A semicircular loop of wire carrying a current I in a magnetic field B.

Solution: (See the given figure)

A segment of wire of length $d\ell$ has a force dF on it of magnitude

$$dF = IB\, d\ell = IBR\, d\theta$$

directed toward the center of the semicircle. The component $dF \cos\theta$ is canceled by the corresponding contribution from the arc segment located at $\pi - \theta$. Only the component $dF \sin\theta$ when summed over the complete semicircle gives a nonzero result. The resultant force is therefore

$$F = \int_0^{\pi} dF \sin\theta$$

$$= IBR \int_0^{\pi} \sin\theta\, d\theta$$

$$= 2IBR.$$

Note that the force experienced by the semicircle is the same as would be experienced by a straight wire of length 2R.

• PROBLEM 9-13

The magnetic flux density B in a region around a straight conducting wire with a current of 500 amperes is given by

$$\bar{B} = \frac{0.0001(yi - xj)}{x + y}$$

The magnetic flux lines represented by this mathematical expression are circles about the z-axis, which coincides with the wire.

Suppose a square conducting loop, with side length equal to 0.2 meter and with a current of 10 amperes, is placed in the region as indicated by the figure. Determine the net force acting on this square loop.

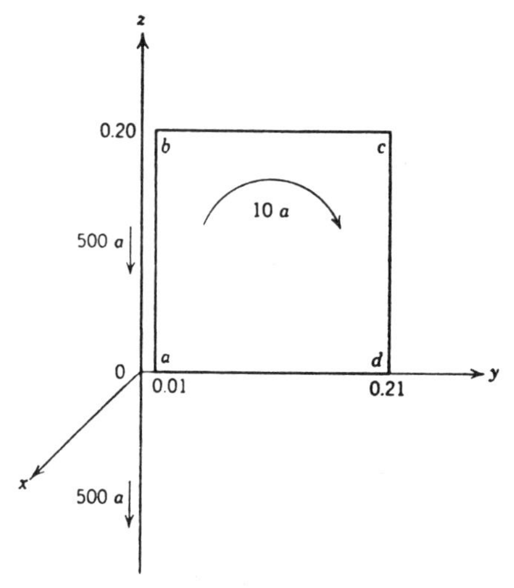

A 500-ampere straight wire and a 10-ampere square loop.

Solution: In the plane of the loop, x = 0 and the 500-a current produces a magnetic flux density $\bar{B} = 0.0001/y\ \bar{i}$. In the figure the flux lines are directed out of the paper, for positive y. It is evident from

$$d\bar{F} = i d\bar{\ell} \times \bar{B} \tag{1}$$

that the specified magnetic field produces forces on the sides of the square loop with each force vector being normal to its side, lying in the plane of the loop, and directed toward the inside of the loop.

Along path ab, y is 0.01 and $\bar{B}$ is $0.01\bar{i}$. The force $\bar{F}_{ab}$ is $Bi\ell\bar{j}$, or $0.01 \times 10 \times 0.2j$, or 0.02j newton.

Along path bc, B varies with y, use Eq. (1). A current element along this path is 10 dy $\bar{j}$, and the force on a current element is $(10\ dy\ \bar{j}) \times (0.0001/y\ \bar{i})$, or $-0.001 \, dy/y \, \bar{k}$. The total force on this side is

$$F_{bc} = -0.001 \int_{0.01}^{0.21} 1/y \, dy \, \bar{k} = -0.00304\bar{k}$$

Along path cd, y is 0.21 and $\bar{B}$ is $0.00048\bar{i}$. The force $\bar{F}_{cd}$ is $-Bi\ell\bar{j}$, or $-0.00096\bar{j}$.

Along path da, a current element is $10 \, dy \, \bar{j}$, with dy negative, and the force on a current element is $-0.001 \, dy/y \, \bar{k}$. For negative dy the limits of the integral are taken with y decreasing from 0.21 to 0.01, and evaluation of the integral gives a force F_{da} of $0.00304\bar{k}$.

The net force acting on the square loop is the vector sum of $\bar{F}_{ab}$, $\bar{F}_{bc}$, $\bar{F}_{cd}$, and $\bar{F}_{da}$. This gives a net force $\bar{F}$ equal to $0.019\bar{j}$ newton. If the loop were rigid but free to move, the magnetic force would move the loop of the figure from left to right. It should be mentioned that the current of the loop produces a magnetic field which is neglected. Although this field contributes to the force on each side of the loop, these force components obviously cancel when the net force is computed.

MAGNETIZATION

• PROBLEM 9-14

Suppose the effect of a magnetic field is to induce a magnetic dipole moment of 10^{-23} A m^2 per atom. Let there be 10^{27} atoms/m^3. Find the surface current density at a surface making an angle of 45° with $\bar{M}$.

Solution: $j_m = \bar{M} \times \hat{n} = Nm \sin 45°$

The tangential component of $\bar{M}$ is

$$(10^{27})(10^{-23})(0.707) = 7070 \text{ A m}^{-1}$$

and this is the surface current density.

10^{27} atoms/m^3 is equivalent to taking 10^9 atoms/m. Then the contribution of each atom to j_m is

$$7.07 \times 10^{-6} \text{ A/atom}$$

This seems like a tremendously large value for the current contribution from a single atom. But while 0.7 μA is a current of macroscopic size, it is several orders of magnitude smaller than the current produced in an atom by one electron circulating with a frequency of 10^{15} Hz.

• PROBLEM 9-15

> Suppose a $\bar{B}$ field is applied to a cube of magnetic material, b m on a side, such that $\bar{M}$ is z directed and varies linearly with x according to $\bar{M} = \bar{a}_z 10x$ A/m, as shown in (a). Find the magnetization current density $\bar{J}_m$ in the material, as well as the surface magnetization current density. Sketch the bound current fields in and on the cube.

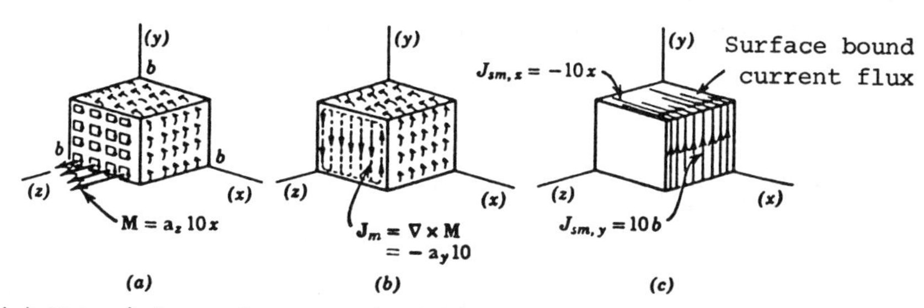

(a) Material sample magnetized linearly with increasing x. (b) Volume magnetization currents produced by transverse variations of M. (c) Surface currents produced by uncanceled segments of bound currents.

<u>Solution</u>: The magnetization current density $\bar{J}_m$ is obtained from

$$\bar{J}_m = \nabla \times \bar{M} = \begin{vmatrix} \bar{a}_x & \bar{a}_y & \bar{a}_z \\ \frac{\partial}{\partial x} & \frac{\partial}{\partial y} & \frac{\partial}{\partial z} \\ 0 & 0 & 10x \end{vmatrix} = -a_y 10 \text{ A/m}^2$$

negative y directed and of constant density as in (b).

The uncancelled segments of the bound currents at the surface of the block constitute a surface density of magnetization currents denoted by $\bar{J}_{sm}$ (A/m). On the end

$x = b$, $\bar{J}_{sm}$ is y directed and has a magnitude equal to that of $\bar{M}$ there; i.e.,

$$\bar{J}_{sm}]_{x=b} = \bar{a}_y M_z]_{x=b} = \bar{a}_y 10b \text{ A/m}$$

while on the top and bottom of the block

$$\bar{J}_{sm}]_{y=b} = -\bar{a}_x M_z]_{y=b} = -\bar{a}_x 10x \text{ A/m}$$

$$\bar{J}_{sm}]_{y=0} = \bar{a}_x M_z]_{y=0} = \bar{a}_x 10x \text{ A/m}$$

No bound currents exist on the end at $x = 0$, since $\bar{M} = 0$ there. These surface effects are shown as flux plots in (c).

• PROBLEM 9-16

Let $\bar{M} = \hat{z}M_0 y$ in Fig. 1. M_0 is a constant, with dimensions $[m^{-2}\ A]$. This shows a rod, of length L into the paper, made of many parallel slabs, the magnetization increasing linearly in the slabs toward the right. Find $\bar{J}_m$ and $\bar{j}_m$.

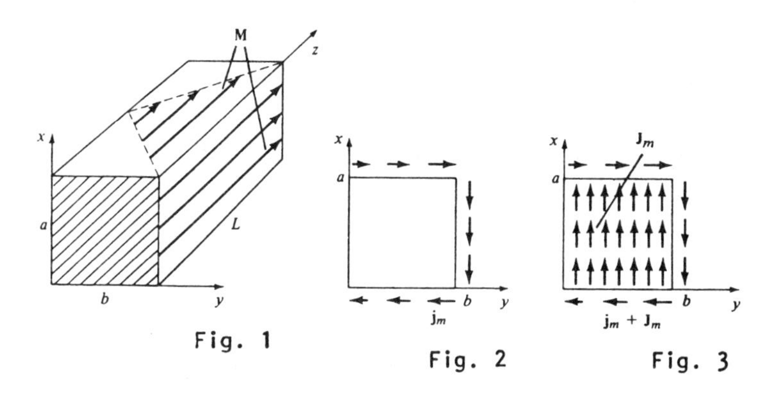

Fig. 1 Fig. 2 Fig. 3

Solution:

$$\bar{j}_m = \bar{M} \times \hat{n} = (\hat{z}M_{0y}) \times \hat{n}$$

Top: $[(\hat{z}M_{0y}) \times \hat{x}]_{x=a} = (\hat{y}M_{0y})_{x=a} = \hat{y}M_{0y}$

Right: $[(\hat{z}M_{0y}) \times \hat{y}]_{y=b} = (-\hat{x}M_{0y})_{y=b} = -\hat{x}M_{0b}$

Bottom: $[(\hat{z}M_{0y}) \times (-\hat{x})]_{x=0} = (-\hat{y}M_{0y})_{x=0} = -\hat{y}M_{0y}$

Left: $[(\hat{z}M_{0y}) \times (-\hat{y})]_{y=0} = (\hat{x}M_{0y})_{y=0} = 0$

Figure 2 shows the surface current around the outside of the rod. $\bar{j}_m$ has an unusual property. Although steady with time, its divergence does not vanish on the top and bottom surfaces.

$$\bar{J}_m = \nabla \times \bar{M} = \hat{x}M_0$$

(The dimensions of $\bar{J}_m$ are $[M_0] = [A\ m^{-2}]$, while $\bar{j}_m$ has dimensions $[A\ m^{-1}]$.) $\bar{J}_m$ has a divergence of zero within the rod. When $\bar{J}_m$ is added to $\bar{j}_m$ the picture of Fig. 3 is obtained. The two contributions on the right side, at $y = b$, do not cancel each other: they have different dimensions so they are different kinds of quantities.

• PROBLEM 9-17

Find the equivalent sheet current density K' for a uniformly magnetized rod 20 cms long, with 10 cm^2 area of cross-section, and a pole strength of 100 amp-meters. Also find the current I required for a 1,000-turn solenoid of the same size to be magnetically equivalent.

Solution: From

$$K' = \frac{m}{v} = \frac{N'IA'}{\ell A} = \frac{NI}{\ell}\frac{nA'}{A} = \frac{NI}{\ell} = K'$$

$$K' = \frac{m}{v} = \frac{Q_m \ell}{v} = \frac{100 \times 0.2}{10^{-3} \times 0.2} = 10^5 \text{ amp/meter}$$

For the solenoid to be equivalent we put $K = K'$, or

$$K = \frac{NI}{\ell} = K' = 10^5$$

from which

$$I = \frac{K'\ell}{N} = \frac{10^5 \times 0.2}{10^3} = 20 \text{ amp}$$

• PROBLEM 9-18

Find the magnetization M_i in the permanent magnet shown in the figure with an air gap of length ℓ_0.

Solution: Applying Ampere's law,

$$\oint_C \bar{H} \cdot d\bar{s} = H_i (\ell - \ell_0) + H_0 \ell_0 = NI$$

where H_i is the magnetic field in the material and

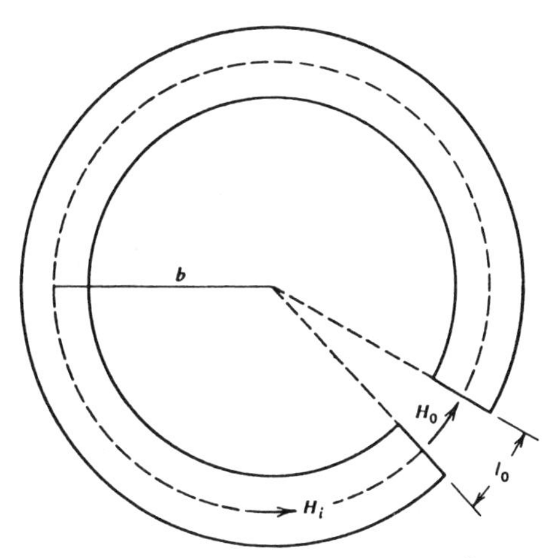

Toroidal magnet with a gap of length l_0.

H_0 the magnetic field value in the gap.

In this case, I = 0. The expression is still valid, however,

$$H_i = \frac{\ell_0 H_0}{(\ell - \ell_0)} \simeq - H_0 \frac{\ell_0}{\ell}$$

showing that the magnetic fields in the two regions are oppositely directed. In the gap, $\bar{M} = 0$ and $H_0 = B_0/\mu_0$ so that $\bar{H}$ and $\bar{B}$ are in the same direction in the gap, but since $B_i = B_0$, $\bar{H}_i$ and $\bar{B}_i$ are oppositely directed within the magnet.

In addition, $B_i = B_0 = \mu_0 H_0 = \mu_0 (H_i + M_i)$ and

$$M_i = \frac{B_i \ell}{\mu_0 (\ell - \ell_0)} \simeq \frac{B_i}{\mu_0} \left(1 + \frac{\ell_0}{\ell}\right)$$

and is in the same direction as B_i, as is to be expected. The change in sign of $\bar{H}$, which occurs at the surface separating the material from the vacuum, results from the discontinuity of $\bar{M}$ at this surface. Since $\bar{M}_i$ and $\bar{H}_i$ are oppositely directed in the material, $\bar{H}_i$ is often described as a "demagnetizing" field.

• **PROBLEM** 9-19

Referring to Fig. 1, a toroidal coil has a radius R and a cross-sectional area $A = \pi r^2$. The coil has a very narrow gap, as shown in the gap detail in Fig. 2a. The coil is made of many turns N of fine insulated wire with a current I. Draw graphs showing the variation of $\bar{B}$, $\bar{M}$, $\bar{H}$, and μ along the line of radius R at the gap (centerline of coil).

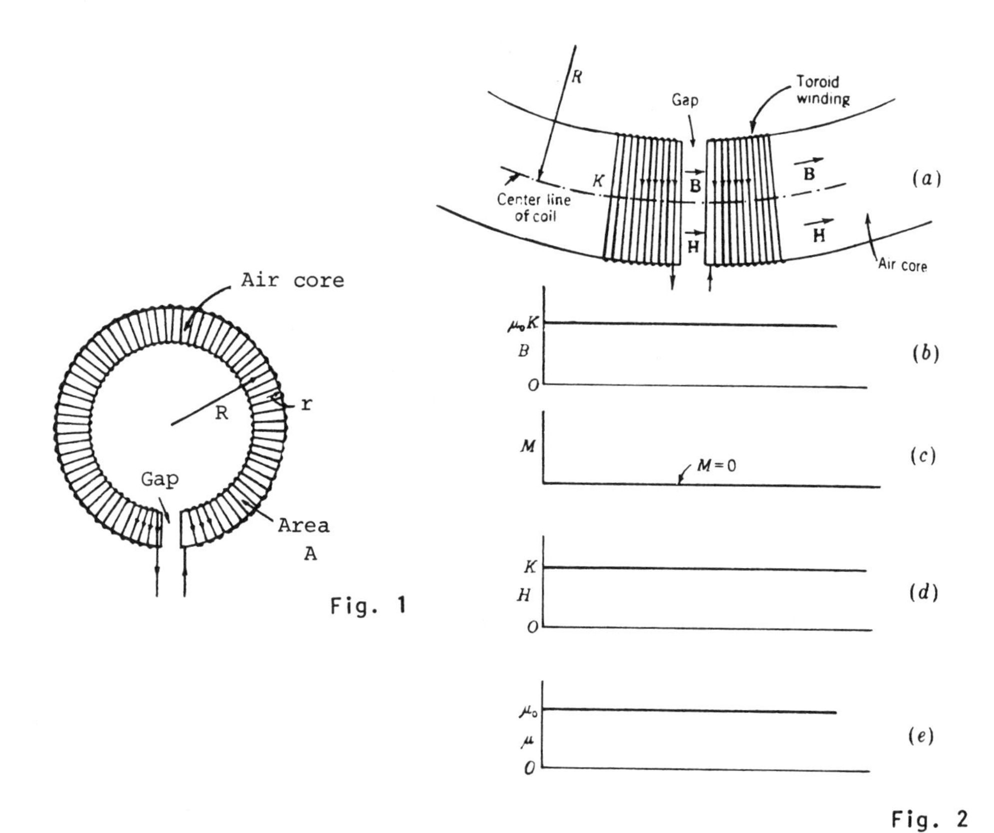

Solution: Neglecting the small effect of the narrow gap, B is substantially uniform around the inside of the entire toroid. From Ampere's law,

$$\oint \bar{B} \cdot d\bar{\ell} = \mu_0 I$$

For this toroid, there are N turns of wire each carrying a current I, therefore

$$\oint B \cdot d\ell = \mu_0 NI$$

$$2\pi RB = \mu_0 NI$$

$$B = \frac{\mu_0 NI}{2\pi R} = \mu_0 K \quad (T)$$

where K is the magnitude of the linear sheet-current density in amperes per meter. A graph of the magnitude B along the centerline of the coil at the gap is shown in Fig. 2b.

No ferromagnetic material is present, so that the magnetization is negligible and $\bar{M} = 0$, as indicated in Fig. 2c. It follows that $\nabla \cdot \bar{M} = 0$ and also $\nabla \cdot \bar{H} = 0$.

Since M = 0,

$$H = \frac{B}{\mu_0} = \frac{\mu_0 K}{\mu_0} = K = \frac{NI}{2\pi R} \quad (\text{A m}^{-1})$$

Therefore, the magnitude of H is constant and equal to the sheet-current density K of the coil winding, as indicated in Fig. 2d. The permeability everywhere is μ_0 (Fig. 2e).

It is to be noted that $\bar{B}$ is continuous and that in this case $\bar{H}$ is also continuous since there is no ferromagnetic material present. Both $\bar{B}$ and $\bar{H}$ have the same direction everywhere in this case.

• PROBLEM 9-20

Suppose an iron ring has wound over it a toroidal coil with a gap in the toroid coinciding with a gap in the ring as shown in Fig. 1 and in the gap detail of fig. 2a. Let the sheet current density for the toroid winding be $K = NI/2\pi R$. Further, let the induced magnetization added to the permanent magnetization in the ring yield a total uniform magnetization (permanent and induced) that is equal in magnitude to 4K. Draw graphs showing the variations of $\bar{B}$, $\bar{M}$, $\bar{H}$, μ, and $\nabla.\bar{H}$ along the centerline of the ring at the gap.

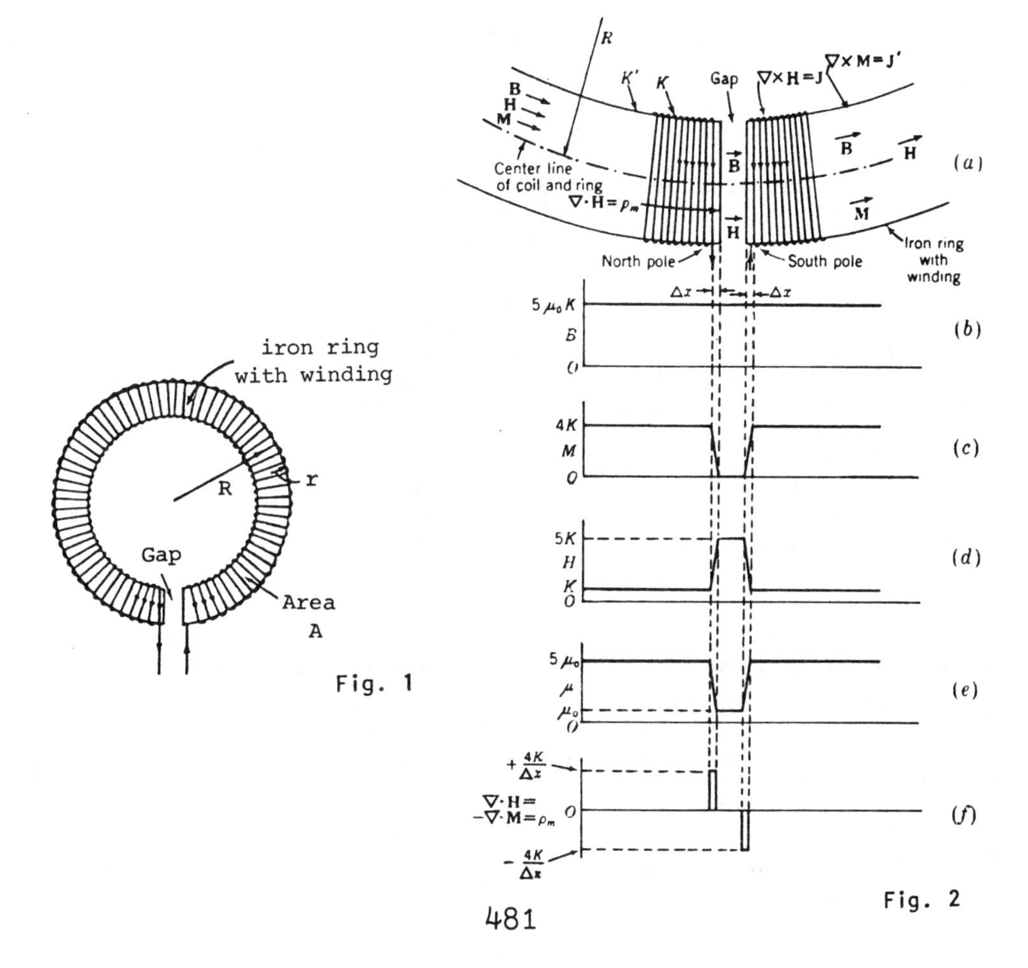

Fig. 1

Fig. 2

Solution: The total magnetization is $M = K' = 4K$. Neglecting the small effect of the narrow gap, the flux density is substantially uniform around the inside of the ring and across the gap. It is given (see fig. 2b) by

$$B = \mu_0(K + K') = 5\mu_0 K \qquad (T)$$

In the ring $M = 4K$, and in the gap $M = 0$, as shown in fig. 2c. It is assumed that M changes linearly over a short distance Δx at the pole faces.

In the gap

$$H = B/\mu_0 = 5K$$

In the ring

$$H = B/\mu_0 - M$$

and so approximately $H = 5K - 4K = K$. The variation of H across the gap is depicted in fig. 2d. In the gap $\mu = \mu_0$. In the ring (see fig. 2e)

$$\mu = \mu_0\left(1 + \frac{M}{H}\right) = \mu_0\left(1 + \frac{4K}{K}\right) = 5\mu_0$$

The divergence of H or pole volume density ρ_m is given by the negative of the divergence of $\bar{M}$. This has a value of $\pm 4K/\Delta x$ over the assumed pole thickness Δx at the pole faces. This is illustrated in fig. 2F. The fact that $\nabla.\bar{H} = \rho_m$ at the pole faces is also indicated in fig. 2a.

Elsewhere $\nabla.\bar{H} = 0$. In this problem, $\bar{B}$ and $\bar{H}$ have the same direction both in the gap and in the ring. In the ring, however, H is weaker than in the gap. In this problem, the toroid has a sheet-current density of $K(A\ m^{-1})$, and the ring has an equivalent sheet-current density around its curved surface of $K' = 4K$ (A/m). Inside a wire of the toroidal coil $\nabla x\bar{H} = \bar{J}$ (A/m^2) as suggested in fig. 2a. Elsewhere $\nabla x\bar{H} = 0$. At the curved surface of the ring $\nabla x\bar{M} = \bar{J}'$ (A/m^2). Elsewhere $\nabla x\bar{M} = 0$.

• PROBLEM 9-21

Consider an iron ring in the shape of a toroidal coil and with a very narrow gap as shown in Figs. 1 and 2a. Assume that the ring has a uniform permanent magnetization $\bar{M}$ that is equal in magnitude to $K = NI/2\pi R$ for a toroidal coil. Draw graphs showing the variation of $\bar{B}$, $\bar{M}$, $\bar{H}$, μ and $\nabla.\bar{H}$ along the centerline of the ring at the gap.

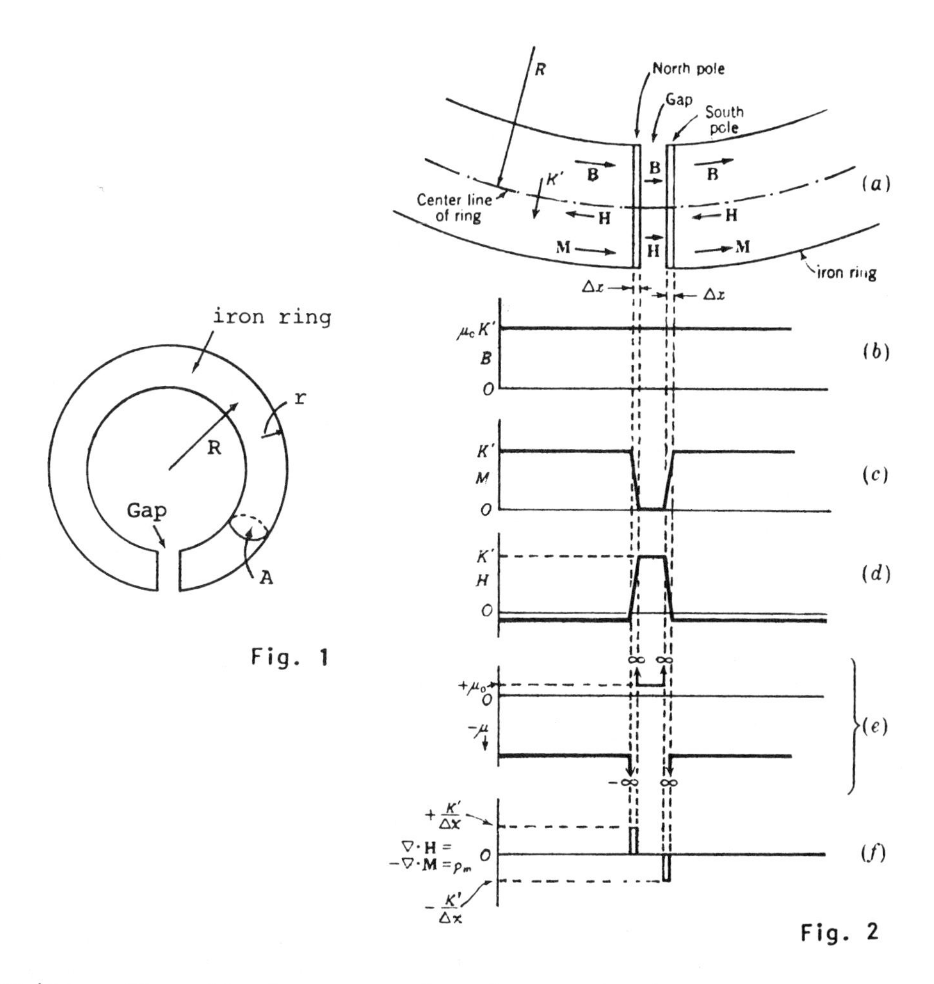

Solution: The ring has a north pole at the left side of the gap and a south pole at the right side. Neglecting the small effect of the narrow gap, $\bar{B}$ is substantially uniform around the interior of the entire ring and also across the gap. It is due entirely to the equivalent sheet-current density K' on the surface of the ring. The magnetization M is equal to the equivalent sheet-current density, for a uniformly magnetized iron ring, K' = M. Thus

$$B = \mu_0 M = \mu_0 K' \qquad (T)$$

where M and K' are, according to the stated conditions, equal to K.

Hence, B is the same as for a toroidal coil of current $iB = K\mu_0 = \mu_0 NI/2\pi R$. Its value at the gap is illustrated in Fig. 2b.

In the ring M = K', but outside the ring and in the gap M = 0. Suppose that the change in M' from zero to K' at the gap occurs over a short distance Δx rather than as a square step function. The graph for M is then as shown in Fig 2c.

Outside the ring and in the gap M = 0; so

$$H = \frac{B}{\mu_0} = K' \qquad (A/m)$$

Inside the ring

$$H = \frac{B}{\mu_0} - M \qquad (A/m)$$

or approximately H = K' - K' = 0. The exact value of H is not zero but is small and negative. The variation of H across the gap is illustrated in Fig. 2d.

The permeability in the ring is large and negative because H is small compared with M and is negative. In the air gap $\mu = \mu_0$. The variation of μ across the gap is suggested in Fig. 2e.

The divergence of $\bar{H}$ equals the negative divergence of $\bar{M}$, and this equals the apparent pole volume density ρ_m in the ring on both sides of the gap. Thus

$$\nabla . \bar{H} = -\nabla . \bar{M} = \rho_m \qquad (A\ m\ m^{-3})$$

This is zero everywhere except at the layers of assumed thickness Δx at the gap. Assuming that $\bar{M}$ changes linearly in magnitude over this thickness, and assuming also that Δx is very small compared with the cross-sectional diameter 2r of the ring, on the centerline

$$\nabla . \bar{M} = \frac{dM_x}{dx} = \frac{\mp K'}{\Delta x} = -\rho_m \quad ,$$

or

$$\nabla . \bar{H} = \frac{\pm K'}{\Delta x} = \rho_m \quad ,$$

where the upper sign in front of K' applies if M decreases and H increases in proceeding across Δx in a positive direction (from left to right). The variation of $\nabla . \bar{H}$ along the centerline is illustrated in Fig. 2f. Hence the pole volume density ρ_m has a value only in the layers of assumed thickness Δx at the sides of the gap. This locates the poles of the ring magnet at the sides of the gap, and for this reason the iron surfaces of the gap are called pole faces.

• PROBLEM 9-22

A long uniform solenoid, as in Fig. 1a, is situated in air and has NI ampere-turns and a length ℓ. A permanently magnetized iron rod (Fig. 1) has the same dimensions as the solenoid and has a uniform magnetization M equal to NI/ℓ for the solenoid. Draw graphs showing the variation of $\bar{B}$, $\bar{M}$, and $\bar{H}$ along the axes of the solenoid and the rod. Also sketch the configuration of the fields for the two cases.

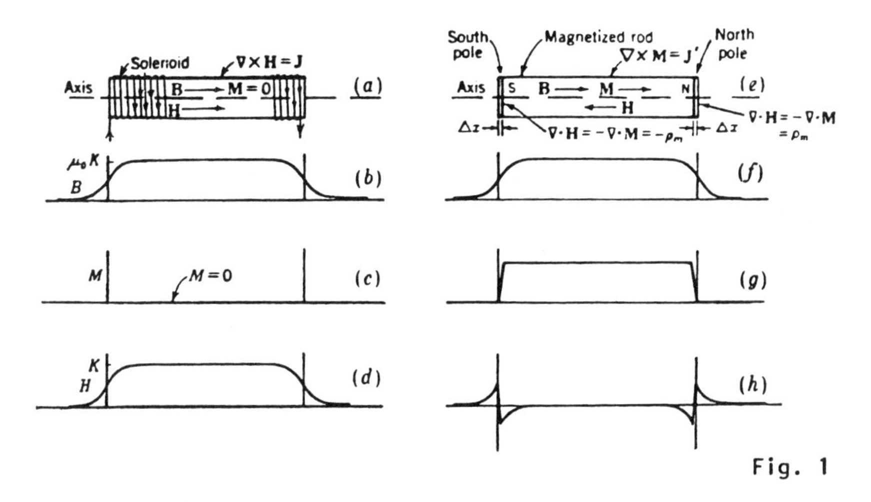

Fig. 1

Solution: Since the rod and solenoid have the same dimensions and $M = K' = K = NI/\ell$, the two are magnetically equivalent. The $\bar{B}$ fields for both are the same everywhere, and the $\bar{H}$ fields for both are the same outside the solenoid and rod. Assuming that the solenoid is long compared with its diameter, the flux density at the center is, from Ampere's law, $\oint \bar{B}.d\bar{\ell} = \mu_0 NI$,

$$B = \mu_0 \frac{NI}{\ell} = \mu_0 K$$

At the ends of the solenoid

$$B = \frac{1}{2} \mu_0 K$$

The variation of $\bar{B}$ along the solenoid axis is shown graphically in Fig. 1b. The variation along the rod axis is the same (Fig. 1f).

For the solenoid case, M = 0 everywhere (Fig. 1c). In the rod the magnetization M is assumed to be uniform, as in Fig. 1g.

For the solenoid case, $\bar{H} = \bar{B}/\mu_0$ everywhere, so that H = K at the center and $H = \frac{1}{2}K$ at the ends. The variation of H along the solenoid axis is shown in Fig. 1d. Outside the rod, H is the same as for the solenoid. Inside the rod $H = (B/\mu_0) - M$, so that the variation is as suggested in Fig. 1h. It is assumed that M changes from 0 to K over a short distance Δx at the ends of the rod. The direction of $\bar{H}$ in the rod is opposite to that for B.

Inside the wires of the solenoid winding $\nabla \times \bar{H} = \bar{J}$, as indicated in Fig. 1a. On the cylindrical surface of the rod, $\nabla \times \bar{M} = \bar{J}'$, as suggested in Fig. 1e. In the solenoid case, $\nabla \cdot \bar{B} = 0$ and $\nabla \cdot \bar{H} = 0$ everywhere. In the rod case, $\nabla \cdot \bar{B} = 0$ everywhere, but $\nabla \cdot \bar{H} = -\nabla \cdot \bar{M} = \rho_m$ at the end faces of the rod.

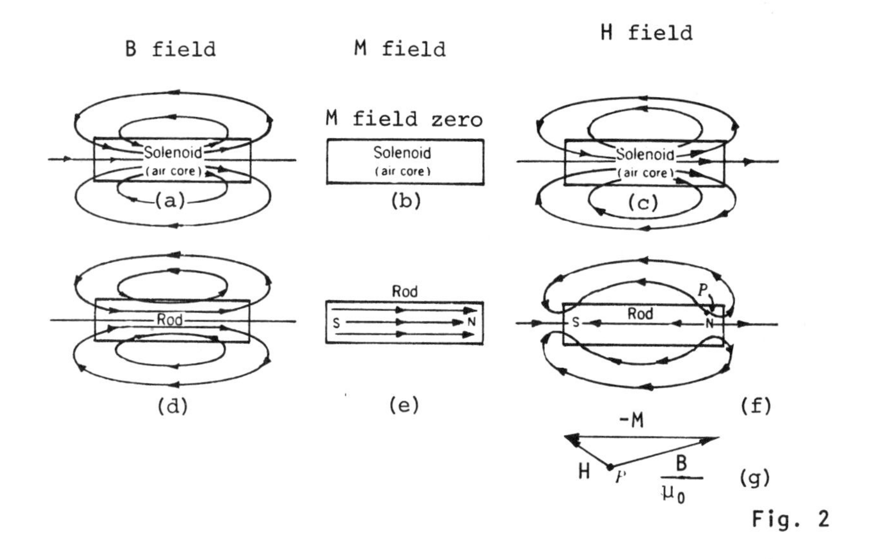

Fig. 2

The $\bar{B}$, $\bar{M}$, and $\bar{H}$ fields for the two cases are sketched in Fig. 2. It is to be noted that inside the rod $\bar{H}$ is directed from the north pole to the south pole. Since $\bar{M}$ and $\bar{B}$ have, in general, different directions in the rod, μ loses its simple scalar significance in this case. Here $\bar{H}$ can be obtained by vector addition, using $\bar{B} = \mu_0(\bar{H} + \bar{M})$. As an example, $\bar{H}$ at the point P in Fig. 2f is obtained by the vector addition of $\bar{B}/\mu_0$ and $-\bar{M}$ as in Fig. 2g.

• PROBLEM 9-23

A line current I of infinite extent is within a cylinder of radius a that has permeability μ, as in the figure. The cylinder is surrounded by free space. What are the $\bar{B}$, $\bar{H}$, and $\bar{M}$ fields everywhere? What is the magnetization current?

Solution: Pick a circular contour of radius r around the current. Using the integral form of Ampere's law, the

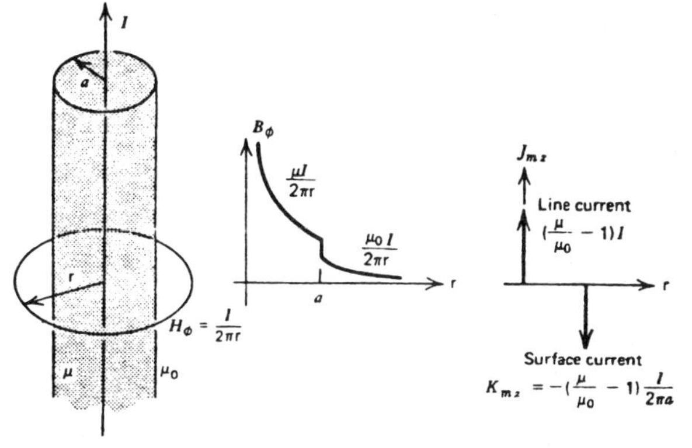

A free line current of infinite extent placed within a permeable cylinder gives rise to a line magnetization current along the axis and an oppositely directed surface magnetization current on the cylinder surface.

H field is of the same form whether inside or outside the cylinder:

$$\oint_L \bar{H} \cdot d\bar{\ell} = H_\phi 2\pi r = I \therefore H_\phi = \frac{I}{2\pi r}$$

The magnetic flux density differs in each region because the permeability differs:

$$B_\phi = \begin{cases} \mu H_\phi = \dfrac{\mu I}{2\pi r}, & 0 < r < a \\[2ex] \mu_0 H_\phi = \dfrac{\mu_0 I}{2\pi r}, & r > a \end{cases}$$

The magnetization is obtained from the relation

$$M = \frac{\bar{B}}{\mu_0} - \bar{H}$$

as

$$M_\phi = \begin{cases} \left(\dfrac{\mu}{\mu_0} - 1\right) H_\phi = \dfrac{\mu - \mu_0}{\mu_0} \dfrac{I}{2\pi r}, & 0 < r < a \\[2ex] 0, & r > a \end{cases}$$

The volume magnetization current can be found using

$$\bar{J}_m = \nabla \times \bar{M} = -\frac{\partial M_\phi}{\partial z} \bar{i}_r + \frac{1}{r} \frac{\partial}{\partial r} (rM_\phi) \bar{i}_z = 0, \quad 0 < r < a$$

There is no bulk magnetization current because there are

no bulk free currents. However, there is a line magnetization current at r = 0 and a surface magnetization current at r = a. They are easily found using Stokes' theorem:

$$\int_S \nabla \times \bar{M} . d\bar{S} = \oint_L \bar{M} . d\bar{\ell} = \int_S \bar{J}_m . d\bar{S}$$

Pick a contour around the center of the cylinder with r < a:

$$M_\phi 2\pi r = \left(\frac{\mu - \mu_0}{\mu_0}\right) I = I_m$$

where I_m is the magnetization line current. The result remains unchanged for any radius r < a as no more current is enclosed since $\bar{J}_m = 0$ for $0 < r < a$. As soon as $r > a$, M_ϕ becomes zero so that the total magnetization current becomes zero. Therefore, at r = a a surface magnetization current must flow whose total current is equal in magnitude but opposite in sign to the line magnetization current:

$$K_{zm} = \frac{-I_m}{2\pi a} = - \frac{(\mu - \mu_0) I}{\mu_0 2\pi a}$$

MAGNETIC BOUNDARY CONDITIONS

• PROBLEM 9-24

A uniform magnetic field of strength B = 1.2 webers/m^2 exists within an iron core ($\mu = 1000\mu_0$) as shown in the figure. If an air gap is cut with the orientation shown, find the magnitude and direction of $\bar{B}$ in the gap.

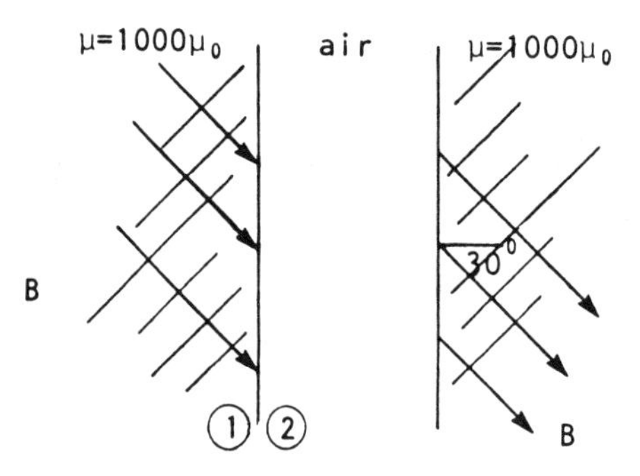

Solution: Using magnetic boundary conditions; surface 1 is in the iron core and surface 2 is in the air gap:

$$B_{n1} = B_{n2} \quad \text{(normal components of } \bar{B} \text{ are equal)}$$

$$H_{t1} = H_{t2} \quad \text{(tangential components of } \bar{H} \text{ are equal)}$$

$$B_{n2} = B_{n1} = B \cos 30° = 1.2\ (0.866) = 1.0\ \text{w/m}^2$$

$$H_{t1} = H_{t2}$$

$$\frac{B_{t1}}{\mu} = \frac{B_{t2}}{\mu_0}$$

$$\therefore\ B_{t2} = \frac{\mu_0}{1000\mu_0} B_{t1}$$

$$B_{t2} = \frac{1}{1000} B_{t1}$$

$$B_{t2} = \frac{1}{1000} (1.2 \sin 30°)$$

$$B_{t2} = 0.00006\ \text{w/m}^2$$

$B \approx 1.0$ since 0.0006 is negligible compared to 1.0 and the angle from the x-axis is approximately equal to 0° since the tangetial component of $\bar{B}$ is so small compared to the normal component.

• PROBLEM 9-25

A coaxial cable consists of a thin center wire carrying current I_0 and a thin outer conductor carrying net current I_0 in the opposite direction. Half the space between the conductors is filled with a magnetic material with permeability μ and the other half with air. Find $\bar{H}$, $\bar{B}$ and μ at all points inside the cable. The figure shows the diagram of the cable.

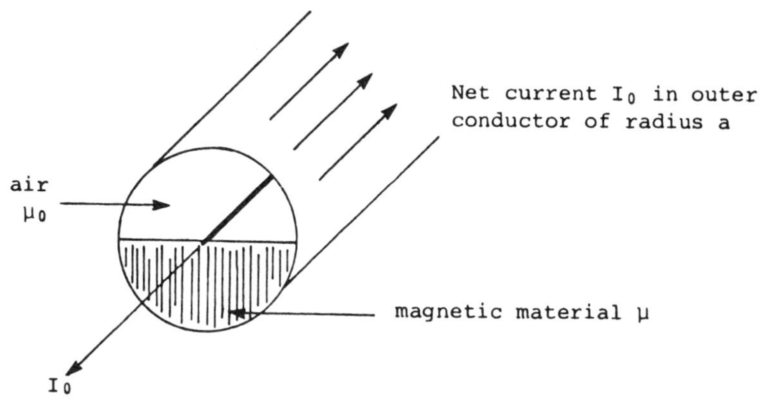

Solution: Using magnetic boundary conditions

$$H_{t1} = H_{t2} = 0$$

since all components of magnetic field are in the radial direction which is normal to the surface between the magnetic material and the air.

$$B_{2n} = B_{1n}$$

Using Ampere's Law,

$$\oint \bar{H} \cdot d\bar{r} = I_{\text{through curve}}$$

For r < a

$$\oint_c \bar{H} \cdot d\bar{r} = I_0$$

$$\oint_c (\bar{H}_{air} + \bar{H}_{mat.}) \cdot d\bar{r} = I_0$$

$$\oint_c \left(\frac{\bar{B}}{\mu_0} + \frac{\bar{B}}{\mu}\right) \cdot d\bar{r} = I_0$$

dr is the differential arc length given by $rd\theta$.

Therefore,

$$\int_0^{\pi} \frac{B}{\mu_0} rd\theta + \int_{\pi}^{2\pi} \frac{B}{\mu} r\, d\theta = I_0$$

$$\frac{\pi Br}{\mu_0} + \frac{\pi Br}{\mu} = I_0$$

Solving for B, we get

$$B = \frac{\mu_0 \mu I}{\mu_0 \pi r + \mu \pi r}$$

For r > a the total current through the curve (circle of radius > a) is zero therefore B = 0.

For r < a

in air,

$$H_a = \frac{B}{\mu_0} = \frac{\mu I}{\mu_0 \pi r + \mu \pi r}$$

In magnetic material

$$H_m = \frac{B}{\mu} = \frac{\mu_0 I}{\mu_0 \pi r + \mu \pi r}$$

In air $\bar{M} = \frac{1}{\mu_0} \bar{B} - \bar{H}_a = \bar{H} - \bar{H} = 0.$

In magnetic material, $\bar{M} = \frac{1}{\mu_0} \bar{B} - \bar{H}_m$

$$M = \frac{\mu I}{\mu_0 \pi r + \mu \pi r} - \frac{\mu_0 I}{\mu_0 \pi r + \mu \pi r}$$

$$M = \frac{I}{\mu_0 \pi r + \mu \pi r} [\mu - \mu_0] = \frac{I\ (\mu - \mu_0)}{\pi r(\mu + \mu_0)}$$

$$M = \frac{I}{\pi r} \left[\frac{\mu - \mu_0}{\mu + \mu_0} \right]$$

• PROBLEM 9-26

Obtain a refractive law for the $\bar{B}$ field at an interface separating two isotropic materials of permeabilities μ_1 and μ_2; i.e., find the relation between the angular deviations from the normal made by B_1 and B_2 at points just to either side of the interface. Refer to the given figure.

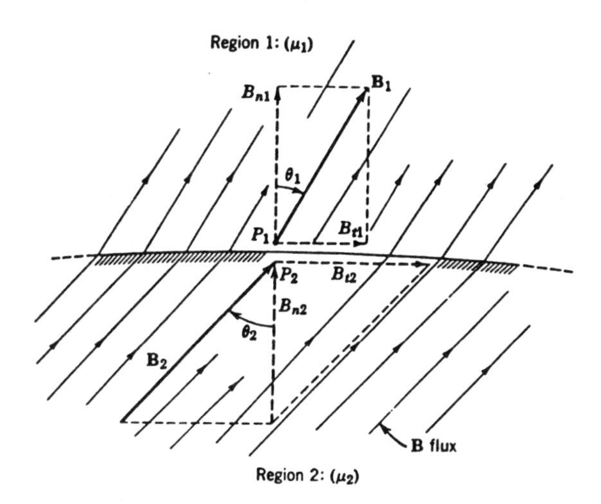

Solution: Assume the total $\bar{B}$ fields tilted from the normal by the angles θ_1 and θ_2 as shown in the figure. The boundary conditions relating the tangential and the normal magnetic field components are $B_{n1} = B_{n2}$ and $H_{t1} = H_{t2}$. The latter can be written

$$\frac{B_{t1}}{\mu_1} = \frac{B_{t2}}{\mu_2} \tag{1}$$

From the geometry of the figure, the tilt-angles obey $\tan\theta_1 = B_{t1}/B_{n1}$ and $\tan\theta_2 = B_{t2}/B_{n2}$, which combine with (1) to yield

$$\tan\theta_2 = \frac{\frac{\mu_2}{\mu_1} B_{t1}}{B_{n1}}$$

Inserting the expression for $\tan\theta_1$,

$$\tan\theta_2 = \frac{\mu_2}{\mu_1}\tan\theta_1$$

• PROBLEM 9-27

Referring to Fig. 1, let medium 1 be air ($\mu_r = 1$) and medium 2 be soft iron with a relative permeability of 7,000. (a) If $\bar{B}$ in the iron is incident normal on the boundary ($\alpha_2 = 0$), find α_1, the angle that the $\bar{B}$ field in the air makes with the normal to the surface. (b) If $\bar{B}$ in the iron is nearly tangent to the surface at an angle $\alpha_2 = 85°$, find α_1.

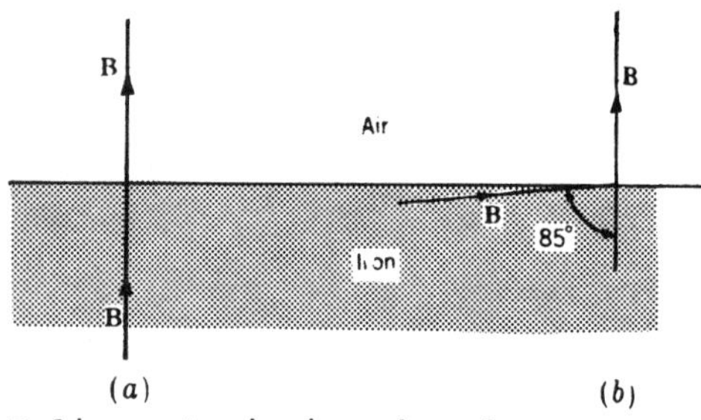

B lines at air-iron boundary.

Fig. 1

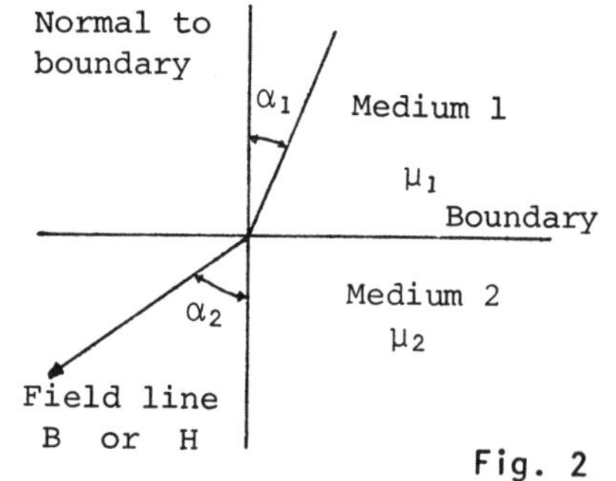

Fig. 2

Boundary between two media of different permeability showing change in direction of magnetic field line.

Solution: (a) From the boundary relations,

$$B_{n1} = B_{n2} \text{ and } H_{t1} = H_{t2} \tag{1}$$

From Fig. 2,

$$B_{n1} = B_1 \cos\alpha_1 \quad \text{and } B_{n2} = B_2 \cos\alpha_2 \tag{2}$$

$$H_{t1} = H_1 \sin\alpha_1 \text{ and } H_{t2} = H_2 \sin\alpha_2 \tag{3}$$

where B_1 = magnitude of $\bar{B}$ in medium 1

B_2 = magnitude of $\bar{B}$ in medium 2

H_1 = magnitude of $\bar{H}$ in medium 1

H_2 = magnitude of $\bar{H}$ in medium 2

Substituting (2) and (3) into (1) and dividing yields

$$\frac{\tan\alpha 1}{\tan\alpha_2} = \frac{\mu_1}{\mu_2} = \frac{\mu_{r1}}{\mu_{r2}} \tag{4}$$

where μ_{r1} = relative permeability of medium 1.

μ_{r2} = relative permeability of medium 2.

In this problem, medium 2 is the iron and medium 1 is the air. Therefore, from eq. (4),

$$\tan\alpha_1 = \frac{\mu_{r1}}{\mu_{r2}} \tan\alpha_2 = \frac{1}{7{,}000} \tan\alpha_2 \tag{5}$$

When $\alpha_2 = 0$, $\alpha_1 = 0$, so that the $\bar{B}$ line in air is also normal to the boundary (see Fig. 1a).

(b) When $\alpha_2 = 85°$, we have, from (5), that $\tan\alpha_1 = 0.0016$, or $\alpha_1 = 0.1°$. Thus, the direction of $\bar{B}$ in air is almost normal to the boundary (within 0.1°) even though its direction in the iron is nearly tangent to the boundary (within 5°) (see Fig. 1b). Accordingly, for many practical purposes the direction of $\bar{B}$ or $\bar{H}$ in air or other medium of low relative permeability may be taken as normal to the boundary of a medium having a high relative permeability.

• PROBLEM 9-28

At a particular instant of time the fields at point 1 in the figure are given by

$$\bar{E}_1 = E_0(3\bar{i}_x + \bar{i}_z)$$

$$\bar{H}_1 = H_0(2\bar{i}_y)$$

where E_0 and H_0 are constants. Find the fields at point 2, lying adjacent to point 1 and on the other side of the interface between media 1 and 2.

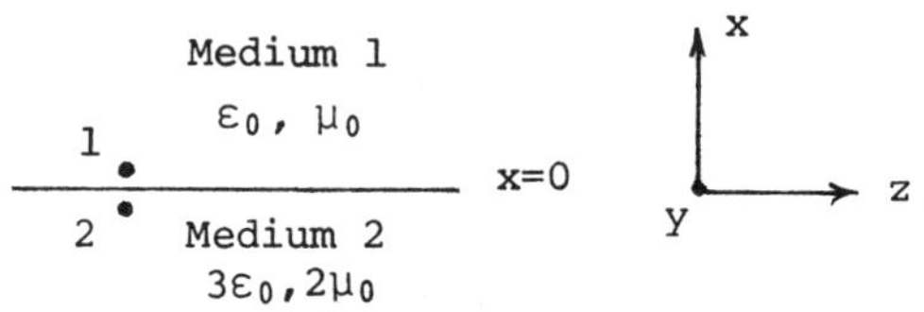

Medium 1
ε_0, μ_0
1
2
$x=0$
Medium 2
$3\varepsilon_0, 2\mu_0$
x
y
z

Solution: From $\bar{i}_n \cdot (\bar{D}_1 - \bar{D}_2) = 0$

$$D_{2x} = D_{1x} = \varepsilon_0(3E_0) = 3\varepsilon_0 E_0$$

$$E_{2x} = \frac{D_{2x}}{3\varepsilon_0} = \frac{3\varepsilon_0 E_0}{3\varepsilon_0} = E_0$$

From $E_{t1} = E_{t2}$

$$E_{2y} = E_{1y} = 0$$

$$E_{2z} = E_{1z} = E_0$$

From $\bar{i}_n \cdot (\bar{B}_1 - \bar{B}_2) = 0$

$$B_{2x} = B_{1x} = \mu_0(0) = 0$$

$$H_{2x} = \frac{B_{2x}}{2\mu_0} = 0$$

From $H_{t1} = H_{t2}$

$$H_{2y} = H_{1y} = 2H_0$$

$$H_{2z} = H_{1z} = 0$$

Thus the required fields at point 2 are found to be

$$\bar{E}_2 = E_0(\bar{i}_x + \bar{i}_z)$$

$$\bar{H}_2 = H_0(2\bar{i}_y)$$

POTENTIAL ENERGY OF MAGNETIC FIELDS

• PROBLEM 9-29

Find the magnetic energy associated with unit length of an infinitely long straight wire of radius a carrying a current I.

Solution: Inside the conductor

$$\bar{B} = \hat{\phi}\left(\frac{\mu_0 I}{2\pi a^2}\right) r$$

Then

$$B^2 = \left(\frac{\mu_0{}^2 I^2}{4\pi^2 a^4}\right) r^2$$

and

$$U = \int \tfrac{1}{2} \; \frac{B^2}{\mu_0} \; d\ell$$

or

$$U_{\ell\,(int)} = \frac{1}{2\mu_0} \int_0^a \left(\frac{\mu_0{}^2 I^2}{4\pi^2 a^4}\right) r^2 \;(2\pi r\; dr) = \frac{\mu_0 I^2}{16\pi}$$

This reveals an interesting feature of the energy distribution: the energy residing within the conductor is independent of the size of the conductor. All infinitely long straight conductors, whether they are thick or thin, have the same internal energy for the same total current.

Outside the conductor

$$\bar{B} = \hat{\Phi}\left(\frac{\mu_0 I}{2\pi}\right) \frac{1}{r}$$

Then

$$B^2 = \left(\frac{\mu_0{}^2 I^2}{4\pi^2}\right) \frac{1}{r^2}$$

and

$$U_{\ell\,(out)} = \frac{1}{2\mu_0} \int_a^\infty \left(\frac{\mu_0{}^2 I^2}{4\pi^2}\right) \frac{1}{r^2} \;(2\pi r\; dr) = \infty$$

The magnetic energy per unit length of an infinitely long conductor is infinite.

• PROBLEM 9-30

Consider a group of rigid current-carrying circuits, none of which extends to infinity, immersed in a medium with linear magnetic properties. Find the energy of this system in terms of H and B.

Solution: The energy of this system is given by

$$U = \frac{1}{2} \sum_{i=1}^{n} I_i \phi_i \tag{1}$$

For the present discussion it is convenient to assume that each circuit consists of only a single loop; then the flux Φ_i may be expressed as

$$\Phi_i = \int_{Si} \bar{B} \cdot \bar{n}\, da = \oint_{Ci} \bar{A} \cdot d\bar{\ell}_i, \tag{2}$$

where A is the local vector potential. Substitution of this result into Eq. (1)

$$U = \frac{1}{2} \sum_{i=1}^{n} I_i \phi_i$$

yields

$$U = \frac{1}{2} \sum_i \oint_{Ci} I_i \bar{A} \cdot d\bar{\ell}_i. \tag{3}$$

Making Eq. (3) somewhat more general, suppose that there are no current circuits defined by wires, but instead each "circuit" is a closed path in the medium (assumed to be conducting) that follows a line of current density. Equation (3) may be made to approximate this situation very closely by choosing a large number of contiguous circuits (C_i), replacing $I_i\, d\ell_i \to J\, dv$, and, finally, by the substitution of

$$\int_V \quad \text{for} \quad \sum_i \oint_{Ci} .$$

Hence

$$U = \frac{1}{2} \int_V \bar{J} \cdot \bar{A}\, dv. \tag{4}$$

The last equation may be further transformed by using the field equation $\nabla \times \bar{H} = \bar{J}$, and the vector identity

$$\nabla \cdot (\bar{A} \times \bar{H}) = \bar{H} \cdot \nabla \times \bar{A} - \bar{A} \cdot \nabla \times \bar{H},$$

whence

$$U = \frac{1}{2}\int_V \bar{H} \cdot \nabla \times \bar{A}\, dv - \frac{1}{2}\int_S \bar{A} \times \bar{H} \cdot \bar{n}\, da, \tag{5}$$

where S is the surface which bounds the volume V. Since, by assumption, none of the current "circuits" extends to infinity, it is convenient to move the surface S out to a very large distance so that all parts of this surface are far from the currents. Of course, the volume of the system must be increased accordingly. Now $\bar{H}$ falls off at least as fast as $1/r^2$, where r is the distance from an origin near the middle of the current distribution to a characteristic point on the surface S; $\bar{A}$ falls off at least as fast as $1/r$; and the surface area is proportional to r^2. Thus the contribution from the surface integral in Eq. (5) falls off as $1/r$ or faster, and if S is moved out to infinity, this contribution vanishes.

By dropping the surface integral in Eq. (5) and extending the volume term to include all space, we obtain

$$U = \frac{1}{2}\int_V \bar{H} \cdot \bar{B}\, dv, \tag{6}$$

since $\bar{B} = \nabla \times \bar{A}$. Equation (6) is restricted to systems containing linear magnetic media, since it was derived from Eq. (1). Also

$$u = \frac{1}{2}\bar{H} \cdot \bar{B}, \tag{7}$$

which, for the case of isotropic, linear, magnetic materials reduces to

$$u = \frac{1}{2}\mu H^2 = \frac{1}{2}\frac{B^2}{\mu}.$$

CHAPTER 10

MAGNETIC CIRCUITS

RELUCTANCE AND PERMEANCE

• PROBLEM 10-1

Find the total reluctance and permeance between the ends of the parallel-connected rectangular iron blocks shown in the figure, assuming that B is uniform in each block and normal to the ends. The permeability of each block is uniform, the value in block 1 being $\mu_1 = 500\mu_0$ and in block 2 being $\mu_2 = 2{,}000\mu_0$.

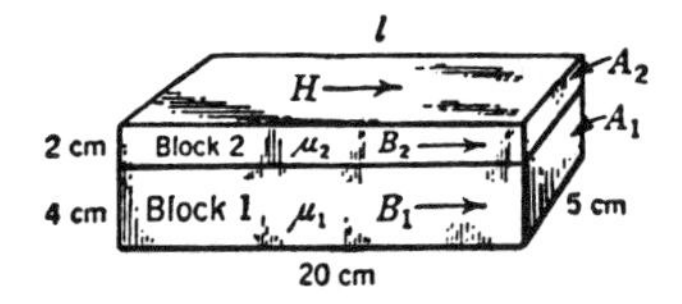

Rectangular iron blocks.

Solution: Since the blocks are in parallel, it is more convenient to calculate the total permeance first. The permeance P_1 of block 1 is

$$P_1 = \frac{\mu_1 A_1}{\ell} = \frac{500 \times 4\pi \times 10^{-7} \times 20 \times 10^{-4}}{0.2}$$

$$= 6.28 \times 10^{-6} \text{ henry}$$

The permeance of block 2 is

$$P_2 = \frac{\mu_2 A_2}{\ell} = \frac{2{,}000 \times 4\pi \times 10^{-7} \times 10 \times 10^{-4}}{0.2}$$

$$= 12.6 \times 10^{-6} \text{ henry}$$

The total permeance equals the sum of the individual permeances; so

$$P_T = P_1 + P_2 = (6.28 + 12.6) \times 10^{-6} = 1.89 \times 10^{-5} \text{ henry}$$

The total relucatance is then given by

$$R_T = \frac{1}{P_T} = \frac{1}{1.89 \times 10^{-5}} = 5.3 \times 10^4 \text{ reciprocal henrys}$$

• PROBLEM 10-2

(a) Referring to the figure 1, the permeability of the block is uniform and has a value $\mu_1 = 500\ \mu_0$ where $\mu_0 = 4\pi \times 10^{-7}$ henry/meter. Assuming that B is uniform throughout the block and normal to the ends, find the reluctance and permeance between ends of the rectangular block.

(b) Now the second block of permeability $\mu_2 = 2{,}000\ \mu_0$, which is connected to the block of part (a) as shown in fig. 2. Find the total reluctance and permeance between the ends of the series connected rectangular blocks assuming B is uniform throughout the blocks and normal to the ends.

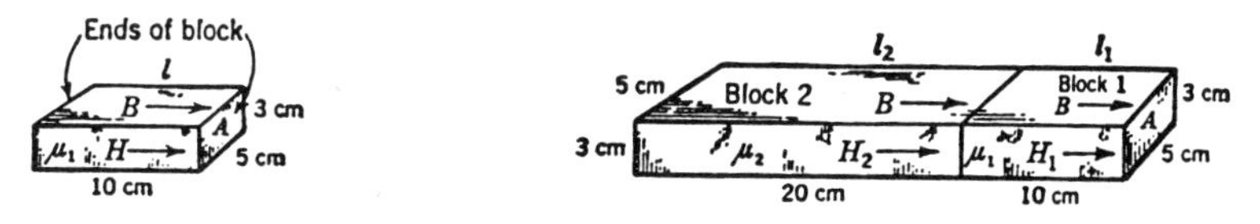

Fig. 1 Fig. 2

Solution: (a) The reluctance R_1 of block 1 is

$$R_1 = \frac{\ell_1}{\mu_1 A} = \frac{0.1}{500 \times 4\pi \times 10^{-7} \times 15 \times 10^{-4}}$$

$$= 1.06 \times 10^5 \text{ reciprocal henrys.}$$

The permeance P is the reciprocal of R; so

$$P_1 = \frac{1}{1.06 \times 10^5} = 9.4 \times 10^{-6} \text{ henry}$$

(b) The reluctance of block 2 is

$$R_2 = \frac{\ell_2}{\mu_2 A}$$

$$= \frac{0.2}{2{,}000 \times 4\pi \times 10^{-7} \times 15 \times 10^{-4}}$$

$$= 0.53 \times 10^5 \text{ reciprocal henrys.}$$

The total reluctance R_T equals the sum of the individual reluctances; so

$$R_T = R_1 + R_2$$

$$= (1.06 + 0.53) \times 10^5 = 1.59 \times 10^5 \text{ reciprocal henrys.}$$

The toal permeance

$$P_T = \frac{1}{R_T} = \frac{1}{1.59 \times 10^5} = 6.3 \times 10^{-6} \text{ henry.}$$

• PROBLEM 10-3

The electromagnet of the figure has an effective closed magnetic path length ℓ, a gap width of g, pole faces of area S, and is wound with n turns. Find the reluctance of the electromagnet and its inductance.

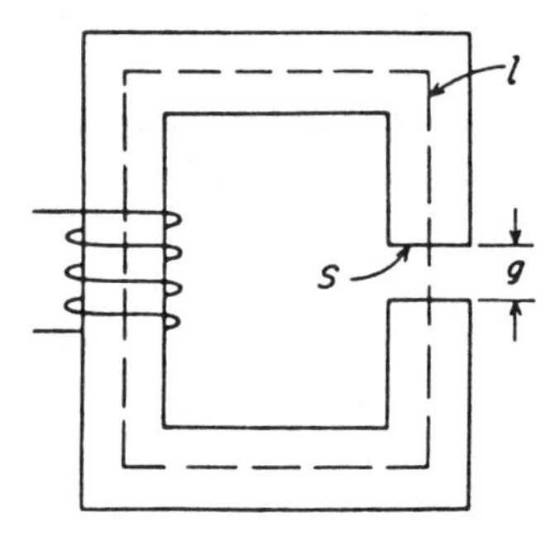

Solution: The reluctance of the core is by definition,

$$R_c = \frac{\ell - g}{K_M \mu_0 S}$$

and of the gap is

$$R_G = \frac{g}{\mu_0 S}$$

so that the total reluctance is

$$R = \frac{\ell - g}{K_M \mu_0 S} + \frac{g}{\mu_0 S}$$

Assume here that the gap width is small compared to the core dimensions, so that the lines of force do not spread out. In addition, assume that the permeability of iron is high, so that $K_M >> 1$, then

$$R = \frac{\ell/K_M + g}{\mu_0 S} = \frac{g}{\mu_0 S}$$

The flux is then

$$\Phi = \frac{nI}{R} = \frac{nI\mu_0 A}{g}$$

and the inductance of the magnet is

$$L = \frac{d\Phi}{dI} = \frac{n\mu_0 A}{g}$$

• PROBLEM 10-4

A toroidal iron core of square cross-section, with a 2 mm air gap and wound with 100 turns, has the dimensions shown. Assume the iron has the constant $\mu = 1000\,\mu_0$. Find (a) the reluctances of the iron path and the air gap and (b) the total flux in the circuit if I = 100 mA.

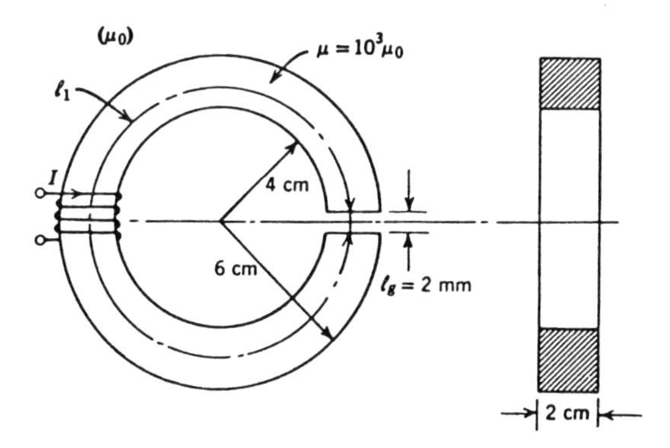

Solution: (a) The reluctance of the iron path, having a median length $\ell_1 \cong 2\pi(0.05) = 0.314$ m and cross-sectional area $A_1 = 4 \times 10^{-4}$ m^2, is

$$R_1 \cong \frac{\ell_1}{\mu_1 A_1} = \frac{0.314 - 0.002}{10^3(4\pi \times 10^{-7})4 \times 10^{-4}} = 0.621 \times 10^6 \text{ H}^{-1}$$

The air gap reluctance, assuming no fringing, becomes

$$R_g = \frac{\ell_g}{\mu_0 A_1} = \frac{0.002}{4\pi \times 10^{-7}(4 \times 10^{-4})} = 3.98 \times 10^6 \text{ H}^{-1}$$

(b) The magnetic flux is given by

$$\psi_m = \frac{nI}{R}$$

i.e., the magnetomotive force nI of the coil divided by the reluctance of the series circuit

$$\psi_m = \frac{nI}{R_1 + R_g} = \frac{10^2(0.1)}{4.6 \times 10^6} = 2.18 \times 10^{-6} \text{ Wb.}$$

With the air gap absent, ψ_m is limited only by the reluctance R_1 of the iron path, becoming $\psi_m = 15.97 \times 10^{-6}$ Wb.

• PROBLEM 10-5

The figure shows a magnetic circuit of square cross section with airgaps, which can be treated as a series parallel combination of reluctances. Find the reluctance of this circuit.

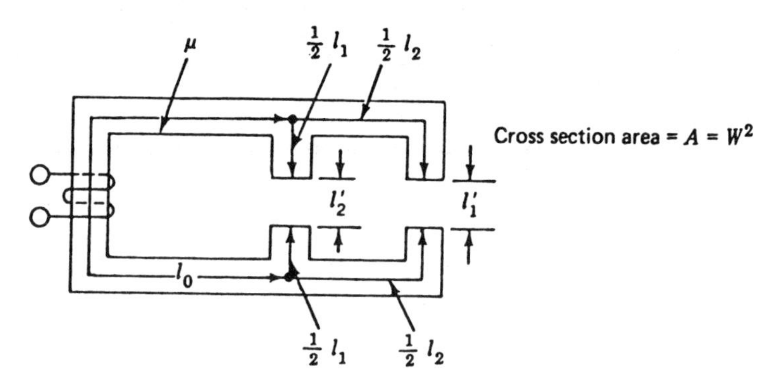

A magnetic circuit with series and parallel reluctance combinations.

Solution: Reluctance is given by

$$R = \frac{NI}{\Phi_S} = R_0 + \frac{(R_1 + R_1')(R_2 + R_2')}{R_1 + R_1' + R_2 + R_2'}$$

where

$$R_0 = \frac{\ell_0}{\mu A}, \quad R_1 = \frac{\ell_1}{\mu A}, \quad R_1' = \frac{\ell_1'}{\mu_0 A}$$

$$R_2 = \frac{\ell_2}{\mu A}, \quad R_2' = \frac{\ell_2'}{\mu_0 A}$$

Note that approximate mean lengths have been used. Because of the fringing at the airgaps the above approximations for R_1' and R_2' are very crude. A better approximation would be obtained if the following rule of thumb is used. Increase each airgap dimension by the length of the airgap, then

$$R_1' = \frac{\ell_1'}{\mu_0 (W + \ell_1')^2}$$

It is important to have a good approximation to the airgap reluctance because the airgap reluctance will often dominate the problem since μ/μ_0 may be of the order of 10^5 or greater.

In some cases, this effect is helpful. The airgap is a linear material, and the dominance of the airgap may make the relationship between Φ_S and I a linear one, even though the magnetic material may be highly nonlinear.

• PROBLEM 10-6

The figure shown illustrates the cross section of an electromagnetic circuit breaker which has circular symmetry about the center axis. The magnetic core and plunger are taken to have the same permeability μ. Let H_m represent the average magnitude of magnetic field intensity in the plunger and central section of the core. Find the reluctance of this magnetic circuit.

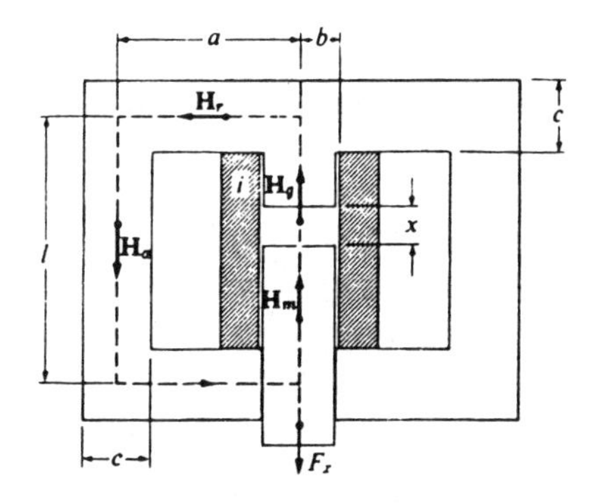

A circuit-breaker electromagnet.

Solution: Define an equivalent path length ℓ_e by means of the equation

$$H_m(\ell_e - x) + H_g x = H_m(\ell - x) + H_g x + \int_b^a H_r dr + H_a \ell$$
$$= iN, \tag{1}$$

which is obtained by applying Ampere's circuital law to the dashed path shown in the figure. The relationship between H_m, H_g, H_r and H_a is obtained from the equations expressing the continuity of magnetic flux:

$$\phi = \mu\pi b^2 H_m = \mu_0 A_g H_g = \mu_0 \pi(b + x)^2 H_g$$
$$= \mu 2\pi rcH_r = \mu\pi[(a + c/2)^2 - (a - c/2)^2]H_a$$
$$= \mu\pi 2acH_a.$$

Solving the above equations for H_g, H_r, and H_a in terms of H_m,

$$H_g = \mu_r\left(\frac{b}{b + x}\right)^2 H_m, \quad H_r = \frac{b^2}{2cr}H_m, \quad H_a = \frac{b^2}{2ac}H_m.$$

Substituting into Eq. (1) yields

$$\ell_e = \ell\left(1 + \frac{1}{2}\frac{b^2}{ac} + \frac{b^2}{2c\ell}\ln\frac{a}{b}\right).$$

The reluctance of the magnetic circuit is

$$R = \frac{\ell_e - x}{\mu\pi b^2} + \frac{x}{\mu_0\pi(b + x)^2}.$$

• PROBLEM 10-7

The figure shows a toroidal core consisting of two different magnetic materials. Find the reluctance of the circuit.

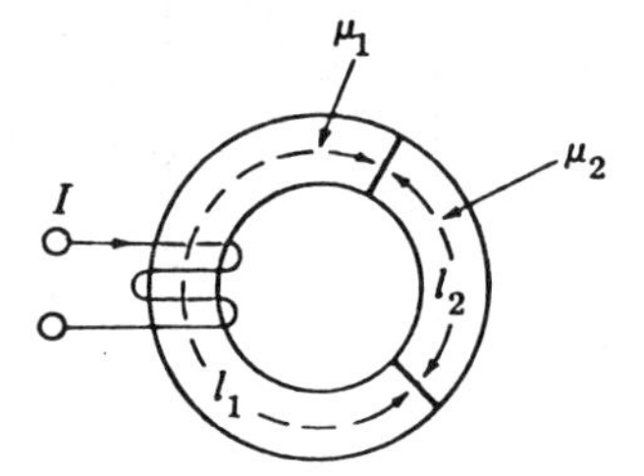

An inhomogeneous magnetic circuit.

Solution: Here use the continuity of normal B, that is, B and the flux are continuous at the boundary between the two materials, but H has a discontinuity.

$$NI = H_1\ell_1 + H_2\ell_2 = \frac{B\ell_1}{\mu_1} + \frac{B\ell_2}{\mu_2}$$

$$R = \frac{NI}{\Phi_s} = \frac{\ell_1}{\mu_1 A} + \frac{\ell_2}{\mu_2 A} = R_1 + R_2$$

where $R_1 = \frac{\ell_1}{\mu_1 A}$ and $R_2 = \frac{\ell_2}{\mu_2 A}$

• PROBLEM 10-8

A magnetic circuit has an air gap of nonuniform separation as suggested in Fig. 1a. The iron has a uniform depth d into the page of 1 meter. The geometry of the gap is identical with the region between ff and gg in the capacitor of fig. 2. Find the permeance of the air gap, neglecting fringing of the field.

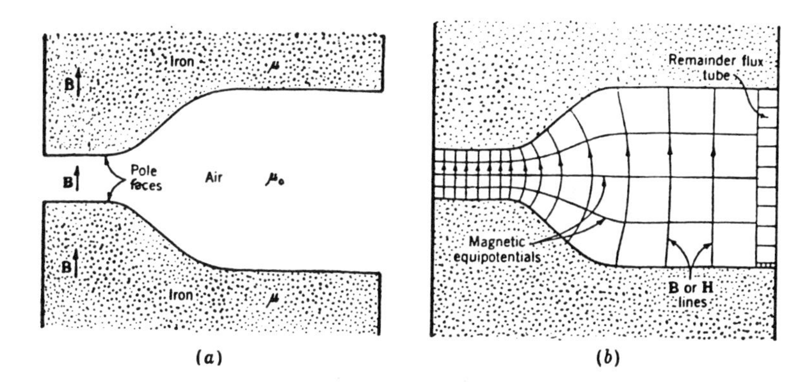

Magnetic field in air gap.

Fig.1

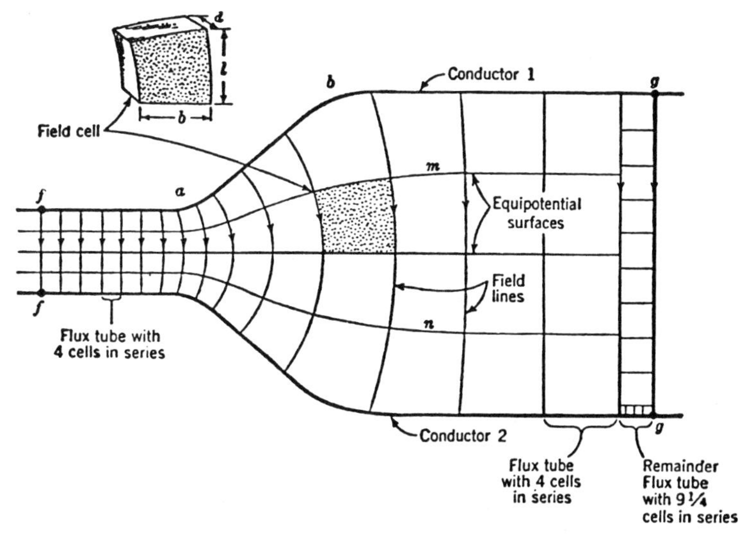

Cross section of two sheet conductors with completed field map. A three-dimensional view of a field cell is also given.

Fig.2

Solution: It may be assumed that the iron permeability is much greater than μ_0 so that the field lines in the gap will be perpendicular to the air-iron boundary, and this boundary can be treated as a magnetic equipotential. Since the geometry of the gap is the same as that for the capacitor in Fig. 2, the field map in Fig. 2 may also serve in the present case, noting that the field lines here are B or H lines and the equipotentials are surfaces of equal magnetic potential U as shown in Fig. 1b.

With the exception of the cells in the remainder flux tube all of the field cells are of the same kind, and the permeance of the air gap is given in terms of cells of the same kind by

$$P = \frac{N}{n} P_0$$

where N = number of field cells (or flux tubes) in parallel (dimensionless)

n = number of field cells in series (dimensionless)

P_0 = permeance of one cell (henrys)

The remainder flux tube has 9¼ cells in series, while the other flux tubes have 4. Hence the remainder tube is

$$\frac{4}{9\frac{1}{4}} = 0.43$$

of the width of a full tube, and N = 15 + 0.43 = 15.43. The total permeance of the gap is then

$$P_T = \frac{15.43}{4} P_0 = 3.86 P_0$$

Since the depth of each cell is 1 meter, the permeance of one cell is

$$P_0 = \mu_0 d = 1.26 \times 1 = 1.26 \ \mu h$$

and the total permeance is

$$P_T = 3.86 \times 1.26 = 4.86 \ \mu h$$

It is assumed in this example that there is no fringing of the field. For an actual gap there would be fringing at the edges, and the actual permeance of the gap would be somewhat larger than given above.

DETERMINATION OF AMPERE-TURNS

• PROBLEM 10-9

If a magnetic circuit contains an air gap 0.08 inch long with an effective cross-sectional area of 3 square inches, find the ampere-turns necessary to produce 300 kilolines across the gap. Refer to the figure.

Solution: All computations will be made in the MKS system of units. Find the reluctance from the defining equation:

$$R = \frac{\ell}{\mu_0 A},$$

$$R = \frac{0.08/39.37}{4\pi \times 10^{-7} \times [3/(39.37)^2]}$$

(since 1 meter = 39.37 inches and $\mu_0 = 4\pi \times 10^{-7}$)

$$= 8.45 \times 10^5 \text{ ampere-turns per weber,}$$

$$\phi = 300{,}000 \times 10^{-8} = 300 \times 10^{-5} \text{ webers.}$$

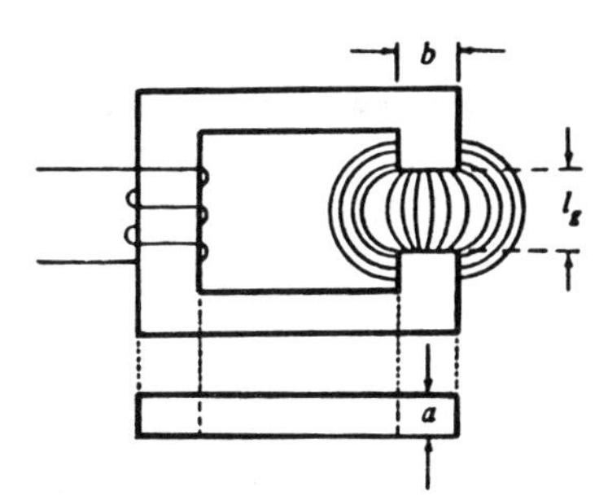

A magnetic circuit with an air gap.

Therefore, the mmf is

$$F = \phi R = 8.45 \times 10^5 \times 300 \times 10^{-5}$$

$$= 2535 \text{ ampere-turns.}$$

It is possible to use a second method, namely:

$$B = \frac{\phi}{A} = \frac{300 \times 10^{-5}}{[3/(39.37)^2]}$$

$$= 1.56 \text{ webers per square meter,}$$

whence

$$H = \frac{B}{\mu_0} = \frac{1.56}{4\pi \times 10^{-7}}$$

$$= 1.24 \times 10^6 \text{ amperes per meter.}$$

Hence

$$F = NI = H\ell$$

$$= 1.24 \times 10^6 \times \frac{0.08}{39.37}$$

$$= 2535 \text{ ampere-turns.}$$

• PROBLEM 10-10

A toroidal type of magnetic core of rectangular cross section has the following parameter values:

$$\ell_m = 20 \text{ cm}, \qquad A_m = (2 \text{ cm})(1 \text{ cm}) = 2 \text{ cm}^2,$$

$$\ell_g = 0.1 \text{ cm}, \qquad \mu_r = 4000.$$

Determine the ampere turns iN required to realize a magnetic flux density in the air gap of magnitude $B_g = 0.1$ Wb/m^2. The air-gap area is taken as $A_g = (2 + 0.1)(1 + 0.1)$ cm^2 = 2.31 cm^2.

Solution: The magnetic flux in the air gap is

$$\phi = B_g A_g = 2.31 \times 10^{-5} \text{ Wb}.$$

Since $\mu_0 = 4\pi \times 10^{-7}$, we find from $R = \frac{I}{\mu A}$ that the numerical value of the core reluctance is given by $R_m = 1.99 \times 10^5$. To eliminate the need for conversion factors, it is convenient to evaluate the ratio

$$\frac{R_g}{R_m} = \mu_r \left(\frac{A_m}{A_g}\right)\left(\frac{\ell_g}{\ell_m}\right) = 17.3.$$

The required number of ampere turns is given by the mmf:

$$v_m = \phi R_m (1 + R_g/R_m) = 84.1 \text{ ampere turns.}$$

Assume that the coil has 200 turns and that the inductance is to be found.

$$L = \phi N^2/iN = 1.1 \times 10^{-2} \text{ H}.$$

• PROBLEM 10-11

Find the number of ampere turns needed to provide a flux density of 1.5 T in the air gap of the magnet shown in Fig. 1, when $\ell_1 = 200$ mm, $\ell_2 = 160$ mm, $\ell_3 = 10$ mm, $A_1 = 0.0016$ m^2, $A_2 = 0.0008$ m^2. The steel used for the magnetic core is that whose magnetization curve is given in Fig. 2.

Solution: It will be assumed that the magnetic flux density is uniform across any cross-section of the core or the air gap. The total reluctance of the circuit is the sum of the reluctances of each section of the core in exactly the same way as the total resistance of an electric circuit is the sum of all the resistances in series in the circuit. Because the core of the circuit changes in cross-section, the flux density will be different in different parts of the circuit and because the magnetization curve is a non-linear relationship, the permeability will also be different in the different parts of the core having different cross-sections. The flux density in the arms of smaller cross-section is the same as that in the air gap. The total flux in the circuit is the product of the flux density and the cross-sectional area:

$$\Psi = 1.5 \times 0.0008 = 0.0012 \text{ Wb.}$$

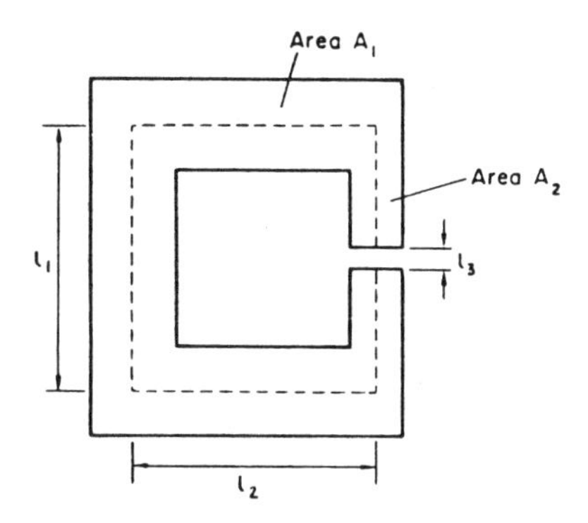

Fig.1

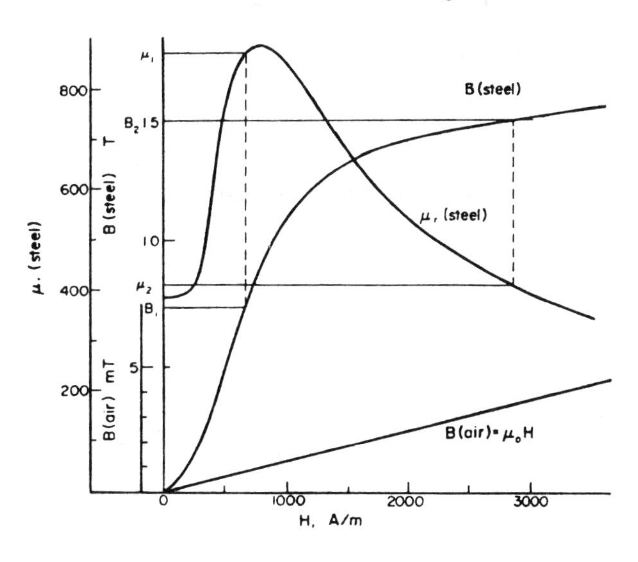

Fig.2

Therefore the flux density in the arms of larger cross-section is given by

$$B = \frac{\Psi}{A}$$

$$B = \frac{0.0012}{0.0016} = 0.75 \text{ T.}$$

It is necessary to turn to the magnetization curve for the steel in order to find the relative permeability of the steel at these flux densities. From the curve it is seen that the two permeabilities are given by

$$\mu_1 = 870\mu_0, \qquad \mu_2 = 410\mu_0 .$$

It will be noticed that the higher value of permeability is associated with the larger area of cross-section and smaller flux density. Therefore the total reluctance of the core and air gap is found from $R_m = \frac{\ell}{\mu A}$,

$$R_m = \frac{(\ell_1 + 2\ell_2)}{\mu_1 A_1} + \frac{(\ell_1 - \ell_3)}{\mu_2 A_2} + \frac{\ell_3}{\mu_0 A_2}$$

$$= \left(\frac{(200 + 320)}{870 \times 1.6} + \frac{(200 - 10)}{410 \times 0.8} + \frac{10}{0.8}\right)\frac{1}{\mu_0} = \frac{13.5}{\mu_0} .$$

The m.m.f. is given by

$$\text{m.m.f.} = NI = \Psi R_m = \frac{0.0012 \times 13.5}{4\pi \times 10^{-7}}$$

$$= 12{,}900 \text{ Ampere turns.}$$

• PROBLEM 10-12

Referring to the iron ring shown in the figure which has a cross-sectional area $A = 10\ \text{cm}^2$, an air gap of width $g = 2$ mm, and a mean length $\ell = 2\pi R = 60$ cm, including the air gap, find the number of ampere-turns required to produce a flux density $B = 1$ weber/meter^2.

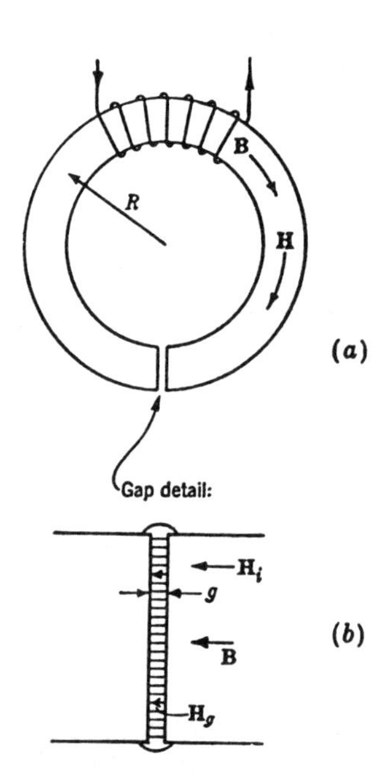

Iron ring with air gap.

Solution: From $\oint \bar{H} \cdot d\bar{\ell} = F = NI$

$$NI = \oint \bar{H} \cdot d\bar{\ell} = H_i(\ell - g) + H_g g \quad (1)$$

where H_i = H field in iron

H_g = H field in gap

From a B-H curve for the iron, H_i = 1,000 amp/meter, and from $\frac{H_g}{H_i} = \mu_r$, H_g is known in terms of H_i. Hence (1) becomes

$$NI = H_i[(\ell - g) + \mu_r g]$$

where μ_r = 795 = relative permeability of iron ring at B = 1 weber/meter^2. Therefore,

$$NI = 1{,}000[(0.6 - 0.002) + 795 \times 0.002]$$

$$= 2{,}188 \text{ amp-turns}$$

The introduction of the narrow air gap makes it necessary to increase the ampere-turns from 600 to 2,188 to maintain the flux density at 1 weber/meter^2.

The above problem may also be solved by calculating the total reluctance of the magnetic circuit. Thus, from (1) we have

$$NI = \frac{\mu A}{\mu A} H_i(\ell - g) + \frac{\mu_0 A}{\mu_0 A} H_g g$$

and

$$NI = BA(R_i + R_g)$$

where $R_i = (\ell - g)/\mu A$ = reluctance of iron part of circuit

$R_g = g/\mu_0 A$ = reluctance of air gap

• PROBLEM 10-13

Consider a transformer core that has a nonuniform cross section as shown in Fig. 1 and assume that the core is made of annealed sheet steel laminations. Find the number of ampere-turns that are required to establish a flux of 0.0006 weber.

Solution: Divide this circuit into two series parts: one very nearly 42 centimeters long with a cross section of 4 square centimeters, and the other 20 centimeters long with a 6-square-centimeter cross section. Changing the dimensions to inches, since the B-H curves shown in Fig. 2 are in English units, we get one magnetic circuit with a mean length of flux path of 16.5 inches and a cross-sectional area of 0.62 square inch, and a series circuit with a mean length of 7.9 inches and cross section of 0.93 square inch. The flux of 0.0006 weber is

$$\phi = 0.0006 \times 10^5 \text{ kilolines} = 60 \text{ kilolines.}$$

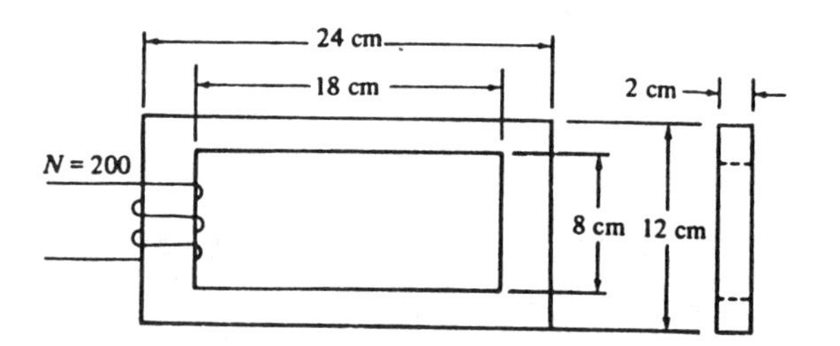

A transformer core with nonuniform cross section

Fig.1

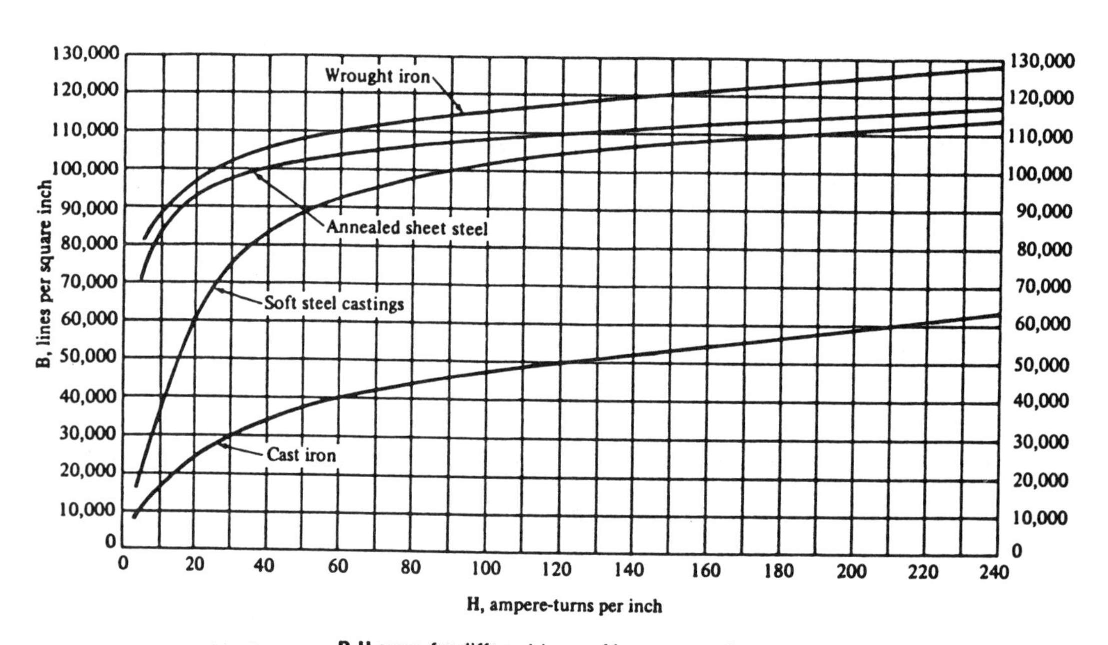

B-H curve for different types of iron commonly used in magnetic circuits

Fig.2

For a flux of 60 kilolines in the magnetic circuit the flux density in the first part is 60/0.62 or 96.8 kilolines per square inch, and in the second part 60/0.93 or 64.8 kilolines per square inch. From the B-H curve for annealed sheet steel the corresponding values of H are 28 ampere-turns per inch and 4 ampere-turns per inch, respectively. The magnetic potential drop for the first part of the magnetic circuit is the product of the magnetizing force by the mean length of the flux path of that circuit; that is,

$$H_1 \ell_1 = 28 \times 16.5$$

$$= 463 \text{ ampere-turns.}$$

The magnetic potential drop for the second part of the circuit is, similarly,

$$H_2 \ell_2 = 7.9 \times 4$$

$$= 32 \text{ ampere-turns.}$$

The mmf must be balanced by the total magnetic potential drop, or

$$F = H_1 \ell_1 + H_2 \ell_2$$

$$= 495 \text{ ampere-turns.}$$

Therefore, to establish a flux of 0.0006 weber in the transformer core requires an mmf of 495 ampere-turns, or 2.48 amperes in the 200 turns.

• PROBLEM 10-14

The dimensions of the magnetic core for a certain transformer are indicated in Figure 1:

Overall:	20 inches by 12 inches
Each window:	4 inches square
Core thickness:	2 inches
Core material:	Annealed sheet steel laminations

Find the mmf supplied to the center leg to establish a flux of 0.00856 weber.

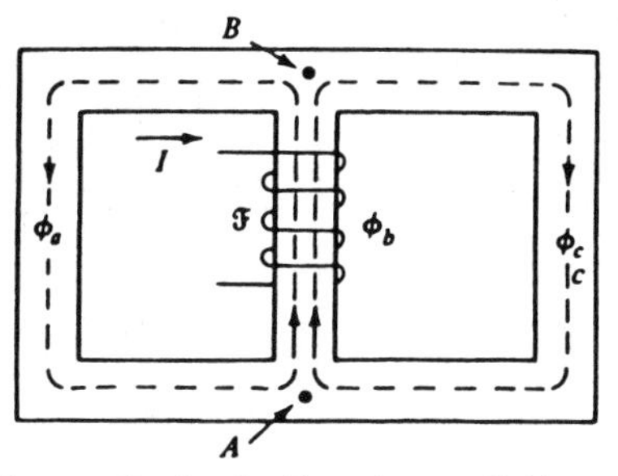

A magnetic circuit with series-parallel branches

Fig.1

Solution: In considering the mean lengths of the flux paths take these lengths as extending from B to A, midway of the core sections.

$$\ell_a = \ell_c = 24 \text{ inches},$$

$$\ell_b = 8 \text{ inches},$$

$$A_a = A_b = A_c = 8 \text{ square inches.}$$

Hence $B_b = 856/8 = 107$ kilolines per square inch, and H_b = 90 ampere-turns per inch from the B-H curve for annealed sheet steel.

From the geometry of the core

$$\phi_a = \phi_c = \frac{\phi_b}{2} = 428 \quad \text{kilolines},$$

$$B_a = B_c = \frac{428}{8} = 53.5 \text{ kilolines per square inch.}$$

Hence $H_a = H_c = 2.5$ ampere-turns per inch. Therefore the total mmf will be

$$F = H_a \ell_a + H_b \ell_b$$

$$= (2.5 \times 24) + (90 \times 8) = 770 \quad \text{ampere-turns.}$$

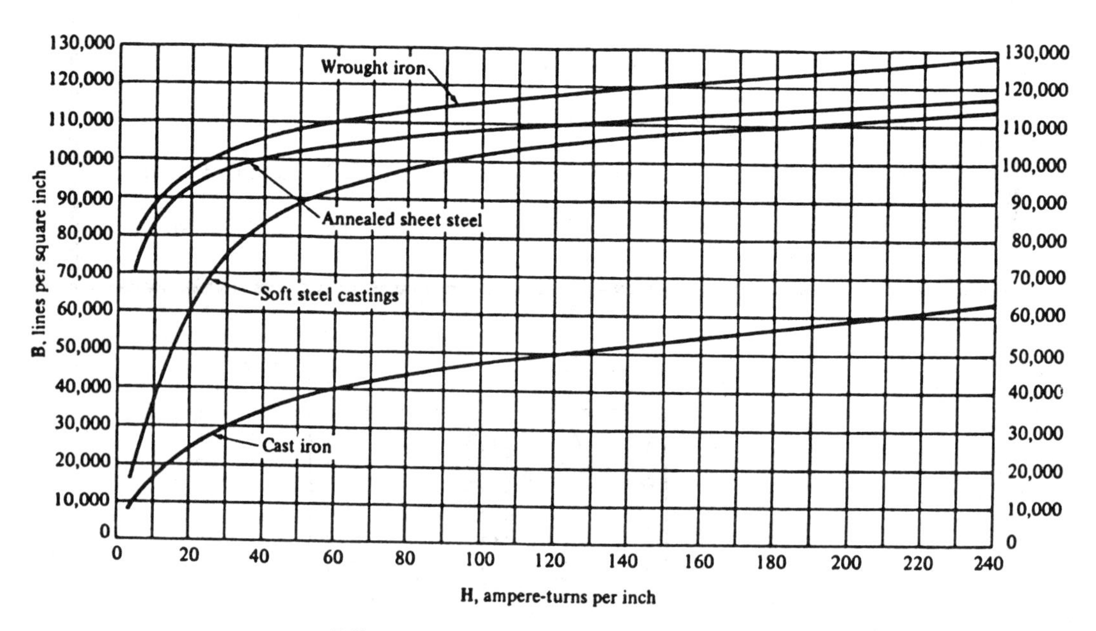

B-H curve for different types of iron commonly used in magnetic circuits

Fig.2

Assume that the magnetic circuit of figure 1 has an air gap 0.1 inch long in leg c. How many ampere-turns are necessary to establish the flux, 0.00856 weber, in leg b. Assume that the core material is made of annealed sheet steel laminations.

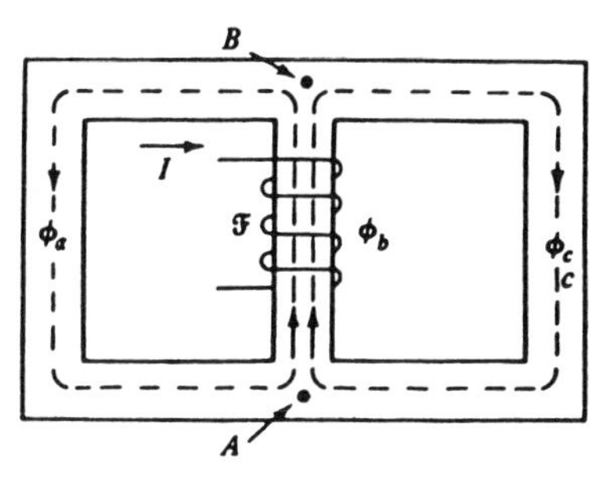

A magnetic circuit with series-parallel branches

Fig.1

Solution: Because the air gap requires an mmf of $H_{air}\ell_{air}$, the magnetic potential drop from B to A is now modified so that

$$U_{BA} = F - H_b\ell_b = H_a\ell_a$$

$$= H_c\ell_c + H_{air}\ell_{air}$$

or $$F = H_b\ell_b + H_c\ell_c + H_{air}\ell_{air}.$$

The flux density in leg b is $B_b = 856/8 = 107$ kilolines per square inch, and the corresponding magnetizing force is 90 ampere-turns per inch. The reluctance of the air gap is large compared to the reluctance of the iron, so the flux in leg c will decrease. Since it is not known how the flux, ϕ_b, will divide, assume a value and use a trial-and-error method. Therefore assume $\phi_c = 214$ kilolines. Hence

$$B_c = \frac{214}{8} = 26.6 \text{ kilolines per square inch.}$$

The corresponding value of the magnetizing force, from the B-H curve, is $H_c = 1.3$ ampere-turns per inch. The magnetic potential drop in the iron for leg c is

$$NI = H_c\ell_c = (1.3)(24) = 31.2 \text{ ampere-turns.}$$

The reluctance for the air gap is now found:

$$R = \frac{\ell}{\mu_0 A} = \frac{[0.1/39.37]}{(4 \times 10^{-7})[(2.1)(4.1)/(39.37)^2]}$$

$$= 3.62 \times 10^5 \text{ ampere-turns per weber.}$$

"Fringing" has been corrected for in the effective area used for the air-gap cross section. The magnetic potential drop across the gap is therefore

$$F = \phi_{air} R = 214 \times 10^{-5} \times 3.62 \times 10^{+5}$$

$$= 775 \text{ ampere-turns.}$$

But the magnetic potential drop from B to A is also

$$H_a \ell_a = H_c \ell_c + H_{air} \ell_{air}$$

$$= 31.2 + 775 = 806.2 \text{ ampere-turns.}$$

Therefore

$$H_a = \frac{806.2}{24} = 33.6 \text{ ampere-turns per inch.}$$

The corresponding value of flux density from the B-H curve is B_a = 79 kilolines per square inch, and

$$\phi_a = B_a A_a = 79 \times 8 = 632 \text{ kilolines.}$$

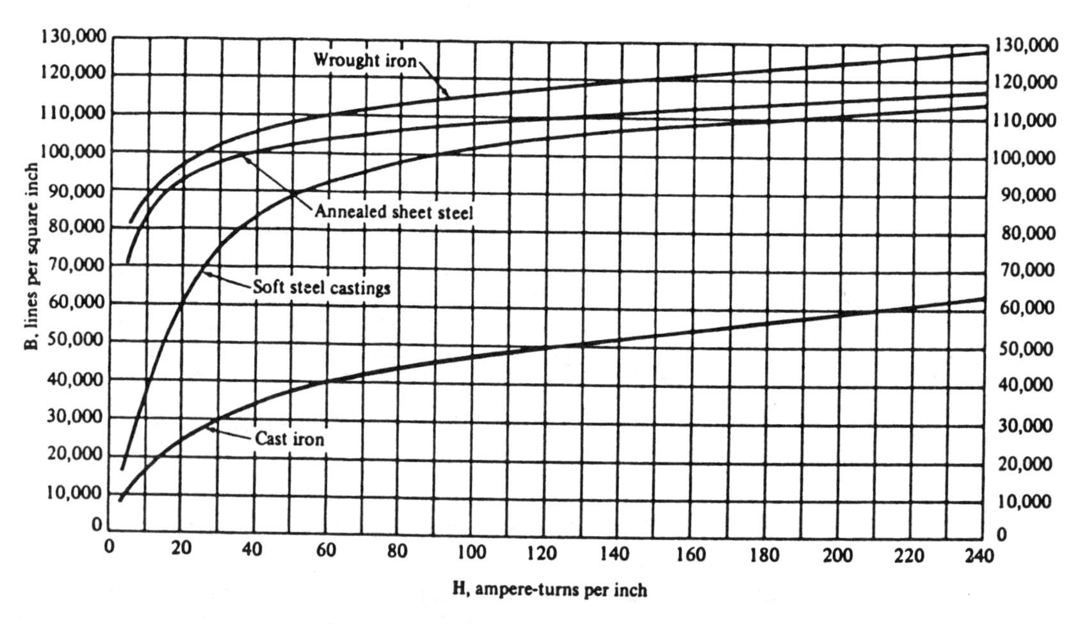

B-H curve for different types of iron commonly used in magnetic circuits

Fig.2

Now check the assumption:

$$\phi_b = \phi_a + \phi_c = 632 + 214 = 846 \text{ kilolines.}$$

The required flux is 846 kilolines.

The first assumption was a judicious guess--namely, that the flux set up in leg c with the air gap would be only one-fourth of the flux without the air gap. If the estimate had been too high or too low, now revise it and proceed as before. The answer is satisfactory from an engineering point of view.

The ampere-turns required are therefore

$$F = H_a \ell_a + H_b \ell_b = 806.2 + (90 \times 8)$$

$$= 1526.2 \text{ ampere-turns.}$$

Note that the total length of the magnetic circuit in leg c is 24 inches and requires only 31 ampere-turns to maintain the desired flux. Yet the air gap with a total length of 1/10 inch, as compared with 24 inches, required 775 ampere-turns (25 times as many) to maintain the same flux. This outcome justifies the assumption that in the "cut-and-try" method of solution for a magnetic circuit with an air gap, a first estimate of the flux density may be found by considering the magnetomotive force to be applied to the air gap alone.

• PROBLEM 10-16

Consider the magnetic circuit illustrated in Fig. 1. The magnetic field intensities and the average path lengths in the various core sections are designated by H_1, H_2, H_3 and l_1, l_2, and l_3, respectively.

(a) Find an equivalent circuit for the circuit in fig. 1 in terms of magnetic fluxes.

(b) Determine the ampere turns required to realize a magnetic flux density in the air gap of magnitude B_g = 0.2 Wb/m², given that

$$\ell_1 = 10 \text{ cm}, \quad A_1 = A_2 = A_3$$

$$\ell_2 = \ell_3 = 25 \text{ cm}, \quad A_1 = 6.25 \text{ cm}^2$$

$$\ell_g = 0.3 \text{ cm}, \quad \mu_r = 4000.$$

Solution: (a) Apply Ampere's circuital law to the two paths C_2 and C_3 shown in the figure and obtain

$$H_1 \ell_1 + H_g \ell_g + H_2 \ell_2 = v_m, \tag{1a}$$

$$H_1 \ell_1 + H_g \ell_g + H_3 \ell_3 = v_m, \tag{1b}$$

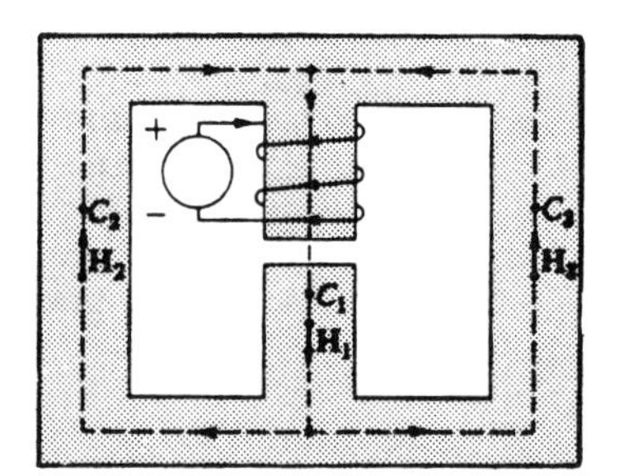

A magnetic circuit with parallel paths for the magnetic flux.

Fig.1

Each term on the left-hand sides of Eqs. (1) represents the magnetic potential difference for the corresponding part of the magnetic circuit. These potential differences will now be expressed in terms of the corresponding fluxes and reluctances. Let ϕ_1 represent the magnetic flux in the center section and ϕ_2 and ϕ_3 represent the magnetic fluxes in the respective outside legs. Then

$$H_k = \phi_1/\mu_k A_k, \qquad k = 1, g,$$

$$H_k = \phi_k/\mu_k A_k, \qquad k = 2, 3.$$

Substituting into Eqs. (1) yields

$$\phi_1 R_1 + \phi_2 R_2 = v_m, \tag{2a}$$

$$\phi_1 R_1 + \phi_3 R_3 = v_m, \tag{2b}$$

where

$$R_1 \triangleq \ell_1/\mu_1 A_1 + \ell_g/\mu_0 A_g, \qquad R_k \triangleq \ell_k/\mu_k A_k, \qquad k = 2,3.$$

Equation (2) in conjunction with the equation of continuity for the magnetic flux, $\phi_1 = \phi_2 = \phi_3$, results in the equivalent circuit illustrated in Fig. 2. Interpret Eqs. (2) by stating that the sum of the magnetic potential differences in a closed loop is equal to the applied mmf.

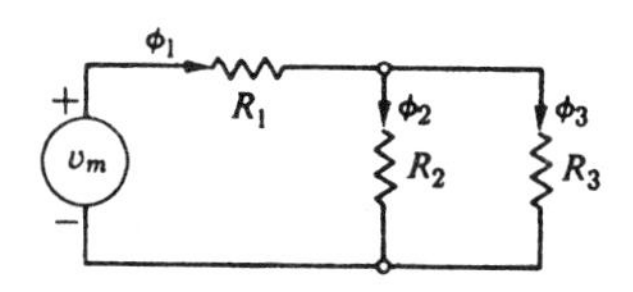

Fig.2

(b) The air-gap area is taken as

$$A_g = (2.5 + 0.3)(2.5 + 0.3)\text{cm}^2 = 7.84 \text{ cm}^2,$$

and the required flux in the center section is $\phi_1 = 1.57 \times 10^{-4}$ Wb. $R_2 = R_3 = 7.96 \times 10^4$ and

$$R_1 = \frac{\ell_1}{\ell_2}\left[1 + \mu_r\left(\frac{A_m}{A_g}\right)\left(\frac{\ell_g}{\ell_1}\right)\right]R_2 = 38.7\ R_2.$$

For the parallel branches the combined permeance is given by $P_c = P_2 + P_3$, and the reluctance is given by the reciprocal of P_c:

$$R_c = \frac{R_2 R_3}{R_2 + R_3} = 0.5\ R_2.$$

The total reluctance is

$$R = R_1 + R_c = 39.2\ R_2,$$

and the number of required ampere turns is given

$$v_m = \phi_1 R = 490 \text{ ampere turns.}$$

• PROBLEM 10-17

The electromagnet of the figure shown has two series-connected 1000-turn coils carrying a steady current I. The dimensions are indicated on the sketch, and the ferromagnetic material is cast iron, with B and H related by the expression

$$B = \frac{0.984H}{1457 + H} \tag{1}$$

Find the current I that produces an air-gap flux density of 0.5 weber/m^2.

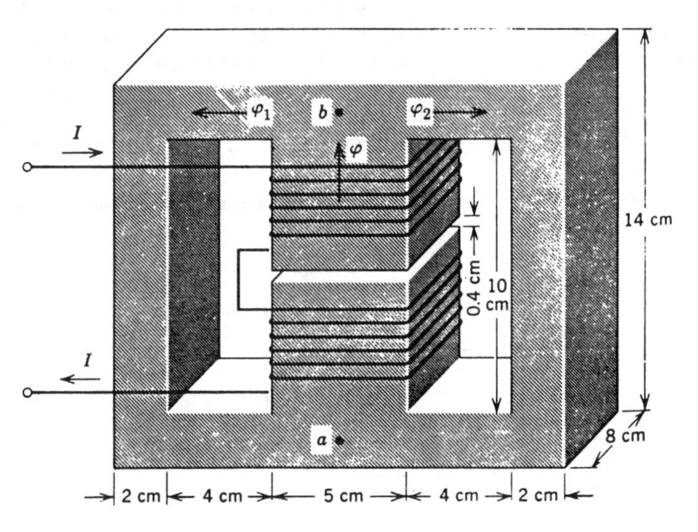

Cast-iron electromagnet.

Solution: In the air gap H is B/μ_0, or 398,000 amperes/m, and the mmf $H\ell$ of the gap is 1592 amperes. The air-gap reluctance $(1/\mu s)$ is 796,000.

In the center leg from a to b the flux density B is 0.5, and H is found from Eq. (1) to be 1505 amp/m. As the path length from a to b, not including the air gap, is 11.6 cm, the mmf $H\ell$ is 175. Thus the total mmf of the center leg from a to b is 1592 + 175, or 1767 amperes. The permeability B/H of the iron of the center leg is 0.000332 henry/m, and the reluctance including that of the air gap is 883,000.

As the magnetic flux density is a solenoidal field, the flux $\phi_1 + \phi_2$ leaving the junction at b equals the flux ϕ that enters the junction. Thus the flux of each side leg is $\frac{1}{2}\phi$, and $\phi_1 = \phi_2 = 0.001$ weber. The flux density is 0.625, and H is found from (1) to be 2537. As the path length is about 24 cm, the mmf from b to a along each side leg is 609 amperes. The permeability B/H is 0.000246, and the reluctance of each leg is 609,000.

Along the closed path of a flux line, the mmf is the sum of the 1767 ampere-turns of the center leg and the 609 ampere-turns of a side leg, or 2376 ampere-turns. This magnetomotive force is given by a current of 1.19 amperes in the 2000 turns of the two coils. Equation (1), which relates B and H, is a fair approximation of a limited range of the magnetization curve of cast iron.

• PROBLEM 10-18

A horse-shoe magnet is formed from a bar of wrought iron 2 ft. long with a section of $1\frac{1}{2}$ sq. in. and a magnetizing coil of 1500 turns is wound over the core. If the magnet is required to lift a load of 200 lb., find the minimum exciting current required, assuming that the two air-gaps of contact with the load are 1 mm. long and each has an area of $1\frac{1}{2}$ sq. in. Take the (relative) permeability of the wrought iron as 700 and the reluctance of the magnetic circuit through the load as 50% of the reluctance of the horse-shoe (excluding the air-gap). Refer to the given figure.

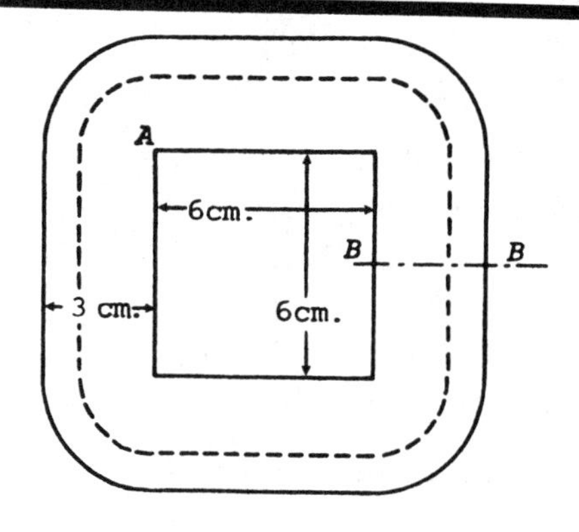

Solution: The force of attraction per gap = 100 lb. = $\frac{100}{0.2247}$ = 445 newtons.

The length of iron: $L = 2$ ft. $= \frac{2}{3.281} = 0.61$ metre.

Length of each gap: $g = 10^{-3}$ metre.

Area of iron and gaps: $A = 1\frac{1}{2}$ sq. in.

$= 9.68 \times 10^{-4}$ sq. metre.

Thus the force per gap, in newtons per sq. metre,

$$= \frac{445 \times 10^4}{9.68} = 4.6 \times 10^5 = \frac{B^2}{2\mu_0},$$

whence

$$B^2 = 1.55, \text{ and } B = 1.075 \text{ Wb/m}^2.$$

(Neglecting leakage, this is the density in both iron and gaps.) H, in the magnet,

$$= \frac{B}{\mu\mu_0} = \frac{1.075 \times 10^7}{4\pi \times 700} = 1220.$$

Thus the ampere-turns for the magnet-iron = 1220 x 0.61

= 746

The ampere-turns for the gaps = $\frac{2Bg}{\mu_0}$ = 1713

The ampere-turns for the load = 50% that for the iron

= 373

Total: 2832

$\therefore$ The exciting current $= \frac{2832}{1500} = 1.89$ amperes.

FLUX PRODUCED BY A GIVEN MMF

• PROBLEM 10-19

Figure 1 shows a toroidal core plus an airgap. The magnetic material consists of medium silicon steel whose characteristics are given in Fig. 2.

(a) Given Φ_S, determine NI, find the current required to establish a magnetic induction field B of 1.5 webers/m^2.

(b) Given NI, determine Φ_s, find the magnetic induction field resulting from 10^4 ampere turns.

(c) Find the magnetic induction field (flux density) resulting from 2×10^4 ampere turns.

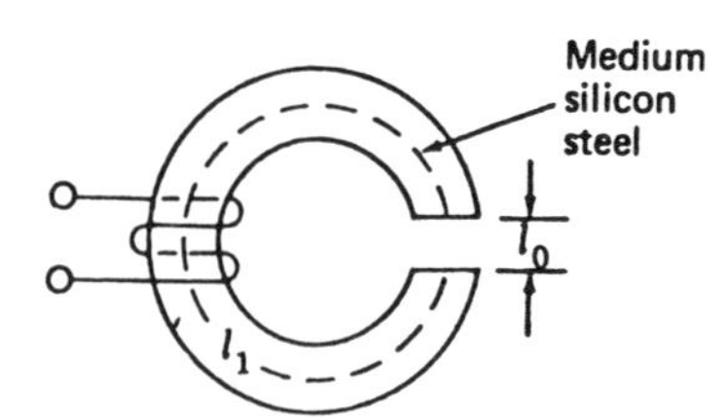

A nonlinear inhomogeneous magnetic circuit problem.

Fig.1

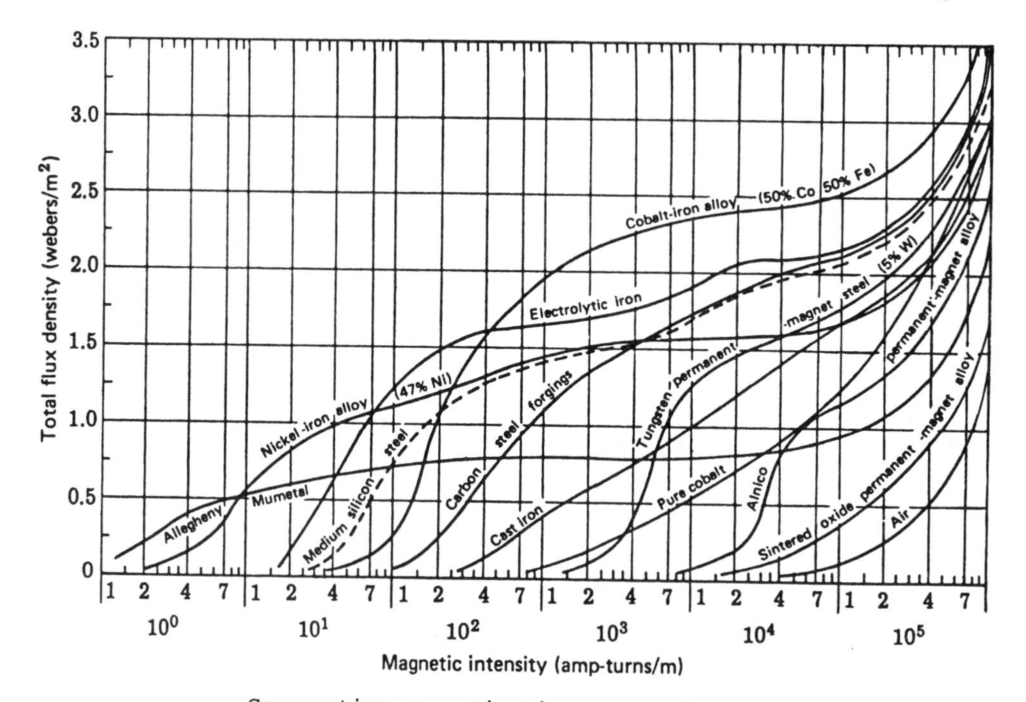

Comparative magnetization curves in MKS units.

Fig.2

Solution: (a) $B = 1.5$ webers/m^2, $H_1 = 3000$ (Fig. 2),

$$H_0 = \frac{B}{\mu_0} = 1.195 \times 10^6$$

$$NI = H_1 \ell_1 + H_0 \ell_0 = 14{,}195 \text{ ampere turns}$$

$$I = 14{,}195/N \text{ amperes}$$

(b) By a trial and error process:

First guess (assume total drop across airgap)

$$NI = 10^4 = H_0 \ell_0$$

$$H_0 = 10^6, \quad B = \mu_0 H_0 = 1.256, \quad H_1 = 230,$$

$$NI = \ell_1 H_1 + \ell_0 H_0 = 10,230 \text{ (close)}.$$

Second guess

$$B = 1.256 \times \frac{10,000}{10,230} = 1.227 \text{ webers/m}^2.$$

Note that (a) was a one-step procedure whereas (b) was a trial and error process, which can also be done in one step, using a graphical procedure. (c) A longer process of trial and error is necessary here (we are now well within the nonlinear region). A series of successive approximations yields

$$B = 1.64 \text{ webers/m}^2.$$

• PROBLEM 10-20

A thin toroid of permanently magnetized material has a uniform magnetic dipole moment (per unit volume) of strength M and oriented parallel to the center of the toroid, as shown

a) Find the direction and magnitude of $\bar{B}$ and $\bar{H}$ at point ① inside the toroid

b) A narrow gap of length ℓ is cut in the material (the material between the dashed lines in the figure is removed). Find B and H at point ① and at point ② in the gap.

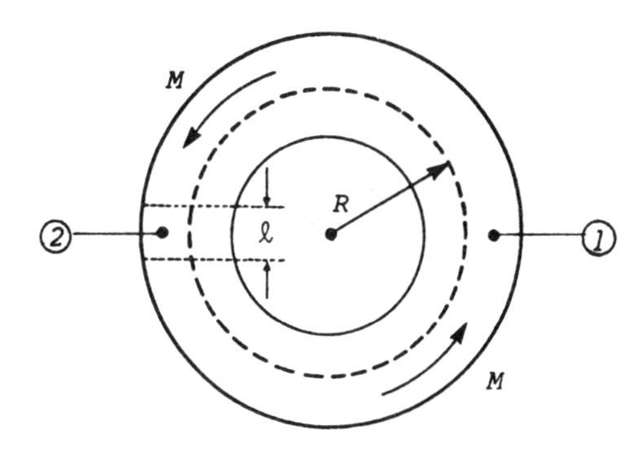

Solution: a) Without the air gap, apply Ampere's Law to find H.

$$\oint_C \bar{H} \cdot d\bar{r} = I_{\text{through curve}}$$

$$H\, 2\pi\, R = I = 0$$

$$\therefore \quad H = 0$$

$$\bar{H} = \frac{1}{\mu_0}\bar{B} - \bar{M}$$

$$\therefore \quad \bar{B} = \mu_0 (\bar{M} + \bar{H})$$

since $\bar{H} = 0$, $\bar{B} = \mu_0 \bar{M}$

b) Again applying Ampere's Law

$$\oint_C \bar{H} \cdot d\bar{r} = I_{\text{through curve}}.$$

$$\oint_C \bar{H}_g \cdot d\bar{r} + \oint \bar{H}_m \cdot d\bar{r} = 0$$

where $\bar{H}_g$ and $\bar{H}_m$ represent the magnetic field intensity in the air and in the magnetic material respectively.

dr for Hg is just the gap distance ℓ since it is small compared to the circumference of the circle and dr for Hm is just the circumference of the circle minus the gap distance ℓ or $2\pi r - \ell$.

Therefore,

$$B \frac{1}{\mu_0}\ell + \left(B \frac{1}{\mu_0} - M\right)(2\pi r - \ell) = 0$$

$$B \frac{1}{\mu_0}\ell + \frac{1}{\mu_0}B(2\pi r) - B \frac{1}{\mu_0}\ell - M2\pi r + M\ell = 0$$

$$\therefore \quad \frac{1}{\mu_0}B\ 2\pi r - M\ 2\pi r + M\ell = 0$$

Solving for B,

$$B = \frac{\mu_0 M\ 2\pi r - M\ell\mu_0}{2\pi r}$$

$$B = \mu_0 M \left(\frac{2\pi r - \ell}{2\pi r}\right)$$ for both the airgap and material.

$$H_g = B \frac{1}{\mu_0} = M\left(\frac{2\pi r - \ell}{2\pi r}\right)$$

$$H_m = B \frac{1}{\mu_0} - M = M\left(\frac{2\pi r - \ell}{2\pi r}\right) - M$$

$$= M\left(1 - \frac{\ell}{2\pi r}\right) - M = M - \frac{\ell M}{2\pi r} - M$$

$$\therefore \quad H_m = -\frac{M\ell}{2\pi r}$$

• PROBLEM 10-21

The toroid shown in the figure is 15.0 cm in mean radius, has an area of cross section of 3.00 cm^2, and the core is of soft iron. The toroid is uniformly wound with 900 closely spaced turns of wire. The iron core is demagnetized initially and, when the switch is closed sending a steady direct conduction current of 0.0292 amp through the wire, the ballistic galvanometer deflection shows a flux change of 10.5×10^{-6} wb.

a) What is the average flux density in the iron?

b) What is the average magnetic field intensity in the iron?

c) What is the permeability of the iron under these conditions?

d) What is its relative permeability?

e) What is the magnetic susceptibility?

f) What is its intensity of magnetization?

g) What is the linear current density?

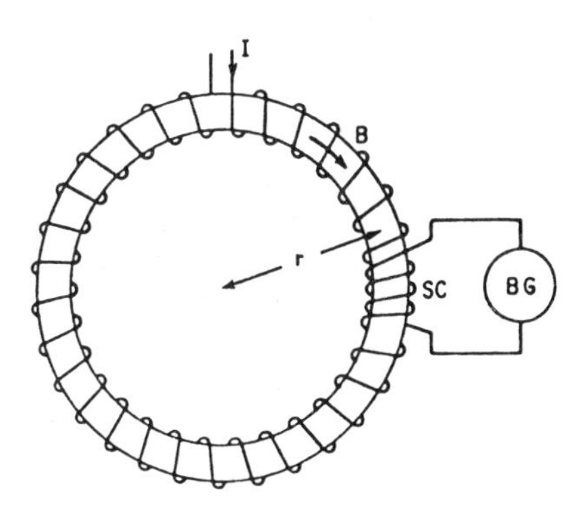

Solution: (a) Since the iron was initially demagnetized the change in flux is equal to the flux, so $B = \Phi/A = 3.50 \times 10^{-2}$ wb/m^2.

(b) For the toroid, H is determined only by the conduction current and geometry of the circuit and not at all by the properties of the core material. However, H is responsible for the magnetization produced in the core material, because H is due to the conduction current, and the conduction current sets up the magnetic field which magnetized the core material. $H = NI/2\pi r = 27.9$ amp.turns/m.

(c) Since B has been computed in (a) from the measured change in flux and H has been computed in (b) from the conduction current and geometry of the circuit, compute μ fi... μ = B/H = 1.26 x 10^{-3} wb/amp m, or μ = 1.26 x 10^{-3} henry/meter.

(d) $\mu_r = \mu/\mu_0 = 1000$.

(e) $\chi = \mu_r - 1 = 999$.

(f) $I = \chi H = 27{,}900$ amp/meter.

(g) $j_a = I = 27{,}900$ amp/m. Compare this with 27.9 amp/meter for the H due to the conduction current in the winding.

• PROBLEM 10-22

An air-core toroid has 500 turns, a cross-sectional area of 6 cm^2, a mean radius of 15 cm, and a coil current of 4A. Find the magnetic field of this toroidal coil using flux densities and check by Ampere's law.

Solution: The magnetic field is confined to the interior of the toroid, and if the closed path of the magnetic circuit is considered along the mean radius, 2,000 A-T are linked,

$$V_{m,source} = 2{,}000 \text{ A-T}$$

Although the field in the toroid is not quite uniform, assume that it is for all practical purposes and calculate the total reluctance of the circuit as

$$R = \frac{L}{\mu S} = \frac{2\pi 0.15}{4\pi 10^{-7} \times 6 \times 10^{-4}}$$

$$= 1.25 \times 10^{9} \text{ A-T/Wb}$$

Thus

$$\Phi = \frac{V_{m,s}}{R} = \frac{2{,}000}{1.25 \times 10^{9}} = 1.6 \times 10^{-6} \text{ Wb}$$

Hence

$$B = \frac{\Phi}{S} = \frac{1.6 \times 10^{-6}}{6 \times 10^{-4}} = 2.67 \times 10^{-3} \text{ Wb/m}^2$$

and finally,

$$H = \frac{B}{\mu} = \frac{2.67 \times 10^{-3}}{4\pi 10^{-7}} = 2{,}120 \text{ A-T/m.}$$

As a check, apply Ampere's circuital law directly in

this symmetrical problem,

$$H_\phi 2\pi r = NI$$

and obtain

$$H_\phi = \frac{NI}{2\pi r} = \frac{500 \times 4}{6.28 \times 0.15} = 2,120 \text{ A/m}$$

at the mean radius.

• PROBLEM 10-23

In the magnetic circuit of the given figure the average path lengths ℓ_1, ℓ_2, and ℓ_3 of the numbered regions are 10 cm, 20 cm, and 0.5 cm, respectively, and each cross-sectional area is 4 cm^2. The relative permeabilities of regions 1 and 2 are 1000 and 800, respectively. For 5000 ampere-turns determine the core flux ϕ and the mmf of each region. Also, find the energy stored in each region, assuming that the permeabilities are independent of the flux density.

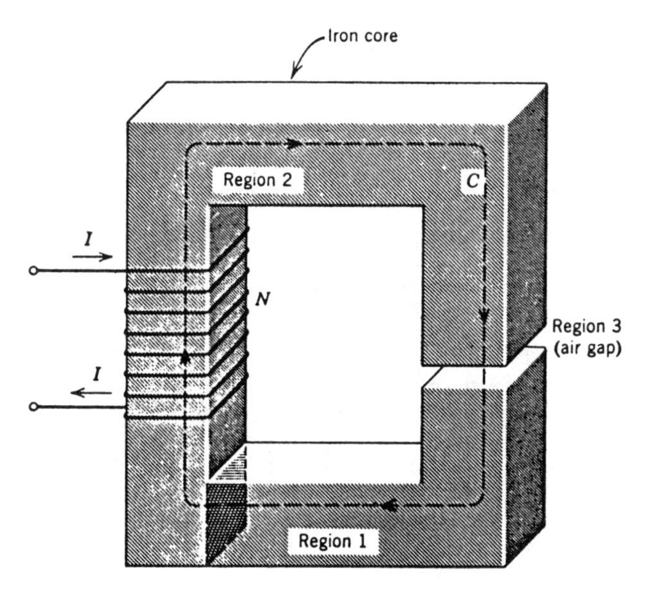

A magnetic circuit.

Solution: The reluctance $\ell_1/(\mu_1 S_1)$ of region 1 is 199,000 ampere-turns per weber, and that of region 2 is 498,000. In the air gap the relative permeability is unity, and even though the path length of the gap is only ½ cm, its reluctance is 9,950,000. Thus the total reluctance is 10,650,000. Dividing 5000 ampere-turns by this reluctance yields a core flux of 0.000469 weber, and the flux density is 1.17 weber/m^2.

The mmf of each region is determined from the product of

the flux and the reluctance of the region. In region 1 the mmf is 93 amperes, and in region 2 it is 234 amperes. The remaining 4673 amperes of the applied mmf appear across the air gap. Note that most of the ampere-turns of the coil are accounted for by the air gap. In problems of this type a rough approximation of the magnetic flux density can be quickly determined by assuming that the mmf of the air gap equals the mmf of the coil.

The energy stored in each region is $(\frac{1}{2}BH)(\ell S)$, for $\frac{1}{2}BH$ is the energy density of a linear medium and the product ℓS is the volume. As $BS = \phi$ and $H\ell$ = mmf, it follows that the stored energy is $\frac{1}{2}\phi$(mmf). The flux is 0.000469 and the mmf's of the regions are 93, 234, and 4673 amperes. Consequently, the stored energies are 0.022, 0.055, and 1.10 joules. Actually, iron is not a linear medium and, therefore, these calculated energies are only approximations. It is interesting to note that most of the energy is stored in the air gap.

• PROBLEM 10-24

(a) The figure shows a circuit with an air gap whose cross section is different from that of the soft-iron yoke. Each winding provides NI/2 ampere-turn. Find the magnetic flux.

(b) If there is a total of 10,000 turns in the two windings, and if I = 1.00 ampere, A_i = 100 centimeters^2, A_g = 50.0 centimeters^2, μ_r = 1,000, L_i = 90.0 centimeters, L_g = 1.00 centimeter, find Φ, B and L.

Solution: (a) This is a general law: reluctances and permeances in a magnetic circuit add in the same way as resistances and conductances in an electric circuit.

Assume that leakage flux is negligible. This will result in quite a large error, as can be seen later.

Applying again Ampere's circuital law to the circuit,

$$NI = H_i L_i + H_g L_g \approx H_g L_g,$$

where the subscript i refers to the iron yoke and g to the air gap. The requirement that the reluctance of the yoke be much smaller than that of the gap makes $H_i L_i << H_g L_g$. The path length L_i in the iron can be taken to be the length measured along the center of the cross section of the yoke.

Now, according to $\nabla \cdot \bar{B} = 0$, the net outward flux of $\bar{B}$ through any closed surface must be equal to zero.

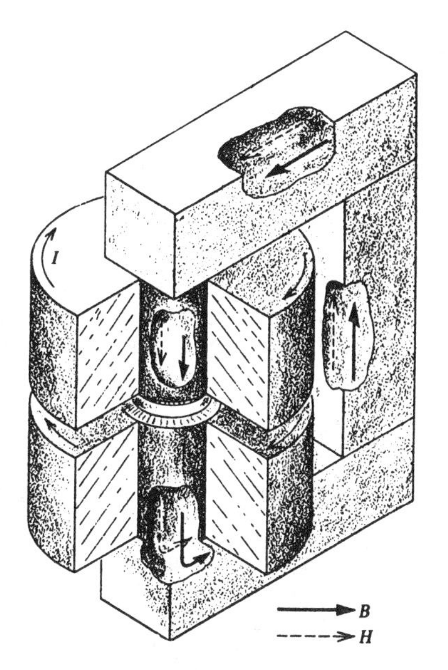

Electromagnet. The coils have been cut out to expose the iron yoke.

Thus, neglecting leakage flux, the flux of $\bar{B}$ must be the same over any cross section of the magnetic circuit and

$$B_i A_i = B_g A_g,$$

where A_i and A_g are respectively the cross sections of the iron yoke and of the air gap.

Combining these two equations,

$$B_g A_g \left[\frac{L_i}{\mu A_i} + \frac{L_g}{\mu_0 A_g} \right] = NI,$$

and the magnetic flux is

$$\Phi = B_g A_g = \frac{NI}{\dfrac{L_i}{\mu A_i} + \dfrac{L_g}{\mu_0 A_g}} \approx \frac{NI\, \mu_0 A_g}{L_g},$$

Here two reluctances are in series, $L_i/\mu A_i$ in the yoke, and $L_g/\mu_0 A_g$ in the gap.

(b) $$\Phi = \frac{10^4}{\dfrac{0.9}{10^3 \times 4\pi \times 10^{-7} \times 10^{-2}} + \dfrac{10^{-2}}{4\pi \times 10^{-7} \times 5 \times 10^{-3}}},$$

$$= 6.3 \times 10^{-3} \text{ weber},$$

$$B = \frac{6.3 \times 10^{-3}}{5 \times 10^{-3}} = 1.3 \text{ teslas} = 1.3 \times 10^4 \text{ gauss},$$

$$L = \frac{6.3 \times 10^{-3}}{1.00} = 6.3 \times 10^{-3} \text{ henry}$$

In this particular case the leakage flux is 70% of the flux in the gap. In other words, the magnetic induction in the gap is not 1.3 teslas, but only 1.3/1.7 = 0.77 tesla.

• PROBLEM 10-25

Figure 1 shows a magnetic circuit constructed of sheet steel. The cross sectionof the core is uniform and measures 2 x 2 in. Compute the number of ampere-turns required to produce a flux of 252 kilolines in the circuit. Assume a stacking factor of 0.9 for the laminations. Use the magnetization curves on fig. 2 and find the flux resulting from an mmf of 270 At.

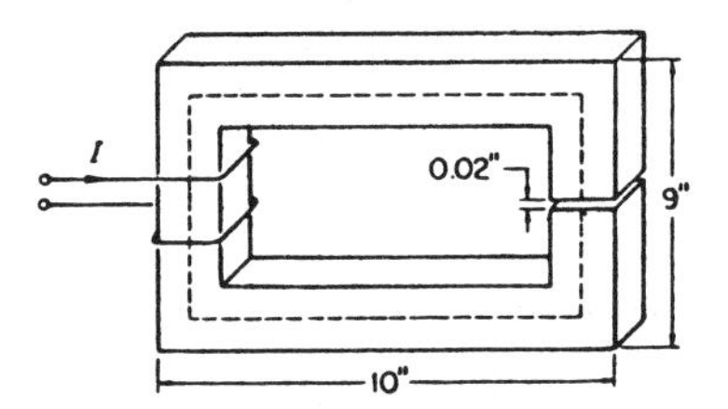

Fig.1

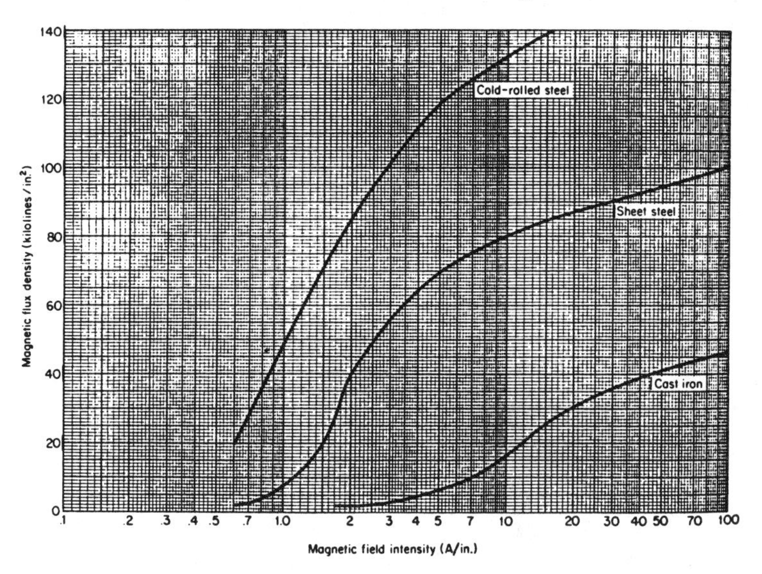

Typical magnetization curves.

Fig.2

Solution: This is a series circuit in which the total mmf is the sum of the mmf in the iron plus the mmf in the air gap. The details of the calculations are

Core:

$$\text{Area} = 2 \times 2 \times 0.9 = 3.6 \text{ in}^2$$

$$\text{Length} = 30 \text{ in.}$$

$$B = 252/3.6 = 70 \text{ kilolines/in.}^2$$

From the graph (Fig. 2):

$$H = 5 \text{ At/in.}$$

$$H\ell = 5 \times 30 = 150 \text{ At.}$$

Air gap:

$$\text{Area} = (2 + 0.02)^2 = 4.08 \text{ in.}^2$$

$$\text{Length} = 0.02 \text{ in.}$$

$$B = 252/4.08 = 61.8 \text{ kilolines/in.}^2$$

$$H = 313 \times 61.8 = 19{,}343 \text{ At/in.}$$

$$H\ell = 19{,}343 \times 0.02 = 387 \text{ At}$$

Therefore the total mmf drop in the magnetic circuit is 150 + 387 = 537 At. And in accordance with the mmf law, NI = 537 At.

The inverse problem of finding the flux resulting from 270 At is considerably more difficult. Because of the nonlinearity of the magnetization curve, it is impossible to deterine the flux by linear-analysis methods. The solution of the problem can be effected, however, by trial and error; that is, the flux in the core is assigned arbitrarily, and the mmf required to produce this flux is found by the previous method. If this mmf agrees with the specified ampere-turns, the solution of the problem is terminated. Otherwise the magnetic flux is assigned different values successively until one is found which requires the specified excitation. Accuracy to within 5 percent is acceptable.

A second method of solution is graphical, and will be illustrated later.

The first trial is such that the assumed flux falls within the range of the magnetization curves. In the problem at hand, a more accurate estimate of flux can be made with the aid of two extreme assumptions:

1. The entire mmf is used in the air gap.

2. The entire mmf is used in the core.

Either of these assumptions leads to a value of flux which is higher than the actual one. Therefore a value lower than either of these should be selected as a first trial.

If the specified 270-At mmf is used in its entirety in the air gap, then

$$H = 270/0.02 = 13{,}500 \text{ At/in.}$$

$$B = 13{,}500/313 = 43.1 \text{ kilolines/in.}^2$$

$$\Phi = 4.08 \times 43.1 = 176 \text{ kilolines}$$

If it is used in the core,

$$H = 270/30 = 90 \text{ At/in.}$$

$$B = 100 \text{ kilolines/in.}^2$$

$$\Phi = 100 \times 3.6 = 360 \text{ kilolines}$$

As a first trial take $\Phi = 170$ kilolines. Next compute the total mmf.

Core:

$$B = 170/3.6 = 47.2 \text{ kilolines/in.}^2$$

$$H = 2.3 \text{ At/in.}$$

$$H\ell = 2.3 \times 30 = 69 \text{ At.}$$

Air gap:

$$B = 170/4.08 = 41.6 \text{ kilolines/in.}^2$$

$$H = 313 \times 41.6 = 13{,}000 \text{ At/in.}$$

$$H\ell = 13{,}000 \times 0.02 = 260 \text{ At}$$

$$\therefore \text{ Total mmf} = 69 + 260 = 329 \text{ At}$$

Since the difference, 329 - 270 = 59, is higher than 5 percent of 270, a second, lower value of Φ must be assumed. Try $\Phi = 140$ kilolines. Then, as before,

Core:

$$B = 140/3.6 = 38.9 \text{ kilolines/in.}^2$$

$$H = 2 \text{ At/in.}$$

$$H\ell = 2 \times 30 = 60 \text{ At.}$$

Air gap:

$$B = 140/4.08 = 34.3 \text{ kilolines/in.}^2$$

$$H = 313 \times 34.3 = 10{,}700 \text{ At/in.}$$

$$H\ell = 10{,}700 \times 0.02 = 214 \text{ At}$$

$$\text{Total mmf} = 214 + 60 = 274 \text{ At.}$$

This value of the total mmf is sufficiently close to the given excitation. So the solution of the problem is now complete.

As previously stated, a direct graphical analysis requiring no guessing is also possible. Refer once again to Fig. 1. Ampere's circuital law requires that, around the magnetic circuit, the algebraic sum of the mmf drops be equal to zero, or

$$\text{mmf}_{core} + \text{mmf}_{air\ gap} = NI$$

From H = 313B (in free air),

$$\text{mmf}_{air\ gap} = H_{air\ gap}\ell_{air\ gap} = (313B_{air\ gap})(0.02)$$

The same flux exists in both the core and in the air gap. However, because of fringing, the flux density in the air is smaller than in the core. More precisely,

$$B_{air\ gap} = \frac{2 \times 2 \times 0.9}{(2 + 0.02)(2 + 0.02)} B_{core} = \frac{3.6}{4.08} B_{core}$$

So

$$\text{mmf}_{air\ gap} = 5.52B_{core}$$

On the other hand,

$$\text{mmf}_{core} = H_{core}\ell_{core} = 30H_{core}$$

Therefore the equation of equilibrium becomes

$$30H_{core} + 5.52B_{core} = 270$$

This is a constraint imposed by the magnetic circuit. A second constraint is imposed by the material of the core in the form of a nonlinear relationship, shown in Fig. 2. The variables of the problem must satisfy both conditions simultaneously. Hence the point of intersection of the proper magnetization curve with the plot of the last equation marks the values of H_{core} and B_{core}, which satisfy both constraints at the same time. If the magnetization curves were plotted on linear scales, the plot of the equation of equilibrium would be a straight line and, hence, easy to draw. In this case, the plot must be made point by point. The student can verify for himself that the two curves cross at B = 38.5 kilolines/in.^{2}, H = 1.95 At/in.

• **PROBLEM 10-26**

Consider the transformer core of annealed sheet steel laminations shown in Fig. 1. Determine the flux that will be set up in this magnetic circuit for a magnetomotive force of 1500 ampere-turns in the concentrated coil.

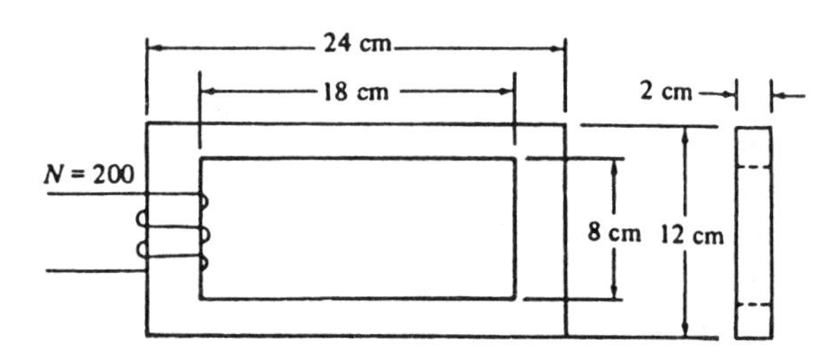

A transformer core with nonuniform cross section

Fig. 1

Solution: Proceed to subdivide the magnetic circuit into two parts, the first approximately 42 centimeters long and 4 square centimeters in cross section, the other 20 centimeters long and 6 square centimeters in cross section. Converting to English units, the first part of the series magnetic circuit has a mean flux-path length of 16.5 inches and a cross section of 0.62 square inch, and the second part has a mean flux-path length of 7.9 inches and a cross section of 0.93 square inch. Now apply the "cut-and-try" method which is summarized for the series-type magnetic circuit as follows:

1. Assume a total flux. Although the value of the flux selected at the first trial depends a great deal on experience, it is advisable not to assume a flux so high that the flux density in the part of the circuit with the smallest cross section is impossible. Also, if one part of the circuit, such as an air gap, will require more ampere-turns than all the other parts of the circuit combined, consider a solution as if this part were the only circuit in the problem.

2. Compute the flux densities for each part of the circuit based on the assumed value of flux.

3. From the B-H curves for the materials used find the corresponding magnetizing force.

4. Find the magnetic potential drop for each part of the circuit by multiplying the magnetizing force for each part by its corresponding mean length of path. Add these potential drops to find the total magnetic potential drop for the entire series circuit.

5. The total magnetic potential drop is then equated

to the mmf required to set up the assumed flux. This mmf is now compared with the available mmf, and a new estimate of the flux is based upon the difference between the computed and available ampere-turns.

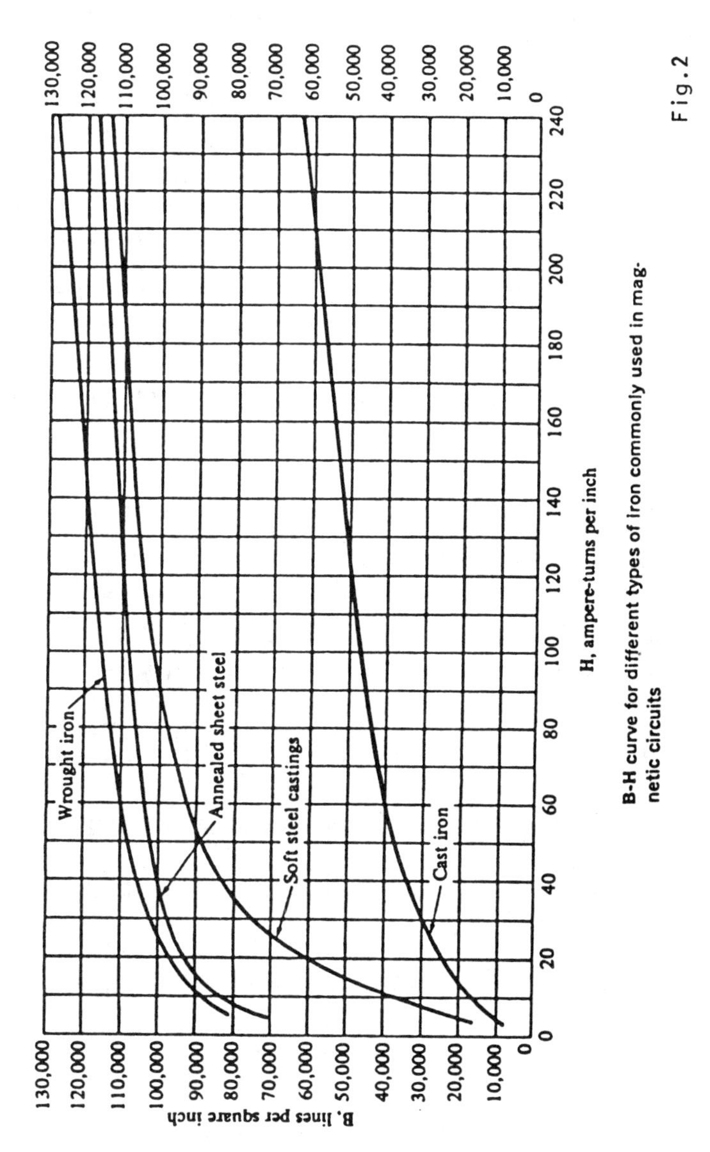

B-H curve for different types of iron commonly used in magnetic circuits

Fig. 2

6. The sequence is repeated with the revised estimate of the flux.

First make a guess at the total flux that will be produced, say 70 kilolines. This guess will enable the computing of the flux densities for each part of the magnetic circuit and, by means of the appropriate B-H curves, the corresponding magnetizing forces. These magnetizing forces are then multiplied by the respective mean lengths of flux path along which they are constant, yielding the magnetic potential drop for each part of the magnetic circuit. The total magnetic potential drop should be equal to the mmf. Usually the first guess will not check, but it will give an indication of a more appropriate guess for the second assumption.

Apply this method to illustrate the process. With the first assumption of 70 kilolines for the flux, the flux density in the first part of the magnetic circuit becomes 70/0.62 or 113 kilolines per square inch, and from the B-H curve for annealed sheet steel the corresponding value of the magnetizing force, H, is 175 ampere-turns

per square inch. The flux density in the second part of the circuit becomes 70/0.93 or 75.5 kilolines per square inch, for which the corresponding value of H is, from the B-H curve, 6.3 ampere-turns per inch. Therefore, the total magnetic potential drop in the magnetic circuit is

$$H_1\ell_1 + H_2\ell_2 = (175)(16.5) + (6.3)(7.9)$$

$$= 2880 + 49.5 = 2929.5 \quad \text{ampere-turns.}$$

The estimate of 70 kilolines is much too high, since such a flux requires nearly 3000 ampere-turns and only 1500 is available. Therefore the first estimate is revised. However, the greatest magnetic potential drop is along the 16.5-inch mean length, and that, moreover, the B-H curve in the region near 113 kilolines per square inch is extremely flat, so that a small reduction in flux density means a very large reduction in magnetizing force. Therefore make a more appropriate guess of 66 kilolines for the flux for the second estimate. Then

$$B_1 = \frac{66}{0.62} = 106 \quad \text{kilolines per square inch,}$$

and the corresponding H_1, from the B-H curve, is H_1 = 80 ampere-turns.

$$B_2 = \frac{66}{0.93} = 71 \quad \text{kilolines per square inch,}$$

and the corresponding H_2, from the B-H curve, is H_2 = 5.5 ampere-turns.

$$F = NI = 80 \times 16.5 + 7.9 \times 5$$

$$= 1320 + 39.5 = 1359.5 \quad \text{ampere-turns.}$$

The available ampere-turns is 1500, and a flux of 66 kilolines requires 1360 ampereturns. Now try 66.5 kilolines for the flux, with a new value for B_1 = 107 kilolines per square inch and a corresponding magnetizing force of H_1 = 90 ampere-turns per inch, resulting in a magnetic potential drop of 90 x 16.5 = 1480 ampere-turns. The value B_2 is now 71.5 kilolines per square inch, having a corresponding magnetizing force of H_2 = 5.1 ampere-turns per inch. Thus the total magnetic potential drop is

$$NI = H_1 \ell_1 + H_2 \ell_2$$

$$= 1480 + 40 = 1520 \quad \text{ampere-turns.}$$

Thus the required flux is greater than 66 kilolines but less than 66.5 kilolines and very nearly equal to 66.5 kilolines. For engineering purposes the answer of 66.5 kilolines is quite good, especially since the hysteresis in the iron will cause an uncertainty greater than the difference between 66 and 66.5 kilolines.

Try No.	ϕ	B_1 $\frac{\phi}{0.62}$	B_2 $\frac{\phi}{0.93}$	H_1 (B-H Curve)	H_2 (B-H Curve)	$(NI)_1$ $16.5\ H_1$	$(NI)_2$ $7.9\ H_2$	Total (NI)
1	70	113	75.5	175	6.3	2880	49.5	2929.5
2	66	106	71	80	5.0	1320	39.5	1360
3	66.5	107	71.5	90	5.5	1480	40.0	1520

In any such trial-and-error method it is advisable to keep an orderly solution in order to avoid retracing any steps that may have gone before. A tabular form for the results of the calculation by the "cut-and-try" method will greatly facilitate the computation.

SELF AND MUTUAL INDUCTANCE

• PROBLEM 10-27

Obtain an expression for the internal inductance of a long wire of circular cross-section. Refer to the figure given.

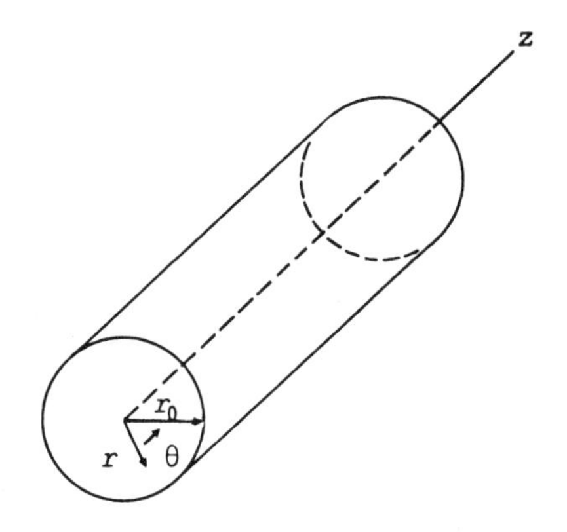

Solution: The inductance is defined as the flux linkages divided by the current causing the flux, i.e.

$$L = \frac{\Lambda}{I} \tag{1}$$

and the number of flux linkages are equal to the magnetic flux Φ, multiplied by the number of times the flux is linked, or

$$\Lambda = N\Phi \tag{2}$$

The flux therefore must be obtained, which can be found from the flux density.

By Ampere's law, the flux density insidethe conductor is found from

$$\oint_{\ell} \bar{H} \cdot d\bar{\ell} = I_{int} = \int_{\ell} \frac{1}{\mu} \bar{B} \cdot d\bar{\ell}$$

Because of symmetry B is independent of I. Therefore

$$\int_{\ell} \frac{1}{\mu} \bar{B} \cdot d\bar{\ell} = \frac{B}{\mu} \int_0^{2\pi r} d\ell = B \frac{2\pi r}{\mu}$$

If the cross-section is small enough to consider the current density J_z = constant, then

$$I_{int} = J_z \; \pi r^2 = I_{total} \left(\frac{r}{r_0}\right)^2$$

since the inner current is equal to the fraction r/r_0 of the total current enclosed by the inner path of radius r.

Therefore, equating (1) to (2)

$$\frac{B}{\mu} 2\pi r = I\left(\frac{r}{r_0}\right)^2$$

giving

$$B = \frac{\mu I}{2\pi r_0^2} \; r.$$

Again, the flux of $d\Phi$ per unit length of conductor is

$$d\Phi = B \; dr(1) = \frac{\mu I}{2\pi} \frac{r}{r_0^2} \; dr.$$

This element of flux is linked only with the inner current I_{int}, which is a $(r/r_0)^2$ part of the total current.

Therefore the element of flux linkage $d\Lambda$ is given by

$$d\Lambda = d\Phi \frac{r^2}{r_0^2} = \frac{\mu I r^3}{2\pi r_0^4} \; dr$$

so that

$$\Lambda = \frac{\mu I}{2\pi r_0^4} \int_0^{r_0} r^3 dr = \frac{\mu I}{8\pi}$$

and (by definition)

$$L_{int} \quad \frac{\Lambda}{I} = \frac{\mu}{8\pi} \; .$$

• PROBLEM 10-28

Find the self-inductance of a long solenoid whose cross-section is shown in the figure.

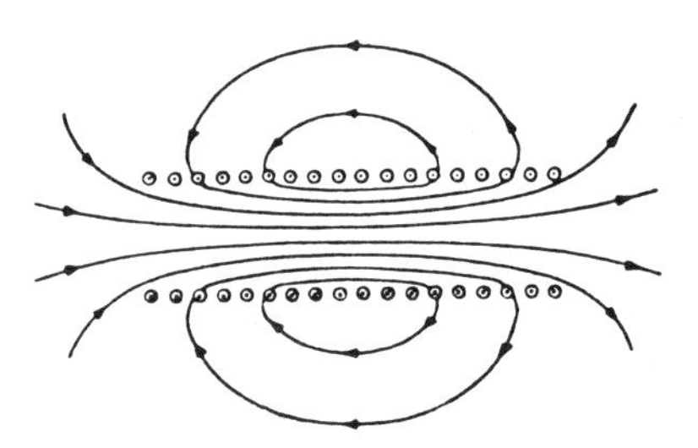

A cross section of a solenoid and its magnetic field.

Solution: The magnetic induction inside a long solenoid, neglecting end effects, is constant, and

$$B = \mu_0 N'I,$$

where N' is the number of turns/meter. Thus

$$\Phi = \frac{\mu_0 NI}{\ell}\pi R^2,$$

where N is the total number of turns, ℓ is the length of the solenoid, and R is its radius. Then

$$L = \frac{N\Phi}{I} = \frac{\mu_0 N^2}{\ell}\pi R^2,$$

$$= \mu_0 N'^2 \ \ell \ \pi R^2.$$

• PROBLEM 10-29

A toroid of length ℓ and rectangular cross-section is shown below. Derive the expression for the inductance in terms of r_2 and r_1. If a very long solenoid is considered, compute the inductance offered by this solenoid. Assume $\mu > \mu_0$.

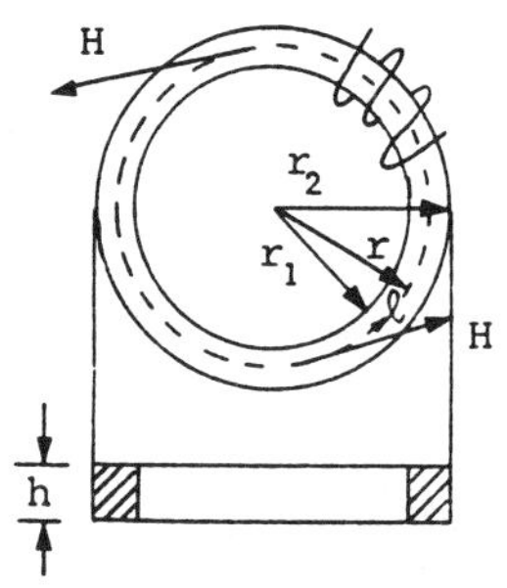

Solution: Now consider a circular loop at a distance of r from the center and let there be a small element of length $d\ell$, now applying Ampere's law:

$$\oint \overline{H} \cdot d\overline{\ell} = H\oint d\ell = H \cdot 2\pi \cdot r = N \cdot I$$

$$\therefore \quad H = \frac{N \cdot I}{2\pi \cdot r}$$

so

$$B = \frac{\mu NI}{2\pi \cdot r}$$

The flux inside the toroid is:

$$\phi = BA = \int_{r_1}^{r_2} B \cdot h \cdot dr = \int_{r_1}^{r_2} \frac{\mu NI \cdot h \cdot dr}{2\pi r}$$ where A = cross-section area

$$= \frac{\mu hNI}{2\pi} \cdot \ln \frac{r_2}{r_1}$$

The flux linkages will be

$$\lambda = N\phi = \frac{\mu hN^2 \cdot I}{2\pi} \cdot \ln \frac{r_2}{r_1}$$

Hence,

$$L = \frac{\lambda}{I} = \frac{\mu hN^2}{2\pi} \cdot \text{In} \frac{r_2}{r_1}$$

If the mean length of toroid equal to ℓ and if the field inside the toroid is uniformly distributed,

$$H = \frac{N \cdot I}{\ell}$$ (i.e., the flux is restricted effectively inside the coil.)

Then, $$\phi = B \cdot A = \mu HA = \frac{\mu N \cdot I \cdot A}{\ell}$$

Hence,

$$L = \frac{\lambda}{I} = \frac{N\phi}{I} = \frac{\mu N^2 A}{\ell}$$

Now, a long solenoid can be considered as a toroid of infinite radius, and hence

$$L = \frac{\mu N^2 \cdot A}{\ell}$$

• PROBLEM 10-30

If the mean radius of a toroidal coil is 10 cm and it has 1500 turns, find the average self-inductance (a) with an air core (b) with an iron core having an average relative incremental permeability of 100.

Solution: (a) The self-inductance is given by

$$L = \frac{N^2 \mu A}{\ell}$$

where for this problem $N = 1500$ turns

$\mu = 4\pi \times 10^{-7}$ henry/meter (air core)

$A = \pi \times (10 \times 10^{-2})^2$

$$\ell = 2\pi \times (10 \times 10^{-2})$$

Therefore

$$L = \frac{(1500)^2 \times 4\pi \times 10^{-7} \times \pi \times (10 \times 10^{-2})^2}{2\pi \times (10 \times 10^{-2})}$$

$$= 141.37 \text{ mH.}$$

(b) Here $\mu = 100 \times 4\pi \times 10^{-7}$ henry/meter.

Therfore

$$L = \frac{(1500)^2 \times 100 \times 4\pi \times 10^{-7} \times \pi \times (10 \times 10^{-2})^2}{2\pi \times (10 \times 10^{-2})}$$

$$= 14.137 \text{ Henrys.}$$

• PROBLEM 10-31

N turns of wire carrying current I are wound around a toroidal form having rectangular cross-section. The magnetic field $B = \phi_0 B(r)$ is in the circumferential direction as shown in Fig. 1. Fig. 2 shows a cross-section of Fig. 1.

a) Using Ampere's law, find B(r) within the coil

b) Find the flux

$$\phi = \iint_{\substack{\text{cross}\\\text{section}}} B(r)\, dA$$

and the inductance $L = N\phi/I$

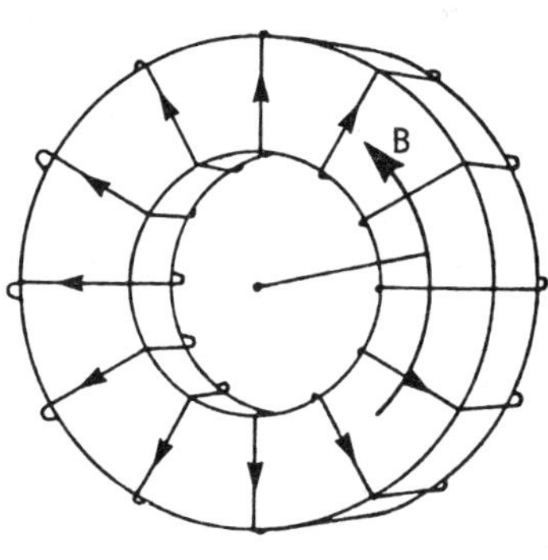

Fig.1

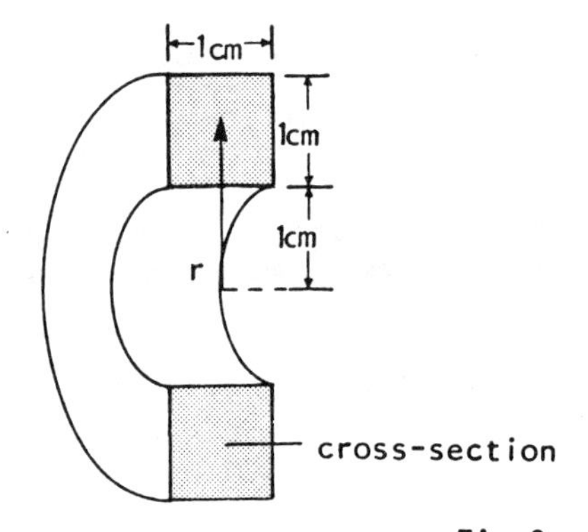

Fig.2

Solution:

a) Using Ampere's Law

$$\oint \bar{B} \cdot d\bar{r} = \mu_0 \text{ I enclosed}$$

The current through curve C is just the number of turns N times the current through the wire I.

$$\oint \bar{B} \cdot d\bar{r} = \mu_0 NI$$

$$2\pi r \; B = \mu_0 NI$$

$$\therefore B = \frac{\mu_0 NI}{2\pi r}$$

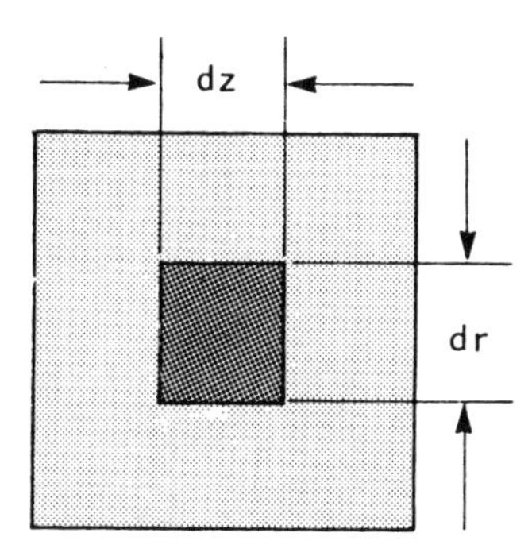

Fig.3

b) Fig. 3 shows the cross-section needed to find the flux. From the diagram the following is obtained,

$$\phi = \int_{0.01}^{0.02} \int_{0.0}^{0.01} \frac{\mu_0 NI}{2\pi r} \, dz \, dr$$

$$\phi = .010 \frac{\mu_0 NI}{2\pi} \ln r \Bigg|_{0.01}^{0.02} = 10^{-2} \frac{\mu_0 NI}{2\pi} \ln 2$$

$$L = \frac{N\phi}{I} = \frac{N \; 10^{-2} \; \mu_0 N}{2\pi} \ln 2$$

$$L = \frac{10^{-2} \; \mu_0 N^2}{2\pi} \ln 2.$$

• PROBLEM 10-32

As shown in Fig. 1 two conducting strips extend normal to the page with a sheet of steady current flowing outward on the upper strip and an equal current flowing inward on the lower strip. The medium in which the strips are located is air. Neglect edge effects. Find the inductance of a 1-meter length of the line.

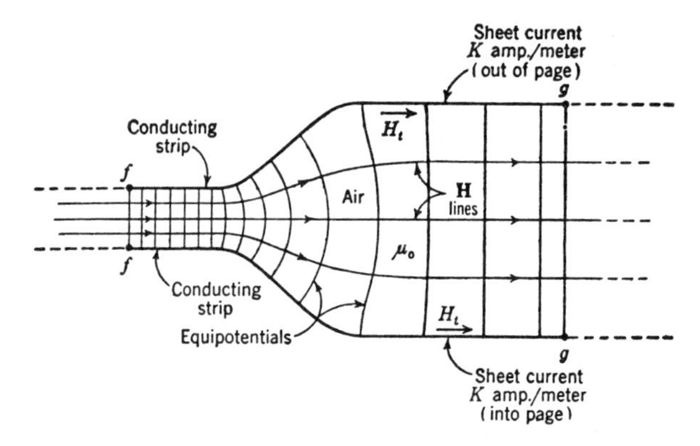

Cross section of strip transmission line.

Fig.1

Solution: Neglecting edge effects, the field map between the strips is identical with that for the iron circuit in Fig. 2.

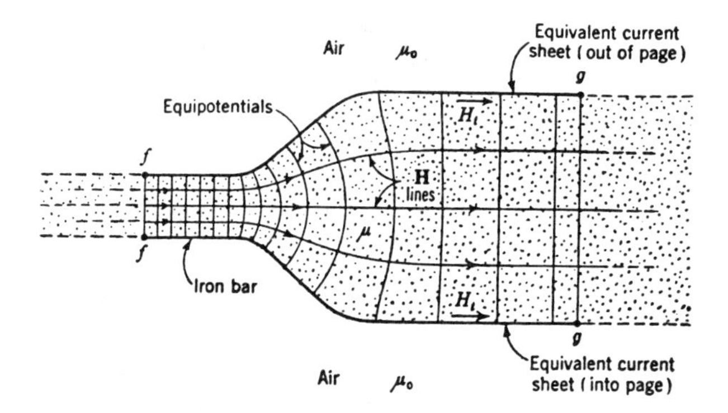

Iron bar of nonuniform cross section with internal field.

Fig.2

If the conducting strips are extended an infinite distance to the left and right, as suggested by the dashed lines in Fig. 1, the field configuration is precisely as indicated. The field between the strips is produced by the currents on the strips. The field in the iron may be regarded as due to an equivalent current sheet at the surfaces of the iron bar normal to the page (Fig. 2).

If each cell in the map is regarded as a strip transmission line with sheet currents along its upper and lower surfaces, the inducatance L_0 for a length d of 1 meter of the single-cell line (normal to the page in Fig. 1) is given by

$$L_0 = \mu_0 d = 1.26 \ \mu h$$

The total inductance L_T of a meter length of the line is then

$$L_T = \frac{4}{15.43} \times 1.26 = 0.326 \ \mu h$$

• PROBLEM 10-33

Find the inductance per unit length for the coaxial transmission line shown in the figure.

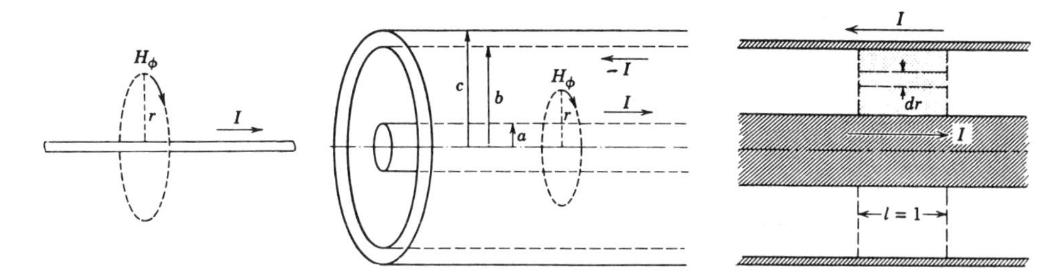

Solution: For a coaxial line as pictured in the figure with axial current I flowing in the inner conductor and returning in the outer, the magnetic field is circumferential and for $a < r < b$, where integration is being done about a circular path of radius r centered on the axis of the wire, and a and b are as shown in the figure, the magnetic field is

$$H_\phi = \frac{I}{2\pi r} \tag{1}$$

For a unit length the magnetic flux between radii a and b is, by integration over the shaded area in the figure

$$\int_S \bar{B} \cdot d\bar{s} = \int_a^b \mu \left(\frac{I}{2\pi r}\right) dr = \frac{\mu I}{2\pi} \ln \frac{b}{a} \tag{2}$$

so from (1) the inductance per unit length is

$$L = \frac{\mu}{2\pi} \ln \frac{b}{a} \text{ henrys/meter.}$$

• PROBLEM 10-34

A double coaxial system consists of a central cylindrical core about which are contained two cylindrical shells of negligible thickness, as shown in the given figure. The relative permeabilities of all regions are unity. The cylinder and the outer shell are assumed to each carry one half of the current +I. The intermediate shell carries the total return current -I. Find the inductance per unit length of the system.

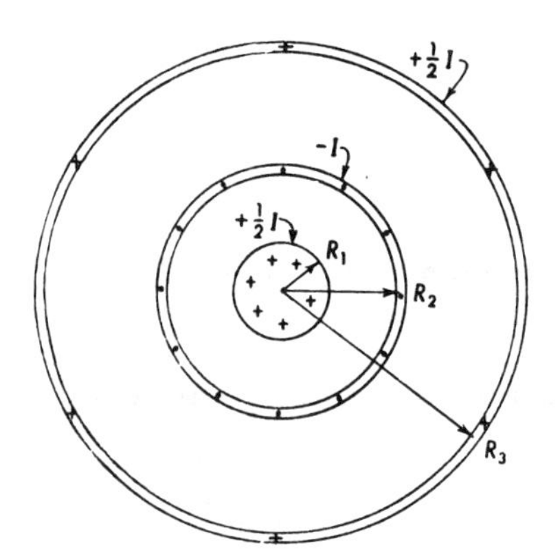

Solution: In the inner cylindrical conductor at a radius of r, the flux density is

$$B = \frac{2 \times 10^{-7}(I_{enclosed})r'}{R^2}$$

where r' is the distance from the center of the cylinder and R is the radius of the cylinder.

$$\therefore B = \frac{2 \times 10^{-7}(I/2)r}{R_1{}^2} = \frac{10^{-7}Ir}{R_1{}^2} .$$

The differential flux in a differential space of thickness dr is

$$d\phi = \frac{10^{-7}Ir}{R_1{}^2} dr,$$

and the differential flux linkage is

$$d\lambda_1 = \tfrac{1}{2} \frac{r^2}{R_1{}^2} d\phi = \frac{10^{-7}Ir^3}{2R_1{}^4} dr.$$

Integrating between 0 and R_1 gives

$$\lambda_1 = \frac{10^{-7}I}{2R_1{}^4} \int_0^{R_1} r^3 \, dr = \frac{10^{-7}I}{8} .$$

In the space were $R_1 \leqq r \leqq R_2$, the flux density is

$$B = \frac{2 \times 10^{-7}(I/2)}{r} = \frac{10^{-7}I}{r} .$$

The differential flux in a differential space of thickness dr is

$$d\phi = \frac{10^{-7}I}{r} dr,$$

and the differential flux linkage is

$$d\lambda_2 = \tfrac{1}{2} d\phi = \frac{10^{-7}I}{2r} dr.$$

Integrating between R_1 and R_2 gives

$$\lambda_2 = \frac{10^{-7}I}{2} \int_{R_1}^{R_2} \frac{dr}{r} = \frac{10^{-7}I}{2} \ln \frac{R_2}{R_1}$$

No appreciable thickness is considered for the current $-I$, and hence there are no appreciable linkages within this current.

In the space where $R_2 \leqq r \leqq R_3$, the flux density is

$$B = \frac{2 \times 10^{-7}(-I/2)}{r} = -\frac{10^{-7}I}{r}$$

The differential flux in a differential space of thickness dr is

$$d\phi = -\frac{10^{-7}I}{r} dr,$$

and the differential flux linkage is

$$d\lambda_3 = (-\tfrac{1}{2})d\phi = \frac{10^{-7}I}{2r} dr.$$

Integrating between R_2 and R_3 gives

$$\lambda_3 = \frac{10^{-7}I}{2} \int_{R_2}^{R_3} \frac{dr}{r} = \frac{10^{-7}I}{2} \ln \frac{R_3}{R_2} .$$

The self-inductance of the system per unit length is

$$\ell = \frac{d(\lambda_1 + \lambda_2 + \lambda_3)}{dI} = \frac{10^{-7}}{8} + \frac{10^{-7}}{2} \ln \frac{R_2}{R_1} + \frac{10^{-7}}{2} \ln \frac{R_3}{R_2}$$

$$= \frac{10^{-7}}{8}\left(1 + 4 \ln \frac{R_3}{R_1}\right).$$

The inductance is independent of the value of R_2.

• PROBLEM 10-35

Calculate the mutual inductance of the coils of fig. 1, assuming a ferromagnetic core with a relative permeability of 5000. The flux ϕ_1 of the left leg from a to b has a mean path length of one meter, and so does the flux ϕ_3 of the right leg. The length of the center leg from a to b is 0.4 m. Each path has a cross-sectional area of 0.01 m^2, and leakage flux is negligible.

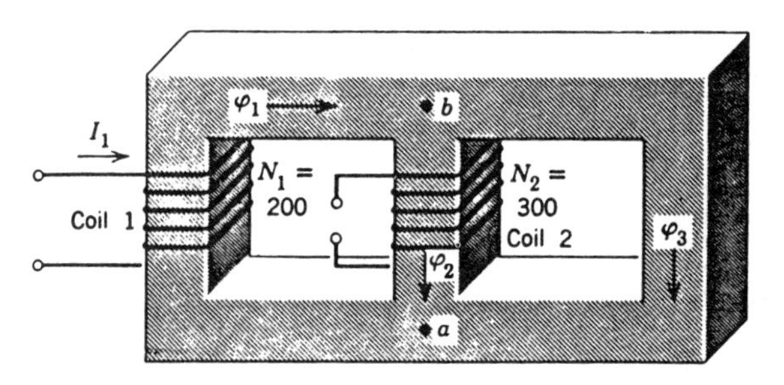

Two coils and a ferromagnetic circuit

Fig.1

Solution: Assume a current I_1 of coil 1 equal to one ampere, with coil 2 open. The flux linkages $N_2\phi_2$ of coil 2 are, therefore, equal to the mutual inductance M. Thus it is needed to determine the flux ϕ_2.

The reluctances R_1, R_2, and R_3 of the three paths are easily calculated from the relation $\ell/(\mu S)$. R_1 and R_3 are 15,900, and R_2 is 6,360. Around the closed magnetic circuit consisting of the left and center legs the mmf N_1I_1 equals $\phi_1R_1 + \phi_2R_2$. Therefore, $200 = 15{,}900\phi_1 + 6{,}360\phi_2$. Around the closed magnetic circuit consisting of the center and right legs the mmf is zero. Consequently, $\phi_2R_2 = \phi_3R_3$, or $6{,}360\phi_2 = 15{,}900\phi_3$. These two equations in terms of the three fluxes, with the aid of the relation $\phi_1 = \phi_2 + \phi_3$, can be solved for ϕ_2, giving $\phi_2 = 0.00699$ weber. $N_2\phi_2$ is 2.1, and the mutual inductance M is 2.1 henrys. The sign of M depends on the positive directions selected for the coil currents.

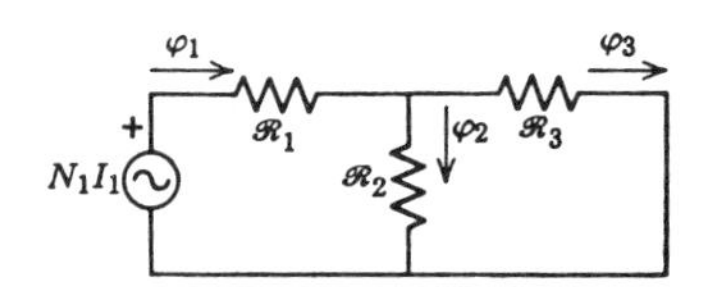

Circuit representation of magnetic-circuit problem, with I_2 of coil 2 equal to zero.

Fig.2

Fig. 2 is a circuit representation of this magnetic-circuit problem, with the reluctances shown as lumped elements.

• PROBLEM 10-36

Find the mutual inductance between the coaxial lines shown in the figure.

Solution: The two coaxial lines shown are inductively

coupled because magnetic flux produced by currents I_1 in line 1 links the center conductor of line 2 in the coupling loop shown. The mutual flux linkage, neglecting flux within the conductors, is given by

$$\psi_{21} = \int_{S_2} \bar{B}_1 \cdot d\bar{S}$$

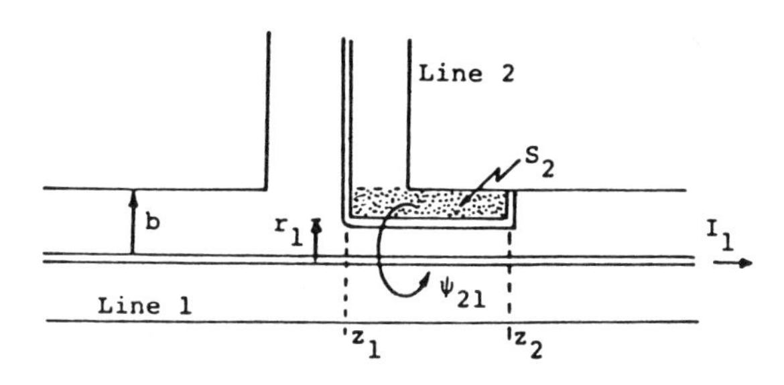

Two coaxial lines with inductive coupling.

The B field for a line of current is given from Ampere's law by

$$B_1 = \frac{\mu_0 I_1}{2\pi r},$$

thus

$$\psi_{21} = \int_{r_1}^{b} \int_{z_1}^{z_2} \frac{\mu_0 I_1}{2\pi r}\, dr\, dz = \frac{\mu_0 I_1}{2\pi}(z_2 - z_1)\ \ln \frac{b}{r_1}$$

Hence the mutual inductance M is given by

$$M = \frac{\psi_{21}}{I_1} = \frac{\mu_0}{2\pi}(z_2 - z_1)\ \ln \frac{b}{r_1}$$

• PROBLEM 10-37

Add a second winding over a solenoid as shown in Fig. 1. Assuming that both windings are long with respect to the common diameter so that end effects may be neglected, calculate the mutual inductance between the two coils.

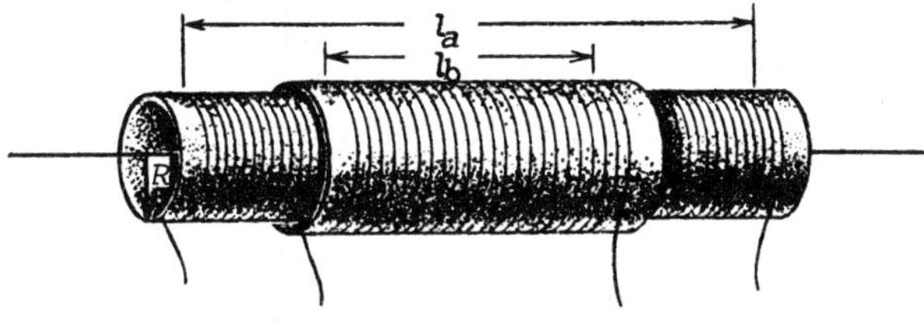

Coaxial solenoids. The two radii are taken to be approximately equal.

Solution: To calculate the mutual inductance between the two coils use the ratio of flux linkage to current. Assume a current I_a in coil a. Then the flux of coil a linking coil b is

$$\Phi_{ab} = \frac{\mu_0 \pi R^2 N_a I_a}{\ell_a},$$

and the mutual inductance is

$$M_{ab} = \frac{N_b \Phi_{ab}}{I_a} = \frac{\mu_0 \pi R^2 N_a N_b}{\ell_a},$$

It is also possible to calculate the mutual inductance by assuming a current I_b in coil b. Then

$$\Phi_{ba} = \frac{\mu_0 \pi R^2 N_b I_b}{\ell_b}.$$

This flux links only $(\ell_b/\ell_a)N_a$ turns of coil a since B falls rapidly to zero beyond the end of a long solenoid, then the mutual inductance is

$$M_{ba} = \frac{\ell_b N_a \Phi_{ba}}{\ell_a I_b} = \frac{\mu_0 \pi R^2 N_a N_b}{\ell_a},$$

as previously.

• **PROBLEM 10-38**

Obtain the expression for the mutual inductance between two circuits consisting of wires bent in the shape of closed circles with a current I_1 passing through circuit 1. (See the figure shown.)

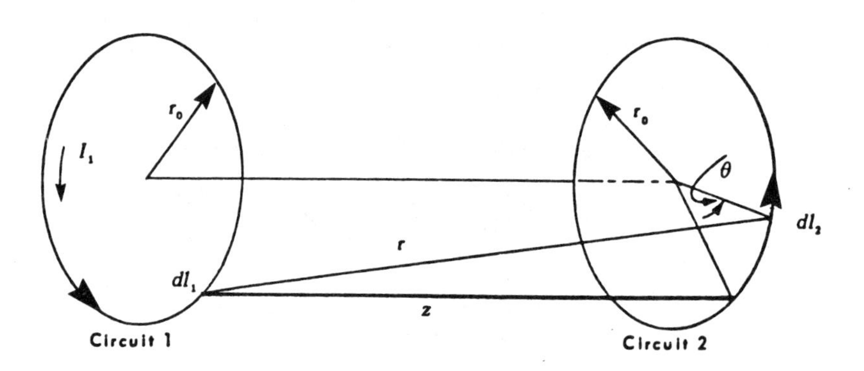

Solution: Evaluate M by using the relation

$$M = \frac{\mu}{4\pi} \oint_1 \oint_2 \frac{d\ell_1 \cdot d\ell_2}{r}$$

First, consider $d\ell_1$ constant and integrate $d\ell_2$ around circuit 2, then integrate around circuit 1. Since

$$d\bar{\ell}_1 \cdot d\bar{\ell}_2 = d\ell_1 \, d\ell_2 \cos\theta$$

and, from the figure.

$$r^2 = z^2 + \left(2 r_0 \sin\frac{\theta}{2}\right)^2$$

$$d\ell_2 = r_0 \, d\theta$$

then

$$\oint_1 d\ell_1 \int_0^{2\pi} \frac{r_0 \cos\theta \, d\theta}{r} =$$

$$\oint_1 d\ell_1 \int_0^{2\pi} \frac{r_0}{\sqrt{z^2 + (2 r_0 \sin\frac{\theta}{2})^2}} \cos\theta \, d\theta$$

To simplify the problem, assume $r \gg 2 r_0$. Then, expanding the coefficient of $\cos\theta \, d\theta$ under the integral sign, and neglecting powers higher than the second,

$$\frac{r_0}{\sqrt{z^2 + \left(2 r_0 \sin\frac{\theta}{2}\right)^2}} = r_0 \left[z^2 + (2 r_0 \sin\frac{\theta}{2})^2\right]^{-\frac{1}{2}}$$

$$= r_0 \left[z^{-1} - \frac{1}{2} z^{-3} \, 4 r_0^2 \sin^2\frac{\theta}{2} + \ldots\right]$$

$$= r_0 \left[\frac{1}{z} - \frac{2 r_0^2 \sin^2 \theta/2}{z^3} + \ldots\right]$$

so that the expression for M now becomes

$$M = \frac{\mu}{4\pi} \oint_1 d\ell_1 \int_0^{2\pi} r_0 \left(\frac{1}{z} - \frac{2 r_0^2 \sin^2\frac{\theta}{2}}{z^3}\right) \cos\theta \, d\theta$$

The integral around circuit 2, in θ is equal to $\pi r_0^3/z^3$ which leads, finally to

$$M = \frac{\pi r_0^3}{z^3} \frac{\mu}{4\pi} \oint_1 d\ell_1 = \frac{\mu r_0^4}{2z^3}$$

(since $\oint_{\ell_1} d\ell_1 = 2\pi r_0$).

A long straight current intersects at right angles a diameter of a circular current, and the plane of the circle makes an acute angle α with the plane through this diameter and the straight current. Show that the coefficient of mutual induction is

$$\mu_0\{c \sec \alpha - (c^2 \sec^2 \alpha - a^2)^{\frac{1}{2}}\}$$

or

$$\mu_0 c(\sec \alpha - \tan \alpha)$$

according as the straight current passes within or without the circle, a being the radius of the circle and c the distance of the straight current from its center. Refer to Fig. 1.

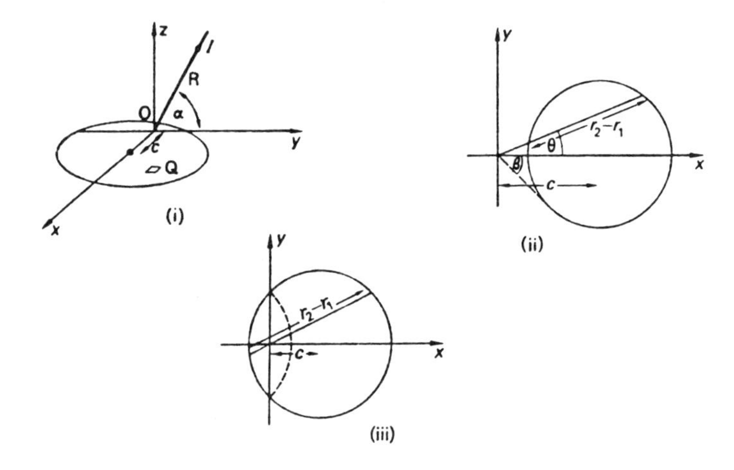

Solution: Choose the axes of coordinates (Fig. 1(i)) so the straight conductor passes through the origin, its direction being given by the unit vector,

$$\bar{e} = \bar{j} \cos \alpha + \bar{k} \sin \alpha.$$

First find the field produced at an arbitrary point Q{x y 0} in the plane of the circle. The vector $\overrightarrow{RQ}$ joining the point R, with position $\ell\bar{e}$ on the conductor, to Q is

$$\bar{i}x + \bar{j}y - \ell(\bar{j} \cos \alpha + \bar{k} \sin \alpha)$$

$$= \bar{i}x + \bar{j}(y - \ell \cos \alpha) - \bar{k}\ell \sin \alpha.$$

This is perpendicular to the conductor if

$$(y - \ell \cos \alpha) \cos \alpha - \ell \sin^2 \alpha = 0, \text{ i.e. } \ell = y \cos \alpha.$$

Hence the vector joining some point R on the conductor to Q such that $\overrightarrow{RQ}$ is perpendicular to $\bar{e}$ is

$$\bar{i}x + \bar{j}y \sin^2 \alpha - \bar{k}y \sin \alpha \cos \alpha.$$

Therefore the field at Q produced by the current I is

$$\bar{B} = \frac{\mu_0 I}{2\pi} \frac{(\bar{j}\cos\alpha + \bar{k}\sin\alpha) \times (\bar{i}x + \bar{j}y \sin^2\alpha - \bar{k}y \sin\alpha\cos\alpha)}{RQ^2}$$

$$= \frac{\mu_0 I}{2\pi} \frac{(-\bar{i}y \sin \alpha + \bar{j}x \sin \alpha - \bar{k}x \cos \alpha)}{x^2 + y^2 \sin^2 \alpha}$$

The flux of $\bar{B}$ through any circuit in $z = 0$ is $N = \iint B_z \, dx \, dy$, where the integral is taken over the area enclosed by the circuit.

$$\therefore M = -\frac{\mu_0}{2\pi} \iint \frac{x \cos \alpha \, dx \, dy}{x^2 + y^2 \sin^2\alpha}.$$

Evaluate the integral in terms of polar coordinates thus:

$$M = -\frac{\mu_0}{2\pi} \cos \alpha \iint \frac{r \cos \theta \, r \, dr \, d\theta}{r^2(\cos^2\theta + \sin^2\alpha \sin^2\theta)}$$

$$= -\frac{\mu_0}{2\pi} \cos \alpha \int \frac{\cos \theta \, d\theta}{\cos^2\theta + \sin^2\alpha \sin^2\theta} \int dr.$$

In the evaluation of the integral distinguish between the cases for which (1) $c > a$ Fig. 1 (ii), and (2) $c < a$, Fig. 1 (iii).

(1) $c > a$; the limits of r are (r_1, r_2) and of θ are $(-\beta, \beta)$, where r_1, r_2 are the (positive) roots of

$$r^2 - 2cr \cos \theta + c^2 = a^2,$$

i.e.

$$r_1 + r_2 = 2c \cos \theta, \quad r_1 r_2 = c^2 - a^2.$$

$$\therefore (r_2 - r_1)^2 = 4(c^2 \cos^2 \theta - c^2 + a^2)$$

$$= 4(a^2 - c^2 \sin^2\theta).$$

Also the angle β is given by $\sin \beta = a/c$.

$$\therefore M = -\frac{\mu_0 \cos\alpha}{2\pi} \int_{-\beta}^{\beta} \frac{\cos \theta \, d\theta}{\cos^2 \theta + \sin^2 \alpha \sin^2 \theta} \int_{r_1}^{r_2} dr.$$

$$= -\frac{2\mu_0 \cos \alpha}{\pi} \int_0^{\beta} \frac{(a^2 - c^2 \sin^2 \theta)^{\frac{1}{2}} \cos \theta \, d\theta}{\cos^2 \theta + \sin^2 \alpha \sin^2 \theta}.$$

By means of the substitution $c \sin \theta = a \sin \phi$, this leads to

$$M = -\mu_0\{c \sec \alpha - (c^2 \sec^2 \alpha - a^2)^{\frac{1}{2}}\}.$$

(2) $c < a$. In this case, Fig. 1 (iii), the flux over the

segment of the circle for which $x < 0$ cancels out the flux for the symmetrical positive segment, so we may take the following limits of integration. Those for r are (r_1, r_2), those for θ are $(-\pi/2, \pi/2)$. In this case r_2 and $-r_1 (r_1 > 0)$ are the roots of

$$r^2 - 2cr\cos\theta + c^2 = a^2,$$

i.e.

$$r^2 - 2cr\cos\theta - (a^2 - c^2) = 0$$

so that

$$r_2 - r_1 = 2c\cos\theta, \qquad r_1 r_2 = a^2 - c^2.$$

Hence

$$M = -\frac{\mu_0 \cos\alpha}{2\pi}\, 2\int_0^{\pi/2} \frac{2c\cos^2\theta\, d\theta}{\sin^2\alpha + \cos^2\alpha\cos^2\theta}.$$

On evaluation this gives

$$M = -\mu_0 c(\sec\alpha - \tan\alpha).$$

• PROBLEM 10-40

The figure shows a tightly wound toroidal coil with N turns and current I_1 and a square coil with current I_2. Calculate the mutual inductance of the coils.

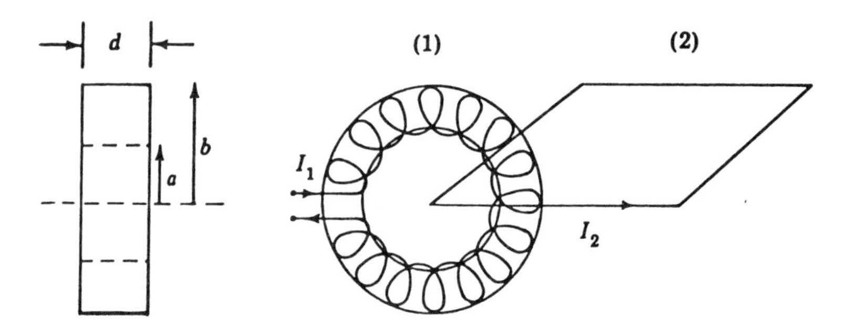

A toroid and a square loop.

Solution: Coil (2) intercepts all of the flux of coil (1) and thus:

$$L_{12} = \Phi_{21}/I_1$$

$$= \frac{1}{I_1}\iint_{S_2} \bar{B}_1 \cdot \bar{n}\, ds = \frac{1}{I_1}\int\int_{S_1} \bar{B}_1 \cdot \bar{n}\, ds$$

From Ampere's law B for a toroid is found to be

$$B = \frac{\mu_0 NI}{2\pi r} \qquad (r < b)$$

$$= 0 \qquad (r > b)$$

$$\frac{1}{I_1}\iint_{S_1} \bar{B}_1 \cdot \bar{n}\, ds = \frac{\mu_0 N I_1}{I_1 2\pi}\int_a^b \frac{Nd}{\bar{r}} \cdot d\bar{r}$$

$$= \frac{\mu_0 N^2 d}{2\pi} \ln r \Big]_a^b$$

$$= \frac{\mu_0 N^2 d}{2\pi} \ln \frac{b}{a}$$

Thus,

$$L_{12} = \frac{\mu_0 N^2 d}{2\pi} \ln (b/a)$$

which is the same as the self inductance of coil (1). In this case, Φ_{21}/I_1 is much easier to evaluate than Φ_{12}/I_2. Note that any coil encircling the toroid as in the figure would have the same mutual inductance.

• PROBLEM 10-41

Calculate the mutual inductance between a long, rectangular primary circuit of width a and a rectangular secondary of height h and length ℓ in the same plane as the primary. The two circuits are shown in the figure.

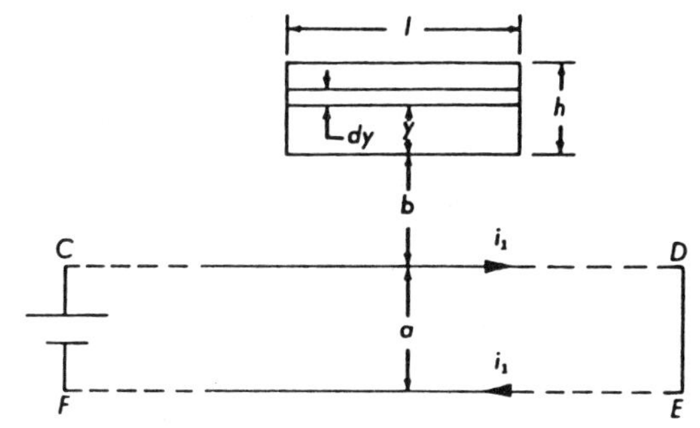

Two inductively coupled thin wire rectangular circuits in a vacuum.

Solution: In order to calculate the mutual inductance it is needed to know ϕ_{21} the flux passing through the secondary due to the current in the primary. The flux density at the point P on the surface S_2 bounded by the secondary is

$$B_1 = \frac{\mu_0 i_1}{2\pi(y + b)} - \frac{\mu_0 i_1}{2\pi(y + a + b)}$$

The first term on the right represents the flux density due to the current in the wire CD and the second term represents the flux density due to the current in the wire EF. The current i_1 in the expression for B_1 is a nonnegative quantity. The positive direction of $\bar{B}_1$ is toward the reader, and choose the same direction to be positive for $dS_2 = \ell\ dy$. Thus, the flux ϕ_{21} is given by

$$\phi_{21} = \int\int \bar{B}_1 . d\bar{S}_2$$

$$= \int_0^h \left[\frac{\mu_0 i_1}{2\pi(y + b)} - \frac{\mu_0 i_1}{2\pi(y + a + b)}\right] \ell\ dy$$

$$= \frac{\mu_0 i_1 \ell}{2\pi}\left[\ln\left(1 + \frac{h}{b}\right) - \ln\left(1 + \frac{h}{a + b}\right)\right]$$

Since h/b is greater than h/(a + b), the flux ϕ_{21} is positive and the direction of circulation of the primary current may be considered to be positive.

The mutual inductance M_{21} between the primary and secondary circuits is

$$M_{21} = \frac{N_2 \phi_{21}}{i_1} = \frac{\mu_0 \ell}{2\pi}\left[\ln\left(1 + \frac{h}{b}\right) - n\left(1 + \frac{h}{a + b}\right)\right]$$

where $N_2 = 1$. If the current in the primary is increasing at the rate di_1/dt, then the induced emf E_2 in the secondary is

$$E_2 = - M_{21} \frac{di_1}{dt}$$

In this case, E_2 is negative and the induced current in the secondary circulates in the negative direction and produces negative flux. This is in accordance with Lenz's law. The induced current in the secondary is in such a direction as to oppose the increase in flux due to the current in the primary.

Show from considerations involving the vector potential $\bar{A}$ that the expression for the mutual inductance of two circuits is the Neumann formula

$$M = \frac{\mu_0}{4\pi} \oint_{C_1} \oint_{C_2} \frac{d\bar{\ell}_1 \cdot d\bar{\ell}_2}{R}$$

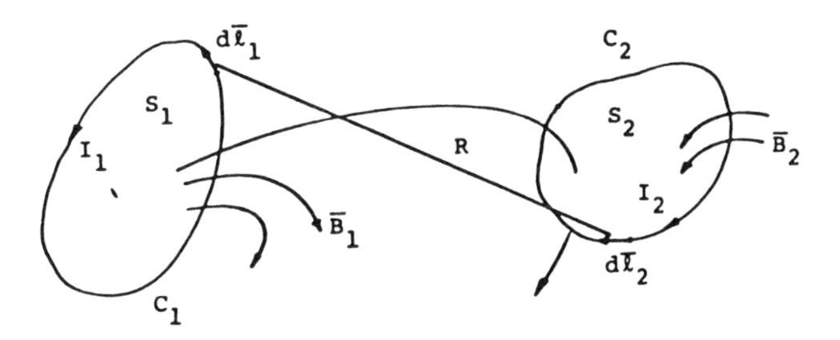

Two circuits with mutual inductive coupling.

Solution: Reference is made to the figure. If the current in circuit 1 is changing with time, there will be a changing magnetic field through circuit 2. As a result, an emf will be induced in circuit 2, as indicated by Faraday's equation,

$$\oint_C \bar{E}.d\bar{\ell} = -\frac{\partial \psi}{\partial t} = -\int_s \frac{\partial \bar{B}}{\partial t} \cdot d\bar{s} \,,$$

$$v_{i2} = -\int_{S_2} \bar{n} \cdot \frac{\partial \bar{B}}{\partial t} dS = -\int_{S_2} \bar{n} \cdot \frac{\partial (\nabla \times \bar{A})}{\partial t} dS$$

By interchanging the space (curl) and time derivatives, and by an application of Stokes' theorem, this expression can be written as a line integral as shown

$$v_{i2} = -\int_{S_2} \bar{n} \cdot \nabla \times \left(\frac{\partial \bar{A}}{\partial t}\right) dS = -\oint_{C_2} \frac{\partial \bar{A}}{\partial t} \cdot d\bar{\ell}$$

where the line integral is taken along circuit C_2. The known form of the vector potential is now included in this expression. This is first written as

$$\bar{A} = \frac{\mu_0}{4\pi} \int_V \frac{\bar{J}_1}{R} dV' = \frac{\mu_0}{4\pi} \oint_{C_1} \frac{I_1}{R} d\bar{\ell}$$

The result by combining the foregoing two formulas is

$$v_{i2} = -\frac{\mu_0}{4\pi} \left(\oint_{C_2} \frac{\partial}{\partial t} \oint_{C_1} \frac{I_1}{R} d\bar{\ell}_1 \cdot d\bar{\ell}_2 \right)$$

In those cases where the current is constant through the circuit (lumped circuit conditions) the current may be taken out from under the integral sign. Note also in this example that only the current changes with time. Therefore the equation may be written as

$$v_{i2} = -\frac{\mu_0}{4\pi} \frac{dI_1}{dt} \oint_{C_2} \oint_{C_1} \frac{d\bar{\ell}_1 \cdot d\bar{\ell}_2}{R}$$

But by definition the coefficient of $-dI/dt$ is the mutual inductance of the two circuits, so that finally

$$M = \frac{\mu_0}{4\pi} \oint_{C_2} \oint_{C_1} \frac{d\bar{\ell}_1 \cdot d\bar{\ell}_2}{R}$$

which is the Neumann formula.

FORCE AND TORQUE IN MAGNETIC CIRCUITS

• **PROBLEM 10-43**

A coil of 60 turns is wound around an iron ring (μ_r = 1000) having a 20-cm diameter and a 10-cm^2 cross section. As indicated in the given figure, the ring contains an air gap of length t = 0.1 mm, which is small enough so that fringing in the air gap may be neglected. Determine the force acting between the pole pieces of the iron ring when the current I = 1A.

Solution: This problem involves a direct application of

$$\bar{F}_m = \oint \left[\mu \bar{H}(\bar{H}.\bar{n}) - \frac{\mu}{2} H^2 \bar{n} \right] da.$$

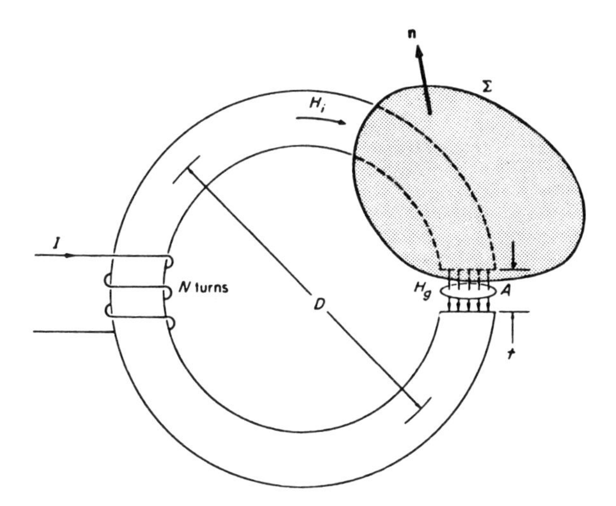

The attracting force of magnets.

The field may be determined by Ampere's Law:

$$\oint \bar{H} \, . \, d\bar{\ell} = NI$$

$$H_i(D\pi - t) + H_g(t) = NI$$

$$\frac{B}{\mu_0 \mu_r}(D\pi - t) + \frac{B}{\mu_0}(t) = NI$$

$$B(D\pi - t) + B\mu_r(t) = NI\mu_0\mu_r$$

$$BD\pi - Bt + B\mu_r(t) = NI\mu_0\mu_r$$

$$B\pi D + B(\mu_r - 1)t = NI\mu_0\mu_r$$

Thus the magnetic flux density in the ring is

$$B = \frac{\mu_0 \mu_r NI}{\pi D + (\mu_r - 1)t} = \frac{4\pi \times 10^{-7} \times 1000 \times 60 \times 1}{0.2\pi + (1000 - 1) \times 10^{-4}} = 0.111 \text{ Wb/m}^2$$

The magnetic field intensities in the iron and in the air gap are $H_i = B/\mu_0\mu_r$ and $H_g = B/\mu_0$, respectively. It is clear that $H_i \ll H_g$, and that the contribution of

H_i to the force expression may therefore be neglected. Accordingly,

$$\bar{F}_m = \int_S \left[\mu_0 \bar{H}_g (\bar{H}_g \cdot \bar{n}) - \frac{\mu_0}{2} H_g^2 \bar{n} \right] da = \int_S \left(\frac{\mu_0 H_g^2}{2} \right) \bar{n}\, da$$

The magnitude of this force is

$$F_m = \frac{\mu_0 H_g^2 A}{2} = \frac{B^2 A}{2\mu_0} = \frac{(0.111)^2 \times 10 \times 10^{-4}}{2 \times 4\pi \times 10^{-7}} = 5 \quad N.$$

and the direction along the normal vector $\bar{n}$. This means that the pole pieces tend to attract each other.

• PROBLEM 10-44

Consider a long solenoid of N turns, and length ℓ carrying current I. A thin iron rod of permeability μ and cross-sectional area A is inserted along the solenoid axis. If the rod is withdrawn until only one-half of its length remains in the solenoid, calculate approximately the force tending to pull it back into place.

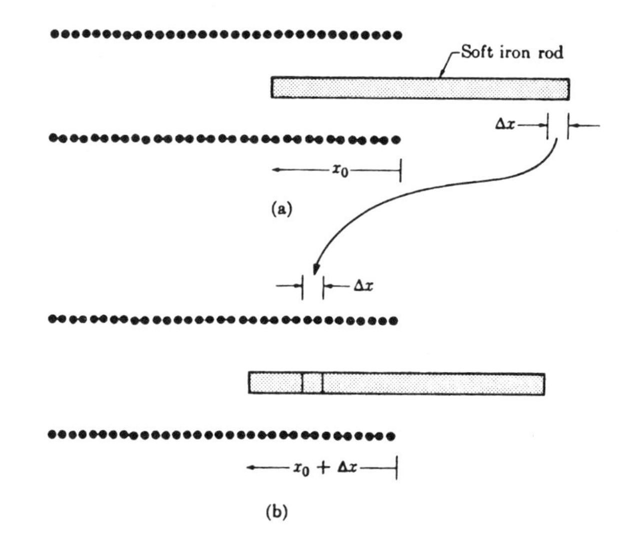

Solution: The magnetic field structure associated with this problem is quite complicated; fortunately, however, it is not necessary to calculate the entire magnetic energy of the system but merely the difference in energy between the two configurations shown in the given figures (a) and (b). The primary field structure (produced by the currents) is relatively uniform in the solenoid. The field structure associated with the magnetized iron rod is complicated, but it moves along with rod. The essential difference between configurations (a) and (b) is that a length Δx from the extreme right-hand end of the rod (outside the field region) is effectively transferred to the uniform field region inside the solenoid, at a place beyond the demagnetizing influence of the magnet pole. Thus

$$W = \frac{1}{2}\int_V \bar{H} \cdot \bar{B}\, dV$$

$$W(x_0 + \Delta x) \approx W(x_0) + \frac{1}{2}\int_{A\Delta x} (\mu - \mu_0)H^2\, dv$$

H for solenoid is found from Ampere's law to be: $H = NI/\ell$. Therefore integrating over the entire volume yields

$$W(x_0 + \Delta x) = W(x_0) + \frac{1}{2}(\mu - \mu_0)\frac{N^2I^2}{\ell^2} A\, \Delta x,$$

and force is the gradient of the magnetic energy or

$$F_x \approx \Delta\left[W(x_0 + \Delta x)\right] = \frac{\partial\left[W(x_0) + \frac{1}{2}(\mu - \mu_0)\frac{N^2I^2}{\ell^2} A\, \Delta x\right]}{\partial x}$$

Thus,

$$F_x \approx \frac{1}{2}(\mu - \mu_0)\frac{N^2I^2A}{\ell^2} .$$

• PROBLEM 10-45

Consider a long solenoid of n turns per unit length, total length ℓ, circular cross section of area S, and carrying a current I circulating as shown in the figure. Also suppose that a cylindrical rod of the same cross section and of permeability μ is inserted into the solenoid a distance z. Assume that the current in the solenoid is kept constant throughout and find the force on the rod.

Solution: A reasonably good solution can be obtained by neglecting the "end effects" that will be associated with the diverging field lines near the ends of the solenoid and rod. Since $|\chi_m| << 1$, the magnetic field will be practically unaffected by the presence of the rod and to a first approximation can be taken as $\bar{H} = nI\bar{z}$. The energy density u_{m_0} in the unoccupied vacuum region of the solenoid of length $\ell - z$ and volume $(\ell - z)S$ as obtained from

$$u_m = \frac{1}{2}\mu H^2$$

is

$u_{m_0} = \frac{1}{2}\mu_0 n^2 I^2$. Similarly, the energy density u_{mM} in the volume zS occupied by the matter will be

$$u_{mM} = \frac{1}{2}\mu n^2 I^2.$$

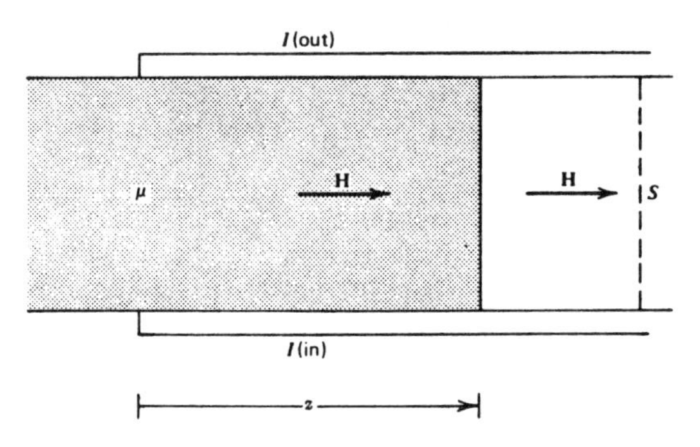

Permeable rod in a long solenoid.

Inserting these constant values into

$$U_m = \int_{\text{all space}} u_m d\tau$$

the total energy of this configuration is

$$U_m(z) = \frac{1}{2}n^2 I^2 S[\mu z + \mu_0(\ell - z)]$$

Since the currents are kept constant, use

$$\bar{F}_m = (\nabla U_m)_I \text{ (constant currents)}$$

to calculate the force:

$$\bar{F}_m = (\nabla U_m)_I = \frac{1}{2}(\mu - \mu_0)n^2 I^2 S\bar{z}$$

$$= \frac{1}{2}\chi_m \mu_0 n^2 I^2 S\bar{z} \tag{1}$$

There are many interesting aspects to this result.

Since everything else in (1) is positive, the sign of $\bar{F}_m$ depends on the sign of χ_m. If the material is paramagnetic so that $\chi_m > 0$, the rod will be attracted into the solenoid, while if it were diamagnetic with a negative susceptibility, it would be repelled. [This is in contrast to the analogous electrostatic result which showed that a dielectric slab would always be attracted in between the capacitor plates; this simply expresses the fact that all electric susceptibilities are positive.] If the material is paramagnetic, $\bar{M}$ will be parallel to $\bar{H}$ by $\bar{M} = \chi_m \bar{H}$ and the associated surface currents $\bar{K}_m$ will be circulating in the same sense as I.

Thus there are two sets of parallel currents and these will attract each other in agreement with the sign found from (1). On the other hand, in a diamagnetic material, $\bar{M}$ will be opposite in direction to $\bar{H}$, its corresponding surface currents will circulate oppositely to the sense of I, and these "unlike" currents will repel each other.

While the force $\bar{F}_m$ is independent of both ℓ and z, it is proportional to the cross-sectional area S; thus introduce a force per unit area $\bar{f}_m$ as

$$\bar{f}_m = \frac{\bar{F}_m}{S} = \frac{1}{2}(\mu - \mu_0)n^2 I^2 \bar{z}$$

This can be put into a more interpretable form by expressing it in terms of the energy densities in the two regions to get

$$\bar{f}_m = (u_{mM} - u_{m_0})\bar{z}$$

In other words, the magnitude of the force per unit area is just equal to the difference in energy densities of the two regions, and its direction is such as to tend to move the material so as to increase the total magnetic energy of the system.

• PROBLEM 10-46

Suppose there are two long ideal solenoids, one of which extends into the other a distance x as illustrated in Figure 1. Assume the windings are thin enough so that their cross-sectional areas are approximately equal to the same value S. Find the force between the solenoids and examine the direction of the force.

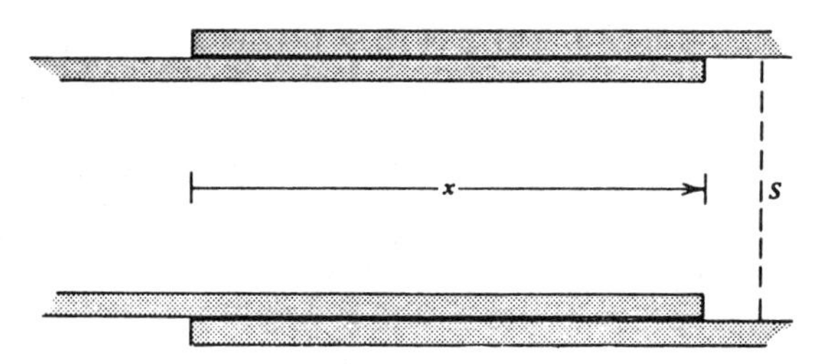

Two interpenetrating long solenoids.

Fig.1

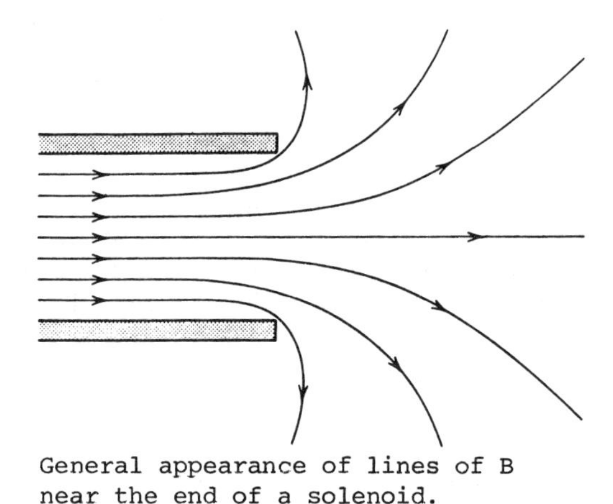

General appearance of lines of B near the end of a solenoid.

Fig.2

Solution: In this case, the solenoids are not infinitely long and it cannot be assumed that the values of $\bar{B}$ produced by them in the overlapping region of interest here are the same as those found in the interior far from the ends. In fact, the lines of $\bar{B}$ near the end tend to diverge quite rapidly from the axial direction as indicated schematically in Figure 2, and thus will not all pass through the distant turns of the other solenoid. Nevertheless, neglect these "end effects," provided that x is already large enough, and find the mutual inductance by considering only the flux contained within the overlapping region. Let the inner solenoid have n turns per unit length and carry a current I; the values for the outer are n' and I'. The mutual inductance can then be obtained using Ampere's law, B can be found and is equal to $B = \mu_0 nI$. If the cross-sectional area of

the outer coil is approximately the same as the inner

coil, then the flux per unit turn of the coil will be $\phi_{turn} = BS = \mu_0 nSI$ so that the total flux through the n' turns will be $\phi = \mu_0 nn'\ell'SI$. Thus the mutual inductance is given by

$$M = \frac{\phi}{I} = \mu_0 nn'\ell'S \text{ but } \ell' = x.$$

Thus

$$M = \mu_0 nn'Sx.$$

Using this in

$$F_{mx} = I\,I'\frac{\partial M}{\partial x}$$

the force is

$$F_{mx} = \mu_0 nn'SII' = \frac{SBB'}{\mu_0} \tag{1}$$

where the last form, expressing it in terms of the inductions produced by each solenoid, was obtained by using

$$B = \mu_0 NI$$

What will be the direction of this force? If I and I' both circulate about their respective solenoids in the same sense, the flux produced in one by the other will be positive and M will be positive. Then $\partial M/\partial x$ will be positive, F_{mx} will be positive, and the inner one will be attracted into the outer one. This is consistent with the qualitative statement that parallel currents attract each other. If I and I' circulate in opposite senses, then the flux produced by one in the other will be negative, according to our sign conventions, making M negative, and then $\partial M/\partial x$ and F_{mx} will both be negative. Therefore, the inner solenoid will be repelled by the outer, again consistent with our qualitative statement that "unlike" currents repel each other. But all of this is also contained in (1) if the currents are assigned relative signs according to their relative sense of circulation. Thus, if they are in the same sense, II' will be positive, as will F_{mx}, while if they are circulating in opposite senses, $II' < 0$ and $F_{mx} < 0$ in agreement with all of the above. Also express this same result quite nicely in terms of the inductions by writing (1) as

$$F_{mx} = \frac{S}{\mu_0} \bar{B} \cdot \bar{B}'$$

which will automatically give the correct sign to F_{mx}.

• PROBLEM 10-47

An electromagnetic linear-stroke motor is often used in control systems to actuate hydraulic valves. The given figure illustrates the cross section of such a device, which has circular symmetry about the center axis. The current i_1 in the center coil sets up flux paths such as C_1 which link all three coils. The current i_2 in coil 2 sets up flux paths such as C_2 which link coil 2. For coil 3 choose $i_3 = i_2$ and take the winding sense opposite to that of coil 2. Derive an expression for F_x.

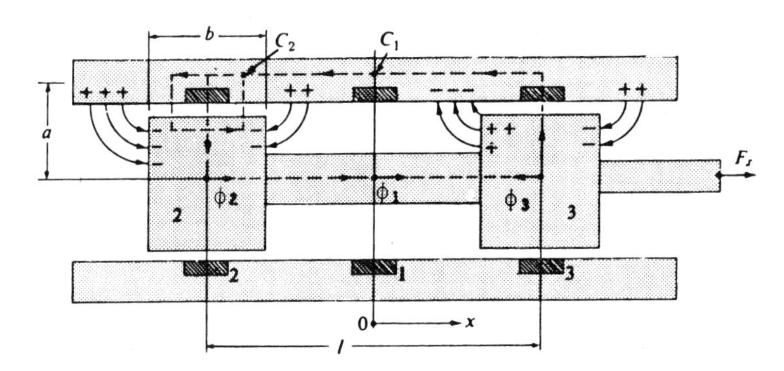

A linear-stroke motor.

Solution: With the polarities as shown in the figure, the magnetic fields of coils 1 and 2 are additive on the left-hand side and subtractive on the right-hand side. The induced magnetic charges (indicated in the figure by the polarity signs) are stronger on the left, resulting in a force of attraction to the left. Section 3 also contributes to the resultant force, and thus for the polarities chosen $F_x < 0$. Now derive an expression for F_x. The magnetic flux linking coil 2 is

$$\phi_2 = \phi_{22} + \phi_{21} ,$$

where ϕ_{22} is the magnetic flux linking coil 2 due to i_2 and ϕ_{21} is the magnetic flux linking coil 2 due to i_1. For small displacements from the center, ϕ_{22} remains constant. Thus, since ϕ_{21} decreases as x increases,

$$d\phi_2 = d\phi_{21} = -B_{g_1} 2\pi a\, dx$$

where $\bar{B}_{g_1}$ is the air-gap magnetic flux density due to i_1. For coil 3

$$d\phi_3 = d\phi_{31} = -B_{g_1} 2\pi a\, dx.$$

The magnetic field energy is

$$W = \frac{1}{2} i_1 N_1 \phi_1 + \frac{1}{2} i_2 N_2 \phi_2 + \frac{1}{2} i_3 N_3 \phi_3$$

Noting that $i_2 = i_3$, $N_2 = N_3$, $F_x = -i_2 N_2 (2\pi a) B_{g_1}$.

In order to determine B_{g_1}, define an equivalent path length ℓ_m by the equation

$$H_m \ell_m + H_g \ell_g = \oint_C \bar{H} \cdot d\bar{r} = i_1 N_1 ,$$

where the mmf is evaluated along the average flux path and where $\bar{H}_m = (1/\mu) \bar{B}_{g_1}$.

$$B_{g_1} = \frac{\mu_0 i_1 N_1}{\ell_g (1 + \ell_m / \mu_r \ell_g)} .$$

Thus

$$F_x = -(i_1 i_2 N_1 N_2) \left(\frac{a}{\ell_g} \right) \frac{2\pi \mu_0}{1 + \ell_m / \mu_r \ell_g} .$$

Note that the direction of the force can be reversed by reversing the direction of i_1 or i_2.

• PROBLEM 10-48

A concentrated coil of N turns of radius R carries a current I_1. At a point on its axis, and distant x from its plane, is situated a coaxial concentrated coil of n turns, radius r and carrying a current I_2. Obtain an expression for the mutual force between the coils, given that r is small so that the variation of field strength across the plane of the second coil (due to I_1) may be neglected. Show that this force is a maximum when x is equal to 1/2R.

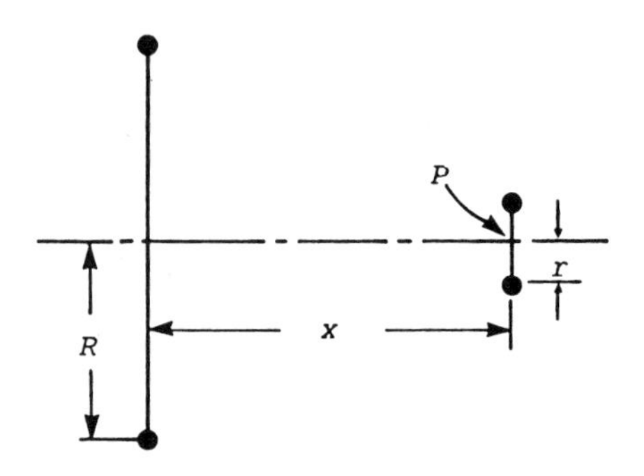

Solution: Let B_0 be the flux-density at P, the center of the small coil, of the field due to the large coil (see figure). Then the flux linking the small coil owing to I_1 in the first is

$$\phi_{21} = \pi r^2 B_0 ,$$

and the mutual inductance M is

$$M = \frac{\phi_{21} n}{I_1} = \frac{\pi r^2 B_0 n}{I_1} .$$

$$B_0 = \mu_0 \frac{I_1 N}{2} \frac{R^2}{(R^2 + x^2)^{3/2}} ,$$

so that

$$M = \frac{\mu_0}{2} (\pi r^2 R^2 nN)(R^2 + x^2)^{-3/2}$$

$$= K(R^2 + x^2)^{-3/2} ,$$

where

$$K = \frac{\mu_0}{2} (\pi r^2 R^2 nN)$$

The mutual force is

$$F = I_1 I_2 \frac{dM}{dx} ,$$

$$\frac{dM}{dx} = - 3Kx(R^2 + x^2)^{-5/2}$$

so that

$$F = 3KI_1I_2 \frac{x}{(R^2 + x^2)^{5/2}}$$

This force is zero both when x = 0 and when x is infinite. At some intermediate value of x it must pass through a maximum. This will be when dF/dx = 0, or

$$\frac{1}{(R^2 + x^2)^{5/2}} - \frac{5x^2}{(R^2 + x^2)^{7/2}} = 0,$$

i.e., when

$$R^2 + x^2 - 5x^2 = 0,$$

or

$$x = \frac{1}{2} R .$$

• PROBLEM 10-49

The accompanying figure illustrates the cross section (taken perpendicular to the axis) of a magnetic device which can supply a mechanical torque proportional to the current. The magnetic field due to the current i induces magnetic surface charges on the rotor and stator surfaces. The force of attraction between the induced magnetic charges results in a torque $\overline{T}$. Determine an expression for T.

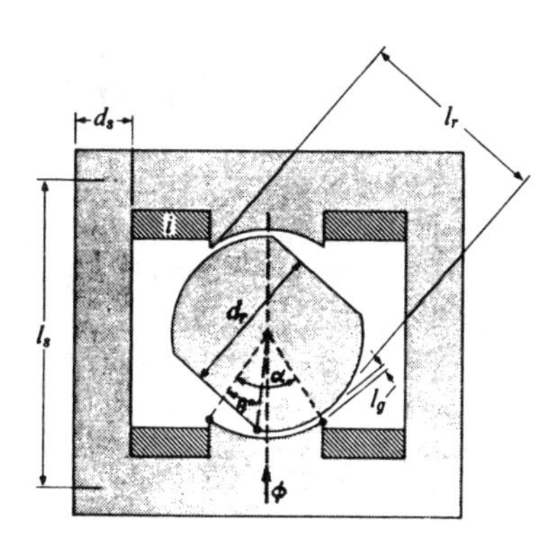

A torque motor.

Solution: Ampere's circuital law applied to the average flux path yields

$$H_r \ell_r + H_g 2\ell_g + H_s \ell_s = iN,$$

where H_r, H_g, and H_s are the average magnitudes for the magnetic field intensities in the rotor, air gap, and stator, respectively. The equation for the continuity of magnetic flux yields

$$\phi = \mu_0 H_g A_g = \mu H_r \ell d_r = 2\mu H_s \ell d_s,$$

where ℓ is the axial length of the rotor. The reluctance is given by $R = R_g + R_r + R_s$, where

$$R_g \triangleq 2\ell_g/\mu_0 A_g \ ,$$

$$R_r \triangleq \ell_r/\mu \ell d_r \ ,$$

$$R_x \triangleq \ell_s /2\mu \ell d_s \ ,$$

and where the air-gap area is $A_g = \ell \ell_r(\alpha - \theta)/2$. When i is taken to be constant, write the magnetic field energy in the form

$$W = \frac{1}{2}\, iN\phi = \frac{1}{2}\left(\frac{iN}{R}\right)^2 .$$

Thus

$$T = \frac{\partial W}{\partial \theta} = -\frac{1}{2}\phi^2 \frac{\partial R}{\partial \theta} = -\frac{1}{4}\phi^2 R_g \frac{\ell_r \ell_g}{A_g}$$

If the relative permeability is very large compared with unity and if ℓ_g/ℓ_r is not excessively small, the reluctance is mainly due to the air gap. Thus

$$\phi \cong iN/R_g \ ,$$

and

$$T \cong -C(iN)^2,$$

where

$$C \triangleq \mu_0 \ell \ell_r / 8\ell_g \ .$$

The torque can be made bidirectional by adding a second pole group to the stator, a group which is situated at right angles with respect to the first pole group. The two stator windings are wound so that the ampere turns are additive for pole group 1, resulting in a torque

$$T = -C(i_1N_1 + i_2N_2)^2,$$

and so that the ampere turns are subtractive for pole group 2. This results in a torque

$$T_2 = C(i_1N_1 - i_2N_2)^2.$$

The resultant torque is then

$$T = T_2 - T_1 = -4C(i_1i_2N_1N_2).$$

• PROBLEM 10-50

Suppose the reluctance of the magnetic circuit of the figure is

$$R = C_1 - C_2 \cos 2\theta$$

with C_1 and C_2 denoting constants. When $\theta = 0$, the reluctance is $C_1 - C_2$, the minimum value of R. When $\theta = 90°$, the reluctance is $C_1 + C_2$, the maximum value of R. If $\theta = \omega t - \delta$ and if the flux ϕ is $\phi_m \cos \omega t$, find the time-average counterclockwise torque.

Solution: When $\omega t = 0$, the flux ϕ is a maximum, and the angular position of the rotor is as shown in Fig. a, with δ being the angle between the two dashed arrows. This angle is called the torque angle.

$$F_x = \frac{dW}{dx} = \frac{\frac{1}{2}\phi di}{dx} + \frac{\frac{1}{2}id\phi}{dx}$$

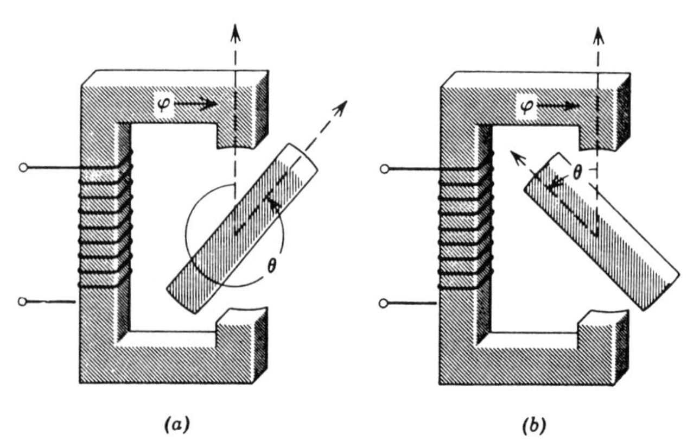

Simple reluctance motor.

If the current is constant

$$F_x = \frac{1}{2} i \frac{d\phi}{dx}$$

But $L = \phi/i$, therefore

$$F_x = \frac{1}{2} i^2 \frac{dL}{dx} .$$

Now $R = Ni/\phi$ and $\phi = \phi/N$, thus

$$F_x = - \frac{1}{2} \phi^2 \frac{dR}{dx} .$$

$$\tau = F_x dx = - \frac{1}{2} \phi^2 \frac{dR}{dx} .$$

The instantaneous torque T is found by substituting the expressions for ϕ and R into

$$T = - \frac{1}{2} \phi^2 \frac{dR}{d\theta}$$

and replacing θ with $\omega t - \delta$. The result is

$$T = -C_2 \phi_m{}^2 \cos^2 \omega t \sin (2\omega t - 2\delta)$$

The time-average torque is easily found by substituting $\frac{1}{2}(1 + \cos 2\omega t)$ for $\cos^2 \omega t$ and by applying another identity to the product of the sine and cosine functions. The resulting time-average torque is

$$\frac{1}{4}C_2\phi_m^{\,2}\ \sin 2\delta\ .$$

It should be noted that this is a maximum when δ is 45°. If there is no load on the motor, the torque angle is practically zero. At maximum load, δ is 45°.

CHAPTER 11

TIME - VARYING FIELDS AND MAXWELL'S EQUATIONS

FARADAY'S LAW

• **PROBLEM 11-1**

Shown in the figure is an electric circuit consisting of a voltmeter and a wire coiled about an iron cylinder. The voltmeter is connected to the ends of the coil by means of sliding contacts, with the tap on the right side sliding along a copper ring to which the coil end is soldered. The current I of the second circuit produces a steady magnetic flux Φ in the iron cylinder. Although the electric circuits are not in the magnetic field, the flux lines pass through the turns of the coils.

When the end of the coil at A is pulled downward, the cylinder rotates on its axis, and the coil unwinds. Suppose the coil is unwound at a rate that causes the magnetic flux over a surface S bounded by this coil and the voltmeter to decrease at a rate of one weber per minute. What does the voltmeter read?

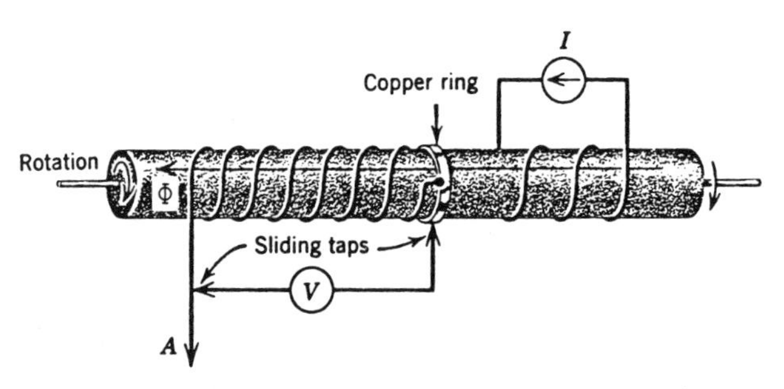

Two coils on an iron core. One is being unwound as the core rotates, and its flux linkages change with time, though Φ is steady.

Solution: The magnetic flux over the surface S is changing with time. However, this change is due neither to a time-changing magnetic field nor to the cutting of flux lines by the path C that bounds the surface S. Therefore, there is no emf, and the voltmeter reads zero. The expression $d\phi/dt$ of emf $= -d\phi/dt$ is zero. An emf around a closed path exists only if a time-changing magnetic flux density B induces a voltage according to the Maxwell-Faraday law, or if there is motional emf.

• PROBLEM 11-2

A small thin circular loop of copper wire is placed in a steadily varying magnetic field $\bar{B}$. Assuming that this magnetic field is being confined to a circular region of radius R', compute the induced electric field $\bar{E}$ in the copper wire for i) r < R' and ii) r > R'.

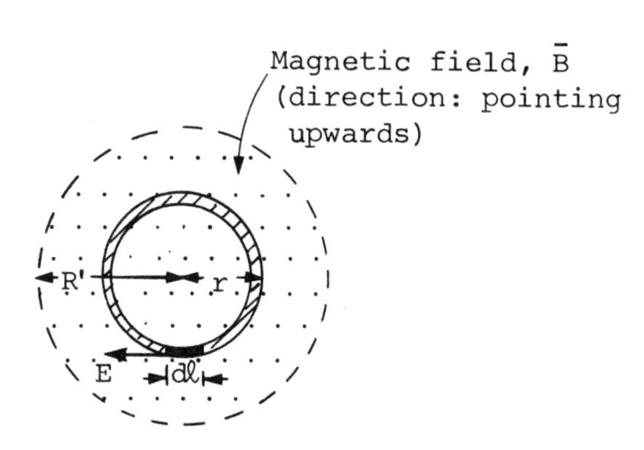

Solution: Let the rate of change of the magnetic field $= \frac{dB}{dt}$.

(a) For r < R':

For this case, the area of the circular loop is $\pi \cdot r^2$.

The flux coming out of this loop at any time =

Φ = Magnetic field × area

$$\therefore \Phi = B \cdot \pi r^2$$

Since the magnetic field is changing at a rate of $\frac{dB}{dt}$ and since the loop is stationary, the induced field is given by

$$\oint \bar{E}_i d\bar{\ell} = -\frac{d\Phi}{dt}$$

$$\therefore E\oint d\ell = -\pi r^2 \cdot \frac{dB}{dt}$$

$$E = -\pi \cdot r^2 \cdot \frac{dB}{dt} \cdot \frac{1}{\oint d\ell}$$

$$= -\pi \cdot r^2 \left(\frac{dB}{dt}\right) \cdot \frac{1}{2\pi \cdot r}$$

$$E = \frac{-r}{2} \frac{dB}{dt}$$

The direction of the $\bar{E}$ is shown in the figure.

(b) For $r > R'$:

$$\Phi = B \cdot \pi \cdot R'^2$$

$$\oint \bar{E}_2 \cdot d\bar{\ell} = -\frac{d\Phi}{dt}$$

$$E_2 = \frac{1}{2\pi r} \cdot \frac{dB}{dt} \cdot \pi \cdot R'^2$$

$$= \frac{R'^2}{r} \cdot \frac{dB}{dt} \cdot \frac{1}{2}$$

$$= \frac{1}{2} \frac{R'^2}{r} \cdot \frac{dB}{dt} \quad \text{(magnitude of the } \bar{E} \text{ field)}$$

• PROBLEM 11-3

A single turn loop is situated in air, with a uniform magnetic field normal to its plane. The area of the loop is 5 meter^2. What is the emf appearing at the terminals of the loop, if the rate of change of flux density is 2 webers/meter^2/sec?

Solution: According to Faraday's law for a loop of N turns, where all turns are linked by the same flux ϕm the emf is

$$E = -N \frac{d\phi m}{dt} \text{ volts}$$

For the given problem $\frac{d\phi m}{dt} = \frac{d(Bm \times Area)}{dt}$

$$= \frac{dBm}{dt} \times \text{area}$$

$$= 2 \times 5 \text{ volts}$$

$$= 10 \text{ volts}$$

Hence the emf appearing at the terminals of the loop is 10 volts.

• PROBLEM 11-4

The circuit shown in the figure 1 has 50 turns and an area of 200 cm^2. It is in a field in which B = 5×10^{-3} weber/m^2, pointing straight into the paper (indicated by the crosses). This field is reduced to one-fifth its original value in 0.2 second.

(a) Compute the average induced emf.

(b) Determine the "direction" of the emf.

(c) The coil has a resistance of 5 ohms. What is the average current in it during the induction period? What total charge flows?

(d) How much energy was required to change the magnetic field by this amount?

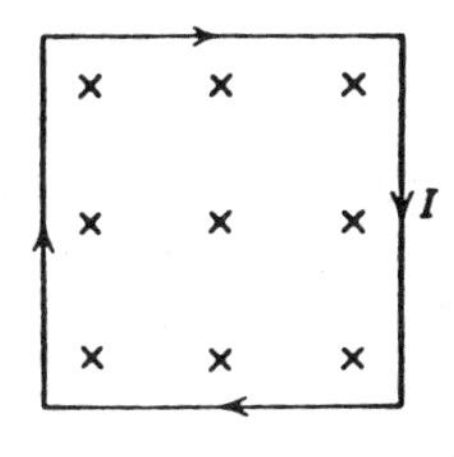

Fig. 1

Solution: (a) The flux is:

$$\phi_0 = (5 \times 10^{-3} \text{ weber/m}^2)(0.02 \text{ m}^2)$$

$$\phi_0 = 10 \times 10^{-5} \text{ weber (at the start).}$$

$$\phi = (1 \times 10^{-3})(0.02) = 2 \times 10^{-5} \text{ weber (at the end).}$$

$$\Delta\phi = -8 \times 10^{-5} \text{ weber.}$$

Using Faraday's Law

$$E = N(\Delta\phi/\Delta t) = 50 \frac{8 \times 10^{-5}}{0.2} = 0.02 \text{ volt.}$$

(b) The original effect is the disappearance of magnetic lines pointing into the paper; the induced current tends to replace these. The right-hand thumb rule shows that the current, and so the emf, must be directed as shown by the arrow.

(c) $E = IR$, or $I = 0.02/5 = 4 \times 10^{-3}$ amp.

$$\Delta q = I\Delta t$$

Thus,

$$\Delta q = (4 \times 10^{-3})(0.2) = 8 \times 10^{-4} \text{ coulomb.}$$

Examination shows that Δq is really independent of the time 0.2 sec, for this time enters into the denominator of E, and so of I, and again in the numerator of Δq. Hence, it is common practice to measure various magnetic effects by measurement of the "induced" charge rather than of the current or emf.

(d) Energy involved in causing the calculated flow of charge is used to change the magnetic field: Energy $= E \cdot \Delta q = (0.02)(8 \times 10^{-4}) = 1.6 \times 10^{-5}$ joule.

• PROBLEM 11-5

An electron is accelerated along a circular orbit by an increasing magnetic flux which treads the orbit centrally. Show that the electron will continue to move in a path of constant radius, if the flux-density at the path is always one-half the mean flux-density enclosed by the path.

Solution: This is the principle of the Betatron. Let R be the radius of the orbit, B the mean flux-density over the orbit and ϕ the flux linking the orbit. Then an e.m.f. is induced around the orbit of magnitude

$$e = -\frac{d\phi}{dt} = -\pi R^2 \frac{dB}{dt}$$

and since the magnetic field is taken to be symmetrical about the axis this means that an electric field E is induced which is concentric with the axis and of the same magnitude at all points of the orbit. The e.m.f. is therefore given by $e = 2\pi RE$, so that

$$E = \frac{R}{2}\frac{dB}{dt}.$$

If B' is the magnetic flux-density at the orbital path and v the velocity of the electron, the latter experiences a radial force qvB', where q is the electronic charge. If m is the electronic mass then

$$qvB' = \frac{mv^2}{R}, \text{ and } B' = \frac{mv}{qR}. \tag{1}$$

The force accelerating the electron is qE and this is equal to the rate of change of momentum. Hence

$$qE = \frac{qR}{2}\frac{dB}{dt} = \frac{d}{dt}(mv). \tag{2}$$

Assume that the electron starts from rest when $B = 0$, and integration of (2) then gives

$$mv = \frac{qRB}{2} \tag{3}$$

so that, from (3) and (1),

$$B = \frac{2mv}{qR} = 2B', \tag{4}$$

which is the required relation between B and B'.

It should be noted that even without assuming the electronic mass, m, to be a constant, for (2) is valid even if m is a function of velocity. Equation (4) therefore still applies if the electron is accelerated to velocities comparable with c, the velocity of light, in which case its mass as used in (2) is given by $m = m_0(1 - v^2/c^2)^{-\frac{1}{2}}$.

• PROBLEM 11-6

Consider the circular wire loop of radius r shown in the figure and assume that the wire loop is being heated in such a way that the radius r is a linear function of the time t, that is,

$$r = vt$$

where v is the constant radial velocity of a point on the loop. Also assume that the magnetic flux density is uniform and that the magnitude of $\overline{B}$ is given by

$$B = B_0(1 + kt)$$

where B_0 and k are constants. The vector $\bar{B}$ is normal to the page and directed toward the reader. Determine the induced emf in the wire loop.

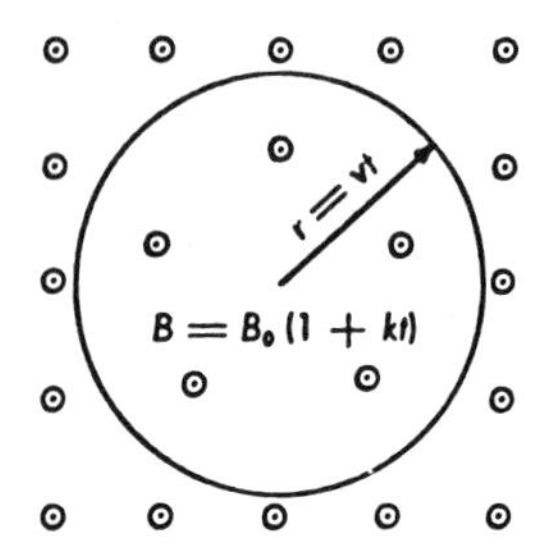

An expanding circular loop in a variable magnetic field.

Solution: The first step in the solution of this problem is to obtain an expression for the flux ϕ. Since

$$\phi = \iint \bar{B} \cdot d\bar{S}$$

and $\bar{B}$ is not a function of the coordinates,

$$\phi = \pi r^2 B$$

where the direction of $d\bar{S}$ was chosen to be parallel to that of $\bar{B}$.

The second step is to find the value of the induced emf from

$$E = -\frac{d\phi}{dt} = -\pi B\left(2r\frac{dr}{dt}\right) - \pi r^2 \frac{dB}{dt}$$

Since

$$\frac{dr}{dt} = v$$

and

$$\frac{dB}{dt} = B_0 k,$$

$$E = -vB(2\pi r) - B_0 k(\pi r^2)$$

Thus, the induced emf consists of two terms. The first depends on the speed v with which elements of the wire loop move through the magnetic field. For this reason, this part of the induced emf is called the motional emf. The second term depends on the time rate of change of the flux density $\bar{B}$. The value of the induced emf may be expressed in general terms as follows:

$$E = \oint \bar{v} \times \bar{B} \cdot d\bar{s} - \iint \frac{\partial \bar{B}}{\partial t} \cdot d\bar{S}$$

where $\oint \bar{v} \times \bar{B} \cdot d\bar{s}$ replaces $-vB(2\pi r)$ and $-\iint \partial\bar{B}/\partial t \cdot d\bar{s}$ replaces $-B_0 k\,(\pi r^2)$. In the case of the wire loop as shown in the figure, since $d\bar{s}$ is directed toward the reader, the vector $d\bar{s}$ is oriented so as to yield counter-clockwise circulation. The cross product, $\bar{v} \times \bar{B}$ is opposite in direction to $d\bar{s}$. Consequently, a minus sign automatically results when the term $\oint \bar{v} \times \bar{B} \cdot d\bar{s}$ is evaluated.

• PROBLEM 11-7

A thin wire is bent into a circle of radius b and placed with its axis concentric with that of a solenoid which carries a time-varying current given by $i = I_0 \sin \omega t$. Find V(t) induced across a small gap left in the conductor, for the two cases of Figure shown: (a) $b > a$ and (b) $b < a$. Include the polarity of V(t) in the answers.

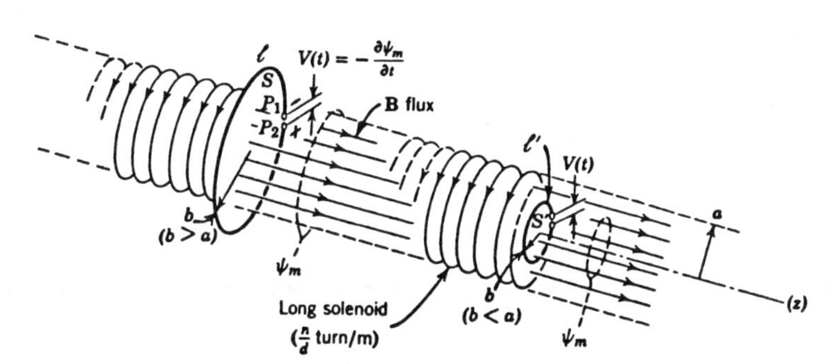

Figure showing open-circuit coils ℓ and ℓ^1 and the induced voltages obtained from time-varying ψ_m.

Solution: (a) Applying Ampere's law yields a result for the B field given by

$$\bar{B}(t) = \bar{a}_z \mu_0 I_0 \left(\frac{n}{d}\right) \sin \omega t.$$

If $b > a$,

$$V(t) = -\frac{d}{dt}\int_S \bar{B} \cdot d\bar{s} = -\frac{d}{dt}\int_S \left[\bar{a}_z \frac{\mu_0 n I_0 \sin \omega t}{d}\right]$$

$$\cdot\ (\bar{a}_z\, ds)$$

$$= -\frac{\omega\mu_0 n\pi a^2 I_0}{d} \cos \omega t \qquad b > a$$

since $\int_S ds = \pi a^2$.

The polarity of V(t) is found by use of right-hand-rule interpretation of the induced voltage law. Assuming, at a given t, that ψ_m through ℓ is increasing in the positive z sense in the Figure, aligning the thumb of the right hand in that direction points the fingers towards the terminal P_2 at the gap, which at that moment is the positive terminal. The presence of the negative sign in the answer, however, requires that the true polarity of V(t) becomes the opposite of the indicated polarity in the Figure, at that instant.

(b) If b < a, the surface S' bounded by the wire ℓ' is smaller than the solenoid cross-section; then

$$V(t) = -\frac{d}{dt}\int_{S'} \bar{B} \cdot d\bar{s} = -\frac{\omega\mu_0 n\pi b^2 I_0}{d} \cos \omega t \qquad b < a$$

• PROBLEM 11-8

Consider the rectangular loop shown in the figure. The width ℓ of the loop is constant, but its length x is increased uniformly with time by moving the sliding conductor at a uniform velocity $\bar{v}$. The flux density $\bar{B}$ is normal to the plane of the loop and its magnitude varies harmonically with time as given by

$$B = B_0 \cos \omega t \tag{1}$$

Find the total emf induced in the loop.

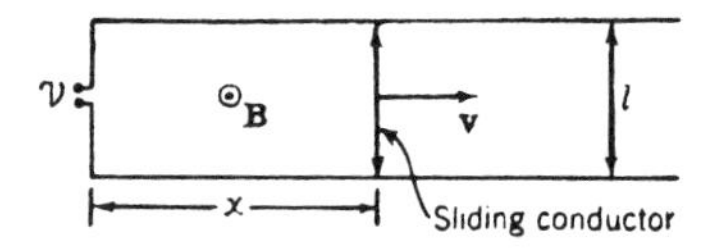

Solution: This is a case involving both motion and a time-changing B. The emf V_m due to the motion is given by

$$V_m = \oint (\bar{v} \times \bar{B}) \cdot d\bar{\ell} = vB\ell = v\ell B_0 \cos \omega t \tag{2}$$

The emf V_t, due to a time-changing B is

$$V_t = -\int_S \frac{\partial \bar{B}}{\partial t} \cdot d\bar{s} = \omega x\ell\, B_0 \sin \omega t \tag{3}$$

By superposition, the total emf V is the sum of the emfs of (2) and (3), or

$$V = \oint (\bar{v} \times \bar{B}) \cdot d\bar{\ell} - \int_S \frac{\partial \bar{B}}{\partial t} \cdot d\bar{s}$$

$$= vB_0 \ \ell \cos \omega t + \omega x B_0 \ \ell \sin \omega t$$

$$= B_0 \ \ell \sqrt{v^2 + (\omega x)^2} \sin (\omega t + \delta) \qquad (4)$$

where $\delta = \tan^{-1}(v/\omega x)$

x = instantaneous length of loop

• PROBLEM 11-9

Referring to Fig. 1 (a), a circular loop described by the equation $x^2 + y^2 = 16$ is located in the x-y plane centered at the origin. The $\bar{B}$ field is described by

$$\bar{B} = \hat{z} 2\sqrt{x^2 + y^2} \quad \cos \omega t \quad (\text{Wb/m}^2)$$

Find the total emf induced in the loop.

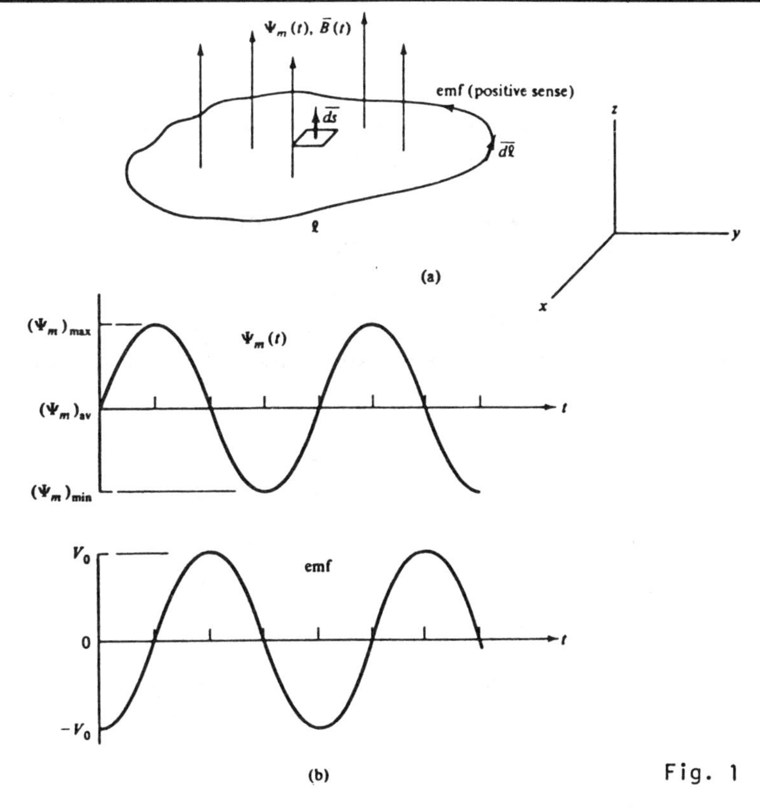

Fig. 1

(a)Time-changing magnetic field linking a closed loop. (b)The time relationship between the time-changing magnetic field and the emf induced in the loop.

Solution: From

$$\text{emf} = -\frac{d}{dt}\left(\int_s \bar{B} \cdot d\bar{s}\right) = -\int_s \frac{\partial \bar{B}}{\partial t} \cdot d\bar{s}$$

$$\text{emf} = -\int_y \int_x -2\omega \sin \omega t \sqrt{x^2 + y^2}\,\hat{z} \cdot \hat{z}\, dx\, dy$$

$$= 2\omega \sin \omega t \iint \sqrt{x^2 + y^2}\, dx\, dy$$

Changing to cylindrical coordinates,

$$\text{emf} = 2\omega \sin \omega t \int_0^{2\pi} \int_0^{4} r_c\, dr_c\, r_c\, d\phi$$

$$= 2 \sin \omega t (2\pi)\, \frac{4^3}{3}$$

$$= \frac{4\omega\pi \times 64}{3} \sin \omega t$$

$$= 268\omega \sin \omega t \quad (V)$$

• PROBLEM 11-10

Consider a fixed rectangular loop of area A, shown in the figure. The flux density $\bar{B}$ is normal to the plane of the loop (outward in the figure) and is uniform over the area of the loop. B varies with respect to time as given by $B = B_0 \cos \omega t$ where B_0 = maximum amplitude of $\bar{B}$ (webers/meter^2), ω = radian frequency, t = time. What is the total emf induced in the loop?

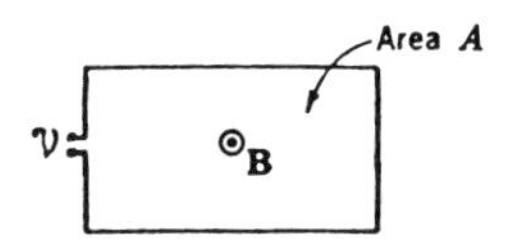

Fixed loop of area A.

Solution: This is a pure case of B change only, there being no motion. Hence, the total emf induced in the loop is

$$V = -\int_s \frac{\partial \bar{B}}{\partial t} \cdot d\bar{s} = A\omega B_0 \sin \omega t \quad \text{volts}$$

This emf appears at the terminals of the loop.

• PROBLEM 11-11

By shaping the pole faces of a cylindrical magnet, as shown in the figure in cross-section, and applying a time varying current to the coils, a magnetic field

$$\bar{B} = -\bar{z}_0 \, B_0 \frac{a}{\sqrt{a^2 + r^2}} \, t$$

can be produced for times $t > 0$. For $t < 0$, the field $B = 0$. Find the electric field induced between the poles for $r < a$ and $r > a$ and all t.

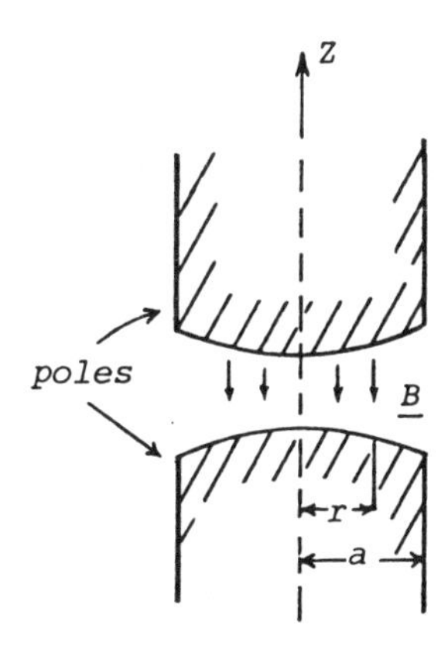

Solution: Since

$$V = \int_S\!\!\int \left(\frac{\partial}{\partial t}\,\bar{B}\right) \cdot \bar{n}\; d\,A \qquad \text{and } V = -\oint \bar{E} \cdot d\bar{r}$$

$$\oint \bar{E} \cdot d\bar{r} = \int_S\!\!\int -\frac{\partial \bar{B}}{\partial t} \cdot \bar{n}\; d\,A \tag{1}$$

$$\frac{\partial \bar{B}}{\partial t} = \frac{\partial}{\partial t}\left(-\bar{z}_0 B_0 \frac{a}{\sqrt{a^2 + r^2}}\, t\right) = -\bar{z}_0 \, B_0 \frac{a}{\sqrt{a^2 + r^2}}$$

Evaluating the right side of equation (1)

$$\int_S\!\!\int -\bar{z}_0 B_0 \frac{a}{\sqrt{a^2 + r^2}} \cdot \bar{n}\; dA \qquad \bar{n} = -\bar{z}_0$$

$$= \int_0^{\hat{R}} \int_0^{2\pi} B_0 \frac{a}{\sqrt{a^2 + r^2}}\, r \; dr \; d\theta = 2\pi \; B_0 \left[a\sqrt{a^2 + r^2}\,\right]_0^{\hat{R}}$$

$$= 2\pi B_0 \; a \; [\; \sqrt{a^2 + \hat{R}^2} - a]$$

where $\hat{R} = R$ for rad $< a$

$\hat{R} = a$ for rad $> a$

It is known that E is a function of radius and time and that it circles around the z axis.

$$\oint \bar{E} \cdot d\bar{r} = \int_0^{\pi} (E(R,t)\bar{a}_{\phi}) \cdot (\bar{a}_{\phi} Ra\theta)$$

$$= 2\pi R\ E(R,t)$$

Equating the right and left sides

$$2\pi R\ E(R,t) = 2\pi\ B_0\ a\ [\sqrt{a^2 + \hat{R}^2} - a]$$

or

$$E(R,t) = \frac{B_0 a}{R} [\sqrt{a^2 + \hat{R}^2} - a]$$

for $t < 0$ $E = 0$

for $t > 0$

$$r < a: \quad E(R,t) = \frac{B_0 a}{R} [\sqrt{a^2 + R^2} - a]$$

$$r > a: \quad E(R,t) = \frac{B_0 a}{R} [\sqrt{za^2} - a]$$

$$E(R,t) = \frac{B_0 a^2 \ (0.4)}{R}$$

• PROBLEM 11-12

A time-varying magnetic field is given by

$$B = \bar{B}_0 \cos \omega t\ \bar{i}_y$$

where B_0 is a constant. Find the induced emf around a rectangular loop in the xz plane as shown in Fig. 1.

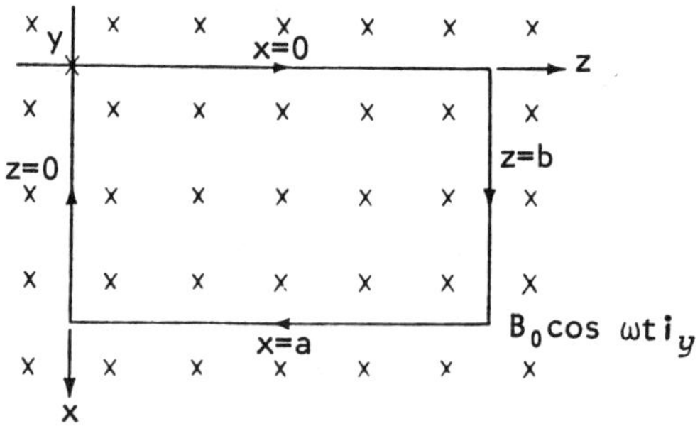

Fig. 1

A rectangular loop in the *xz* plane situated in a time-varying magnetic field.

Solution: The magnetic flux enclosed by the loop and directed into the paper is given by

$$\psi = \int_S \bar{B} \cdot d\bar{S} = \int_{z=0}^{b}\int_{x=0}^{a} B_0 \cos \omega t \; \bar{i}_y \cdot dx\, dz\, \bar{i}_y$$

$$= B_0 \cos \omega t \int_{z=0}^{b}\int_{x=0}^{a} dx\, dz = ab\, B_0 \cos \omega t$$

The induced emf in the clockwise sense is then given by

$$\oint_C \bar{E} \cdot d\bar{\ell} = - \frac{d}{dt} \int_S \bar{B} \cdot d\bar{S}$$

$$= - \frac{d}{dt}[abB_0 \cos \omega t] = ab\, B_0\omega \sin \omega t$$

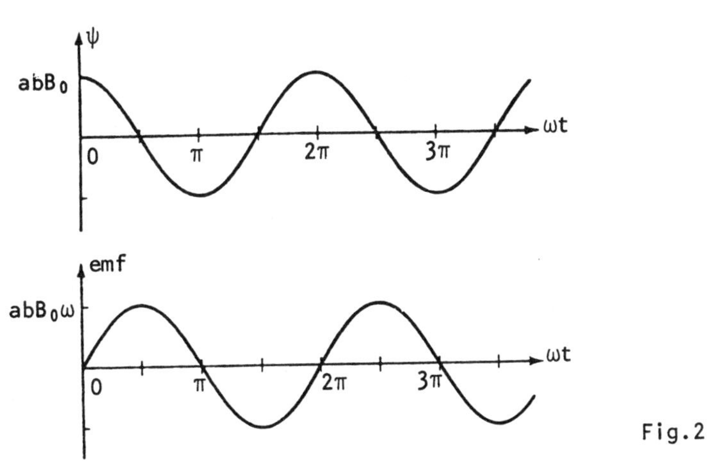

Time variations of magnetic flux ψ enclosed by the loop of Fig.1, and the resulting induced emf around the loop.

The time variations of the magnetic flux enclosed by the loop and the induced emf around the loop are shown in Fig. 2. It can be seen that when the magnetic flux enclosed by the loop is decreasing with time, the induced emf is positive, thereby producing a clockwise current if the loop were a wire. This polarity of the current gives rise to a magnetic field directed into the paper inside the loop and hence acts to increase the magnetic flux enclosed by the loop. When the magnetic flux enclosed by the loop is increasing with time, the induced emf is negative, thereby producing a counterclockwise current around the loop. This polarity of the current gives rise to a magnetic field directed out of the paper inside the loop and hence acts to decrease the magnetic flux enclosed by the loop. These observations are consistent with Lenz's law.

• PROBLEM 11-13

Consider a long, straight wire connected with flexible leads to a voltmeter and executing simple harmonic motion in a sinusoidally-varying magnetic field (see figure). The z-axis is the equilibrium position of the wire, which vibrates in the y, z-plane with a velocity given by

$$\bar{v} = \bar{j} v_m \cos \omega t \tag{1}$$

where v_m is the maximum value of the magnitude of $\bar{v}$. The time-variation of the induction can be expressed by

$$\bar{B} = \bar{i}\, B_m \cos \omega t \tag{2}$$

where B_m is the corresponding maximum, or amplitude, of $\bar{B}$.

Use Faraday's law to find the voltage produced by this combination.

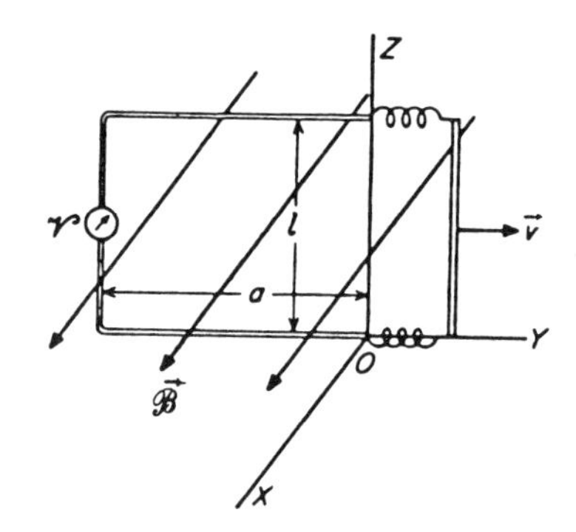

A wire vibrating in an alternating magnetic field

Solution: Faraday's law

$$\oint \bar{E} \cdot d\bar{\ell} = \int \left[\frac{\partial \bar{B}}{\partial t} - \text{curl } (\bar{v} \times \bar{B}) \right] \cdot d\bar{s}$$

Applying Stoke's theorem to the left side.

$$\oint \bar{E} \cdot d\bar{\ell} = \int \frac{\partial \bar{B}}{\partial t} \cdot d\bar{S} - \oint \bar{v} \times \bar{B} \cdot d\bar{\ell} \tag{3}$$

First find the contribution to the output voltage due to the motion by assuming that $\bar{B}$ is "frozen" at some instant t at the value given by eq. (2). There is a contribution to the second integral on the right of eq. (3) only along the moving wire (assuming that the flexible leads contribute nothing) and this integral becomes

$$\oint \bar{v} \times \bar{B} \cdot d\bar{\ell} = \int_0^{\ell} \bar{j}v_m \cos \omega t \times \bar{i}\, B_m \cos \omega t \cdot \bar{k}\, d\ell$$

$$= -v_m B_m \ell \cos^2 \omega t = -v_m B_m \ell \left(\frac{1 + \cos 2\omega t}{2}\right)$$

Similarly, find the contribution due to the variation of B by assuming the wire to be "frozen" at the position shown, so that the first term becomes

$$\int \frac{\partial \bar{B}}{\partial t} \cdot d\bar{S} = \int_{-a}^{y} (-\bar{i}B_m\, \omega \sin \omega t) \cdot (\bar{i}\ell\, dy)$$

$$= -B_m \omega \ell \sin \omega t [y + a]$$

By eq. (1)

$$y = \frac{v_m}{\omega} \sin \omega t$$

since

$$v = \frac{dy}{dt} = v_m \cos \omega t$$

Hence,

$$\int \frac{\partial \bar{B}}{\partial t} \cdot d\bar{S} = -B_m\, \omega \ell a \sin \omega t - B_m \ell v_m \left(\frac{1 - \cos 2\omega t}{2}\right)$$

Combining these two results,

$$\oint \bar{E} \cdot d\bar{\ell} = -B_m \omega\, \ell a \sin \omega t + B_m\, \ell v_m \cos 2\omega t$$

so that the output contains a fundamental and a second harmonic term.

• PROBLEM 11-14

A wire pendulum swings with velocity at the tip given by

$$V_0 = \omega d \cos \omega t$$

where d is the maximum horizontal displacement. Also, $\omega = 2\pi/T$ where T = 4 sec, is the period at the earth's surface. Fig. 1 shows this pendulum. If B = 0.25 webers/m^2 out of the paper, d = 0.1m and R = 3m, find the terminal voltage V(t) as a function of time.

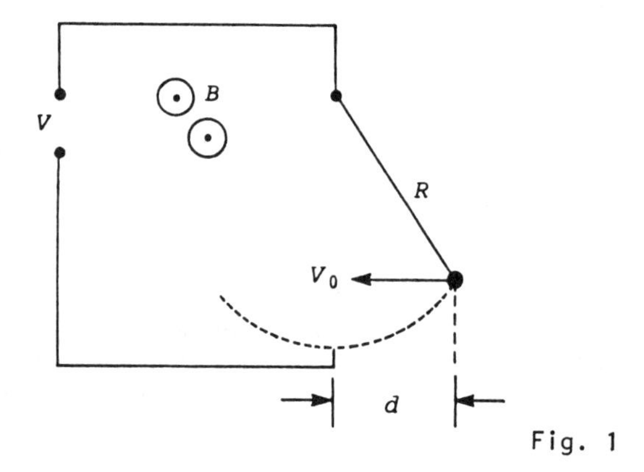

Fig. 1

Solution: The problem can be solved using Faraday's Law. Use

$$+V = -\frac{d}{dt}\iint \bar{B}\cdot\bar{n}\,dA$$

$$= \frac{d}{dt}[B\cdot \text{Area}(t)]$$

$$= \frac{d}{dt}\{B[\text{constant area} + \text{changing area}]\}$$

$$= \frac{d}{dt}\{B[\text{constant} + \tfrac{1}{2}R\,x\,(t)]\}$$

Derivative of a constant equals zero and $x(t) = d \sin \omega t$

$$\therefore\ V = +\frac{d}{dt}\{B\,\tfrac{1}{2}Rd \sin \omega t\} = \frac{BRd}{2}\frac{d}{dt}(\sin \omega t)$$

$$V = \frac{BR}{2}\,\omega d \cos \omega t$$

$$V = \frac{0.25(3)(0.1)(\frac{2\pi}{4})}{2} \cos \frac{2\pi}{4}t$$

$$V = \frac{0.15\pi}{8} \cos \frac{\pi}{2}\,t$$

$$V = 0.01875\pi \cos \frac{\pi}{2}\,t$$

• PROBLEM 11-15

A rectangular loop of wire with three sides fixed and the fourth side movable is situated in a plane perpendicular to a uniform magnetic field $\bar{B} = B_0 \bar{i}_z$, as illustrated in the figure. The movable side consists of a conducting bar moving with a velocity v_0 in the y direction. Find the emf induced in the loop.

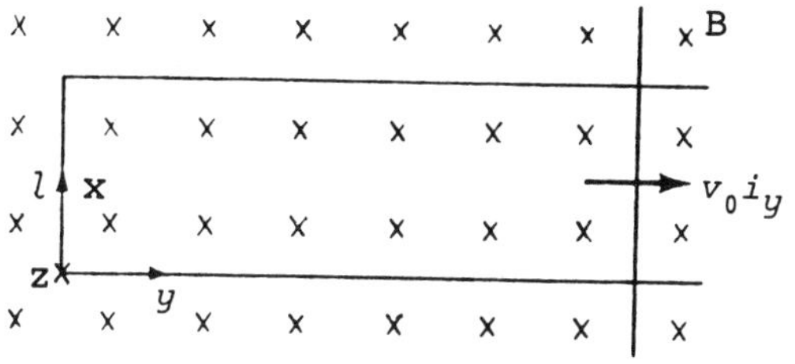

A rectangular loop of wire with a movable side situated in a uniform magnetic field.

Solution: Letting the position of the movable side at any time t be $y_0 + v_0 t$, the magnetic flux enclosed by the loop and directed into the paper is

$$\psi = (\text{area of the loop}) B_0$$

$$= \ell(y_0 + v_0 t) B_0.$$

The emf induced in the loop in the clockwise sense is then given by

$$\oint \bar{E} \cdot d\bar{\ell} = -\frac{d}{dt}\psi$$

$$= -\frac{d}{dt}[\ell(y_0 + v_0 t) B_0]$$

$$= -B_0 \ell v_0$$

Thus if the bar is moving to the right, the induced emf produces a current in the counterclockwise sense. Note that this polarity of the current is such that it gives rise to a magnetic field directed out of the paper inside the loop. The flux of this magnetic field is in opposition to the flux of the original magnetic field and hence tends to decrease it. This observation is in accordance with "Lenz's law", which states that the induced emf is such that it acts to oppose the change in the magnetic flux producing it. The minus sign on the right side of Faraday's law ensures that Lenz's law is always satisfied.

• PROBLEM 11-16

Consider a nonexpanding circuit moving through a field which has both spatial and temporal variations. (Refer to the accompanying figure). Find the induced voltage around the loop.

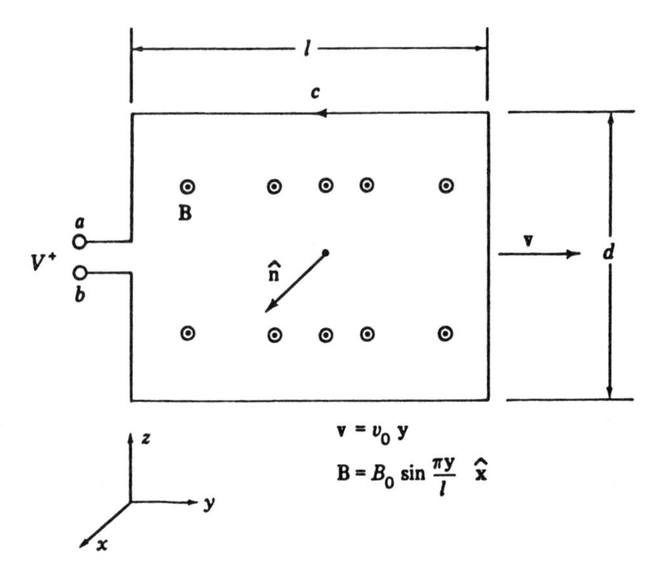

A moving, non-expanding circuit.

Solution: Using

$$V_{ab} = -\frac{d}{dt} \iint_s \bar{B} \cdot \bar{n}\, ds,$$

$$\therefore V_{ab} = -\frac{d}{dt} \int_0^d \int_{v_0 t}^{v_0 t+\ell} B_0 \sin \frac{\pi y}{\ell}\, dy\, dz$$

$$= +\frac{d}{dt} \frac{B_0 d\ell}{\pi} \left[\cos \frac{\pi}{\ell}(v_0 t + \ell) - \cos \frac{\pi v_0 t}{\ell}\right]$$

$$= 2v_0 B_0 d \sin\left(\frac{\pi}{\ell} v_0 t\right)$$

Alternatively, using

$$V_{ab} = \oint_c \bar{V} \times \bar{B} \cdot d\bar{\ell}$$

$$V_{ab} = -\iint_s \underbrace{\frac{\partial \bar{B}}{\partial t}}_{0} \cdot \bar{n}\, ds + \oint_c \bar{v} \times \bar{B} \cdot d\ell$$

The first term is zero; consider only voltages induced by the Lorentz force. Clearly, only the right-hand and left-hand sides of the circuit contribute. The voltage induced across the left-hand portion is $v_0 B_0 d \sin \pi/\ell\, v_0 t$; the voltage induced across the right-hand portion is equal in magnitude and opposite in sign because of the $\sin \pi y/d$ variation of the field. Therefore the total voltage around the loop is

$$V_{ab} = 2v_0 B_0 d \sin \frac{\pi}{\ell} v_0 t$$

The long solenoid of Figure 1(a), and 1(b), showing its cross section, carries a suitably slowly time-varying current $i = I_0 \sin \omega t$. Determine from Ampere's law the quasi-static magnetic flux density developed inside the coil of radius a, and then use Faraday's law to find the induced electric intensity field both inside and outside the coil.

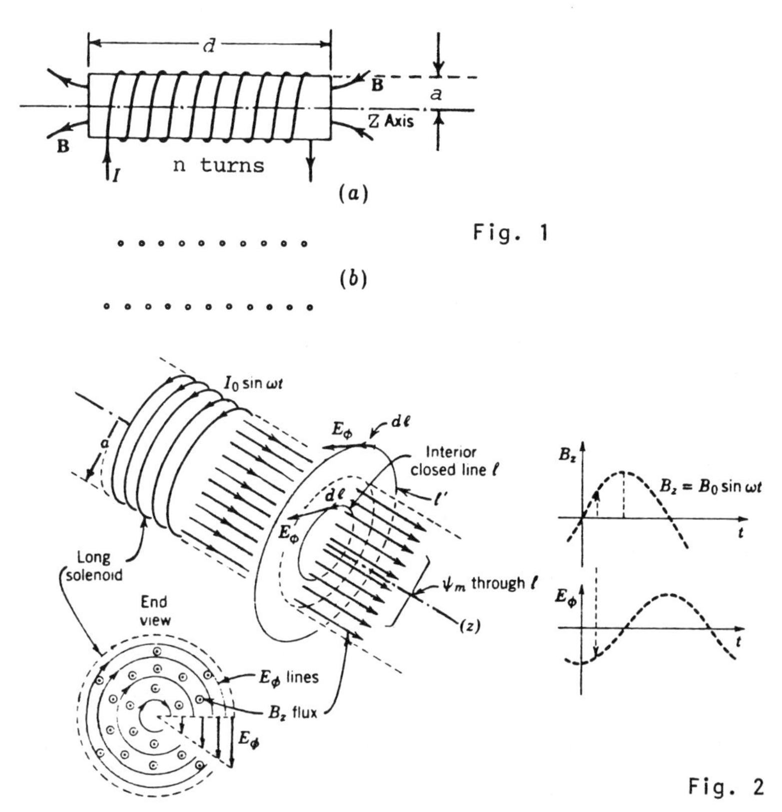

Showing the assumed integration path ℓ used for finding the induced E field of a solenoid, and the resulting E field.

Solution: From Ampere's law, the magnetic flux density inside the long solenoid carrying a static current I is found to be

$$B = \frac{\mu_0 n I}{d} a_z .$$

Thus, the solenoid current $I_0 \sin \omega t$ will to a first-order approximation provide the quasi-static magnetic flux density

$$\bar{B}(t) = \bar{a}_z \mu_0 I_0 \left(\frac{n}{d}\right) \sin \omega t = \bar{a}_z B_0 \sin \omega t$$

in which $B_0 = \mu_0 n I_0 / d$, the amplitude of $\bar{B}$. This assumption is reasonably accurate for an angular frequency ω

which is not too large. The electric field $\bar{E}$ induced by this time-varying $\bar{B}$ field is found by means of Faraday's law

$$\oint_{\ell} \bar{E} \cdot d\bar{\ell} = -\frac{\partial}{\partial t} \oint_{s} \bar{B} \cdot d\bar{s}$$

the line integral of which is first taken around the symmetric path ℓ of radius ρ inside the coil, as shown in Figure 2. Faraday's law becomes

$$\oint_{\ell} (\bar{a}_{\phi} E_{\phi}) \cdot \bar{a}_{\phi} d\ell = -\frac{d}{dt}\int_{s} (\bar{a}_z B_0 \sin \omega t) \cdot \bar{a}_z \, ds$$

in which, from the circular symmetry, E_{ϕ} must be a constant on ℓ. Thus

$$E_{\phi} \oint_{\ell} d\ell = -\omega B_0 \cos \omega t \int_{s} ds \tag{1}$$

but $\oint_{\ell} d\ell = 2\pi\rho$ and $\int_{s} ds = \pi\rho^2$, so that

$$E_{\phi} = -\frac{\omega B_0 \rho}{2} \cos \omega t \qquad \rho < a$$

is the first-order solution for the electric field intensity generated by the time-varying magnetic flux of the solenoid. Observe that E_{ϕ} varies in direct proportion to ρ, as shown in Figure 2. The negative sign in the result implies that as the net flux ψ_m through S is increasing (in the positive z direction), the sense of the induced $\bar{E}$ field is negative ϕ directed. This is symbolized in the time diagram of Figure 2.

Upon applying Faraday's law to the closed line ℓ' exterior to the coil the same formula as eq. (1) applies but now $\oint_{\ell} d\ell = 2\pi\rho$ and $\int_{s} ds = \pi a^2$, so that

$$E_{\phi} = \frac{-\omega B_0 a^2}{2\rho} \cos \omega t \qquad \rho > a$$

The electric field generated outside the long solenoid by the time-varying magnetic flux, therefore, varies inversely with respect to ρ. Both answers are directly proportional to ω because they are governed by the time rate of change of the net magnetic flux intercepted by the surface.

• PROBLEM 11-18

The z-component of the magnetic field in the air gap of a circularly symmetric magnet and in the neighborhood of the magnet axis varies sinusoidally with time and is distributed according to the law

$$H_z = H_0 \left[1 - \left(\frac{r}{a}\right)^2\right] \cos \omega t; \text{ for } r << a$$

where r is the radial coordinate in a cylindrical coordinate system coincident with the magnet axis. Find the electric field in the symmetry plane of the gap as produced according to Faraday's law. Note that the H-field in the symmetry plane has only a z-component, as shown in Fig. 1.

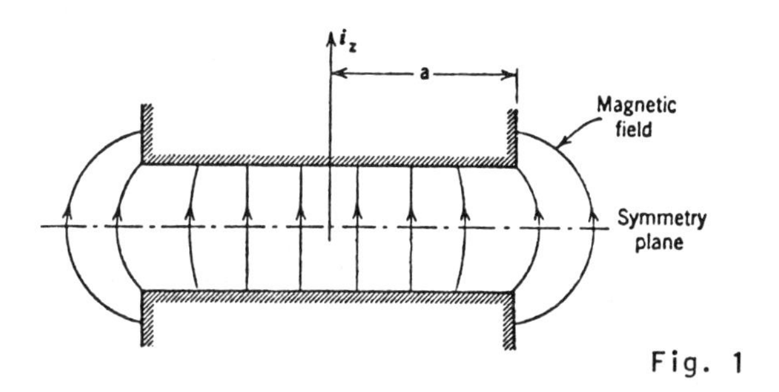

Fig. 1

Solution: Describe all field components in a cylindrical coordinate system with its z-axis coincident with the axis of symmetry of the magnetic field. Since the H-field in the symmetry plane is given completely, whereas off the symmetry plane it is not given directly, it is most convenient to apply Faraday's law

$$\oint_C E_t ds = -\frac{d}{dt}\int_S \mu_0 H_n da \tag{1}$$

to a circle lying in the symmetry plane and centered on the axis. The contour integral in Eq. 1 computed around the circle is

$$\oint_C E_t ds = \int_0^{2\pi} E_\phi r \, d\phi \tag{2}$$

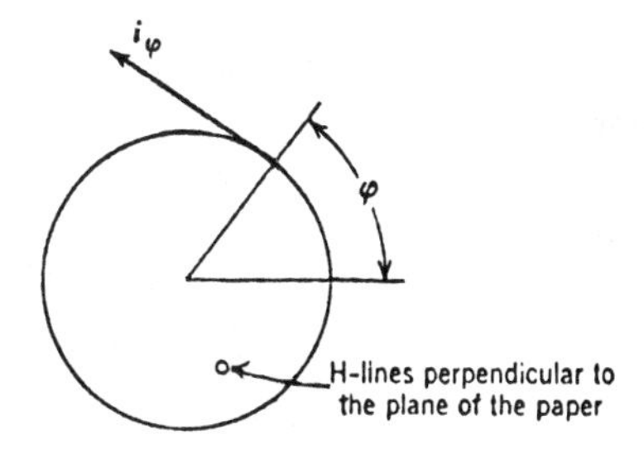

Fig. 2

The surface integral in Eq. 1 has to be evaluated over the area of the circle. The quantity H_n is the component of H normal to the plane of the circle (Fig. 2); thus

$$H_n = H_z \quad , \tag{3}$$

therefore

$$\int_S H_n \, da = \int_S H_z \, da = \int_{r=0}^{r} \int_{\phi=0}^{2\pi} H_z(r,\phi) r \, dr \, d\phi$$

$$= \int_{r=0}^{r=r} \int_{\phi=0}^{2\pi} H_0 \left[1 - \left(\frac{r}{a}\right)^2\right] \cos \omega t \; r \, dr \, d\phi \tag{4}$$

$$= H_0 \cos \omega t \; 2\pi \int_{r=0}^{r=r} \left[1 - \left(\frac{r}{a}\right)^2\right] r \, dr$$

$$= H_0 \cos \omega t \; \pi r^2 \left(1 - \frac{1}{2} \frac{r^2}{a^2}\right)$$

According to Eq. 1, μ_0 times the negative time derivative of the foregoing surface integral must be equal to the contour integral.

$$r \int_0^{2\pi} E_\phi d\phi = +\mu_0 H_0 \omega \sin \omega t \; \pi r^2 \left[1 - \frac{1}{2}\left(\frac{r}{a}\right)^2\right] \tag{5}$$

The integration on the left-hand side of Eq. 5 can be carried out directly and gives

$$E_\phi = \omega \mu_0 H_0 \sin \omega t \; \frac{r}{2}\left[1 - \frac{1}{2}\left(\frac{r}{a}\right)^2\right] \tag{6}$$

Thus, the magnetic field produces an electric field that is 90° out of time phase with the magnetic field.

• PROBLEM 11-19

A rigid, rectangular conducting loop with the dimensions a and b is located between the poles of a permanent magnet as shown. Let $B = a_z B_0$, constant as shown over the left portion of the loop, and assume the loop is pulled to the right at a constant velocity $\bar{v} = \bar{a}_x v_0$. Find (a) the emf induced around the loop, (b) the direction of the current caused to flow in the loop, (c) the force on the wire resulting from the current flow, and (d) the magnitude and polarity of the open-circuit voltage V(t) appearing at a gap in the wire at P shown.

Solution: (a) The sense of the line integration is assumed counterclockwise looking from the front, as in (b) of the accompanying figure. The emf induced about the loop is found by use of

$$\oint_{\ell(t)} \bar{E} \cdot d\bar{\ell} = \oint_{\ell(t)} (\bar{v} \times \bar{B}) \cdot d\bar{\ell}$$

On P_1 $(\bar{v} \times \bar{B}) \cdot d\bar{\ell} = [(\bar{a}_x v_0) \times (\bar{a}_z B_0)] \cdot \bar{a}_y\, dy =$ $-v_0 B_0\, dy$, obtaining

$$\oint_{\ell(t)} \bar{E} \cdot d\bar{\ell} = \int_{y=b}^{0} (-v_0 B_0)\, dy = v_0 B_0 b \tag{1}$$

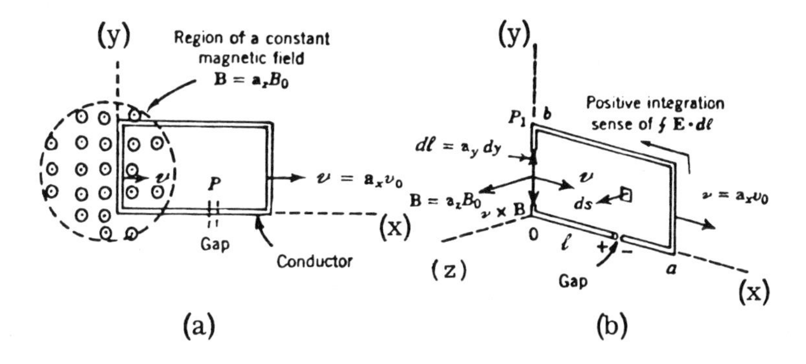

(a) Moving wire loop in a constant magnetic field.
(b) Geometry showing assumed line-integration sense.

(b) The positive sign of (1) denotes that the induced emf about ℓ is in the same sense (counterclockwise) as the direction of integration. The result (1) therefore causes a current to flow in the same direction.

(c) The force acting on the wire carrying the current $\bar{I}$ immersed in the field B is obtainable from $\bar{F}_B = q(\bar{v} \times \bar{B})$. The force on a differential charge $dq = \rho_v dv$ being $dF_B = dq(\bar{v} \times \bar{B})$, and with $\rho_v v$ the current density $\bar{J}$ in the wire, one obtains

$$d\bar{F}_B = \bar{J} \times \bar{B}\, dv \tag{2}$$

The product $\bar{J}\, dv$ defines a volume current element $\bar{I}\, d\ell$ in a thin wire, so (2) becomes

$$d\bar{F}_B = \bar{I}\, d\ell \times \bar{B} \tag{3}$$

Integrating (3) over the length $0P_1$ of the wire obtains the total force

$$\bar{F}_B = \int_{\ell} \bar{I}\, d\ell \times \bar{B} = \int_{y=b}^{0} I(\bar{a}_y\, dy) \times (\bar{a}_z B_0)$$

$$= -\bar{a}_x B_0 I b \tag{4}$$

in which the integration in the direction of the

current produces the proper vector sense of the force. $\overline{F}_B$ is a force to the left in the figure, opposing the motion of the wire.

(d) A small gap at P in the wire renders the loop open circuited, reducing I to zero and yielding $V(t) = \oint_\ell \overline{E} \cdot d\overline{\ell} = v_0 B_0 b$ across the gap. The polarity is determined by the direction of $\overline{v} \times \overline{B}$, directed around the loop towards the positive terminal of the gap as in (b) of the figure.

• PROBLEM 11-20

Shown in the figures is a circular metallic loop having a radius of one meter. Suppose a time-varying magnetic field exists in the region such that the magnetic flux density at points of the circular area is approximately

$$\overline{B} = 10^{-8}\ (1 - r^2/4 + r^4/64)\ \sin 3 \times 10^8 t\ \overline{k}$$

with r denoting the distance from the center of the loop and $\overline{k}$ representing a unit vector directed out of the paper. Find the counterclockwise voltage drop around the loop.

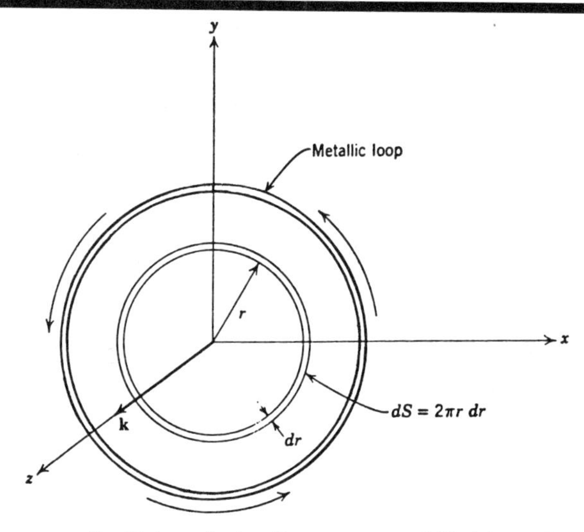

Metallic loop of unit radius in a magnetic field of density $\mathbf{B} = B_z\mathbf{k}$.

Solution: Let the positive side of the plane surface S bounded by the path C of the loop be the side facing the reader. By the right-hand integration rule the line integral of $\overline{E}$ is taken in the counterclockwise direction. Thus the counterclockwise voltage drop v is found by differentiating $\overline{B}$ with respect to time and substituting into the Maxwell-Faraday equation. Noting that the unit vector k and each differential vector area dS have the same direction,

$$v = -\int_S \frac{\partial \overline{B}}{\partial t} \cdot d\overline{s}$$

$$v = -\int_S 3(1 - r^2/4 + r^4/64) \cos 3 \times 10^8 t \; dS$$

As the integrand is a function of the space coordinate r, a convenient differential area is that between two circles of radii r and r + dr. This area is $2\pi r\, dr$, and the limits of r are zero and one. Consequently,

$$v = -6\pi \cos 3 \times 10^8 t \int_0^1 (r - r^3/4 + r^5/64)\, dr$$

Evaluation of the integral yields a counterclockwise voltage drop v equal to $-8.3 \cos 3 \times 10^8 t$ volts.

The electric field lines in the plane area are circles concentric with the metallic loop. Because of the circular symmetry, the field intensity equals the voltage drop divided by the circumference of the loop. Therefore, at points on the conductor the counterclockwise electric field intensity is $-1.32 \cos 3 \times 10^8 t$ volts/m.

• PROBLEM 11-21

Shown in the figure is a slot cut in a copper plate. A sinusoidal voltage v, equal to $10 \sin 10^9 t$ volts, is applied across the center of the slot as indicated. Such a center-fed slot is frequently used as an antenna at very high frequencies. For all practical purposes the copper plate behaves like a perfect conductor at such high frequencies. Find the magnetic flux passing through the left half of the slot, the right half of the slot, and the entire slot.

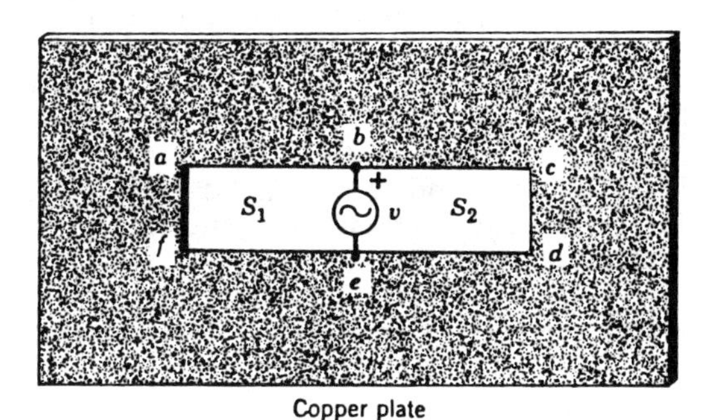

Copper plate

A slot antenna

Solution: Let the plane surface of the slot be divided into surfaces S_1 and S_2 as indicated, with the positive sides of these surfaces selected as the sides facing the reader. Application of the Maxwell-Faraday law to the surface S_1 and its contour C_1 counterclockwise around S_1 gives

$$\oint_{C_1} \bar{E} \cdot d\bar{\ell} = -\frac{\partial \Phi_1}{\partial t}$$

with Φ_1 denoting the magnetic flux out of the paper over the surface S_1. The voltage drop along the path bafe just inside the conducting material is negligibly small, for the copper plate of this example can be regarded as a perfect conductor. Electric fields do not exist in perfect conductors. A finite electric field in a conductor with no resistance would result in the absurdity of an infinite current. It follows that the voltage drop around the closed path C_1 equals the voltage drop along that portion of the path from e to b, and this voltage drop is -v. Therefore, -v equals $-\partial\Phi_1/\partial t$, or

$$-10 \sin 10^9 t = -\partial\Phi_1/\partial t$$

Integration with respect to time gives

$$\Phi_1 = -10^{-8} \cos 10^9 t \text{ weber}$$

The constant of integration is zero if the magnetic field has no steady component.

The voltage drop counterclockwise around the path C_2 that bounds the surface S_2 is obviously equal to v. Therefore, v equals $-\Phi_2$, with Φ_2 denoting the magnetic flux out of the paper over the surface S_2. Deduce that Φ_2 is the negative of Φ_1, or

$$\Phi_2 = 10^{-8} \cos 10^9 t \text{ weber}$$

The total magnetic flux over the area of the slot is the sum of Φ_1 and Φ_2, and this sum is zero. Thus the total flux is zero at every instant of time.

• PROBLEM 11-22

A wavefront of an electromagnetic wave in a perfect dielectric is shown in the figure. The wavefront propagates in a direction normal to its surface with velocity c. Ahead of the wavefront there are no fields. Immediately behind the wavefront the $\bar{E}$ and $\bar{H}$ fields are parallel to the surface and normal to one another. The crosses of the figure indicate magnetic field lines directed into the paper. Show that $E/H = \mu c$, with μ denoting the permeability of the medium.

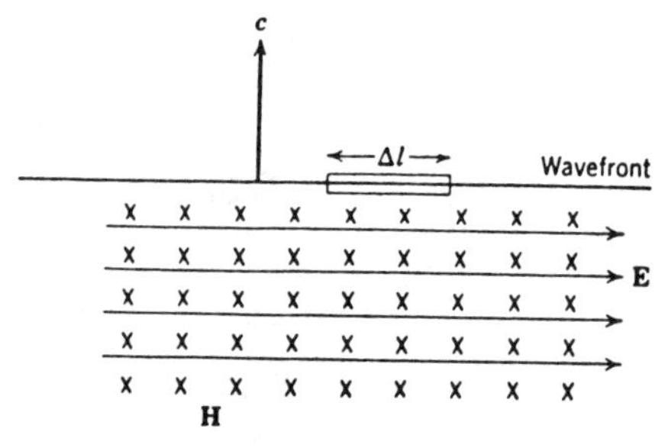

Cross section of the wavefront of an electromagnetic wave.

Solution: Apply the Maxwell-Faraday law to the very small rectangle constructed about a portion of the wavefront, as illustrated in the figure. The wavefront is passing through the stationary rectangle at the instant under consideration. The clockwise cirulation of $\bar{E}$ around the closed path C of the rectangle is obviously equal to $E\ \Delta\ell$, with $\Delta\ell$ denoting the length of the rectangle. This must equal $-\Phi$, with the positive side of the surface S of the rectangle being the side facing the reader. Select the back side of the rectangle as the positive side, then $E\ \Delta\ell = +\partial\Phi/\partial t$.

The time rate of increase of the magnetic flux out of the back side of the rectangle is $Bc\ \Delta\ell$, because $c\ \Delta\ell$ represents the rate of increase of the surface area occupied by the flux density $\bar{B}$. Therefore, $E\ \Delta\ell = Bc\ \Delta\ell$, or $E = Bc$, or $E/H = \mu c$.

It is easily shown that $\bar{E}$ and $\bar{H}$ must be normal to one another at the surface of a wavefront. Suppose that there is a component of $\bar{H}$ in the direction of $\bar{E}$. Then the magnetomotive force around the path C of the rectangle of the figure has value. This violates the Maxwell-Ampere law, since there is neither convection nor displacement current over the surface S of the rectangle. Therefore, $\bar{H}$ cannot have a component in the direction of $\bar{E}$.

• PROBLEM 11-23

Two thin, parallel bars 0.1 m long, spaced 0.1 m apart move with equal, constant velocity of 10 m/s in perfect contact with resistanceless rails, as indicated in Fig. 1.

The resistance of each bar is 0.01 Ω. The z-component of a static magnetic field varies as $B_z = 10^3x^2y$ tesla in the region between the rails. An ideal voltmeter is connected across the rails. Analyze this problem, and for the instant that bar A is at x = 0.1 m, deduce such quantities as the power transferred to the bars as work, the net force on the bars, the force on each bar, and the reading of the voltmeter.

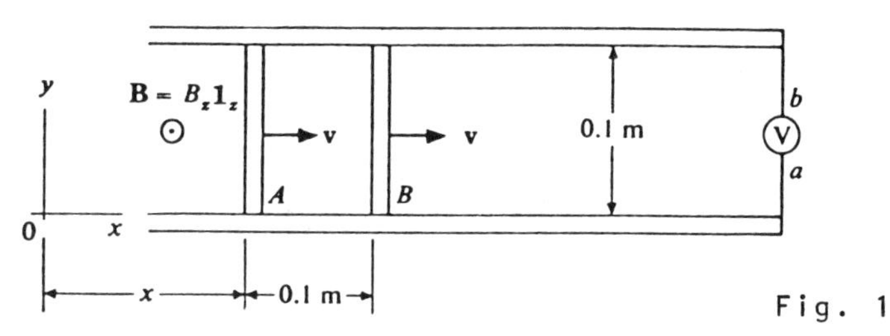

Fig. 1

Two bars A and B moving with equal constant velocity along resistanceless rails in a static magnetic field.

Solution: Since the magnetic field is nonuniform in x, and $\bar{v}$ is constant, there will obviously be a net induced emf in the loop formed by the bars and the rails. Also, it can be surmised that the emf will vary with time, but since the loop is small, quasistationary conditions will apply. Neglecting the time-varying magnetic field due to the induced current, the current will be given by

$$I = \frac{E}{R_{bars}} = \frac{E}{0.02} = 50\ E$$

The magnitude and polarity of E may be determined in several ways. First apply

$$\Lambda = \int_S \bar{B} \cdot d\bar{s} = \oint_\ell \bar{A} \cdot d\bar{\ell}$$

Looking into the paper and taking $d\bar{S}$ in the direction of vision, a general expression for the flux linkage in the loop is

$$\Lambda_\ell = 10^3 \int_S (x^2 y \bar{i}_z) \cdot (-dx\ dy\ \bar{i}_z)$$

$$= -10^3 \int_x^{x+0.1} dx \int_0^{0.1} dy\ x^2 y$$

$$\Lambda_\ell = -\frac{10x^3}{6}\Bigg|_x^{x+0.1} = -\frac{(3x^2 + 0.3x + 0.01)}{6}$$

where x is the location of bar A. Thus, for a clockwise integration of $\bar{v} \times \bar{B}$, the induced emf is

$$E = -\frac{\partial \Lambda_\ell}{\partial t} = \frac{1}{6}(6x + 0.3)\frac{dx}{dt} = (x + 0.05)\frac{dx}{dt}$$

For bar A at $x = 0.1$ and $v = dx/dt = 10$,

$$E = 1.5\ V$$

and

$$I = 75\ A \quad \text{(clockwise in the loop)}$$

The net force on the two bars may be computed from

$$F = \frac{I^2 R}{v} = \frac{(75)^2 \times 0.02}{10} = 11.25\ N$$

The power transferred to the bars is

$$P = Fv = I^2 R = 112.5\ W.$$

The individual electromagnetic forces on each bar may be computed from

$$F_i = I_i \int_{\ell_i} d\,\bar{\ell}_i \times \bar{B}_i .$$

On bar A,

$$\bar{F}_A = 75 \times 10^3 \int_0^{0.1} dy\, \bar{i}_y \times 0.01y\, \bar{i}_z$$

$$= \frac{750y^2}{2} \bar{i}_x \Big|_0^{0.1} = 3.75\, \bar{i}_x \text{ N}$$

On bar B,

$$\bar{F}_B = 75 \times 10^3 \int_{0.1}^{0} dy\, \bar{i}_y \times 0.04y\, \bar{i}_z$$

$$= 1500y^2\, \bar{i}_x \Big|_{0.1}^{0} = -15.0\, \bar{i}_x \text{ N}$$

The net electromagnetic force $\bar{F}$ is

$$\bar{F} = \bar{F}_A + \bar{F}_B = -11.25\, \bar{i}_x \text{ N}$$

which checks the previous calculation. Note that the electromagnetic force opposes the velocity and the applied force.

In order to determine the voltmeter reading it will be necessary to determine the induced emf in at least one bar, preferably both, as a check on previous calculations. Using

$$\varepsilon = \int_{\ell} \bar{v} \times \bar{B} \cdot d\bar{\ell}$$

$$E_A = 10^4 \int_0^{0.1} 0.01y\, dy = 50y^2 \Big|_0^{0.1} = 0.50 \text{ V.}$$

$$E_B = 10^4 \int_0^{0.1} 0.04y\, dy = 200y^2 \Big|_0^{0.1} = 2.00 \text{ V.}$$

From inspection of $\bar{v} \times \bar{B}$ in each case, the polarities are as indicated in Fig. 2. The IR drops in the bars are

$$IR_A = IR_B = 75 \times 0.01 = 0.75 \text{ V}$$

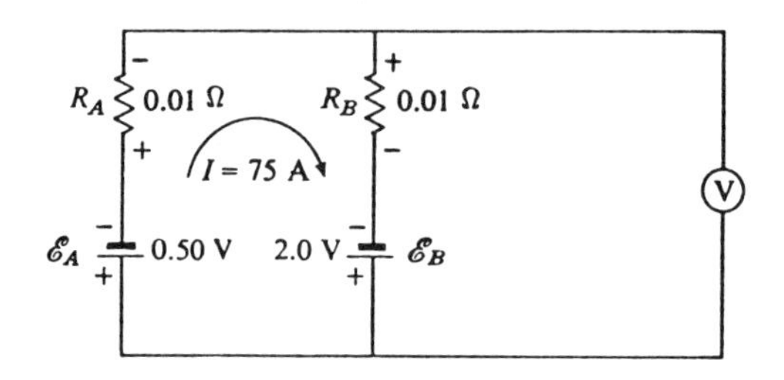

Fig. 2

Equivalent circuit for Fig. 1 when $v=10$, $B_z=10^3x^2y$ and $x=0.1$.

Therefore the voltmeter reading will be 1.25 V. If it is a polarity-sensitive instrument, the terminal marked (+) must be at the bottom to get an upscale reading.

MAXWELL'S EQUATIONS

• PROBLEM 11-24

Given an electric field

$$\bar{E} = \bar{j}A \cos \omega \left(t - \frac{z}{c}\right)$$

Determine the time-dependent magnetic intensity $\bar{H}$ in free space.

Solution: Use Maxwell's Equation

$$\bar{\nabla} \times \bar{E} = -\frac{\partial \bar{B}}{\partial t} = -\mu_0 \frac{\partial \bar{H}}{\partial t}$$

where $\bar{B}$ is the magnetic induction and μ_0 is the permeability of free space expanding the $\bar{\nabla} \times \bar{E}$ in rectangular coordinates

$$\bar{i}\left(\frac{\partial E_z}{\partial y} - \frac{\partial E_y}{\partial z}\right) + \bar{j}\left(\frac{\partial E_x}{\partial z} - \frac{\partial E_z}{\partial x}\right) + \bar{k}\left(\frac{\partial E_y}{\partial x} - \frac{\partial E_x}{\partial y}\right) = -\mu_0 \frac{\partial \bar{H}}{\partial t}$$

$$E_z = E_x = 0, \; E_y = f(t, z) \text{ only}$$

$$-\bar{i}\frac{\partial E_y}{\partial z} = -\mu \partial \frac{\partial \bar{H}}{\partial t}$$

$$\frac{\partial E_y}{\partial z} = \frac{\partial}{\partial z}\left[A \cos \omega \left(t - \frac{z}{c}\right)\right] = \frac{\omega}{c} A \sin \omega \left(t - \frac{z}{c}\right)$$

$$H_x = \frac{\omega A}{c\mu_0} \int \sin \omega \left(t - \frac{z}{c}\right) dt = - \frac{A}{c\mu_0} \cos \omega \left(t - \frac{z}{c}\right) + C_1$$

$$\bar{H} = - \frac{\bar{i}A}{c\mu_0} \cos \omega \left(t - \frac{z}{c}\right)$$

The constant C_1 can be set equal to zero since fields which are constant in time do not influence the time varying part.

• PROBLEM 11-25

Consider a simple magnetic field which increases exponentially with time,

$$\bar{B} = B_0 e^{bt} \bar{a}_z$$

where B_0 is constant. Find the electric field produced by this varying $\bar{B}$ field.

Solution: Choose a circular path of radius a in the z = 0 plane, along which E_ϕ must be constant by symmetry, then from

$$\text{emf} = \oint \bar{E} \cdot d\bar{\ell} = -\int_S \frac{\partial \bar{B}}{\partial t} \cdot d\bar{s}$$

$$\text{emf} = 2\pi a E_\phi = -bB_0 e^{bt} \pi a^2$$

The emf around this closed path is $-bB_0 e^{bt}\pi a^2$. It is proportion to a^2, since the magnetic flux density is uniform and the flux passing through the surface at any instant is proportional to the area. The emf is evidently the same for any other path in the z = 0 plane enclosing the same area.

Now replace a by r, the electric field intensity at any point is

$$\bar{E} = -\tfrac{1}{2} bB_0 e^{bt} r \bar{a}_\phi$$

Now attempt to obtain the same answer from

$$(\nabla \times \bar{E}) = - \frac{\partial \bar{B}}{\partial t}$$

which becomes

$$(\nabla \times \bar{E})_z = -bB_0 e^{bt} = \frac{1}{r} \frac{\partial(rE_\phi)}{\partial r}$$

Multiplying by r and integrating from 0 to r (treating t as a constant, since the derivative is a partial derivative),

$$-\tfrac{1}{2}bB_0 e^{bt} r^2 = rE_\phi + K$$

or

$$\bar{E} = -\tfrac{1}{2}bB_0 e^{bt} r\bar{a}_\phi$$

once again.

• PROBLEM 11-26

Consider an infinitely long cylindrical region containing a $\bar{B}$ field given in cylindrical coordinates by

$$\bar{B} = \begin{cases} B_0 \cos(\omega t + \alpha)\bar{z} & (\rho \le a) \\ 0 & (\rho > a) \end{cases} \quad (1)$$

where B_0 = const. In other words, $\bar{B}$ is spatially constant over the area of the circle but harmonically oscillating in time; visualize this as being produced by an infinite ideal solenoid with an alternating current in the windings. Find the induced $\bar{E}$ field due to this alternating $\bar{B}$ field and plot the result.

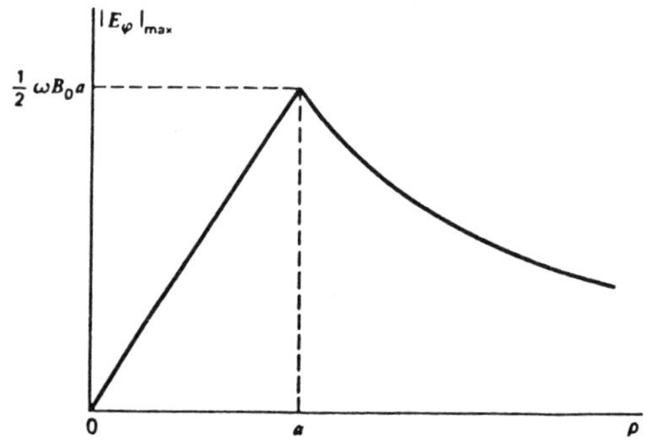

The induced electric field as a function of distance from the axis of a cylinder of radius *a* containing an alternating induction.

Solution: $\nabla \times \bar{E} = -\frac{\partial \bar{B}}{\partial t} = \omega B_0 \sin(\omega t + \alpha)\bar{z}$. The cylindrical symmetry of this problem leads to the expectation that $\bar{E}$ will lie in the xy plane and have the form $\bar{E} = E_\phi(\rho)\,\bar{\phi}$, that is, be tangent to circles of radius ρ. Accordingly, choose such a circle as a path of integration, and then, for any ρ,

$$\oint_C \bar{E} \cdot d\bar{s} = \oint_C E_\phi \bar{\phi} \cdot \rho d\phi \bar{\phi} \quad = 2\pi\rho E_\phi$$

$$= \int_S (\nabla \times \bar{E}) \cdot d\bar{a} = \omega B_0 \sin(\omega t + \alpha) \int da_z \qquad (2)$$

with the use of Stoke's theorem. The area integral equals $\pi\rho^2$ if $\rho \leq a$, and has the constant value πa^2 if $\rho > a$ because of (1). Substituting these into (2),

$$E_\phi = \frac{1}{2}\omega B_0 \rho \sin(\omega t + \alpha) \qquad (\rho \leq a)$$

$$E_\phi = \frac{1}{2}\omega B_0 \left(\frac{a^2}{\rho}\right) \sin(\omega t + \alpha) \qquad (\rho > a)$$

The figure shows the maximum value of $|E_\phi|$, that is, its amplitude, as a function of ρ. As in the last example, E_ϕ is 90° out of phase with $\bar{B}$, depending as it does on the rate of change of $\bar{B}$, rather than on its absolute value. Induced electric fields produced in this general way are the basis of operation of the charged particle accelerator known as the betatron.

• PROBLEM 11-27

Given a magnetic field in free space where there is neither charge nor current density ($\rho = J = 0$),

$$\bar{B} = \bar{i}\, a \sin(\omega t - nx) + \bar{j}\, any \cos(\omega t - nx)$$

where a, n, and ω are constants.

Use a Maxwell equation to derive the time-dependent part $\bar{E}$ of the electric field.

Solution: The Ampere-Maxwell equation is (1)

$$\nabla \times \bar{H} = \bar{J} + \frac{\partial \bar{D}}{\partial t} \qquad (1)$$

where $\bar{J} = 0$ in this problem.

In free space $\bar{D} = \varepsilon_0 \bar{E}$, $\bar{B} = \mu_0 \bar{H}$. Hence (1) becomes

$$\frac{1}{\mu_0} \nabla \times \bar{B} = \varepsilon_0 \frac{\partial \bar{E}}{\partial t}$$

$$\nabla \times \bar{B} = \mu_0 \varepsilon_0 \frac{\partial \bar{E}}{\partial t} \tag{2}$$

$$\nabla \times \bar{B} = \bar{i}\left(\frac{\partial B_z}{\partial y} - \frac{\partial B_y}{\partial z}\right) + \bar{j}\left(\frac{\partial B_x}{\partial z} - \frac{\partial B_z}{\partial x}\right)$$

$$+ \bar{k}\left(\frac{\partial B_y}{\partial x} - \frac{\partial B_x}{\partial y}\right); \quad B_z = 0$$

$B_y \neq f(z)$, hence $\dfrac{\partial B_y}{\partial z} = 0$

$B_x \neq f(z)$, hence $\dfrac{\partial B_x}{\partial z} = 0$

$$\therefore \nabla \times \bar{B} = \bar{k}\left(\frac{\partial B_y}{\partial x} - \frac{\partial B_x}{\partial y}\right), \quad B_x \neq f(y) \quad \frac{\partial B_x}{\partial y} = 0$$

$$\frac{\partial B_y}{\partial x} = an^2 y \sin(\omega t - nx)$$

$$\nabla \times \bar{B} = \bar{k}\, a\, n^2 y \sin(\omega t - nx) = \mu_0 \varepsilon_0 \frac{\partial \bar{E}}{\partial t}$$

$$\therefore \frac{\partial \bar{E}}{\partial t} = \frac{\bar{k}}{\mu_0 \varepsilon_0} [an^2 y \sin(\omega t - nx)]$$

$$\bar{E} = \frac{\bar{k}}{\mu_0 \varepsilon_0} \frac{an^2 y}{\omega} \int_0^t \sin(\omega t - nx)\, \omega dt$$

$$\bar{E} = -\frac{\bar{k} a\, n^2 y}{\mu_0 \varepsilon_0 \omega} \cos(\omega t - nx)$$

• PROBLEM 11-28

Determine the field of two equal and opposite distributions of charge on the surfaces of two parallel conducting plates by using Maxwell's equation. For simplicity assume that the plates are in free space.

Make the further simplifying assumption that fringing may be neglected.

Solution: In mathematical language, this means that, with reference to the given figure, $\partial/\partial y$ and $\partial/\partial z$ are equal to zero.

Let a and b be the dimensions of each plate, and d their distance of separation. Let ρ_s be the density of surface charge on the top plate, and $-\rho_s$ the density on the lower plate. Since the charges are stationary, the resulting field is independent of time, and $\bar{J} = 0$. The differential laws at a point not on the plates, consequently, reduce to the simpler forms

$$\nabla \times \bar{E} = 0 \tag{1}$$

$$\nabla \times \bar{H} = 0 \tag{2}$$

$$\nabla \cdot \bar{B} = 0 \tag{3}$$

$$\nabla \cdot \bar{D} = 0 \tag{4}$$

Eliminate two of the four field vectors by introducing the constitutive relations

$$\bar{D} = \varepsilon_0 \bar{E} \qquad \bar{B} = \mu_0 \bar{H}$$

Since ε_0 and μ_0 are both constants, Eqs. (3) and (4) may be replaced, respectively, by

$$\nabla \cdot \bar{H} = 0$$

$$\nabla \cdot \bar{E} = 0$$

Thus $\bar{E}$ must satisfy the pair

$$\nabla \times \bar{E} = 0 \qquad \nabla \cdot \bar{E} = 0$$

and H must satisfy the similar pair

$$\nabla \times \bar{H} = 0 \qquad \nabla \cdot \bar{H} = 0$$

Since the equations in $\bar{E}$ and $\bar{H}$ are uncoupled, deal with the $\bar{E}$ field first.

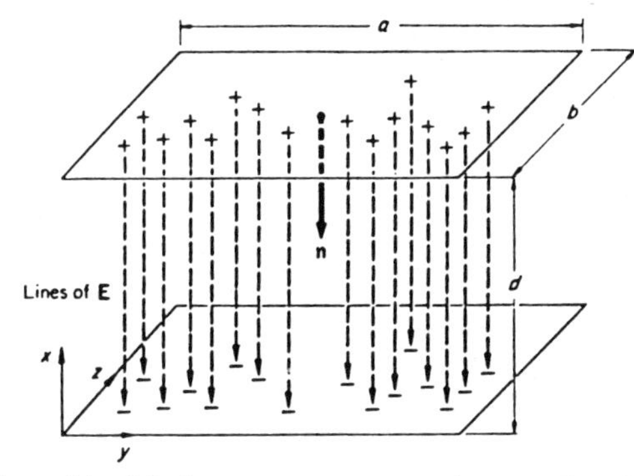

The field between two charged plates.
Uniform charge distribution.

Using the definitions of curl and divergence, and remembering the hypothesis $\partial/\partial y = \partial/\partial z = 0$, expand the field equations in $\bar{E}$ to obtain

$$\text{From } \nabla \times \bar{E} = 0: \qquad -\frac{\partial E_z}{\partial x}\bar{a}_y + \frac{\partial E_y}{\partial x}\bar{a}_z = 0 \tag{5}$$

From $\nabla \cdot \bar{E} = 0$: $$\frac{\partial E_x}{\partial x} = 0 \qquad (6)$$

From Eq. (5) it is clear that both E_z and E_y must be constants independent of x, y, and z. However, the boundary conditions require that they should both vanish on the conducting plates; otherwise the continuity of tangential $\bar{E}$ would require the existence of an $\bar{E}$ field within the conducting plates. This $\bar{E}$ field in turn would exert a Lorentz force on the charges, causing them to move, contrary to the assumed stationary character of the charge distribution. The condition may be satisfied only if

$$E_y = E_z = 0$$

On the other hand, from Eq. (6) it can be concluded that

$$E_x = \text{constant } C$$

To evaluate the constant C, calculate

$$D_x = \varepsilon_0 E_x = \varepsilon_0 C$$

and recall the boundary condition

$$\bar{n} \cdot (\bar{D}_2 - \bar{D}_1) = \rho_s \qquad (7)$$

Consider the upper plate, which is denoted medium 1. If medium 2 is the space between the plates, the unit normal to the upper plate will point downward, as shown in the given figure. Since $\bar{E}_1$ is zero, $\bar{D}_1$ in Eq. (7) is zero. With $\bar{n} = -\bar{a}_x$, Eq. (7) gives

$$-\bar{a}_x \cdot D_x \bar{a}_x = \rho_s$$

Therefore

$$D_x = -\rho_s$$

The same result would be obtained by repeating the procedure for the lower plate. Therefore

$$C = -\frac{\rho_s}{\varepsilon_0}$$

and at any point between the plates

$$\bar{E} = C\bar{a}_x = -\frac{\rho_s}{\varepsilon_0}\bar{a}_x$$

The same formal procedure can be used to obtain the $\bar{H}$ field. The student will verify that, since there are no currents anywhere, the result is $\bar{H} = 0$, everywhere.

DISPLACEMENT CURRENT

• **PROBLEM 11-29**

A parallel-plate capacitor consists of two circular plates of radius R = 10.0 cm. Suppose that the capacitor is being charged at a uniform rate so that the electric field between the plates changes at the constant rate $dE/dt = 10^{13} V \cdot m^{-1} \cdot sec^{-1}$. Find the displacement current for the capacitor. Derive an expression for the magnitude B of the induced magnetic field at a distance r from the center of the capacitor in a direction parallel to the plates. Evaluate B at r = R.

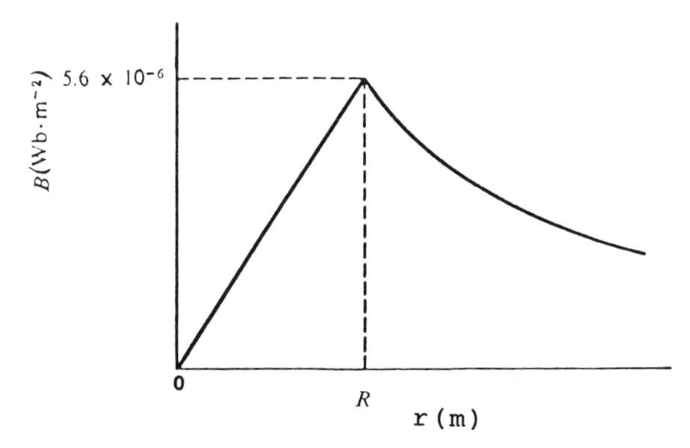

Variation of the induced magnetic field between the plates of the capacitor due to the changing electric field.

Solution: The total displacement current I_D is

$$I_D = \varepsilon_0 \frac{\partial E}{\partial t} \pi R^2$$

$$= 8.9 \times 10^{-12} \times 10^{13} \times 3.14 \times (0.1)^2$$

$$= 2.8 \text{ A}.$$

From Ampere's law,

$$\oint \bar{B} \cdot d\bar{\ell} = \mu_0 \varepsilon_0 \int \frac{\partial \bar{E}}{\partial t} \cdot d\bar{S}.$$

For $r \le R$

$$B(2\pi r) = \mu_0 \varepsilon_0 \frac{dE}{dt} (\pi r^2)$$

or

$$B = \frac{\mu_0 \varepsilon_0}{2} r \frac{dE}{dt} .$$

For $r \geq R$

$$B(2\pi r) = \mu_0 \varepsilon_0 \frac{dE}{dt}(\pi R^2)$$

or

$$B = \frac{\mu_0 \varepsilon_0}{2} \frac{R^2}{r} \frac{dE}{dt} .$$

The variation of the induced magnetic field with r is shown in the figure. The magnitude of the field at r = R is

$$B = \frac{\mu_0 \varepsilon_0}{2} R \frac{dE}{dt}$$

$$= \frac{1}{2} \times 4\pi \times 10^{-7} \times 8.9 \times 10^{-12} \times 0.1 \times 10^{13}$$

$$= 5.6 \times 10^{-6} \text{ Wb} \cdot \text{m}^{-2}.$$

Note that even though the displacement current is reasonably large, it produces only a small magnetic field.

• PROBLEM 11-30

(a) Illustrated in the figure is a conducting rod with a circular cross section of radius 2.1 centimeters. The current i varies sinusoidally with time at a frequency of 1590 cycles per second, and its positive direction is specified by the arrow on the illustration. Inside the wire the drift current density $J_z\overline{k}$ varies with respect to the distance r from the center of the wire because of a phenomenon known as skin effect. The current density J_z is given by the approximate expression

$$J_z = 0.001e^{\pi r} \sin(\omega t + \pi r) \text{ amperes/cm}^2$$

with $\omega = 10^4$ and with r measured in centimeters. Determine the current i as a function of time.

(b) The rod in part (a) is made of brass having a conductivity of 1.57×10^7 mhos per meter. Determine the displacement current i_d as a function of time. Assume $\varepsilon_r = 1$.

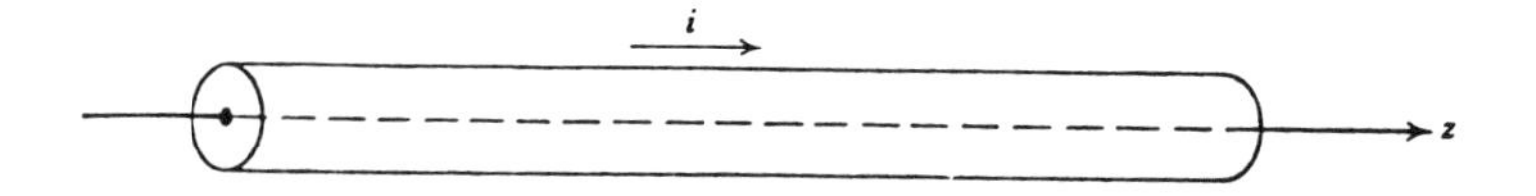

Cylindrical conductor

Solution: As J_z depends only on the distance r from the center of the wire and time t, it is clear that a differential area of a cross-sectional surface can be taken as the area between two circles of radii r and r + dr. This is $2\pi r\,dr$ square centimeters. The current arrow on the illustration specifies the positive side of the surface as the side facing in the positive z-direction. Therefore, the vector differential area is $2\pi r\,dr\,\bar{k}$, and $\bar{J} \cdot d\bar{S}$ equals $J_z 2\pi r\,dr$. The current is

$$i = \int_0^{2.1} J_z 2\pi r\,dr$$

Substitution for J_z gives

$$i = 0.002\pi \int_0^{2.1} re^{\pi r} \sin(\omega t + \pi r)dr$$

A change of variable from r to x, with $x = \pi r$, yields the equation

$$i = \frac{0.002}{\pi} \int_0^{2.1\pi} xe^x \sin(\omega t + x)dx$$

By integrating by parts one can readily show that

$$\int xe^x \sin(\omega t + x)\,dx = [\tfrac{1}{2}e^x][(x \cos x + x\sin x - \sin x)$$

$$x \sin \omega t + (x \sin x - x \cos x + \cos x)$$

$$x \cos \omega t] \qquad (1)$$

Utilizing Eq. (1), the current i is

$$i = 1.87 \sin \omega t - 0.769 \cos \omega t$$

This can be expressed as

$$i = 2.024 \sin(\omega t - 22.3°) \text{ amperes}$$

by application of the trigonometric identity $A \sin \omega t + B \cos \omega t = C \sin(\omega t + \theta)$ where $C = \sqrt{A^2 + B^2}$ and $\theta = \arctan B/A$. The sign of the current is alternately positive and negative as time varies.

(b) The displacement current density $\bar{J}_d$ is $\dot{\bar{D}}$, or $\varepsilon\dot{\bar{E}}$. As $\bar{E} = \bar{J}/\sigma$, it follows that $\bar{J}_d$ equals $(\varepsilon/\sigma)\dot{\bar{J}}$. The ratio ε/σ is $8.854 \times 10^{-12}/1.57 \times 10^7$, or 5.64×10^{-19} second, and

$$\bar{J}_d = 5.64 \times 10^{-19}\dot{J}_z\bar{k}$$

The drift current density J_z is given in part (a). Its partial time derivative equals J_z multiplied by ω, or 10^4, with the angle of the sine function increased by 90°. Thus

$$J_d = 5.64 \times 10^{-18} e^{\pi r} \sin(\omega t + 90° + \pi r) \text{ ampere/cm}^2$$

with r measured in centimeters. Comparison of this with the expression for the drift current density J_z reveals that J_d can be found from J_z by multiplying J_z by 5.64×10^{-15} and by increasing its angle by 90°. Deduce that the displacement current can be found from the drift current i by multiplying i by 5.64×10^{-15} and by adding 90° to its angle. This gives

$$i_d = 1.14 \times 10^{-14} \sin(\omega t + 67.7°) \text{ ampere}$$

It should be noted that the displacement current is negligible compared with the drift current. This is true in all good conductors.

GENERATORS

• PROBLEM 11-31

A coil of N turns and area A is rotated in a uniform magnetic field B, at an angular velocity ω_0 rad/sec as shown in Fig. 1. Calculate the emf generated, using the Farady induction law.

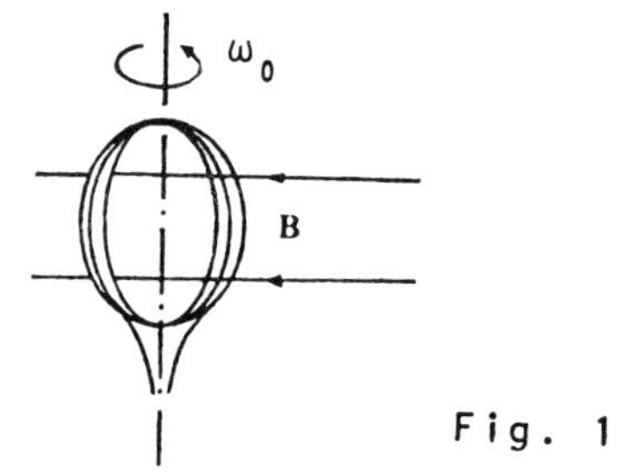

Fig. 1

A simple generator: a rotating coil in an external magnetic field.

Solution: The flux linkage threading the coil at an angle θ between the plane of the coil and the direction of B is given by

$$N\phi = NAB \sin\theta = NAB \sin\omega_0 t$$

where $\omega_0 t$ is the value of θ for a particular time t.

Then

$$E = - N \frac{d\phi}{dt} = -\omega_0 NAB \cos\omega_0 t$$

Thus this generator gives a sinusoidal emf of amplitude $\omega_0 NAB$. The source of this emf is the mechanical work done in rotating the coil. This is zero (in a frictionless system) if no current flows and increases linearly with increasing current.

• PROBLEM 11-32

The rectangular loop in Fig. 1, of length a and width b, is rotating with uniform angular velocity ω about the y-axis. The entire loop lies in a uniform, constant B-field, parallel to the z-axis. Calculate the motional emf in the loop from

$$E = \int_b^a \bar{E}_n \cdot d\bar{s} = E_n \ell = vB\ell.$$

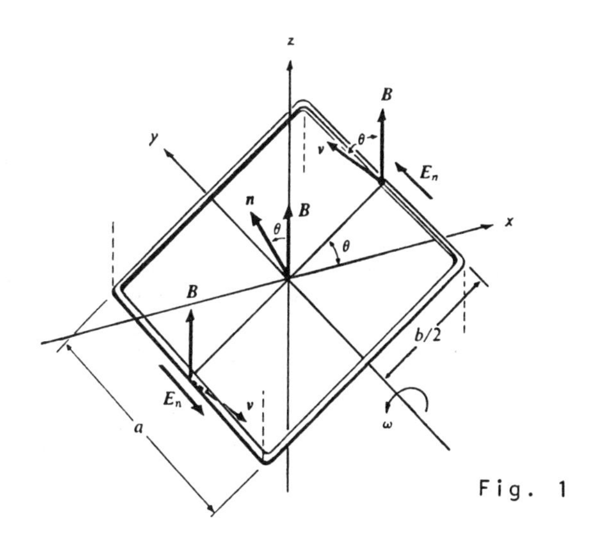

Fig. 1

Rectangular loop rotating with constant angular velocity in a uniform magnetic field.

Solution: The velocity v of the sides of the loop of length a is

$$v = \omega \frac{b}{2}.$$

The direction of the motional field $\bar{E}_n$ $(= \bar{v} \times \bar{B})$ in each of the sides of length a is shown in the diagram.

Its magnitude is

$$E_n = vB \sin \theta = \omega \frac{b}{2} B \sin \theta.$$

The motional fields in the other two sides of the loop are transverse to these sides and contribute nothing to the emf. The line integral of $\bar{E}_n$ around the loop reduces to $2E_n a$, so

$$E = \oint \bar{E}_n \cdot d\bar{s} = 2E_n a = \omega(ab)\, B \sin \theta.$$

The product ab equals the area A of the loop, and if the loop lies in the xy-plane at t = 0, then $\theta = \omega t$. Hence

$$E = \omega AB \sin \omega t. \tag{1}$$

The emf therefore varies sinusoidally with the time. The maximum emf E_m, which occurs when $\sin \omega t = 1$, is

$$E_m = \omega AB,$$

so write Eq. (1) as

$$E = E_m \sin \omega t. \tag{2}$$

The rotating loop is the prototype of the alternating current generator.

The rotating loop in Fig. 1 can be utilized as the source in an external circuit by making connections to slip rings S, S, which rotate with the loop as shown in Fig. 2 (a). Stationary brushes bearing against the rings are connected to the output terminals a and b. The instantaneous terminal voltage v_{ab}, on open circuit, equals the instantaneous emf. Figure 2 (b) is a graph of $\bar{v}_{ab}$ as a function of time.

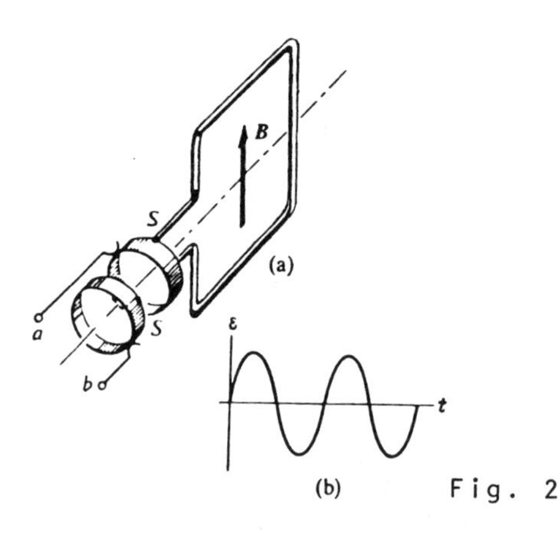

Fig. 2

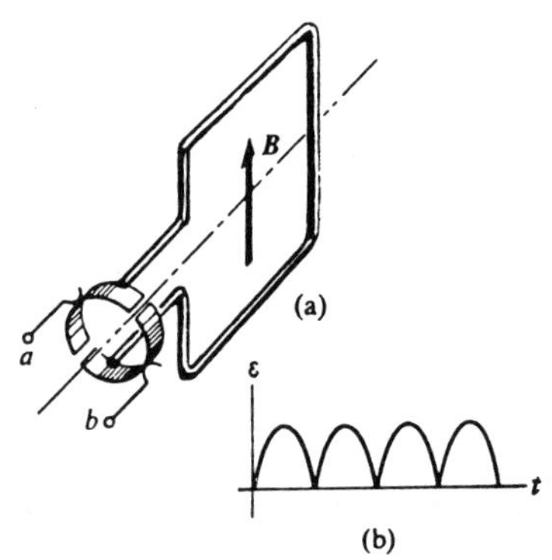

Fig. 3

A terminal voltage that always has the same sign, although it fluctuates in magnitude, can be obtained by connecting the loop to a split ring or commutator, as in Fig. 3(a). At the position in which the emf reverses, the connections to the external circuit are interchanged. Figure 3(b) is a graph of the terminal voltage, and the device is the prototype of a dc generator.

Now repeat the problem using Faraday's Law,

$$E = -\frac{d\Phi}{dt}$$

The flux through the loop equals that through its projected area on the xy-plane (shaded in Fig. 1).

$$\Phi = \bar{B} \cdot \bar{A} = BA \cos \theta = BA \cos \omega t.$$

Then

$$\dot{\Phi} = \frac{d\Phi}{dt} = -\omega BA \sin \omega t$$

and

$$E = -\dot{\Phi} = \omega BA \sin \omega t.$$

in agreement with the previous result.

Note that the maximum value of E occurs when $\theta = 90°$ and the flux through the loop is zero, and that $E = 0$ when $\theta = 0$ and the flux is a maximum. That is, the emf depends not on the flux through the loop, but on its rate of change.

• PROBLEM 11-33

The simple device sketched in Fig. 1, exhibits the principle of operation of a mechanical generator. A metal rod can be slid along a rigid wire frame which is in a uniform magnetic field $\bar{B}$, as shown. What is the rate of energy dissipation and the work rate? Assume that the flux is negative (into the paper).

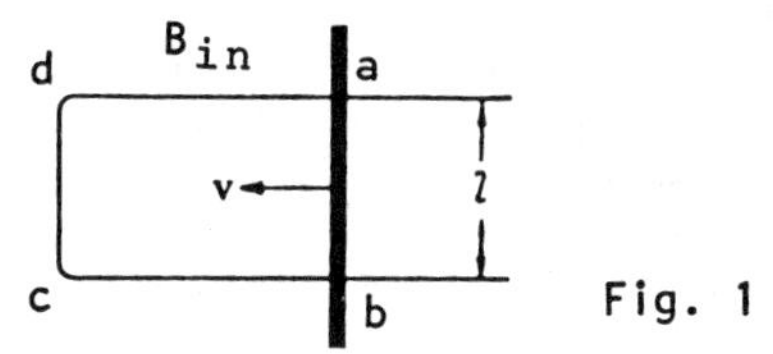

Sliding wire on a stationary loop in a field B.

Solution: While the rod of length ℓ is being moved with velocity v, the area within the loop increases at a rate $v\ell$. If the field intensity in the region is B, the time rate of change of flux within the loop is $d\phi/dt = v\ell B$. Once the time variation of ϕ inside the loop is found, the Faraday law can be used to find the induced emf.

$$E = -\frac{d\phi}{dt} = |v\ell B| \quad \text{volts}$$

Since ϕ itself is negative (away from the viewer) and decreasing with time, $d\phi/dt$ is positive; so the emf is negative, or clockwise.

To calculate the rate of electric energy dissipation into heat compared with the mechanical work rate is easy to do if we first determine the current which results from the emf around the circuit. The current is given by

$$i = \frac{E}{R} = \frac{v\ell B}{R} \quad \text{amp}$$

where R is the resistance around the loop.

Electric energy is being dissipated in the resistance at a rate

$$P = Ei = \frac{v^2\ell^2B^2}{R} \quad \text{watts}$$

The source of this energy is the mechanical work being done in moving the rod. The force on the rod carrying a current i in a field B is $F = Bi\ell$, so that the rate at which mechanical work being done is

$$\frac{dW}{dt} = Fv = v\ell Bi = \frac{v^2\ell^2B^2}{R} \quad \text{watts}$$

Comparison of the work rate and power dissipation in the resistance of the loop gives the identity expected on the basis of energy conservation.

• PROBLEM 11-34

(a) The figure shows a cross-section of a homopolar generator. B is a copper cylinder which rotates in a radial field between two cylindrical pole-pieces A and C, of which A is the north pole. The circuit is completed by means of two fixed brushes sliding on the cylinder, one at each of its ends. The mean flux-density of the field over the thickness of the cylinder wall is 50,000 maxwells (lines) per sq. inch, the mean radius of the cylinder is 5 in., its axial length 20 in., and it rotates at 2000 r.p.m. Find the magnitude and direction of the e.m.f. which appears between the brushes.

(b) Suppose that it is possible to make the whole of the external circuit (that is, that part extending from brush to brush through the voltmeter) rotate in a clockwise direction at a speed of 1000 r.p.m. Find the magnitude and direction of the e.m.f. in the circuit

(1) with the cylinder at rest.

(2) when the cylinder rotates clockwise at 2000 r.p.m.

(3) when the cylinder rotates counter-clockwise at 1000 r.p.m.

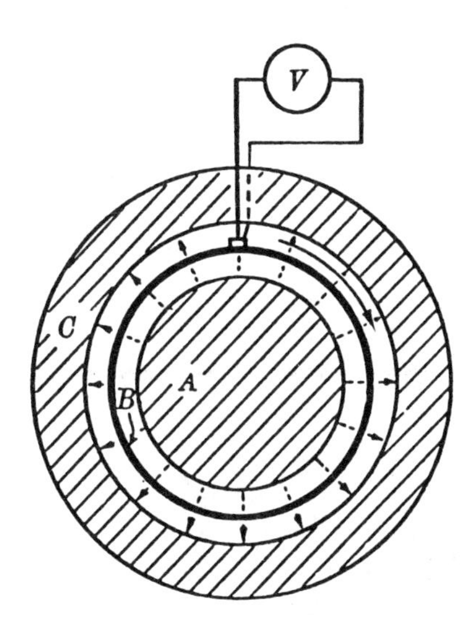

Homopolar generator

Solution: (a) From the data it is clear that the e.m.f. measured by the voltmeter is induced entirely in the rotating cylinder.

The peripheral speed of the cylinder, relative to the fixed voltmeter, is

$$v = \frac{2000}{60} \times 2\pi \times 5 \times 0.0254 = 26.6 \text{ meters per sec.},$$

$$L = 20 \times 0.0254 = 0.508 \text{ meter},$$

$$B = \frac{50{,}000}{(2.54)^2} \text{ gauss} = 0.775 \text{ weber per sq. meter;}$$

hence

$$e = BLv = 0.775 \times 0.508 \times 26.6 = 10.47 \text{ volts.}$$

The direction of this e.m.f. is, in the rotating cylinder, towards the reader.

(b) (1) The cylinder rotates at 1000 r.p.m. in a counter-clockwise direction with respect to the voltmeter. Hence

$$v = 13.3 \text{ meters per sec.}$$

and e = 5.23 volts in a direction opposite to that in part (a).

(2) The cylinder now rotates at 1000 r.p.m. in a clockwise direction with respect to the voltmeter. Hence e = 5.23 volts in the same direction as in (a).

(3) The cylinder rotates counter-clockwise at 2000 r.p.m. with respect to the voltmeter, so that e = 10.47 volts, in a direction opposite to that of (a).

• PROBLEM 11-35

A disk of radius R, shown in the figure, lies in the xz-plane and rotates with uniform angular velocity ω about the y-axis. The disk is in a uniform, constant $\bar{B}$-field parallel to the y-axis. Calculate the motional emf between the center and the rim.

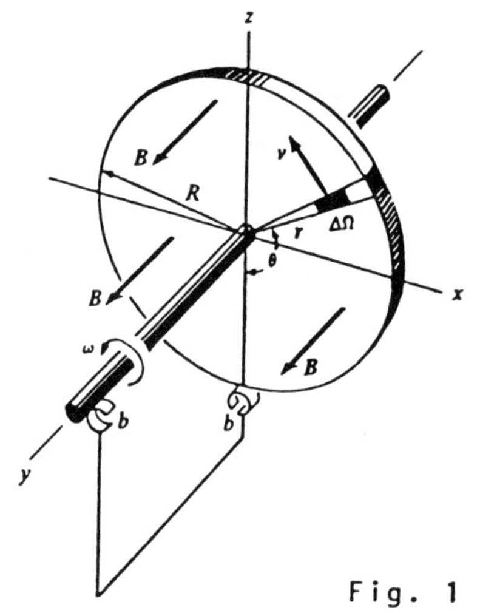

Fig. 1

Solution: Consider a short portion of a narrow radial segment of the disk, of length dr. Its velocity is $v = \omega r$, and since $\bar{v}$ is at right angles to $\bar{B}$, the motional field E_n in the segment is

$$E_n = vB = \omega rB.$$

The direction of $\bar{E}_n$ is radially outward.

The emf between center and rim is

$$E = \int_0^R \bar{E}_n \cdot d\bar{r} = \omega B \int_0^R r\, dr = \omega BR^2/2.$$

All of the radial segments of the disk are in parallel, so the emf between center and rim equals that in any radial segment. The entire disk can therefore be considered a source for which the emf between center and rim equals $\omega BR^2/2$. The source can be included in a closed circuit by completing the circuit through sliding contacts of brushes b, b.

To compute the emf from Faraday's law, $E = -\frac{d\phi}{dt}$, consider the circuit to be the periphery of the shaded areas in Fig. 1. The rectangular portion in the yz-plane is fixed. The area of the shaded section in the xz-plane is $\frac{1}{2}R^2\theta$, and the flux through it is

$$\Phi = \tfrac{1}{2}BR^2\theta.$$

As the disk rotates, the shaded area increases. In a time dt, the angle θ increases by $d\theta = \omega\ dt$. The flux increases by

$$d\Phi = \tfrac{1}{2}BR^2\ d\theta = \tfrac{1}{2}BR^2\ \omega\ dt,$$

and the induced emf is

$$E = \frac{d\Phi}{dt} = \tfrac{1}{2}Br^2\omega,$$

in agreement with the previous result.

• PROBLEM 11-36

A direct current generator is constructed by having a metal cart travel around a set of perfectly conducting rails forming a large circle. The rails are L m apart and there is a uniform field B_0 normal to their plane (see fig. 1). The cart has a mass m; it is driven by a rocket engine having a constant thrust F_0; and a resistor R acts as an electrical load. Find the voltage as a function of time, and also the value of the voltage after the generator has reached steady-state conditions.

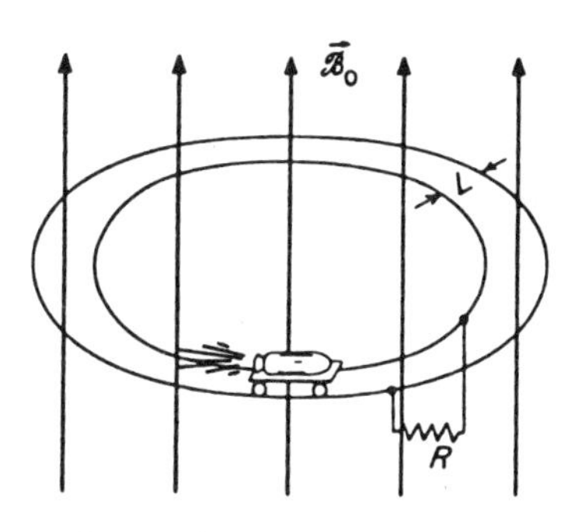

Fig. 1

A rocket generator.

Solution: By $F = I \oint d\ell \times B$, the force on a current element L m in length moving with velocity v at right angles to a field B_0 is

$$F = I \int_0^L B_0 \, d\ell = B_0 I L$$

Now the current is

$$I = \frac{V}{R}$$

where V is the induced voltage, and this in turn is given by

$$V = B_0 vL$$

so that

$$F = B_0^2 L^2 \frac{v}{R}$$

The equation of motion of the cart is then

$$m \frac{dv}{dt} = F_0 - \frac{B_0^2 L^2 v}{R}$$

or

$$m\frac{dv}{dt} + \frac{B_0^2 L^2 v}{R} = F_0$$

To solve this equation, separate variables in the following fashion

$$\frac{dv}{(F_0/m) - (B_0^2 L^2/mR)v} = dt$$

Introducing the abbreviations

$$a = \frac{F_0}{m}, \quad b = \frac{B_0^2 L^2}{mR}$$

and integrating,

$$\int \frac{dv}{a - bv} = -\frac{1}{b} \ln (a - bv) = t + C$$

where C is a constant of integration or

$$a - bv = e^{-b(t+C)}$$

To evaluate C, using the initial conditions specified by v = 0 at t = 0. Then

$$a = e^{-bC}$$

and

$$a - bv = e^{-bt}e^{-bC} = e^{-bt}a$$

so that

$$v = \frac{a}{b}(1 - e^{-bt})$$

or

$$v = \frac{F_0 R}{B_0^2 L^2}\ (1 - e^{-t/(mR/B_0^2 L^2)})$$

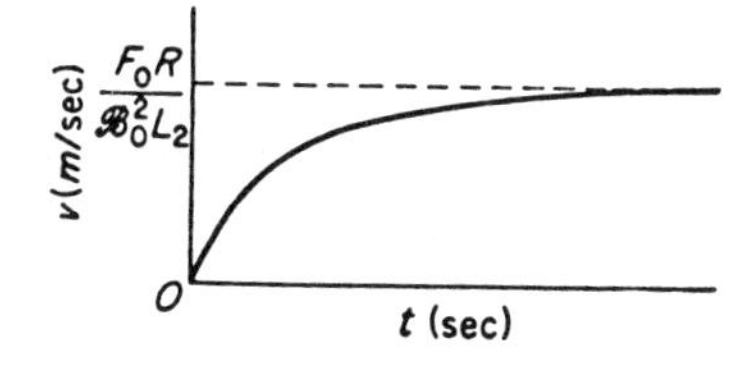

Fig. 2

Velocity as a function of time for the rocket generator.

This result shows that the velocity rises from its initial value of v = 0 and approaches its steady-state value asymptotically (Fig. 2). The quantity $(mR/B_0^2 L^2)$ is the time constant for this process, and plays the same role as the time constant RC for the charging of a capacitor C through a resistance R. The induced voltage is then

$$V = \frac{F_0 R}{B_0 L}\ (1 - e^{-(B_0^2 L^2/mR)t})$$

and this reaches a steady-state value of $F_0 R/B_0 L$ as the time t approaches large values.

CHAPTER 12

PLANE WAVES

ENERGY AND THE POYNTING VECTOR

• **PROBLEM 12-1**

If 5 watts/meter^2 is the Poynting vector of a plane wave traveling in free space, find its average energy density.

Solution:

The Poynting vector, energy density, and the velocity are related by

$$\frac{\text{Poynting vector}}{\text{Velocity}} = \text{Energy density}$$

i.e. $\frac{S_{av}}{\upsilon} = W_{av}$

Here υ (in free space) $= 3 \times 10^8$ meters/sec

$S_{av} = 5$ watts/meter^2 (given)

$$\therefore W_{av} = \frac{5}{3 \times 10^8} = \frac{5 \times 10^{-8}}{3} \text{ Joule/meter}^3$$

or $\quad W_{av} = \frac{1}{6}$ erg/meter^3

• **PROBLEM 12-2**

Consider a straight wire of radius r_0 oriented along the Z-axis carrying a steady current I. Determine the total power entering a unit length of wire.

Solution: First find the electric and magnetic fields

of this configuration. The magnetic field at the surface of the wire is

$$\bar{H} = \frac{I}{2\pi r_0} \bar{a}_\phi$$

The electric field is in the Z-direction and it is equal to

$$\bar{E} = \frac{\bar{J}}{\sigma} = \frac{J\bar{a}_z}{\sigma} = \frac{I}{\pi(r_0)^2 \sigma} \bar{a}_z$$

Hence the Poynting vector at the surface of the wire is equal to

$$\bar{S} = \bar{E}_z \times \bar{H}_\phi = - \frac{I^2}{2\pi^2 (r_0)^3 \sigma} \bar{a}_r$$

The total power entering a unit length of wire is

$$\oint_s \bar{S} \cdot d\bar{s} = \frac{I^2}{2\pi^2 (r_0)^3 \sigma} \cdot 2\pi r_0 = I^2 \frac{1}{\pi(r_0)^2 \sigma} = I^2 R$$

which indicates that the field supplies the energy to balance out the I^2R heating losses in the wire. The contribution to the surface integral is only from the cylindrical part of the surface; the contribution from the top and the bottom surfaces is zero since $\bar{E}_z \times \bar{H}_\phi$ is perpendicular to $d\bar{s}$.

• PROBLEM 12-3

Care must be employed when sinusoidal quantities multiply each other if the quantities are represented by complex exponential functions, i.e., by either the real or the imaginary parts of these functions. Calculate the Poynting vector for a linearly polarized plane wave in free space.

Solution: For a linearly polarized wave in free space,

$$\bar{E} = \hat{y} E_0 \cos (\omega t - kx)$$

$$\bar{B} = \hat{z} B_0 \cos (\omega t - kx)$$

Then

$$\bar{S} = \left(\frac{1}{\mu_0}\right) \bar{E} \times \bar{B} = \hat{x} \frac{1}{\mu_0} E_0 B_0 \cos^2 (\omega t - kx)$$

At x = 0 say, the instantaneous energy flux will be fluctuating but unidirectional:

$$\bar{S}\ (x = 0,\ t) = \hat{x}\ \frac{1}{\mu_0} E_0 B_0 \cos^2 \omega t$$

while the energy flow averaged over any number of whole cycles will be

$$\langle \bar{S} \rangle = \hat{x}\ \frac{1}{2\mu_0}\ E_0 B_0$$

This is the answer obtained when the calculation is made using the trigonometric functions.

Now employ the exponential functions. At x = 0 say:

$$\bar{E} = \hat{y} E_0 e^{i\omega t}$$

$$\bar{B} = \hat{z} B_0 e^{i\omega t}$$

It is no longer correct to write $\bar{S} = (1/\mu_0)\ \bar{E} \times \bar{B}$ for this implies

$$\bar{S} = \left(\frac{1}{\mu_0}\right) R(\bar{E} \times \bar{B}) = \frac{1}{\mu_0}\ R\ [(\hat{y}E_0\ e^{i\omega t}) \times (\hat{z}B_0 e^{i\omega t})]$$

$$= \frac{1}{\mu_0} R\left(\hat{x} E_0 B_0 e^{2i\omega t}\right) = \hat{x}\ \frac{1}{\mu_0}\ E_0 B_0 \cos 2\omega t$$

Not only is this instantaneous value wrong, but the average value is zero instead of $\hat{x}\ (1/\mu_0)(E_0/\sqrt{2})(B_0/\sqrt{2})$.

Suppose $\bar{S} = (1/\mu_0)\ E \times B^*$ is tried, where the asterisk means complex conjugate. This really stands for

$$\bar{S} = \frac{1}{\mu_0}\ R(\bar{E} \times \bar{B}^*) = \frac{1}{\mu_0}\ R[(\hat{y}E_0 e^{i\omega t}) \times (\hat{z}B_0 e^{-i\omega t})]$$

$$= \frac{1}{\mu_0}\ R(\hat{x}E_0 B_0)$$

This, also, does not yield the correct value for the instanteous value of $\bar{S}$. (A similar result would be obtained with $(1/\mu_0)\ \bar{E}^* \times \bar{B}$.) But, except for a factor of ½, it gives the correct value for the average power. It is common practice, therefore, to use the form

$$\langle \bar{S} \rangle = R\left(\frac{1}{2\mu_0}\ \bar{E} \times \bar{B}^*\right)$$

for the time-averaged value of the Poynting vector.

• PROBLEM 12-4

The approximate radiation fields of a certain antenna are $H_\phi = (1/r) \sin\theta \cos(\omega t - \beta r)$ and $E_\theta = 377 H_\phi$. Determine the energy flow in watts out of the volume surrounded by the spherical surface S of radius r, with center at the origin.

Solution: $\bar{E} \times \bar{H} = E_\theta H_\phi \bar{a}_r$, because $\bar{a}_\theta \times \bar{a}_\phi = \bar{a}_r$. Therefore,

$$\bar{E} \times \bar{H} = (377/r^2) \sin^2 \theta \cos^2 (\omega t - \beta r)\bar{a}_r$$

$$p = \oint_S (\bar{E} \times \bar{H}) \cdot d\bar{S} \qquad (1)$$

gives the power out of the surface S. A differential vector surface area $d\bar{S}$ of this spherical surface is $r^2 \sin\theta \, d\theta \, d\phi \, \bar{a}_r$. Substitution for $\bar{E} \times \bar{H}$ and $d\bar{S}$ into Eq. (1), with the limits of θ being 0 and π and the limits of ϕ being 0 and 2π, yields

$$p = 377 \cos^2 (\phi t - \beta r) \int_0^{2\pi}\int_0^{\pi} \sin^3 \theta \, d\theta \, d\phi$$

Evaluation of the integrals gives $p = 3160 \cos^2 (\omega t - \beta r)$ watts. As time varies, the cosine squared function varies between zero and unity, with an average value of one-half. Therefore, the time-average power is 1580 watts, and this is the radiated power.

• PROBLEM 12-5

From $-\oint \bar{P} \cdot d\bar{s} = \int_V \bar{J} \cdot \bar{E} \, dv, \qquad (1)$

evaluate the total power-flux entering the closed surface S embracing a length ℓ of a long, round wire carrying a direct current I as in (a) of the accompanying figure. Compare the result with the volume integral of (1).

Solution: The closed surface S is noted in (b). The Poynting vector P on the peripheral surface $\rho = a$ is obtained from the known $\bar{E}$ and $\bar{H}$ fields, $\bar{H}$ being given from Ampere's law by $\bar{H} = \bar{a}_\phi \frac{I}{2\pi\rho}$ while $\bar{E}$ is obtained from the current density $J_z = I/A$ combined with $J = \sigma E$.

$$\bar{E} = \bar{a}_z E_z = \bar{a}_z J_z/\sigma = \bar{a}_z I/\sigma A$$

$$\bar{H} = \bar{a}_\phi H_\phi = \bar{a}_\phi I/2\pi a$$

The Poynting vector at $\rho = a$ on S is

$$\bar{P} = \bar{E} \times \bar{H} = \left(\frac{\bar{a}_z I}{\sigma A}\right) \times \left(\frac{\bar{a}_\phi I}{2\pi a}\right) = -\bar{a}_\rho \frac{I^2}{2\pi a A \sigma}$$

As seen in (b), $\bar{P}$ on the end caps contributes nothing to the inward power-flux, making the total inward power-flux over S

$$p \equiv -\oint_S \overline{P} \cdot d\overline{s} = -\int_{z=0}^{\ell}\int_{\phi=0}^{2\pi}\left(-\overline{a}_\rho \frac{I^2}{2\pi a A\sigma}\right) \cdot \overline{a}_\rho \rho \; d\phi dz$$

$$= \frac{I^2 2\pi a \ell}{2\pi a A\sigma} = I^2 \frac{\ell}{\sigma A} = I^2 R \text{ Watts.}$$

a result expressed in terms of the resistance of the wire.

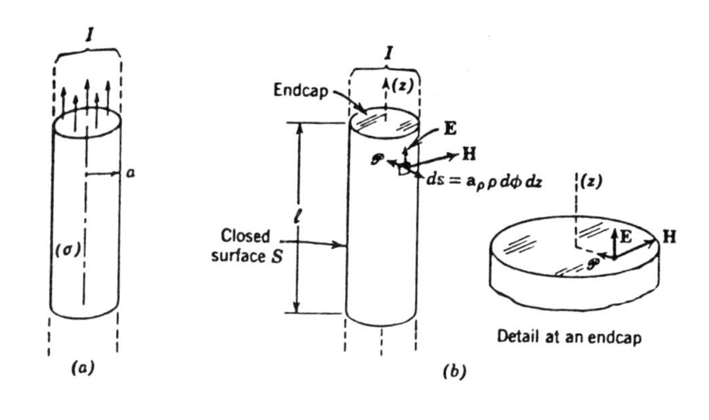

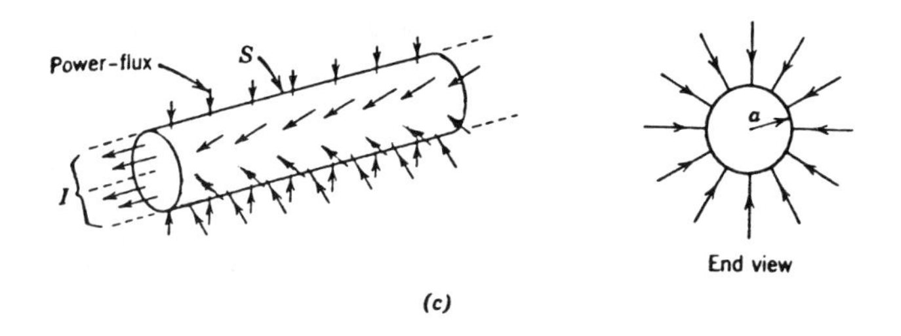

(a) Long, round wire carrying a static current I.
(b) The E and H fields on the surface S.
(c) Inward power-flux associated with direct current flow in a wire.

From (1), the result I^2R is also obtainable from the volume integral of $\overline{J} \cdot \overline{E}$ taken throughout the interior of S. Thus

$$\int_V \overline{J} \cdot \overline{E} \, dv = \int_V (\sigma\overline{E}) \cdot \overline{E} \, dv = \int_V \sigma E_z^2 \, dv$$

$$= \int_{z=0}^{\ell}\int_{\phi=0}^{2\pi}\int_{\rho=0}^{a} \sigma \left(\frac{I}{\sigma A}\right)^2 \rho d\rho \; d\phi \; dz$$

integrating to I^2R as expected. The positive sign accounts for the actual inward sense of the power-flux p over S, as noted in (c).

• PROBLEM 12-6

Use Poynting's theorem to find the rate at which energy is increasing within a very long solenoid of radius a wound on a core of permeability μ and show that the energy contained in a length ℓ is equal to $\frac{1}{2} Li^2$ (see the figure), exclusive of the energy in the wire.

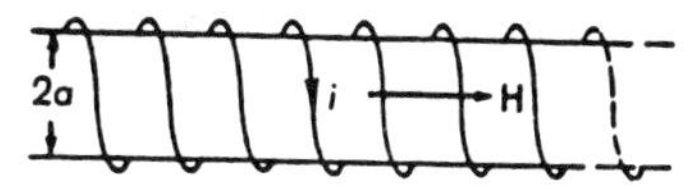

A long solenoid of radius *a* wound on a core of permeability μ.

Solution: By using the Ampere circuital law,

$$H = ni \quad \text{and} \quad B = \mu ni$$

where n is the number of turns per unit length and i is the conduction current (the only important current). Now, using the Faraday law for a circular path of radius a

$$\oint \bar{E} \cdot d\bar{s} = 2\pi aE = -\iint \frac{\partial \bar{B}}{\partial t} \cdot d\bar{S} = -\mu n \frac{di}{dt} (\pi a^2)$$

Therefore, the magnitude of $\bar{E}$ at the surface of the core is

$$E = \frac{1}{2} \mu na \frac{di}{dt}$$

and the magnitude of $\bar{E} \times \bar{H}$ is

$$\frac{1}{2} \mu n^2 ai \frac{di}{dt}$$

This is the power flowing in through unit area of the surface of the solenoid. The total power flowing into a volume of length ℓ is

$$\frac{dW}{dt} = \left(\frac{1}{2} \mu n^2 ai \frac{di}{dt}\right) (2\pi a\ell)$$

$$= \mu n^2 \pi a^2 \ell i \frac{di}{dt}$$

The total energy stored within the volume is obtained by integration

$$W = \frac{1}{2} (\mu n^2 \pi a^2 \ell) i^2$$

Since $\mu n^2 \pi a^2 \ell$ is the self-inductance of a length ℓ of the solenoid,

$$W = \frac{1}{2} Li^2$$

Consider a portion of a long straight cylindrical conductor of length ℓ and radius a as shown in the figure. Assume a constant current in the z direction, which is distributed uniformly across the cross section so that $\bar{J}_f = J_f\hat{z}$ = const. Find the poynting vector S at the surface just outside of the conductor.

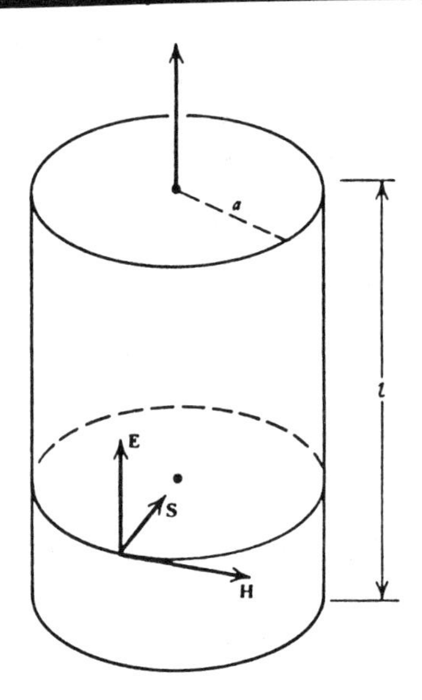

Fields and Poynting vector for a cylinder carrying a constant current.

Solution: The electric field just outside the cylinder is equal to the constant value inside and given by $E = J_f/\sigma$ making it parallel to the axis as shown. The value of $\bar{B}$ outside the cylinder is found from Ampere's law to be

$$\frac{\mu_0 I}{2\pi a}\hat{\phi},$$

so that, in this case, $\bar{H}$ just outside will be $\bar{H} = \bar{B}/\mu_0$, or

$$H = \frac{I}{2\pi a}\hat{\phi} = \frac{1}{2}J_f a\hat{\phi}$$

since $J_f = I/\pi a^2$. Substituting these expressions into $\bar{S} = \bar{E} \times \bar{H}$, and solving

$$\bar{S} = \frac{J_f}{\sigma}\hat{z} \times \frac{1}{2}J_f a\hat{\phi} = -\frac{J_f^2 a}{2\sigma}\hat{\rho} \tag{1}$$

It is clear from the figure, that $\bar{S}$ is normal to the surface and directed radially inward so that there is a steady flow of energy into the conductor as S is constant. Now the element of area $d\bar{a}$ is directed outward from the surface; also, $\bar{S}$ is parallel to the ends of the cylinder. Therefore, the total rate at which energy is flowing into the volume is given by

$$-\oint \bar{S} \cdot d\bar{a} = S\int da = S(2\pi a\ell) = \frac{J_f^{\,2}}{\sigma}(\pi a^2 \ell)$$

$$= \frac{J_f^{\,2}}{\sigma}(\text{volume}) \qquad (2)$$

with the use of (1), and where $\pi a^2 \ell$ is the volume of the conductor. Compare this with

$$w = \frac{J_f^{\,2}}{\sigma}$$

and see that (2) says that the total rate at which energy is flowing into the conductor is exactly equal to the rate at which energy is being dissipated into heat within the volume. This is exactly what is required for the steady-state situation assumed. The ultimate source of this energy is some device, like a battery, which maintains the potential difference between the ends of the conductor by continuously doing work on the charges as they pass through the complete circuit. The description here involving $\bar{S}$ depicts the energy as being trasferred from the source by means of the fields that it establishes throughout space as a result of the charge and current distributions within it. Finally, the energy flow passes normally through the surface of the conductor in just the right amount to be transformed into heat. Thus the interpretation given to the Poynting vector provides a complete and internally consistent description of the processes involved in this case.

• PROBLEM 12-8

A straight, isolated, copper wire of conductivity σ and diameter d carries a steady current I as shown in the figure. Integrate the Poynting vector over the entire boundary of a piece of wire of length L, and compare to the Joule heat developed in this section.

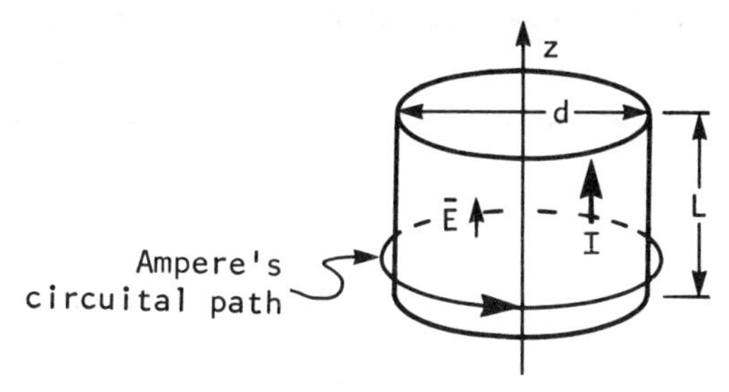

Solution: Use cylindrical coordinates.
To find $\bar{H}$, magnetic intensity, use Ampere's circuital law

$$\oint \bar{H} \cdot d\bar{\ell} = I = \int \bar{J} \cdot d\bar{a} = \bar{J} \cdot \int d\bar{a} = J\left(\frac{\pi d^2}{4}\right)$$

$$H_\theta \oint d\ell = H_\theta (\pi d) = J \frac{\pi d^2}{4}$$

$$\bar{H} = \frac{\hat{\theta} J d}{4}$$

The Poynting vector is $\bar{S} = \bar{E} \times \bar{H}$, $\bar{E} = \hat{k} E_z$

$$\bar{S} = \hat{k} E_z \times \hat{\theta} H_\theta = \hat{r} \frac{E_z j_z d}{4} = \bar{S}$$

In accordance with Ohm's law

$$\bar{E} = \frac{\bar{J}}{\sigma}, \; E_z = \frac{J_z}{\sigma}$$

$$\bar{S} = \frac{\hat{r} J_z{}^2 d}{4\sigma}$$

$$\int_{\substack{\text{curved} \\ \text{surface}}} \bar{S} \cdot d\bar{a} = \frac{J_z{}^2 d}{4\sigma} \int \hat{r} \cdot \hat{r} \, da = \frac{J_z{}^2 d}{4\sigma} (\pi d L)$$

$$\therefore \int \bar{S} \cdot d\bar{a} = \frac{J_z{}^2}{\sigma}\left(\frac{\pi d^2}{4}\right) L = \frac{J^2}{\sigma} \text{ (volume)}$$

$$\text{joule heat/unit volume} = \bar{J} \cdot \bar{E} = \sigma \bar{E} \cdot \bar{E}$$

$$= \sigma E^2 = \frac{\sigma J^2}{\sigma^2} = \frac{J^2}{\sigma}$$

Total joule heat developed in cylinder $= \int \bar{S} \cdot d\bar{a}$

• PROBLEM 12-9

Find the energy relations for a plane wave.

Solution: First consider the case $\sigma = 0$.

Substitute the equation

$$\bar{H} = \frac{\hat{k} \times \bar{E}}{\mu v} = \frac{\hat{k} \times \bar{E}}{z} \text{ into}$$

$$\langle \bar{S} \rangle = \tfrac{1}{2} \text{Re} (\bar{E}_c \times \bar{H}_c^*).$$

Using

$$z = \left(\frac{\mu}{\varepsilon}\right)^{\frac{1}{2}}, \quad \bar{A} \times (\bar{B} \times \bar{C}) = \bar{B}(\bar{A} \cdot \bar{C}) - \bar{C}(\bar{A} \cdot \bar{B})$$

and $\hat{k} \cdot \bar{E} = 0$,

$$\langle\bar{S}\rangle = \frac{1}{2}\left(\frac{\varepsilon}{\mu}\right)^{\frac{1}{2}} \mathrm{Re}[\bar{E} \times (\hat{k} \times \bar{E}^*)] = \frac{1}{2}\left(\frac{\varepsilon}{\mu}\right)^{\frac{1}{2}} (\bar{E} \cdot \bar{E}^*)\hat{k} \qquad (1)$$

since $\bar{E} \cdot \bar{E}^*$ is real. Now use

$$E = E_0 e^{i(k \cdot r - \omega t)}, \qquad B = B_0 e^{i(k \cdot r - \omega t)}$$

and

$$\bar{H} = \frac{\hat{k} \times \bar{E}}{z},$$

then (1) can also be written as

$$\langle\bar{S}\rangle = \frac{1}{2}\left(\frac{\varepsilon}{\mu}\right)^{\frac{1}{2}} |\bar{E}_0|^2\hat{k} = \frac{1}{2}\left(\frac{\mu}{\varepsilon}\right)^{\frac{1}{2}} |\bar{H}_0|^2\hat{k} \qquad (2)$$

Since $\bar{E} \cdot \bar{E}^* = \bar{E}_0 \cdot \bar{E}_0^* = |E_0|^2$, and so on for $\bar{H}$. Note from this result that not only is the energy flow in the direction of propagation $\hat{k}$, as already known, but also it is proportional to the square of the amplitude of either $\bar{E}$ or $\bar{H}$.

In the same way, the energy densities are

$$\langle u_e\rangle = \frac{1}{4}\varepsilon\bar{E} \cdot \bar{E}^* = \frac{1}{4}\varepsilon|\bar{E}_0|^2 = \frac{1}{4}\mu|\bar{H}_0|^2 = \langle u_m\rangle \qquad (3)$$

so that the average energy densities are equal. The total average energy density then becomes

$$\langle u\rangle = \langle u_e\rangle + \langle u_m\rangle = \frac{1}{2}\varepsilon|\bar{E}_0|^2 = \frac{1}{2}\mu|\bar{H}_0|^2 \qquad (4)$$

which makes it possible to write eq. (2) as

$$\langle\bar{S}\rangle = \frac{\langle u\rangle}{\sqrt{\mu\varepsilon}}\hat{k} = \langle u\rangle v\hat{k} = \langle u\rangle\bar{v} \qquad (5)$$

with the help of $v = \dfrac{1}{\sqrt{\mu\varepsilon}}$

Thus the average energy current is the product of the average energy density and the velocity of the wave. This is in accord with the analogous result for current density $\bar{J} = \rho\bar{v}$.

In a conducting medium where $\sigma \neq 0$ use

$$\bar{H} = \frac{|\kappa|}{\mu\omega}e^{i\Omega}\,\hat{k} \times \bar{E}$$

and, when this is substituted into

$$\langle\bar{S}\rangle = \frac{1}{2}\text{Re}\ (\bar{E}_c \times \bar{H}_c^*)$$

$$\langle\bar{S}\rangle = \frac{|\kappa|}{2\mu\omega}\text{Re}\quad \bar{E} \times (\hat{k} \times \bar{E}^*)\ e^{-i\Omega}$$

$$= \frac{|k|\cos\Omega}{2\mu\omega}\ e^{-2\beta\zeta}|\bar{E}_0|^2\hat{k}$$

$$= \frac{\alpha e^{-2\beta\zeta}}{2\mu\omega}\ |\bar{E}_0|^2\hat{k}$$

$$= \frac{e^{-2\beta\zeta}}{2\mu v}\ |\bar{E}_0|^2\ \hat{k} \tag{6}$$

with the use of

$$e^{iu} = \cos u + i \sin u,\ e^{-iu} = \cos u - i \sin u,$$

$$\alpha = |\kappa| \cos \Omega,\quad \beta = |\kappa| \sin \Omega,\ v = \frac{\omega}{\alpha},$$

$$E = E_0{}^{i(\kappa\cdot r-\omega t)},\ B = B_0\ e^{i(\kappa . r-\omega t)},$$

$$\kappa = \kappa\hat{n} = \kappa\ \hat{k},\quad \zeta = \hat{n}\cdot r,\ \text{and}\ \kappa = (\alpha \pm i\beta).$$

and where ζ now measures the distance in the direction of propagation. Similarly, the average total energy density becomes

$$\langle u\rangle = \frac{\alpha^2}{2\mu\omega^2}e^{-2\beta\zeta}|\bar{E}_0|^2$$

$$= \frac{e^{-2\beta\zeta}}{2\mu v^2}|\bar{E}_0|^2 \tag{7}$$

with the additional use of $\alpha^2 - \beta^2 = \omega^2\ \mu\varepsilon$

and $\alpha = |\kappa| \cos \Omega$ and $\beta = |\kappa| \sin \Omega$.

Note that in this case $\langle\bar{S}\rangle = \langle\mu\rangle v\hat{k}$ just as given by (5) for the nonconducting medium.

Both $\langle\bar{S}\rangle$ and $\langle u\rangle$ are seen to be proportional to $e^{-2\beta\zeta}$ so that they decrease with twice the attenuation factor of the fields; this is a result of the fact that they both are proportional to the square of the amplitudes. This energy is lost because of the resistive heating of the material arising from its conductivity.

• PROBLEM 12-10

Given the plane wave defined by

$$\bar{E} = \bar{a}_x E_x^+(z,t) = \bar{a}_x\ E_m^+ \cos\ (\omega t - \beta_0 z)$$

and

$$\bar{H} = \bar{a}_y H_y^+ (z,t) = \bar{a}_y \frac{E_m^+}{\eta_0} \cos (\omega t - \beta_0 z),$$

determine the net power flux p(t) entering a closed, box-shaped surface S having dimensions as in the accompanying figure. Show that the time rate of increase of the electromagnetic energy within the volume of the box provides the same answer.

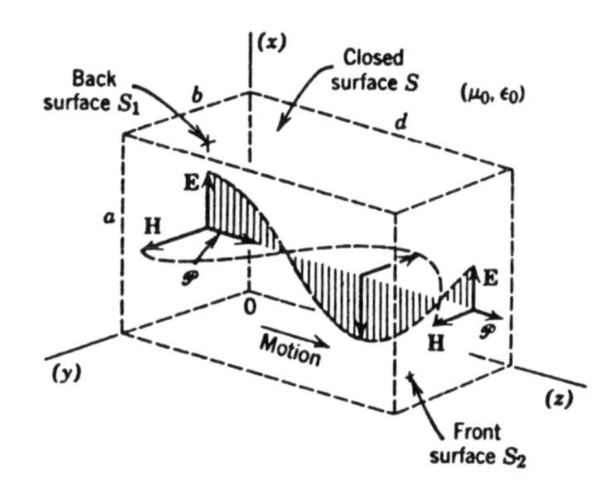

Solution: Because $\bar{P}$ is everywhere z directed, the only contributions to power-flux entering the box are on the ends S_1 and S_2 shown, so

$$p(t) = -\oint_S \bar{P} \cdot d\bar{s} = \frac{\partial}{\partial t} \int_V \left[\frac{\mu_0 H^2}{2} + \frac{\varepsilon_0 E^2}{2}\right] dv \tag{1}$$

yields

$$p_1(t) \equiv -\int_{S_1} \bar{P} \cdot d\bar{s}$$

$$= -\int_{y=0}^{b} \int_{x=0}^{a} \left[\bar{a}_z \frac{(E_m^+)^2}{\eta_0} \cos^2(\omega t - \beta_0 z)\right]_{z=0} \cdot (-\bar{a}_z \, dx \, dy)$$

$$= \frac{\left(E_m^+\right)^2}{\eta_0} ab \cos^2 \omega t$$

$$p_2(t) \equiv -\int_{S_2} \bar{P} \cdot d\bar{s} = -\frac{(E_m^+)^2}{\eta_0} ab \cos^2 (\omega t - \beta_0 d)$$

The net power-flux entering S is therefore

$$-\oint_S \bar{P} \cdot d\bar{s} \equiv \bar{P}(t) = p_1(t) + p_2(t)$$

$$= \frac{(E_m^+)^2}{\eta_0} ab[\cos^2 \omega t - \cos^2 (\omega t - \beta_0 d)]$$

$$= \frac{(E_m^+)^2}{2\eta_0} ab[\cos 2\omega t - \cos 2(\omega t - \beta_0 d)]W \qquad (2)$$

the last being obtained by use of $\cos^2 \theta = 1/2 + (1/2) \cos 2\theta$.

Equivalently, if the right side of eq. (1) is integrated throughout the volume of the box, eq. (2) should again be obtained. Substituting the values of E and H into the integral yields

$$\frac{\partial}{\partial t} \int_V \left[\frac{\mu_0 H^2}{2} + \frac{\varepsilon_0 E^2}{2}\right] dv$$

$$= \frac{\partial}{\partial t}\left\{\frac{\mu_0 (E_m^+)^2}{4\eta_0^2} \int_0^d \int_0^b \int_0^a [1 + \cos 2(\omega t - \beta_0 z)] dx\, dy\, dz\right.$$

$$\left. + \frac{\varepsilon_0 (E_m^+)^2}{4} \int_0^d \int_0^b \int_0^a [1 + \cos 2(\omega t - \beta_0 z)] dx\, dy\, dz\right\}$$

$$= \frac{\partial}{\partial t}\left\{\frac{\varepsilon_0 (E_m^+)^2}{2} \int_0^d \int_0^b \int_0^a [1 + \cos 2(\omega t - \beta_0 z)] dx\, dy\, dz\right\}$$

$$= \frac{\omega\varepsilon_0 (E_m^+)^2 ab}{2\beta_0} [-\cos 2(\omega t - \beta_0 z)]_0^d$$

$$= \frac{(E_m^+)^2 ab}{2\eta_0} [\cos 2\omega t - \cos 2(\omega t - \beta_0 d)]$$

agreeing with eq. (2) as expected.

NORMAL INCIDENCE

• PROBLEM 12-11

A plane 1,000 Mc/sec traveling wave in air with peak electric field intensity of 1 volt/meter is incident normally on a large copper sheet. Find the average power absorbed by the sheet per square meter of area.

Solution: First let the intrinsic impedance of copper be calculated at 1,000 Mc/sec, as follows

$$\dot{Z}_e = \frac{1 + j}{\sqrt{2}} 376.7 \sqrt{\frac{\mu_r}{\varepsilon_r}} \sqrt{\frac{\omega\varepsilon}{\sigma}}$$

For copper $\mu_r = \varepsilon_r = 1$ and $\sigma = 5.8 \times 10^7$ mhos/meter. Hence the real part of $\dot{Z}_c$ is

$$\text{Re } \dot{Z}_c = (\cos 45°)(376.7)\sqrt{\frac{2\pi \times 10^9 \times 8.85 \times 10^{-12}}{5.8 \times 10^7}}$$

$$= 8.2 \times 10^{-3} \text{ohm}$$

Next find the value of H_0 at the sheet (tangent to the surface). This is very nearly double H for the incident wave. Thus

$$H_0 = 2\frac{E}{Z} = \frac{2 \times 1}{376.7} \text{ amp/meter}$$

From $S_{av} = \frac{1}{2}\text{Re}\left(\dot{H}_z \dot{H}_z^* \dot{z}_c\right) = \frac{1}{2}\left|\dot{H}_z\right|^2 \text{Re}\dot{z}_c = \frac{1}{2}H_0^2 \text{Re}\dot{z}_c$ watts/m²

the average power per square meter into the sheet is then

$$S_{av} = \frac{1}{2}\left(\frac{2}{376.7}\right)^2 8.2 \times 10^{-3} = 1.16 \times 10^{-7} \text{ watt/meter}^2$$

• PROBLEM 12-12

A thin lossless dielectric with permittivity ε and permeability μ is coated onto the interface between two infinite half-spaces of lossless media with respective properties (ε_1, μ_1) and (ε_2, μ_2), as shown in the figure. What coating parameters ε and μ and thickness d will allow all the time-average power from region 1 to be transmitted through the coating to region 2? Such coatings are applied to optical components such as lenses to minimize unwanted reflections and to maximize the transmitted light intensity.

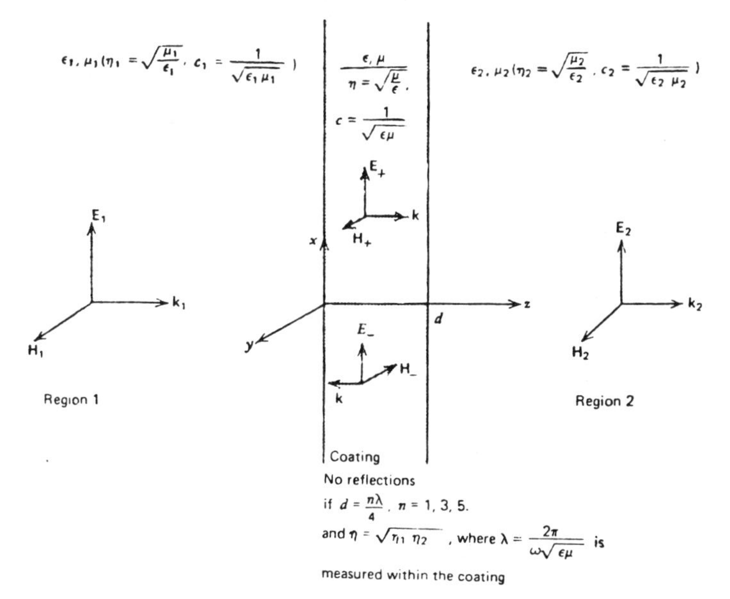

A suitable dielectric coating applied on the interface of discontinuity between differing media can eliminate reflections at a given frequency

Solution: For all the incident power to be transmitted into region 2 there can be no reflected field in region 1, although there are oppositely traveling waves in the coating due to the reflection at the second interface. Region 2 only has positively z-directed power flow. The fields in each region are thus of the following form:

Region 1

$$\bar{E}_1 = \text{Re}[\hat{E}_1 e^{j(\omega t - k_1 z)} \bar{i}_x], \qquad k_1 = \omega/c_1 = \omega\sqrt{\varepsilon_1 \mu_1}$$

$$\bar{H}_1 = \text{Re}\left[\frac{\hat{E}_1}{\eta_1} e^{j(\omega t - k_1 z)} \bar{i}_y\right], \qquad \eta_1 = \frac{\mu_1}{\varepsilon_1}$$

Coating

$$\bar{E}_+ = \text{Re}[\hat{E}_+ e^{j(\omega t - kz)} \bar{i}_x], \qquad k = \omega/c = \omega\sqrt{\varepsilon\mu}$$

$$\bar{H}_+ = \text{Re}\left[\frac{E_+}{\eta} e^{j(\omega t - kz)} \bar{i}_y\right], \qquad \eta = \sqrt{\frac{\mu}{\varepsilon}}$$

$$\bar{E}_- = \text{Re}[\hat{E}_- e^{j(\omega t + kz)} \bar{i}_x]$$

$$\bar{H}_- = \text{Re}\left[-\frac{\hat{E}_-}{\eta} e^{j(\omega t + kz)} \bar{i}_y\right]$$

Region 2

$$\bar{E}_2 = \text{Re}[\hat{E}_2 e^{j(\omega t - k_2 z)} \bar{i}_x], \qquad k_2 = \omega/c_2 = \omega\sqrt{\varepsilon_2 \mu_2}$$

$$\bar{H}_2 = \text{Re}\left[\frac{\hat{E}_2}{\eta_2} e^{j(\omega t - k_2 z)} \bar{i}_y\right], \qquad \eta_2 = \sqrt{\frac{\mu_2}{\varepsilon_2}}$$

Continuity of tangential $\bar{E}$ and $\bar{H}$ at $z = 0$ and $z = d$ requires

$$\hat{E}_1 = \hat{E}_+ + \hat{E}_-, \qquad \frac{\hat{E}_1}{\eta_1} = \frac{\hat{E}_+ - \hat{E}_-}{\eta}$$

$$\hat{E}_+ e^{-jkd} + \hat{E}_- e^{+jkd} = \hat{E}_2 e^{-jk_2 d}$$

$$\frac{\hat{E}_+ e^{-jkd} - \hat{E}_- e^{+jkd}}{\eta} = \frac{\hat{E}_2 e^{-jk_2 d}}{\eta_2}$$

Each of these amplitudes in terms of $\hat{E}_1$ is then

$$\hat{E}_+ = \frac{1}{2}\left(1 + \frac{\eta}{\eta_1}\right)\hat{E}_1$$

$$\hat{E}_- = \frac{1}{2}\left(1 - \frac{\eta}{\eta_1}\right)\hat{E}_1$$

$$\hat{E}_2 = e^{jk_2 d}[\hat{E}_+ e^{-jkd} - \hat{E}_- e^{+jkd}]$$

Solving this last relation self-consistently requires that

$$\hat{E}_+ e^{-jkd}\left(1 - \frac{\eta_2}{\eta}\right) + \hat{E}_- e^{jkd}\left(1 + \frac{\eta_2}{\eta}\right) = 0$$

Writing $\hat{E}_+$ and $\hat{E}_-$ in terms of $\hat{E}_1$ yields

$$\left(1 + \frac{\eta}{\eta_1}\right)\left(1 - \frac{\eta_2}{\eta}\right) + e^{2jkd}\left(1 + \frac{\eta_2}{\eta}\right)\left(1 - \frac{\eta}{\eta_1}\right) = 0$$

Since this relation is complex, the real and imaginary parts must separately be satisfied. For the imaginary part to be zero requires that the coating thickness d be an integral number of quarter wavelengths measured within the coating,

$$2kd = n\pi \Rightarrow d = n\lambda/4, \qquad n = 1, 2, 3 \ldots$$

The real part then requires

$$\left(1 + \frac{\eta}{\eta_1}\right)\left(1 - \frac{\eta_2}{\eta}\right) \pm \left(1 + \frac{\eta_2}{\eta}\right)\left(1 - \frac{\eta}{\eta_1}\right) = 0 \begin{cases} n \text{ even} \\ n \text{ odd} \end{cases}$$

For the upper sign where d is a multiple of half-wavelengths the only solution is

$$\eta_2 = \eta_1 \qquad (d = n\lambda/4, \quad n = 2, 4, 6, \ldots)$$

which requires that media 1 and 2 be the same so that the coating serves no purpose. If regions 1 and 2 have differing wave impedances, the lower sign where d is an odd integer number of quarter wavelengths must be used so that

$$\eta^2 = \eta_1\eta_2 \Rightarrow \eta = \sqrt{\eta_1\eta_2} \qquad (d = n\lambda/4, n = 1, 3, 5, \ldots)$$

Thus, if the coating is a quarter wavelength thick as measured within the coating, or any odd integer multiple of this thickness with its wave impedance equal to the geometrical average of the impedances in each adjacent region, all the time-average power flow in region 1 passes through the coating into region 2:

$$\langle S_z \rangle = \frac{1}{2}\frac{|\hat{E}_2|^2}{\eta_1} = \frac{1}{2}\frac{|\hat{E}_2|^2}{\eta_2}$$

$$= \frac{1}{2}\text{Re}\left[(\hat{E}_+ e^{-jkz} + \hat{E}_- e^{+jkz})\frac{(\hat{E}_+^* e^{+jkz} - \hat{E}_-^* e^{-jkz})}{\eta}\right]$$

$$= \frac{1}{2\eta}(|\hat{E}_+|^2 - |\hat{E}_-|^2)$$

Note that for a given coating thickness d, there is no reflection only at select frequencies corresponding to wavelengths $d = n\lambda/4$, $n = 1, 3, 5, \ldots$. For a narrow band of wavelengths about these select wavelengths, reflections are small. The magnetic permeability of coatings and of the glass used in optical components are usually that of free space while the permittivities differ. The permittivity of the coating ε is then picked so that

$$\varepsilon = \sqrt{\varepsilon_2 \varepsilon_0}$$

and with a thickness corresponding to the central range of the wavelengths of interest (often in the visible).

• PROBLEM 12-13

A uniform plane wave is normally incident in air upon a slab of plastic with the parameters shown, a quarter wave thick at the operating frequency f = 1 MHz. The x polarized wave has the amplitude

$$\hat{E}^{+}_{m1} = 100e^{j0°} \text{ V/m}.$$

Utilize the concepts of reflection coefficient and total field impedance to find the remaining wave amplitudes.

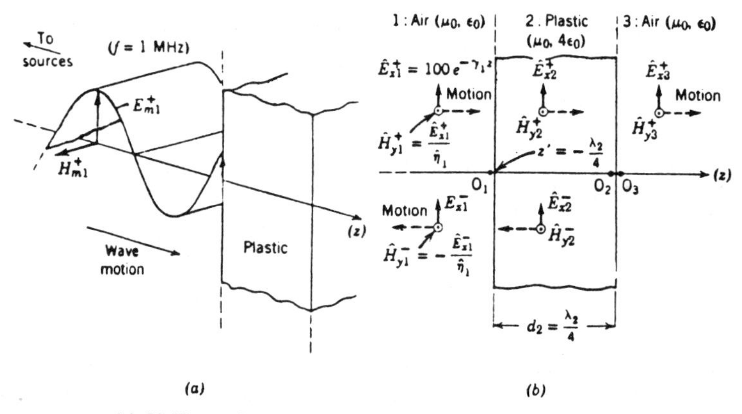

(a) Uniform plane wave normally incident on a plastic slab. (b) Side view showing wave components in the regions.

Solution: To obviate carrying cumbersome phase terms across the interfaces, assume separate z origins, 0_1, 0_2, and 0_3 shown in (b) of the figure. The wave amplitudes are referred to these origins. First, values of $\hat{\eta}$ for each region are found by

$$\hat{\eta}_1 = \hat{\eta}_3 = \sqrt{\mu_0/\varepsilon_0} = 120\pi \ \Omega;$$

in the plastic slab,

$$\hat{\eta}_2 = \sqrt{\mu_0/4\varepsilon_0} = 60\pi \ \Omega.$$

The propagation constants $\gamma = \alpha + j\beta$ are computed from

$$\alpha = \frac{\omega\sqrt{\mu\varepsilon}}{\sqrt{2}}\left[\sqrt{1+\left(\frac{\sigma}{\omega\varepsilon}\right)^2} - 1\right]^{\frac{1}{2}}$$

and (1)

$$\beta = \frac{\omega\sqrt{\mu\varepsilon}}{\sqrt{2}}\left[\sqrt{1+\left(\frac{\sigma}{\omega\varepsilon}\right)^2} + 1\right]^{\frac{1}{2}} ;$$

thus, in lossless region 2,

$$\gamma_2 = j\beta_2 = j\omega\sqrt{\mu_0(4\varepsilon_0)} \qquad (=j2\pi/\lambda_2)$$

Then finding the complex wave amplitudes proceeds as follows:

(a) Begin in region 3, containing no reflected wave. $\hat{\Gamma}_3'(z)$, from

$$\hat{\Gamma}_3(z) = \frac{\hat{E}_m^-}{\hat{E}_m^+} e^{2\gamma z} \tag{2}$$

is therein zero, yielding the total field impedance from

$$\hat{Z}(z) = \frac{\hat{E}_x(z)}{\hat{H}_y(z)} = \hat{\eta}\,\frac{1+\hat{\Gamma}(z)}{1-\hat{\Gamma}(z)} \quad \Omega \tag{3}$$

$$\hat{Z}_3(z) = \hat{\eta}_3(1+0)/(1-0) = \hat{\eta}_3 = 120\pi \ \Omega.$$

By

$$\hat{Z}(a+) = \hat{Z}(a-) , \tag{4}$$

the total field impedance $\hat{Z}_2(0)$ just inside region 2 has the same value, i.e.,

$$\hat{Z}_2(0) = \hat{Z}_3(0) = 120\pi \ \Omega.$$

(b) By use of $$\hat{\Gamma}(z) = \frac{\hat{Z}(z) - \hat{\eta}}{\hat{Z}(z) + \hat{\eta}} \tag{5}$$

$\hat{\Gamma}_2$ at 0_2 in region 2 becomes

$$\hat{\Gamma}_2(0) = \frac{\hat{Z}_2(0) - \hat{\eta}_2}{\hat{Z}_2(0) + \hat{\eta}_2} = \frac{120\pi - 60\pi}{120\pi + 60\pi} = \frac{1}{3}$$

The equation

$$\hat{\Gamma}(z') = \hat{\Gamma}(z)e^{2\gamma(z'-z)} \tag{6}$$

is employed to translate $\hat{\Gamma}_2(0)$ to the value $\hat{\Gamma}_2(-d_2)$ at the input plane of region 2. With

$$z' = -d_2 = -\lambda_2/4$$

and

$$\gamma_2 = j\beta_2 = j2\pi/\lambda_2$$

$$\hat{\Gamma}_2 \left(-\frac{\lambda_2}{4}\right) = \hat{\Gamma}_2(0)e^{2\gamma_2 [(-\lambda_2/4)-0]}$$

$$= \hat{\Gamma}_2(0)e^{2(j2\pi/\lambda_2)(-\lambda_2/4)} = \frac{1}{3}e^{-j\pi}$$

$$= -\frac{1}{3}$$

(c) Steps (a) and (b) are repeated to find $\hat{Z}$ and $\hat{\Gamma}$ in the next region to the left. First, the use of (3) at $z = -d_2$ in region 2 yields

$$\hat{Z}_2(-d_2) = \hat{\eta}_2 \frac{1 + \hat{\Gamma}_2(-d_2)}{1 - \hat{\Gamma}_2(-d_2)} = 60\pi \frac{1 + \left(-\frac{1}{3}\right)}{1 - \left(-\frac{1}{3}\right)} = 30\pi\ \Omega$$

which from the continuity relation (4) yields

$$\hat{Z}_2(-d) = 30\pi\ \Omega = \hat{Z}_1(0).$$

The reflection coefficient at the output plane of region 1, from (5), is

$$\hat{\Gamma}_1(0) = \frac{\hat{Z}_1(0) - \hat{\eta}_1}{\hat{Z}_1(0) + \hat{\eta}_1} = \frac{30\pi - 120\pi}{30\pi + 120\pi} = -\frac{3}{5}$$

The reflected wave amplitude $\hat{E}_{m1}^-$ is now obtained, using the definition (2) of reflection coefficient. Applied at z = 0 in region 1, given

$$\hat{E}_{m1}^+ = 100e^{j0°},$$

it yields

$$\hat{E}_{m1}^- = \hat{E}_{m1}^+ \hat{\Gamma}_1(z)e^{-2\gamma_1 z}]_{z=0} = (100^{j0°})\left(-\frac{3}{5}\right) = -60 \text{ V/m}$$

Then the total electric field in air region 1 is

$$E_{x1}^-(z) = E_{m1}^+ e^{-\gamma_0 z} + E_{m1}^- e^{\gamma_1 z}$$

$$= 100e^{-j\beta_0 z} - 60e^{j\beta_0 z} \text{ V/m}$$

The total field is obtained by use of

$$\hat{H}_y(z) = \frac{\hat{E}_m^+}{\hat{\eta}} e^{-\gamma z} [1 - \hat{\Gamma}(z)] \tag{7}$$

$$\hat{H}_{y1}(z) = \frac{100}{120\pi} e^{-j\beta_0 z} - \frac{(-60)}{120\pi} e^{j\beta_0 z}$$

$$= 0.266\, e^{-j\beta_0 z} + 0.159 e^{j\beta_0 z} \text{ A/m}$$

(d) The rest of the problem concerns finding $\hat{E}_{m2}^+$, $\hat{E}_{m2}^-$, and $\hat{E}_{m3}^+$. For example, E_{m2}^+ is obtained by specializing

$$\hat{E}_{x1}(z) \text{ to } z = 0$$

at the interface, whence

$$\hat{E}_{x1}(0) = 100 - 60 = 40 \text{ V/m} = \hat{E}_{x2}(-d_2)$$

in which the last equality is evident from the continuity condition.

The total electric field in region z is

$$\hat{E}_{x2}(z) = \hat{E}^{+}_{m2}\, e^{-\gamma_2 z}[1 + \hat{\Gamma}_2(z)],$$

but at $z = -d_2$, all quantities are known except E^{+}_{m2}; solving for it yields

$$\hat{E}^{+}_{m2} = \hat{E}_{x2}(z) \left.\frac{e^{\gamma_2 z}}{1 + \hat{\Gamma}_2(z)}\right]_{z = -d_2}$$

$$= 40 \frac{e^{(j2\pi/\gamma_2)(-d_2)}}{1 - (-\frac{1}{3})} = 60e^{-j(\pi/2)} = -j60 \text{ V/m}$$

Then applying (2) to region 2, $\hat{E}^{-}_{m2}$ is obtained from $\hat{E}^{+}_{m2}$ just found. A similar procedure applied at the second interface yields $\hat{E}^{+}_{m3}$ completing the problem.

• PROBLEM 12-14

A plane wave is normally incident from medium 1 on an interface with medium 2, where $\varepsilon'_{r1} = 2$, $\sigma_1 = 0.2(\mho/m)$, $\varepsilon'_{r2} = 4$, and $\sigma_2 = 0.1(\mho/m)$. Find γ_1, γ_2, η_1, η_2, and the amplitudes E^{-}_{x10}, H^{+}_{y10}, H^{-}_{y10}, E^{+}_{x20}, H^{+}_{y20}, in terms of E^{+}_{x10} at f = 10(GHz).

Find the values of the phasors $\vec{E}^{-}_{x10}$, $\vec{H}^{+}_{y10}$, $\vec{H}^{-}_{y10}$, $\vec{E}^{+}_{x20}$, and $\vec{H}^{+}_{y20}$ relative to the phasor $\vec{E}^{+}_{x10}$.

Solution: Find $\dfrac{\sigma_1}{\omega\varepsilon'_1}$ and $\dfrac{\sigma_2}{\omega\varepsilon'_2}$.

$$\frac{\sigma_1}{\omega\varepsilon'_1} = \frac{0.2}{20\pi(10^9)(2)\left(\frac{10^{-9}}{36\pi}\right)} = 0.18$$

$$\frac{\sigma_2}{\omega\varepsilon'_2} = \frac{0.1}{20\pi(10^9)(4)\left(\frac{10^{-9}}{36\pi}\right)} = 0.05$$

TABLE 1 Comparisons of Velocity, Impedance, and the Propagation Constant for Free Space, Lossless Material, and Conducting or Lossy material

	Velocity	Intrinsic Impedance	Propagation Constant
Free space	$c = \frac{1}{\sqrt{\mu_0\epsilon_0}} = \frac{\omega}{\beta} \approx 3 \times 10^8$ (m s^{-1})	$\eta = \sqrt{\frac{\mu_0}{\epsilon_0}} \approx 120\pi\,(\Omega) \approx 377\,(\Omega)$	$\gamma = j\beta = j\omega\sqrt{\mu_0\epsilon_0}$ (m^{-1})
Lossless dielectric	$U = \frac{1}{\sqrt{\mu\epsilon}} = \frac{c}{\sqrt{\mu_r\epsilon_r}} = \frac{\omega}{\beta}$ (m s^{-1})	$\eta = \sqrt{\frac{\mu}{\epsilon}} = \sqrt{\frac{\mu_0\mu_r}{\epsilon_0\epsilon_r}} \approx 120\pi\sqrt{\frac{\mu_r}{\epsilon_r}}\,(\Omega)$	$\gamma = j\beta = j\omega\sqrt{\mu_0\epsilon_0\mu_r\epsilon_r}$ (m^{-1})
Lossy dielectric, $\epsilon = \epsilon' - j\frac{\sigma}{\omega}$	$U = \frac{\omega}{\beta}$ (m s^{-1})	$\eta = \frac{\sqrt{\mu/\epsilon'}}{\sqrt{1 - j\frac{\sigma}{\omega\epsilon'}}}\,(\Omega)$	$\gamma = \alpha + j\beta = j\omega\sqrt{\mu\epsilon'}\left(\sqrt{1 - j\frac{\sigma}{\omega\epsilon'}}\right)$ (m^{-1})
Slightly lossy dielectric, $\frac{\sigma}{\omega\epsilon'} \ll 1$	$U = \frac{\omega}{\beta}$ (m s^{-1})	$\eta \approx \sqrt{\frac{\mu}{\epsilon'}}\left(1 + j\frac{\sigma}{2\omega\epsilon'}\right)(\Omega)$	$\gamma = \alpha + j\beta \approx j\omega\sqrt{\mu\epsilon'}\left(1 - j\frac{\sigma}{2\omega\epsilon'}\right)$ (m^{-1})
Good conductor $\frac{\sigma}{\omega\epsilon'} \gg 1$	$U = \frac{\omega}{\beta}$ (m s^{-1})	$\eta \approx \sqrt{\frac{\omega\mu}{\sigma}}\angle 45°\,\Omega$	$\gamma = \alpha + j\beta \approx \sqrt{\pi f\mu\sigma}(1 - j)$ (m^{-1})

From Table 1,

$$\eta_1 = \frac{\sqrt{\mu_0/\varepsilon'_1}}{\sqrt{1 - j0.18}} = \frac{377/\sqrt{2}}{\sqrt{1 - j0.18}} = 264.46\ \angle 5.10°\ (\Omega)$$

$$= (263.41 + j23.52)\ (\Omega)$$

$$\eta_2 = \frac{\sqrt{\mu_0/\varepsilon'_2}}{\sqrt{1 - j0.05}} = \frac{377/\sqrt{4}}{\sqrt{1 - j0.05}} = 188.26\ \angle 1.43°\ (\Omega)$$

$$= (188.20 + j4.69)\ (\Omega)$$

Also from Table 1,

$$\gamma_1 = j\omega\sqrt{\mu\varepsilon_1'}\ \sqrt{1 - j0.18}$$

$$= j20\pi(10^9)\frac{\sqrt{2}}{3 \times 10^8}\ 1.01\ \angle{-5.10°}$$

$$= 299\ \angle 84.9°$$

$$= (26.6 + j297.97)\quad (m^{-1})$$

$$= \alpha_1 + j\beta_1$$

and

$$\gamma_2 = j20\pi(10^9)\ \frac{\sqrt{4}}{3 \times 10^8}\ 1.0012\ \angle{-1.43°}$$

$$= 419.4\ \angle 88.57°$$

$$= (10.46 + j419.25)\ (m^{-1})$$

$$= \alpha_2 + j\beta_2$$

ρ and τ must be found to compute the amplitudes

$$\rho = \frac{\eta_2 - \eta_1}{\eta_2 + \eta_1} = \frac{188.20 + j4.69 - 263.41 - j23.52}{188.20 + j4.69 + 263.41 + j23.52}$$

$$= 0.17\ \angle{-169.52°}$$

$$= -0.1685 - j0.0312$$

From

$$\tau = \frac{2\eta_2}{\eta_2 + \eta_1}$$

$$\tau = 1 + \rho = 1 - 0.1685 - j0.0312$$

$$= 0.8315 - j0.0312 = 0.8321\ \angle{-2.15°}$$

Then

$$E_{x10}^{-} = |\rho|E_{x10}^{+} = 0.17E_{x10}^{+} (Vm^{-1})$$

$$H_{y10}^{+} = \frac{E_{x10}^{+}}{|\eta_1|} = \frac{E_{x10}^{+}}{|264.46|} = 0.0038E_{x10}^{+} \quad (Am^{-1})$$

$$H_{y10}^{-} = |\rho|H_{y10}^{+} = (0.17)(0.0038E_{x10}^{+}) = 0.00064E_{x10}^{+} \quad (Am^{-1})$$

$$E_{x20}^{+} = |\tau|E_{x10}^{+} = 0.8321E_{x10}^{+} \quad (Vm^{-1})$$

$$H_{y20}^{+} = \frac{E_{x20}^{+}}{|\eta_2|} = \frac{0.8321E_{x10}^{+}}{188.26} = 0.00442E_{x10}^{+} \quad (Am^{-1})$$

The magnitudes of the phasors have been determined and the angles can be found using equations (1) through (6):

$$E_{x1}^{+} = E_{x10}^{+}e^{-\alpha_1 z} \cos(\omega t - \beta_1 z) \tag{1}$$

$$H_{y1}^{+} = \frac{E_{x10}^{+}}{|\eta_1|} e^{-\alpha_1 z} \cos(\omega t - \beta_1 - \theta_{\eta_1}) \tag{2}$$

$$E_{x1}^{-} = |\rho|E_{x10}^{+}e^{\alpha_1 z} \cos(\omega t + \beta_1 z + \theta_{\rho}) \tag{3}$$

$$H_{y1}^{-} = \frac{|\rho|}{|\eta_1|} E_{x10}^{+}e^{\alpha_1 z} \cos(\omega t + \beta_1 z - \theta_{\eta_1} + \theta_{\rho} + \pi) \tag{4}$$

$$E_{x2}^{+} = |\tau|E_{x10}^{+}e^{-\alpha_2 z} \cos(\omega t - \beta_2 z + \theta_{\tau}) \tag{5}$$

$$H_{y2}^{+} = \frac{|\tau|}{|\eta_2|} E_{x10}^{+}e^{-\alpha_2 z} \cos(\omega t - \beta_2 z - \theta_{\eta_2} + \theta_{\tau}) \tag{6}$$

where

$$\eta_1 = |\eta_1| \angle\theta_{\eta_1}, \quad \eta_2 = |\eta_2| \angle\theta_{\eta_2}, \quad \rho = |\rho| \angle\theta_{\rho},$$

and $\tau = |\tau| \angle\theta_{\tau}$. Note also

$$E_{x10}^{+} = |\vec{E}_{x10}^{+}|, \text{ etc.}$$

From the previous statements and eq. (2)

$$\frac{\vec{H}^{+}_{y10}}{\vec{E}^{+}_{x10}} = 0.0038 \quad \angle -\theta_{\eta_1} \quad (\Omega)$$

$$= 0.0038 \quad \angle -5.10° \quad (\Omega)$$

From eq. (3)

$$\frac{\vec{E}^{-}_{x10}}{\vec{E}^{+}_{x10}} = 0.17 \quad \angle \theta_{\rho}$$

$$= 0.17 \quad \angle -169.52°$$

From eq. (4)

$$\frac{\vec{H}^{-}_{y10}}{\vec{E}^{+}_{x10}} = 0.00064 \quad \angle (-\theta_{\eta_1} + \theta_{\rho} + 180°)$$

$$= 0.00064 \quad \angle 5.38° \quad (\Omega^{-1})$$

From eq. (5)

$$\frac{\vec{E}^{+}_{x20}}{\vec{E}^{+}_{x10}} = 0.8321 \quad \angle \theta_{\tau}$$

$$= 0.8321 \quad \angle -2.15°$$

From eq. (6)

$$\frac{\vec{H}^{+}_{y20}}{\vec{E}^{+}_{x10}} = 0.00442 \quad \angle (-\theta_{\eta_2} + \theta_{\tau})$$

$$= 0.00442 \quad \angle -3.58° \quad (\Omega^{-1})$$

The complex values of these phasors are plotted in Fig.1,

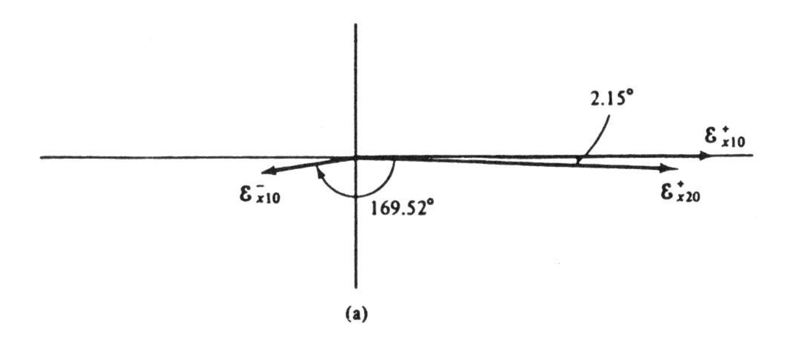

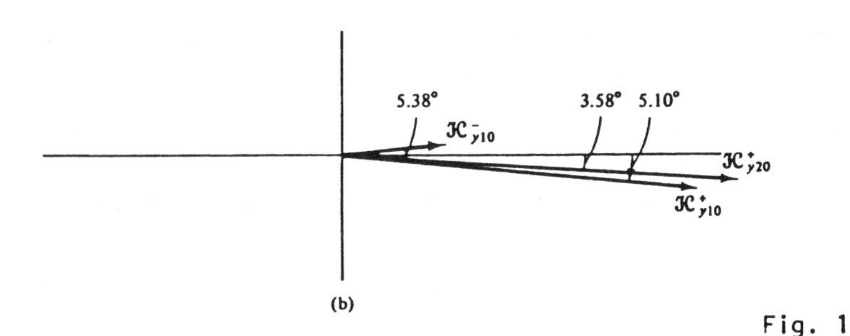

Fig. 1

where the angle of

$$\vec{E}^+_{x10}$$

is used as a reference for all other phasors.

BOUNDARY CONDITIONS

• PROBLEM 12-15

Polystyrene has a relative permittivity of 2.7. If a wave is incident at an angle of $\theta_i = 30°$ from air onto polystyrene, calculate the angle of transmission θ_t. Interchange polystyrene and air and repeat the calculation.

Solution: From air onto polystyrene, $\varepsilon_1 = \varepsilon_0$, $\mu_1 = \mu_0$, $\varepsilon_2 = 2.7\varepsilon_0$, $\mu_2 = \mu_0$. From

$$\sin\theta_t = \frac{\eta_1}{\eta_2}\sin\theta_i$$

$$\sin\theta_t = \sqrt{\frac{1}{2.7}}\,(0.5) = 0.304$$

$$\theta_t = 17.7°$$

From polystyrene onto air, $\varepsilon_1 = 2.7\varepsilon_0$, $\mu_1 = \mu_0$, $\varepsilon_2 = \varepsilon_0$, $\mu_2 = \mu_0$.

$$\sin\theta_t = \sqrt{2.7}\,(0.5) = 0.822$$

$$\theta_t = 55.2°$$

• PROBLEM 12-16

Consider a plane boundary between two media of permeability μ_1 and μ_2 as in the figure. Find the relation between the angles α_1 and α_2. Assume that the media are isotropic with $\vec{B}$ and $\vec{H}$ in the same direction.

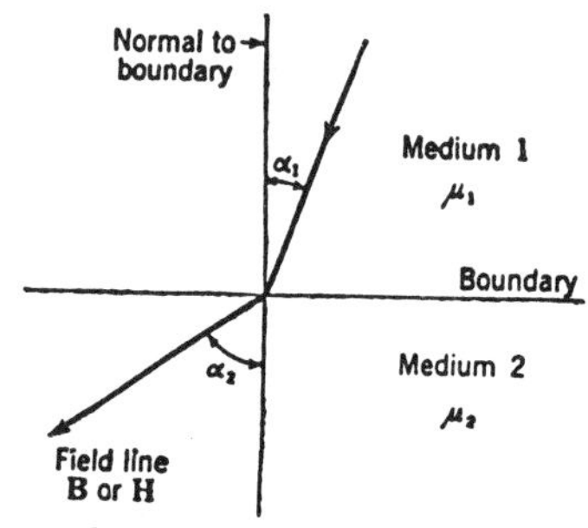

Boundary between two media of different permeability showing change in direction of magnetic field line.

Solution: From the boundary relations,

$$B_{n1} = B_{n2}$$

and (1)

$$H_{t1} = H_{t2}$$

From the figure,

$$B_{n1} = B_1 \cos\alpha_1$$

and (2)

$$B_{n2} = B_2 \cos\alpha_2$$

and

$$H_{t1} = H_1 \sin \alpha_1$$

and (3)

$$H_{t2} = H_2 \sin \alpha_2$$

where

B_1 = magnitude of $\bar{B}$ in medium 1

B_2 = magnitude of $\bar{B}$ in medium 2

H_1 = magnitude of $\bar{H}$ in medium 1

H_2 = magnitude of $\bar{H}$ in medium 2

Substituting (2) and (3) into (1) and dividing yields

$$\frac{\tan \alpha_1}{\tan \alpha_2} = \frac{\mu_1}{\mu_2} = \frac{\mu_{r1}}{\mu_{r2}} \tag{4}$$

where

μ_{r1} = relative permeability of medium 1 (dimensionless)

μ_{r2} = relative permeability of medium 2 (dimensionless)

Equation (4) gives the relation between the angles α_1 and α_2 for B and H lines at the boundary between two media.

• PROBLEM 12-17

Distilled water is a dielectric having the constants $\varepsilon_r = 81$, $\mu_r = 1$. If a wave is incident from water onto a water-air interface, calculate the critical angle. If the incident $E_1 = 1\ \text{V m}^{-1}$ and the incident angle is 45°, calculate the magnitude of the field strength in the air (a) at the interface and (b) $\lambda/4$ away from the interface.

Solution: The critical angle is defined as

$$\theta_{ic} = \sin^{-1}\sqrt{\frac{\varepsilon_2}{\varepsilon_1}}$$

Therefore,

$$\theta_{ic} = \sin^{-1}\sqrt{\frac{1}{81}} = 6.38°$$

For $\theta_i = 45°$, $\sin\theta_t = \frac{\eta_1}{\eta_2}\sin\theta_i$ and

$$\cos\theta_t = \sqrt{1 - \sin^2\theta_t}$$

Therefore,

$$\sin\theta_t = \sqrt{81}\,(0.707) = 6.36$$

and

$$\cos\theta_t = \sqrt{1 - (6.36)^2} = j6.28$$

so that

$$\alpha = \omega\sqrt{\mu_2\varepsilon_2}\sqrt{\frac{\varepsilon_1}{\varepsilon_2}\sin^2\theta_i - 1}$$

giving

$$\alpha = \frac{2\pi}{\lambda_0}6.28 = \frac{39.49}{\lambda_0} \qquad (\text{Np m}^{-1})$$

and

$$\tau_1 = 1 + \rho_1 = 1 + \frac{z_2\cos\theta_i - z_1\cos\theta_t}{z_2\cos\theta_i + z_1\cos\theta_t}$$

giving

$$\tau_1 = 1 + \rho_1 = 1 + \frac{0.707 - \sqrt{\frac{1}{81} - 0.5}}{0.707 + \sqrt{\frac{1}{81} - 0.5}}$$

$$= 1.42 \;\angle -44.60°.$$

Therefore, the magnitude of the field strength is:

(a) at the interface: $|E_t| = 1.42\ \text{Vm}^{-1}$

(b) $\lambda/4$ away from the interface:

$$|E_t| = 1.42\exp\left(-\frac{39.49}{\lambda_0}\frac{\lambda_0}{4}\right) = 73.2\ \mu\ \text{Vm}^{-1}$$

indicating that the surface wave is very tightly bound to the water surface.

• **PROBLEM 12-18**

Consider a perfect dielectric medium z < 0 bounded by a perfect conductor z > 0, as shown in the figure. Let the fields in the dielectric medium be given by the superposition of (+) and (-) uniform plane waves propagating normal to the conductor surface, that is,

$$\bar{E} = E_1 \cos(\omega t - \beta z)\,\bar{i}_x + E_2 \cos(\omega t + \beta z)\,\bar{i}_x$$

$$\bar{H} = \frac{E_1}{\eta} \cos(\omega t - \beta z)\,\bar{i}_y - \frac{E_2}{\eta} \cos(\omega t + \beta z)\,\bar{i}_y$$

where

$$\beta = \omega\sqrt{\mu\varepsilon}$$

and

$$\eta = \sqrt{\mu/\varepsilon}.$$

Investigate the relationship between E_2 and E_1.

Also, find the total electric and magnetic fields in the dielectric and the current density at the surface between the dielectric and the conductor.

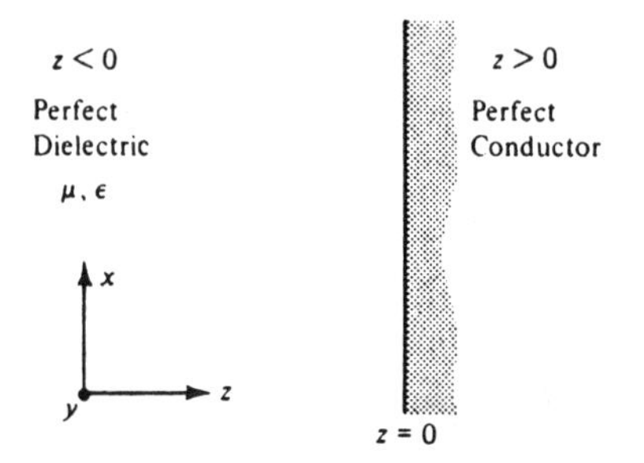

A perfect dielectric medium bounded by a perfect conductor.

Solution: Since E_x is tangential to the perfect conductor surface, the boundary condition for the tangential component of E given by

$$\bar{i}_n \times \bar{E} = 0$$

requires that

$$[E_x]_{z=0} = 0$$

or

$$[E_1 \cos(\omega t - \beta z) + E_2 \cos(\omega t + \beta z)]_{z=0} = 0$$

$$E_1 \cos \omega t + E_2 \cos \omega t = 0 \quad \text{for all } t$$

Thus the required relationship is

$$E_2 = -E_1$$

Proceeding further, the total electric field in the dielectric is given by

$$\bar{E} = E_1 \cos(\omega t - \beta z)\bar{i}_x - E_1 \cos(\omega t + \beta z)\bar{i}_x$$

$$= 2E_1 \sin \omega t \sin \beta z \, \bar{i}_x$$

and the total magnetic field in the dielectric is given by

$$\bar{H} = \frac{E_1}{\eta} \cos(\omega t - \beta z)\bar{i}_y + \frac{E_1}{\eta} \cos(\omega t + \beta z)\bar{i}_y$$

$$= \frac{2E_1}{\eta} \cos \omega t \cos \beta z \, \bar{i}_y$$

Now, from the boundary condition for the tangential compontent of $\bar{H}$ given by

$$\bar{i}_n \times \bar{H} = \bar{J}_S$$

$$[\bar{J}_S]_{z=0} = \bar{i}_n \times [\bar{H}]_{z=0} = -\bar{i}_z \times [\bar{H}]_{z=0}$$

$$= -\bar{i}_z \times \frac{2E_1}{\eta} \cos \omega t \, \bar{i}_y$$

$$= \frac{2E_1}{\eta} \cos \omega t \, \bar{i}_x$$

• PROBLEM 12-19

A uniform plane wave is described by the electric and magnetic fields

$$\bar{E} = \bar{a}_x E_x^+(z,t) = \bar{a}_x E_m^+ \cos(\omega t - \beta_0 z)$$

$$\bar{H} = \bar{a}_y H_y^+(z,t) = \bar{a}_y \frac{E_m^+}{\eta_0} \cos(\omega t - \beta_0 z)$$

and propagates in air between two perfectly conducting, parallel plates of great extent as in (a). The inner surfaces of the plates are located at x = 0 and x = a. Obtain expressions for (a) the surface charge field and (b) the surface currents on the two conductors.

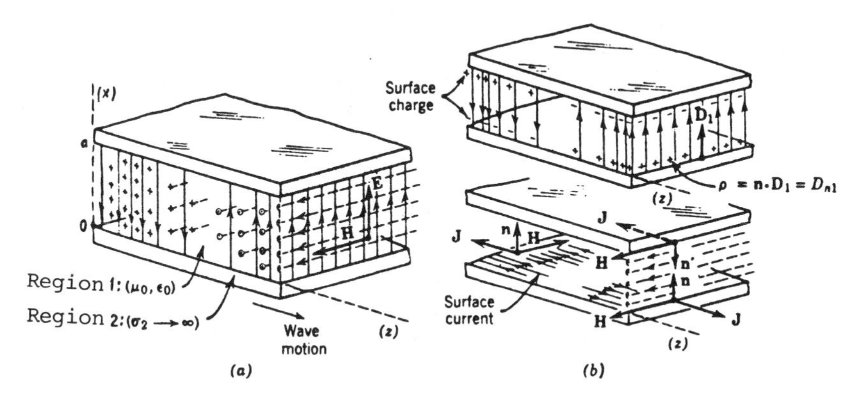

(a) Parallel- plate system supporting a uniform plane wave field. (b) Charge and current distribution on conductor inner surfaces.

Solution: (a)The given $\bar{E}$ is everywhere normal to the plates at x = 0 and x = a, satisfying the boundary condition of $D_{n1} = \rho_s$. The surface charge distributions thus become

$$\rho_s = \bar{n} \cdot \bar{D}_1 = \varepsilon_0 \bar{n} \cdot \bar{E}_1 = \varepsilon_0 \bar{a}_x \cdot \bar{a}_x E_x^+$$

$$= \varepsilon_0 E_m^+ \cos(\omega t - \beta_0 z) \quad x = 0$$

$$\rho_s = \bar{n}' \cdot \bar{D}_1 = -\varepsilon_0 \bar{a}_x \cdot \bar{a}_x E_x^+ = -\varepsilon_0 E_m^+ \cos(\omega t - \beta_0 z)$$

$$x = a$$

implying that $\bar{E}$ lines emerge from positive charges and terminate on negative ones.

(b) The given $\bar{H}$ must be everywhere tangential to the perfect conductors at x = 0 and x = a, to satisfy boundary conditions for magnetic media, yielding there

$$\bar{J}_s = \bar{n} \times \bar{H}_1 = \bar{a}_x \times \bar{a}_y H_y^+ = \bar{a}_z \frac{E_m^+}{\eta_0} \cos(\omega t - \beta_0 z) \text{ at } x = 0$$

$$\bar{J}_s = \bar{n}' \times \bar{H}_1 = -\bar{a}_x \times \bar{a}_y H_y^+ = -\bar{a}_z \frac{E_m^+}{\eta_0} \cos(\omega t - \beta_0 z) \text{ at } x = a$$

It is seen that, in any fixed z plane, current flows in opposite z directions in the two conductors.

• PROBLEM 12-20

Suppose that a uniform plane wave in a medium with constants ε_1, μ_1, σ_1 is incident normally upon a second medium of infinite depth having the constants ε_2, μ_2, σ_2, as shown in the figure. Find the reflection and transmission coefficients for the electric and magnetic field. Find the same when an electromagnetic wave in air is incident normally upon a copper sheet at 1 MHz frequency.

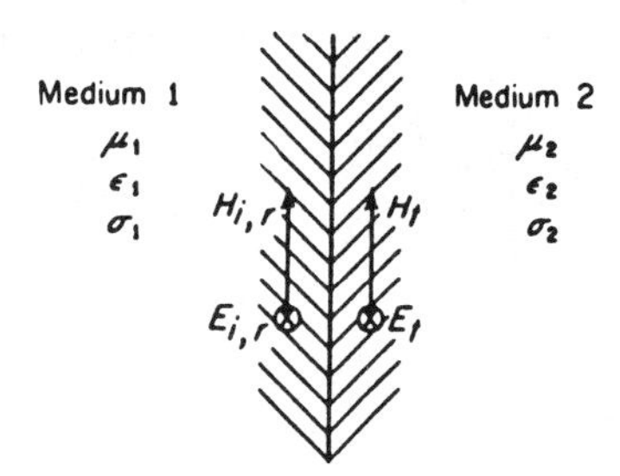

Reflection and transmission at a boundary between two conductive media.

Solution: At the boundary, continuity of the tangential field components demands that

$$E_i + E_r = E_t$$

and

$$H_i + H_r = H_t$$

The latter equation may be modified by noting that

$$\eta_1 H_i = E_i, \qquad \eta_1 H_r = -E_r, \qquad \eta_2 H_t = E_t$$

Thus if E_i is regarded as known, we have two equations in the two unknowns E_r and E_t. These equations may be expressed as

$$E_r - E_t = -E_i$$

$$\frac{E_r}{\eta_1} + \frac{E_t}{\eta_2} = \frac{E_i}{\eta_1}$$

and their solution yields the reflection and transmission coefficients

$$\frac{E_r}{E_i} = \frac{\eta_2 - \eta_1}{\eta_2 + \eta_1} \tag{1}$$

$$\frac{E_t}{E_i} = \frac{2\eta_2}{\eta_2 + \eta_1} \tag{2}$$

The reflection and transmission coefficients for the magnetic field are

$$\frac{H_r}{H_i} = \frac{\eta_1 - \eta_2}{\eta_2 + \eta_1} \tag{3}$$

$$\frac{H_t}{H_i} = \frac{2\eta_1}{\eta_2 + \eta_1} \tag{4}$$

For the given problem assume

$$\mu_1 = \mu_v \qquad \mu_2 = \mu_v$$

$$\varepsilon_1 = \varepsilon_v \qquad \varepsilon_2 = \varepsilon_v$$

$$\sigma_1 = 0 \qquad \sigma_2 = 5.8 \times 10^7 \text{ mhos/m}$$

so that

$$\eta_1 = \sqrt{\frac{\mu_v}{\varepsilon_v}} = 377 \text{ ohms}$$

$$\eta_2 = \sqrt{\frac{j2\pi \times 10^6 \times 4\pi \times 10^{-7}}{5.8 \times 10^7 + j2\pi \times 10^6 \times 8.854 \times 10^{-12}}}$$

$$= 0.000369 \ \angle 45° \text{ ohms}$$

Then the ratio of reflected to incident electric field strengths is

$$\frac{E_r}{E_i} = \frac{3.69 \times 10^{-4} \angle 45° - 377}{3.69 \times 10^{-4} \angle 45° + 377}$$

$$= -0.9999986 \angle -0.000079°$$

$$= -\frac{H_r}{H_i}$$

It is seen that differences between these reflection coefficients for copper and the coefficients of minus and plus unity, which would be obtained for a perfect reflector, are indeed negligible. For most practical purposes, copper can be considered a perfect reflector of radio waves.

The relative strengths of the transmitted fields for this case are

$$\frac{E_t}{E_i} = \frac{7.38 \times 10^{-4} \angle 45°}{3.69 \times 10^{-4} \angle 45° + 377} = +0.00000196 \angle 45°$$

$$\frac{H_t}{H_i} = \frac{2 \times 377}{377 + 3.69 \times 10^{-4} \angle 45°} = 1.9999986 \angle -0.00004°$$

PLANE WAVES IN CONDUCTING AND DIELECTRIC MEDIA

• PROBLEM 12-21

Copper has a conductivity $\sigma = 5.8 \times 10^7$ mhos/meter, and $\mu = 1.26 \times 10^{-6}$ Henry/meter. If a plane wave at a frequency of 200 Hz is normally incident on copper find the depth of penetration of this wave.

Solution: The depth of penetration is given by

$$\delta = \frac{1}{\sqrt{f\pi\mu\sigma}}$$

where f = frequency, μ = permeability and σ = conductivity.

Substituting the given values

$$\delta = \frac{1}{\sqrt{200 \times \pi \times 1.26 \times 10^{-6} \times 5.8 \times 10^{7}}}$$

$$= 4.66 \times 10^{-3} \text{ meter}$$

• PROBLEM 12-22

Find the depth of penetration for copper at 1,100 and 10,000 mHz. What difference is found for different frequencies?

Solution: The depth of penetration is given by

$$\delta = \frac{1}{\alpha} = \sqrt{\frac{2}{\omega\mu\sigma}}$$

where

$$\omega = 2\pi \times f \text{ radians}$$

$$\mu = 4\pi \times 10^{-7} \text{ henry/m}$$

$$\sigma = 58 \times 10^{6} \text{ mhos/m}$$

Now

$$\delta(\text{at 1 mHz}) = \sqrt{\frac{2}{(2\pi \times 10^{6})(4\pi \times 10^{-7})(58 \times 10^{6})}}$$

$$= 0.006 \text{ mm}$$

$$\delta(\text{at 100mHz}) = 0.0066 \text{ mm}$$

$$\delta(\text{at 10,000 mHz}) \cong 0.000 \text{ mm}$$

And we find that as the frequency is increased the depth of penetration becomes smaller.

• PROBLEM 12-23

Given a lossless medium with $\varepsilon_r = 10$ and $\mu_r = 5$ find its impedence.

Solution: The impedance of this medium is given by

$$Z = \sqrt{\mu/\varepsilon}$$

where $\mu = \mu_o\mu_r$ and $\varepsilon = \varepsilon_o\varepsilon_r$

Therefore

$$Z = \sqrt{5 \times 4\pi \times 10^{-7}/(10 \times 10^{-9}/36\pi)}$$

$$= 266 \text{ ohms}$$

• PROBLEM 12-24

If the magnitude of H in a plane wave is 1 amp/meter, find the magnitude of E for a plane wave in free space.

Solution: It is known that E, H and Z (intrinsic impedance) are related by

$$\frac{E}{H} = Z = \sqrt{\frac{\mu}{\varepsilon}}$$

For air

$$Z = \sqrt{\frac{\mu_0}{\varepsilon_0}} = 376.7 \text{ ohms}$$

Hence in the given case

$$E = H \times Z = 1 \times 376 \cdot 7$$

$$= 376.7 \text{ volt/meter.}$$

• PROBLEM 12-25

(a) Paraffin has a relative permittivity $\varepsilon_r = 2.1$. Find the index of refraction for paraffin and also the phase velocity of a wave in an unbounded medium of paraffin.

(b) Distilled water has the constants $\sigma \simeq 0$, $\varepsilon_r = 81$, $\mu_r = 1$. Find η and v.

Solution: (a) The index of refraction

$$\eta = \sqrt{2.1} = 1.45$$

$$v = \frac{c}{\sqrt{2.1}} = 2.07 \times 10^8 \text{ meters/sec}$$

(b)

$$\eta = \sqrt{81} = 9$$

$$v = \frac{c}{\sqrt{81}} = 0.111c = 3.33 \times 10^7 \text{ meters/sec}$$

• PROBLEM 12-26

The x and y components of an elliptically polarized wave in air are

$$E_x = 3 \sin(\omega t - \beta z) \text{ volts/meter}$$

$$E_y = 6 \sin(\omega t - \beta z + 75°) \text{ volts/meter}$$

What is the power per unit area conveyed by the wave?

Solution: The wave is traveling in the positive z direction. The average power per unit area is equal to the average Poynting vector, which has a magnitude

$$S_{av} = \frac{1}{2}\frac{E^2}{Z} = \frac{1}{2}\frac{E_1{}^2 + E_2{}^2}{Z}$$

From the stated conditions the amplitude E_1 = 3 volts/meter, and the amplitude E_2 = 6 volts/meter. Also for air Z = 376.7 ohms. Hence

$$S_{av} = \frac{1}{2}\frac{3^2 + 6^2}{376.7} = \frac{1}{2}\frac{45}{376.7} \simeq 0.06 \text{ watt/meter}^2$$

• PROBLEM 12-27

Find the magnitude of E for a plane wave in free space, if the magnitude of H for this wave is 5 amp/meter.

Solution: It is known that E, H and Z (intrinsic impedance) are related by

$$\frac{E}{H} = Z = \sqrt{\frac{\mu}{\varepsilon}}$$

For air $\mu = \mu_0 = 4\pi \times 10^{-7}$ henry/meter

$$\varepsilon = \varepsilon_0 = \frac{10^{-9}}{36\pi} \text{ farad/meter}$$

Therefore

$$\frac{E}{H} = Z = \sqrt{4\pi \times 10^{-7} \times 10^{9} \times 36\pi}$$

$$= 376.7 \text{ ohms}$$

In the given problem

$$E = H \times Z = 5 \times 376.7 = 1883 \text{ volts/meter}$$

• PROBLEM 12-28

A certain medium has a relative permittivity of 25 and a relative permeability of unity. Evaluate (a) the phase velocity of a wave in this medium; (b) the index of refraction of the medium.

Solution: (a) The phase velocity of the wave in a medium with relative permittivity ε_r and relative permeability μ_r is given by

$$V = 1/\sqrt{\varepsilon_o \varepsilon_r \mu_o \mu_r}$$

where ε_o and μ_o are permittivity and permeability of air.

Hence for the given problem

$$V = 1/\sqrt{(25 \times 10^{-9}/36\pi) \times 1 \times 4\pi \times 10^{-7}} =$$

$$0.6 \times 10^{8} \text{ meters/sec}$$

(b) The index of refraction is

$$\eta = \varepsilon_r \mu_r$$

$$= \sqrt{25 \times 1} = 5$$

• PROBLEM 12-29

> The amplitude of a plane wave in a medium with $\mu_r = 1$ and $\varepsilon_r = 4$ is $H_o = 5$ amp/meter. Find (a) the average Poynting vector, (b) the maximum energy density for this plane wave.

Solution: It is known that

$$\frac{E}{H} = \sqrt{\frac{\mu}{\varepsilon}}$$

so $$E = H\sqrt{\frac{\mu}{\varepsilon}} = 5 \times \sqrt{\frac{4\pi \times 10^{-7} \times 36\pi}{4 \times 10^{-9}}} = 300\pi$$

$$= 942 \text{ volts/meter}$$

The average poynting vector is given by

$S_{av} = \frac{E^2}{2R}$ and here $R = \sqrt{\frac{\mu}{\varepsilon}} = 188.5$ ohms.

$$\therefore\ S_{AV} = \frac{E^2}{2R} = 942 \times 942/(2 \times 188.5) = 2354 \text{ W/m}^2$$

(b) The maximum energy density of this wave is given by $W_{peak} = \varepsilon E_0^{\ 2}$

$$= \frac{4}{36\pi} \times 10^{-9} \times 300\pi \times 300\pi$$

$$= 3.1 \times 10^{-4} \text{ joule/meter}^2$$

• PROBLEM 12-30

> A plane wave with an electric field intensity of $100\sqrt{\pi}$ V/m is traveling in a medium with $\varepsilon = 10^{-9}/36\pi$. Find the magnetic and total energy densities.

Solution: For a plane traveling wave the electric and magnetic energy densities are equal. Therefore it is sufficient to evaluate electric energy density which is

$$\omega_e = \frac{1}{2} \varepsilon E^2$$

$$= \frac{1}{2} \times (10^{-9}/36\pi) \times 100 \times 100 \times \pi$$

$$= 1.38 \times 10^{-7} \text{ joules/meter}^3$$

Therefore the magnetic energy density is 2×10^{-7} joules/meter^3.

The total energy density is

$$\omega = \omega_e + \omega_h = 2.7 \times 10^{-7} \text{ joules/meter}^3$$

• PROBLEM 12-31

What is Hagen-Rubens relation?

Solution: For normal incidence from air on a conducting medium, with $\hat{n}_1 = 1$, $\hat{n}_2 = n + ik$, the reflectance is

$$R_n = \frac{(n - 1)^2 + k^2}{(n + 1)^2 + k^2} .$$

Since all the transmitted energy is eventually absorbed in a semi-infinite conducting medium, the absorptance is defined as

$$A = 1 - R.$$

For normal incidence

$$A_n = \frac{4n}{(n + 1)^2 + k^2} .$$

The absorptance is small (high reflectance) if $n \ll 1$, or $n \gg 1$, or $k \gg 1$. When $\hat{n} \cong k \gg 1$(with $K_i = g/\varepsilon_0\omega \gg 1$), where K_i and K_r denote magnitudes of incident and reflected waves and

$$K_i \gg |K_r|, \; n \cong k \cong \sqrt{K_i/2} ,$$

$$A_n \cong 2/k \ll 1.$$

In this case

$$k \cong \sqrt{K_i/2} = \sqrt{g/2\varepsilon_0\omega},$$

so that

$$A_n \cong 2\sqrt{2\varepsilon_0\omega/g}.$$

This is called the Hagen-Rubens relation; it should hold for moderately good conductors in the microwave region and below, and for metals into the infrared, with g the d-c conductivity. With the same values used in calculating the skin depth, we find that for silver at

$$f = 10^{10} s^{-1} \text{ (3 cm wavelength)},$$

and

$$g = 3 \times 10^7 \text{ mhos/m},$$

$$A_n = 2\sqrt{2(8.854 \times 10^{-12})(2\pi \times 10^{10})/3 \times 10^7} =$$

$$= 3.9 \times 10^{-4},$$

$$R_n = 0.9996.$$

For sea water at $f = 6 \times 10^4 \ s^{-1}$

$$A_n = 25 \times 10^{-4},$$

$$R_n = 0.9975.$$

• PROBLEM 12-32

If E = 5 volts/meter is the amplitude of an incident plane wave which is reflected at normal incidence from a boundary surface and E_1 is the amplitude of the reflected wave, find:

(a) Under what conditions is there a pure standing wave?

(b) The standing wave ratio when E_1 = 1/10 volt/meter.

(c) The value of E_{max} and E_{min} when E_1 = 3/5 volt/meter.

Solution: (a) There is a pure standing wave when $E_1 = -E_0$ i.e. $E_1 = -5$ volt/meter

(b) When $E_1 = \frac{1}{10}$ v/m, the standing wave ratio is

$$VSWR = \frac{E_0 + E_1}{E_0 - E_1} = \frac{5 + 0.1}{5 - 0.1} = 1.04$$

(c) It is known that

$$VSWR = \frac{E_0 + E_1}{E_0 - E_1} = \frac{v_{max}}{v_{min}}$$

so

$$E_{max} = E_0 + E_1 = 5 + 3/5 = 5.6 \text{ volts}$$

$$E_{min} = E_0 - E_1 = 5 - 3/5 = 4.4 \text{ volts}$$

• PROBLEM 12-33

A plane 1-GHz traveling wave in air with peak electric field intensity of 1 v/m is incident normally on a large copper sheet. Find the average power absorbed by the sheet per square meter of area.

Solution: First let the intrinsic impedance of copper be calculated at 1 GHz.

$$\dot{\bar{Z}}_c = \frac{1 - j}{\sqrt{2}} 376.7 \sqrt{\frac{\mu_r}{\varepsilon_r}} \sqrt{\frac{\omega\varepsilon}{\sigma}}$$

For copper, $\mu_r = \varepsilon_r = 1$ and $\sigma = 58 \text{ M m}^{-1}$. Hence the real part of $\dot{Z}_c$ is

$$\text{Re}\dot{\bar{Z}}_c = (\cos 45°)(376.7)\sqrt{\frac{2\pi \times 10^9 \times 8.85 \times 10^{-12}}{5.8 \times 10^7}}$$

$$= 8.2 \text{ m}\Omega$$

Next find the value of H_0 at the sheet (tangential to

the surface). This is very nearly double H for the incident wave.

Thus

$$H_0 = 2\,\frac{E}{Z} = \frac{2 \times 1}{376.7}\ \text{A/M}$$

From

$$S_{AV} = \frac{1}{2}\,H_0{}^2 \text{Re}\bar{\dot{Z}}_c$$

$$S_{AV} = \frac{1}{2}\left(\frac{2}{376.7}\right)^2 8.2 \times 10^{-3} = 116\text{nW/m}^2$$

• PROBLEM 12-34

Consider a plane wave in air impinging normally on a relatively large, thick slab of polyethylene, as shown in the given figure. Let the peak value of the incident electric field intensity be 10 V/m. Find the transmitted and reflected power densities.

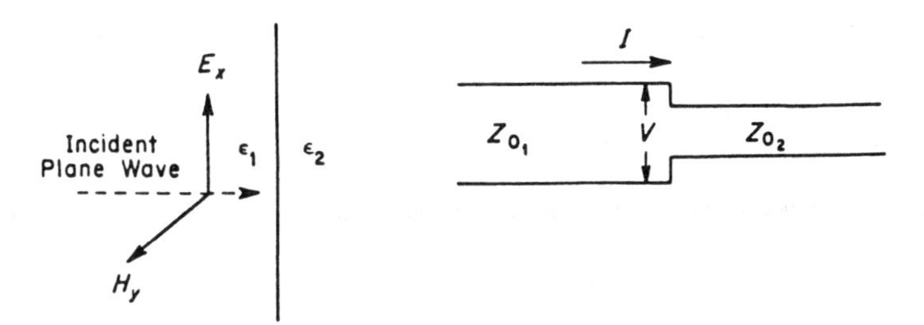

Analogous plane-wave and transmission-line discontinuities.

Solution: First the intrinsic impedances for air and polyethylene are found to be

$$\eta_{air} = \sqrt{\frac{\mu_0}{\varepsilon_0}} = \sqrt{\frac{4\pi \times 10^{-7}}{(1/36\pi) \times 10^{-9}}} = 377\ \text{ohms}$$

$$\eta_{polyethylene} = \sqrt{\frac{\mu_0}{\varepsilon_r \varepsilon_0}} = \frac{1}{\sqrt{\varepsilon_r}}\sqrt{\frac{\mu_0}{\varepsilon_0}} = \frac{377}{1.5}\ \text{ohms}$$

These intrinsic impedances correspond to the characteristic impedances Z_{01} and Z_{02} in the transmission-line analogy shown in the given figure. The reflection coefficient at the boundary is then

$$K = \frac{\eta_{polyethylene} - \eta_{air}}{\eta_{polyethylene} + \eta_{air}} = \frac{377/1.5 - 377}{377/1.5 + 377} = -\frac{1}{5}$$

The minus sign indicates 180-degree phase difference between incident and reflected waves at the boundary. The reflected wave then has a peak amplitude of

$$|E^-| = |K||E^+| = (0.2)(10) = 2 \text{ V/m}$$

$$P_{incident} = \frac{(10)^2}{(2)(377)} = 0.133 \text{ W/m}^2$$

$$P_{reflected} = \frac{(2)^2}{(2)(377)} = 0.0053 \text{ W/m}^2$$

$$P_{transmitted} = P_{incident} - P_{reflected} = 0.128 \text{ W/m}^2$$

• PROBLEM 12-35

Referring to the given figure, let medium 1 be air ($\mu_r = 1$) and medium 2 be soft iron with a relative permeability of 7,000.

(a) If $\bar{B}$ in the iron is incident normally on the boundary ($\alpha_2 = 0$), find α_1.

(b) If B in the iron is nearly tangent to the surface at an angle $\alpha_2 = 85°$, find α_1.

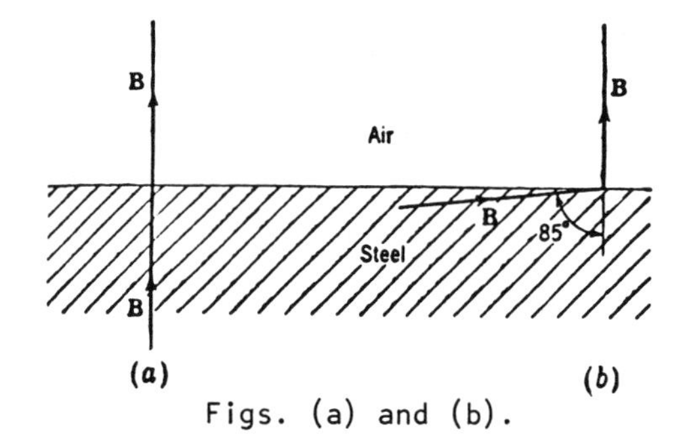

Figs. (a) and (b).

Solution: (a) From

$$\frac{\tan \alpha_1}{\tan \alpha_2} = \frac{\mu_1}{\mu_2} = \frac{\mu_{r_1}}{\mu_{r_2}}$$

$$\tan \alpha_1 = \frac{\mu_{r_1}}{\mu_{r_2}} \tan \alpha_2 = \frac{1}{7,000} \tan \alpha_2 \qquad (1)$$

When $\alpha_2 = 0$, $\alpha_1 = 0$, so that the $\bar{B}$ line in air is also normal to the boundary (see Fig. a).

(b) When $\alpha_2 = 85°$, from (1) $\tan \alpha_1 = 0.0016$ or $\alpha_1 = 0.1°$. Thus, the direction of $\bar{B}$ in air is almost normal to the boundary (within $\frac{1}{10}°$) even though its direction in the iron is nearly tangent to the boundary (within 5°) (see Fig. b). Accordingly, for many practical purposes the direction of B or H in air or other medium of low relative permeability may be taken as normal to the boundary of a medium having a high relative permeability.

• PROBLEM 12-36

For sea water with $\gamma = 5$ mhos/meter and $\varepsilon_r = 80$, find the distance a radio signal can be transmitted at 25 kcps and 25 mcps, if the range is taken to be the distance at which 90 per cent of the wave amplitude is attenuated.

Solution: Since

$$E = \bar{E}e^{-\rho x} = \bar{E}e^{-(\alpha+j\beta)x}$$

The attenuation is therefore

$$e^{-\alpha x} = 0.1$$

or taking ln of both sides

$$-\alpha x = \ln 0.1 = -2.30$$

so that

$$x = \frac{2.30}{\alpha}$$

$$\alpha = \text{real part of } \rho = \omega\sqrt{\frac{\mu\varepsilon}{2}\left(\sqrt{\frac{\gamma^2}{\omega^2\varepsilon^2} + 1} - 1\right)}$$

The proof is shown below:

Since

$$\rho = \sqrt{j\omega\mu(\gamma + j\omega\varepsilon)} = \alpha + j\beta$$

then

$$\rho^2 = j\omega\mu\gamma - \omega^2\mu\varepsilon = \alpha^2 - \beta^2 + 2j\alpha\beta$$

Equating reals and imaginaries

$$-\omega^2\mu\varepsilon = \alpha^2 - \beta^2$$

$$2\alpha\beta = \gamma\omega\mu$$

$$\alpha\beta = \frac{\gamma\omega\mu}{2}$$

or

$$\alpha = \frac{\gamma\omega\mu}{2\beta} \quad ; \qquad \beta = \frac{\gamma\omega\mu}{2\alpha}$$

$$-\omega^2\mu\varepsilon = \alpha^2 - \beta^2 = \alpha^2 - \frac{\gamma^2\omega^2\mu^2}{4\alpha^2}$$

or

$$-4\alpha^2\omega^2\mu\varepsilon = 4\alpha^4 - \gamma^2\omega^2\mu^2$$

Transpose

$$4\alpha^4 + 4\alpha^2\omega^2\mu\varepsilon - \gamma^2\omega^2\mu^2 = 0$$

and divide by 4

$$\alpha^4 + \alpha^2\omega^2\mu\varepsilon - \frac{\gamma^2\omega^2\mu^2}{4} = 0$$

Completing the square

$$\alpha^4 + \alpha^2\omega^2\mu\varepsilon + \left(\frac{\omega^2\mu\varepsilon}{2}\right)^2 = \frac{\gamma^2\omega^2\mu^2}{4} + \left(\frac{\omega^2\mu\varepsilon}{2}\right)^2$$

or

$$\left(\alpha^2 + \frac{\omega^2\mu\varepsilon}{2}\right)^2 = \frac{\gamma^2\omega^2\mu^2}{4} + \frac{\omega^2\mu^2\varepsilon^2}{4}$$

or

$$\left(\alpha^2 + \frac{\omega^2\mu\varepsilon}{2}\right)^2 = \omega^4 \frac{\mu^2\varepsilon^2}{4}\left(\frac{\gamma^2}{\omega^2\varepsilon^2} + 1\right)$$

Take square root of both sides

$$\alpha^2 = \omega^2 \frac{\mu\varepsilon}{2} \sqrt{\frac{\gamma^2}{\omega^2\varepsilon^2} + 1} - \frac{\omega^2\mu\varepsilon}{2}$$

$$\alpha^2 = \frac{\omega^2\mu\varepsilon}{2} \left(\sqrt{\frac{\gamma^2}{\omega^2\varepsilon^2} + 1} - 1 \right)$$

therefore

$$\alpha = \omega \sqrt{\frac{\mu\varepsilon}{2} \left(\sqrt{\frac{\gamma^2}{\omega^2\varepsilon^2} + 1} - 1 \right)}$$

Substituting given values

for 25 kcps:

$$\alpha = 2\pi(25)10^3 \left[\frac{1.26(10^{-6})80(8.85)10^{-12}}{2} \cdot \left(\sqrt{\frac{25}{4\pi^2(25)^2 10^6 [80(8.85)10^{-12}]^2} + 1} - 1 \right) \right]^{\frac{1}{2}}$$

$\alpha = .715$

For 25 mcps:

$$\alpha = 2\pi(25)10^6 \left[\frac{1.26(10^{-6})80(8.85)10^{-12}}{2} \cdot \left(\sqrt{\frac{25}{4\pi^2(25)^2 10^{12} [80(8.85)10^{-12}]^2} + 1} - 1 \right) \right]^{\frac{1}{2}}$$

$= 22.6$

Thus for 25 kcps

$$x = \frac{2.3}{.715} = 3.22 \text{ meters}$$

For 25 mcps

$$x = \frac{2.3}{22.6} = .104 \text{ meters} \cong 10 \text{ cm}$$

• PROBLEM 12-37

a) Paraffin has a relative permittivity $\varepsilon_r = 2.1$. Find the index of refraction for paraffin and also the phase velocity of a wave in an unbounded medium of paraffin.

b) Distilled water has the constants $\varepsilon_r = 81$, $\mu_r = 1$. Find the index of refraction and the phase velocity.

Solution: a) The index of refraction is the reciprocal of the relative phase velocity p. That is,

$$\eta = \frac{1}{p} = \frac{1}{v/c} = \frac{c}{v} = \sqrt{\mu_r \varepsilon_r}$$

For a nonferrous material such as this one, μ_r is very nearly unity so that

$$\eta = \sqrt{\varepsilon_r}$$

The index of refraction

$$\eta = \sqrt{2.1} = 1.45$$

The phase velocity

$$v = \frac{c}{\sqrt{2.1}} = 207 \text{ µm/s}$$

b) $\eta = \sqrt{81} = 9$

$$v = c/\sqrt{81} = 0.111c = 33.3 \text{ µm/s}.$$

The index of refraction given for water is the value at low frequencies ($f \to 0$). At light frequencies, say for sodium light ($\lambda = 5893 \text{ A}^0$), the index of refraction is observed to be about 1.33 instead of 9 as calculated on the basis of the relative permittivity. The explanation for the difference is that the permittivity ε is not a constant but is a function of frequency. At zero frequency $\varepsilon_r = 81$, but at light frequencies $\varepsilon_r = (1.33)^2 = 1.77$. The index of refraction and permittivity of many other substances also vary as a function of the frequency.

• **PROBLEM 12-38**

Consider a plane wave in air impinging obliquely on a large, thick slab of polyethylene. The polarization of the wave is in the plane of incidence, as shown in the figure. Determine the reflected and transmitted waves as a function of the angle of incidence, θ_1 which is 60°. The incident wave has a peak value of 10 V/m.

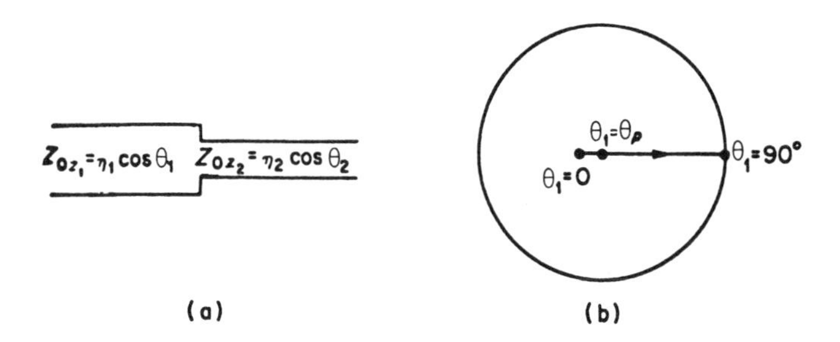

Solution: First construct the transmission-line analogy shown in the (a) portion of the accompanying figure. It can be seen now that the reflection coefficient will vary as shown in the (b) portion as the incident angle is varied from 0 to 90 degrees. This figure will be recognized as a Smith chart with the impedance contours suppressed. When the incident angle is zero, this is just the normal incidence case and K = - 1/5. As θ_1 is increased, K moves to the right along the real axis and ends up at unity for θ_1 = 90 degrees. Note that there will be a value of θ_1 for which the reflection coefficient is zero. No reflection will occur for this value of θ_1, which is known as the "polarizing angle."

The polarizing angle is found by equating Z_{0z_1} to Z_{0z_2}. For the case of $\bar{E}$ in the plane of incidence and $\mu_1 = \mu_2$,

$$\theta_p = \tan^{-1}\sqrt{\frac{\varepsilon_2}{\varepsilon_1}}$$

Now, for any given incident angle θ_1, the refracted angle θ_2 may be computed from Snell's law which is

$$\frac{\sin\theta_1}{\sin\theta_2} = \sqrt{\frac{\mu_2\varepsilon_2}{\mu_1\varepsilon_1}}$$

Then the resultant component waves may be evaluated.

$$\theta_2 = \sin^{-1}\left(\sqrt{\frac{1}{2.25}}\sin 60°\right) = 35.3°$$

and

$$Z_{0z1} = (377)(\cos 60°) = 188.5 \text{ ohms}$$

$$Z_{0z2} = \left(\frac{377}{\sqrt{2.25}}\right)(\cos 35.3°) = 205 \text{ ohms}$$

Therefore

$$K = \frac{205 - 188.5}{205 + 188.5} = 0.042$$

and

$$|E^+_{x1}| = (10)(\cos 60°) = 5 \text{ V/m}$$

$$|E^-_x| = |K||E^+_{x1}| = 0.21 \text{ V/m}$$

$$|E^+_{x2}| = |1 + K||E^+_{x1}| = 5.21 \text{ V/m}$$

The complete wave picture may now be reconstructed from the tangential components and the known directions of propagation. That is,

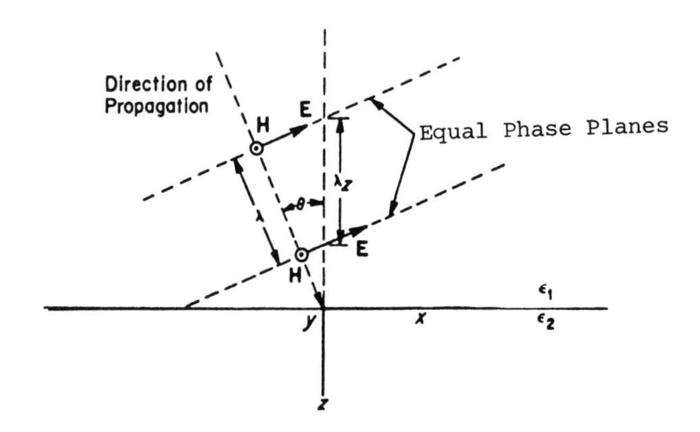

Plane wave with electric field in plane of incidence.

$$|\bar{E}^+_1| = (5)(\sec 60°) = 10 \text{ V/m}$$

$$|\bar{E}^-_1| = (0.21)(\sec 60°) = 0.42 \text{ V/m}$$

$$|\bar{E}^+_2| = (5.21)(\sec 35.3°) = 6.38 \text{ V/m}$$

An x polarized wave arrives from the left at f = 1 MHz with an amplitude $\hat{E}^{+}_{m1} = 100\, e^{j0°}$ V/m. It is incident upon a lossless slab an eighth of a wavelength thick, backed with a quarter wave lossy slab, with parameters as shown in the diagram. Find the total magnetic and electric fields in region 1.

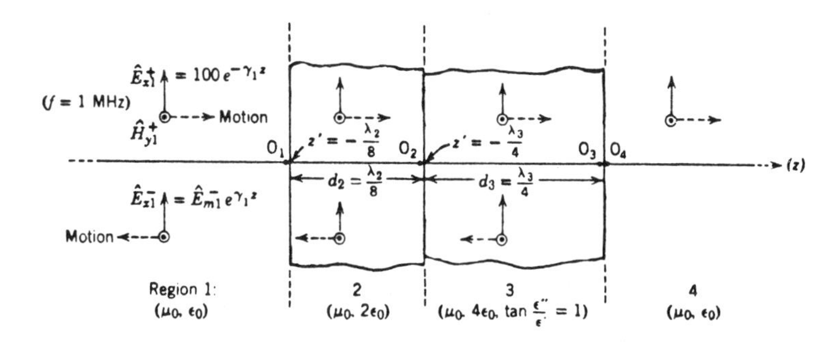

Solution: The origins assumed for the four regions are noted in the diagram. A tabulation of α, β, γ, and $\hat{\eta}$ obtained for the regions using

$$\alpha = \frac{\omega\sqrt{\mu\varepsilon}}{\sqrt{2}}\left[\sqrt{1 + \left(\frac{\sigma}{\omega\varepsilon}\right)^2} - 1\right]^{\frac{1}{2}},$$

$$\beta = \frac{\omega\sqrt{\mu\varepsilon}}{\sqrt{2}}\left[\sqrt{1 + \left(\frac{\sigma}{\omega\varepsilon}\right)^2} + 1\right]^{\frac{1}{2}}, \tag{1}$$

and

$$\eta = \sqrt{\mu/\varepsilon}$$

is given here:

Region	μ_r	ϵ_r	$\frac{\epsilon''}{\epsilon'}$	$\alpha\ (m^{-1})$	$\beta\ (m^{-1})$	λ (m)	$\hat{\eta}\ (\Omega)$
1	1	1	0	0	0.0209	300	377
2	1	2	0	0	0.0296	212	266
3	1	4	1	0.0191	0.0461	136	$159e^{j22.5°}$
4	1	1	0	0	0.0209	300	377

(a) Beginning in region 4 which contains no reflection, $\hat{\Gamma}_4(0)$ is zero from

$$\hat{\Gamma}(z) = \frac{\hat{E}_m^-}{\hat{E}_m^+} e^{2\gamma z} \tag{2}$$

yielding, from

$$\hat{Z}(z) = \hat{\eta} \frac{1 + \hat{\Gamma}(z)}{1 - \hat{\Gamma}(z)} \ \Omega , \tag{3}$$

$\hat{Z}_4(0) = \hat{\eta}_4 = 120\pi \ \Omega = \hat{Z}_3(0)$.

(b) Inserting the latter into

$$\hat{\Gamma}(z) = \frac{\hat{Z}(z) - \hat{\eta}}{\hat{Z}(z) + \hat{\eta}} \tag{4}$$

yields

$$\hat{\Gamma}_3(0) = \frac{377 - 159e^{j22.5°}}{477 + 159e^{j22.5°}} = 0.451e^{-j21.4°}$$

Now

$$\hat{\Gamma}(z') = \hat{\Gamma}(z)e^{2\gamma(z'-z)} \tag{5}$$

$\therefore$ At the input plane $z = -d_3 = -\lambda_3/4$,

$$\hat{\Gamma}_3(-d_3) = \hat{\Gamma}_3(0)e^{2\gamma_3(-d_3-0)} = \hat{\Gamma}_3(0)e^{-2\alpha_3 d_3}e^{-j2\beta_3 d_3}$$

$$= (0.451e^{-j21.4°})(e^{-2(0.0191)34})e^{-j180°}$$

$$= 0.1233e^{-j201.4°}$$

(c) Steps (a) and (b) are repeated to find $\hat{Z}$ and $\hat{\Gamma}$ at the output plane of region 2. Thus (3) yields

$$\hat{Z}_3(-d_3) = \hat{\eta}_3 \frac{1 + \hat{\Gamma}_3(-d_3)}{1 - \hat{\Gamma}_3(-d_3)}$$

$$= 159e^{j22.5} \ \frac{1 + 0.1233e^{-j201.4°}}{1 - 0.1233e^{-j201.4°}}$$

$$= 126.2e^{j27.9°}\ \Omega$$

and from

$$\hat{Z}(a^+) = \hat{Z}(a^-) \quad , \tag{6}$$

$$\hat{Z}_3(-d_3) = \hat{Z}_2(0) \quad ,$$

yielding from

$$\hat{\Gamma}(z) = \frac{\hat{Z}(z) - \hat{\eta}}{\hat{Z}(z) + \hat{\eta}} \tag{4}$$

$$\hat{\Gamma}_2(0) = \frac{\hat{Z}_2(0) - \hat{\eta}_2}{\hat{Z}_2(0) + \hat{\eta}_2} = \frac{126.2e^{j27.9°} - 266}{126.2e^{j27.9°} + 266} = 0.434e^{j150.3°}$$

The latter transforms, by the use of (5) at the input plane $z = -d_2$

$$= \Gamma_2(-d_2) = \hat{\Gamma}_2(0)e^{2\gamma_2(-d_2-0)}$$

$$= (0.434e^{j150.3°})e^{2(j2\pi/\lambda_2)(-\lambda_2/8)} = 0.434e^{j60.3°}$$

(d) The total field impedance there, from (3), is

$$\hat{Z}_2(-d_2) = \hat{\eta}_2\frac{1 + \hat{\Gamma}_2(-d_2)}{1 - \hat{\Gamma}_2(-d_2)} = 266\ \frac{1 + 0.434e^{j60.3°}}{1 - 0.434e^{j60.3°}}$$

$$= 390e^{j42.9°}\ \Omega$$

which, by continuity across the interface, yields

$$\hat{Z}_1(0) = 390e^{j42.9^\circ}\,\Omega .$$

From (4)

$$\hat{\Gamma}_1(0) = \frac{\hat{Z}_1(0) - \hat{\eta}_1}{\hat{Z}_1(0) + \hat{\eta}_1} = \frac{390e^{j42.9^\circ} - 377}{390e^{j42.9^\circ} + 377} = 0.393e^{j87.1^\circ}$$

The reflected wave amplitude is obtained using (2); applying it at z = 0 yields $\hat{E}^-_{m_1} = \hat{E}^+_{m_1}\hat{\Gamma}_1(0) = (100)(0.393e^{j87.1^\circ}) = 39.3^{j87.1^\circ}$, whence the total fields in region 1 become,

$$\hat{E}_{x_1}(z) = 100e^{j\beta_1 z} + 39.3e^{j(\beta_1 z + 87.1^\circ)}\text{V/m}$$

$$\hat{H}_{y_1}(z) = \frac{100}{377}e^{-j\beta_1 z} - \frac{39.3}{377}e^{j(\beta_1 z+87.1^\circ)}\text{A/m}$$

• PROBLEM 12-40

A homogeneous non-magnetic insulating slab, of dielectric constant K, whose surfaces are parallel to the x-y plane, is moving with uniform velocity w along the x-axis. Show that the velocity of the wave in the insulator, as measured by a stationary observer, is given by

$$v = v_0 + \left(1 - \frac{1}{K}\right) w,$$

where $v_0 = c/\sqrt{K}$, the velocity of the wave when the insulator is at rest, and provided that w^2/c^2 may be neglected.

Solution: Consider a point inside the moving insulator, and stationary relative to the observer who measures w. Let the electric and magnetic fields at this point be E and B, along the z- and y-axes respectively. Let E_1 be the electric field just outside the surfaces of the insulating slab and E_2 the component due to charges displaced inside the insulator, so that

$$E = E_1 + E_2. \tag{1}$$

Let B_1 be the component of B due to the changing electric field (displacement current), and B_2 the component due to the motion of the displaced charges in the insulator (convection current), so that

$$B = B_1 + B_2. \tag{2}$$

The "inducing" electric intensity acting upon the structure of the moving insulator is

$$E_0 = E_1 + (w \times B_1) = E_1 + wB_1$$

(relative to the moving insulator, B_2 does not exist), so that from

$$E = -(1 - 1/K)E_0$$

$$E_2 = -\left(1 - \frac{1}{K}\right)(E_1 + wB_1), \tag{3}$$

and (1) becomes

$$E = \frac{E_1}{K} - \left(1 - \frac{1}{K}\right) wB_1, \tag{4}$$

Then,

$$\frac{\partial E}{\partial x} = \frac{\partial B}{\partial t}, \tag{5}$$

while the reciprocal relation

$$\frac{\partial B_y}{\partial x} = \frac{\mu K}{C^2}\frac{\partial E_z}{\partial t}, \tag{a}$$

becomes

$$\frac{\partial B_1}{\partial x} = \frac{1}{c^2}\frac{\partial E_1}{\partial t}. \tag{6}$$

[The component E_2 causes B_2, and does not contribute to B_1. In the case of a stationary insulator, $E_1 = KE$, and (6) is then identical with (a).]

The component B_2 of the magnetic field, is, from

$$B = \frac{V \times E}{C^2},$$

$$B_2 = \frac{w \times E_2}{c^2} = \frac{\left(1 - \frac{1}{K}\right) wE_1}{c^2} \tag{7}$$

neglecting w^2/c^2. Hence, from (2) and (5),

$$\frac{\partial E}{\partial x} = \frac{\partial B_1}{\partial t} + \frac{\left(1 - \frac{1}{K}\right) w}{c^2} \frac{\partial E_1}{\partial t}$$

or

$$\frac{\partial E}{\partial x} = \frac{\partial B_1}{\partial t} + \frac{(K - 1)w}{c^2} \frac{\partial E}{\partial t}, \tag{8}$$

from (4) and again neglecting w^2/c^2; while, from (4) and (6),

$$\frac{\partial B_1}{\partial x} = \frac{K}{c^2} \frac{\partial E}{\partial t} + \frac{(K - 1)w}{c^2} \frac{\partial B_1}{\partial t}. \tag{9}$$

If the solution of (8) and (9) is of the form

$$E = \alpha \sin p(x - vt),$$

$$B_1 = \beta \sin p(x - vt),$$

then (8) and (9) reduce to

$$-\beta v = \alpha\left\{1 + \frac{(K - 1)wv}{c^2}\right\}, \tag{10}$$

$$-\frac{K}{c^2} \alpha v = \beta\left\{1 + \frac{(K - 1)wv}{c^2}\right\}. \tag{11}$$

Multiplying these equations, and taking the square root:

$$\sqrt{K} \frac{v}{c} = 1 + \frac{(K - 1)wv}{c^2}$$

or

$$\frac{1}{v} = \frac{\sqrt{K}}{c} - \frac{(K - 1)}{c^2} w.$$

Now $c/\sqrt{K} = v_0$, the wave velocity when $w = 0$, and $c^2 = Kv_0^2$, so that

$$v = \frac{v_0}{1 - \left(\frac{K-1}{K}\right)\frac{w}{v_0}}$$

$$= v_0 + 1 - \frac{1}{K} \ w, \tag{12}$$

neglecting w^2/v_0^2.

Further, by substituting this result in (10) and (11),

$$B_1 = -\frac{E}{v_0} \ . \tag{13}$$

To find B, the resultant magnetic field, from (4),

$$E_1 = KE + (K - 1) \ wB_{1_0}$$

$$= E\left\{K - (K - 1)\frac{w}{v_0}\right\}, \tag{14}$$

from (13). And from (7) and (14), neglecting w^2/c^2,

$$B_2 = \frac{(K - 1)wE}{c^2} ,$$

so that

$$B = B_1 + B_2 = -E\left\{\frac{1}{v_0} - \frac{(K - 1)w}{c^2}\right\}$$

$$= -\frac{E}{v} \ . \tag{15}$$

• **PROBLEM 12-41**

A right circularly polarized wave (RCP) is incident at an angle of 45° from air onto (a) a perfect conductor and (b) polystyrene ($\varepsilon_r = 2.7$). What is the polarization state of the reflected wave for these two cases?

Solution: (a) When medium 1 is air and medium 2 is a

perfect conductor,

$$\rho_{||} = 1 \ \angle 180^\circ \qquad \rho_\perp = 1 \ \angle 180^\circ$$

For a right circularly polarized wave, $\gamma_i = 45^\circ$, $\delta_i = -90^\circ$, so that from

$$\delta_r = \delta_i + \pi + (\phi_\perp - \phi_{||}),$$

$$\delta_r = -90^\circ + 180^\circ + (180^\circ - 180^\circ) = 90^\circ$$

and from

$$\gamma_r = \tan^{-1}\left(\frac{|\rho_{||}|}{|\rho_\perp|} \times \tan \gamma_i\right),$$

$$\gamma_r = \tan^{-1}\left(\frac{1}{1} \times 1\right) = 45^\circ.$$

Therefore the reflected wave is left circularly polarized (LCP), as shown in figure 1a. Conversely, if the incident wave had been LCP, the reflected wave would be RCP.

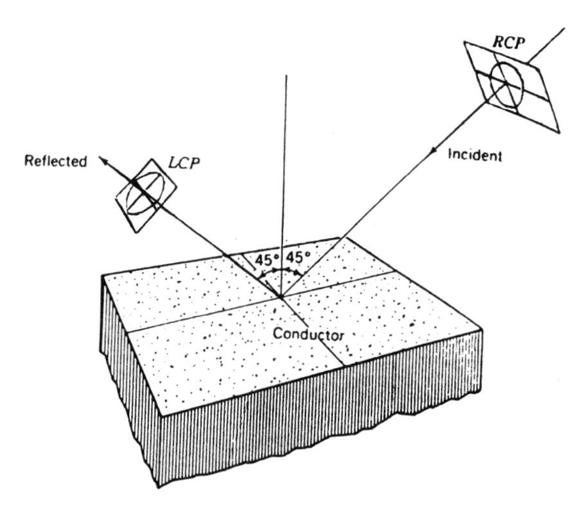

Right circularly polarized (RCP) wave incident on a perfect conductor. The reflected wave is left circularly polarized (LCP). No wave is transmitted.

Fig. 1a

(b) When medium 1 is air and medium 2 is polystyrene, then from

$$\rho_{||} = \frac{-\varepsilon_r \cos\theta_i + \sqrt{\varepsilon_r - \sin^2\theta_i}}{\varepsilon_r \cos\theta_i + \sqrt{\varepsilon_r - \sin^2\theta_i}} \; ; \; \varepsilon_r = \varepsilon_2/\varepsilon_1$$

$$\rho_{||} = \frac{-2.7(0.707) + \sqrt{2.7 - 0.5}}{2.7(0.707) + \sqrt{2.7 - 0.5}} = 0.126 \quad \angle 180^\circ$$

and from

$$\rho_{\perp} = \frac{\cos\theta_i - \sqrt{\varepsilon_r - \sin^2\theta_i}}{\cos\theta_i + \sqrt{\varepsilon_r - \sin^2\theta_i}}$$

$$\rho_{\perp} = \frac{0.707 - \sqrt{2.7 - 0.5}}{0.707 + \sqrt{2.7 - 0.5}} = 0.354 \quad \angle 180^\circ$$

Therefore,

$$\delta_r = -90^\circ + 180^\circ + (180^\circ - 180^\circ) = 90^\circ$$

$$\gamma_r = \tan^{-1} \frac{0.126}{0.354} = 19.6^\circ$$

Substituting these values into

$$\tan 2\tau_r = \tan 2\gamma_r \cos \delta_r \,,$$

$$\sin 2\varepsilon_r = \sin 2\gamma_r \sin \delta_r \,,$$

and

$$\cos 2\gamma_r = \cos 2\varepsilon_r \cos 2\tau_r$$

yields

$$\tau_r = 0^\circ.$$

$$\varepsilon_r = \cot^{-1}(\pm AR_r)$$

yields

$$AR_r = 2.81.$$

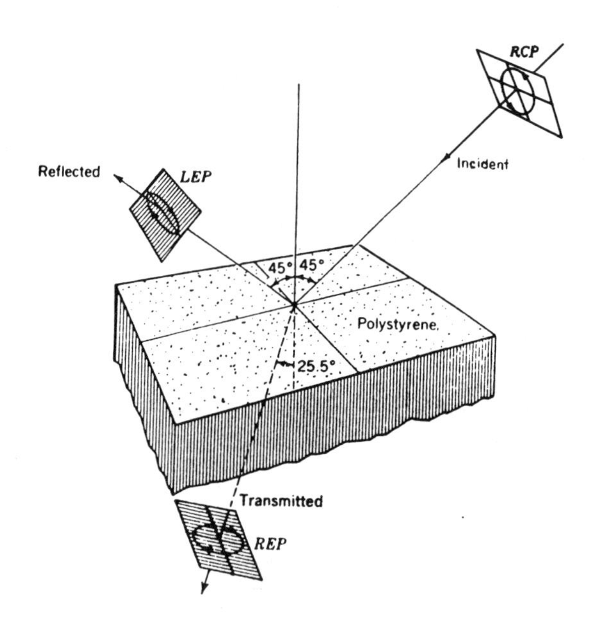

Right circularly polarized (RCP) wave incident on dielectric slab of polystyrene. The reflected wave is left elliptically polarized (LEP) with major axis of the polarization ellipse horizontal and the transmitted wave right elliptically polarized (REP) with major axis of the polarization ellipse in the (vertical) plane of incidence. Fig. 1b

If the wave is right elliptically polarized, the minus sign is used; if it is left elliptically polarized, the plus sign is used in the equation

$$\varepsilon_r = \cot^{-1}(\mp AR_r).$$

Thus the reflected wave is left elliptically polarized (LEP), as shown in Fig. 1b.

• PROBLEM 12-42

Suppose a uniform plane wave with the amplitude $1000e^{j0°}$ V/m propagates in the +z direction at $f = 10^8$ Hz in a conductive region having the constants $\mu = \mu_0$, $\varepsilon = 4\varepsilon_0$, $\sigma/\omega\varepsilon = 1$.

(a) Find β, α, and $\hat{\eta}$ for the wave.

(b) Find the associated $\bar{H}$ field, and sketch the wave along the z axis at t = 0.

(c) Find the depth of penetration, the wavelength, and the phase velocity. Compare λ and v_p with their values in a lossless ($\sigma = 0$) region having the same μ and ε values. Assume only E_x and H_y components for the wave.

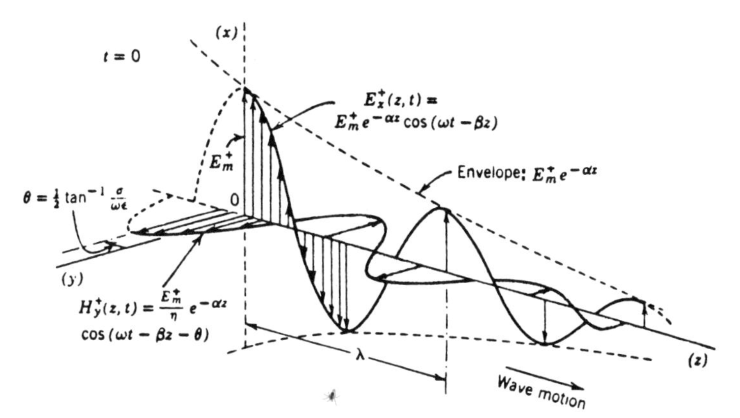

Positive z traveling fields of a uniform plane wave in a conductive region, shown at t = 0.

Solution: (a) The attenuation and phase factors are given by

$$\alpha = \frac{\omega\sqrt{\mu\varepsilon}}{\sqrt{2}}\left[\sqrt{1 + \left(\frac{\sigma}{\omega\varepsilon}\right)^2} - 1\right]^{\frac{1}{2}}$$

and

$$\beta = \frac{\omega\sqrt{\mu\varepsilon}}{\sqrt{2}}\left[\sqrt{1 + \left(\frac{\sigma}{\omega\varepsilon}\right)^2} + 1\right]^{\frac{1}{2}}$$

Thus,

$$\alpha = \frac{\omega\sqrt{\mu\varepsilon}}{\sqrt{2}}\left[\sqrt{1 + \left(\frac{\sigma}{\omega\varepsilon}\right)^2} - 1\right]^{\frac{1}{2}}$$

$$= \frac{\omega\sqrt{\mu_0 4\varepsilon_0}}{\sqrt{2}} \quad [\sqrt{1 + 1^2} - 1]^{\frac{1}{2}}$$

$$= \frac{2\omega}{\sqrt{2}c}\ [0.414]^{\frac{1}{2}} = 1.90 \text{ Np/m} \qquad (1)$$

$$\beta = \frac{2\omega}{\sqrt{2}c}\ [2.414]^{\frac{1}{2}} = 4.58 \text{ rad/m} \qquad (2)$$

The propagation constant is therefore

$$\gamma = 1.9 + j4.58\text{m}^{-1}.$$

The complex wave impedance is given by

$$\eta = \frac{\sqrt{\frac{\mu}{\varepsilon}}}{\sqrt{1 - j\frac{\sigma}{\omega\varepsilon}}} = \frac{\sqrt{\frac{\mu_0}{4\varepsilon_0}}}{\sqrt{1 - j1}} = \frac{\frac{1}{2}\sqrt{\frac{\mu_0}{\varepsilon_0}}}{[1 + 1^2]^{\frac{1}{4}}} \angle\theta$$

$$\theta = \frac{1}{2}\tan^{-1}\left(\frac{1}{1}\right) = \frac{\pi}{8}$$

Therefore,

$$\hat{\eta} = \frac{\frac{1}{2}\sqrt{\frac{\mu_0}{\varepsilon_0}}}{[1 + 1^2]^{\frac{1}{4}}} e^{j(1/2)\text{arc tan } 1}$$

$$= \frac{60\pi}{1.19} e^{j(1/2)(\pi/4)} = 159e^{j(\pi/8)} \ \Omega \qquad (3)$$

(b) The $\hat{H}$ field is found by use of

$$\frac{\hat{E}_x^+(z)}{\hat{H}_y^+(z)} = \sqrt{\frac{\mu}{\varepsilon - j\frac{\sigma}{\omega}}} \equiv \hat{\eta} = \frac{-E_x^-(z)}{H_y^-(z)}$$

$$\hat{H}_y^+(z) = \frac{\hat{E}_x^+(z)}{\hat{\eta}} = \frac{1000e^{-\alpha z}e^{-j\beta z}}{159e^{j\pi/8}}$$

$$= 6.29e^{-1.9z}e^{-j[4.58z+(\pi/8)]} \text{A/m} \qquad (4)$$

to yield the real-time expressions

$$E_x^+(z,t) = \text{Re}[\hat{E}_x^+(z)e^{j\omega t}] = \text{Re}[1000e^{-\alpha z}e^{-j\beta z}e^{j\omega t}]$$

$$= 1000e^{-1.9z}\cos(\omega t - 4.58z) \text{ V/m} \qquad (5)$$

$$H_y^+(z,t) = 6.29e^{-1.9z}\cos[\omega t - 4.58z - (\pi/8)] \text{A/m} \qquad (6)$$

(c) The depth of penetration is found using (1):

$$\delta = \alpha^{-1} = 0.52 \text{ m},$$

the distance the wave must travel to diminish to e^{-1} (or 38.8%) of any reference value.

The wavelength is obtained using the value of β

$$\lambda = \frac{2\pi}{\beta} = \frac{2\pi}{4.58} = 1.37 \text{ m}$$

comparing with that for a lossless region $(\mu_0, 4\varepsilon_0)$ as follows

$$\lambda^{(0)} = \frac{2\pi}{\beta^{(0)}} = \frac{2\pi}{2\pi f\sqrt{\mu_0 4\varepsilon_0}} = \frac{c}{2f} = \frac{3 \times 10^8}{2(10^8)} = 1.5 \text{ m}$$

The effect of finite conductivity is thus to foreshorten the wavelength.

The phase velocity in the conductive region is

$$v_p = \frac{\omega}{\beta} = \frac{2\pi(10^8)}{4.58} = 1.37 \times 10^8 \text{ m/sec}$$

which compares with that in the lossless region as follows:

$$v_p^{(0)} = \frac{\omega}{\beta^{(0)}} = \frac{\omega}{\omega\sqrt{\mu_o 4\varepsilon_o}} = \frac{c}{2} \cong 1.5 \times 10^8 \text{ m/sec}$$

Conduction thus serves to slow down v_p. The foregoing numerical results may be added to the figure to provide a picture of the wave motion in the conductive region.

• PROBLEM 12-43

A 1-Mc/sec (300 meters wavelength) plane wave traveling in a normally dispersive, lossless medium has a phase velocity at this frequency of 3×10^8 meters/sec. The phase velocity as a function of wavelength is given by

$$v = k\sqrt{\lambda}$$

where k = constant. Find the group velocity.

Solution: From

$$u = v - \frac{\lambda dv}{d\lambda}$$

the group velocity is

$$u = v - \lambda\frac{dv}{d\lambda} = v - \frac{k}{2}\sqrt{\lambda}$$

or

$$u = v\left(1 - \frac{1}{2}\right)$$

Hence

$$u = \frac{v}{2} = 1.5 \times 10^{8} \text{ meters/sec}$$

• PROBLEM 12-44

Find the 10 percent and 60 percent depth of penetrations of an electromagnetic wave of amplitude

$$E_y = E_0 e^{-x/\delta}$$

(E_0 is initial amplitude) traveling in the x-direction.

Solution:

$$E_y = E_0 e^{-x/\delta}$$

For ten percent depth of penetration

$$E_0 e^{-x/\delta} = 0.1E_0$$

or

$$e^{-x/\delta} = 0.1$$

Taking natural logarithms on both sides

$$-x/\delta = -2.3$$

or

$$x = 2.3\delta$$

which is the 10percent depth of penetration

Similarly for 60 percent depth of penetration

$$e^{-x/\delta} = 0.6$$

or

$$x = 0.51\delta$$

is the 60 percent depth of penetration.

• PROBLEM 12-45

The conductivity of silver is $g = 3 \times 10^7$ mhos/m at microwave frequencies. Find the skin depth at 10^{10}Hz. Also calculate the frequency at which skin depth in sea water is one meter.

Solution: The skin depth is given by the formula

$$\delta = \sqrt{\frac{2}{\mu_0 \omega g}}$$

where

$$\mu_0 = 4\pi \times 10^{-7}$$

$$\omega = 2\pi \times 10^{10}$$

So for silver the skin depth is

$$\delta = \sqrt{\frac{2}{(2\pi \times 10^{10})(3 \times 10^7)(4\pi \times 10^{-7})}} = 9.2 \times 10^{-5} \text{ cm.}$$

Now given skin depth = 1 meter for sea water.

For sea water $\mu = \mu_0$ and $g \simeq 4.3$ mhos/m. The expression for the frequency corresponding to a given skin depth δ is

$$\omega = \frac{2}{g\mu_0\delta^2} = \frac{2}{4.3 \times 4\pi \times 10^{-7}\delta^2} \text{ s}^{-1}$$

$$= \frac{3.70 \times 10^5}{\delta^2} \; s^{-1}$$

which yields

$$f = 58.6 \times 10^3 \text{ Hz}$$

PLANE WAVES IN FREE SPACE

• PROBLEM 12-46

If the conductivity and relative permittivity of a medium are 0.3 mho/meter and 60 respectively, assuming that $\mu_r = 1$, does the medium behave like a conductor or a dielectric at a frequency of: (a) 100 KHz, (b) 3000MHz?

Solution: At 100 KHz

$$\frac{\sigma}{\omega\varepsilon} = \frac{0.3}{2 \times \pi \times 100 \times 10^3 \times 60 \times 10^{-9}/36\pi} = 900$$

Since

$$\frac{\sigma}{\omega\varepsilon} > 100$$

the medium behaves like a good conductor.

At 3×10^3 MHz

$$\frac{\sigma}{\omega\varepsilon} = \frac{0.3}{2 \times \pi \times 3 \times 10^3 \times 10^6 \times 60 \times 10^{-9}/36\pi} = 0.03$$

Since

$$\frac{\sigma}{\omega\varepsilon} < 100$$

the medium behaves like a dielectric.

• PROBLEM 12-47

A 10 MHz plane wave that is traveling in free space has a peak amplitude $E_0 = 50\mu V/m$. Determine the following: (a) the time average electric energy density of the wave, (b) the peak total energy density, (c) the average Poynting vector, (d) the peak Poynting vector, (e) the energy contained in a cube 10 km on a side.

Solution:

(a) Electric energy density $= \langle\frac{\varepsilon_0 E^2}{2}\rangle$

$$= \frac{10^{-9}}{36\pi} \frac{(50 \times 10^{-6})^2}{4} = 0.552 \times 10^{-22} \text{ Joule/m}^3$$

(b) Total peak energy density is

$$\varepsilon_0 E^2 = 2.208 \times 10^{-22} \text{Joule/m}^3,$$

since

$$\frac{1}{2} \varepsilon_0 E^2 = \frac{1}{2} \mu_0 H^2$$

(c)

$$\langle S_c \rangle = \sqrt{\frac{\varepsilon_0}{\mu_0}} \frac{E^2}{2} = \frac{E^2}{2Z_0} = \frac{(50 \times 10^{-6})^2}{2 \times 120\pi}$$

$$= 3.66 \times 10^{-14} \text{ Watt/m}^2$$

$$\left(Z_0 = \sqrt{\frac{\mu_0}{\varepsilon_0}} = \text{characteristic impedance; } H = \frac{\varepsilon_0}{\mu_0} E\right)$$

(d) $S_{peak} = 2\langle S_c \rangle = 7.32 \times 10^{-14} \text{ Watt/m}^2$

(e) $\langle U \rangle = \frac{\varepsilon_0 E^2}{2} (\text{vol}) = 1.104 \times 10^{-22} \times (10^4)^3$

$$= 1.104 \times 10^{-10} \text{ Joule, avg.}$$

• PROBLEM 12-48

The intensity of solar radiation at a point in space that is at the mean distance of the earth from the sun is 1.35×10^3 J . m^{-2} . sec^{-1}. If radiation from the sun were all at one wavelength, what would be the amplitude of electromagnetic waves from the sun at the position of the earth?

Solution: The vector $\bar{E} \times \bar{B}/\mu_0$ gives the energy flow in terms of the instantaneous values of $\bar{E}$ and $\bar{B}$. Since both $\bar{E}$ and $\bar{B}$ vary sinusoidally with time, their average values are $(2)^{-\frac{1}{2}}\bar{E}_0$ and $(2)^{-\frac{1}{2}}\bar{B}_0$, respectively, where $\bar{E}_0$ and $\bar{B}_0$ are the amplitudes of the electric and magnetic components of the wave. Therefore, the average value of the energy flow is

$$\frac{1}{2} E_0 \times \frac{B_0}{\mu_0} \text{ J} \cdot \text{m}^{-2} \cdot \text{sec}^{-1}.$$

Since $E_0 = cB_0$, the magnitude of the average energy flow is

$$\frac{1}{2} E_0 \frac{B_0}{\mu_0} = \frac{1}{2} \frac{E_0^2}{\mu_0 c} = 1.35 \times 10^3 \text{ J} \cdot \text{m}^{-2} \cdot \text{sec}^{-1}.$$

Therefore,

$$E_0^2 = 2(4\pi \times 10^{-7})(3.00 \times 10^8)(1.35 \times 10^3)$$

$$= 1.02 \times 10^6$$

and

$$E_0 = 1.01 \times 10^3 \text{ V} \cdot \text{m}^{-1}$$

is the amplitude of the electric component of the wave. The amplitude of the magnetic component is

$$B_0 = \frac{E_0}{c} = \frac{1.01 \times 10^3}{3.00 \times 10^8} = 3.37 \times 10^{-6} \text{ Wb} \cdot \text{m}^{-2}.$$

• PROBLEM 12-49

A 10-GHz plane wave traveling in free space has an amplitude $E_x = 1$ V/m.

(a) Find the phase velocity, the wavelength, and the propagation constant.

(b) Determine the characteristic impedance of the medium.

(c) Find the amplitude and direction of the magnetic field intensity.

(d) Repeat part (a) if the wave is traveling in a lossless, unbounded medium, having a permeability the same as free space, but permittivity four times that of free space.

Solution: (a) For free space

$$v_p = c = \frac{1}{\sqrt{\mu_0 \varepsilon_0}} = \frac{1}{\sqrt{(4\pi \times 10^{-7})(8.854 \times 10^{-12})}}$$

$$\cong 3 \times 10^8 \text{ m/s}$$

so that

$$\lambda = \frac{c}{f} = \frac{3 \times 10^8}{10 \times 10^9} = 3 \times 10^{-2} \text{m} = 3 \text{ cm}$$

and

$$\beta = \frac{2\pi}{\lambda} = \frac{2\pi}{3} = 2.093 \text{ rad/cm}$$

(b) The characteristic (intrinsic) impedance of free space is

$$\eta = \sqrt{\frac{\mu_0}{\varepsilon_0}} = \sqrt{\frac{4\pi \times 10^{-7}}{8.854 \times 10^{-12}}} = 377 \ \Omega$$

(c)

$$H_y^+ = \frac{E_x^+}{\eta} = \frac{1}{377} = 2.65 \times 10^{-3} \text{ A/m}$$

if the wave is traveling in the +z direction, or

$$H_y^- = -\frac{E_x^+}{\eta} = -\frac{1}{377} = -2.65 \times 10^{-3} \text{ A/m}$$

if the wave is traveling in the -z direction. Here the minus sign indicates a reversal in space direction.

(d) For a dielectric medium with $\varepsilon_r = 4$,

$$v_p = \frac{1}{\sqrt{\mu_0 \varepsilon_0 \varepsilon_r}} = \frac{1}{2\sqrt{\mu_o \varepsilon_o}} = 1.5 \times 10^8 \text{ m/s}$$

so that

$$\lambda = \frac{v_p}{f} = \frac{1.5 \times 10^8}{10 \times 10^9} = 1.5 \text{ cm}$$

and

$$\beta = \frac{2\pi}{\lambda} = \frac{2\pi}{1.5} = 4.186 \text{ rad/cm}$$

• PROBLEM 12-50

A vertically polarized electro-magnetic wave has a wavelength of 360 meters. The amplitude of its electric field is 0.01 volt per meter. Find (a) the frequency, (b) the amplitude of the magnetic field component, (c) the amplitude of the e.m.f. induced in a vertical single wire of height 50 meters, (d) the amplitude of the e.m.f. induced in a rectangular loop whose vertical sides are 10 meters long and whose horizontal sides are 60 meters long, its plane being parallel to the direction of propagation, and (e) the mean value of the energy carried, per cubic meter. Assume that the medium is free space.

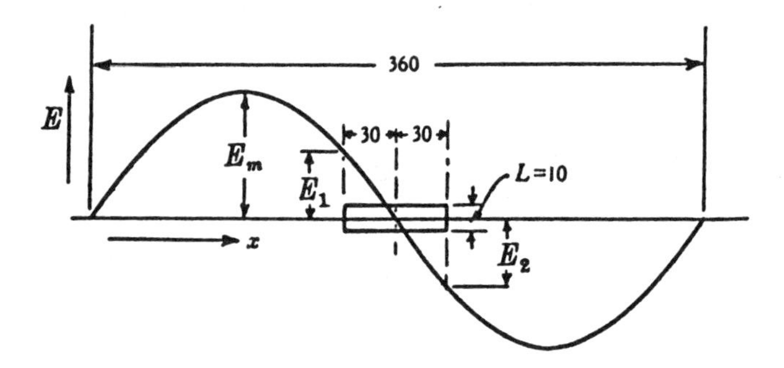

Solution: (a)

$$f = \frac{c}{\lambda} = \frac{3 \times 10^8}{360} = 8.34 \times 10^5 = 834 \text{ kilo-cycles per sec.}$$

(b)

$$B = \frac{E}{c} = \frac{0.01}{3 \times 10^8} = 3.33 \times 10^{-11} \text{ Wb/m}^2 = 3.33 \times 10^{-7} \text{ gauss.}$$

(c) $e = EL = 0.01 \times 50 = 0.5$ volt.

(d) The e.m.f. in the loop will be a maximum when the field at the center of the loop is zero (see the figure). The field E_1 at one vertical side will be

$$E_1 = E_{max} \sin 30° = 0.005 \text{ volt per meter,}$$

and the field at the other vertical side

$$E_2 = -0.005 \text{ volt per meter},$$

so that the total e.m.f. in the loop will be

$$e = L(E_1 - E_2) = 0.1 \text{ volt}.$$

(e) The maximum value of the energy per cubic meter is

$$W_m = \varepsilon_0 E^2 \text{ joules}$$

$$= 8.854 \times 10^{-12} \times 10^{-4}$$

$$= 8.854 \times 10^{-16} \text{ joules}.$$

The mean value, since $\bar{E}$ varies sinusoidally, is half the maximum. That is,

$$W = \frac{1}{2} W_m = 4.427 \times 10^{-16} \text{ joules per cu. meter}.$$

• PROBLEM 12-51

Suppose a uniform plane wave in empty space has the electric field

$$\hat{E}(z) = \bar{a}_x \, 1000 e^{-j\beta_0 z} \text{ V/m} \qquad (1)$$

its frequency being 20 MHz. (a) What is its direction of travel? Its amplitude? Its vector direction in space? (b) Find the associated $\hat{B}$ field and the equivalent $\hat{H}$ field. (c) Express $\hat{E}$, $\hat{B}$, and $\hat{H}$ in real-time form. (d) Find the phase factor β_0, the phase velocity, and the wavelength of this electromagnetic wave.

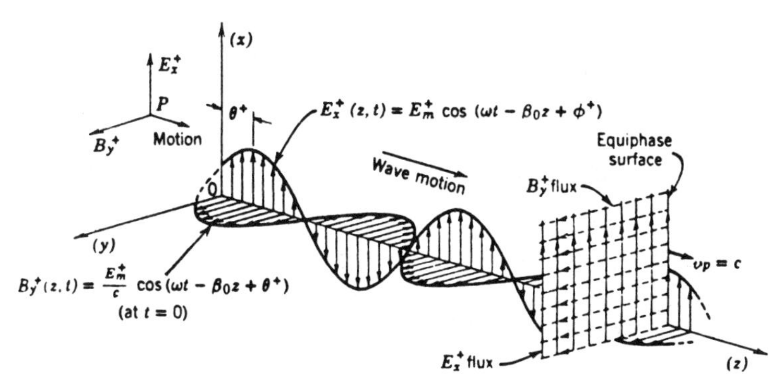

Vector plot of the fields of a uniform plane wave along the z axis. Note the typical plane equiphase surface, depicting fluxes of E_x^+ and B_y^+.

Solution: (a) A comparison of (1) with

$$\hat{E}_x(z) = \hat{E}_m^+ e^{-j\beta_0 z} + \hat{E}_m^- e^{j\beta_0 z}$$

or

$$\hat{E}_{x(z)} = \hat{E}_x^+ (z) + \hat{E}_x^- (z)$$

reveals a positive z traveling wave, whence the symbolism $\hat{E}(z) = \bar{a}_x\hat{E}_x^+(z)$. The given figure shows a sketch of the traveling wave. The real amplitude is $\hat{E}_m^+ = 1000$ V/m with the vector field x directed in space.

(b) Using

$$\frac{\hat{E}_x^+(z)}{\hat{B}_y^+(z)} = \frac{1}{c},$$

yields

$$\hat{B}_y^+(z) = \frac{\hat{E}_x^+(z)}{c} = \frac{1000}{c}e^{-j\beta_0 z} = 3.33 \times 10^{-6}e^{-j\beta_0 z} \text{ Wb/m}^2$$

The use of

$$\frac{\hat{E}_x^+(z)}{\hat{H}_x^+(z)} = \sqrt{\frac{\mu_0}{\varepsilon_0}} \equiv \eta_0 \cong 120\pi$$

gives the magnetic intensity

$$\hat{H}_y^+(z) = \frac{\hat{E}_x^+(z)}{\eta_0} = \frac{1000}{120\pi} e^{-i\beta_0 z} = 2.65e^{-j\beta_0 z} \text{ A/m}$$

(c) The real-time fields are obtained from

$$E(u_1, u_2, u_3, t) = R_e[\hat{E}(u_1, u_2, u_3)e^{j\omega t}]$$

and

$$B(u_1, u_2, u_3, t) = R_e[\hat{B}(u_1, u_2, u_3)e^{j\omega t}]$$

by taking the real part after multiplication by $e^{j\omega t}$.

$$E_x^+(z, t) = \text{Re}[1000e^{-j\beta_0 z}e^{j\omega t}] = 1000 \cos (\omega t - \beta_0 z)\text{V/m}$$

$$B_y^+(z, t) = 3.33 \times 10^{-6} \cos(\omega t - \beta_0 z) \text{ Wb/m}^2 \text{(or T)}$$

$$H_y^+(z, t) = 2.65 \cos(\omega t - \beta_0 z)\text{A/m}$$

(d) $$\beta_0 \equiv \omega\sqrt{\mu_0\varepsilon_0} = \frac{\omega}{c} = \frac{2\pi(20 \times 10^6)}{3 \times 10^8} = 0.42 \text{ rad/m}$$

$$v_p = \frac{\omega}{\beta_0} = \sqrt{\mu_0 \varepsilon_0} = c = 3 \times 10^8 \text{ m/sec}$$

$$\lambda_0 = \frac{2\pi}{\beta_0} = \frac{c}{f} = \frac{3 \times 10^8}{20 \times 10^6} = 15\text{m}$$

PLANE WAVES AND CURRENT DENSITY

• PROBLEM 12-52

Assume a plane wave with E = 1 volt/meter and a frequency of 300×10^6 cps moving in free space, impinges on a thick copper sheet located perpendicularly to the direction of propagation. Find

(1) E at the plate surface

(2) H at the same location

(3) δ the depth of penetration

(4) J_c (the conduction current density at a depth of 10^{-2} mm)

(5) The surface impedance, $\sqrt{j\omega\mu/\gamma}$

Assume $\gamma = 5.8 \times 10^7$ mhos/m, $\varepsilon = \varepsilon_v$, $\mu = \mu_v$.

Solution: For (1) first find

$$\frac{E_r}{E_i} = \frac{2\rho_2}{\rho_2 + \rho_1}$$

Now $\rho_1 = 377$ ohms (for air or vacuum)

$$\rho_2 = \sqrt{\frac{j\omega\mu_v}{\gamma + j\omega\varepsilon_v}} = \left[\frac{j(2\pi)300(10^6)4\pi(10^{-7})}{5.8(10^7) + j2\pi(300)10^6(8.85)10^{-12}}\right]^{\frac{1}{2}}$$

$$= \left[\frac{j2370}{5.8(10^7) + \underbrace{j167(10^{-4})}_{\text{(negligible)}}}\right]^{\frac{1}{2}}$$

$$= 6.39(10^{-3}) \angle 45°$$

Substituting in [1]

$$\frac{\bar{E}_{refr}}{1} = \frac{2(6.39)10^{-3}}{377} \angle 45°$$

$$= 3.39(10^{-8}) \angle 45°$$

(2)

$$\frac{H_{ref}}{H_{inc}} = \frac{2\rho_1}{\rho_1 + \rho_2}$$

But

$$\frac{E_i}{H_i} = \rho_1$$

or

$$H_i = \frac{E_i}{\rho_1} = \frac{1}{377}$$

Therefore

$$\bar{H}_{ref} = \frac{2(377)E_i}{[377 + \underbrace{2.05(10^{-3})}_{\text{(negligible)}} \angle 45°](377)}$$

$$= 5.31(10^{-3})\bar{E}_i \text{ amp/m}$$

(3)

$$\delta = \sqrt{\frac{2}{\omega\mu\gamma}} = \left[\frac{2}{2\pi(300)10^6(4\pi)10^{-7}(5.8)10^7}\right]^{1/2}$$

$$= \sqrt{14.57}\ (10^{-6})\text{m} = 3.81(10^{-3})\text{mm}$$

(4)

$$\bar{J}_c = \gamma\bar{E}_{refr}$$

$$= (5.8)10^7(3.39)10^{-5} \angle 45°$$

$$= 1910 \text{ amp/m}^2 \angle 45°$$

At a distance x below the surface

$$J = J_c\ e^{-px}$$

$$p = \sqrt{j\omega\mu\gamma}$$

$$= [2\pi(300)10^6(4\pi)10^{-7}(5.8)10^7]^{1/2} \angle 45°$$

$$= [13.73(10^{10})]^{\frac{1}{2}} \angle 45°$$

$$= 3.7(10^5) \angle 45°$$

$$\text{Re}[p] = \alpha = \frac{3.7(10^5)}{\sqrt{2}} = 2.62(10^5)$$

When $x = .01(10^{-3})\text{m}$

$$\bar{J} = \bar{J}_c e^{-2.62(10^5)0.01(10^{-3})}$$

$$= 1910(0.0729) \angle 45°$$

$$= 144 \angle 45° \text{ amp/m}^2$$

(5) The surface impedance is $\sqrt{j\omega\mu\gamma}$

Substituting the values of Ω, μ, and γ,

$$\rho_2 = 2.05(10^{-3}) \angle 45°$$

• PROBLEM 12-53

A plane wave having a frequency of 1590 MHz is traveling in a medium for which $\mu_r = \varepsilon_r = 1$ and $\sigma = 0.1$ mho/m. If the rms electric field intensity of the wave is 10 Volt/m, determine: (a) the conduction current density, (b) the displacement current density, (c) the total current density.

Solution: Given that $E_0 = 10$ Volt/m, then $E = 10\sqrt{2} \sin \omega t$. Then

(a) $J_c = \sigma E = 0.1 \times 10 = 1 \text{ Amp/m}^2$

(b) Since $\frac{\partial E}{\partial t} = 10\sqrt{2}\ \omega \cos \omega t = 10\sqrt{2}\ \omega \sin(\omega t + \frac{\pi}{2})$

$$J_d = \frac{\partial D}{\partial t} = \varepsilon \frac{\partial E}{\partial t} = \omega\varepsilon E_0 \sin(\omega t + \frac{\pi}{2})$$

$$= 2\pi \times 1.59 \times 10^9 \times \frac{10^{-9}}{36\pi} \times 10 \sin(\omega t + \frac{\pi}{2})$$

$$= 0.883 \sin(\omega t + \frac{\pi}{2}) \text{ Amp/m}^2$$

(c) The phasor representation of the total current is

given simply as

$$\vec{J}_t = J_c + J_d = (1 + j0.88) \text{ Amp/m}^2$$

• PROBLEM 12-54

A brass rod of conductivity 1.57×10^5 mhos/cm has a radius of 2.1 cm. A cylindrical coordinate system is located so that the z-axis is along the axis of the rod. If the rod carries a current of $2.024 \sin 10^4 t$ amperes in the positive z-direction, determine the current density J_z as a function of the radial distance r in centimeters.

Solution: At the given frequency the skin depth δ of brass is found from

$$\delta = \frac{1}{\sqrt{\pi f \mu \sigma}}$$

to be 0.318 cm, or approximately $1/\pi$ cm. The radius is about seven skin depths and, therefore, the approximate current density can be found by treating the rod as a semi-infinite solid. As the distance into the conductor is 2.1 - r cm, this should be substituted for x in

$$J_z = J_0 e^{-x/\delta} \sin(\omega t - x/\delta).$$

Also, a constant should be added to the phase of J_z, because the current i, which equals the surface integral of J_z, has a phase angle of zero. It follows that

$$J_z = J_1 e^{r/\delta} \sin(\omega t + r/\delta + \theta) \text{ amperes/cm}^2$$

with J_1 and θ to be determined.

The current i is $\int J_z 2\pi r \, dr$ with the limits of zero and 2.1 cm. This can be evaluated by integrating by parts, giving

$$i = 2024 J_1 \sin(\omega t + \theta - 22.3°)$$

Comparison with the given current shows that $J_1 = 0.001$ and $\theta = 22.3°$. Substitution for J_1, θ, ω, and δ into the expression for J_z gives

$$J_z = 0.001 e^{\pi r} \sin(10^4 t + \pi r + 22.3°) \text{ amperes/cm}^2$$

Note that the sign of J_z at any fixed instant depends on the distance r.

CHAPTER 13

TRANSMISSION LINES

EQUATIONS OF TRANSMISSION LINES

• PROBLEM 13-1

At a frequency of 1590 Hz find (a) the characteristic impedance; (b) the phase velocity of wave propagation for a transmission line with the following constants $R = 10^{-2}$ ohm/meter, $G = 10^{-6}$ mho/meter, $L = 10^{-6}$ Henry/meter, $C = 10^{-9}$ Farad/meter.

Solution: (a) The characteristic impedance of the line is given by

$$Z_0 = \sqrt{\frac{R + j\omega L}{G + j\omega C}}$$

$$Z_0 = \sqrt{\frac{10^{-2} + j \times 2\pi \times 1590 \times 10^{-6}}{10^{-6} + j \times 2\pi \times 1590 \times 10^{-9}}}$$

$$Z_0 = 37.5 \underline{/-20°} \ \Omega.$$

(b) The phase velocity of wave propagation on the line is given by

$$\upsilon = \frac{\omega}{\text{Im}\sqrt{zy}}$$

$$\upsilon = 2\pi \times 1590/\sqrt{j \times 2\pi \times 1590 \times 10^{-6}}\sqrt{j \times 2\pi \times 1590 \times 10^{-6}}$$

$$= 2.96 \times 10^{7} \text{ meters/sec}$$

• PROBLEM 13-2

At a frequency of 4000 kilocycles a parallel-wire transmission line has the following parameters: $R = 0.025$ ohm/m, $L = 2\ \mu h/m$, $G = 0$, $C = 5.56\ \mu\mu f/m$. The line is 100 meters long, terminated in a resistance of 300 ohms. Find the efficiency of transmission, defined as the ratio of the power delivered to the load to the power supplied to the input terminals.

Solution: The angular frequency ω is $8\pi \times 10^6$, and ωL is 50.3 ohms. The ratio $R/\omega L$ equals 0.0005. From

$$Z_0 \simeq \sqrt{\frac{L}{C}}\left[1 + \frac{j}{2\omega}\left(\frac{G}{C} - \frac{R}{L}\right)\right]$$

note that the characteristic impedance is approximately equal to $\sqrt{L/C}$, or 600 ohms. By

$$\alpha = \frac{1}{2}(R\sqrt{C/L} + G\sqrt{L/C})$$

the attenuation constant α is approximately 0.000021. The exponential of $-\alpha\ell$, with ℓ equal to 100 meters, is 0.998. Thus the attenuation is very small, and the line is nearly lossless.

If I denotes the rms line current, the power lost in the differential length dz is $I^2R\ dz$, and the power delivered to the load R_r if $I_r{}^2R_r$, with I_r denoting the rms load current. In order to determine the efficiency of transmission it is needed to find the line losses in terms of the load current I_r.

As $V_r = I_rR_r$, it follows from $I = I_r \cos\beta(\ell - z) + j(V_r/ Z_0) \sin\beta(\ell - z)$ that

$$I^2 = I_r^2[\cos^2 \beta y + (R_r/R_0)^2 \sin^2 \beta y]$$

with y denoting the distance from the receiving end. The line loss can be expressed as $\int_0^\ell I^2R\ dy$. Substitution for I^2, with R_r/R_0 replaced by ½, gives

$$P_{lost} = I_r{}^2R \int_0^{100} (\cos^2 \beta y + \tfrac{1}{4} \sin^2 \beta y)\ dy$$

The constant R is 0.025, and the constant β is $\omega\sqrt{LC}$, or 0.0838. Making these substitutions and evaluating the integral, the power loss is found to be $1.51I_r{}^2$.

The power to the load is $300I_r{}^2$. Therefore, the efficiency of transmission is

$$\frac{P_{load}}{P_{load} + P_{loss}} = \frac{300I_r^2}{300I_r^2 + 1.51I_r^2} = 99.5\%$$

• PROBLEM 13-3

Solve the problem of transmission through a slab which has a dielectric constant of 4 and a thickness d.

Solution: For $\varepsilon_2 = 4\varepsilon_0$,

$$\eta_2 = \sqrt{\frac{\mu_0}{4\varepsilon_0}} = \frac{\eta_0}{2}$$

and from

$$\eta_2 \frac{1 + \Gamma_2}{1 - \Gamma_2} = \eta_0$$

$\Gamma_2 = 1/3$. Inserting this value and $\eta_2 = \eta_0/2$ into

$$\eta_0 \frac{1 + \Gamma_1}{1 - \Gamma_1} = \eta_2 \frac{1 + \Gamma_2 e^{-j2\beta_2 d}}{1 - \Gamma_2 e^{-j2\beta_2 d}}$$

yields

$$\frac{1 + \Gamma_1}{1 - \Gamma_1} = \frac{1}{2}\left(\frac{1 + 1/3e^{-j2\beta_2 d}}{1 - 1/3e^{-j2\beta_2 d}}\right)$$

from which it is obvious that Γ_1 is complex unless $e^{-j2\beta_2 d}$ is real. Since solution for the complex case is really complex, examine the real case first.

Note that $e^{-j2\beta_2 d}$ is real when $2\beta_2 d = n\pi$, and it becomes +1 when $2\beta_2 d = n\pi$, with n even, and -1 when $2\beta_2 d = n\pi$, with n odd. For n even,

$$\frac{1 + \Gamma_1}{1 - \Gamma_1} = \frac{1}{2}\left(\frac{1 + 1/3}{1 - 1/3}\right) = 1$$

and $\Gamma_1 = 0$. Thus $E_r/E_i = \Gamma_1 = 0$, and hence $E_r = 0$; there is no reflection when n is even. Since $\beta_2 = 2\pi/\lambda_2$, $2\beta_2 d = n\pi$ gives

$$d = n\frac{\lambda_2}{4}, \; n = 0, 2, 4, \ldots$$

Thus there is 100 percent transmission if the slab is an integral number of half wavelengths thick.

Similarly, for n odd,

$$\frac{1 + \Gamma_1}{1 - \Gamma_1} = \frac{1}{2}\left(\frac{1 - 1/3}{1 + 1/3}\right) = \frac{1}{4}$$

and $\Gamma_1 = -3/5$.

From Poynting's theorem, the ratio of the reflected to incident power is

$$\frac{E_r^{\ 2}}{E_i^{\ 2}} = \left(\frac{3}{5}\right)^2 = \frac{9}{25} \quad \text{or} \quad 36 \text{ percent}$$

Hence 36 percent of the power is reflected and 64 percent of it is transmitted for n odd, which means for $d = n(\lambda_2/4)$, with n odd.

A careful examination of the equation for thickness d other than an integral number of quarter wavelengths reveals that the power transmitted varies smoothly from 64 percent to 100 percent as the thickness varies. Additionally, the complex nature of Γ_1 means that the reflected wave is not in time phase with the incident wave at the interface.

• PROBLEM 13-4

A pulse of voltage is applied to the line of Fig. 1 by throwing the switch to position 2 at $t = 0$, and returning the switch to position 1 at $t = t_1$, where t_1 is less than the time T it takes a wave to travel the length of the line. (The short circuit is provided so the sending end impedance is zero regardless of the voltage.) Analyze the waves on the line.

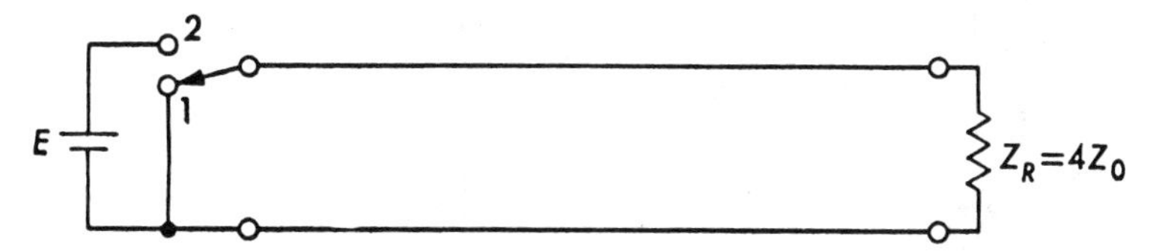

Terminated line arranged for transmission of a pulse of voltage.

Fig. 1

Solution: The voltage wave is shown in fig. 2a, for several values of t, and the current wave is shown in fig. 2b. The current is shown in units of E/Z_0.

When the wave reaches the receiving end, it is reflected with a reflection coefficient k_R:

$$k_R = \frac{Z_R - Z_0}{Z_R + Z_0} = \frac{4Z_0 - Z_0}{4Z_0 - Z_0} = \frac{3}{5}$$

The current wave is reflected in the ratio

$$\frac{i^-}{i^+} = -\frac{3}{5}$$

The reflected waves of voltage and current, as well as the total voltage and current, are also shown in Fig. 2. It can be noted that v_R/i_R is always equal to $4Z_0$.

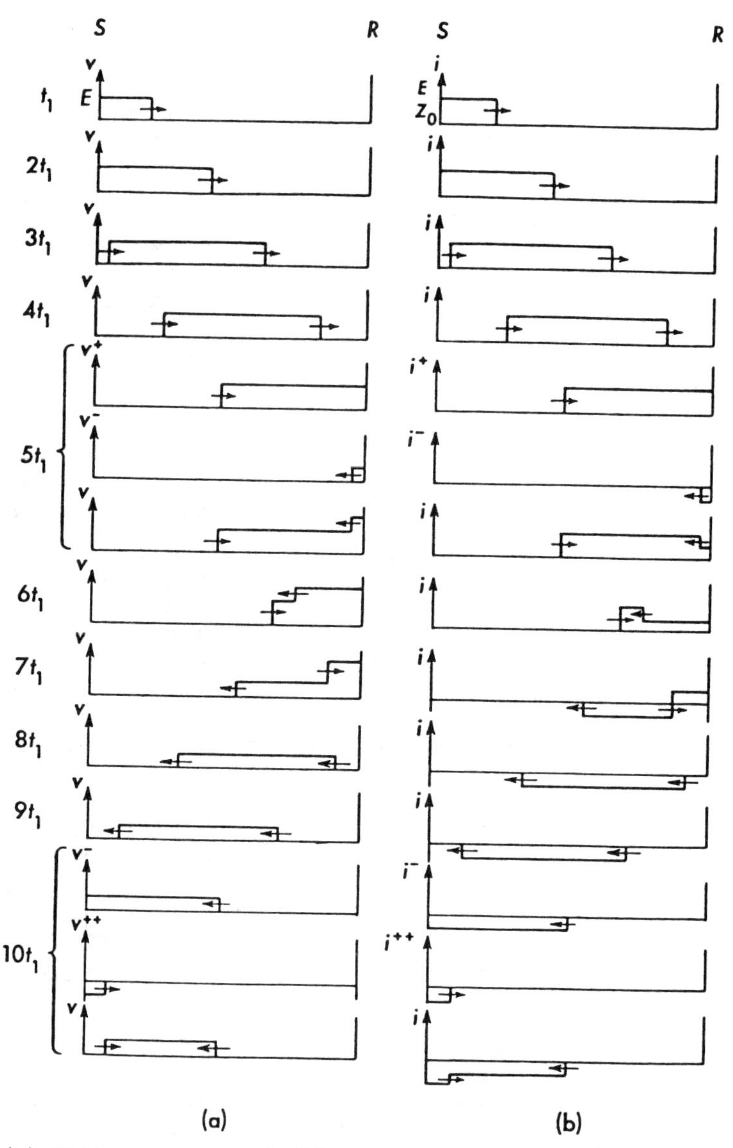

(a) Voltage waves and (b) current waves, on line of Fig. 1 at various instants of time. Arrows indicate directions of motion of wavefronts.

Fig. 2

When the reflected waves return to the sending end, they "see" a short circuit, so they are again reflected. The reflection coefficient at the sending end is -1 for voltage and -1 for current. Observe carefully that the incident wave is now the backward traveling wave, and the reflected wave travels in the forward direction. Thus

$$k_S = \frac{v^{++}}{v^-} = -\frac{i^{++}}{i^-}$$

where v^{++}, i^{++} now represent the (second) reflected wave traveling in the forward direction: This wave is also shown in Fig. 2.

In the foregoing example the reflection at the sending end gave rise to no difficulty because the battery had been disconnected before the wave returned. Even if the battery is not disconnected, however, it has no effect on reflections, except for its internal resistance Z_S, because the reflected and rereflected waves may be considered separately from the original incident wave and are merely superimposed upon it.

• PROBLEM 13-5

Let both Z_S and Z_L in Fig.1a be pure resistances. In particular, let $Z_S = 150\ \Omega$, $Z_L = 50\ \Omega$. Also, let $Z_0 = 50\ \Omega$. Assume that the line length is 800 m, and that the velocity of propagation on the line is 200 m/μs. At t = 0, a single pulse of 16 V and 1-μs duration is transmitted by the generator. Discuss the pulse behavior.

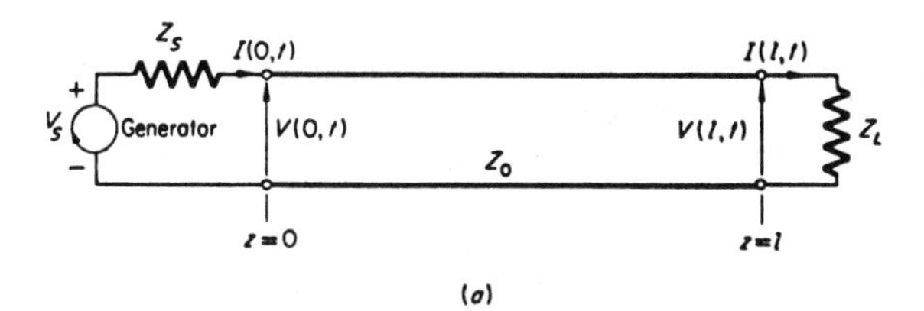

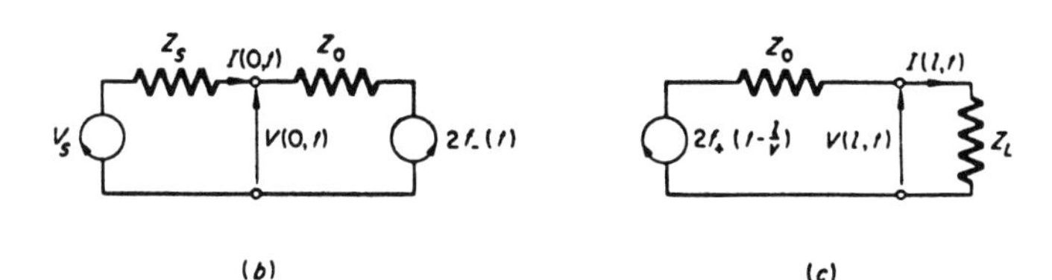

Circuit characterization of a lossless transmission line. (a) A finite line; (b) sending-end equivalent circuit; (c) receiving-end equivalent circuit.

Fig. 1

Solution: At the initial instant, t = 0, the line is completely at rest, and both the voltage and the current are zero for $0 < z \leq \ell$. More specifically, $f_-(0) = 0$, so that, in Fig. 1b, the generator feeds a series combination of two resistances, Z_S and Z_0. Hence, by simple voltage division, the pulse amplitude at the input terminals to the line is

$$V(0,t) = \frac{16 \times 50}{150 + 50} = 4 \text{ V} \qquad 0 < t < 1 \ \mu s$$

This 4-V pulse arrives at the load 800/200 = 4 μs later.

From Fig. 1c, the load voltage $V(\ell,t)$ is seen to be equal to $f_+(t - \ell/v)$ when $Z_L = Z_0$. Additionally, the voltage equation in

$$V(\ell,t) = f_+(t - \frac{\ell}{v}) + f_-(t + \frac{\ell}{v})$$

and $$I(\ell,t) = \frac{1}{Z_0} [f_+(t - \frac{\ell}{v}) - f_-(t + \frac{\ell}{v})]$$

predicts $f_-(t + \ell/v) \equiv 0$ when $V(\ell,t) = f_+(t - \ell/v)$. Therefore, in this case, there is no reflection, and all the pulse power is absorbed in the load. [Notice that the reflection coefficient at the load is $\Gamma_L = (50 - 50)/(50 + 50) = 0$, which also means that there is no reflection. However, we cannot use this approach when reactive elements are included in the load.]

Then, at the load, we have a pulse of 4 V magnitude, which begins at t = 4 μs and lasts for 1 μs.

• PROBLEM 13-6

> Consider the coaxial transmission line shown in the figure composed of two perfectly conducting concentric cylinders of radii a and b enclosing a linear medium with permittivity ε and permeability μ. Assume the voltage difference between the cylinders is v with the inner cylinder carrying a total current i. Find the transmission line equations for this coaxial cable.

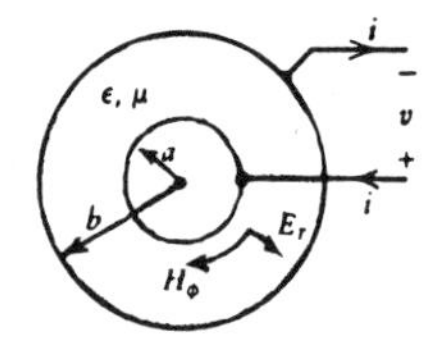

Coaxial cable

Solution: Solve for the transverse dependence of the fields as if the problem were static, independent of time. If the voltage difference between cylinders is v with the inner cylinder carrying a total current i, the static fields are

$$E_r = \frac{v}{r \ln (b/a)}, \qquad H_\phi = \frac{i}{2\pi r}$$

The surface charge per unit length q and magnetic flux per unit length λ are

$$q = \varepsilon E_r (r = a) 2\pi a = \frac{2\pi\varepsilon v}{\ln (b/a)}$$

$$\lambda = \int_a^b \mu H_\phi \, dr = \frac{\mu i}{2\pi} \ln \frac{b}{a}$$

so that the capacitance and inductance per unit length of this structure are

$$C = \frac{q}{v} = \frac{2\pi\varepsilon}{\ln (b/a)}, \qquad L = \frac{\lambda}{i} = \frac{\mu}{2\pi} \ln \frac{b}{a}$$

where, as required

$$LC = \varepsilon\mu$$

Substituting E_r and H_ϕ into

$$\frac{\partial}{\partial z}(i_z \times H_t) = \varepsilon \frac{\partial E_b}{\partial t}$$

yields the following transmission line equations:

$$\frac{\partial E_r}{\partial z} = -\mu \frac{\partial H_\phi}{\partial t} \Rightarrow \frac{\partial v}{\partial z} = -L \frac{\partial i}{\partial t}$$

$$\frac{\partial H_\phi}{\partial z} = -\varepsilon \frac{\partial E_r}{\partial t} \Rightarrow \frac{\partial i}{\partial z} = -C \frac{\partial v}{\partial t}$$

• PROBLEM 13-7

A plane electromagnetic wave is normally incident on a plane semi-infinite dielectric (see the figure shown). The standing-wave ratio in the free-space medium 1 is r = 1.73 and there is an average rate of flow of energy in the z-direction of 1 W/m^2. Determine the relative permittivity of the dielectric, the maximum and minimum electric field intensities in medium 1, and the electric field intensity in medium 2.

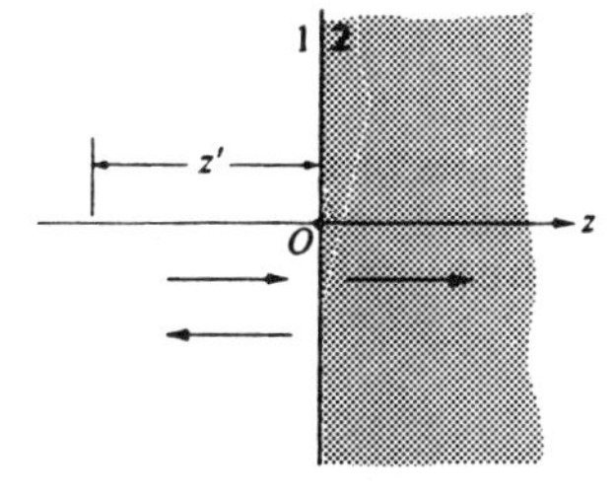

A plane electromagnetic wave normally incident on a semi-infinite dielectric.

Solution: Represent medium 1 by a transmission line analog. This is terminated at $z = 0$ in the characteristic impedance of medium 2, which is considered to be the load. Thus

$$Z_\ell = Z_0/\sqrt{\varepsilon_r} \; .$$

Use $\quad Z = Z_0 \dfrac{Z_\ell \cos \beta_z' + j\, Z_0 \sin \beta_z'}{Z_0 \cos \beta_z' + j\, Z_\ell \sin \beta_z'}$

and

$$\Gamma = \frac{|V-|}{|V_+|} = |\Gamma|, \quad z = r\, Z_0,$$

and see that at the first maximum of $|V|$

$$\beta z' = \pi/2, \qquad Z = Z_0^2/Z_\ell = Z_0\sqrt{\varepsilon_r} = rZ_0,$$

and that at the first minimum of $|V|$ away from the load

$$\beta z' = \pi, \qquad Z = Z_\ell = Z_0/\sqrt{\varepsilon_r} = Z_0/r.$$

Thus $\varepsilon_r = r^2 = 3$ and

$$\Gamma_\ell = \frac{1 - \sqrt{\varepsilon_r}}{1 + \sqrt{\varepsilon_r}} = -0.268.$$

$$P = \frac{|V_+|^2}{2Z_0}\,(1 - |\Gamma|^2)$$

yields the average rate of flow of energy per unit area:

$$\frac{|V_+|^2}{2Z_0}\,[1 - |\Gamma|^2] = 1 \text{ W/m}^2.$$

Since $Z_0 = 120\pi$ ohms, $|V_+| = 28.5$ V. Thus

$$|V_-| = |\Gamma V_+| = 7.6 \text{ V},$$

and in medium 1

$$|E|_{max} = 36.1 \text{ V/m}, \qquad |E|_{min} = 20.9 \text{ V/m}.$$

At $z = 0$, a minimum of $|E|$ exists and it follows from the continuity of the tangential component of $\bar{E}$ that in medium 2, $|E| = 20.9$ V/m. Note that the rate of flow of energy per unit area in medium 2 is

$$P = \frac{1}{2}\,\frac{|E|^2}{Z_\ell} = 1 \text{ W/m}^2,$$

as required by the conservation of energy.

• **PROBLEM 13-8**

A 100-mile telephone line has a series resistance of 4 ohms/mile, an inductance of 3 mh/mile, a leakage conductance of 1 μmho/mile, and a shunt capacitance of 0.015 μf/mile, at an angular frequency $\omega = 5000$. At the sending end there is a generator supplying 100 volts peak, at 5000 radians per second, in series with a resistance of 300 ohms. The load at the receiving end consists of a 200-ohm resistor. Find the voltage and current as functions of z, and calculate their values at the midpoint of the line.

Solution:

$$Z = R + j\omega L = 4 + j15 = 15.53 \underline{/75.1^\circ}$$

$$Y = G + j\omega C = (1 + j75)10^{-6} = 75 \times 10^{-6} \underline{/82.15^\circ}$$

$$\gamma = \sqrt{ZY} = 0.0342 \underline{/82.15^\circ} = 0.00466 + j0.0338$$

$$Z_0 = \sqrt{Z/Y} = 455 \underline{/-7.1^\circ} = 452 - j56.1$$

The load impedance $Z_r = 200$. The reflection coefficient $\rho = \dfrac{Z_r - Z_0}{Z_r + Z_0}$ is found to be $-0.396 \underline{/-7.6^\circ}$.

Calculation of the input impedance Z_i requires evaluation of $e^{-2\gamma \ell}$. As $\ell = 100$ this exponential becomes

$$e^{-2\gamma\ell} = e^{-0.932} e^{-j6.76} = 0.394 \underline{/-6.76} = 0.394 \underline{/-27.3^\circ}$$

The product $\rho e^{-2\gamma\ell}$ becomes $-0.128 + j0.0890$. Calculation of Z_i yields $Z_i = 353 \underline{/3.3^\circ} = 353 + j20.1$.

The sending-end current $\overline{I}_s$ is $\overline{V}_g/(Z_g + Z_i)$. The impedance Z_g is 300 ohms resistive, and $\overline{V}_g = 100 \underline{/0^\circ}$. Therefore, $\overline{I}_s = 100/(300 + 353 + j20.1) = 0.153 \underline{/-1.8^\circ}$. The sending-end voltage $\overline{V}_s$ is $I_s Z_i$, or $54 \underline{/1.5^\circ}$, or $54 + j1.41$.

The coefficients $\overline{V}_1$ and $\overline{V}_2$ are

$$\overline{V}_1 = \frac{1}{2}(\overline{V}_s + I_s Z_0) = 61.4 \underline{/-4.35^\circ}$$

$$\overline{V}_2 = \frac{1}{2}(\overline{V}_s - I_s Z_0) = -9.58 \underline{/-39.35^\circ}$$

Consequently, the line voltage $\overline{V}$ and the line current $\overline{I}$, as functions of z, are

$$\overline{V} = (61.4 \underline{/-4.35^\circ}) e^{-\gamma z} - (9.58 \underline{/-39.35^\circ}) e^{\gamma z}$$

$$\bar{I} = (0.135\ \underline{/2.75°})e^{-\gamma z} + (0.021\ \underline{/-32.3°})e^{\gamma z}$$

with $\gamma = 0.00466 + j0.0338$.

At the midpoint of the line, z is 50, and γz is 0.233 + j1.69. The exponential of $-\gamma z$ becomes $e^{-0.233}e^{-j1.69}$, or 0.791 $\underline{/-96.8°}$. The exponential of γz becomes 1.262 $\underline{/96.8°}$. Therefore,

$$\bar{V} = (61.4\ \underline{/-4.35°})(0.791\ \underline{/-96.8°})$$

$$- (9.58\ \underline{/-39.35°})(1.262\ \underline{/96.8°})$$

Evaluation gives $\bar{V} = 60.0\ \underline{/-105.4°}$. This is the phasor voltage, maximum value, at z = 50 miles. Similarly, the current is found to be $\bar{I} = 0.0829\ \underline{/-87.4°}$. The instantaneous voltage and current at the midway point are

$$v = 60 \sin (5000t - 105.4°)$$

$$i = 0.0829 \sin (5000t - 87.4°)$$

The complex impedance at the midway point, looking toward the load, is the ratio of $\bar{V}$ to $\bar{I}$. This gives an impedance of 725 $\underline{/-18°}$.

The voltage and current are readily determined at any point z, although the calculations are tedious. The values of $\bar{V}$ and $\bar{I}$ at the sending end, at the 50-mile point, and at the receiving end are given below.

$\bar{V}_S = 54\ \underline{/1.5°}$ $\qquad$ $\bar{I}_S = 0.153\ \underline{/-1.8°}$

$\bar{V}_{50} = 60\ \underline{/-105.4°}$ $\qquad$ $\bar{I}_{50} = 0.0829\ \underline{/-87.4°}$

$\bar{V}_r = 23.4\ \underline{/-193°}$ $\qquad$ $\bar{I}_r = 0.117\ \underline{/-193°}$

If the line were terminated in its characteristic impedance, there would be no reflected wave, and the voltage and current would decrease exponentially with respect to the axial coordinate z. However, in this problem there is reflection. The voltage rises and then falls as z increases, and the current falls and then rises. The sum of the incident and reflected waves yields this standing-wave effect.

• PROBLEM 13-9

Referring to the terminated transmission line with short-circuited stub shown in Fig. 1, the load $Z_L = 150 + j50$ ohms. The line and stubs have a characteristic impedance $Z_0 = R_0 = 100$ ohms. Find values for d_1 and d_2 such that there is no reflected wave at A (SWR = 1).

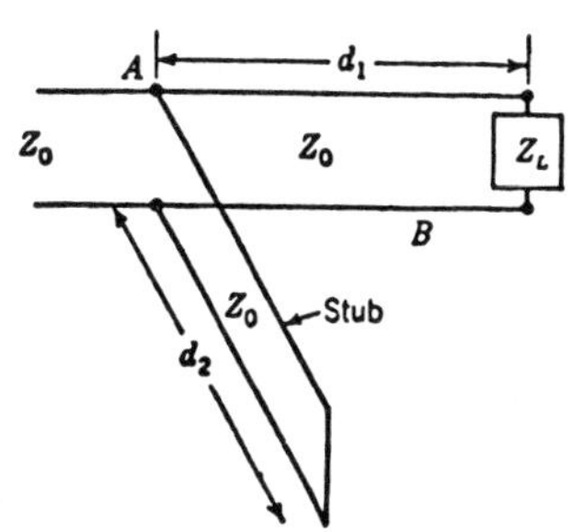

Terminated transmission line with single matching stub. Both the stub position (d_1) and its length (d_2) are adjustable.

Fig. 1

Solution: The normalized value of the load is

$$Z_n = \frac{Z_L}{R_0} = \frac{150 + j50}{100} = 1.5 + j0.5$$

The chart is then entered at the point 1.5 + j0.5 as indicated by P_1 in Fig. 2. For clarity most of the rectangular and circular coordinate lines are omitted in this figure. Point P_1 is on the SWR circle for which SWR = 1.77. Hence the SWR at B is 1.77. Now, moving along the SWR = 1.77 circle away from the load (clockwise), we proceed to the point P_2. This is just ¼ wavelength (90 electrical degrees) from the point P_3 that lies on the SWR = 1.77 circle at $R_n = 1$. At P_2, which is 0.194 wavelength from the load, the normalized impedance is 0.78 - j0.41. Moving ¼ wavelength farther on the chart gives the impedance ¼ wavelength farther from the load or the admittance at the same location (0.194 wavelength from the load).

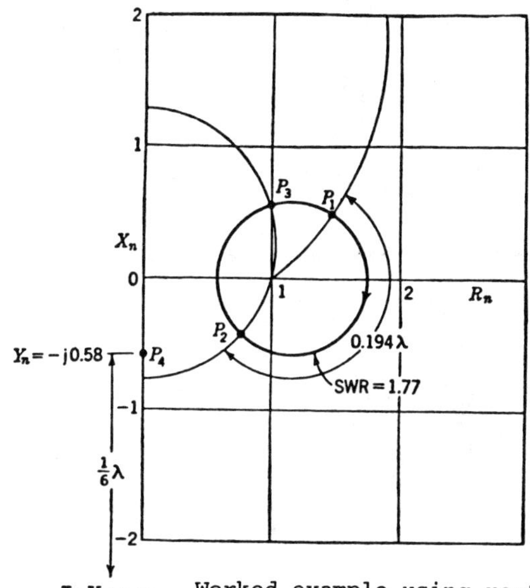

Fig. 2

Worked example using rectangular impedance chart.

Since the stub is connected in parallel to the line, it is convenient to deal in admittances. To do this, the chart is now considered to be an admittance chart, the point P_3 giving the normalized admittance at a distance of 0.194 wavelength from the load as

$$Y_n = 1.0 + j0.58$$

The actual admittance is $1/R_0$ times this value, or

$$0.01 + j0.0058 \text{ mhos}$$

For there to be no reflection at A (Fig. 1) requires that the stub present a normalized admittance to the line of $-j0.58$, so that the resultant $Y_n = 1.0 + j0$ and, hence, the impedance looking to the right at the junction is $100 + j0$ ohms. A normalized admittance $Y_n = -j0.58$ (pure susceptance) is indicated at P_4, and note that the distance required from a short circuit ($Y_n = \pm\infty$) to obtain this value is $\frac{1}{6}$ wavelength. Thus the required stub length

$$d_2 = \frac{1}{6} = 0.167 \text{ wavelength}$$

The required distance of the stub from the load as obtained above is

$$d_1 = 0.194 \text{ wavelength}$$

• PROBLEM 13-10

Consider the terminated line with two short-circuited stubs portrayed in Fig. 1. The position at which the stubs connect to the line is fixed, as shown, but the stub lengths, d_1 and d_2, are adjustable. This kind of arrangement is called a double-stub tuner. The load Z_L = $50 + j100$ ohms. The line and stubs have a charcteristic impedance $Z_0 = R_0 = 100$ ohms. Find the shortest values of d_1 and d_2 such that there is no reflected wave at A.

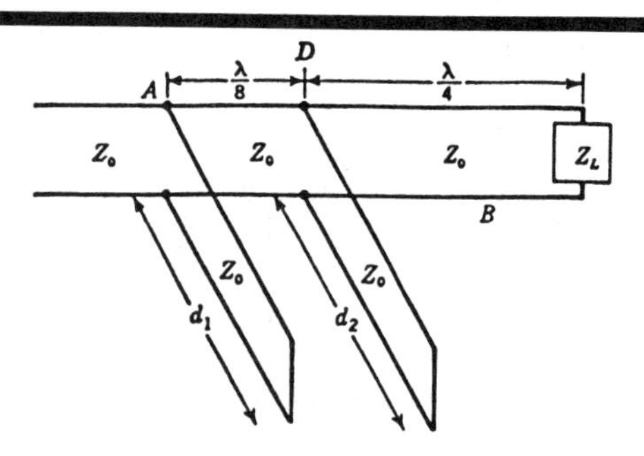

Double-stub tuner with short-circuited stubs. Fig. 1

Solution: The normalized value of the load impedance is

$$Z_n = \frac{50 + j100}{100} = 0.5 + j1.0$$

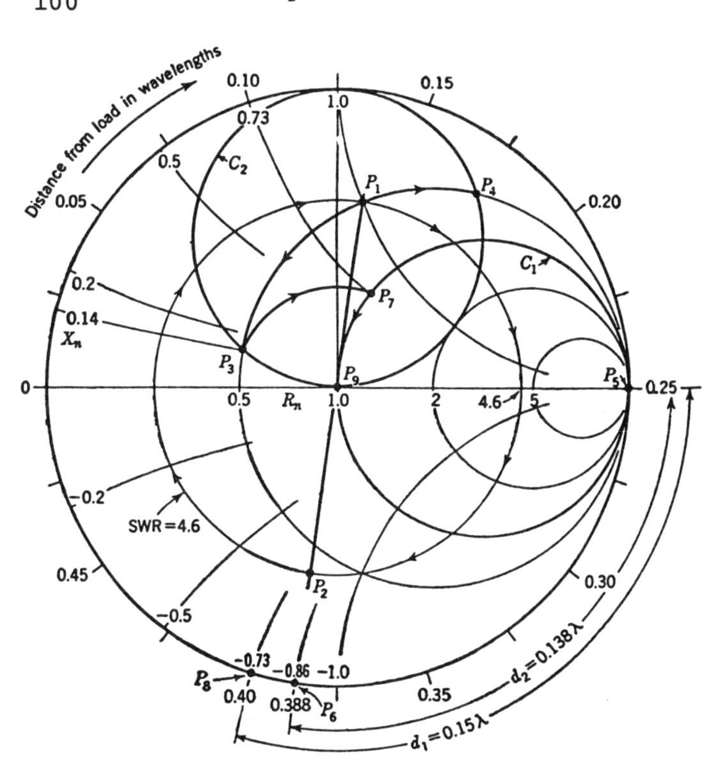

Worked example using Smith chart. Fig. 2

The chart (Fig. 2) is entered at this normalized impedance as indicated by the point P_1. Constructing a SWR curve through P_1, we note that the SWR at B (Fig. 1) is 4.6. Next, constructing the diametric line through P_1, locate P_2 halfway around the constant SWR circle from P_1. Thus, the normalized load admittance is 0.4-j0.8. Now, moving clockwise along the constant SWR circle from P_2 a distance of $\frac{1}{4}$ wavelength away from the load (toward the generator), arrive back at P_1. Thus at the point D the normalized admittance of the main line (looking toward the load) is 0.5 + j1.0. Since the reflection at A must be zero, the admittance of the main line at A (without the stub of length d_1 connected) must fall on the circle marked C_1 (Fig. 2). Therefore, at the junction of the the stub of length d_2 the admittance must fall on this circle rotated back (counterclockwise) $\frac{1}{4}$ wavelength to the position indicated by the circle marked C_2.

The admittance added by the stub of length d_2 will cause the total admittance to move from P_1 along a constant conductance line. In order to end up on the circle C_2, move either to the left, arriving at P_3, or to the right, arriving at P_4. Moving to P_3 results in shorter stubs; so make the stub of such length as to bring the total admittance to P_3. This requires a stub admittance (pure susceptance) of

$$Y_n = -j(1.0 - 0.14) = -j0.86.$$

A short-circuited stub has an infinite SWR so that the admittance at points along the stub are on the circle at the periphery of the chart. At the short circuit the admittance is infinite (point P_5). Therefore, in order to present a value

$$Y_n = -j0.86 \text{ (point } P_6\text{)}$$

the stub length must be given by

$$d_2 = 0.388 - 0.25 = 0.138 \text{ wavelength}$$

Next, moving along the constant SWR curve from P_3 to P_7, the line admittance at A is $Y_n = 1.0 + j0.73$. Hence a stub admittance of $Y_n = -j0.73$ is required in order to make the total normalized admittance at A equal to 1.0 + j0, and therefore the actual impedance at A equal to 100 + j0 ohms. A value

$$Y_n = -j0.73$$

falls at point P_8. Therefore the length of the stub is given by

$$d_1 = 0.40 - 0.25 = 0.15 \text{ wavelength}$$

Connecting this stub brings the total admittance (or impedance) to the center of the chart (point P_9).

To summarize, the required stub lengths are

$$d_1 = 0.15 \text{ wavelength}$$

$$d_2 = 0.138 \text{ wavelength}$$

If the movement had been to P_4 instead of to P_3, one would have ended up with longer stubs, namely,

$$d_1 = 0.443 \text{ wavelength}$$

$$d_2 = 0.364 \text{ wavelength}$$

• **PROBLEM 13-11**

The circuit of Fig. 1 is in steady-state condition just before the switch is closed, shorting out part of the load resistance. Draw the distance-time plots for voltage and current, and plot V_L, I_L versus time after t = 0.

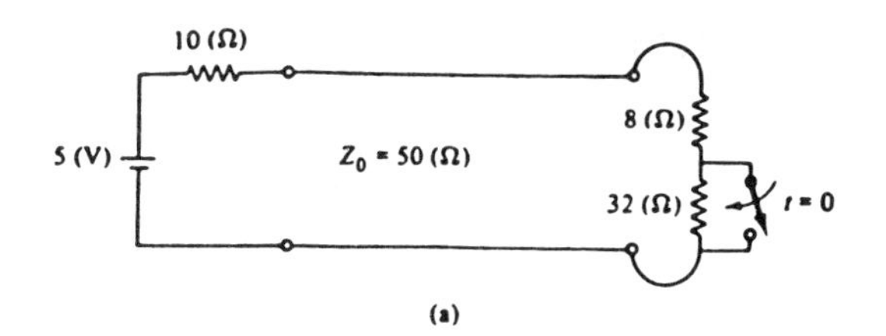

(a)

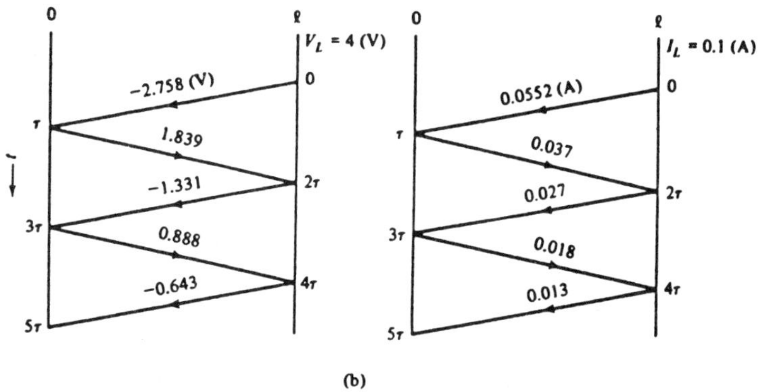

(b)

Fig. 1

Solution: At t = 0-,

$$V_L = V^+ + V^- = V^+(1 + \rho_L) = 5\frac{40}{50} = 4 \quad (V)$$

$$\rho_L = Z_L - Z_0/Z_L + Z_0$$

$$\rho_L = \frac{40 - 50}{40 + 50} = -\frac{1}{9} = -0.1111$$

Then

$$V^+ = \frac{4V}{\frac{8}{9}} = 4.5 \quad (V)$$

and

$$V^- = -\frac{1}{9}(4.5\ V) = -0.5 \quad (V)$$

Prior to the closing of the switch, the load voltage can be considered as composed of an incident voltage of 4.5(V) and a reflected voltage of -0.5 (V). Closing the switch cannot instantaneously change the incident voltage, but the reflected voltage does change instantly to maintain boundary conditions. Then, at t = 0+,

$$V^+ = \text{same} = 4.5 \text{ (V)}$$

$$\rho_L = \frac{8 - 50}{8 + 50} = -0.724$$

$$V^- = -0.724(4.5) = -3.258 \text{ (V)}$$

Thus

$$\Delta V^- = -3.258 - (-0.5) = -2.758 \text{ (V)}$$

and

$$V_L(t = 0+) = 4.5 - 3.258 = 1.242 = (4.0 - 2.758) \text{ (V)}$$

$$= [V_L(t = 0-) + \Delta V^-] \text{ (V)}$$

The change in reflected current can readily be calculated from ΔV^- and Z_0. Since voltage and current are reflected with opposite polarity,

$$\Delta I_L = \Delta I^- = \frac{\Delta V^-}{Z_0} = \frac{2.758}{50} = 0.0552 \quad \text{(A)}$$

and

$$I_L(t = 0+) = I_L(t = 0-) + \Delta I_L^- = \frac{5}{10 + 40} + 0.0552$$

$$= 0.1552 \text{ (A)}$$

To calculate load voltage and current as a function of time, ρ_G must be known. Hence,

$$\rho_G = \frac{Z_G - Z_0}{Z_G + Z_0}$$

$$\rho_G = \frac{10 - 50}{10 + 50} = -0.667 = -\frac{2}{3}$$

Now write V_L and I_L as a function of time.

$$V_L = \{4.0 + 2.758[-u(t) + \frac{2}{3}u(t - 2\tau)$$

$$- 0.724\left(\frac{2}{3}\right)u\ (t - 2\tau)$$

$$+ 0.724\left(\frac{2}{3}\right)^2 u(t - 4\tau) - (0.724)^2\left(\frac{2}{3}\right)^2 u(t - 4\tau)$$

$$+ \ldots]\} \text{ (V)}$$

$$= [4.0 - 2.758u(t) + 0.507u(t - 2\tau)$$

$$+ 0.245u(t - 4\tau)$$

$$+ \ldots] \quad (V)$$

and

$$I_L = \{0.1 + 0.0552[u(t) + \frac{2}{3}u(t - 2\tau)$$

$$+ 0.724\left(\frac{2}{3}\right) u(t - 2\tau)$$

$$+ 0.724\left(\frac{2}{3}\right)^2 u(t - 4\tau) + (0.724)^2 \left(\frac{2}{3}\right)^2 u(t - 4\tau)$$

$$+ \ldots]\} \quad (A)$$

$$= [0.1 + 0.0552u(t) + 0.063u(t - 2\tau)$$

$$+ 0.031u(t - 4\tau)$$

$$+ \ldots] \quad (A)$$

Both V_L and I_L are plotted in Fig. 2.

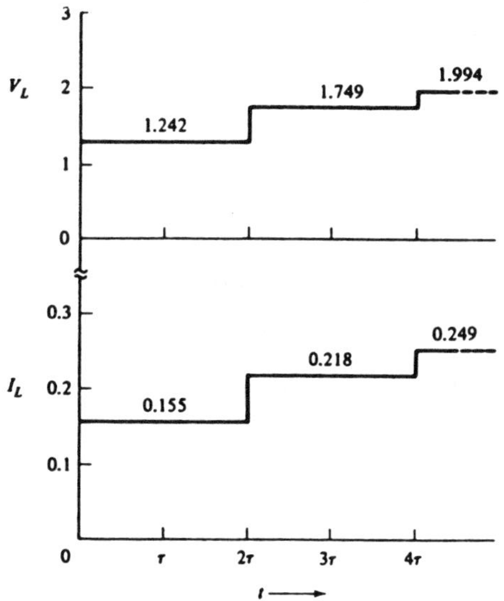

Fig. 2

• PROBLEM 13-12

Determine, using both the exact and approximate expressions, the attenuation and phase shift constants, and the real and imaginary parts of Z_0, for a parallel-wire transmission line consisting of two No. 8 AWG copper wires spaced 12 in. apart.

Solution: Reference to wire tables discloses that R = 6.634 ohms/mile (both conductors). For a parallel-wire transmission line,

$$L = \frac{\mu}{4\pi} + \frac{\mu}{\pi} \ln \frac{S}{r} \text{ henry/meter}$$

so (assuming $\mu = \mu_0$)

$$L = \left(10^{-7} + 4 \times 10^{-7} \ln \frac{12}{0.0643}\right) \times 1609 \frac{\text{meters}}{\text{mile}}$$

(henry/mile)

$$= 0.00016 + 0.00335 = 0.00351 \text{ henry/mile}$$

(The internal inductance is less than 5 percent of the external inductance.)

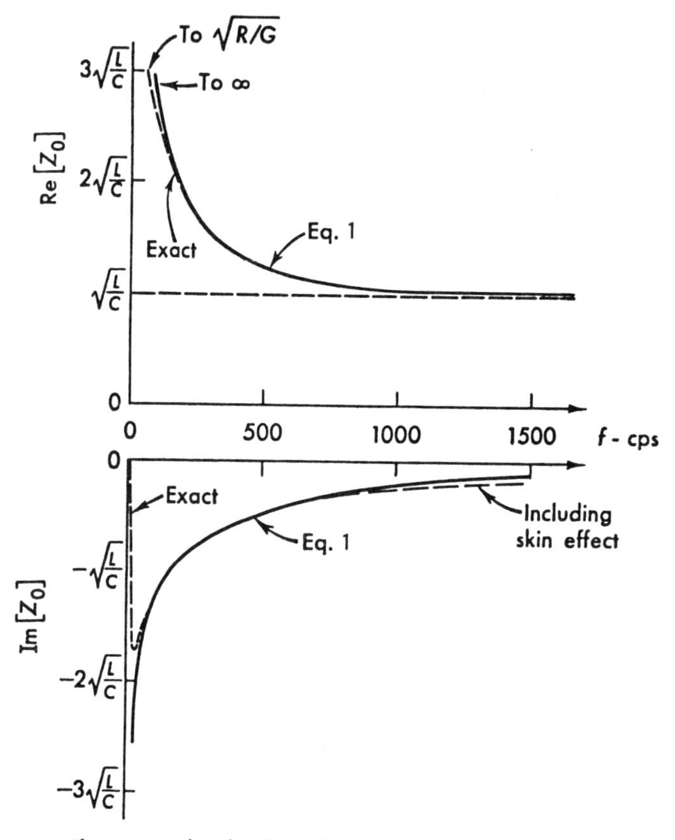

Characteristic impedance of a lossy line as a function of frequency. Fig. 1

Although no explicit expression has been obtained for capacitance, note that

$$L_e C = c^2 = \mu\varepsilon$$

(This equation is true only to the extent that the high-frequency inductance equals the d-c inductance.) Whence, in free space or air

$$C = \frac{\mu_0 \varepsilon_0}{L} = \frac{\mu_0 \varepsilon_0}{(\mu_0/\pi)\ \ln\ (S/r)} = \frac{\pi \varepsilon_0}{\ln\ (S/r)}$$

$$= \frac{10^{-9}}{36\ \ln\ (S/r)} \quad \text{farad/meter}$$

$$= \frac{10^{-9}}{36\ \ln\ (12/0.0643)} \times 1609 \text{ meters/mile}$$

$$= 0.00851\ \mu\text{f/mile}$$

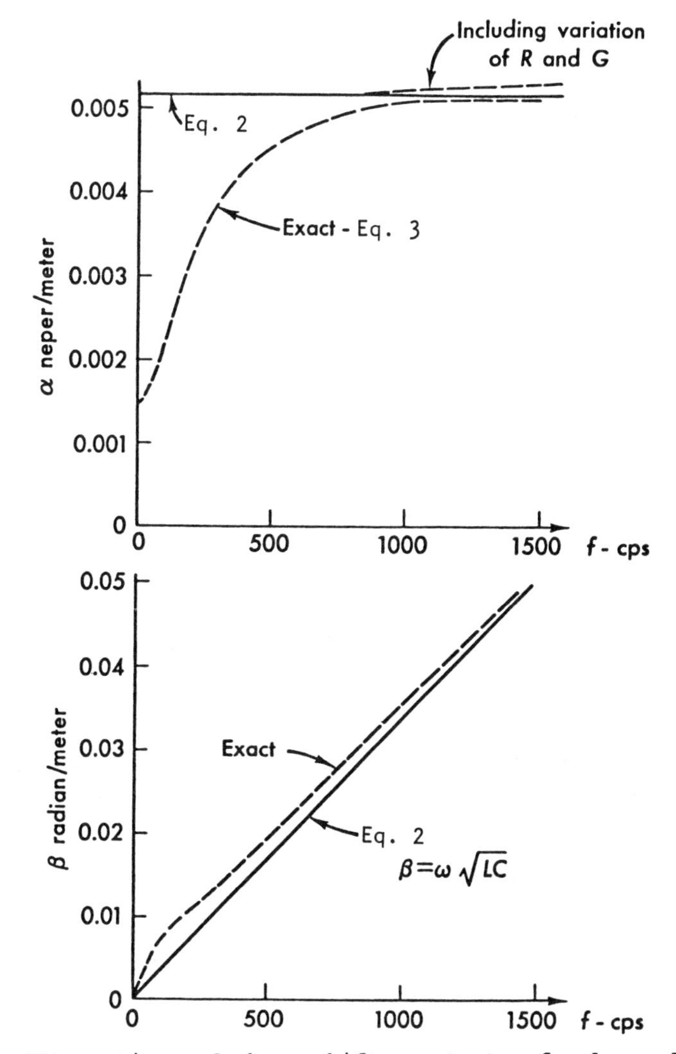

Attenuation and phase shift constants of a lossy line as a function of frequency. Fig. 2

The value of G must be measured experimentally, as it depends primarily on the leakage conductance of the insulators supporting the line. It also exhibits considerable frequency dependence, varying almost linearly with frequency. Take as an average value 0.3×10^{-6} mho/mile, which holds for dry insulators at 1000 cps.

For this line R/L = 1890 and G/C = 35.3, so expect the approximate formula for Z_0,

$$Z_0 \approx \sqrt{\frac{L}{C}}\sqrt{1 + \frac{R}{j\omega L}} \tag{1}$$

to hold at all but the lowest frequencies. The values of the resistive and reactive components of Z_0 are shown in Fig. 1.

In this figure also are shown the values at low frequencies given by the exact expressions and the values at high frequencies, including the result of skin effect.

In Fig. 2 are plotted α and β, which are the real and imaginary parts of γ. The results of the approximate expression

$$\gamma \approx \frac{G}{2}\sqrt{\frac{L}{C}} + \frac{R}{2}\sqrt{\frac{C}{L}} + j\omega\sqrt{LC} \tag{2}$$

are shown, as well as the exact values from

$$\gamma = \sqrt{(R + j\omega L)(G + j\omega C)} = \alpha + j\beta \tag{3}$$

and a better computation which includes the variation of R and G with frequency.

INPUT IMPEDANCES

• PROBLEM 13-13

Consider a space where $\varepsilon_r = 4$; $\mu_r = 1$, $\gamma = 0$. A wave defined by the expression $E_y = 2 \cos(\omega t - \beta x)$ is moving through this medium. Find v if $\omega = 2\pi \times 10^6$ cycles per second. Also compute β, λ, H_z and the wave impedance $z = E_y/H_z$.

Solution: Since

$$v = \frac{1}{\sqrt{\varepsilon\mu}} = \frac{c}{\sqrt{\varepsilon_r\mu_r}}$$

then

$$v = \frac{3 \times 10^8}{\sqrt{4 \times 1}} \text{ m/sec} = 1.5 \times 10^8 \text{ m/sec}$$

$$\beta = \frac{2\pi}{\lambda} = \frac{2\pi f}{v} = \frac{2\pi \times 10^6}{1.5 \times 10^8} = 4.19 \times 10^{-2} \text{ radians/meter}$$

and

$$\lambda = \frac{2\pi}{\beta} = \frac{2\pi}{4.19 \times 10^{-2}} = 150 \text{ meters}$$

By Maxwell's equation I: curl $H_0 = j\omega\varepsilon E_0$,

$$E_y = \rho H_z, \text{ where } \rho = \sqrt{\frac{\mu}{\varepsilon}}$$

Therefore, since $\rho = \sqrt{\frac{1}{4}}$ (377)   (free space)

$$H_z = \frac{2}{377} \times 2 \cos (\omega t - \beta x)$$

$$= .0106 \cos (\omega t - \beta x)$$

$$= \frac{E_y}{H_z} = \rho = \frac{2}{.0106} = 188 \text{ ohms}$$

• PROBLEM 13-14

Find the characteristic impedance (or resistance) of the lossless coaxial line shown in the Figure. The line is air-filled.

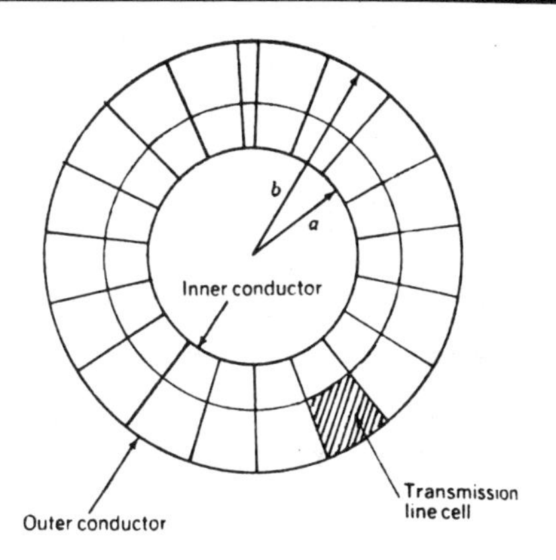

Coaxial transmission line with 18.3 transmission-line cells in parallel and 2 in series.

Solution: Dividing the space between the conductors into curvilinear squares or cells by graphical field mapping, there are a total of 18.3 squares in parallel and 2 in series.

The characteristic impedance of each cell is 376.7 Ω. Hence, from

$$Z_0 = \frac{N}{n} Z'_0$$

where N = number of cells in series

n = number of cells in parallel

Z'_0 = characteristic impedance of one cell $(=\sqrt{\mu/\varepsilon})$

$$= \sqrt{\frac{\mu_0}{\varepsilon_0}} = 376.7\Omega \text{ (for air-filled coaxial line).}$$

the characteristic impedance of the line of the figure is

$$Z_0 = \frac{N}{n} Z'_0 = \frac{2}{18.3} 376.7 = 41.2\Omega$$

If a cross section of this line is drawn to scale with conducting paint on a sheet of space paper 41.2 Ω would be measured directly on an ohmmeter connected between the inner and outer conductors.

• PROBLEM 13-15

An $\bar{E}$ field given by $\hat{x}\ 100e^{-\gamma z}$ (V/m) is traveling through a material $[\varepsilon'_r = 4,\ \sigma = 0.1\ (\mho m^{-1}),\ \mu = \mu_0]$ and the frequency is 2.45 (GHz). Find α and β and the decibels per meter attenuation in the material. Calculate the intrinsic impedance and then calculate both the impedance and propagation constant using the low loss approximations of

$$\eta_{\sigma/\omega\varepsilon' << 1} \approx \sqrt{\frac{\mu}{\varepsilon'}}\left(1 + \frac{1}{2}j\frac{\sigma}{\omega\varepsilon'}\right)$$

and

$$\gamma_{\sigma/\omega\varepsilon' << 1} \approx \frac{1}{2}\sigma\sqrt{\frac{\mu}{\varepsilon'}} + j\omega\sqrt{\mu\varepsilon'}\ .$$

Compare the results.

Solution: From $\gamma = j\omega\sqrt{\mu\varepsilon'}\ \sqrt{1 - j\frac{\sigma}{\omega\varepsilon'}}$

$$\gamma = j2\pi(2.45)(10^9)\sqrt{4\mu_0\varepsilon_0}\ \sqrt{1 - j\frac{0.1}{4\omega\varepsilon_0}}$$

$$= j1.539(10^{10})\ \frac{\sqrt{4}}{3 \times 10^8}\ (1.008\ \ \angle -5.197°)$$

$$= 103.44\ \ \angle 84.8°$$

$$= 9.37 + j103 = \alpha + j\beta \quad (m^{-1})$$

$$\frac{|E_x|_{z=1}}{|E_x|_{z=0}} = e^{-\alpha(1)} = e^{-9.37}$$

The dB attenuation is equal to

$$10 \log_{10} \frac{|E_x|_0^2}{|E_x|_1^2} = 20 \log_{10} e^{9.37}$$

$$= (20 \log_{10} e)(9.37)$$

$$= 81.39 \left(\frac{dB}{m}\right)$$

Then

$$\alpha = 9.37 \left(\frac{\text{nepers}}{m}\right)$$

$$\beta = 103 \left(\frac{\text{radians}}{m}\right)$$

$$\text{Attenuation} = 8139 \left(\frac{db}{m}\right)$$

At a depth equal to $1/\alpha$, the wave is attenuated to $e^{=1}$ times its initial value. This depth is referred to as the skin depth.

Next calculate $\sigma/\omega\varepsilon'$.

$$\frac{\sigma}{\omega\varepsilon'} = \frac{0.1}{2\pi(2.45 \times 10^9)4(8.854 \times 10^{-12})} = 0.1834$$

$$\sqrt{1 - j\frac{\sigma}{\omega\varepsilon'}} = \sqrt{1 - j0.1834} = 1.0083 \quad \underline{/- 5.197°}$$

$$\eta = \frac{\sqrt{\mu/\varepsilon'}}{\sqrt{1 - j\frac{\sigma}{\omega\varepsilon'}}} = \frac{377/\sqrt{4}}{1.0083 \quad \underline{/- 5.197°}} = 186.9 \quad \underline{/5.197°}$$

$$= (186.2 + j16.93) \quad (\Omega)$$

By low loss approximation,

$$\gamma \approx \frac{1}{2}\sigma\sqrt{\frac{\mu}{\varepsilon'}} + j\omega\sqrt{\mu\varepsilon'}$$

$$= \frac{1}{2}(0.1)\frac{377}{\sqrt{4}} + j2\pi(2.45 \times 10^9)\frac{\sqrt{4}}{3 \times 10^8}$$

$$= 9.43 + j102.6 = (\alpha + j\beta) \quad (m^{-1})$$

compared to

$$9.37 + j103$$

from before,

$$\eta_{\text{low loss}} \approx \sqrt{\frac{\mu}{\varepsilon'}} + \frac{1}{2}j\frac{\sigma}{\omega\varepsilon'}\sqrt{\frac{\mu}{\varepsilon'}}$$

$$= \frac{377}{4} + \frac{1}{2} j0.1834 \frac{377}{\sqrt{4}}$$

$$= (188.5 + j17.29) \quad (\Omega)$$

compared to

$$(186.2 + j16.93) \quad (\Omega)$$

calculated by the exact method.

• PROBLEM 13-16

(a) The air-filled coaxial line in Fig. 1 has a radius ratio b/a = 2. Find its characteristic impedance. (b) Find the characteristic impedance for a two-wire line as shown in Fig. 2.

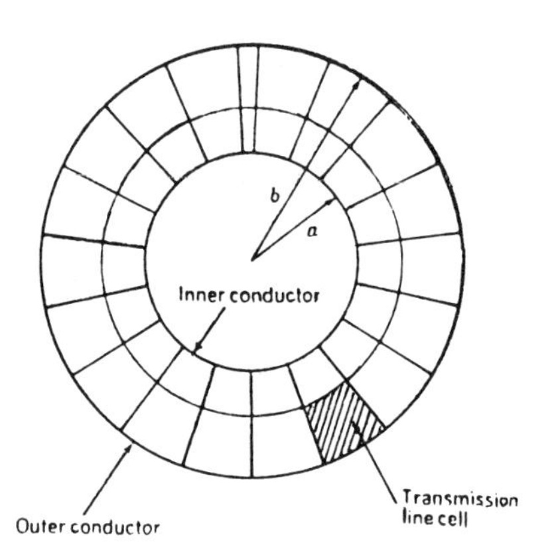

FIGURE 1
Coaxial transmission line with 18.3 transmission-line cells in parallel and 2 in series.

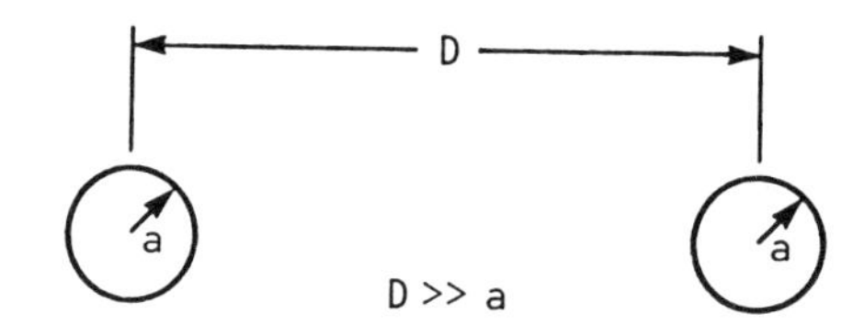

Fig. 2 Two wire transmission line.

Solution: The characteristic impedance of a coaxial line is given by

$$Z_0 = \frac{1}{2\pi}\sqrt{\frac{\mu}{\varepsilon}} \ln \frac{b}{a} = 0.367 \sqrt{\frac{\mu}{\varepsilon}} \log \frac{b}{a}$$

If there is no ferromagnetic material present, $\mu = \mu_0$ and

$$Z_0 = \frac{138}{\sqrt{\varepsilon_r}} \log \frac{b}{a} \qquad (\Omega)$$

where ε_r = relative permittivity of medium filling line, a = outside radius of inner conductor,

b = inside radius of outer conductor

For an air-filled line $\varepsilon_r = 1$, and

$$Z_0 = 138 \log \frac{b}{a} \quad (\Omega) \tag{1}$$

From (1)

$$Z_0 = 138 \log 2 = 41.4 \ \Omega$$

In a similar way, the characteristic impedance can be obtained for the two-wire line. Thus, if $D >> a$,

$$Z_0 = \frac{1}{\pi}\sqrt{\frac{\mu}{\varepsilon}} \ln \frac{D}{a} = 0.73\sqrt{\frac{\mu}{\varepsilon}} \log \frac{D}{a} \quad (\Omega) \tag{2}$$

If there is no ferromagnetic material present, $\mu = \mu_0$ and (2) reduces to

$$Z_0 = \frac{276}{\sqrt{\varepsilon_r}} \log \frac{D}{a} \quad (\Omega) \tag{3}$$

where ε_r = relative permittivity of medium

D = center-to-center spacing

a = radius of conductor (in same units as D)

If the medium is air, $\varepsilon_r = 1$ and (3) becomes

$$Z_0 = 276 \log \frac{D}{a} \quad (\Omega) \tag{4}$$

The characteristic impedances are summarized in the Table.

TABLE Characteristic impedance of coaxial and two-wire lines

Type of line	Characteristic impedance Ω
Coaxial (filled with medium of relative permittivity ε_r)	$Z_0 = \frac{138}{\sqrt{\varepsilon_r}} \log \frac{b}{a}$
Coaxial (air-filled)	$Z_0 = 138 \log \frac{b}{a}$
Two-wire (in medium of relative permittivity ε_r) ($D \gg a$)	$Z_0 = \frac{276}{\sqrt{\varepsilon_r}} \log \frac{D}{a}$
Two-wire (in air) ($D \gg a$)	$Z_0 = 276 \log \frac{D}{a}$

SMITH CHART

• PROBLEM 13-17

Given that, at a specified point on a transmission line, the line impedance is 50 + j50 Ω. The characteristic impedance of the line is 50 Ω. What is the line admittance at this point?

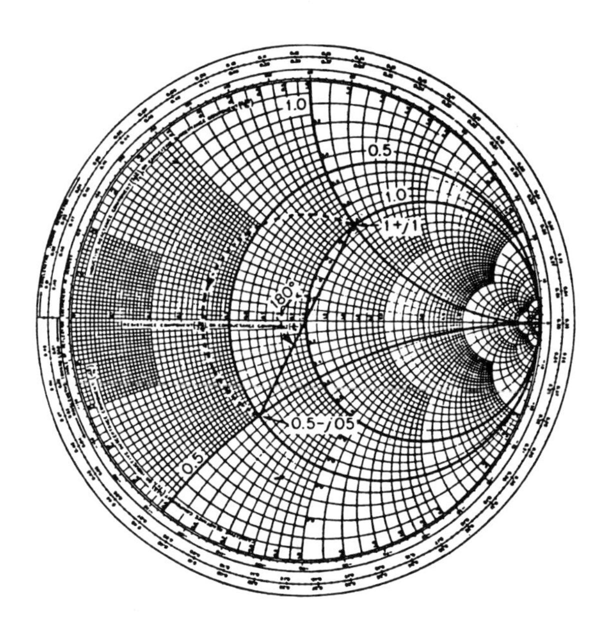

Solution: By direct calculation

$$Y = \frac{1}{Z} = \frac{1}{50 + j50} = (10 - j10) \times 10^{-3} \mho$$

By use of the Smith chart (refer to the figure shown)

$$Z = Z_0 (r + jx) = 50(1 + j1)$$

Locate the Smith chart impedance coordinate point 1 + j1. Draw the circle $\Gamma_0 e^{-j\beta s}$ through 1 + j1. At 180° along this circle from 1 + j1 read the Smith chart admittance coordinates 0.5 - j0.5. Now

$$Y = Y_0 (g + jb) = 1/50(0.5 - j0.5) = (10 - j10) \times 10^{-3} \mho$$

The steps are illustrated in the figure.

Although the Smith chart procedure seems more involved than direct calculation, it actually facilitates most calculations of this type.

• **PROBLEM 13-18**

Given a generator connected to a passive load by means of a transmission line whose characteristic impedance is 50 Ω. At a reference position s = 0 it is found that the line impedance, $Z(0) = V(0)/I(0)$, is given by

$$Z(0) = 100 + j50$$

What will the line impedance be 0.2λ toward the generator from the reference position?

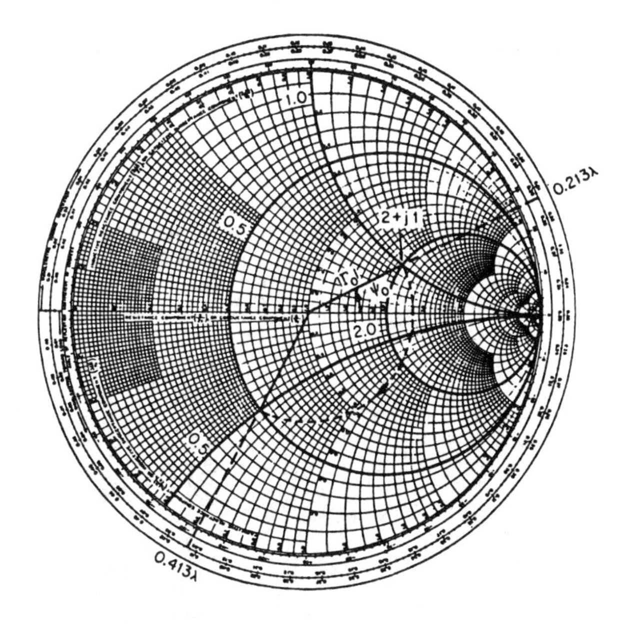

Solution: First express the problem in Smith chart coordinates by writing

$$z_n(0) = \frac{Z(0)}{Z_0} = \frac{100 + j50}{50} = 2 + j1$$

Locate the Smith chart point 2 + j1 as shown in the figure. Note that the reference point is at 0.213λ "toward the generator," or $\psi_0 = 26.2°$, and that the origin for the two scales is not the same. The value 0.213λ has no significance in this problem, except that 0.2λ toward the generator from this reference point is at 0.213λ + 0.2λ = 0.413λ "toward the generator." ψ at this point is -117.8°.

Next use a compass to draw the (circular) locus of $\Gamma_0 e^{-j2\beta s}$; read the Smith chart coordinates at the point on this circle corresponding to s = 0.2λ, that is, at 0.413λ "toward the generator." Obtain

$$z_n(s = 0.2\lambda) = 0.5 - j0.5$$

Hence $Z(0.2\lambda) = (0.5 - j0.5)50 = 25 - j25\ \Omega$

is the line impedance 0.2λ toward the generator from the point where the line impedance is $100 + j50$.

• PROBLEM 13-19

Given that the VSWR is $S = 3$, that the first minimum of $|V(s)|$ is 5 cm from the load, and that the distance between successive minima is 20 cm, while the characteristic impedance of the line is 50Ω. Find the load impedance.

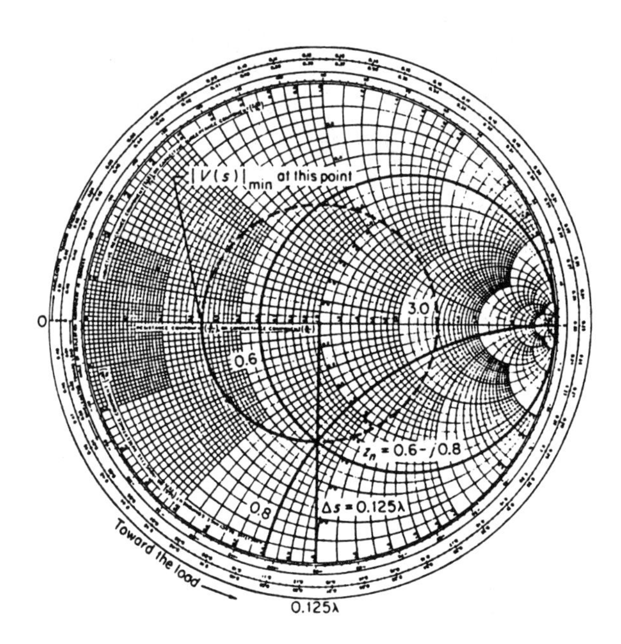

Solution: The wavelength is 40 cm. The load is 5/40 $= 0.125\lambda$ "toward the load" from the first voltage minimum. With reference to the given figure, locate the point $r_{min} = 1/S = 1/3$ and use a compass to draw the circle $\Gamma_0 e^{-j2\beta s}$. Draw the line $\Delta s = 0.125\lambda$ "toward the load." At the intersection of the circle and the line read the Smith chart coordinates

$$z_n(0) = r + jx = 0.6 - j0.8$$

Multiply by Z_0 and obtain

$$Z(0) = Z_L = Z_0 z_n(0) = 50(0.6 - j0.8) = 30 - j40\ \Omega$$

• PROBLEM 13-20

Find, using the Smith chart, the reciprocal of 1.2 - j0.8.

Solution: First enter 1.2 - j0.8 on the Smith chart (figure shown); then swing k through 180°. The result, read directly, is

$$y = 0.575 + j0.390$$

It is rather unlikely that the average engineer will l regularly equipped with Smith charts, so that in ordinary problems this method of inversion is not very helpful. However, this is a useful step in many matching problems.

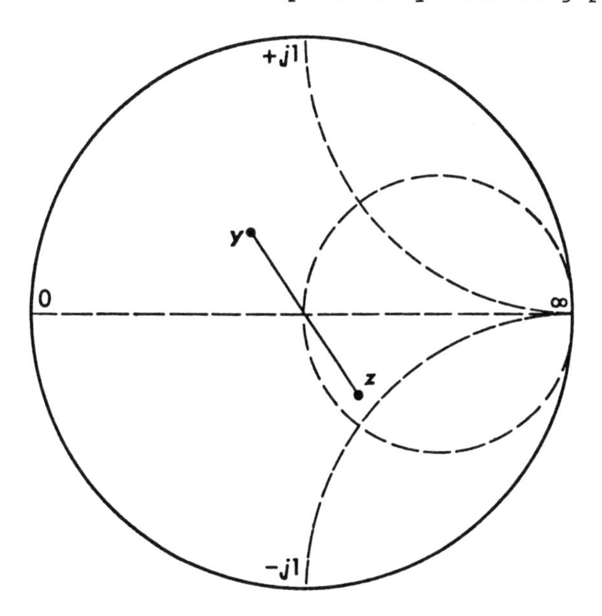

• PROBLEM 13-21

The load impedance on a 50-Ohm line is

$$z_L = 50(1 + j)$$

What is the admittance of the load? First use direct computation, then check by use of the Smith chart.

Solution: By direct computation,

$$Y_L = \frac{1}{Z_L} = \frac{1}{50(1 + j)} = \frac{(1 - j)}{100}$$

To use the Smith chart, find the normalized impedance at A in the Figure:

$$Z_{nL} = 1 + j$$

The normalized admittance that is the reciprocal of the normalized impedance is found by locating the impedance a distance $\lambda/4$ away from the load end at B:

$$Y_{nL} = 0.5(1 - j) \rightarrow Y_L = Y_n Y_0 = (1 - j)/100$$

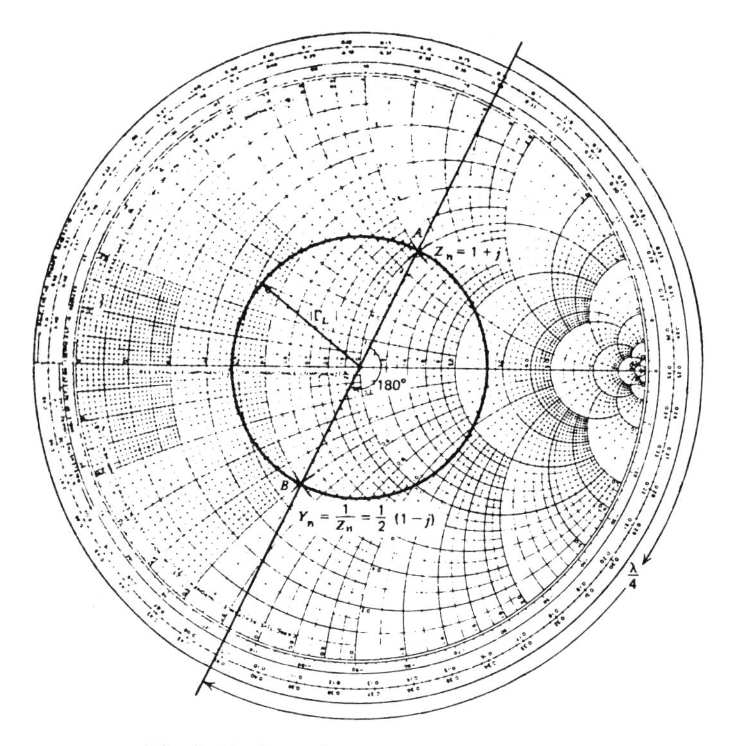

The Smith chart offers a convenient way to find the reciprocal of a complex number using the property that the normalized impedance reflected back by a quarter wavelength inverts. Thus, the normalized admittance is found by locating the normalized impedance and rotating this point by 180° about the constant $|\Gamma_L|$ circle.

Note that the point B is just 180° away from A on the constant $|\Gamma_L|$ circle. For more complicated loads the Smith chart is a convenient way to find the reciprocal of a complex number.

• PROBLEM 13-22

Consider the load impedance $Z_L = 25 + j50\Omega$, terminating a $50\,\Omega$ line. If the line is 60 cm long and the operating frequency is such that the wavelength on the line is 2m, find Z_{in}, by use of the Smith chart.

Solution: The normalized load impedance is $Z = 0.5 + j1$ as marked at A on Fig. 1; $\Gamma = 0.62e^{j1.45} = 0.62\underline{/83°}$. If the line is 60 cm long and the operating frequency is such

The Smith chart contains the constant-r circles and constant-x circles, an auxiliary radial scale to determine $|\Gamma|$, and an angular scale on the circumference for measuring ϕ.

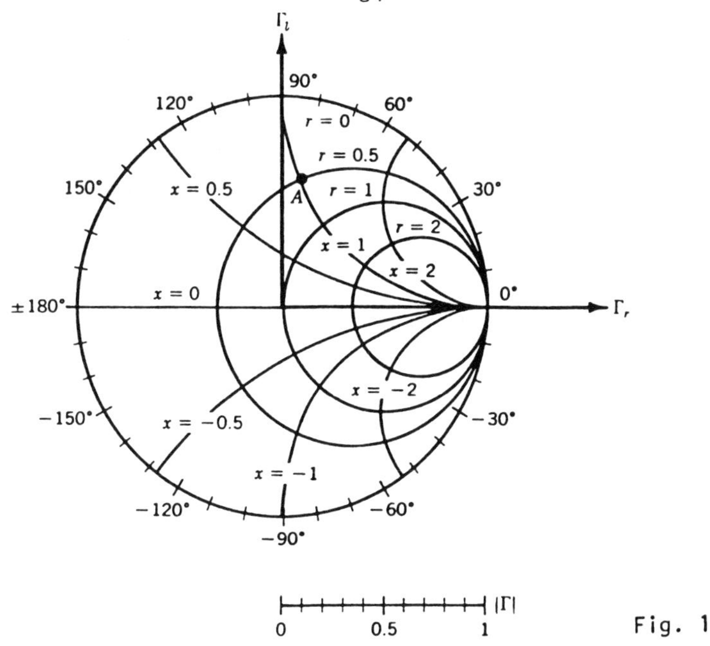

Fig. 1

The normalized input impedance produced by a normalized load impedance $z = 0.5 + j1$ on a line 0.3λ long is $z_{in} = 0.28 - j0.40$.

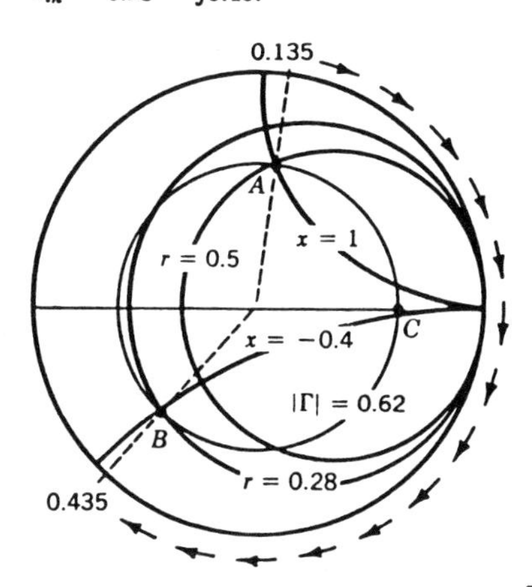

Fig. 2

that the wavelength on the line is 2 m, then $\beta\ell = 360° \times 60/200 = 108°$. To find z_{in}, therefore, move $2 \times 108° = 216°$ in a clockwise direction about the $|\Gamma| = 0.62$ circle. Although this could be accomplished with a compass and protractor, it is easier to use the wtg (wavelength toward generator) scale. A straight line drawn from the origin through A intersects the wtg scale at 0.135, as shown on Fig. 2. Since 60 cm is 0.300 wavelength, find z_{in} on the $|\Gamma| = 0.62$ circle opposite a wtg reading of 0.135 + 0.300 = 0.435. This construction is shown in Fig. 2, and the point locating the input impedance is marked B. The normalized input impedance is read as 0.28 - j0.40, and thus Z_{in} = 14 - j20. A more accurate analytical calculation gives Z_{in} = 13.7 - j20.2.

• PROBLEM 13-23

Find the sending-end impedance of a 50-mile telephone line for which

$$\bar{Z}_0 = 650 - j100 \text{ ohms}$$

$$\alpha = 0.00503 \text{ neper/mile}$$

$$\beta = 0.0350 \text{ radian/mile}$$

when Z_R is 1000 + j0 ohms.

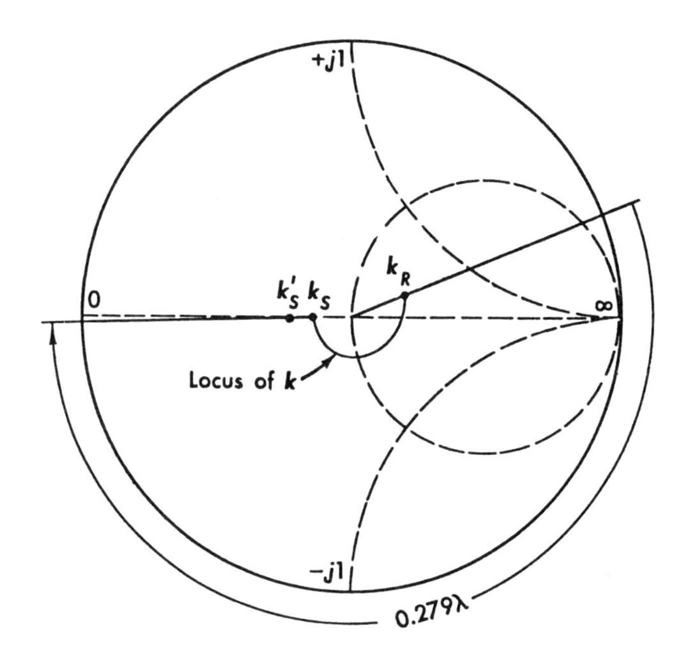

Solution: The normalized receiving-end impedance is

$$\bar{z}_R = \frac{1000 + j0}{650 - j100} = 1.502 + j0.232$$

(Observe that $\bar{z}_R$ is not real, even though $\bar{Z}_R$ is resistive.) Entering this value on the Smith chart (figure shown)

$$\bar{k}_R = 0.223 \angle 21.5°$$

The length of the line in wavelengths is

$$\frac{\ell}{\lambda} = \frac{\beta\ell}{2\pi} = \frac{0.0350 \times 50}{2\pi} = 0.2785$$

and rotating $\bar{k}_R$ through this distance transforms $\bar{k}$ to the point $\bar{k}'_S$. This is not the sending-end impedance, however, as the effect of attenuation as well as of phase shift must be included. From

$$k(d) = k_R e^{-2\alpha d} e^{-j2\beta d}$$

$$k_S = k_R \varepsilon^{-2\alpha \ell}$$

where the k's are the magnitudes of the $\bar{k}$'s.

Thus

$$k_S = 0.223E^{-2x0.00503x50} = 0.135$$

and

$$\bar{k}_S = 0.135 \angle -179.2°$$

whence

$$\bar{z}_S = 0.76 - j0.01$$

or

$$\bar{Z}_S = (0.76 - j0.01)(650 - j100) = 493 - j83$$

(If many calculations involving lossy lines are to be made, they can be facilitated through the use of an auxiliary scale for k, marked in units of $-2\alpha d$.)

• PROBLEM 13-24

(a) Suppose that a 58-cm length of lossy line which is known to be less than $\lambda/2$ long is open-circuited at one end and that the input impedance Z(s) at the other end is measured to be

$$\frac{Z(s)}{Z_0} = 0.20 + j0.25$$

Find α and β.

(b) Assume that this same transmission line is terminated with a load impedance such that

$$\frac{Z(0)}{Z_0} = 1.0 + j2.0$$

and that $Z_0 = 50 + j5\Omega$. Find the input impedance.

Solution: (a) Let d be the length of the line. Then

$$\frac{Z(0)}{Z_0} \to \infty \quad \text{at open end}$$

$$\frac{Z(s)}{Z_0} = 0.20 + j0.25 \quad \text{at input}$$

$$\frac{1 + 1e^{-2\alpha d}e^{-j2\beta d}}{1 - 1e^{-2\alpha d}e^{-j2\beta d}} = 0.20 + j0.25$$

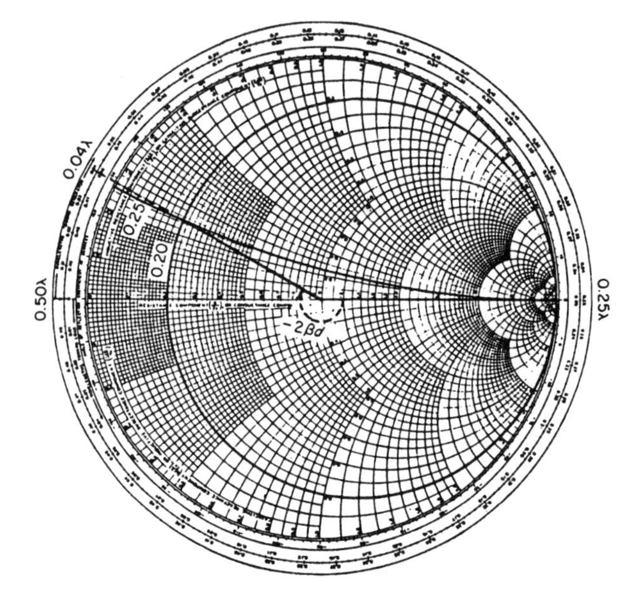

Fig. 1

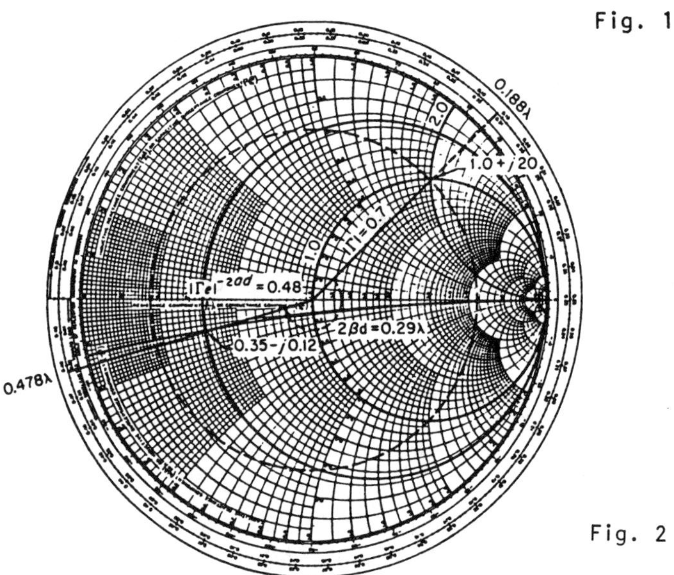

Fig. 2

Smith chart calculations for a lossy transmission line.

As shown in Fig. 1, locate the point 0.2 + j0.25 and draw the line from the center to this point. By measurement, its length is 0.685 of the chart's outer radius, which means that $|\Gamma(d)| = 0.685$. Extending this line to the scale around the circumference, d = 0.04λ + 0.25λ = 0.29λ.

From these measurements and the length d = 0.58 m, obtain $e^{-2\alpha d} = 0.685$, or $2\alpha d = 0.375$, and

$$\alpha = \frac{0.375}{1.16} = 0.323 \text{ Np/m}$$

Actually, since it is the factor $2\alpha d$ which appears in the exponential, and since d is frequently specified in wavelengths, it is convenient at times to calculate

$$2\alpha = \frac{0.375}{0.29} = 1.295 \text{ Np/wavelength}$$

(b) Refer to Fig. 2. Locate the point 1.0 + j2.0 as shown. Rotate 0.29λ "toward the generator" and draw the radial line shown. Then $Z(0.29\lambda)/Z_0$ lies on this line. Using a compass and a linear scale, find $|\Gamma| = 0.7$. Using $e^{-2\alpha d} = 0.685$, obtain $|\Gamma| e^{-2\alpha d} = 0.48$. Again, using a compass and the linear scale, locate the point $|\Gamma| e^{-2\alpha d} e^{-j2\beta d}$ along the previously located radial line, as shown in Fig. 2. Read

$$\frac{Z(0.29\lambda)}{Z_0} = 0.35 - j0.12$$

and since $Z_0 = 50 + j5$, calculate

$$Z(0.29\lambda) = (50 + j5)(0.35 - j0.12) = 18.1 - j4.25\Omega.$$

• PROBLEM 13-25

Suppose that the transmission line of figure 1 has a load impedance $Z_L = 100 - j50$ ohms, with $Z_0 = 50$ ohms. The point P of Fig. 2 then corresponds to $r_L = 2$ and $jx_L = j1$; Q corresponds to $y_L = 0.4 + j2.0$, the load admittance. Q occurs at 0.037λ toward the generator on the chart. Find the position, x, where a short-circuited stub, having the normalized $y_{stub} = 0 + jb_{stub}$, should be put in parallel with the line input admittance y = g + jb such that

$$y_{stub} + y = g + j(b + b_{stub}) = 1 + j0$$

Further, find the length of the stub required.

Solution: In Fig. 2 the circle with O as center and OQ as radius gives the admittance at all the points along the line with this particular load. There are two points where the normalized conductance is unity. At R_1 normalized input admittance is g + jb = 1 + j1; this corresponds to a distance 0.162λ toward the generator. At R_2 g + jb = 1 - j1, at a distance 0.338λ toward the generator. R_1 gives a point on the line whose distance from the load is $0.162\lambda - 0.037\lambda = 0.125\lambda$. R_2 gives a point on the line distant $0.338\lambda - 0.037\lambda = 0.301\lambda$ from the load. Take R_1, the one closer to the load, as the place to put the stub.

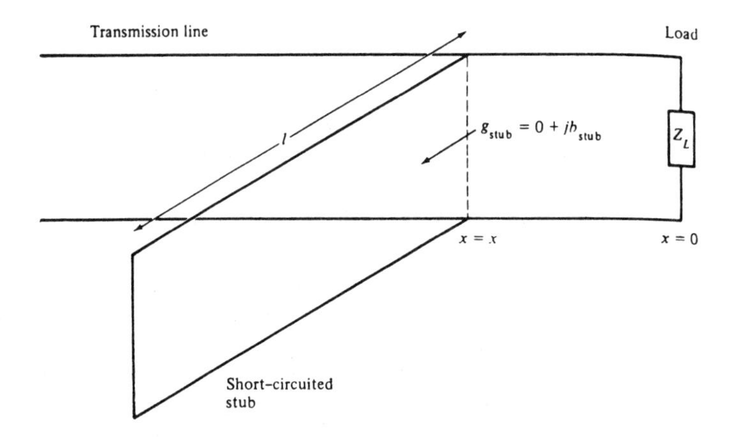

Fig. 1

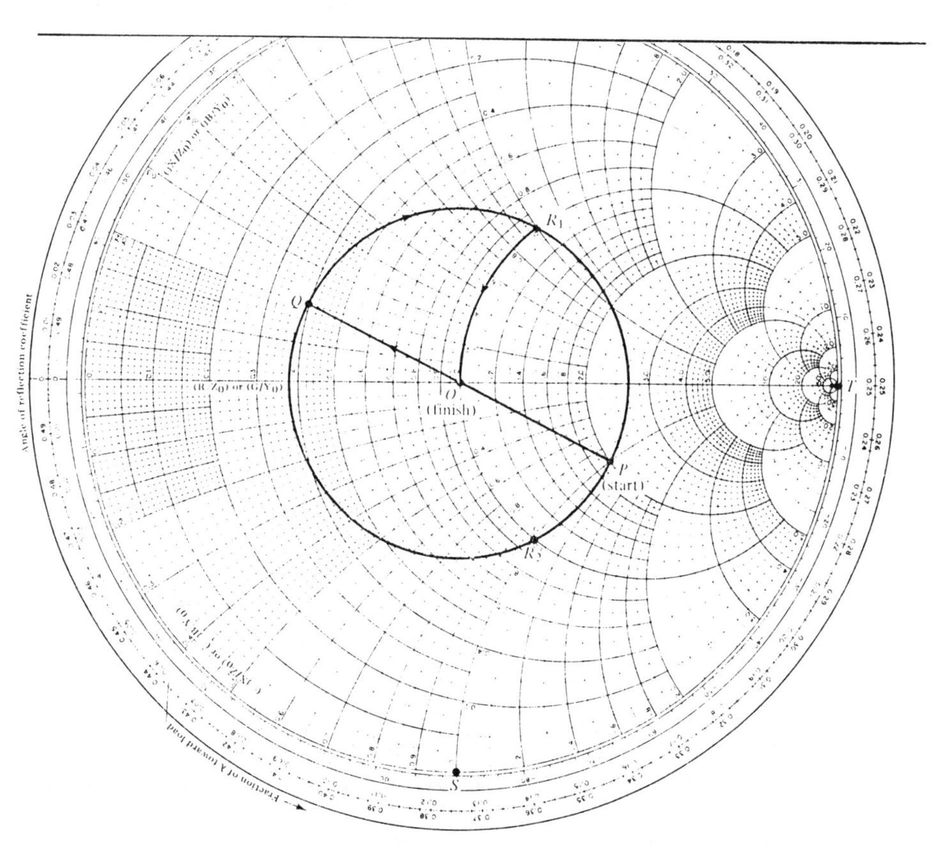

Fig. 2

At $x = 0.125\lambda$ the normalized input admittance of the line and load to the right is $1 + j1$. So at $x = 0.125\lambda$, a short-circuited stub whose normalized input admittance is $y_{stub} = -j1$ is needed for then the line, looking to the right toward the parallel combination, will see $(1 - j1) + (0 + j1) = 1 + j0$ for the total input admittance.

The point S on the chart having $y_{stub} = 0 - j1$ occurs at 0.375λ toward the generator; this gives the input admittance of the stub. The point T, on the chart with $y = 0 - j\infty$ at 0.25λ, corresponds to the short-circuited end of the stub. So the stub must be $0.375\lambda - 0.25\lambda = 0.125\lambda$ long.

• PROBLEM 13-26

Figure 1 shows the configuration of a typical problem that may be met in practice. Say a generator feeds an antenna by means of a coaxial transmission line 1.72 m long; for measurement purposes a slotted section has been inserted between the generator and the transmission line, and is tied to the line by means of a connector. The line, the connector, and the slotted section all have a common characteristic impedance of 50Ω. A minimum in the standing wave on the slotted section is observed 9 cm from the connector. The generator frequency is 750 MHz, and a voltage standing wave ratio of 3 is obtained along the slotted section. What impedance does the antenna present to the line at this frequency?

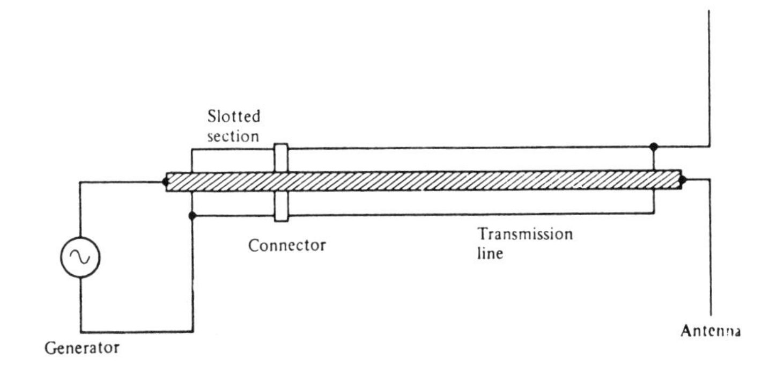

Fig. 1

Solution:
The magnitude of the reflection coefficient is

$$|\Gamma| = \frac{VSWR - 1}{VSWR + 1} = \frac{3 - 1}{3 + 1} = 0.5$$

The wavelength is

$$\lambda = \frac{c}{f} = \frac{3 \times 10^8}{750 \times 10^6} = 0.4m$$

The first voltage minimum, 0.09 m to the left of the connector, corresponds to $\frac{0.09}{0.4}$ of a wavelength, 0.225λ from the connector.

On the Smith chart of Fig. 2 the circle about the center with a radius equal to 0.5 that of the radius of the maximum circle gives the points having $|\Gamma| = 0.5$. On any such circle the point a, where it crosses the horizontal axis to the left of center, corresponds to a voltage minimum on the line, while the point b on the chart corresponds to a voltage maximum on the line. To show this,

$$V = V_1 e^{-j\beta z} + V_2 e^{j\beta z}$$

$$= V_1 e^{-j\beta(\ell - x)} \left[1 + \left(\frac{V_2}{V_1} e^{j2\beta\ell} \right) e^{-j2\beta x} \right]$$

so $$|V| = |V_1| (|1 + \Gamma_L e^{-j2\beta x}|) = |V_1| (|1 + \Gamma|)$$

The length of the line on the Smith chart from the extreme left point on the chart to any point on the circle is $|1 + \Gamma|$, and this has a minimum value for point α and a maximum value for point b.

Thus the value of ψ on any Smith chart for a point that lies at a voltage minimum on the line is $\psi = 180°$, i.e., at horizontal left. Similarly, a voltage maximum on the Smith chart lies along the $\psi = 0°$ direction. In the present case the voltage minimum on the slotted section of Fig. 1 lies at point a on the Smith chart of Fig. 2. The input to the transmission line proper in Fig. 1, at the connector, is 0.225λ toward the load, so on the Smith chart going from point a, 0.225λ toward the load along the circle of constant Γ, arriving at point P on the chart. Then read the $z_{in} = r_{in} + jx_{in}$ coordinates giving the normalized input impedance of the transmission line at the connector.

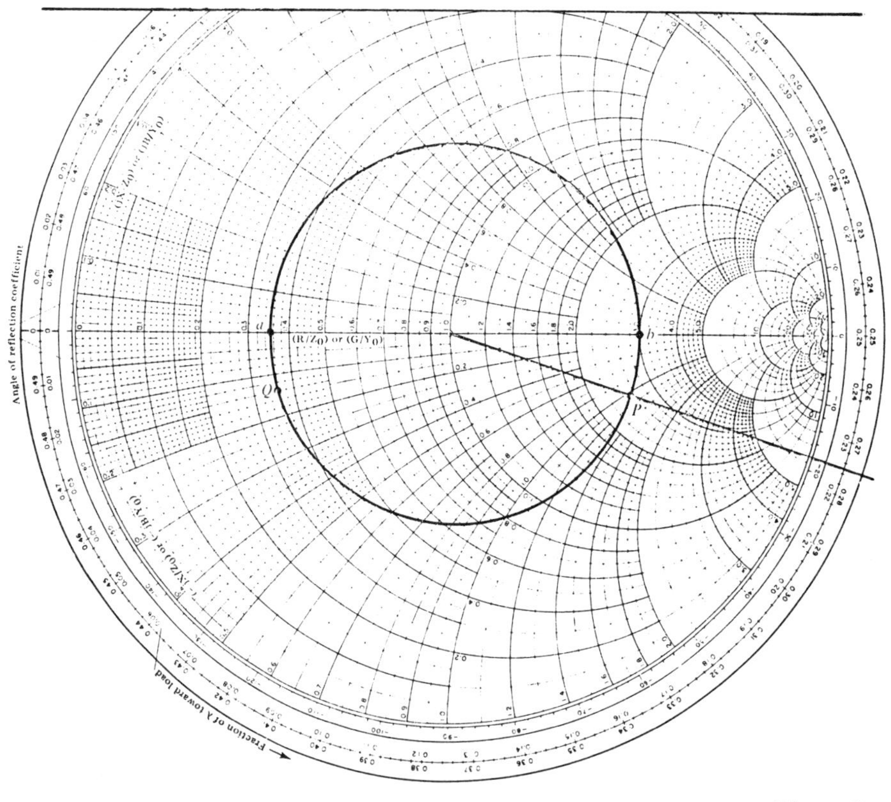

Fig. 2

Since the point P on the chart may be read either as z_{in} or y_{in} it is necessary to remember one fact from the text: if the load is capacitive then a minimum lies closest to the load, while if the load is inductive then a maximum lies closest. Here a minimum lies closest, so the load is capacitive. The point P has a negative reactance or susceptance associated with it, so the former must be used. P, here, gives an impedance.

So in this case at P on the Smith chart, read $r_{in} = 2.5 - jx_{in} = -j1.0$; thus Z_{in} is $50(2.5 - j1.0) = 125 - j50$ ohms for the input impedance to the transmission line at this frequency.

It is desired to know the input impedance at the antenna rather than at the input to the line. With $\lambda = 0.4$ m, the length of the line is 1.72/0.4 or 4.3 wavelengths long. Subtract an integral number of wavelengths from this, leaving 0.3λ. On the Smith chart then move along the constant VSWR circle, already drawn, another 0.3λ toward the load. The radius must then indicate $0.225\lambda + 0.3\lambda = 0.525\lambda$ at the rim of the diagram.

The calibration stops at 0.5λ, however, so go to 0.025λ, giving the point Q. Here $r_{in} = 0.34$ and $-jx_{in} = -j0.14$. The antenna impedance is thus

$$50(0.34 - j0.14) = 17 - j7\Omega$$

• PROBLEM 13-27

Find the sending-end impedance of a lossless line 10 m long having a characteristic impedance of 470 ohms (resistive), and terminated in an impedance of 320 - j175 ohms. The wavelength at the frequency employed is 1.80 m.

Solution: The first step is to obtain the normalized receiving-end impedance. This is

$$\bar{z} = \frac{320 - j175}{470} = 0.681 - j0.372 \text{ ohms}$$

Entering this on the Smith chart (figure shown)

$$\bar{k}_R = 0.281 \underline{/-118.2°}$$

(the angle is read directly; the magnitude must be measured with an auxiliary scale).

The line is 10/1.80 = 5.56 wavelengths long. Thus

$$\bar{k}_S = \bar{k}_R \varepsilon^{-j4\pi 5.56}$$

Since the standing wave pattern is repeated every half wavelength, this is equivalent to

$$\bar{k}_S = \bar{k}_R \varepsilon^{-j4\pi x0.06}$$

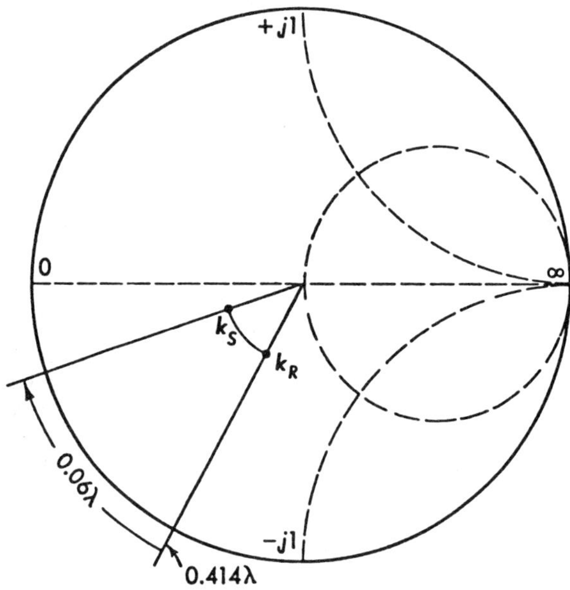

The Smith chart is calibrated directly in wavelength, so the only number which needs concern is 0.06. At a point corresponding to $\bar{k}_R$ read 0.414λ. Adding 0.06 to this yields 0.474λ. Swinging $\bar{k}_R$ through the arc, 0.06λ yields $\bar{k}_S$ (Fig.). This is

$$\bar{k}_S = 0.281 \underline{/-161.3°}$$

and, from the chart,

$$z_S = 0.58 - j0.12 \text{ ohms}$$

Observe that it was never actually necessary to know the value of $\bar{k}$, explicitly.

• PROBLEM 13-28

(a) Find the load impedance on a line for which $\bar{Z}_0 = 500$ ohms, on which the standing wave ratio is 2.3, and on which minima occur at 1.23 m and 0.37 m from the load.

(b) Find the proper location and value of a shunt admittance to match the load described in (a).

Solution: (a) Since minima occur every half-wavelength

$$\lambda = 2(1.23 - 0.37) = 1.72 \text{ m}$$

Thus the minimum is 0.37/1.72 = 0.215 wavelengths from the load.

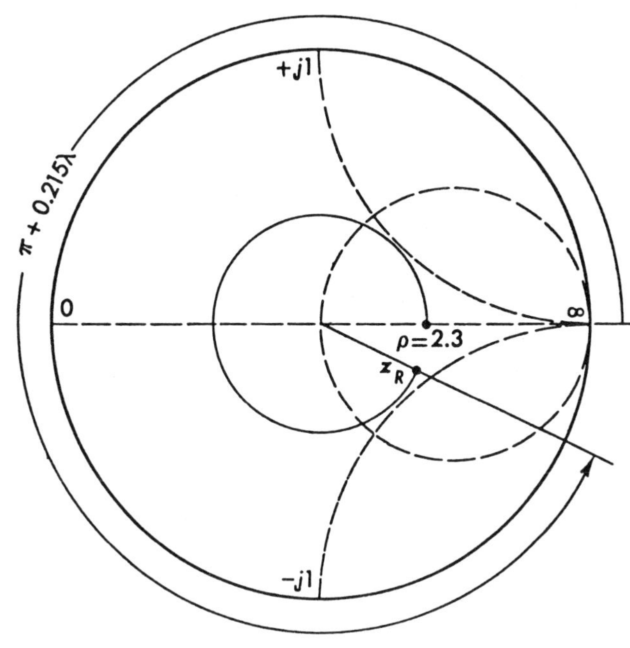

Fig. 1

(b) It was determined in (a)

$$\bar{z}_R = 1.91 - j0.76 \text{ ohms}$$

This may be converted to $\bar{y}_R$ by using the inversion procedure. The steps of finding $\bar{z}_R$ and then inverting it are not really necessary, however. Since $\bar{z}_R$ was determined by first finding $\bar{k}_R$ and rotating this through d_{min}/λ (0.215 wavelength) plus 180°, it is only necessary, in order to find $\bar{y}_R$, to rotate $\bar{k}_R$ through d_{min}/λ (plus nothing). Such a procedure yields (Fig. 2)

$$\bar{y}_R = 0.448 + j0.180$$

The circle $\bar{k}_R$ = constant intersects the circle of unit real part of $\bar{y}$ in two places:

$$\bar{y} = 1 + j0.85 \quad \text{at} \quad d = 0.122\lambda$$

$$\bar{y} = 1 - j0.85 \quad \text{at} \quad d = 0.308\lambda$$

In the first position an inductive susceptance of -0.85 is required, or in the second a capacitive susceptance of +0.85 is required. Since Z_0 is 500 ohms, $\bar{Y}_0 = 1/\bar{Z}_0 =$

0.002 mho, and the required susceptances become -0.0017 mho or +0.0017 mho, respectively.

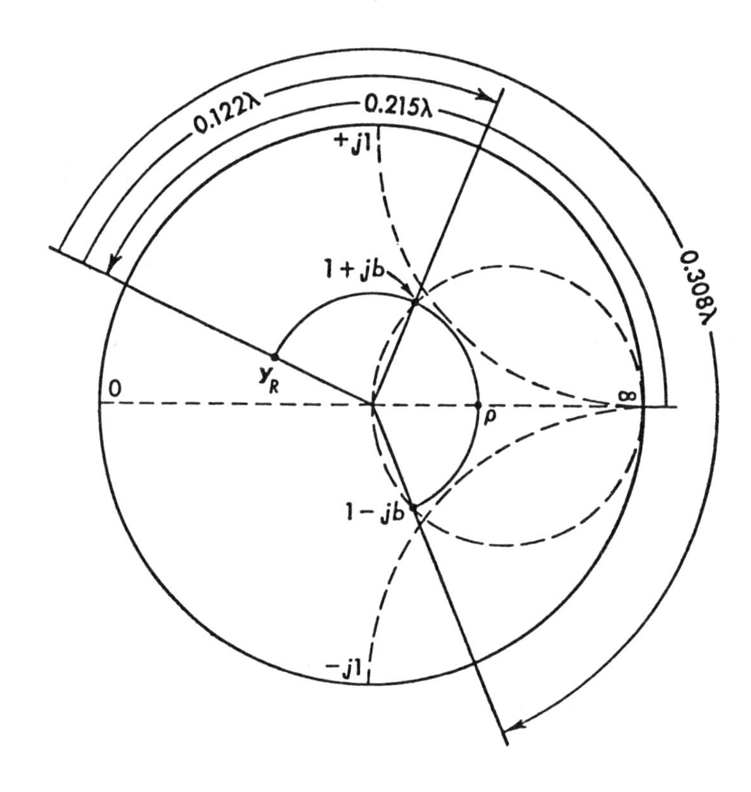

Fig. 2

The value of k_R is found (implicitly) by entering ρ at z = 2.3 + j0 ohms (fig. 1). Rotating this through 0.215 wavelengths plus 180° gives $\bar{k}_R$. The corresponding value of $\bar{z}_R$ is

$$\bar{z}_R = 1.91 - j0.76 \text{ ohms}$$

or

$$\bar{Z}_R = (1.91 - j0.76)500 = 955 - j380 \text{ ohms}$$

(It is seen that it was never necessary to know explicitly the value of $\bar{k}$.)

A transmission line of characteristic impedance 50 ohms is terminated by a load impedance $\overline{Z}_R$ = (15 - j20) ohms. Find the following quantities by using the Smith chart.

(1) Reflection coefficient at the load.

(2) VSWR on the line.

(3) Distance of the first voltage minimum of the standing wave pattern from the load.

(4) Line impedance at d = 0.05λ.

(5) Line admittance at d = 0.05λ.

(6) Location nearest to the load at which the real part of the line admittance is equal to the line characteristic admittance.

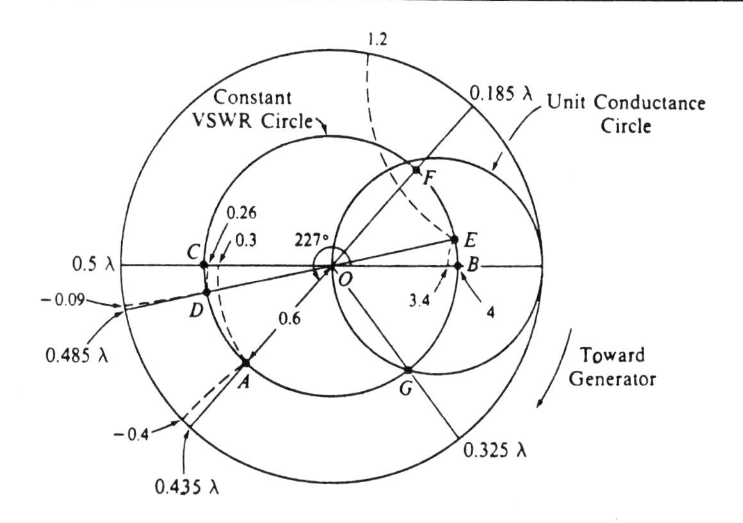

Solution: Proceed with the solution of the problem in the following step-by-step manner with reference to the figure shown.

(a) Find the nomalized load impedance.

$$\overline{z}_R = \frac{\overline{Z}_R}{Z_0} = \frac{15 - j20}{50} = 0.3 - j0.4$$

(b) Locate the normalized load impedance on the Smith chart at the intersection of the 0.3 constant normalized resistance circle and -0.4 constant normalized reactance circle (point A).

(c) Locating point A actually amounts to computing the reflection coefficient at the load since the Smith chart is a transformation in the $\overline{\Gamma}$ plane. The magnitude of the reflection coefficient is the distance from the center (O) of the Smith chart (origin of the $\bar{\Gamma}$ plane) to the point A based on a radius of unity for the outermost circle. For this example, $|\bar{\Gamma}(0)| = 0.6$. The phase angle of $\bar{\Gamma}(0)$ is the angle measured from the horizontal axis to the right of O (positive real axis in the $\bar{\Gamma}$ plane) to the line OA in the counter-clockwise direction. This angle is indicated on the chart along its circumference. For this example, $\underline{/\bar{\Gamma}(0)} = 227°$. Thus

$$\bar{\Gamma}(0) = 0.63^{j227°}$$

(d) To find the VSWR, recall that at the location of a voltage maximum, the line impedance is purely real and maximum. Denoting this impedance as R_{max},

$$R_{max} = \frac{V_{max}}{I_{min}} = \frac{|\bar{V}^+|(1 + |\bar{\Gamma}|)}{(|\bar{V}^+|/Z_0)(1 - |\bar{\Gamma}|)} = Z_0(\text{VSWR})$$

Thus the normalized value of R_{max} is equal to the VSWR. Therefore move along the line to the location of the voltage maximum, which involves going around the constant $|\bar{\Gamma}|$ circle to the point on the positive real axis. To do this on the Smith chart, draw a circle passing through A and with center at O. This circle is known as the "constant VSWR circle" since for points on this circle, $|\bar{\Gamma}|$ and hence VSWR = $(1 + |\bar{\Gamma}|)/(1 - |\bar{\Gamma}|)$ is a constant. Impedance values along this circle are normalized line impedances as seen moving along the line. In particular, since point B (the intersection of the constant VSWR circle with the horizontal axis to the right of O) corresponds to voltage maximum, the normalized impedance value at point B which is purely real and maximum, is equal to the VSWR. Thus, for this example, VSWR = 4.

(e) Just as point B represents the position of a voltage maximum on the line, point C (intersection of the constant VSWR circle with the horizontal axis to the left of O, i.e., the negative real axis of the $\bar{\Gamma}$ plane) represents the location of a voltage minimum. Hence, to find the distance of the first voltage minimum from the load, move along the constant VSWR circle starting at point A (load impedance) towards the generator (clockwise direction on the chart) to reach point C. Distance moved along the constant VSWR circle in this process can be determined by recognizing that one complete revolution around the chart ($\bar{\Gamma}$-plane diagram) constitutes movement on the line by 0.5λ. However, it is not necessary to compute in this manner since distance scales in terms of λ are provided along the periphery of the chart for movement in both directions. For this example, the distance from the load to the first voltage minimum = $(0.5 - 0.435)\lambda = 0.065\lambda$. Conversely, if the VSWR and the location of the voltage minimum are specified, then find the load impedance following the above procedures in reverse.

(f) To find the line impedance at $d = 0.05\lambda$, start at point A and move along the constant VSWR circle towards the generator (in the clockwise direction) by a distance of 0.05λ to reach point D. Thus, from the coordinates corresponding to point D, the normalized line impedance at d = 0.05λ is (0.26 - j0.09) and hence the line impedance at d = 0.05λ is 50(0.26 - j0.09) or (13 - j4.5) ohms.

(g) To find the line admittance at d = 0.05λ, recall that

$$[\bar{Z}(d)]\left[\bar{Z}(d + \frac{\lambda}{4})\right] = Z_0^2$$

so that

$$[\bar{z}(d)]\left[\bar{z}(d + \frac{\lambda}{4})\right] = 1$$

or

$$\bar{y}(d) = \bar{z}(d + \frac{\lambda}{4}) \tag{1}$$

Thus the normalized line admittance at a point D is the same as the normalized line impedance at a distance λ/4 from it. Hence, to find $\bar{y}(0.05\lambda)$, start at point D and move along the constant VSWR circle by a distance λ/4 to reach point E (note that this point is diametrically opposite to point D) and read its coordinates. This gives $\bar{y}(0.05\lambda) = (3.4 + j1.2)$. Then $\bar{Y}(0.05\lambda) = \bar{y}(0.05\lambda) \times Y_0 = (3.4 + j1.2) \times 1/50 = (0.068 + j0.024)$ mhos.

(h) Relationship (1) permits the use of the Smith chart as an admittance chart instead of an impedance chart. In other words, if it is desired to find the normalized line admittance $\bar{y}(Q)$ at a point Q on the line, knowing the normalized line admittance $\bar{y}(P)$ at another point on the line, simply locate $\bar{y}(P)$ by entering the chart at coordinates equal to its real and imaginary parts and then moving along the constant VSWR circle by the amount of the distance from P to Q in the proper direction to obtain the coordinates equal to the real and imaginary parts of $\bar{y}(Q)$. Thus it is not necessary first to locate $\bar{z}(P)$ diametrically opposite to $\bar{y}(P)$ on the constant VSWR circle, then move along the constant VSWR circle to locate $\bar{z}(Q)$ and then find $\bar{y}(Q)$ diametrically opposite to $\bar{z}(Q)$. To find the location nearest to the load at which the real part of the line admittance is equal to the line characteristic admittance, first locate $\bar{y}(0)$ at point F diametrically opposite to point A which corresponds to $\bar{z}(0)$. Then move along the constant VSWR circle towards the generator to reach point G on the circle corresponding to constant real part equal to unity. This is called the "unit conductance curve." Distance moved from F to G is read off the chart as (0.325 - 0.185)λ = 0.14λ. This is the distance closest to the load at which the real part of the normalized line admittance is equal to unity and hence the real part of the line admittance is equal to the line characteristic admittance.

• PROBLEM 13-30

Find the ratio of reflected power to incident power, for a 75-ohm load on a 50-ohm line.

Suppose a 50-ohm load is used instead of the 75-ohm load, what does it signify for an incident power of 50W?

Solution: (a) The relation between incident power, reflected power and the reflection coefficient KL is given by

$$\frac{\text{Reflected Power}}{\text{Incident Power}} = \frac{P_r}{P_i} = |KL|^2$$

For the given problem $KL = \frac{75 - 50}{75 + 50} = 0.2$ which leads to

$$\frac{P_r}{P_i} = |KL|^2 = (0.2)^2 = 0.04.$$

i.e., only 4 percent of the power is reflected from the load.

(b) If $R_L = 75$ ohms

$$\frac{P_r}{P_i} = 0.04$$

$$\therefore P_r = P_i \times 0.04 = 50 \times 0.04 = 2W$$

i.e. reflected power = 2W

∴ The source supplies 50-2 = 48W to 75 ohm load.

Now if $R_L = 50$ ohm then $KL = \frac{R_L - R_O}{R_L + R_O} = 0$ and $P_r = 0$ consequently.

∴ The source supplies all the 50W power to the load R_L.

• PROBLEM 13-31

Design a transparent window for a radar antenna. This problem arises because it is necessary to protect antennas from the weather by appropriate covers or radomes. Assume that the antenna is off to the left in free space, $z < -L$. Region 1 lies between $z = -L$ and 0, where a slab of perfect dielectric is placed, making it as thin as possible to keep the assumption of zero losses valid. To the right in region 2, $z > 0$, is the free-space region into which the radar signal is sent.

Solution: In order to avoid any reflection of power back into the antenna, or in order to match the antenna to the outside world, set $\eta_{in} = 377$. Since $\eta_2 = 377$, from

$$\eta_{in} = \eta_1 \frac{\eta_2 + j\eta_1 \tan \beta_1 L}{\eta_1 + j\eta_2 \tan \beta_1 L}$$

$$377 = \eta_1 \frac{377 + j\eta_1 \tan \beta_1 L}{\eta_1 + j377 \tan \beta_1 L}$$

Multiplying out,

$$j377^2 \tan \beta_1 L = j\eta_1{}^2 \tan \beta_1 L$$

Since $\eta_1 < 377$ for all nonmagnetic materials, this equation can be satisfied only by selecting $\beta_1 L = n\pi$. The thinnest radome is obtained when $\beta_1 L = \pi$, or $L = \lambda_1/2$. Thus, if the operating frequency is 10,000 MHz, select a lightweight low-loss plastic for which $\varepsilon_{R1} = 2.25$ and use a thickness

$$L = \frac{\lambda_1}{2} = \frac{U_1}{2f_1} = \frac{3 \times 10^8}{2\sqrt{2.25}\ 10^{10}} = 10^{-2}\text{m} \qquad \text{(or 1 cm)}$$

If the radome were 0.5m thick, it could be shown that $\eta_{in} = 167.5\Omega$ and 14.8 percent of the incident power would be reflected.

• PROBLEM 13-32

A 50-Ω transmission line is connected to a load impedance $Z_L = 25 - j50$. Find the position and length of the short-circuited stub required to match the line.

Solution: Since $Z_O = 50\Omega$, the normalized load impedance is

$$z_L = 0.5 - j1.0$$

From the Smith chart (figure) find

$$y_L = 0.4 + j0.8$$

at 0.115λ "toward the generator" (or $\phi_0 = 97°$), and

$$y(s) = 1.0 + j1.6$$

at 0.179λ "toward the generator," giving $s = (0.179 - 0.115)\lambda$, or 0.064λ, as the matching point nearest the load.

Notice that there is a second matching point, where

$$y(s) = 1.0 - j1.6$$

at 0.321λ "toward the generator." This point is (0.321 - 0.115)λ, or 0.206λ from the load.

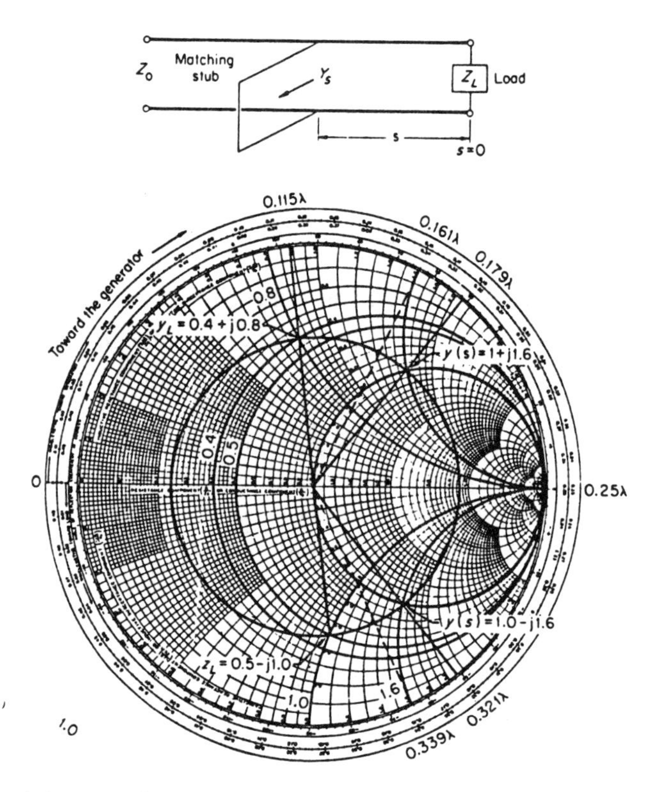

The first matching point requires a stub whose normalized susceptance b is -1.6. Reading around the outer circle from -j∞, the length of the short-circuited stub with this value of b is (0.339 - 0.250)λ, or 0.089λ long.

The second matching point would require a stub whose normalized susceptance b is +1.6. This stub would have a length of 0.161λ if it were open-circuited or a length of (0.161 + 0.250)λ = 0.411λ if it were short-circuited.

• PROBLEM 13-33

The load impedance $Z_L = 50(1 + j)$ on a 50-Ohm line is to be matched by a double-stub tuner of $\frac{3}{8}\lambda$ spacing. What stub lengths ℓ_1 and ℓ_2 are necessary?

Solution: The normalized load impedance $Z_{nL} = 1 + j$ corresponds to a normalized load admittance:

$$Y_{nL} = 0.5(1 - j)$$

Then the two solutions for Y_a lie on the intersection of the circle shown in Figure 1a with the r = 0.5 circle:

$$Y_{a1} = 0.5 - 0.14j$$

$$Y_{a2} = 0.5 - 1.85j$$

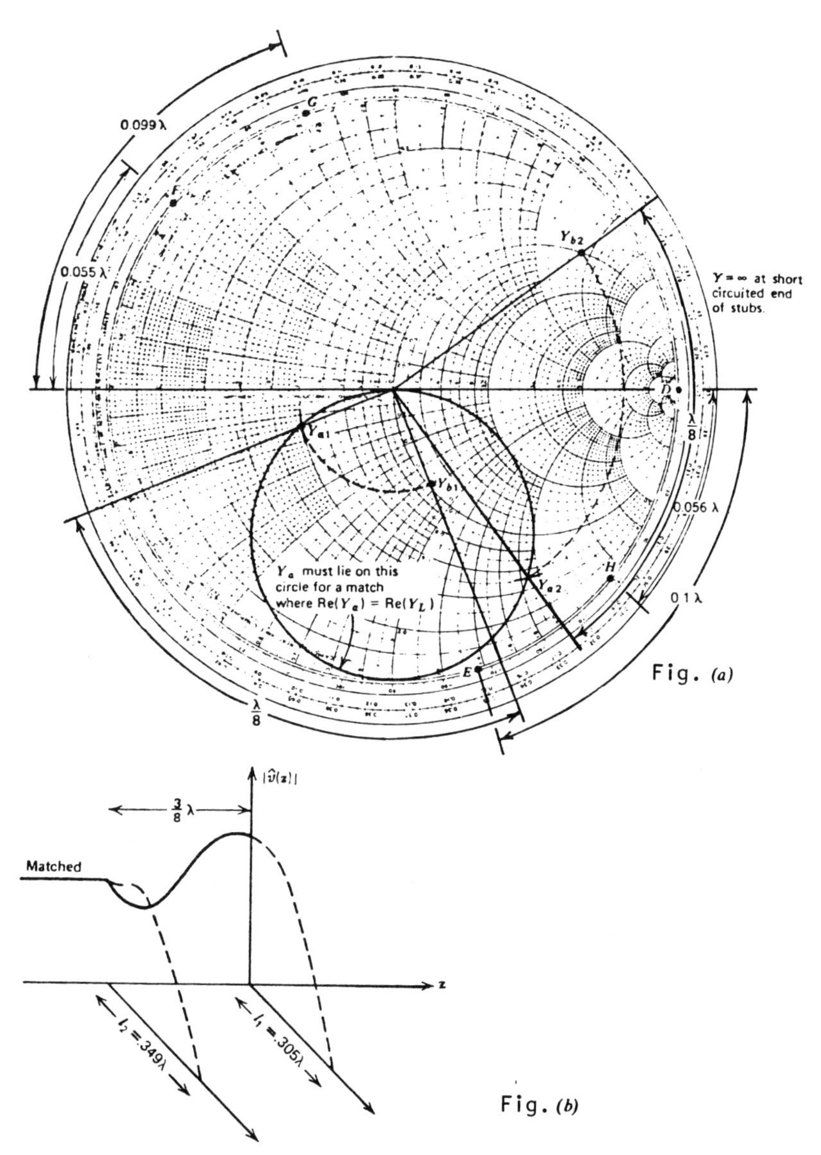

(a) The Smith chart construction for a double-stub tuner of $\frac{3}{8}\lambda$ spacing with $Z_{nL} = 1 + j$. (b) The voltage standing wave pattern.

Then find Y_1 by solving for the imaginary part of the equation $Y_a = Y_1 + Y_L$:

$$Y_1 = j\,\text{Im}\,(Y_a - Y_L) = \begin{cases} 0.36j \rightarrow \ell_1 = 0.305\lambda & \text{(F)} \\ -1.35j \rightarrow \ell_1 = 0.1\lambda & \text{(E)} \end{cases}$$

By rotating the Y_a solutions by $\frac{3}{8}\lambda$ back to the generator (270° clockwise, which is equivalent to 90° counterclockwise), their intersection with the r = 1 circle gives the solutions for Y_b as

$$Y_{b1} = 1.0 - 0.72j$$

$$Y_{b2} = 1.0 + 2.7j$$

This requires Y_2 to be

$$Y_2 = -j\,\text{Im}\,(Y_b) = \begin{cases} 0.72j \rightarrow \ell_2 = 0.349\lambda & \text{(G)} \\ -2.7j \rightarrow \ell_2 = 0.056\lambda & \text{(H)} \end{cases}$$

The voltage standing wave pattern along the line and stubs is shown in Figure 1b. Note the continuity of voltage at the junctions. The actual stub lengths can be those listed plus any integer multiple of $\lambda/2$.

• PROBLEM 13-34

Given that

$$Y_L = (8 + j8) \times 10^{-3}\,\mho$$

$$Y_0 = 20 \times 10^{-3}\,\mho$$

Use one short-circuited stub in parallel with the load and one short-circuited stub in parallel with the line at d = 0.22λ; match the line and find the length of the required stubs.

Solution: With reference to the given figure:

1. Draw the rotated circle.
2. Locate $y_L = 0.4 + j0.4$.
3. Follow the $g_L = 0.4$ curve to the point $y(0) = 0.4 + j0.7$, where it intersects the rotated circle.
4. Calculate $b_1 = 0.7 - 0.4 = 0.3$
5. From $y(0) = 0.4 + j0.7$, follow the $|\Gamma_0|$ = constant circle a distance $2\beta d = 0.22\lambda$ "toward the generator" to the point $y(d)$ on the matching circle.
6. Read $y(d) = 1 - j1.45$.

7. The required normalized susceptance of the matching stub at s = d is 1.45.

The corresponding short-circuited stub lengths are

For $b_1 = 0.3$: $\ell_1 = 0.297\lambda$

For $b_2 = 1.45$: $\ell_2 = 0.404\lambda$

There are two points of intersection. The entire procedure is valid for either point of intersection, but it generally leads to different matching-stub lengths.

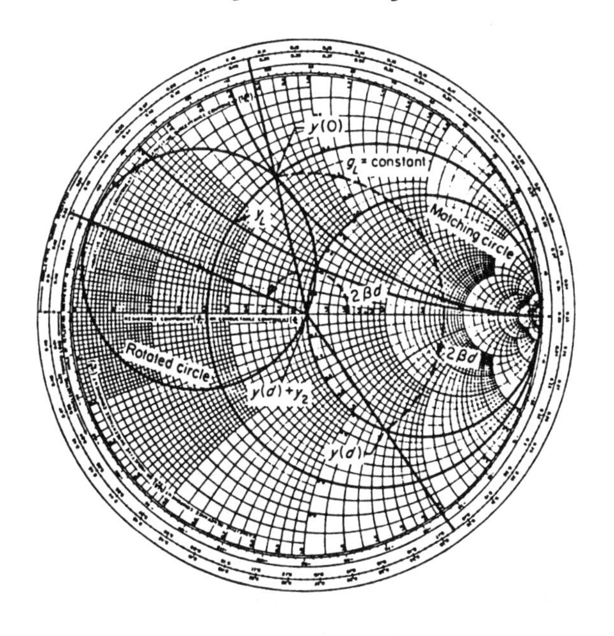

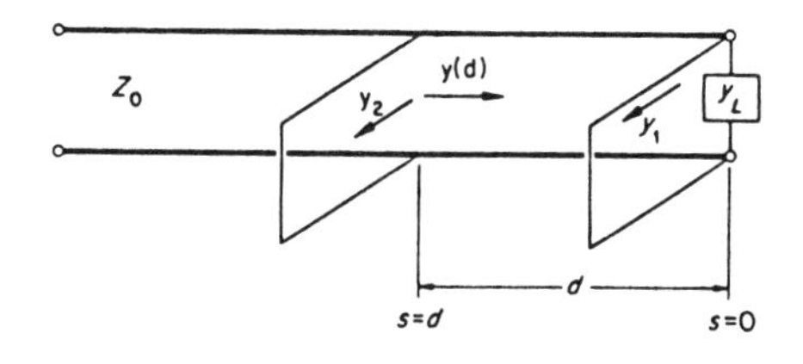

• PROBLEM 13-35

A transmission line has its first two voltage minima at 1.25 and 2.77 m from the load. The VSWR is 2.1. What should be the length and location of a matching stub?

Solution: The wavelength is

$$\lambda = 2(2.77 - 1.25) = 3.04 \text{ m}$$

Hence the first null is 1.25/3.04 = 0.411 wavelength from

the load. Entering the VSWR (2.1 + j0) on the impedance scale of the Smith chart (Fig. 1) and rotating $\bar{k}$ through 0.411 wavelength yield $\bar{y}_R$. (By omitting the π of $\theta = \pi + 2\beta d_{min}$, $-\bar{k}_R$ is found, rather than $\bar{k}$, which yields $\bar{y}_R$ directly.) Thus:

$$\bar{y}_R = 1.07 - j0.78$$

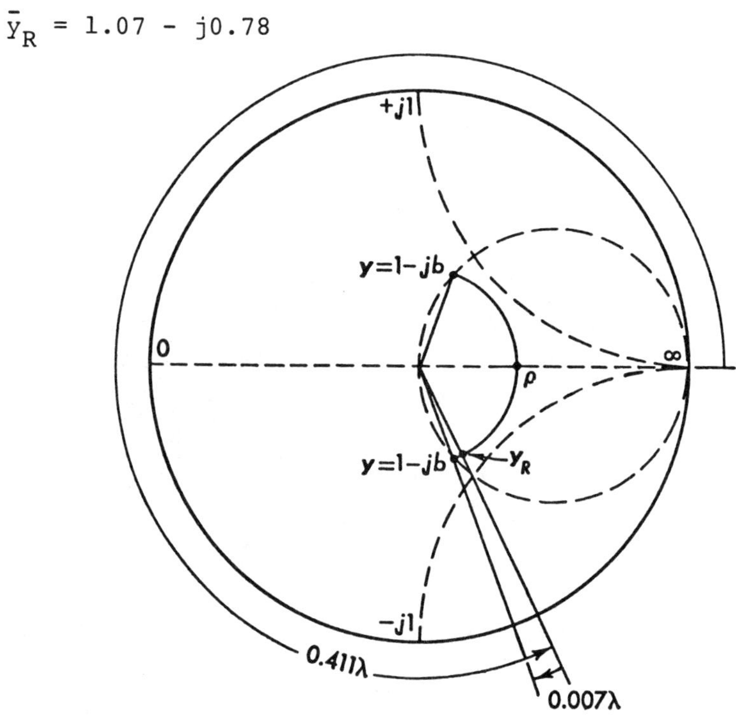

Fig. 1

The circle of constant $\bar{k}$ intersects the unit real-part circle in two places, corresponding to

$$\bar{y} = 1.00 - j0.75$$

$$y = 1.00 + j0.75$$

at distances

$$d = 0.007\lambda$$

$$d = 0.315\lambda$$

from the load, respectively.

Although matching can be accomplished at either of these points, it is preferably done nearer the load because otherwise a voltage maximum would occur between the matching stub and the load, which would mitigate some of the advantages of matching. Thus, a shunt susceptance of +j0.75 is needed.

A short circuit appears on the Smith chart at the origin of the z-coordinates, or its admittance is at the point at infinity. It is transformed into an admittance of

0 + j0.75 by a line 0.352λ long (Fig. 2).

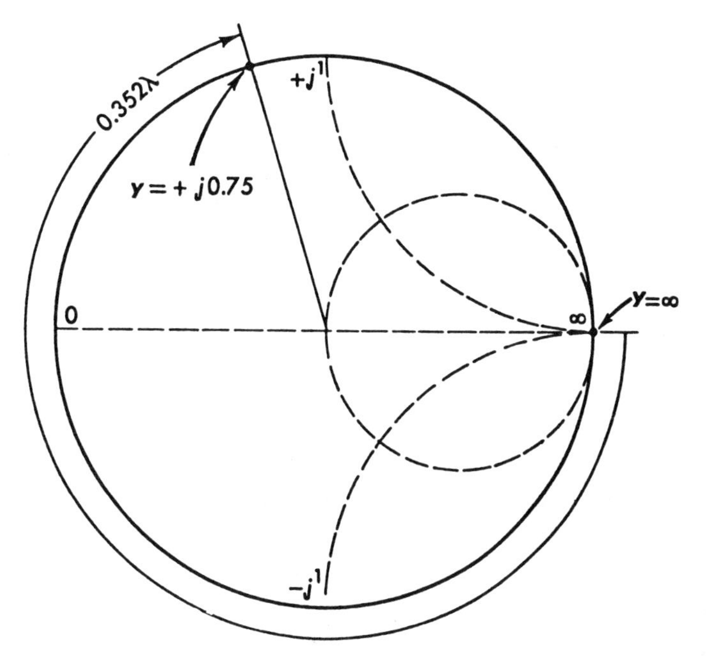

Fig. 2

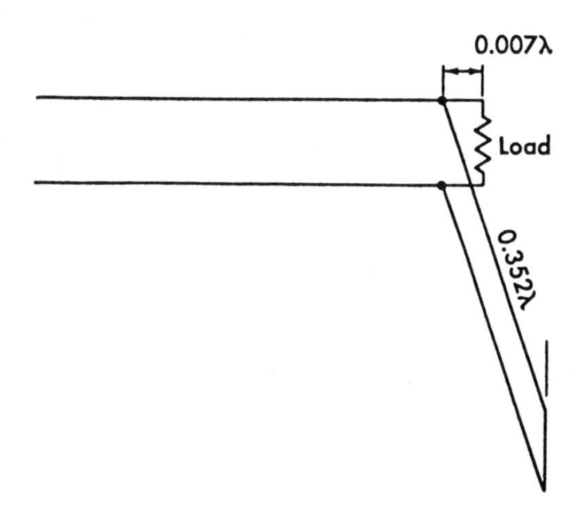

Fig. 3

The final arrangement of load and matching stub is shown in Fig. 3.

It may be difficult, mechanically, to place the proper matching stub so close (2 cm) to the load. In such a case it may be more desirable to add some susceptance to the load before attempting matching.

REFLECTION COEFFICIENT

• PROBLEM 13-36

Find the input impedance of a 50-ohm line terminated in +j50 ohms, for a line length such that $\beta d = \frac{\pi}{1}$ radian.

Solution: The reflection coefficient is $(j1 - 1)/(j1 + 1) = j1$.
Now the voltage reflection coefficient at the input-end of the line is $K_{in} = K_L e^{-j2\beta d}$ and in this case it is $(j1)(e^{-j\pi}) = -j1$. And from this the input impedance is found to be $50(1 - j1)(1 + j1) = -j50$ ohms.

• PROBLEM 13-37

A 50 V pulse is incident on a 30 ohm load in a line with characteristic resistance of 50 ohms. Find the following a) Incident Current Pulse, b) Reflection Coefficient, c) Reflected Voltage Pulse, d) Reflected Current Pulse, e) Net voltage and current at the load.

Solution: a) The incident current pulse is $\frac{V^+(0,t)}{R_0} = \frac{50}{50} = 1$ Ampere.

b) The reflection coefficient $K_L = \frac{R_L - R_0}{R_L + R_0} = \frac{30 - 50}{30 + 50} = -0.25$.

c) Reflected voltage pusle $= V^+(0,t) \times KL = 50 \times 0.25 = 12.5V$.

d) Reflected current pulse $= I^+(0,t) \times (-KL) = 1 \times (-0.25) = 0.25$ Amperes.

e) Net voltage at the load $= 50 - 12.5 = 37.5V$.

Net current at the load $= 1 + 0.25 = 1.25$ Amperes.

• PROBLEM 13-38

An electromagnetic wave impinges normally on a metallic sheet. Compare the reflection coefficients for copper and iron if $\gamma = 5.8 \times 10^7$ mho/m for copper, 1×10^6 mho/m for iron, $\mu_r = \mu_v$ for copper and $\mu_r = 1000\ \mu_v$ for iron. Take frequency to be 1 megacycle.

Solution:

$$\Gamma = \frac{\rho_2 - \rho_1}{\rho_1 + \rho_2} \qquad \text{where } \rho_1 \text{ is for air, } \rho_2 \text{ is for the metal}$$

$$\rho_1 = 377 \text{ ohms}$$

$$\rho_2 \text{ for copper} = \sqrt{\frac{j\omega\mu}{\gamma + j\omega\mu}} = \sqrt{\frac{j\,2\pi(10^6)\;4\pi(10^{-7})}{5.8(10^7) + \cancel{j2\pi(10^6)\;4\pi \times 10^{-7}}}} \quad \text{(negligible)}$$

$$= j\left[\frac{79.2(10^{-8})}{5.8}\right]^{\frac{1}{2}}$$

$$= 3.69(10^{-4})\angle 45° = 2.61(10^{-4}) + j2.61(10^{-4})$$

$$\rho_2 \text{ for iron} = \sqrt{\frac{j2\pi(10^6)4\pi(10^{-7})\;10^3}{10^6 + \cancel{j2\pi(10^6)4\pi(10^{-7})\;10^3}}} \quad \text{(negligible)}$$

$$= \left[j\,\frac{7900}{10^6}\right]^{\frac{1}{2}} = 89(10^{-3})\angle 45°$$

$$= 0.089\angle 45° = 0.063 + j0.063$$

Therefore

$$\Gamma_{iron} = \frac{.089\angle 45° - 377\angle 0}{.089\angle 45° + 377\angle 0} \cong \frac{-376.91}{377.089}$$

$$\cong .99$$

$$\Gamma_{copper} = \frac{2.61(10^{-4}) + j2.61(10^{-4}) - 377}{377 + 2.61(10^{-4}) + j2.61(10^{-4})}$$

$$= 1$$

Therefore

$\Gamma_{copper} = 1$ (almost perfect reflector)

$\Gamma_{iron} = .99$ (not quite, but very good reflector)

• PROBLEM 13-39

The VSWR (voltage standing wave ratio) on a 50-Ohm (characteristic impedance) transmission line is 2. The distance between successive voltage minima is 40 cm while the distance from the load to the first minima is 10 cm. What is the reflection coefficient and load impedance?

Solution: The following are given:

VSWR = 2

$$kd_{min} = \frac{2\pi(10)}{2(40)} = \frac{\pi}{4}$$

The reflection coefficient is given from

$$\Gamma_L = |\Gamma_L| e^{j\phi}$$

where $\phi = 2kd_{min} - \pi$ and $|\Gamma_L| = \frac{VSWR - 1}{VSWR + 1}$

Therefore $\phi = 2\left(\frac{\pi}{4}\right) - \pi = -\frac{\pi}{2}$ and $|\Gamma_L| = \frac{2 - 1}{2 + 1} = \frac{1}{3}$. Thus,

$$\Gamma_L = \frac{1}{3} e^{-j\pi/2} = \frac{-j}{3}$$

while the load impedance is found from

$$Z_L = \frac{Z_0[1 - jVSWR \tan kd_{min}]}{[VSWR - j \tan kd_{min}]}$$

as

$$Z_L = \frac{50(1 - 2j)}{2 - j}$$

$$= 40 - 30j \text{ ohm}$$

• PROBLEM 13-40

Find a suitable coating for lenses in order to minimize reflection. At normal incidence, the situation is that shown in the figure.

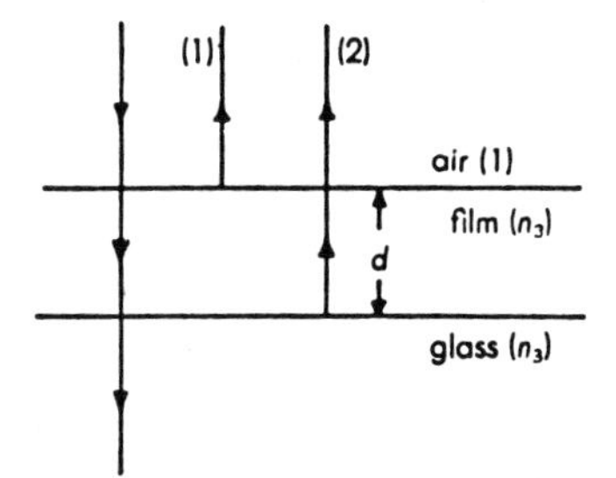

Transmitted and reflected rays at normal incidence on a coated lens.

Solution: The design of a suitable coating for glass involves the determination of the proper index of refraction of the film and the determination of the correct thickness. As a first step, make the reflection coefficients the same at both boundaries. The equality of the reflection coefficients requires that

$$\frac{n_2 - 1}{n_2 + 1} = \frac{n_3 - n_2}{n_3 + n_2}$$

or

$$n_2 = \sqrt{n_3} \tag{1}$$

Equation (1) implies that the index of refraction of the film is the geometric mean of the refractive indexes of air and glass.

In order to have destructive interference, wave (2) in the figure must be 180 degrees out of phase with wave (1). Since both wave (1) and wave (2) are reflected at boundaries of optically denser mediums, both waves suffer a reversal in phase due to reflection. Therefore, in order for waves (1) and (2) to interfere destructively, the distance 2 d must be equal to an odd number of half wavelengths (in the film) -- that is,

$$2d = m\left(\frac{\lambda}{2n_2}\right)$$

where m = 1, 3, 5, etc.

The crucial question now is whether the amplitude of the resultant reflected electric field intensity equals zero. First, realize that wave (2) in the figure consists of an infinite series of waves. Thus, to find the amplitude of wave (2) in the figure, find the sum of the following series:

$$- b\tau_{12}\tau_{21}A - b^3\tau_{12}\tau_{21}A - b^5\tau_{12}\tau_{21}A - \ldots$$

where, of course, $\tau_{12}\tau_{21} = 1 - b^2$. The negative signs indicate that the phases of all of the returning waves represented by the terms in the series are opposite to the phase of wave (1) in the figure. The path of a given wave in the series differs from that of a preceding wave by a half wavelength (in the film), but there is also one additional reflection at the boundary of an optically denser medium. All of the waves represented by the terms in the series, therefore, have the same phase. The sum of the series is

$$-b(1 - b^2)A(1 + b^2 + b^4 + \ldots)$$

$$= -b(1 - b^2)A(1 - b^2)^{-1} = -bA$$

The amplitude of wave (1) in the figure is bA. Consequently, the amplitude of the resultant reflected electric field intensity is zero. The film makes the glass nonreflecting for the particular wavelength λ (in air).

• PROBLEM 13-41

Assume a two-wire 300-Ω line (Z_0 = 300 Ω), such as the lead-in wire from the antenna to a television or FM receiver. The circuit is shown in the figure. The line is 2 m long and the dielectric constant is such that the velocity on the line is 2.5 x 10^8 m/s. The line is terminated with a receiver having an input resistance of 300 Ω and has an input source with an internal impedance of 300 Ω and a crest amplitude of 60 V at a frequency of 100 MHz.

a) Determine the reflection coefficient, the standing wave ratio, the input impedance, the input and load currents, the voltage of the load and the power supplied by the source to the load.

b) Assume a second receiver with an input resistance of 300 Ω placed in parallel with the first receiver. Repeat the calculations in part (a).

c) The load is now replaced by a capacitor with impedance, $Z_L = -j\ 300\Omega$. Repeat part (a) for this case.

A transmission line that is matched at each end produces no reflections and thus delivers maximum power to the load.

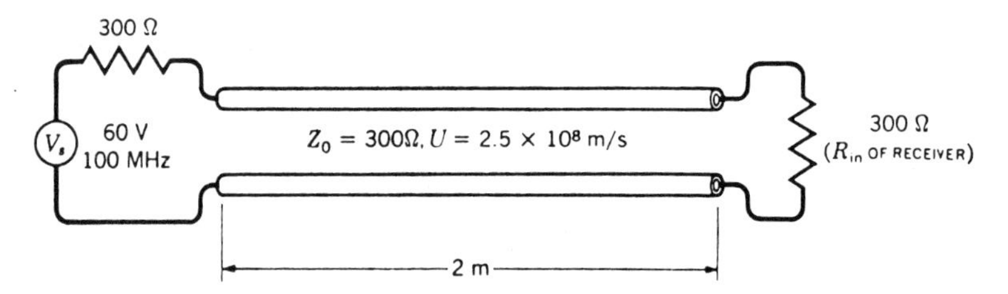

Solution: (a) Since the load impedance is equal to the characteristic impedance, the line is matched; the reflection coefficient is zero, and the standing-wave ratio is unity. For the given velocity and frequency the wavelength on the line is 2.5 m and the phase constant is 0.8π rad/m; the attenuation constant is zero. The electrical length of the line $\beta\ell$ is $2 \times 0.8\pi$, or 1.6π rad. This length may also be expressed as 288°, or 0.8 wavelength.

The input impedance offered to the voltage source is 300 Ω, and since the internal impedance of the source is 300 Ω, the voltage at the input to the line is half of 60 V, or 30 V. The source is matched to the line and delivers the maximum available power to the line. Since there is no reflection and no attenuation, the voltage at the load is also 30-V crest amplitude, but it is delayed in phase by 1.6π rad. Thus

$$V_{in} = 30 \cos 2\pi\ 10^8 t$$

whereas

$$V_L = 30 \cos (\ 2\pi 10^8 t - 1.6\pi)$$

The input current is

$$I_{in} = \frac{V_{in}}{300} = 0.1 \cos 2\pi\ 10^8 t$$

while the load current is

$$I_L = 0.1 \cos (2\pi 10^8 t - 1.6\pi)$$

The average power delivered to the input of the line by the source is equal to that delivered to the load by the line,

$$P_{in} = P_L = \tfrac{1}{2} \times 30 \times 0.1 = 1.5 \text{ W}$$

(b) The load impedance is now 150 Ω, the reflection coefficient is

$$\Gamma = \frac{Z_L - Z_0}{Z_L + Z_0}$$

$$\Gamma = \frac{150 - 300}{150 + 300} = -\frac{1}{3}$$

and the standing-wave ratio on the line is

$$s = \frac{1 + |\Gamma|}{1 - |\Gamma|}$$

$$s = \frac{1 + \frac{1}{3}}{1 - \frac{1}{3}} = 2$$

The input impedance is no longer 300 Ω, but is now

$$Z_{in} = Z_0 \frac{Z_L + jZ_0 \tan \beta\ell}{Z_0 + jZ_L \tan \beta\ell} = 300 \frac{150 + j300 \tan 228°}{300 + j150 \tan 288°}$$

$$= 510\underline{/-23.8°} = 466 - j206 \quad \Omega$$

which is a capacitive impedance. The current flowing through the source is thus

$$I_{s,in} = \frac{60}{766 - j206} = 0.0756\underline{/15°} \quad \text{A}$$

and the power supplied to the line by the source is

$$P_{in} = \frac{1}{2} \times (0.0756)^2 \times 466 = 1.333 \text{ W}$$

Since there are no losses in the line, 1.333 W must also be delivered to the load. Note that this is less than the 1.50 W which were delivered to a matched load; moreover, this power must divide equally between two receivers, and thus each receiver now receives only 0.667 W. Since the input impedance of each receiver is 300Ω, the voltage across the receiver is easily found as

$$0.667 = \frac{1}{2}\frac{|V_{Ls}|^2}{300}$$

$$|V_{Ls}| = 20 \text{ V}$$

in comparison with the 30 V obtained across the single load.

(c) No average power can be delivered to the load. As a consequence, the reflection coefficient is

$$\Gamma = \frac{-j300 - 300}{-j300 + 300} = -j1 = 1\underline{/-90°}$$

and the reflected wave is equal in amplitude to the incident wave. Hence the standing-wave ratio is

$$s = \frac{1 + |-j1|}{1 - |-j1|} = \infty$$

and the input impedance is a pure reactance,

$$Z_{in} = 300\,\frac{-j300 + j300 \tan 288°}{300 + j(-j300) \tan 288°} = j153 \text{ ohms.}$$

to which no average power can be delivered.

CHAPTER 14

WAVE GUIDES AND ANTENNAS

CUTOFF FREQUENCIES FOR TE AND TM MODES

• PROBLEM 14-1

A square wave guide of side 2b and a circular wave guide of diameter 2d have the same cross-sections. If the cut-off wave length of the square guide is 4b, find the cutoff wave length of the circular wave guide.

Solution: The square and circular guides have the same cross-sectional areas

$$4b^2 = \pi\left(\frac{2d}{2}\right)^2$$

$$d = 1.13b$$

Since $\lambda_{oc} = 2b$ for the square wave guide, is obtained as the cutoff wave length for the circular guide

$$\lambda_{oc} = \frac{2d}{1.13} = 1.77d$$

• PROBLEM 14-2

Find the cutoff frequency of the first higher-order mode for an air-dielectric 50-ohm coaxial line whose inner conductor has a radius of 2 mm.

Solution: The characteristic impedance of the line specifies the outer radius b through the equation

$$Z_0 = 60 \ln \frac{b}{a}$$

Therefore

$$b = ae^{Z_0/60} = (2 \times 10^{-3})e^{5/6} = 4.6 \text{ mm}$$

The approximate cutoff frequency of the first of the higher-order modes may now be computed from

$$f_c \simeq \frac{\text{free velocity}}{2 \times \text{width}} = \frac{v}{\pi(a+b)}$$

as

$$f_c \simeq \frac{3 \times 10^8}{\pi(2 + 4.6)10^{-3}} = 14.5 \text{ GHz.}$$

• PROBLEM 14-3

Determine the lowest four cutoff frequencies referred to the cutoff frequency of the dominant mode for three cases of rectangular waveguide dimensions: (i) b/a = 1, (ii) b/a = 1/2, and (iii) b/a = 1/3. Given a = 3 cm, then find the propagating mode(s) for f = 9000 MHz for each of the three cases.

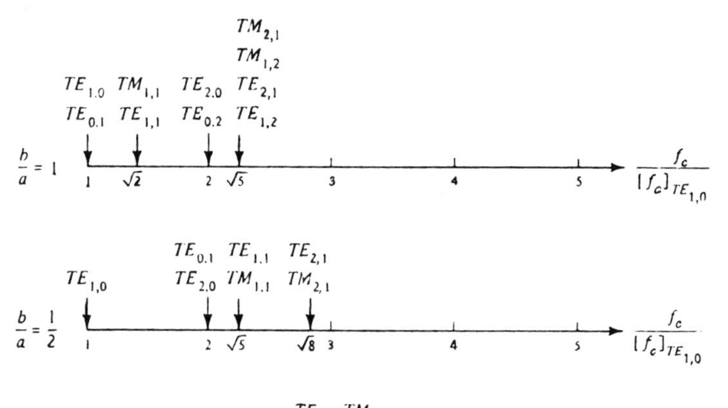

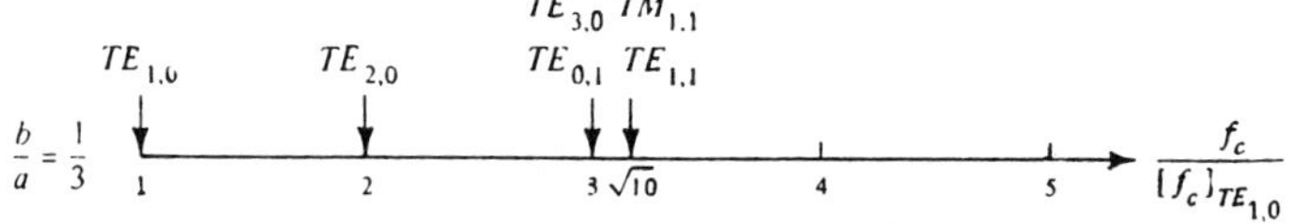

Lowest four cutoff frequencies referred to the cutoff frequency of the dominant mode for three cases of rectangular waveguide dimensions.

Solution: The expression for the cutoff wavelength for a $TE_{m,n}$ mode where m = 0, 1, 2, 3, ... and n = 0, 1, 2, 3, ... but not both m and n equal to zero and for a

$TM_{m,n}$ mode where m = 1, 2, 3, ... and n = 1, 2, 3, ... is given by

$$\lambda_c = \frac{1}{\sqrt{(m/2a)^2 + (n/2b)^2}}$$

The corresponding expression for the cutoff frequency is

$$f_c = \frac{v_p}{\lambda_c} = \frac{1}{\sqrt{\mu\varepsilon}}\sqrt{\left(\frac{m}{2a}\right)^2 + \left(\frac{n}{2b}\right)^2}$$

$$= \frac{1}{2a\sqrt{\mu\varepsilon}}\sqrt{m^2 + \left(n\,\frac{a}{b}\right)^2}$$

The cutoff frequency of the dominant mode $TE_{1,0}$ is $1/2a\sqrt{\mu\varepsilon}$. Hence

$$\frac{f_c}{[f_c]_{TE_{1,0}}} = \sqrt{m^2 + \left(n\,\frac{a}{b}\right)^2}$$

By assigning different pairs of values for m and n, the lowest four values of $f_c/[f_c]_{TE_{1,0}}$ can be computed for each of the three specified values of b/a. These computed values and the corresponding modes are shown in the figure.

For a = 3 cm, and assuming free space for the dielectric in the waveguide,

$$[f_c]_{TE_{1,0}} = \frac{1}{2a\sqrt{\mu\varepsilon}} = \frac{3 \times 10^8}{2 \times 0.03} = 5000 \text{ MHz}$$

Hence for a signal of frequency f = 9000 MHz, all the modes for which $f_c/[f_c]_{TE_{1,0}}$ is less than 1.8 propagate. From the figure, these are

$TE_{1,0}$, $TE_{0,1}$, $TM_{1,1}$, $TE_{1,1}$ for b/a = 1

$TE_{1,0}$ for b/a = 1/2

$TE_{1,0}$ for b/a = 1/3

It can be seen from the figure that for b/a $\leq$ 1/2, the

second lowest cutoff frequency that corresponds to that of the $TE_{2,0}$ mode is twice the cutoff frequency of the dominant mode $TE_{1,0}$. For this reason, the dimension b of a rectangular waveguide is generally chosen to be less than or equal to a/2 in order to achieve single-mode transmission over a complete octave (factor of two) range of frequencies.

• PROBLEM 14-4

For a rectangular wave guide with width and height a and b respectively, find the cutoff wave length and frequency for TE_{11} and TE_{10} modes. What is the significance of TE_{10} mode?

Solution: The cutoff wave length and frequency for $TE_{m,n}$ mode of propagation are given by

$$(\lambda_c)_{m,n} = \frac{2}{\sqrt{(m/a)^2 + (n/b)^2}}$$

$$(f_c)_{m,n} = \frac{1}{2\sqrt{\mu\varepsilon}} \sqrt{(m/a)^2 + (n/b)^2}$$

So for TE_{11}:

$$(\lambda_c)_{1,1} = \frac{2}{\sqrt{(1/a)^2 + (1/b)^2}} = \frac{2ab}{\sqrt{a^2 + b^2}}$$

$$(f_c)_{1,1} = \frac{1}{2\sqrt{\mu\varepsilon}} \sqrt{(1/a)^2 + (1/b)^2}$$

$$= \frac{\sqrt{a^2 + b^2}}{2ab\sqrt{\mu\varepsilon}}$$

For $TE_{1,0}$:

$$(\lambda_c)_{1,0} = \frac{2}{\sqrt{(1/a)^2}} = 2a$$

$$(f_c)_{1,0} = \frac{1}{2\sqrt{\mu\varepsilon}} \sqrt{(1/a)^2} = \frac{1}{2a\sqrt{\mu\varepsilon}}$$

Here for TE_{10} from $\lambda_c = 2a$, the cutoff frequency is that frequency for which the width of the guide is a half wave length. It does not depend at all on the other dimensions. This TE_{10} mode is frequently referred to as the dominant mode of the rectangular guide.

• PROBLEM 14-5

An air filled hollow rectangular wave guide has cross-sectional dimensions $z_1 = 10$ cm, $y_1 = 6$ cm. Find the cutoff frequencies for the following modes: TM_{10}, TM_{20}, TM_{11}, TM_{21}.

Solution: For a TM wave to pass through a hollow rectangular wave guide, m and n should be integers and not equal to zero. So TM_{10} and TM_{20} mode of waves will not pass through.

For a TM_{mn} mode wave the cutoff frequency is given by

$$f_c = \frac{1}{2\sqrt{\mu\varepsilon}} \sqrt{\left(\frac{m}{z_1}\right)^2 + \left(\frac{n}{y_1}\right)^2}$$

In this problem since the hollow cylinder is filled with air

$$\varepsilon = \frac{10^{-9}}{36\pi} \text{ farad/meter,}$$

$$\mu = 4\pi \times 10^{-7} \text{ henry/meter}$$

∴ For TM_{11}

$$f_c = \frac{1}{2 \times \sqrt{4\pi \times 10^{-7} \times \frac{10^{-9}}{36\pi}}} \times \sqrt{\left(\frac{1}{10}\right)^2 + \left(\frac{1}{6}\right)^2} = 2920 \text{ MHz}$$

For TM_{21}

$$f_c = \frac{1}{2 \times \sqrt{4\pi \times 10^{-7} \times \frac{10^{-9}}{36\pi}}} \times \sqrt{\left(\frac{2}{10}\right)^2 + \left(\frac{1}{6}\right)^2} = 3907 \text{ MHz}$$

• PROBLEM 14-6

The dimensions of a rectangular cavity resonator with air dielectric are a = 4 cm, b = 2 cm, and d = 4 cm as shown in the figure. Determine the three lowest frequencies of oscillation and specify the mode(s) of oscillation, transverse with resepct to the z direction for each frequency.

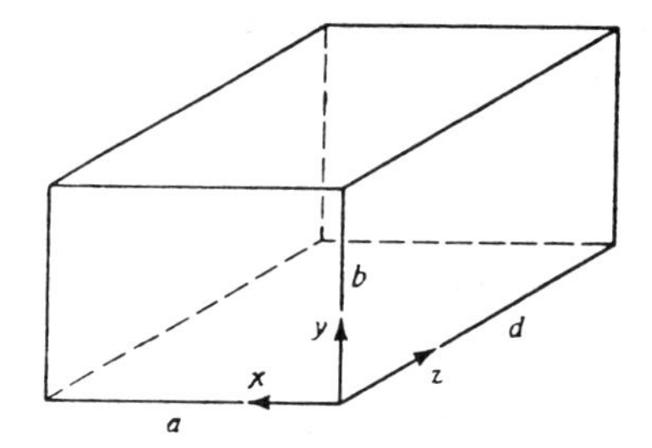

A rectangular cavity resonator.

Solution: By substituting $\mu = \mu_0$, $\varepsilon = \varepsilon_0$, and the given dimensions for a, b, and d in

$$f_{osc} = \frac{v_p}{\lambda_{osc}} = \frac{1}{\sqrt{\mu\varepsilon}}\sqrt{\left(\frac{m}{2a}\right)^2 + \left(\frac{n}{2b}\right)^2 + \left(\frac{\ell}{2d}\right)^2} ,$$

the following is obtained:

$$f_{osc} = 3 \times 10^8 \sqrt{\left(\frac{m}{0.08}\right)^2 + \left(\frac{n}{0.04}\right)^2 + \left(\frac{\ell}{0.08}\right)^2}$$

$$= 3750 \sqrt{m^2 + 4n^2 + \ell^2} \text{ MHz}$$

By assigning combinations of integer values for m, n, and ℓ and recalling that both m and n must be nonzero for TM modes, the three lowest frequencies of oscillation are found to be

$3750 \times \sqrt{2} = 5303$ MHz for $TE_{1,0,1}$ mode

$3750 \times \sqrt{5} = 8385$ MHz for $TE_{0,1,1}$, $TE_{2,0,1}$, and $TE_{1,0,2}$ modes

$3750 \times \sqrt{6} = 9186$ MHz for $TE_{1,1,1}$ and $TM_{1,1,1}$ modes

• PROBLEM 14-7

Assume the spacing a, between the plates of a parallel-plate waveguide to be 5 cm. Investigate the propagating $TE_{m,0}$ modes for f = 10,000 MHz.

Solution: The cutoff wavelengths for $TE_{m,0}$ modes are given by

$$\lambda_c = \frac{2a}{m} = \frac{10}{m}\ \text{cm} = \frac{0.1}{m}\ \text{m}$$

The result is independent of the dielectric between the plates. If the medium between the plates is free space, then the cutoff frequencies for the $TE_{m,0}$ modes are

$$f_c = \frac{3 \times 10^8}{\lambda_c} = \frac{3 \times 10^8}{0.1/m} = 3m \times 10^9\ \text{Hz}$$

For f = 10,000 MHz = 10^{10} Hz, the propagating modes are $TE_{1,0}$($f_c = 3 \times 10^9$ Hz), $TE_{2,0}$($f_c = 6 \times 10^9$ Hz), and $TE_{3,0}$($f_c = 9 \times 10^9$ Hz).

For each propagating mode, θ, λ_g, and v_{pz} can be found by using

$$\cos\theta = \lambda/\lambda_c = f_c/f,\ \lambda_g = 2\pi/\beta\sin\theta = \lambda/\sqrt{1 - (\lambda/\lambda_c)^2}$$

$$= \lambda/\sqrt{1 - (f_c/f)^2}\ ,$$

and

$$v_{pz} = \frac{\omega}{p\sin\theta} = \frac{v_p}{\sin\theta} = \frac{v_p}{\sqrt{1 - (\lambda/\lambda_c)^2}} = \frac{v_p}{\sqrt{1 - (f_c/f)^2}}$$

respectively. Values of these quantities are listed in the following table:

Mode	λ_c, cm	f_c, MHz	θ, deg	λ_g, cm	v_{pz}, m/s
$TE_{1,0}$	10	3000	72.54	3.145	3.145×10^8
$TE_{2,0}$	5	6000	53.13	3.75	3.75×10^8
$TE_{3,0}$	3.33	9000	25.84	6.882	6.882×10^8

• PROBLEM 14-8

Design a rectangular waveguide which, at 10 GHz, will operate in the TE_{10} mode with 25 percent safety factor ($f \geq 1.25f_c$) when the interior of the guide is filled with air. It is required that the mode with the next higher cutoff will operate at 25 percent below its cut-off frequency.

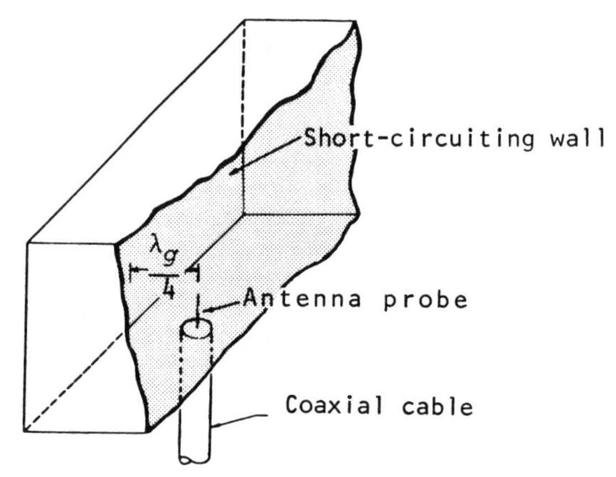

Excitation method for TE_{10} mode.

Solution: For the TE_{10} mode, $f_c = c/2a$. Therefore the lower bound for the dimension a is

$$\frac{3 \times 10^{10}}{2 \times 10^{10}} = 1.5 \text{ cm}$$

The upper bound is found from the condition

$$f \geq 1.25\ f_c.$$

Thus

$$a \leq \frac{1.25c}{2f} = \frac{1.25 \times 3 \times 10^{10}}{2 \times 10^{10}} = 1.875 \text{ cm}$$

so that a must be chosen such that $1.5 \leq a \leq 1.875$.

The wave with the next higher cutoff frequency is the TE_{01} mode. By analogy, $f'_c = c/2b$, and

$$f \le 0.75f'_c = 0.75 \frac{c}{2b}$$

from which the choice of b must be such that

$$b \le \frac{0.75c}{2f} = \frac{0.75 \times 3 \times 10^{10}}{2 \times 10^{10}} = 1.125 \text{ cm}$$

If the upper bounds a = 1.875 cm, b = 1.125 cm is chosen, the cutoff frequency of the next higher mode, the TE_{11} mode, will be 15.52 GHz, well beyond the operating frequency. Therefore wave propagation will be confined to the TE_{10} mode.

One method of exciting the TE_{10} mode is shown in the given figure.

• PROBLEM 14-9

Determine all the modes that can be transmitted in a rectangular waveguide with a cross-section of 0.04 × 0.07 m. Assume the guide excited at $3(10^9)$ cps and at $6(10^9)$ cps.

Solution: For air dielectric

$$f_{co} = \frac{c}{2\pi}\sqrt{\left(\frac{m\pi}{a}\right)^2 + \left(\frac{n\pi}{b}\right)^2}$$

No modes can be propagated whose cut-off frequencies are below f_{co} for the guide dimensions. Therefore calling $3(10^9) = f_{co}$,

$$3(10^9) > \frac{3(10^{10})}{2\pi}\sqrt{\left(\frac{m\pi}{7}\right)^2 + \left(\frac{n\pi}{4}\right)^2}$$

or

$$\sqrt{\left(\frac{m\pi}{7}\right)^2 + \left(\frac{n\pi}{4}\right)^2} < \frac{(2\pi)3(10^9)}{3(10^{10})}$$

Square both sides

$$\frac{m^2}{49} + \frac{n^2}{16} < (0.2)^2 = 0.04$$

and

$$16\ m^2 + 49\ n^2 < 49(16)(0.04)$$

or

$$16\ m^2 + 49\ n^2 < 31.4$$

If m = 1, n = 0, propagation can take place. Therefore TE_{10} is the only possible mode.

TM waves require TM_{11} minimum, therefore no TM waves can be supported at 3000 mcps.

At $6(10^9)$ cps, $16\ m^2 + 49\ n^2 < 125.5$.

m	*n*
1	0
0	1
1	1
2	0
2	1

Therefore, the possible modes are TE_{10}, TE_{01}, TE_{11}, TE_{20}

and TM_{11}, TM_{20}, TM_{21}

• PROBLEM 14-10

A waveguide with rectangular cross-section has dimensions of 2 cm high by 4 cm wide as shown in figure 1. Assuming TE operation, good conductivity of the walls, and the dielectric to be air, compute: (a) mode, (b) λ_g, (c) v_f, (d) v_g, if the guide is operated at a frequency 20% higher than f_{co}.

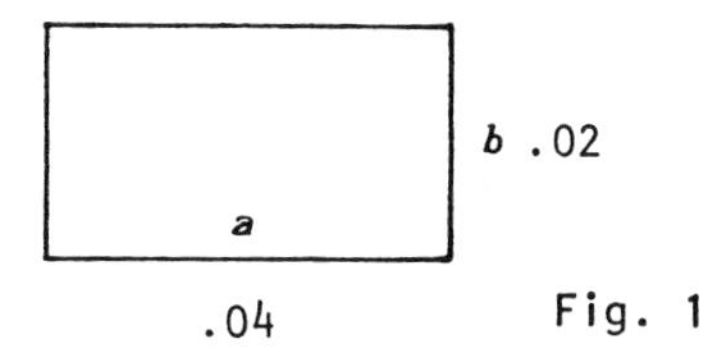

Fig. 1

Solution: From

$$\lambda_{co} = \frac{2}{\sqrt{\left(\frac{m}{a}\right)^2 + \left(\frac{n}{b}\right)^2}}$$

where for the TE_{10} mode

$m = 1$ and $n = 0$:

(a)

$$\lambda_{co} = \frac{2}{\sqrt{(1/a)^2}} = 2a = .08 \text{ m}$$

$$f = \frac{c}{\lambda_{co}} = \frac{3(10^8)}{.08} = 3750 \text{ mcps}$$

20% above 3750 = 4500 mcps

$$\lambda = \frac{3(10^8)}{4.5(10^9)} = 6.67 \text{ cm}$$

For the TE_{01} mode

$$\lambda_{co} = \frac{2}{\sqrt{(1/b)^2}} = 2b = .04 \text{ m} = 4 \text{ cm}$$

$f_{co} = 7500$ mcps, so operation must be confined to TE_{10} mode.

(b)

$$\lambda_g = \frac{\lambda}{\sqrt{1 - (\lambda/\lambda_{co})^2}}$$

$$= \frac{6.67}{\sqrt{1 - (6.67/8)^2}} = 12 \text{ cm}$$

(c)

$$\lambda_g = 2a \tan\theta; \ \tan\theta = \frac{0.12}{0.08} = 1.5$$

$$\therefore \qquad \theta = 56.4°$$

$$\cos\theta = .553$$

Therefore using v_f = phase velocity = $\frac{v}{\cos\theta}$ (see Fig. 2)

$$v_f = \frac{3(10^8)}{.553} = 5.40(10^8) \text{ meters per sec.}$$

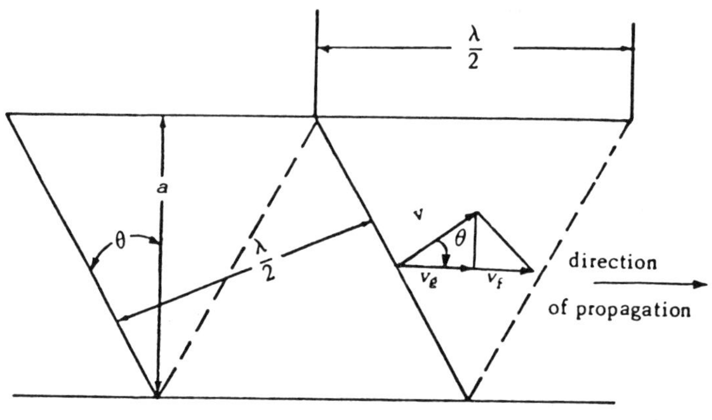

Fig. 2

(d)

$$v_g = v \cos\theta = 3(10^8).553 = 1.66(10^8) \text{ meters}$$

per sec.

• PROBLEM 14-11

The cross-sectional dimensions of an air filled hollow rectangular conducting wave guide, are $y_1 = 6$ cm and $z_1 = 10$ cm.

(a) Find the cutoff frequencies for the following modes: TEM, TE_{10}, TE_{20}, TE_{01}, TE_{11}, and TE_{21}.

(b) For TE_{10} mode, find the phase velocity at a frequency of twice its cutoff frequency.

Solution: It is known that

$$\frac{1}{\sqrt{\mu\varepsilon}} = \frac{1}{\sqrt{4\pi \times 10^{-7} \times \frac{1}{36\pi} \times 10^{-9}}} = 3 \times 10^8$$

$\therefore$ TEM will not Pass

For $TE_{m,n}$ mode, the cutoff frequency is given by

$$f_c = \frac{1}{2\sqrt{\mu\varepsilon}}\sqrt{\left(\frac{m}{z_1}\right)^2 + \left(\frac{n}{y_1}\right)^2}$$

$\therefore$ For TE_{10}

$$f_c = \frac{3 \times 10^8}{2}\sqrt{\left(\frac{1}{10}\right)^2} \qquad = 1500 \text{ MHz}$$

For TE_{20}

$$f_c = \frac{3 \times 10^8}{2}\sqrt{\left(\frac{2}{10}\right)^2} \qquad = 3000 \text{ MHz}$$

For TE_{01}

$$f_c = \frac{3 \times 10^8}{2}\sqrt{\left(\frac{1}{6}\right)^2} \qquad = 2500 \text{ MHz}$$

For TE_{11}

$$f_c = \frac{3 \times 10^8}{2}\sqrt{\left(\frac{1}{6}\right)^2 + \left(\frac{1}{10}\right)^2} \qquad = 2920 \text{ MHz}$$

For TE_{21}

$$f_c = \frac{3 \times 10^8}{2}\sqrt{\left(\frac{1}{6}\right)^2 + \left(\frac{2}{10}\right)^2} \qquad = 3905 \text{ MHz}$$

(b) The Phase Velocity is given by

$$v = \frac{v_0}{\sqrt{1 - \frac{\lambda_0}{\lambda_c}^2}}$$

where

$$v_0 = \frac{1}{\sqrt{\mu\varepsilon}} = 3 \times 10^8 \text{ m/sec}$$

λ_0 = Wave length of the traveling wave

λ_c = Cutoff wave length

Since $f_0 = 2f_c$ (given)

$\therefore \quad \lambda_0 = 1/2\ \lambda_c$

$\therefore$ Eq. (1) becomes

$$v = \frac{3 \times 10^8}{\sqrt{1 - \left(\frac{1}{2}\right)^2}} = 3.46 \times 10^8 \text{ meters/sec.}$$

• PROBLEM 14-12

Consider the parallel-plate waveguide discontinuity shown in the figure. For $TE_{1,0}$ waves of frequency f = 5000 MHz incident on the junction from the free space side, find the reflection and transmission coefficients.

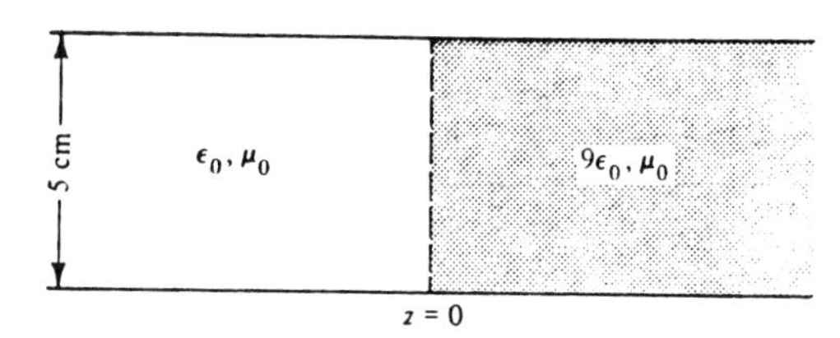

Solution: For the $TE_{1,0}$ mode, λ_c = 2a = 10 cm, independent of the dielectric. For f = 5000 MHz,

λ_1 = wavelength on the free space side

$$= \frac{3 \times 10^8}{5 \times 10^9} = 6 \text{ cm}$$

λ_2 = wavelength on the dielectric side

$$= \frac{3 \times 10^8}{\sqrt{9} \times 5 \times 10^9} = \frac{6}{3} = 2 \text{ cm}$$

Since $\lambda < \lambda_c$ in both sections, $TE_{1,0}$ mode propagates in both sections. Thus the intrinsic impedance of the free space side and the intrinsic impedance of the dielectric side respectively are

$$\eta_{g_1} = \frac{\eta_1}{\sqrt{1 - (\lambda_1/\lambda_c)^2}} = \frac{120\pi}{\sqrt{1 - (6/10)^2}} = 471.24 \text{ ohms}$$

$$\eta_{g_2} = \frac{\eta_2}{\sqrt{1 - (\lambda_2/\lambda_c)^2}} = \frac{120\pi/\sqrt{9}}{\sqrt{1 - (2/10)^2}} = \frac{40\pi}{\sqrt{1 - 0.04}}$$

$$= 128.25 \text{ ohms}$$

The reflection coefficient is given by

$$\Gamma = \frac{\eta_{g_2} - \eta_{g_1}}{\eta_{g_2} + \eta_{g_1}} = \frac{128.25 - 471.24}{128.25 + 471.24} = -0.572$$

$$\tau = 1 + \Gamma = 1 - 0.572 = 0.428$$

For $f = 4000$ MHz, the result would be $\Gamma = -0.629$ and $\tau = 0.371$.

PROPAGATION AND ATTENUATION CONSTANTS

• PROBLEM 14-13

Assume a parallel plane waveguide consisting of two sheets of good conductor separated by 0.1 m and operated in the TE_1 mode. Find the propagation constant p at frequencies of 10^8, 10^9, and 10^{10} (10G) cycles per second. Does propagation take place? If the same guide is excited in the TE_2 mode, repeat the calculations.

Solution: By

$$p = \sqrt{\left(\frac{m\pi}{a}\right)^2 - \omega^2\mu\varepsilon}$$

with $m = 1$

$$p = \sqrt{\left(\frac{\pi}{a}\right)^2 - 4\pi^2 f^2 \mu\varepsilon} = \pi\sqrt{\left(\frac{1}{a}\right)^2 - \mu\varepsilon(2f)^2}$$

$$= \pi\sqrt{\left(\frac{1}{a}\right)^2 - \left(\frac{2f}{c}\right)^2}$$

(a) At $f = 10^8$ cps

$$p = \pi\sqrt{\left(\frac{1}{0.1}\right)^2 - \left(\frac{2(10^8)}{3(10^8)}\right)^2} = \pi\sqrt{100 - 4/9}$$

$= 31.3$ nepers/meter

$\bar{p}$ in this case is real, which means attenuation and no phase shift. Therefore no propagation at 10^8 cps.

(b) At $f = 10^9$ cps

$$p = \pi\sqrt{100 - \left[\frac{20}{3}\left(\frac{10^8}{10^8}\right)^2\right]} = \pi\sqrt{100 - 44.4}$$

$= 23.5$ nepers/meter

$\bar{p}$ is again real, therefore no propagation.

(c) At $f = 10^{10}$ cps

$$p = \pi\sqrt{100 - \left(\frac{200}{3}\right)^2} = \pi\sqrt{-44.5(10^2) + 100}$$

$= j207$ radians/meter

Here p is imaginary, therefore $\beta = 207$ radians/meter and propagation takes place.

Now for the TE_2 mode, (here m = 2),

$$p = \sqrt{\left(\frac{2\pi}{a}\right)^2 - (2\pi f)^2\mu\varepsilon} = 2\pi\sqrt{\left(\frac{1}{a}\right)^2 - \left(\frac{f}{c}\right)^2}$$

(a)

$$p = 2\pi\sqrt{100 - \left[\frac{10^8}{3(10^8)}\right]^2} = 62.8 \text{ nepers/meter}$$

Here p is real, therefore no propagation.

(b)

$$p = 2\pi\sqrt{100 - \left[\frac{10^9}{3(10^8)}\right]^2} = 2\pi\sqrt{100 - 11.1}$$

$= 59.1$ nepers/meter

This is still below cut-off, therefore no propagation.

(c)

$$p = 2\pi\sqrt{100 - \left[\frac{10^{10}}{3(10^8)}\right]^2} = 2\pi\sqrt{100 - (33.3)^2}$$

$$= j199 \text{ radians/meter}$$

Propagation takes place since quantity is imaginary.

• PROBLEM 14-14

A rectangular waveguide, 5 cm by 2.5 cm, has an air dielectric. At an excitation frequency of 4000 megacycles, determine the propagation constants of the 10, 01, 20, 11, 02, and 55 modes. Assume perfect conductors.

Solution: From

$$\gamma = j\omega\sqrt{\mu\varepsilon}\,\sqrt{1 - (f_c/f)^2}$$

the propagation constant γ is determined to be

$$\gamma = j83.8\sqrt{1 - (f_c/4000)^2}$$

with f_c in megacycles. From

$$f_c = \left(\frac{1}{2}\,/\sqrt{\mu\varepsilon}\right)\sqrt{(m/a)^2 + (n/b)^2}$$

$$f_c = 3000\sqrt{m^2 + 4n^2} \text{ megacycles}$$

The cut-off frequencies of the 10, 01, 20, 11, 02, and 55 modes are found to be 3000, 6000, 6000, 6720, 12,000, and 33,500 megacycles, respectively. As the excitation frequency is 4000 megacycles, only the 10 mode will propagate without attenuation.

Using the calculated cut-off frequencies, the propagation constants are determined to be j55.4 for the 10 mode, 93.8 for the 01 and 20 modes, 113 for the 11 mode, 237 for the 02 mode, and 697 for the 55 mode. The propagation constants for all modes except the 10 mode are positive real numbers. When a wave propagates through a distance z_1 such that $\alpha z_1 = 1$, the amplitude decreases to 1/e, or 36.8%, of its original value. For the 01 and 20 modes the distance z_1 is 1.07cm, and for the 55 mode the distance is 0.14cm. All modes except the TE_{10} wave are very rapidly attenuated.

• PROBLEM 14-15

A common air filled rectangular waveguide has the interior dimensions a = 0.9 in. and b = 0.4 in. (2.29cm x 1.02cm), the so-called X-band guide. (a) Find the cutoff frequency of the lowest-order, nontrivial TM mode. (b) At a source frequency that is twice the cutoff value of (a), determine the propagation constant for this mode. Also obtain the wavelength in the guide, the phase velocity, and the intrinsic wave impedance. (c) Repeat (b), assuming $f = f_c/2$.

Solution: (a) From

$$f_{c,mn} = \frac{1}{2\pi\sqrt{\mu\varepsilon}}\left[\left(\frac{m\pi}{a}\right)^2 + \left(\frac{m\pi}{b}\right)^2\right]^{\frac{1}{2}},$$

it is seen that the cutoff frequency has its lowest value for TM modes if m = 1 and n = 1, the smallest integers producing nontrivial fields. Thus for the TM_{11} mode, the given dimensions yield

$$f_{c,11} = \frac{1}{2\pi\sqrt{\mu_0\varepsilon_0}}\left[\left(\frac{\pi}{a}\right)^2 + \left(\frac{\pi}{b}\right)^2\right]^{\frac{1}{2}}$$

$$= \frac{3\times 10^8}{2}\left[\frac{1}{(0.0229)^2} + \frac{1}{(0.0102)^2}\right]^{\frac{1}{2}}$$

$$= 16{,}100 \text{ MHz}$$

The TM_{11} mode will thus propagate in this guide if its frequency exceeds 16,100 MHz. Below this frequency, the mode is evanescent.

(b) At f = 32,200 MHz,

$$\gamma_{m,n} = j\beta_{mn} = j\omega\sqrt{\mu\varepsilon}\sqrt{1 - \left(\frac{f_{c,mn}}{f}\right)^2} \qquad f > f_{c,mn}$$

yields

$$\beta_{11} = \beta^{(0)}\sqrt{1 - \left(\frac{f_{c,11}}{f}\right)^2} = \frac{2\pi(32.2\times 10^9)}{3\times 10^8}\sqrt{1 - \left(\frac{1}{2}\right)^2}$$

$$= 585 \text{ rad/m}$$

In free space, $\lambda^{(0)} = c/f = 3 \times 10^8/32.2 \times 10^9$

$$= 0.933 \text{ cm},$$

so, $$\lambda_{11} = \frac{\lambda^{(0)}}{\sqrt{1 - \left(\frac{f_{c,11}}{f}\right)^2}} = \frac{0.933}{\sqrt{1 - \left(\frac{1}{2}\right)^2}} = \frac{0.933}{0.866} = 1.076 \text{ cm}$$

while the phase velocity and intrinsic wave impedance from

$$v_{p,mn} = \frac{\omega}{\beta_{mn}} = \frac{v_p(0)}{\sqrt{1 - \left(\frac{f_{c,mn}}{f}\right)^2}}$$

where

$$v_p(0) = (\mu\varepsilon)^{\frac{1}{2}}$$

and $$\hat{\eta}_{TM,\,mn} = \frac{j\beta_{mn}}{j\omega f} = \eta(0) \sqrt{1 - \left(\frac{f_{c,m,n}}{f}\right)^2}$$

where $$\eta(0) = \sqrt{\frac{\mu}{\varepsilon}}$$

are

$$v_{p,11} = \frac{3 \times 10^8}{0.866} = 3.46 \times 10^8 \text{ m/sec}$$

$$\hat{\eta}_{TM,11} = 377(0.866) = 326\ \Omega$$

(c) At f = 8.05 GHz,

$$\alpha_{mn} \equiv \omega\sqrt{\mu\varepsilon} \sqrt{\left(\frac{f_{c,mn}}{f}\right)^2 - 1} \qquad f < f_{c,mn}$$

obtains

$$\alpha_{11} = \beta^{(0)} \sqrt{\left(\frac{f_{c,11}}{f}\right)^2 - 1} = \frac{2\pi(8.05 \times 10^9)}{3 \times 10^8} \sqrt{(2)^2 - 1}$$

$$= 291 \text{ Np/m}$$

Below $f_{c,11}$, wavelength and phase velocity are undefined, in view of evanescent fields, but below cutoff

$$\hat{\eta}_{TM,11} = -j\eta^{(0)} \sqrt{\left(\frac{f_{c,11}}{f}\right)^2 - 1} = -j377\sqrt{(2)^2 - 1}$$

$$= -j653\ \Omega$$

• PROBLEM 14-16

Find the attentuation per meter along a wave guide for an applied wave length $\lambda_0 = 2$ meters, if the cutoff wave length of the guide is $\lambda_{oc} = 20$ cm.

Solution: The attenuation per meter along the guide for an applied wave length λ_0 of 1 meter is given by

$$\alpha = \frac{2\pi}{\lambda_0} \sqrt{(\lambda_0/\lambda_{oc})^2 - 1} \quad \text{nepers/meter}$$

Therefore

$$\alpha = \frac{2\pi}{1} \sqrt{(2/0.2)^2 - 1}$$

$$= 20\pi \text{ nepers/meter}$$

or

$$20\pi \times 8.68 = 545 \text{ db/meter}$$

• PROBLEM 14-17

Consider an air-filled rectangular waveguide whose a and b dimensions are 0.9 and 0.4 in., respectively. Find (a) the cutoff frequency of the dominant mode and also that of the next higher-order mode, and (b) the attentuation factor associated with the next higher-order mode at a frequency midway between the two cutoff frequencies.

Solution: The cutoff frequency for the dominant mode can be computed from

$$\omega_c = \frac{1}{\sqrt{\mu\varepsilon}} \sqrt{\frac{m\pi}{a^2} + \frac{n\pi}{b^2}}$$

Letting m = 1 and n = 0 gives for a result

$$f_c = \frac{1}{2\pi\sqrt{\mu\varepsilon}}\sqrt{\left(\frac{\pi}{a}\right)^2 + (0)^2} = \frac{3 \times 10^8}{(2)(0.9)(2.54 \times 10^{-2})}$$

$$= 6.56 \text{ GHz}$$

After trying a few values of m and n, it can be seen that the next higher-order mode is the TE_{20} mode. Its cutoff frequency is

$$f_c = \frac{1}{2\pi\sqrt{\mu\varepsilon}}\sqrt{\left(\frac{2\pi}{a}\right)^2 + (0)^2} = 13.1 \text{ GHz} \tag{1}$$

Thus, within the frequency range from 6.56 to 13.1 GHz, only the TE_{10} mode can propagate in the ordinary sense of the word.

Looking next at the propagation constant for the TE_{20} mode, one finds from

$$\gamma = j\beta = j\sqrt{\omega^2\mu\varepsilon - \omega_c{}^2\mu\varepsilon} \tag{2}$$

and

$$\beta = \omega\sqrt{\mu\varepsilon}\sqrt{1 - \left(\frac{\omega_c}{\omega}\right)^2} \tag{3}$$

that it is (for a mid-range frequency of 9.84 GHz)

$$\gamma = j\beta = j\omega\sqrt{\mu\varepsilon}\sqrt{1 - \left(\frac{\omega_c}{\omega}\right)^2}$$

$$= j\,\frac{(2\pi)(9.84 \times 10^9)}{3 \times 10^8}\sqrt{1 - \left(\frac{13.1}{9.84}\right)^2} = 181 \text{ Np/m} .$$

This will be recognized as a sizeable attenuation factor —over 1500 dB/m. Furthermore, looking at the higher-order modes beyond the TE_{20} one, one finds that they are attenuated at even a higher rate. This can be seen by rearranging Eq. (2) and (3).

as

$$\gamma = j\omega\sqrt{\mu\varepsilon}\sqrt{1 - \left(\frac{\omega_c}{\omega}\right)^2} = \omega_c\sqrt{\mu\varepsilon}\sqrt{1 - \left(\frac{\omega}{\omega_c}\right)^2}$$

and noting that each successive higher-order mode has a higher cutoff frequency.

• PROBLEM 14-18

Find the attenuation constant for a 300-MHz TEM wave in an infinite parallel-plane transmission line with a spacing between planes of 100 mm. The planes or walls are made of copper, and the medium between the planes is air.

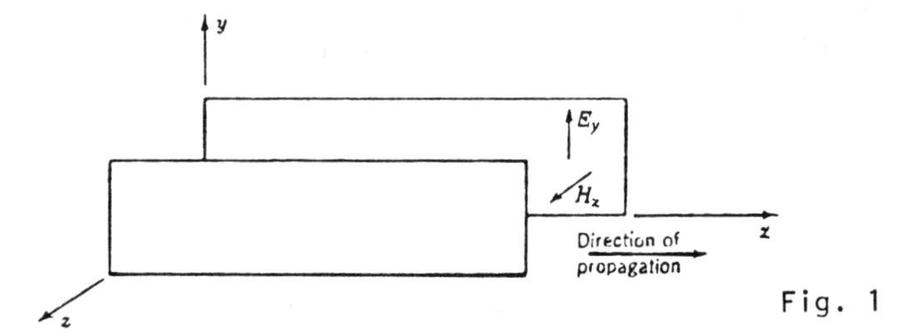

Fig. 1

Transmission system of two conducting planes parallel to the xy plane. The planes are assumed to be infinite in extent (infinite parallel-plane transmission line).

Solution: For a TEM wave the transverse impedance equals the intrinsic impedance: so

$$\alpha = \frac{\text{Re } Z_c \int |H_{t_1}|^2 d\ell}{2 \text{ Re } Z_{yz} \iint |H_{t_1}|^2 \, ds}$$

becomes

$$\alpha = \frac{2 \text{ Re } Z_c \int_0^{y_1} |H_{t_1}|^2 dy}{2 \text{ Re } Z_d \int_0^{y_1}\int_0^{z_1} |H_{t_2}|^2 \, dzdy} \tag{1}$$

where

y_1 = arbitrary distance along conducting wall (see figure 1)

z_1 = spacing between walls, m

Re Z_c = real part of intrinsic impedance of conducting walls, Ω

Re Z_d = real part of intrinsic impedance dielectric medium between walls (Z_d is entirely real for lossless medium

The integral with H_{t_1} involves power lost in one wall of the line. The total power loss in both walls is twice this; hence the factor 2 in the numerator. For a TEM wave H is everywhere parallel to the walls and normal to the direction of propagation, so that both H_{t_1} and H_{t_2} are perpendicular to the page instead of as suggested in Fig. 2. It follows that

$$|H_{t_1}| = |H_{t_2}| = \text{a constant.}$$

Attenuation α in decibels and nepers per meter as a function of frequency for various modes in an air-filled infinite-parallel-plane waveguide of copper with 100-mm spacing between planes.

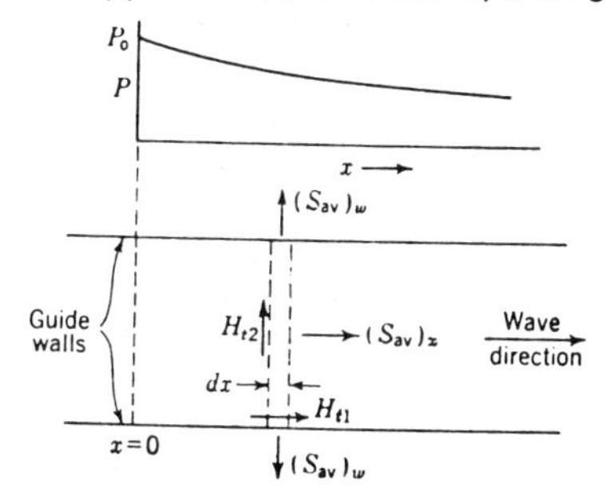

Fig. 2

Power lost in walls of waveguide results in attenuation.

Hence, (1) reduces to

$$\alpha = \frac{\text{Re } Z_c y_1}{\text{Re } Z_d \; y_1 z_1} = \frac{\text{Re } Z_c}{z_1 \text{ Re } Z_d}$$

For copper at 300 MHz, Re Z_c = 4.55 mΩ, while for air Re Z_d = 376.7 Ω. Therefore

$$\alpha = 1.2 \times 10^{-4} \text{ Np m}^{-1}$$

or $\alpha = 1.04 \times 10^{-3} \text{ dB m}^{-1}$

Thus, the attenuation amounts to about 1 dB km^{-1}.

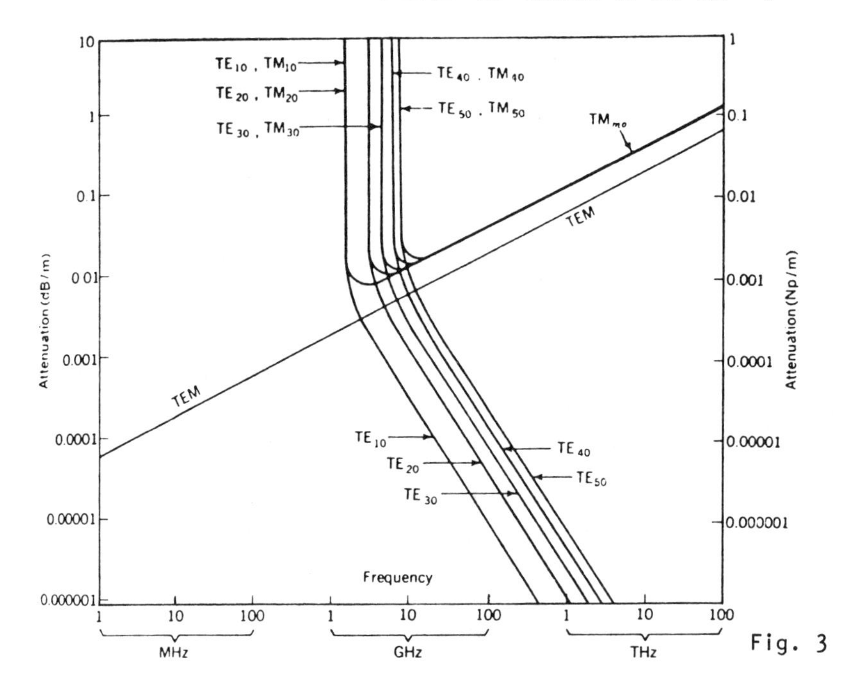

Fig. 3

The attenuation in decibels per meter (and in nepers per meter) as a function of frequency for a number of modes in an air-filled infinite-parallel-plane guide of copper is shown in Fig. 3. The spacing between planes is 100 mm. Note that there is no cutoff for the TEM mode and attenuation decreases for this mode with decreasing frequency. However, at 100 GHz the TE_{10} mode has the lowest attenuation although the TE_{20} and higher-order modes can also be transmitted. The TE_{20} and higher-order modes all have higher attenuation than the TE_{10} mode.

• PROBLEM 14-19

Find the attenuation constant for the TE_{nr} mode in a circular cylindrical waveguide of radius a at frequencies above cutoff. Determine this attenuation in decibels per meter for the TE_{01} and TE_{11} modes in a 50.8-mm-diameter copper pipe.

Solution: For a TE_{nr} mode, the following equations:

$$H_\phi = \frac{n\gamma H_0}{k^2 r} \sin(n\phi)\, J_n(kr) e^{(j\omega t - \gamma z)} \qquad (1)$$

$$H_z = H_0 \cos(n\phi)\, J_n(kr) e^{(j\omega t - \gamma z)} \qquad (2)$$

yield

$$|H_{t1}|^2 = |H_z(r = a)|^2 + |H_\phi(r = a)|^2$$

$$= |H_0|^2 J_n^2(ka)\left[\cos^2(n\phi) + \frac{|\gamma|^2 n^2}{k^4 a^2}\sin^2(n\phi)\right] \qquad (3)$$

Also, the following equation:

$$H_r = \frac{-\gamma H_0}{k^2}\cos(n\phi)\frac{d}{dr}\frac{J_n(kr)}{} e^{(j\omega t - \gamma z)}$$

combined with equation (1) yields

$$|H_{t2}|^2 = |H_r|^2 + |H_\phi|^2$$

$$= \frac{|H_0|^2|\gamma|^2}{k^4}\left\{\cos^2(n\phi)\left[\frac{dJ_n(kr)}{dr}\right]^2\right.$$

$$\left. + \frac{n^2}{r^2}\sin^2(n\phi)\ J_n^2(kr)\right\} \qquad (4)$$

Substituting (3) and (4) into

$$H_z = H_o \cos(n\phi)\ J_n(kr)$$

gives

$$\alpha = \frac{k^4 \mathrm{Re}Z_c\ J_n^2(ka)}{2\beta^2 \mathrm{Re}Z_{r\phi}} \times \frac{A}{B}$$

where

$$A = \int_0^{2\pi}\cos^2 n\phi(ad\phi) + \frac{\beta^2 n^2}{k^4 a^2}\int_0^{2\pi}\sin^2 n\phi(ad\phi)$$

$$B = \int_0^{2\pi}\int_0^{a}\left[\cos^2 n\phi\left(\frac{dJ_n(kr)}{dr}\right)^2 + \frac{n^2}{r^2}\sin^2 n\phi \times J_n^2(kr)\right] r\, dr d\phi$$

Note that

$$\int_0^{2\pi}\cos^2 n\phi d\phi = \begin{cases} 2\pi & n = 0 \\ \pi & n \neq 0 \end{cases}$$

and

$$\int_0^{2\pi} \sin^2 n\phi d\phi = \begin{cases} 0 & n = 0 \\ \pi & n \neq 0 \end{cases}$$

Thus for TE_{nr} modes ($n = 0,1,2,3...$)

$$\alpha = \frac{k^4 \mathrm{Re}Z_c a}{2\beta^2 \mathrm{Re}Z_{r\phi}} \frac{J_n^2(ka)\pi[1 + (\beta^2 n^2/k^4 a^2)]}{\pi \int_0^a \left\{ \left[\frac{dJ_n(kr)}{dr}\right]^2 + \frac{n^2}{r^2} J_n^2(kr) \right\} r\, dr} \tag{5}$$

Evaluating (5) and referring to Tables 1 and 2 for $\mathrm{Re}Z_{r\phi}$ gives

$$\alpha = \frac{\mathrm{Re}Z_c}{a\mathrm{Re}Z_d\sqrt{1 - (f_c/f)^2}} \left[\left(\frac{f_c}{f}\right)^2 + \frac{n^2}{(k'_{nr})^2 - n^2} \right] \quad (\mathrm{Npm}^{-1}) \tag{6}$$

where $\mathrm{Re}Z_c = \sqrt{\pi\mu/\sigma}\sqrt{f} = 2.63 \times 10^{-7}\sqrt{f}(\Omega)$ for copper and $\mathrm{Re}Z_d = 377\Omega$ for an air-filled waveguide. The value of K'_{nr} can be found from Table 3.

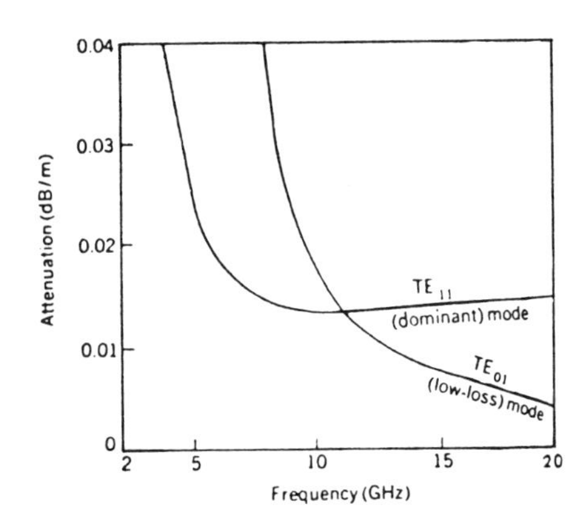

Fig. 1

Attenuation α in decibels per meter as a function of frequency for TE_{01} (low-loss) and TE_{11} (dominant) modes in a circular copper waveguide of 50.8 mm inside Diameter.

Evaluating (6) for TE_{01} and TE_{11} modes yields the attenuation curves of Fig. 1. Note that for the TE_{01} (low-loss) mode the attenuation decreases monotonically as the frequency increases. This trend also occurs for any TE_{0r} mode. The TE_{11} (dominant) mode has a minimum at $f/f_c = 3.1$. A minimum at this f/f_c ratio also occurs with all TE_{nr} modes for which $n = 0$.

Parameters of rectangular and cylindrical waveguides are summarized in Table 2.

Table 1 RELATIONS FOR TE_{mn} MODES IN HOLLOW RECTANGULAR WAVEGUIDES

Name of relation	Relation
Cutoff frequency	$f_c = \dfrac{1}{2\sqrt{\mu\varepsilon}}\sqrt{\left(\dfrac{n}{y_1}\right)^2 + \left(\dfrac{m}{\varepsilon_1}\right)^2}$ (Hz)
Cutoff wavelength	$\lambda_{oc} = \dfrac{2}{\sqrt{(n/y_1)^2 + (m/z_1)^2}}$ (m)
Wavelength in guide	$\lambda_g = \dfrac{\lambda_0}{\sqrt{1 - (\lambda_0/\lambda_{oc})^2}}$ (m)
Phase velocity	$v_p = \dfrac{v_0}{\sqrt{1 - (n\lambda_0/2y_1)^2 - (m\lambda_0/2z_1)^2}}$ $= \dfrac{v_0}{\sqrt{1 - (\lambda_0/\lambda_{oc})^2}}$ $= \dfrac{v_0}{\sqrt{1 - (f_c/f)^2}}$ (ms^{-1}) where $v_0 = 1/\sqrt{\mu\varepsilon}$
Transverse-wave impedance	$Z_{yz} = \dfrac{Z_d}{\sqrt{1 - (n\lambda_0/2y_1)^2 - (m\lambda_0/2z_1)^2}}$ $= \dfrac{Z_d}{\sqrt{1 - (\lambda_0/\lambda_{oc})^2}}$ $= \dfrac{Z_d}{\sqrt{1 - (f_c/f)^2}}$ (Ω) where $Z_d = \sqrt{\mu/\varepsilon}$

Table 2

RECTANGULAR AND CIRCULAR CYLINDRICAL WAVEGUIDE PARAMETERS

Symbol	Name	Equation (rectangular or cylindrical)	
α	Attenuation constant:		
	Below cutoff frequency	$\alpha = \frac{2\pi}{\lambda 0}\sqrt{\left(\frac{\lambda_0}{\lambda_{oc}}\right)^2 - 1}$	(Npm^{-1})
	Above cutoff frequency	$\alpha = \frac{ReZ_c \int \lvert H_{t1}\rvert^2 d\ell}{2ReZ_{yz}\iint \lvert H_{t2}\rvert^2 ds}$	(Npm^{-1})
β	Phase Constant	$\beta = \sqrt{\left(\frac{2\pi}{\lambda_o}\right)^2 - k2}$	$(radm^{-1})$
v_p	Phase Velocity	$v_p = \frac{\omega}{\beta} = \frac{r_0}{\sqrt{1 - (\lambda_o/\lambda_{oc})^2}}$	(ms^{-1})
Z_{ys}, $Z_{r\phi}$	Transverse-wave impedance	Rectangular $Z_{ys} = \frac{Z_d}{\sqrt{1 - (\lambda_o/\lambda_{oc})^2}}$ Cylindrical $Z_{r\phi} = \frac{Z_d}{\sqrt{1 - (\lambda_0/\lambda_{oc})}}$ (Ω)	TE_{nr} mode
λ_{oc}	Cutoff wavelength	$\lambda_{oc} = \frac{2}{\sqrt{(n/y_1)^2 + (m/z_1)^2}}$ TE_{mn} or TM_{mn} mode $\lambda_{oc} = \frac{2\pi r_0}{k'_{nr}}$ or $\frac{2\pi r_0}{k_{nr}}$ (m) TE_{nr} mode (k'_{nr}), TM_{nr} mode (k_{nr})	

γ = propagation constant = $\alpha + j\beta$.

Table 3 CYLINDRICAL WAVEGUIDE MODES

Mode designation	Eigenvalues		Cutoff wavelength, λ_{oc}
	k'_{nr}	k_{nr}	
TM_{01}		2.405	$2.61r_0$
TE_{01} (low loss)	3.832		$1.64r_0$
TM_{02}		5.520	$1.14r_0$
TE_{02}	7.016		$0.89r_0$
TE_{11} (dominant)	1.840		$3.41r_0$
TM_{11}		3.832	$1.64r_0$
TE_{12}	5.330		$1.18r_0$
TM_{12}		7.016	$0.89r_0$
TE_{21}	3.054		$2.06r_0$
TM_{21}		5.135	$1.22r_0$
TE_{22}	6.706		$0.94r_0$
TE_{31}	4.201		$1.49r_0$
TM_{31}		6.379	$0.98r_0$
TE_{41}	5.318		$1.18r_0$
TM_{41}		7.588	$0.83r_0$
TE_{51}	6.416		$0.98r_0$

FIELD COMPONENTS IN WAVE-GUIDES

• PROBLEM 14-20

Consider a rectangular waveguide which is 2 cm by 1 cm (a and b dimensions, respectively) and is air-filled. Find (a) the mode with the lowest cutoff frequency; (b) the phase velocity, phase-shift constant, wavelength within the guide, and characteristic wave impedance; (c) the explicit expressions in phasor form for the six field components.

Solution: (a) The cutoff frequency for any particular mode is given by

$$\omega_c = \frac{1}{\sqrt{\mu\varepsilon}}\sqrt{\frac{m\pi}{a^2} + \frac{n\pi}{b^2}} \ .$$

Note that one cannot let either m or n be zero without having a trivial situation for which all components of E and H are zero. Therefore, the lowest possible value for ω_c is obtained when both m and n are unity. This corresponds to the TM_{11} mode, and its cutoff frequency is

$$\omega_c \text{ (for } TM_{11} \text{ mode)} = \frac{1}{\sqrt{\mu\varepsilon}} \sqrt{\left(\frac{\pi}{a}\right)^2 + \left(\frac{\pi}{b}\right)^2}$$

$$= 105.5 \times 10^9 \text{ rad/s}$$

or

$$f_c = \frac{\omega_c}{2\pi} = 16.8 \text{ GHz}$$

(b) For a system operating in the TM_{11} mode at a frequency of $1.5\omega_c$, the phase velocity, phase-shift constant, wavelength within the guide, and characteristic wave impedance can be found from

$$\gamma = j\beta = j\sqrt{\omega^2\mu\varepsilon - \omega^2{}_c\mu\varepsilon} \tag{1}$$

$$\beta = \omega\sqrt{\mu\varepsilon}\sqrt{1 - \left(\frac{\omega_c}{\omega}\right)^2} \tag{2}$$

$$v_p = \frac{\omega}{\beta} = \frac{V}{\sqrt{1 - (\omega_c/\omega)^2}} \tag{3}$$

$$\lambda_g = \frac{V_p}{f} = \frac{\lambda}{\sqrt{1 - (\omega_c/\omega)^2}} \tag{4}$$

$$Z_0(T_m) = \frac{\gamma}{j\omega\varepsilon} = \eta\sqrt{1 - \left(\frac{\omega_c}{\omega}\right)^2} \tag{5}$$

They are

$$v_p = (3 \times 10^8) \frac{1}{\sqrt{1 - (1/1.5)^2}} = 4.03 \times 10^8 \text{ m/s}$$

$$\beta = \frac{(1.5)(105.5 \times 10^9)}{4.03 \times 10^8} = 392 \text{ rad/m}$$

$$\lambda_g = \frac{4.03 \times 10^8}{(1.5)(16.8 \times 10^9)} = 0.016 \text{ m}$$

$$Z_{0(TM_{11})} = 377\sqrt{1 - (1/1.5)^2} = 280 \text{ ohms}$$

(c) Finally, the explicit expressions (in phasor form) for all six field components are obtained by directly substituting into the Eqs.

$$E'_z = E_0 \sin \frac{m\pi x}{a} \sin \frac{n\pi y}{b}$$

$$H'_z = 0$$

$$H'_x = j \frac{\omega\varepsilon n\pi E_0}{bk_c^2} \sin \frac{m\pi x}{a} \cos \frac{n\pi y}{b}$$

$$H'_y = -j \frac{\omega\varepsilon m\pi E_0}{ak_c^2} \cos \frac{m\pi x}{a} \sin \frac{n\pi y}{b}$$

$$E'_x = -\frac{\gamma m\pi E_0}{ak_c^2} \cos \frac{m\pi x}{a} \sin \frac{n\pi y}{b}$$

$$E'_y = -\frac{\gamma n\pi E_0}{bk_c^2} \sin \frac{m\pi x}{a} \cos \frac{n\pi y}{b}$$

where

$$k_c^2 = \gamma^2 + \omega^2\mu\varepsilon = \left(\frac{m\pi}{a}\right)^2 + \left(\frac{n\pi}{b}\right)^2$$

and multiplying by $e^{-j\beta z}$ in order to reinsert the z-dependence. Assuming the peak value of E_z to be E_0, the result is

$$E_z = E_0 \sin (50\pi x) \sin (100\pi y)\, e^{-j\beta_z}$$

$$H_z = 0$$

$$H_x = j(3.57 \times 10^{-3})E_0 \sin (50\pi x) \cos (100\pi y)\, e^{-j\beta_z}$$

$$H_y = -j(1.79 \times 10^{-3})E_0 \cos(50\pi x) \sin(100\pi y) e^{-j\beta_z}$$

$$E_x = -j0.50E_0 \cos(50\pi x) \sin(100\pi y) e^{-j\beta_z}$$

$$E_y = -j1.0E_0 \sin(50\pi x) \cos(100\pi y) e^{-j\beta_z}$$

• PROBLEM 14-21

In the free-space dielectric of a rectangular waveguide the axial magnetic field H_z is zero, and the axial electric field E_z is

$$E_z = 10{,}000 \sin 20\pi x \sin 40\pi y\, e^{-0.02z}$$

$$\times \cos(18\pi \times 10^9 t - 40\pi z + 30°)$$

Find H_x as a function of the space coordinates and time, and evaluate H_x at $(x, y, z, t) = (0.01, 0.01, 2, 0)$.

Solution: Letting $\gamma = 0.02 + j40\pi$ and $\omega = 18\pi \times 10^9$, E_z in complex-exponential form is

$$E_z = 10{,}000 \sin 20\pi x \sin 40\pi y\, e^{j30°} e^{j\omega t - \gamma z}$$

with the real part understood. Thus $\bar{E}_z$ is

$$\bar{E}_z = 10{,}000 \sin 20\pi x \sin 40\pi y\, e^{j30°}$$

Differentiating with respect to y gives

$$\partial\bar{E}_z/\partial y = 400{,}000\pi \sin 20\pi x \cos 40\pi y\, e^{j30°}$$

$$H_x = \frac{\sigma + j\omega\varepsilon}{k_c^{\,2}} \frac{\partial \varepsilon_z}{\partial y} - \frac{\gamma}{k_c} \frac{\partial H_z}{\partial x}$$

will be used, with σ and $\bar{H}_z$ set equal to zero. The partial derivative of $\bar{E}_z$ with respect to y has been determined. This must be multiplied by $j\omega\varepsilon_0/k_c^{\,2}$. As

$\gamma^2 = -1600\pi^2$ and $\omega^2\mu_0\varepsilon_0 = 3600\pi^2$, the sum $\gamma^2 + \omega^2\mu_0\varepsilon_0$ gives $k_c^{\,2} = 2000\pi^2$. Multiplying

$\partial\bar{E}_z/\partial y$ by $j\omega\varepsilon_0/k_c{}^2$

gives

$$\bar{H}_x = j31.8 \sin 20\pi x \cos 40\pi y \; e^{j30°}$$

or

$$\bar{H}_x = 31.8 \sin 20\pi x \cos 40\pi y \; \underline{/120°}$$

Inserting the exponential of $j\omega t - \gamma z$ and dropping the imaginary part of the result

$$H_x = 31.8 \sin 20\pi x \cos 40\pi y \; e^{-0.02z} \cos(\omega t - 40\pi z + 120°)$$

At (x, y, z, t) = (0.01, 0.01, 2, 0) the expression for H_x becomes

$$H_x = 31.8 \sin 0.2\pi \cos 0.4\pi \; e^{-0.04} \cos(-80\pi + 120°)$$

Evaluation gives $H_x = -2.77$ amperes per meter.

• PROBLEM 14-22

If $\bar{E}_z = 1000 \sin 20\pi x \sin 25\pi y$ and $\bar{H}_z = j5 \cos 20\pi x \cos 25\pi y$, determine E_x as a function of the space coordinates and time. The frequency is 5000 megacycles, and the dielectric has a dielectric constant of 4 and a conductivity of 0.0001. Assume perfect conductors and cosinusoidal time variations.

Solution: It is evident that $m\pi/a = 20\pi$ and $n\pi/b = 25\pi$. Therefore, $k_c{}^2$ is $(20\pi)^2 + (25\pi)^2$, or 10,120. As $\omega = \pi 10^{10}$, $\mu = \mu_0$, $\varepsilon = 4\varepsilon_0$, and $\sigma = 0.0001$,

$$\gamma = \sqrt{-\omega^2\mu\varepsilon + (m\pi/a)^2 + (n\pi/b)^2 + j\omega\mu_\sigma}$$

becomes

$$\gamma = \sqrt{-33{,}880 + j3.96} = 0.011 + j184 \cdot$$

$\bar{E}_x$ has one component due to $\bar{E}_z$ and a second component due to $\bar{H}_z$. These components are

$$\bar{E}'_x = -\frac{\gamma}{k_c^2}\frac{\partial \bar{E}_z}{\partial x} \qquad \bar{E}''_x = -\frac{j\omega\mu}{k_c^2}\frac{\partial \bar{H}_z}{\partial y}$$

with $\bar{E}_x = \bar{E}'_x + \bar{E}''_x$. The constant $-\gamma/k_c^2 = -j0.0182$. Multiplying $\partial \bar{E}_z/\partial x$ by this constant gives

$$\bar{E}'_x = -j1140 \cos 20\pi x \sin 25\pi y$$

The constant $-j\omega\mu/k_c^2 = -j3.90$. Multiplying $\partial \bar{H}_z/\partial y$ by this gives

$$\bar{E}''_x = -1530 \cos 20\pi x \sin 25\pi y$$

The sum of $\bar{E}'_x$ and $\bar{E}''_x$ is

$$\bar{E}_x = -(1530 + j1140) \cos 20\pi x \sin 25\pi y$$

Multiplying $\bar{E}_x$ by $\exp(j\omega t - \gamma z)$, with $\gamma = 0.011 + j184$, and selecting the real part of the resulting expression, obtain

$$E_x = 1910 \cos 20\pi x \sin 25\pi y\, e^{-0.011z} \times \cos(\omega t - 184z - 143.3°)$$

The amplitude of E_x is $1910 \cos 20\pi x \sin 25\pi y\, e^{-0.011z}$, and the phase is $(\omega t - 184z - 143.3°)$. The amplitude decreases exponentially with z. The wave velocity ω/β is 1.71×10^8 m/sec. This is greater than the normal velocity of light in this dielectric, this velocity being 1.5×10^8 m/sec.

• PROBLEM 14-23

For the uniform plane wave defined by

$$\bar{E} = [\bar{a}_x + E_y\bar{a}_y + (2 + j5)\bar{a}_z]e^{-j2.3(-0.6x+0.8y)+j\omega t}$$

$$\bar{H} = (H_x\bar{a}_x + H_y\bar{a}_y + H_z\bar{a}_z)e^{-j2.3(-0.6x+0.8y)+j\omega t}$$

where H_x, H_y, H_z are all independent of x, y, and z, determine

(a) The components E_y, H_x, H_y, H_z, assuming that $\mu = \mu_0$ and $\varepsilon = \varepsilon_0$

(b) The frequency and corresponding wavelength

(c) The equation of the surface of constant phase

(d) The state of polarization of the wave

Solution: (a) By comparison with

$$\bar{E} = (E_1 e^{-\gamma n \cdot r} + E_2 e^{\gamma n \cdot r}) e^{j\omega t}$$

it is clear that

$$\gamma \bar{n} \cdot \bar{r} = j2.3(-0.6x + 0.8y) = j2.3\bar{n} \cdot \bar{r}$$

where

$$\bar{n} = -0.6\bar{a}_x + 0.8\bar{a}_y$$

and since $\gamma = \alpha + j\beta$, $\alpha = 0$, and $\beta = 2.3$. Since $\bar{n} \cdot \bar{E} = 0$, it follows that

$$-0.6 + 0.8E_y = 0$$

and

$$E_y = 0.75$$

Now $-j\gamma = \beta = \omega\sqrt{\mu_0 \varepsilon_0}$; hence

$$\bar{H} = \frac{\beta}{\omega\mu_0} \bar{n} \times \bar{E} = \frac{1}{\eta} \bar{n} \times \bar{E} = \frac{1}{377} \begin{vmatrix} \bar{a}_x & \bar{a}_y & \bar{a}_z \\ -0.6 & 0.8 & 0 \\ 1 & 0.75 & 2 + j5 \end{vmatrix}$$

so that

$$H_x = \frac{0.8(2 + j5)}{377} = (4.24 + j10.6) \times 10^{-3}$$

$$H_y = \frac{0.6(2 + j5)}{377} = (3.18 + j7.95) \times 10^{-3}$$

$$H_z = -\frac{0.6 \times 0.75 + 0.8}{377} = 3.31 \times 10^{-3}$$

(b) The wavelength is determined as

$$\lambda = \frac{2\pi}{\beta} = 2.73 \text{ m}$$

and the frequency as

$$f = \frac{c}{\lambda} = \frac{3 \times 10^8}{2.73} = 1.1 \times 10^8 \text{ Hz}$$

(c) The equation of the surface of constant phase

$$\bar{n} \cdot \bar{r} = -0.6x + 0.8y = \text{constant}$$

obviously defines a plane which is parallel to the z axis.

(d) Separating the real and imaginary parts of the magnitude of E,

$$\bar{E}_r = \bar{a}_x + 0.75\bar{a}_y + 2\bar{a}_z$$

$$\bar{E}_i = 5\bar{a}_z$$

Clearly, $\bar{E}_r$ and $\bar{E}_i$ are neither collinear nor mutually orthogonal in space. Therefore the polarization of the wave cannot possibly be linear nor circular; it must be elliptical.

• PROBLEM 14-24

(a) For an air-filled waveguide whose inside dimensions are 3.0(in.) × 1.5(in.), find the cut-off frequency and cut-off wavelength for the TE_{10} mode. (b) For the same waveguide, calculate the fields in terms of an arbitrary constant H_{z0} for f = 2.45 (GHz) operating in the TE mode. Refer to the figure shown.

(b) The waveguide has a λ_g of 0.2(m) as determined by slotted line measurements. Find the frequency, phase velocity, and group velocity.

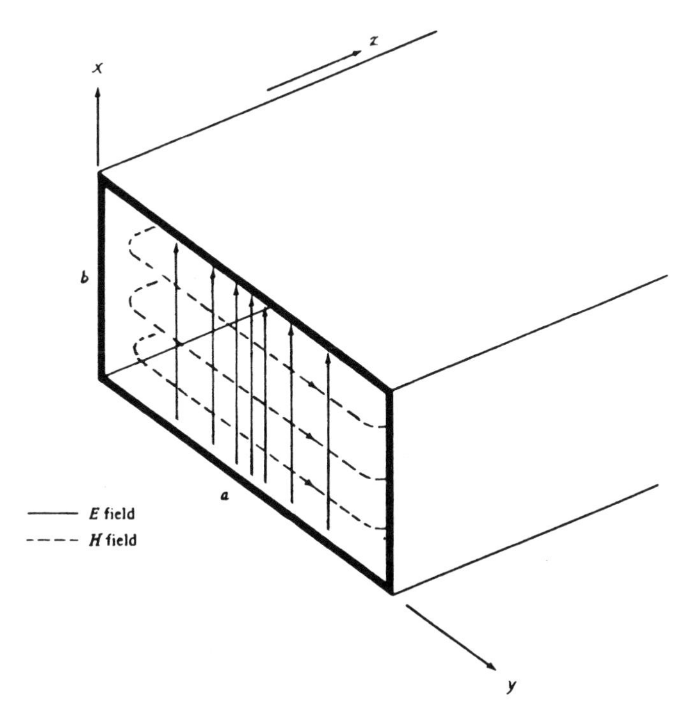

Illustration of the transverse field distribution in a waveguide cross section operating in the TE_{10} mode.

Solution: For the TE_{10} mode, m = 1, n = 0.

(a) $a = 3.0(\text{in.}) = 0.0764(\text{m})$

$b = 1.5(\text{in.}) = 0.0381\ (\text{m})$

$\lambda_c = 2a = 0.1524(\text{m})$

$$f_c = \frac{c}{\lambda_c} = \frac{3 \times 10^8\ (\text{m/sec})}{0.1524(\text{m})} = 1.97(\text{GHz})$$

(b) Since $f > f_c$, and $\gamma = j\beta$,

$$\beta = \sqrt{\omega^2\mu\varepsilon - \frac{\pi^2}{a^2}}$$

$$= \sqrt{\left(\frac{2\pi(2.45 \times 10^9)}{3 \times 10^8}\right)^2 - \left(\frac{\pi}{0.0762}\right)^2}$$

$$= 30.5 \ (m^{-1})$$

$$k^2 = k_x^2 + k_y^2 = 0 + \left(\frac{\pi}{a}\right)^2 = 1700$$

$$\frac{m\pi}{a} = \frac{(1)\pi}{0.0762} = 41.23$$

$$\frac{\gamma\pi}{a} = j\frac{\beta\pi}{a} = j(30.5)(41.23) = j1258$$

$$j\omega\mu\frac{\pi}{a} = j2\pi(2.45 \times 10^9)(4\pi \times 10^{-7})(41.23)$$

$$= j7.97 \times 10^5$$

Then

$$H_z = H_{z_0} \cos 41.23ye^{-j30.5z}$$

$$H_y = \left(\frac{j1258H_{z0} \sin 41.23y}{1700}\right) e^{-j30.5z}$$

$$= j0.74H_{z0} \sin 41.23ye^{-j30.5z}$$

$$H_x = 0$$

$$E_x = \left(\frac{j7.97 \times 10^5 H_{z0} \sin 41.23y}{1700}\right) e^{-j30.5z}$$

$$= j468H_{z0} \sin 41.23ye^{-30.5z}$$

$$E_y = 0$$

(c) $$\lambda_g = 0.2 = \frac{2\pi}{\beta}$$

$$\beta = \frac{2\pi}{0.2} = 10\pi = \sqrt{\omega^2\mu\varepsilon - \frac{\pi^2}{a^2}}$$

$$\omega^2\mu\varepsilon = \beta^2 + \frac{\pi^2}{a^2} = \left(\frac{2\pi}{\lambda_g}\right)^2 + \left(\frac{\pi}{a}\right)^2$$

$$f = \frac{1}{2\sqrt{\mu\varepsilon}} \sqrt{\left(\frac{2}{\lambda_g}\right)^2 + \left(\frac{1}{a}\right)^2}$$

$$= 1.5 \times 10^8 \sqrt{\left(\frac{2}{0.2}\right)^2 + \left(\frac{1}{0.076}\right)^2}$$

$$= 2.47 \text{ (GHz)}$$

$$v_p = \frac{U}{\sqrt{1 - \left(\frac{f_c}{f}\right)^2}} = \frac{3 \times 10^8}{\sqrt{1 - \left(\frac{1.97}{2.47}\right)^2}}$$

$$= 4.97 \times 10^8 \text{ (ms}^{-1}\text{)}$$

$$v_g = \frac{U^2}{v_p} = \frac{(3 \times 10^8)^2}{4.97 \times 10^8} = 1.81 \times 10^8 \text{ (ms}^{-1}\text{)}$$

• PROBLEM 14-25

A TE_{10} wave is excited at 4000 megacycles at z = 0 in a rectangular waveguide with transverse dimensions 5 by 2.5 centimeters. The dielectric has a conductivity of 2×10^{-6} mho/m, the copper has a conductivity of 5.8×10^7, and the permittivity and permeability are everywhere the same as for free space. Assume that the excitation is such that B_1 in

$$\bar{E}_y = -jB_1 \sin \pi x/a$$

equals j1000. Calculate the phase and group velocities, the power flow, and the copper and dielectric losses per unit length. Also, determine the fields as functions of the space coordinates and time.

Solution: As the cut-off frequency is $1/(2a\sqrt{\mu\varepsilon})$, or 3000 mc, the excitation frequency is above cut-off. The wavelength is 7.5 cm, and the cut-off wavelength is 10 cm.

The phase and group velocities, calcuated from

$$v_p = \frac{1/\sqrt{\mu\varepsilon}}{\sqrt{1 - (\lambda/2a)^2}}$$

and

$$v_g = (1/\sqrt{\mu\varepsilon})\ \sqrt{1 - (\lambda/2a)^2}$$

are 4.54×10^8 m/sec and 1.99×10^8 m/sec, respectively. The phase constant β is determined from

$$\gamma = (j2\pi/\lambda)\sqrt{1 - (\lambda/2a)^2}$$

to be 55.4 radians per meter.

In order to calculate the attenuation constant find the surface resistivity of copper. From

$$R_s = \sqrt{\pi f \mu/\sigma}$$

R_s = 0.0165 ohm. Utilizing

$$\alpha_c = \frac{R_s}{\sqrt{\mu/\varepsilon}}\left[\frac{1}{b} + \frac{2}{a}\left(\frac{f_c}{f}\right)^2\right] \Big/ \sqrt{1 - \left(\frac{f_c}{f}\right)^2}$$

and

$$\alpha_d = \frac{\sigma\sqrt{\mu/\varepsilon}}{\sqrt{1 - (f_c/f)^2}}$$

α_c = 0.00413 and α_d = 0.00057. Therefore, α = 0.0047 neper/m.

From $B_1 = 2f\mu aB = j1000$, $B = j1.99$.

$$P = \frac{ab}{4}\sqrt{\frac{\mu}{\varepsilon}}\ \left(\frac{f}{f_c}\right)^2\sqrt{1 - \left(\frac{f_c}{f}\right)^2}\ B^2$$

can now be employed to determine the power flow. The calculated power P is 548 milliwatts, and because of attenuation, this must be multiplied by the exponential of $-2\alpha z$. Therefore

$$P = 548e^{-0.0094z} \text{ millwatts}$$

At z = 0, the power is 548 mw, and at z = 10 meters the power is 498 mw. The difference is due to copper and dielectric losses.

The copper loss P_c per unit length equals $2\alpha_c P$, or

$4.53e^{-0.0094z}$ mw/m. This could also have been determined from

$$P_c = R_s[b + \frac{1}{2}a(f/f_c)^2]B^2$$

The dielectric loss P_d per unit length is $2\alpha_d P$, or $0.62e^{-0.0094z}$ mw/m. At z = 0 the copper loss is 4.53 mw/m, the dielectric loss is 0.62 mw/m, and the total loss is 5.15 mw/m. At z = 10 the total loss is 4.7 mw/m.

The constant $B_2 = (2a/\lambda)\sqrt{1 - (\lambda/2a)^2}\, B = j1.75$.

The Equations

$$H_z = B \cos \pi x/a,\ E_y = -jB_1 \sin \pi x/a, \text{ and}$$

$$H_x = jB_2 \sin\pi x/a$$

become

$$\bar{H}_z = j1.99 \cos 20\pi x$$

$$\bar{E}_y = 1000 \sin 20\pi x$$

$$\bar{H}_x = -1.75 \sin 20\pi x$$

Multiply by $\exp(j\omega t - \gamma z)$; with $\gamma = 0.0047 + j55.4$, and accept the real part only,

$$H_z = 1.99 \cos 20\pi x\ e^{-0.0047z} \cos(\omega t - 54.4z + 90°)$$

$$E_y = 1000 \sin 20\pi x\ e^{-0.0047z} \cos(\omega t - 55.4z)$$

$$H_x = 1.75 \sin 20\pi x\ e^{-0.0047z} \cos(\omega t - 55.4z + 180°)$$

• **PROBLEM 14-26**

Consider the rectangular resonator shown in the figure. The field components for this structure are

$$E_x = E_z = H_y = 0$$

$$E_y = 2jE_0 \sin\frac{\pi z}{d} \sin\frac{\pi x}{a}$$

$$H_x = \frac{2E_0\sqrt{1 - (\omega_c/\omega_r)^2}}{\eta} \cos\frac{\pi z}{d} \sin\frac{\pi x}{a}$$

$$H_z = \frac{-2E_0\omega_c}{\eta\omega_r} \sin\frac{\pi z}{d} \cos\frac{\pi x}{a}$$

Determine the total energy stored in the electric field and the total energy stored in the magnetic field and prove that these quantities are indeed equal.

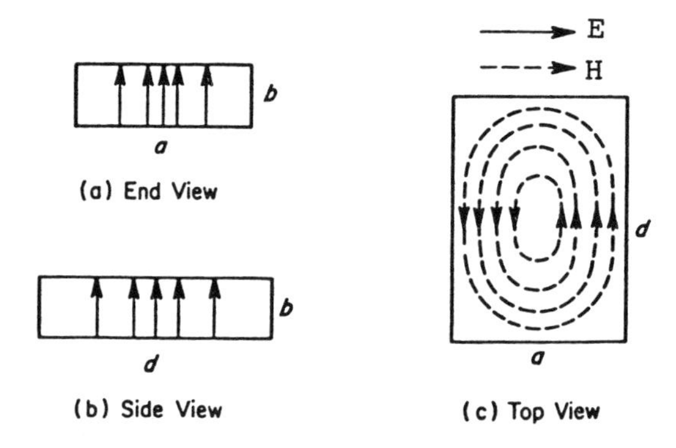

Field configuration for TE_{101} mode.

Solution: The total time-averaged energy stored in the electric field is

$$\frac{1}{2}\iiint \frac{1}{2}\varepsilon|\bar{E}|^2 dv = \int_0^d \int_0^b \int_0^a \frac{\varepsilon_0}{4}|E_0|^2 \sin^2\frac{\pi z}{d}$$

$$\sin^2\frac{\pi x}{a}\, dx\, dy\, dz = \varepsilon_0 \frac{|E_0|^2}{4} abd$$

The total time-averaged energy stored in the magnetic field is

$$\frac{1}{2}\iiint \frac{1}{2}\mu|\bar{H}|^2\, dv = \int_0^d\int_0^b\int_0^a \frac{\mu_0}{4}\frac{4|E_0|^2}{\eta^2}$$

$$\times\left\{\left[1-\left(\frac{\omega_c}{\omega_r}\right)^2\right]\cos^2\frac{\pi z}{d}\sin^2\frac{\pi x}{a} + \left(\frac{\omega_c}{\omega_r}\right)^2\sin^2\frac{\pi z}{d}\cos^2\frac{\pi x}{a}\right\} dx\, dy\, dz$$

$$= \frac{4\mu_0}{4}\frac{|E_0|^2}{\eta^2}\left[1-\left(\frac{\omega_c}{\omega_r}\right)^2+\left(\frac{\omega_c}{\omega_r}\right)^2\right]\frac{abd}{4}$$

$$= \frac{\mu_0|E_0|^2}{4\mu_0/\varepsilon_0}abd$$

$$= \frac{\varepsilon_0|E_0|^2}{4}\,abd$$

thus proving the conclusion arrived at in

$$\iiint_V \mu|\bar{H}|^2\, dv = \iiint \varepsilon|\bar{E}|^2\, dv$$

that the total time-averaged energy stored in the electric field is equal to the total time-averaged energy stored in the magnetic field of the resonant cavity.

ABSORBED AND TRANSMITTED POWER

• PROBLEM 14-27

An air-filled, X-band, rectangular waveguide carries a positive z traveling TE_{10} mode at f = 9 GHz. (a) Find the phase constant, wavelength, phase velocity, and intrinsic wave impedance associated with this mode at the given frequency. (b) If E_y^+ has the amplitude 10^4 V/m, determine the amplitudes of H_x^+ and H_z^+. What time-average power flux is transmitted through every cross-sectional surface of the waveguide by this mode?

Solution: (a) At 9GHz, the wavelength in unbounded free space is $\lambda^{(0)} \doteq v_p^{(0)}/f = (3 \times 10^8)/(9 \times 10^9) = 3.33$ cm.

With a = 0.9 in., the ratio $f_{c,10}/f$ given by $\frac{f_{c,10}}{f} = \frac{v_p^{(0)}}{2af} = \frac{\lambda^{(0)}}{2a} = 0.729$, while $\beta^{(0)} = \omega\sqrt{\mu_o\varepsilon_o} = 2\pi/\lambda^{(0)} = 60\pi$ rad/m. By use of $\gamma_{10} = j\beta_{10} = j\beta^{(0)}\sqrt{1-\left[\frac{\lambda^{(0)}}{2a}\right]^2}$ rad/m

$f > f_{c,10}$,

the phase constant becomes $\beta_{10} = \beta^{(0)}\sqrt{1 - \left[\frac{\lambda^{(0)}}{2a}\right]^2}$

$= 60\pi\sqrt{1 - (0.720)^2} = 60\pi(0.683) = 128.8 \text{ m}^{-1}$.

Thus, the following equations:

$$\lambda_{10} = \frac{\lambda^{(0)}}{\sqrt{1 - \left[\frac{\lambda^{(0)}}{2a}\right]^2}}, \quad v_{p,10} = \frac{v_p^{(0)}}{\sqrt{1 - \left[\frac{\lambda^{(0)}}{2a}\right]^2}}$$

and $$\hat{\eta}_{TE,10} = \frac{\eta^{(0)}}{\sqrt{1 - \left[\frac{\lambda^{(0)}}{2a}\right]^2}} \qquad f > f_{c,10}$$

yield

$$\lambda_{10} = \frac{3.33}{0.683} = 4.88 \text{ cm}$$

$$v_{p,10} = \frac{3 \times 10^8}{0.683} = 4.39 \times 10^8 \text{ m/sec}$$

$$\hat{\eta}_{TE,10} = \frac{120\pi}{0.683} = 552\,\Omega$$

(b) With $\hat{E}^+_{y,10} = 10^4$ V/m, the remaining amplitudes, from

$$\hat{E}^\pm_y(x) = \hat{E}^\pm_{y,10} \sin\left(\frac{\pi}{a}x\right)$$

$$\hat{H}^\pm_x(x) = \mp \frac{\hat{E}^\pm_{y,10}}{\hat{\eta}_{TE,10}} \sin\left(\frac{\pi}{a}x\right) = \hat{H}^\pm_{x,10} \sin\left(\frac{\pi}{a}x\right)$$

and

$$\hat{H}_z^{\pm}(x) = j\frac{\hat{E}^{\pm}_{y,10}\lambda^{(0)}}{2\eta^{(0)}a}\cos\left(\frac{\pi}{a}x\right) = \hat{H}^{\pm}_{z,10}\cos\left(\frac{\pi}{a}x\right),$$

are

$$\hat{H}^{+}_{x,10} = -\frac{\hat{E}^{+}_{y,10}}{\hat{\eta}_{TE,10}} = -\frac{10^4}{552} = -18.1 \text{ A/m}$$

$$\hat{H}^{+}_{z,10} = j\frac{\hat{E}^{+}_{y,10}\lambda^{(0)}}{2\eta^{(0)}a} = \frac{j10^4(0.033)}{2(120\pi)0.0229} = j19.3$$

$$= 19.3e^{j90°} \quad \text{A/m}$$

The time-average power-flux transmitted through any cross-section is obtained using $P_{av} = \int_s (1/2)\ \text{Re}\ [\hat{E} \times \hat{H}^*]\cdot ds$, in which $\hat{E} = \bar{a}_y\hat{E}^{+}_y e^{(-j\beta_{10}z)}$, $\hat{H} = [\bar{a}_x\hat{H}^{+}_z + \bar{a}_z\hat{H}^{+}_z]e^{(-j\beta_{10}z)}$, and $\hat{E}^{+}_y$, $\hat{H}^{+}_x$, and $\hat{H}^{+}_z$ are supplied. Then

$$P_{av} = \int_{y=0}^{b}\int_{x=0}^{a} \tfrac{1}{2}\text{Re}\left\{\bar{a}_z\frac{(\hat{E}^{+}_{y,10})(\hat{E}^{+}_{y,10})^*}{\hat{\eta}^*_{TE,10}}\right.$$

$$\left. \times \sin^2\frac{\pi}{a}x(e^{-j\beta_{10}z})(e^{-j\beta_{10}z})^*\right\}\cdot\bar{a}_z dx\, dy$$

$$= \frac{|\hat{E}^{+}_{y,10}|^2}{2\eta_{TE,10}}b\int_0^a \sin^2\frac{\pi}{a}x dx = \frac{|\hat{E}^{+}_{y,10}|^2}{4\eta_{TE,10}}ab$$

With $\hat{E}^{+}_{y,10} = 10^4$ V/m, $\eta_{TE,10} = 552\Omega$, $a = 0.0229$ m, and $b = 0.0102$ m, the time-average transmitted power becomes $P_{av} = 10.6$W.

• PROBLEM 14-28

Compute the power transmitted (average) and the peak electric field across the b dimension (height) of a waveguide that has a width of .07 m and height of .035 m. Operation is at the TE_{10} mode at a frequency of 3000 mcps with a peak magnetic field intensity of 10 amp/m at the guide center. Assume negligible attenuation.

Solution: The specific impedance Z_{wg} is

$$Z_{wg} = \frac{E_y}{H_x} = \frac{j\omega\mu}{p}$$

If there are no losses, then $\alpha = 0$ and $p = j\beta$, leading to

$$Z_{wg} = \frac{\omega\mu}{\beta}$$

But $\beta = \frac{2\pi}{\lambda_g}$, where λ_g is the wavelength in guide, so that

$$Z_{wg} = \frac{\omega\mu\lambda_g}{2\pi} = f\mu\lambda_g$$

Now since c, the velocity of light = $f\lambda$, with λ being the wavelength outside of guide, substitute for f, giving

$$Z_{wg} = \frac{c}{\lambda}\,\mu\lambda_g$$

Also

$$c = \sqrt{\frac{1}{\mu_r\mu_v} \times \frac{1}{\varepsilon_r\varepsilon_v}} = \sqrt{\frac{1}{\mu_r\varepsilon_r}}\sqrt{\frac{1}{\mu_v\varepsilon_v}}$$

or

$$Z_{wg} = \sqrt{\frac{1}{\mu_r\varepsilon_r}} \times \mu_r\mu_v\lambda_g\sqrt{\frac{1}{\mu_v\varepsilon_v}}$$

$$= \frac{\sqrt{\mu_v\mu_r}}{\sqrt{\varepsilon_v\varepsilon_r}}\,\frac{\lambda_g}{\lambda}$$

But $\sqrt{\frac{\mu_v}{\varepsilon_v}} = 377$ ohms, the impedance of free space.

Therefore

$$Z_{wg} = \sqrt{\frac{\mu_r}{\varepsilon_r}} \times \frac{\lambda_g}{\lambda} \times 377 \text{ ohms}$$

$$Z_{wg} = \sqrt{\frac{\mu_r}{\varepsilon_r}} \frac{\lambda_g}{\lambda} \times 377$$

which for air becomes

$$Z_{wg} = \frac{\lambda_g}{\lambda} \times 377$$

From

$$\lambda_{co} \text{ (for air)} = \frac{2}{\sqrt{\left(\frac{m}{a}\right)^2 + \left(\frac{n}{b}\right)^2}}$$

$$\lambda_{co} = 2a = 0.14 \text{ m}$$

$$\lambda = \frac{3(10^8)}{3(10^9)} = 0.1 \text{ m}$$

and

$$\lambda_g = \frac{\lambda}{\sqrt{1 - (\lambda/\lambda_{co})^2}} = \frac{0.1}{\sqrt{1 - (0.1/0.14)^2}}$$

$$= \frac{0.1}{0.69} = 0.145 \text{ m}$$

Therefore

$$Z_{wg} = \frac{0.145}{0.1} (377) = 545 \text{ ohms.}$$

Total power = $\frac{1}{4} Z_{wg} H_m^2$ (area of guide)

$$= \tfrac{1}{4}(545)(10)^2(0.07)(0.035) = 34.7 \text{ watts}$$

Peak field across small dimension = $H_m Z_{wg} = 10(545)$

$$= 5450 \frac{\text{volts}}{\text{meter}} .$$

• PROBLEM 14-29

At 9 GHz the inside dimensions of the standard waveguide used in practice are 0.9 by 0.4 in. Find the maximum power that can be transmitted in the TE mode, assuming that the air dielectric will break down when the electric field intensity exceeds 3 × 10 V/m.

Solution: From

$$E_y = -j\,\frac{\omega\mu a}{\pi}\,C\,\sin\frac{\pi x}{a}\,e^{j\beta z}$$

the maximum value of the transverse electric field intensity occurs at $\hat{x} = a/2$ and is equal to

$$\frac{\omega\mu a C}{\pi} = \frac{2\pi \times 9 \times 10^9 \times 4\pi \times 10^{-7} \times 0.9 \times 0.0254 \times C}{\pi} \quad \text{V/m}$$

This quantity is not to exceed 3×10^6 V/m. Therefore the maximum value of the constant C is 5.79×10^3 A/m.

On the other hand,

$$\lambda_c = 2a = 2 \times 0.9 \times 0.0254 = 0.0457 \text{ m}$$

$$\lambda = \frac{c}{f} = \frac{3 \times 10^8}{9 \times 10^9} = 0.0333 \text{ m}$$

and

$$\sqrt{\frac{\mu}{\varepsilon}} = 377 \quad \Omega$$

Substituting these quantities in the following equation

$$P_T = \frac{ab}{4}\sqrt{\frac{\mu}{\varepsilon}}\left(\frac{\lambda_c}{\lambda}\right)^2 C^2\sqrt{1 - \left(\frac{\lambda}{\lambda_c}\right)^2}$$

$$= \frac{0.9 \times 0.4 \times (0.0254)^2}{4}\,(377)\left(\frac{0.0457}{0.0333}\right)^2$$

$$\times \quad (5.79 \times 10^3)^2\sqrt{1 - \left(\frac{0.0333}{0.0457}\right)^2}$$

$$= 0.944 \times 10^6 \text{ W}$$

as the maximum amount of power that can be transmitted without causing breakdown of the air dielectric.

• PROBLEM 14-30

Consider a guide consisting of a pair of highly conducting plates. For a plane-parallel wave moving in the z direction and having a frequency of 10^8 cps, determine the power absorbed per square meter in the plates. Assume E = 1 volt/m, $\gamma = 5.8(10^7)$ mhos/m.

Solution: This wave will have two components, E_x and H_y and a small component E_z due to the currents in the walls that are induced by H_y. In air

$$\frac{E_x}{H_y} = \rho_{air} = 377 \text{ ohms};$$

$$H_y = \frac{E_x}{\rho_{air}} = \frac{1}{377} = 2.65(10^{-3}) \text{ amp/m.}$$

In metal

$$\rho_m = \sqrt{\frac{\omega\mu}{\gamma}} \angle 45°$$

$$= \left[\frac{2\pi(10^8)4\pi(10^{-7})}{5.8(10^7)}\right]^{\frac{1}{2}} \angle 45°$$

$$= \sqrt{13.6}\ (10^{-3}) \angle 45°$$

$$= 3.69(10^{-3}) \angle 45° \quad \text{ohms/m}^2$$

$$E_z = \rho_m H_v = 2.65(10^{-3})3.69(10^{-3}) \angle 45°$$

$$E_z = 9.78(10^{-6}) \angle 45°$$

$$P = \frac{1}{\sqrt{2}} \frac{|E_z|^2}{\rho_m} = \frac{[9.78(10^{-6})]^2}{1.414(3.69)10^{-3}} = 18.3(10^{-9}) \text{ watts/m}^2$$

$$P = \tfrac{1}{2}(E_z H_y) = \frac{|E_x||H_y|}{\sqrt{2}}$$

$$= 9.78(10^{-6})2.65(10^{-3}).707 = 18.4(10^{-9}) \quad \text{watts/m}^2$$

• PROBLEM 14-31

Determine the wave impedance of a particular load in an X-band waveguide (0.9 by 0.4 in.) by means of a slotted section of waveguide. With a shorting plane placed over the load end of the guide, adjacent nulls are found to be 2.44 cm apart. When the short is replaced with the load, the standing-wave ratio is found to be 2.0, and the nulls have shifted toward the load by 0.81 cm. (1) What is the frequency of operation? (2) What is the reflection coefficient and wave impedance at the load? (3) If the incident power is 10 mW, what is the net power delivered to the load?

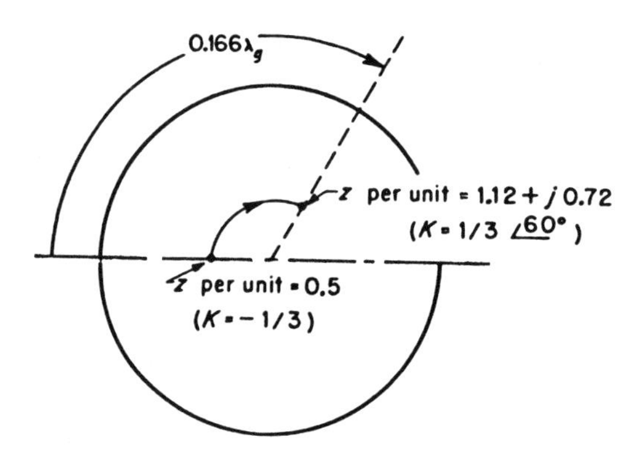

Smith chart for the given problem.

Solution: It is fundamental in wave theory that nulls appear at half-wavelength intervals, and therefore

$$\lambda_g = (2)(2.44) = 4.88 \text{ cm}$$

Now, noting that the free wavelength λ is just the free velocity v divided by the frequency f, the expression for λ_g which is

$$\frac{\lambda_g}{\sqrt{1 - \left(\frac{\omega_c}{\omega}\right)^2}}$$

can be written as

$$\lambda_g = \frac{v}{f} \frac{1}{\sqrt{1 - (f_c/f)^2}}$$

This equation can be readily solved for the frequency, and the result is

$$f = \sqrt{f_c^2 + \left(\frac{v}{\lambda_g}\right)^2}$$

The frequency may now be computed as (recalling that f_c for X-band waveguide is 6.56 GHz)

$$f = \sqrt{(6.56)^2 + \left(\frac{0.3}{0.0488}\right)^2} \times 10^9 = 9.0 \text{ GHz}$$

The reflection coefficient and per-unit wave impedance can be found with the aid of a Smith chart, just as in transmission theory. By referring to the figure, entering the chart at a point of minimum impedance and move back toward the source to a point known to be some multiple of a half-wavelength from the load. This takes place on a constant-radius circle corresponding to a standing-wave ratio of 2, and the distance moved is $(0.81/4.88)\lambda_g = 0.166\lambda_g$. The resulting reflection coefficient and wave impedance are

$$\text{Reflection coefficient} = 0.333 \angle 60°$$

$$\text{Per-unit wave impedance} = 1.12 + j0.72$$

The ohmic value of the wave impedance could be obtained, of course, by multiplying the per-unit value by the characteristic impedance. However, this is not normally of primary interest, so this will not be done.

Finally, if the incident power is 10 mW, then the net power delivered to the load is

$$P_{load} = P_{incident} - P_{reflected}$$

$$= 10(1-|K|^2) = 8.9 \text{ mW}$$

CHARACTERISTICS OF ANTENNAS

• PROBLEM 14-32

A satellite power station in a synchronous orbit is beaming microwave power to earth. If the beamwidth of the transmitting antenna on the satellite is 0.1°, and the distance from the earth's surface is 36,000 (km), what is the size of the spot illuminated by the antenna on the earth's surface? Assume a circular spot or beam area as shown in the figure.

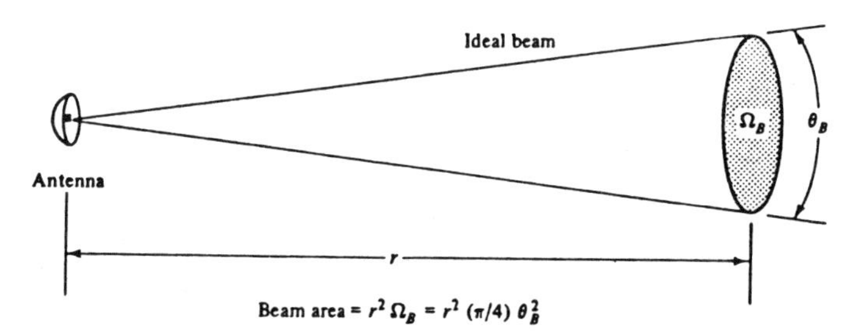

Illustration of the concept of a beam solid angle for an ideal antenna.

Solution: The beamwidth in radians is

$$\theta_B = \frac{\pi}{180°} \times 0.1° = 1.745 \times 10^{-3} \text{ (radian)}$$

and from

$$\Omega_B = \frac{\pi}{4}\,\theta_B^{\,2}$$

$$\Omega_B = \pi/4 \times (1.745 \times 10^{-3})^2$$

$$= 2.39 \times 10^{-6} \text{ (steradian)}$$

Then the area of the spot is $r^2\Omega_B$, or

$$A_{spot} = 2.39 \times 10^{-6}(36{,}000 \times 10^3)^2 = 3.10 \times 10^9 \quad (m^2)$$

• PROBLEM 14-33

An electromagnetic wave of 40KHz is radiated by an antenna to reach a point 60 Km away from it. Find the time taken by this wave to reach the point.

Solution: The time taken by the wave is

$$T = \varepsilon / c$$

where

ε = distance,

$c = 3 \times 10^8$ meters/sec

which is the velocity of light (and of the wave)

Therefore

$$T = \frac{60 \times 1000}{3 \times 10^8} = 2 \times 10^{-4} \text{ seconds}$$

$$= 0.2 \text{ msec.}$$

• PROBLEM 14-34

Find the radiation resistance of a dipole antenna $\frac{1}{10}$ wavelength long.

Solution: From

$$R = 20(\beta\ell)^2 = 80\pi^2 \frac{\ell}{\lambda} \quad \text{ohms}$$

$$R = 80\pi^2 \left(\frac{1}{10}\right)^2 = 7.9 \text{ ohms}$$

The radiation resistance of antennas other than the short dipole can be calculated as above provided the far field is known as a function of angle. From the power equation $P = \frac{1}{2} I_0^2 R$,

$$R = \frac{2P}{I_0^2}$$

and

$$P = \frac{1}{2} \int_s \text{Re} H_\phi H_\phi^* Z ds = \frac{1}{2} \int_s |H_\phi|^2 \text{Re } Z \, ds$$

the radiation resistance at the terminals of an antenna is given by

$$R = \frac{120\pi}{I_0^2} \int_s |H|^2 ds \quad \text{ohms}$$

where $|H|$ = amplitude of far H field (amp/meter)

I_0 = amplitude of terminal current (amp)

• PROBLEM 14-35

Find the radiation resistance of a single-turn circular loop with a circumference of $\frac{1}{4}$ wavelength.

Solution: The area of the loop is

$$A = \pi a^2 = \pi \left(\frac{\lambda}{8\pi}\right)^2 = 0.00497\lambda^2$$

Hence the radiation resistance is

$$R = 31,171 \, (0.00497)^2 = 0.77 \text{ ohm}$$

This is a small value of radiation resistance. However, if the resistance of the loop due to the resistivity of the wire is small compared with the radiation resistance, the loop may be an efficient radiator. As the frequency is decreased (circumference in wavelengths less), the radiation resistance rapidly reduces to such a small value that for practical purposes radiation is negligible.

• PROBLEM 14-36

Calculate the directivity of a short dipole.

Solution: From

Radiation Intensity = U = $S_r r^2$ and

$$P = \frac{1}{2}\int_s \text{Re}H_\phi H_\phi^* Z ds \quad = \frac{1}{2}\int_s |H_\phi|^2 \text{ Re}Z ds \ ,$$

$$U = S_r r^2 = 60\pi |H_\phi|^2 r^2 \tag{1}$$

where

$$|H_\phi| = \frac{\omega I_0 \ell \sin\theta}{4\pi c r} \tag{2}$$

Substituting this in (1),

$$U = 60\pi \left(\frac{\omega I_0 \ell}{4\pi c}\right)^2 \sin^2\theta \tag{3}$$

and the maximum value of U is given by

$$U_m = 60\pi \left(\frac{\omega I_0 \ell}{4\pi c}\right)^2 \tag{4}$$

Introducing (3) and (4) into $D = \dfrac{4\pi U_m}{\iint U d\Omega}$ yields for the directivity

$$D = \frac{4\pi}{\int_0^{2\pi}\int_0^{\pi} \sin^3\theta \, d\theta \, d\phi} = \frac{4\pi}{\frac{8}{3}\pi} = \frac{3}{2}$$

Hence, the directivity of a short dipole is $\frac{3}{2}$. That is, the maximum radiation intensity is 1.5 times as much as if the power were radiated uniformly in all directions.

• PROBLEM 14-37

Suppose an antenna has a power input of 40π W and an efficiency of 98 percent. If the radiation intensity has been found to have a maximum value of 200 W/unit solid angle, find the directivity and gain of the antenna.

Solution: The average power per unit solid angle, U_{avg} is given by

$$P_{avg} = \frac{P_{rad}}{4\pi} = \frac{40\pi(0.98)}{4\pi} = 9.8 \text{ watts/steradian}$$

where P_{rad} is the power of the antenna's radiation.

Since the directivity, D, is defined as

$$\text{Directivity} = \frac{\text{Radiation intensity per unit solid angle}}{\text{Average power per unit solid angle}} \ ,$$

$$D = \frac{200 \text{ watt/sr}}{9.8 \text{ watt/sr}} = 20.41$$

or in decibels,

$$D = 10 \log_{10}(20.41) = 13.10 \text{ dB}$$

To find the gain G, recall that for an antenna

$$G = \frac{\text{radiation intensity in a given direction}}{\text{power density of an isotropic radiator}}.$$

For the isotropic radiator, the radiation intensity U is:

$$U = \frac{P_{in}}{4\pi} = 10 \text{ watt/steradian.}$$

Hence,

$$G = \frac{200 \text{ W/sr}}{10 \text{ W/sr}} = 20$$

or in decibels,

$$G = 10 \log_{10}(20) = 13.01 \text{ dB}$$

• PROBLEM 14-38

Find the gain of an antenna for a half-wave dipole.

Solution: The gain for a half-wave dipole is given by

$$g = \frac{4\pi r^2 P_r}{W}$$

where $P_r = \frac{15 I_m^2}{\pi r^2} \left\{ \frac{\cos [(\pi/2) \cos \theta]}{\sin \theta} \right\}^2$ watts/meter^2

and $W = I_m^2 R_r / 2$

Generally gain is given at its maximum value. For a half-wave dipole the direction for maximum is

$$\theta = \pi/2, \; R_r = 73.09 \text{ ohms}$$

So $$g_{max} = 4\pi r^2 \times \frac{15 I_m^2}{\pi r^2} \left\{ \frac{\cos [(\pi/2) \cos 90°]}{\sin 90\%} \right\}^2 \times \frac{2}{I_m^2 \times 73.09} \simeq 1.64$$

• PROBLEM 14-39

The directivity of an antenna is 50 and the antenna operates at a wavelength of 4 meters. What is its maximum effective aperture?

Solution: The maximum effective aperture of an antenna, A_{em}, is given by

$$A_{em} = \frac{D\lambda^2}{4\pi}$$ where D = directivity

λ = radiation wavelength

$$A_{em} = \frac{50 \times 16}{4\pi} = 63.6 \text{ meter}^2$$

• PROBLEM 14-40

A short dipole has a radiation resistance $R_r = \sqrt{\mu_0/\varepsilon_0}\,\frac{(\beta\ell)^2}{6\pi}$ ohms. Find A_{em} the maximum effective aperture of this dipole.

Solution: The maximum effective aperture is given by

$$A_{em} = V^2/4SR_r \quad (1)$$

where

V = rms voltage induced by the passing wave

$= E\ell$ volts

S = Poynting vector of the incident wave

$= E^2/Z_0$ watts/meter^2

where $Z_0 = \sqrt{\mu_0/\varepsilon_0}$ ohms.

Substituting these values and the given radiation resistance in (1)

$$A_{em} = \frac{V^2}{4SR_r} = \frac{E^2\ell^2}{4\frac{E^2}{Z_0}\left[\frac{\sqrt{\mu_0/\varepsilon_0}\;\beta^2\ell^2}{6\pi}\right]} = \frac{E^2\ell^2 Z_0 6\pi}{E^2\ell^2 Z_0 4\beta^2}$$

$$= \frac{6\pi}{4\beta^2} = \frac{6\pi}{4\left(\frac{2\pi}{\lambda}\right)^2} = \frac{6\lambda^2}{16\pi}$$

$$= \frac{3}{8\pi}\lambda^2 = 0.119\,\lambda^2 \text{ square meters.}$$

Hence the required maximum effective aperture is $0.119\lambda^2$ square meters.

Note that the maximum effective aperture is solely wavelength dependent. Hence, since λ is proportional to frequency, the maximum effective aperture can also be said to be strictly frequency dependent.

• PROBLEM 14-41

For a short dipole antenna, evaluate the effective aperture A_{em}.

Solution: $A_{em} = \frac{\upsilon^2}{4SR_r}$

where υ = emf induced in the short dipole

R_r = Radiation Resistance

S = Poynting vector

The emf induced in the short dipole is a maximum when the dipole is parallel to the incident electric field E. Hence

$$\upsilon = E\ell \text{ volts}$$

The Poynting vector

$$S = \frac{E^2}{Z_0} \text{ watts/meter}^2$$

where Z_0 = intrinsic impedance of medium (air or vacuum) $(= \sqrt{\mu_0/\varepsilon_0})$. The radiation resistance is

$$R_r = \sqrt{\frac{\mu_0}{\varepsilon_0}} \frac{(\beta\ell)^2}{6\pi} \quad \text{ohms}$$

Substituting these values for υ, S, and R_r into the maximum effective aperture of a short dipole is

$$A_{em} = \frac{3}{8\pi} \lambda^2 = 0.119\lambda^2$$

Thus, regardless of how small the dipole is, it can collect power over an aperture of 0.119 wavelength^2 and deliver it to its terminal impedance or load. It is assumed here that the dipole is lossless. However, in practice, losses are present due to the finite conductivity of the dipole conductor so that the actual effective aperture is less than A_{em}.

• PROBLEM 14-42

For point-to-point communication at the higher frequencies the desired radiation pattern is a single narrow lobe or beam. To obtain such a characteristic (at least approximately) a multi-element linear array is usually used. An array is linear when the elements of the array are spaced equally along a straight line. In a uniform linear array the elements are fed with currents of equal magnitude and having a uniform progressive phase shift along the line. The pattern of such an array can be obtained by adding vectorially the field strengths due to each of the elements. For a uniform array of non-directional elements the field strength would be

$$E_T = E_0 \left| 1 + e^{j\Psi} + e^{j2\Psi} + e^{j3\Psi} + \ldots + e^{j(n-1)\Psi} \right| \quad (1)$$

where

$$\Psi = \beta d \cos \phi + \alpha$$

and α is the progressive phase shift between elements. (α is the angle by which the current in any element leads the current in the preceding element.)

Compute the pattern of linear array.

Solution: Eq. (1) is viewed as a geometric progression and written in the form

$$\frac{E_T}{E_0} = \left| \frac{1 - e^{jn\psi}}{1 - e^{j\psi}} \right| = \left| \frac{\sin \frac{n\psi}{2}}{\sin \frac{\psi}{2}} \right| \qquad (2)$$

The maximum value of this expression is n and occurs when $\psi = 0$. This is the principal maximum of the array. Since $\psi = \beta d \cos \phi + \alpha$ the principal maximum occurs when

$$\cos \phi = - \frac{\alpha}{\beta d}$$

For a broadside array the maximum radiation occurs perpendicular to the line of the array at $\phi = 90$ degrees, so $\alpha = 0$ degrees. For an end-fire array the maximum radiation is along the line of the array at $\phi = 0$, so $\alpha = - \beta d$ for this case.

The expression (2) is zero when

$$\frac{n\psi}{2} = \pm k\pi \quad , \; k = 1,2,3,\ldots$$

These are the nulls of the pattern. Secondary maxima occur approximately midway between the nulls, when the numerator of expression (2) is a maximum, that is when

$$\frac{n\psi}{2} = \pm (2m + 1) \frac{\pi}{2} \qquad m = 1,2,3,\ldots$$

The first secondary maximum occurs when

$$\frac{\psi}{2} = \frac{+3\pi}{2n}$$

(note that $\psi/2 = \pi/2n$ does not give a maximum). The amplitude of the first secondary lobe is

$$\left|\frac{1}{\sin(\psi/2)}\right| = \left|\frac{1}{\sin(3\pi/2n)}\right|$$

$$\approx \frac{2n}{3\pi} \quad \text{for large n}$$

The amplitude of the principal maximum was n so the amplitude ratio of first secondary maximum to principal maximum is $2/3\pi = 0.212$. This means that the first secondary maximum is about 13.5 db below the principal maximum, and this ratio is independent of the number of elements in the uniform array, as long as the number is large.

RADIATED AND ABSORBED POWER OF ANTENNAS

• PROBLEM 14-43

If S_R is the power radiated by a transmitting antenna and S_T is the power (maximum) in the load of a receiving antenna, show that

$$\frac{S_R}{S_T} = \frac{A_{Tem}\ A_{Rem}}{(d\lambda)^2}$$

where A_{Tem} and A_{Rem} are the maximum effective apertures of transmitting and receiving antennas, d is the distance between the two antennas and λ is the wave length. Both the antennas are situated in free space.

Solution: The useful power picked up by the receiving antenna is given by the product of effective aperture and average Poynting vector in the oncoming wave

$$S_R = A_{Rem}\ P_{AVERAGE} \tag{1}$$

The power density at the receiver is the power density of an isotropic radiator ($S_T/4\pi d^2$) times the transmitting antenna gain.

$$S_R = \frac{S_T\ A_{Rem}\ G_T}{4\pi d^2} \tag{2}$$

where S_T = power transmitted,

$$G_T = 4\pi\ A_{Tem}/\lambda^2 = \text{antenna gain}, \tag{3}$$

and

A_{Rem} = effective aperture of receiver.

Substituting (3) in (2) gives,

$$\frac{S_R}{S_T} = \frac{A_{Rem}\quad 4\pi A_{Tem}}{x^2 \cdot 4\pi d^2} = \frac{A_{Tem}\ A_{Rem}}{(d\lambda)^2}$$

Hence the proof.

• PROBLEM 14-44

A communication link between two $\lambda/2$ dipole antennas is established in a free space environment. If the transmitter delivers 1 kW of power to the transmitting antenna and the transmitter gain G_T is 1.64, how much power will be received by a receiver connected to the receiving dipole 500 km from the transmitter if the frequency is 200 MHz? Assume that the path between dipoles is normal to each dipole, and the dipoles are perfectly aligned. See Fig. (1)

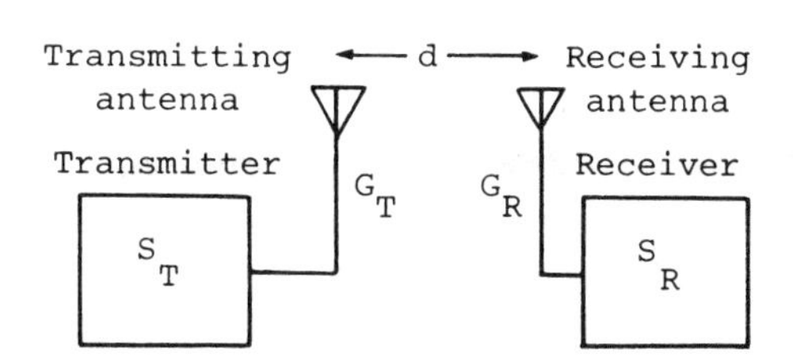

Figure 1.

Solution: For a $\lambda/2$ dipole antenna with transmitter gain $G_T = 1.64$, the effective aperture of the transmitting antenna $\left(\text{from } G_T = \frac{4\pi\ A_{TE}}{\lambda^2}\right)$ is given as

$$A_{TE} = \frac{G_T\lambda^2}{4\pi}$$

$$= \frac{1.64\ \left(\frac{3 \times 10^8}{2 \times 10^8}\right)^2}{4\pi}$$

$$= 0.294\ m^2$$

$$= A_{RE} = \text{effective receiver area}$$

This must be true since the antennas are identical in type and differ only in their use. Thus,

$$G_T = G_R = 1.64$$

$$A_{TE} = A_{RE} = 0.294 \text{ m}^2$$

The power received, S_R , is given by:

$$S_R = \frac{S_T\ G_T\ A_{RE}}{4\pi d^2} = \frac{S_T\ G_T\ G_R}{(4\pi\ d/\lambda)^2}$$

$$= \frac{(10^3)(1.64)(0.294)}{4\pi(5 \times 10^5)^2}$$

Thus,

$$S_R = 1.5329 \times 10^{-10} \text{ watts}$$

• **PROBLEM 14-45**

For free-space transmission, derive the basic transmission path loss, L. Also, determine the basic transmission loss between a ground based antenna and an antenna on a aircraft at distances of 1, 10, 100 and 200 m

(a) at a frequency of 300 MHz.

(b) at a frequency of 3 GHz.

<u>Solution</u>: For a power S_T radiated from an isotropic antenna the power density at a distance d is $S_T/4\pi d^2$. If the transmitting antenna has a gain G_T in the desired direction the power density is increased to $G_T S_T/4\pi d^2$. The effective area of the receiving antenna is $A_{RE} = \lambda^2 G_R/4\pi$ so the power received will be

$$S_R = \frac{G_T S_R}{4\pi d^2}\ \frac{\lambda^2 G_R}{4\pi} = \frac{\lambda^2 G_T G_R}{(4\pi d)^2} S_T \quad \text{in watts.}$$

Hence, the ratio of received to transmitted power is

$$\frac{S_R}{S_T} = \frac{\lambda^2 G_R G_T}{(4\pi d)^2} \quad .$$

The basic transmission path loss, L, is defined as the reciprocal of this ratio, expressed in decibels, for transmission between isotropic antennas ($G_T = G_R = 1$). That is:

$$L = 10 \log_{10} \frac{(4\pi d)^2}{\lambda^2} = 10 \log_{10} \left(4\pi \frac{d}{\lambda}\right)^2$$

Part (A) At a frequency of f = 300 MHz

For distance = 1 meter:

$$L = 10 \log_{10} \left[4\pi \frac{1 \text{ m}}{\left(\dfrac{3\text{x}10^8 \text{msec}^{-1}}{3\text{x}10^8 \text{Hz}}\right)} \right]^2$$

$= +21.984$ dB

For distance = 10 meters:

$$L = 10 \log_{10} \left[4\pi \frac{10 \text{ m}}{\left(\dfrac{3\text{x}10^8 \text{msec}^{-1}}{3\text{x}10^8 \text{Hz}}\right)} \right]^2$$

$= +41.984$ dB

For distance = 100 meters:

$$L = 10 \log_{10} \left[4\pi \frac{100 \text{ m}}{\left(\dfrac{3\text{x}10^8 \text{msec}^{-1}}{3\text{x}10^8 \text{Hz}}\right)} \right]^2$$

$= +61.984$ dB

For distance = 200 meters:

$$L = 10 \text{ Log}_{10} \left[4\pi \frac{200 \text{ m}}{\left(\dfrac{3\text{x}10^8 \text{msec}^{-1}}{3\text{x}10^8 \text{Hz}}\right)} \right]^2$$

$= +68.005$ dB

Part (B) At a frequency of 3 GHz

For distance = 1 meter:

$$L = 10 \text{ Log}_{10} \left[4\pi \frac{1 \text{ m}}{\left(\dfrac{3\text{x}10^8 \text{msec}^{-1}}{3\text{x}10^9 \text{Hz}}\right)} \right]^2$$

$= +41.984$ dB

For distance = 10 meters:

$$L = 10 \operatorname{Log}_{10} \left[4\pi \frac{10 \text{ m}}{\left(\dfrac{3\text{x}10^8 \text{msec}^{-1}}{3\text{x}10^9 \text{Hz}} \right)} \right]^2$$

$$= +61.984 \text{ dB}$$

For distance = 100 meters:

$$L = 10 \operatorname{Log}_{10} \left[4\pi \frac{100 \text{ m}}{\left(\dfrac{3\text{x}10^8 \text{msec}^{-1}}{3\text{x}10^9 \text{Hz}} \right)} \right]^2$$

$$= +81.984 \text{ dB}$$

For distance = 200 meters:

$$L = 10 \operatorname{Log}_{10} \left[4\pi \frac{200 \text{ m}}{\left(\dfrac{3\text{x}10^8 \text{msec}^{-1}}{3\text{x}10^9 \text{Hz}} \right)} \right]^2$$

$$= +88.005 \text{ dB}$$

The homogram of transmission path loss for free space to calculate L, where d = 10 meters and f = GHz.

• PROBLEM 14-46

For the two-antenna system shown in the figure, one antenna is transmitting a 1 GHz signal 30 meters above moist earth. The receiving antenna is 16 km away and 10 meters above moist earth. If the relative permitivity of moist earth is $\varepsilon_r = 12$ and the conductivity σ = 0.02 mho/meter, show that the signal is almost perfectly reflected by the earth to the receiver.

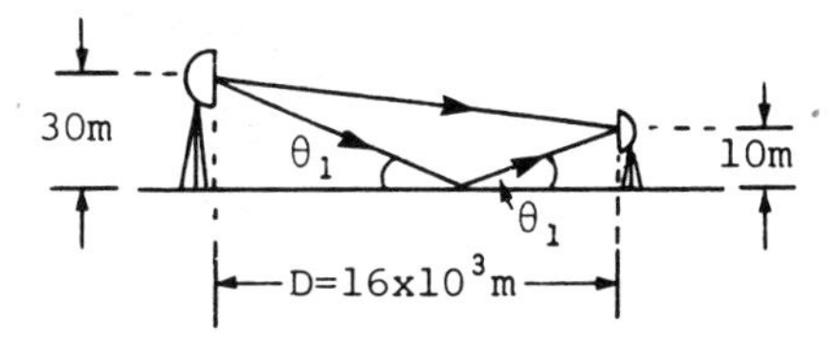

Solution: When the distance between the transmitter and receiver is small enough for the earth to be considered flat, the energy transmitted may reach the receiver by a direct path between the antennas or by a route involving reflection from the earth's surface. From electromagnetic theory, the reflection coefficient Γ is given as

$$\Gamma = \frac{\eta_2 \sin \theta_1 - \eta_1 \sin \theta_2}{\eta_2 \sin \theta_1 + \eta_1 \sin \theta_2}$$

and

$$\eta = \sqrt{\frac{\mu}{\left(\varepsilon + \dfrac{\sigma}{j\omega}\right)}}$$

where η = characteristic impedance of the medium. (consider air medium #1, moist earth medium #2)

θ_1 = angle of incidence of the energy

θ_2 = angle of refraction

σ = conductivity of the medium

$\varepsilon = \varepsilon_o \varepsilon_r$ = electric permitivity

$\varepsilon_o = \frac{1}{36\pi} \times 10^{-9}$ F/m

ε_r = relative permitivity of the medium

$\mu = \mu_o \mu_r$ = magnetic permeability

$\mu_o = 4\pi \times 10^{-7}$ H/m

$\mu_r \simeq 1$ for non-magnetic materials.

Thus, for air (or free space) $\varepsilon = 1$, $\sigma = 0$

$$\eta_1 = \sqrt{\frac{4\pi \times 10^{-7}}{\varepsilon_o + \frac{(0)}{j\omega}}} = \sqrt{\frac{4\pi \times 10^{-7}}{\frac{1}{36\pi} \times 10^{-9}}} = 120\pi \simeq 377 \text{ ohms}$$

for moist earth, $\varepsilon_r = 12$, $\sigma = 0.02$

$$\eta_2 = \sqrt{\frac{4\pi \times 10^{-7}}{\left(\frac{12}{36\pi} \times 10^{-19} + \frac{0.02}{j(2\pi \times 10^9)}\right)}} = \sqrt{\frac{4\pi \times 10^{-7}}{1.06 \times 10^{-10} - j2.812 \times 10^{-23}}}$$

Because the imaginary part is over ten orders of magnitude less than the real part, we can approximate to get

$$\eta_2 \simeq \sqrt{\frac{4\pi \times 10^{-7}}{1.06 \times 10^{-10}}} = 108.881 \text{ ohms}$$

$\therefore\ \eta_2 \simeq 109$ ohms

From the figure, $\theta_1 = \tan^{-1} \dfrac{(30 + 10)\text{m}}{16 \times 10^3 \text{ m}} = 0.0025$ radians

at the reflection point, the angle of incidence is so near grazing that:

$\tan \theta_1 \simeq \sin \theta_1 \simeq \theta_1 = 0.0025$ radians.

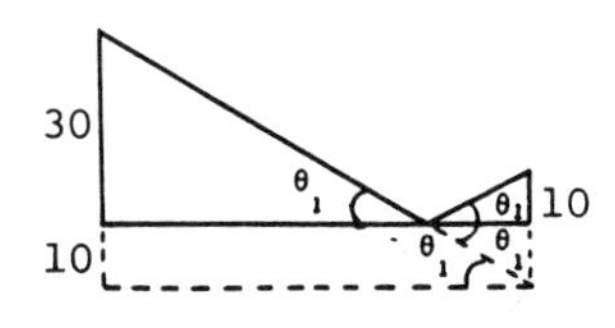

Thus, $\cos \theta_1 \simeq 1.$

Now, to find θ_2, use Snell's law:

$$\frac{\cos \theta_1}{\cos \theta_2} = \sqrt{\frac{\varepsilon_2}{\varepsilon_1}} = \sqrt{\frac{\varepsilon_o \varepsilon_{r_2}}{\varepsilon_o \varepsilon_{r_1}}} = \sqrt{12} = 2\sqrt{3} = 3.464$$

Hence,

$$\cos \theta_2 = \frac{1}{3.464}$$

or $\theta_2 = 73.22^o$

Thus, $\sin \theta_2 = 0.957$

This gives

$$\Gamma = \frac{\text{Reflected field}}{\text{Incident field}} = \frac{109(0.0025) - 120\pi(0.957)}{109(0.0025) + 120\pi(0.957)}$$

$$= -0.9985$$

$$\simeq -1$$

It is thus obvious that moist earth behaves as a perfect reflector.

When the transmission distance is great compared with the antenna heights, reflection is practically complete and takes place with a reversal of phase as indicated by the negative sign. The two waves do not cancel out in spite of the phase reversal because the direct and indirect paths have different lengths.

• PROBLEM 14-47

Using the field-cell concept, calculate the inductance and also capacitance per unit length of the coaxial transmission line shown in cross section in the figure.

The line is air-filled.

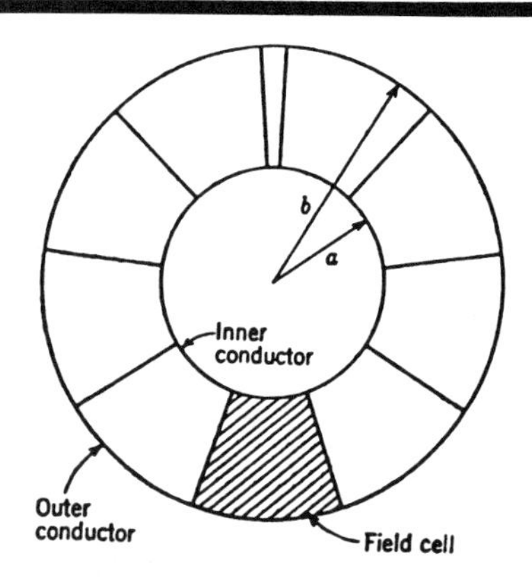

Coaxial transmission line divided into 9.15 field-cell lines in parallel.

Solution: The inductance per unit length of the coaxial line is given by

$$\frac{L}{d} = \frac{1}{n}\frac{L_0}{d} = \frac{\mu_0}{n} \qquad \text{henrys/meter}$$

where

L_0/d = inductance per unit length of transmission-line cell

n = number of line cells in parallel

μ_0 = permeability of air = $4\pi \times 10^{-7}$ henry/meter (= 1.26 μh/meter)

Dividing the space between the coaxial conductors into curvilinear squares, 9.15 line cells in parallel are obtained. Thus

$$\frac{L}{d} = \frac{1.26}{9.15} = 0.138 \ \mu\text{h/meter}$$

Note that the radius of the outer conductor is twice the radius of the inner so that from

$$\frac{L}{d} = 0.46 \log b/a \tag{1}$$

which is the same result as obtained above. The equation $\frac{L}{d} = 0.46 \log b/a$ is exact for this case. The accuracy of the cell method depends on the accuracy of construction of the curvilinear squares. However, the cell method (or field-mapping method) is applicable to conductor configurations that might be very difficult to handle mathematically.

The capacitance per unit length of the coaxial line of the figure is given by

$$\frac{C}{d} = n\frac{C_0}{d} = n\varepsilon_0 \qquad \text{farads/meter}$$

where C_0/d = capacitance per unit length of line cell

n = number of line cells in parallel

ε_0 = permittivity of air = 8.85 $\mu\mu$f/meter

Thus

$$\frac{C}{d} = 9.15 \times 8.85 = 81 \ \mu\mu f/\text{meter}$$

Using the exact relation

$$\frac{C}{d} = \frac{24.2}{\log(b/a)} \ \mu\mu f/\text{meter}$$

obtain

$$\frac{C}{d} = \frac{24.2}{\log 2} = 81 \quad \mu\mu f/\text{meter}$$

which is the same as obtained by the cell method.

SECTION II

SUMMARY OF ELECTROMAGNETIC PROPAGATION IN CONDUCTING MEDIA

Martin B. Kraichman*

List of Principal Symbols

$\mathbf{A}$	magnetic vector potential
A	end point of finite length electric antenna
$\mathbf{B}$	magnetic flux density
B	end point of finite length electric antenna
$\mathbf{D}$	electric displacement
$\mathbf{E}$	electric field intensity
$\mathbf{E}_0$	impressed electric field intensity; complex vector amplitude of incident electric field intensity in Fresnel's equations
$\mathbf{E}_1$, $\mathbf{E}_2$	complex vector amplitudes of refracted and reflected electric field intensities, respectively, in Fresnel's equations
e	base of natural logarithms
$\mathbf{F}$	electric vector potential
$F(n)$	surface wave attenuation factor defined by (3.33)
$\mathbf{H}$	magnetic field intensity
$\mathbf{H}_0$, $\mathbf{H}_1$, $\mathbf{H}_2$	complex vector amplitudes of incident, refracted, and reflected magnetic field intensities, respectively, in Fresnel's equations
HED	horizontal electric dipole
HMD	horizontal magnetic dipole

* U. S. White Oak Laboratory, Silver Spring, Maryland

h source height or depth measured from the interface of a homogeneous, semi-infinite, conducting medium

I electric current

$I_0(z)$, $I_1(z)$ modified Bessel functions of the first kind, of orders zero and one respectively, with argument z; also abbreviated as I_0 and I_1

$\mathbf{J}$ electric current density

$\mathbf{J}_0$ impressed electric current density

j imaginary $\sqrt{-1}$

$K_0(z)$, $K_1(z)$ modified Bessel functions of the second kind, of orders zero and one, respectively, with argument z; also abbreviated as K_0 and K_1

L wavelength of ripple in conductor surface; length of electric antenna; inductance per unit length of transmission line

ln natural logarithm

$\mathbf{M}_0$ impressed magnetic dipole moment per unit volume

m magnetic dipole moment in units of ampere-meter^2

n Sommerfeld numerical distance defined by (3.34)

$\mathbf{P}_0$ impressed electric dipole moment per unit volume

p propagation function defined in (2.44); electric current moment in units of ampere-meter

Q layered medium stratification factor defined by (2.72) to (2.74)

Q_E, Q_H layered medium stratification factors defined by (3.65) and (3.66), respectively

Q'_E, Q'_H layered medium stratification factors similar to Q_E and Q_H respectively but with the source and field points interchanged

R the distance $\left(\rho^2 + z^2\right)^{1/2}$ in section 3.2; transmission line resistance per unit length

R' distance $\left(\rho^2 + h^2\right)^{1/2}$

R_1, R_2 distances defined by (3.13) and (3.14) respectively; distances defined in section 4.2.1

r radial spherical coordinate; ratio of distances in section 3.2.2.3; distance $\left(y^2 + z^2\right)^{1/2}$ in section 4.2

TE, TM transverse electric and transverse magnetic, respectively

t time variable

VED vertical electric dipole

VMD vertical magnetic dipole

x, y, z rectangular coordinates

ρ, ϕ, z cylindrical coordinates

r, θ, ϕ spherical coordinates

α	attenuation constant, real part of propagation constant
β	phase constant, imaginary part of propagation constant
γ	propagation constant $\alpha + j\beta$
γ_0	propagation constant of free space
γ_1, γ_2	propagation constants of a homogeneous conducting half-space; propagation constants of the top and second layers respectively of a stratified conducting half-space
γ_e	effective propagation constant of a stratified conducting half-space
δ	skin depth in a conducting medium
δ_1	skin depth in the top layer of a stratified conducting half-space
ϵ	permittivity of a medium
η	intrinsic impedance of a medium
$\theta_0, \theta_1, \theta_2$	angles defined in figure 2.1; also θ_0 defined in figure 2.2
μ	permeability of a medium
π	constant 3.14159...
$\mathbf{\Pi}$	electric Hertz vector
$\mathbf{\Pi}^*$	magnetic Hertz vector
ρ	free electric charge density; radial cylindrical coordinate; field reflection coefficient; smallest radius of curvature of a conducting surface
σ	conductivity of a medium
σ_1, σ_2	conductivities of a homogeneous half-space; conductivities of the top and second layers, respectively, of a stratified medium
τ	field transmission coefficient
ϕ	phase angle; angular cylindrical coordinate
ψ	scalar magnetic potential; angle defined in figure 2.2; scalar field function defined by (2.82)
ω	angular frequency

Chapter 1

BASIC EQUATIONS AND THEOREMS

This chapter presents the basic equations and theorems of electromagnetic theory with emphasis on propagation in conducting media. Rationalized mks units are used throughout this book.

1.1 MAXWELL'S EQUATIONS

Maxwell's equations for an electromagnetic field are:

$$\nabla \times \mathbf{E} = -\frac{\partial \mathbf{B}}{\partial t} \tag{1.1}$$

$$\nabla \times \mathbf{H} = \mathbf{J} + \frac{\partial \mathbf{D}}{\partial t} \tag{1.2}$$

$$\nabla \cdot \mathbf{B} = 0 \tag{1.3}$$

$$\nabla \cdot \mathbf{D} = \rho \tag{1.4}$$

in which ρ is the free charge density and $\mathbf{J}$ is the electric current density, which includes both induced conduction current and impressed or source current from batteries and generators. Convection current should also be included in $\mathbf{J}$, although in this book such current will not be considered. In dielectric media, the electric intensity $\mathbf{E}$ is related to the displacement $\mathbf{D}$ by

$$\mathbf{D} = \epsilon_0 \mathbf{E} + \mathbf{P} \tag{1.5}$$

where ϵ_0 = permittivity of free space = $(1/36\pi) \times 10^{-9}$ farad/meter and
$\mathbf{P}$ = polarization or electric dipole moment/unit volume

In magnetic materials the magnetic flux density $\mathbf{B}$ is related to the magnetic intensity $\mathbf{H}$ by

$$\mathbf{B} = \mu_0(\mathbf{H} + \mathbf{M}) \tag{1.6}$$

where μ_0 = permeability of free space = $4\pi \times 10^{-7}$ henry/meter and
$\mathbf{M}$ = magnetization or magnetic dipole moment/unit volume

Some authors relate $\mathbf{B}$ to $\mathbf{H}$ by

$$\mathbf{B} = \mu_0 \mathbf{H} + \mathbf{M}' \tag{1.7}$$

in which case $\mathbf{M}'$ has the units of $\mathbf{B}$ (webers/meter^2) instead of $\mathbf{H}$ (ampere/meter). Although this definition allows a symmetry between (1.5) and (1.7), it results in units of magnetic moment, weber-meter, which differ from the more commonly used loop dipole moment, ampere-meter^2.

In a linear medium, **P** and **M** (excluding permanent polarization and magnetization) are proportional to **E** and **H**, respectively, and consequently (1.5) and (1.6) may be written

$$\mathbf{D} = \epsilon \mathbf{E} \tag{1.8}$$

$$\mathbf{B} = \mu \mathbf{H} \tag{1.9}$$

where ϵ and μ are respectively the permittivity and permeability of the medium. These parameters may be expressed as

$$\epsilon = \epsilon_r \epsilon_0 \tag{1.10}$$

$$\mu = \mu_r \mu_0 \tag{1.11}$$

in which ϵ_r is the relative permittivity or dielectric constant and μ_r is the relative permeability.

A statement of the conservation of charge,

$$\nabla \cdot \mathbf{J} = -\frac{\partial \rho}{\partial t} \tag{1.12}$$

can be derived from (1.2) and (1.4). For a linear conducting medium that obeys Ohm's law,

$$\mathbf{J} = \sigma(\mathbf{E} + \mathbf{E}_0) \tag{1.13}$$

where σ is the conductivity of the medium, **E** is the induced electric field, and $\mathbf{E}_0$ is the impressed electric field. By combining this equation with (1.8) and (1.12) an expression for the decay of charge in a conducting medium in the absence of an impressed current may be written as

$$\rho = \rho_0 e^{-\frac{\sigma}{\epsilon} t} \tag{1.14}$$

with the constant of integration ρ_0 being equal to the charge density at the time $t = 0$. Equation (1.14) shows that there can be no permanent distribution of free charge within a homogeneous medium with a nonzero conductivity and that the decay of the charge density is independent of any existing field. The time

$$\tau = \frac{\epsilon}{\sigma} \tag{1.15}$$

required for the charge at any point to decay to $1/e$ of its original value is called the relaxation time. In all but the poorest conductors, τ is exceedingly small.

1.2 AUXILIARY POTENTIALS

Except in very simple cases, the fields cannot conveniently be solved directly from Maxwell's equations. Accordingly, it is desirable to introduce auxiliary potentials by means of which second-order partial differential equations may be derived and solved as boundary value problems. The fields may then be obtained from the potentials by differentiation.

1.2.1 Magnetic and Electric Potentials

Expressing the fields in terms of a magnetic vector potential **A** and an electric scalar potential Φ, which are consistent with Maxwell's equations, results in

$$\mathbf{E} = -\nabla\Phi - \frac{\partial \mathbf{A}}{\partial t} \tag{1.16}$$

$$\mathbf{B} = \nabla \times \mathbf{A} \tag{1.17}$$

Substituting the above expressions for the field into Maxwell's equations results in a pair of coupled second-order partial differential equations. Because the vector potential **A** is not uniquely defined until its divergence is specified, the divergence may conveniently be chosen to uncouple the differential equations. Thus choosing

$$\nabla \cdot \mathbf{A} = -\mu\epsilon \frac{\partial \Phi}{\partial t} - \mu\sigma\Phi \tag{1.18}$$

the resulting differential equations become

$$\nabla^2 \mathbf{A} - \mu\epsilon \frac{\partial^2 \mathbf{A}}{\partial t^2} - \mu\sigma \frac{\partial \mathbf{A}}{\partial t} = -\mu \mathbf{J}_0 \tag{1.19}$$

$$\nabla^2 \Phi - \mu\epsilon \frac{\partial^2 \Phi}{\partial t^2} - \mu\sigma \frac{\partial \Phi}{\partial t} = -\frac{\rho}{\epsilon} \tag{1.20}$$

These symmetrical inhomogeneous differential equations are the propagation equations for a conducting medium. Here $\mathbf{J}_0$ represents that part of the current density produced by impressed electromotive forces. The induced conduction current density in the medium is represented by the first time derivative terms containing σ and the displacement current density by the second time derivative terms containing ϵ. In those cases in which the displacement current is negligible compared with the conduction current, the resulting differential equations are characteristic of a diffusion process.

In a region where $\rho = 0$, a possible solution of (1.20) is $\Phi = 0$. This means that $\nabla \cdot \mathbf{A} = 0$ and from the conservation of charge that $\nabla \cdot \mathbf{J}_0 = 0$. The electromagnetic field can then be determined exclusively by the vector potential as

$$\mathbf{E} = -\frac{\partial \mathbf{A}}{\partial t} \tag{1.21}$$

and

$$\mathbf{B} = \nabla \times \mathbf{A} \tag{1.22}$$

The vector potential **A** may be determined by solving (1.19) in which the impressed current density has zero divergence. The field in a region that contains no sources (ρ=0, $\mathbf{J}_0$=0) may also be given by (1.21) and (1.22). It should also be noted that where there are no sources, the $\nabla \cdot \mathbf{D} = 0$, and the electromagnetic field can alternatively be expressed in terms of an electric vector potential **F** and a scalar magnetic potential ψ such that

$$\mathbf{D} = -\nabla \times \mathbf{F} \tag{1.23}$$

and

$$\mathbf{H} = -\nabla\psi - \frac{\partial \mathbf{F}}{\partial t} - \frac{\sigma}{\epsilon}\mathbf{F} \tag{1.24}$$

The potentials ψ and **F** are then related by

$$\nabla \cdot \mathbf{F} = -\mu\epsilon \frac{\partial \psi}{\partial t} \tag{1.25}$$

Without loss of generality, the scalar potential ψ may be set equal to zero and the electromagnetic field may then be found exclusively in terms of the vector potential $\mathbf{F}$. Thus

$$\mathbf{D} = -\nabla \times \mathbf{F} \tag{1.26}$$

$$\mathbf{H} = -\frac{\partial \mathbf{F}}{\partial t} - \frac{\sigma}{\epsilon}\mathbf{F} \tag{1.27}$$

where $\mathbf{F}$ is the solution to

$$\nabla^2 \mathbf{F} - \mu\epsilon\frac{\partial^2 \mathbf{F}}{\partial t^2} - \mu\sigma\frac{\partial \mathbf{F}}{\partial t} = 0 \tag{1.28}$$

and satisfies $\nabla \cdot \mathbf{F} = 0$.

1.2.2 Hertz Vector Potential

It is possible to define an electromagnetic field in terms of a single vector function or Hertz vector for which the impressed sources are either electric or magnetic dipole distributions. The Hertz vector potentials and the propagation equations for such sources in a conducting medium are most conveniently expressed for harmonic time variation and will be discussed in the next section.

1.3 HARMONIC TIME VARIATION

The results presented in the subsequent sections of this handbook will be stated for a single-frequency sinusoidal time variation. A particular pulse shape or time dependence may then be expressed by means of a Fourier series or integral. The complex exponential $e^{j\omega t}$ is used to represent a harmonic time variation of angular frequency ω. Thus, a harmonic electric field may be written as $\mathbf{E}_a e^{j(\omega t - \phi)}$, where $\mathbf{E}_a$ is the real vector amplitude and ϕ is the phase. This may also be written as $\mathbf{E}e^{j\omega t}$ with $\mathbf{E} = \mathbf{E}_a e^{-j\phi}$. Because the factor $e^{j\omega t}$ appears in each term of an equation, it is usually suppressed. It is understood, however, that physical entities must finally be represented by the real part of the result.

1.3.1 Maxwell's Equations

Maxwell's equations for the harmonic case may be written

$$\nabla \times \mathbf{E} = -j\omega\mathbf{B} \tag{1.29}$$

$$\nabla \times \mathbf{H} = \mathbf{J}_0 + (\sigma + j\omega\epsilon)\mathbf{E} \tag{1.30}$$

$$\nabla \cdot \mathbf{B} = 0 \tag{1.31}$$

$$\nabla \cdot \mathbf{D} = \rho \tag{1.32}$$

in which the first derivative is indicated by the factor $j\omega$ and the electric current density $\mathbf{J}$ has been separated into an impressed term $\mathbf{J}_0$ and an induced conduction current term $\sigma\mathbf{E}$. Equation (1.30) may also be written as

$$\nabla \times \mathbf{H} = \mathbf{J}_0 + j\omega\epsilon^* \mathbf{E} \tag{1.33}$$

in which the complex permittivity ϵ^* is defined as

$$\epsilon^* = \epsilon\left(1 + \frac{\sigma}{j\omega\epsilon}\right) \tag{1.34}$$

The equations for harmonic fields derived for nonconducting media may be made valid for conducting media by replacing the real permittivity ϵ by the complex permittivity ϵ^*. This in effect introduces a conduction current into the equations.

1.3.2 Magnetic and Electric Potentials

The equations given in the previous sections for a conducting medium may be stated for harmonic variation by replacing $\partial/\partial t$ by $j\omega$ and $\partial^2/\partial t^2$ by $-\omega^2$. Thus, for example, (1.19) becomes

$$\nabla^2 \mathbf{A} - \gamma^2 \mathbf{A} = -\mu \mathbf{J}_0 \tag{1.35}$$

where

$$\gamma^2 = -\omega^2 \mu\epsilon + j\omega\mu\sigma \tag{1.36}$$

and γ is the propagation constant of the conducting medium. After $\mathbf{A}$ is determined by solving (1.35) subject to boundary conditions, the fields may be obtained from

$$\mathbf{E} = \frac{j\omega}{\gamma^2} \nabla\nabla \cdot \mathbf{A} - j\omega \mathbf{A} \tag{1.37}$$

$$\mathbf{B} = \nabla \times \mathbf{A} \tag{1.38}$$

where $\nabla\Phi$ in (1.16) has been replaced by its equivalent in terms of $\mathbf{A}$ from (1.18). Therefore, for harmonic time variation, the electromagnetic field may be expressed explicitly in terms of the magnetic vector potential alone.

1.3.3 Hertz Vector Potentials

The electromagnetic field of an impressed electric dipole moment distribution $\mathbf{P}_0$ in a conducting medium may be expressed in terms of an electric Hertz vector Π as

$$\mathbf{E} = \nabla\nabla \cdot \Pi - \gamma^2 \Pi \tag{1.39}$$

$$\mathbf{B} = \frac{\gamma^2}{j\omega} \nabla \times \Pi \tag{1.40}$$

where

$$\mathbf{D} = \epsilon \mathbf{E} + \mathbf{P}_0 \tag{1.41}$$

and Π satisfies

$$\nabla^2 \Pi - \gamma^2 \Pi = \frac{\omega^2 \mu}{\gamma^2} \mathbf{P}_0 \tag{1.42}$$

A comparison of (1.39) with (1.37) reveals that

$$\mathbf{A} = \frac{\gamma^2}{j\omega} \Pi \tag{1.43}$$

which shows that the magnetic vector potential and the electric Hertz vector are only trivially different for harmonic time variation.

For an impressed magnetic dipole distribution $\mathbf{M}_0$, the electromagnetic field may be expressed in terms of a magnetic Hertz vector Π^* as

$$\mathbf{E} = -j\omega\mu\nabla \times \Pi^* \tag{1.44}$$

$$\mathbf{H} = \nabla\nabla \cdot \Pi^* - \gamma^2 \Pi^* \tag{1.45}$$

where

$$\mathbf{B} = \mu(\mathbf{H} + \mathbf{M}_0) \tag{1.46}$$

and Π^* satisfies

$$\nabla^2 \Pi^* - \gamma^2 \Pi^* = -\mathbf{M}_0 \tag{1.47}$$

If both electric and magnetic sources are present, the electromagnetic field may be obtained by a superposition of the solutions for each kind of source.

1.4 PARTICULAR SOLUTIONS FOR AN UNBOUNDED HOMOGENEOUS REGION WITH SOURCES

The well-known solution of (1.35) for a unit point source in an unbounded homogeneous conducting region (Green's function) is

$$G(\mathbf{r}|\mathbf{r}') = \frac{\mu}{4\pi} \frac{e^{-\gamma|\mathbf{r}-\mathbf{r}'|}}{|\mathbf{r}-\mathbf{r}'|} \tag{1.48}$$

which represents a diverging wave when the positive root for γ is chosen. In the above, the primed quantities refer to source coordinates, and the unprimed represent field coordinates. If the impressed sources are all within a finite distance from the coordinate origin, then by means of superposition the magnetic vector potential may be written

$$\mathbf{A}(\mathbf{r}) = \frac{\mu}{4\pi} \int \mathbf{J}_0(\mathbf{r}') \frac{e^{-\gamma|\mathbf{r}-\mathbf{r}'|}}{|\mathbf{r}-\mathbf{r}'|} dv' \tag{1.49}$$

This relationship shows that the contributions from the various volume elements dv' do not add with their original phase but are affected by an additional phase lag $\gamma|\mathbf{r}-\mathbf{r}'|$. Because the propagation constant in a conducting medium is complex, this phase lag not only represents a retardation but an attenuation as well.

Expressions similar to (1.49) exist for the Hertz vectors and are given by

$$\Pi(\mathbf{r}) = \frac{-\omega^2\mu}{4\pi\gamma^2} \int \mathbf{P}_0(\mathbf{r}') \frac{e^{-\gamma|\mathbf{r}-\mathbf{r}'|}}{|\mathbf{r}-\mathbf{r}'|} dv' \tag{1.50}$$

and

$$\Pi^*(\mathbf{r}) = \frac{1}{4\pi} \int \mathbf{M}_0(\mathbf{r}') \frac{e^{-\gamma|\mathbf{r}-\mathbf{r}'|}}{|\mathbf{r}-\mathbf{r}'|} dv' \tag{1.51}$$

1.5 POYNTING VECTOR

With the aid of well-known vector theorems, the relation between the surface and volume integrals

$$\int_S (\mathbf{E}\times\mathbf{H})\cdot d\mathbf{S}+\int_v \mathbf{E}\cdot\mathbf{J}\,dv=-\frac{\partial}{\partial t}\int_v\left(\frac{\epsilon}{2}E^2+\frac{\mu}{2}H^2\right)dv \tag{1.52}$$

may be derived from (1.1) and (1.2). Replacing $\mathbf{E}$ in the second integral on the left in (1.52) with its equivalent from (1.13) results in

$$-\int_S (\mathbf{E}\times\mathbf{H})\cdot d\mathbf{S}=\int_v \frac{J^2}{\sigma}dv+\frac{\partial}{\partial t}\int_v\left(\frac{\epsilon}{2}E^2+\frac{\mu}{2}H^2\right)dv-\int_v \mathbf{E}_0\cdot\mathbf{J}\,dv \tag{1.53}$$

The Poynting vector $\mathbf{P}$ is defined as

$$\mathbf{P}=\mathbf{E}\times\mathbf{H}\quad \text{watts/meter}^2 \tag{1.54}$$

and has the nature of a power flux. Thus (1.53) states that within a volume v bounded by a closed surface S, the power entering is equal to the Joule heat loss plus the increase in the energy stored in the electric and magnetic fields minus the work done on the system by impressed electromotive forces.

Although the total energy flow through a closed surface is represented correctly by (1.53), the intensity of energy flow at a point given by (1.54) is not necessarily correct since any vector integrating to zero over S may be added to $\mathbf{P}$ without affecting the total flow.

The time average of the Poynting vector for harmonic variation using complex exponential notation is given by

$$\mathbf{P}_{\text{av}}=\tfrac{1}{2}\text{Re}\,(\mathbf{E}\times\tilde{\mathbf{H}}) \tag{1.55}$$

where Re stands for the real part of the quantity in parentheses and $\tilde{\mathbf{H}}$ is the complex conjugate of $\mathbf{H}$.

1.6 RECIPROCITY THEOREM

Consider two sets of impressed sources $\mathbf{J}_0^a$, $\mathbf{M}_0^a$ and $\mathbf{J}_0^b$, $\mathbf{M}_0^b$ with their respective fields $\mathbf{E}^a$, $\mathbf{H}^a$ and $\mathbf{E}^b$, $\mathbf{H}^b$ all at the same frequency. From the vector identity for the divergence of $(\mathbf{E}^a\times\mathbf{H}^b-\mathbf{E}^b\times\mathbf{H}^a)$ together with Maxwell's equations, Ohm's law, and the constitutive relations, it follows that

$$\nabla\cdot\left(\mathbf{E}^a\times\mathbf{H}^b-\mathbf{E}^b\times\mathbf{H}^a\right)=\left(\mathbf{J}_0^a\cdot\mathbf{E}^b-j\omega\mu\mathbf{M}_0^a\cdot\mathbf{H}^b\right)-\left(\mathbf{J}_0^b\cdot\mathbf{E}^a-j\omega\mu\mathbf{M}_0^b\cdot\mathbf{H}^a\right) \tag{1.56}$$

For source-free points this reduces to

$$\nabla\cdot\left(\mathbf{E}^a\times\mathbf{H}^b-\mathbf{E}^b\times\mathbf{H}^a\right)=0 \tag{1.57}$$

which is called the Lorentz reciprocity theorem. From the divergence theorem, the integral form of the Lorentz reciprocity theorem is

$$\int_S (\mathbf{E}^a\times\mathbf{H}^b-\mathbf{E}^b\times\mathbf{H}^a)\cdot d\mathbf{S}=0 \tag{1.58}$$

for a source-free region bounded by a surface S. In a region containing sources of finite extent, the integration of (1.56) over all space (bounding surface S infinitely remote) results in the reciprocity theorem*

$$\int_{\infty}\left(\mathbf{J}_0^a\cdot\mathbf{E}^b - j\omega\mu\mathbf{M}_0^a\cdot\mathbf{H}^b\right)dv = \int_{\infty}\left(\mathbf{J}_0^b\cdot\mathbf{E}^a - j\omega\mu\mathbf{M}_0^b\cdot\mathbf{H}^a\right)dv \tag{1.59}$$

Equation (1.59) also applies to regions of finite extent whenever (1.58) is satisfied. For example, fields in a region bounded by a perfect conductor satisfy (1.58) and hence the relation (1.59) applies. Both reciprocity theorems are valid for any linear, isotropic medium (including inhomogeneous media) obeying Ohm's law.

The application of (1.59) to two-antenna systems with impressed currents leads to the well-known result that *the voltage at antenna* b *from an impressed current at antenna* a *is equal to the voltage at antenna* a *from the same impressed current at antenna* b.

1.7 BOUNDARY CONDITIONS

The following well-known boundary conditions hold at the interface between two media, neither of which is perfectly conducting:

$$(\mathbf{E}_2 - \mathbf{E}_1)\times\mathbf{n} = 0 \tag{1.60}$$

$$(\mathbf{H}_2 - \mathbf{H}_1)\times\mathbf{n} = 0 \tag{1.61}$$

$$(\mathbf{D}_2 - \mathbf{D}_1)\cdot\mathbf{n} = \rho_s \tag{1.62}$$

$$(\mathbf{B}_2 - \mathbf{B}_1)\cdot\mathbf{n} = 0 \tag{1.63}$$

In the above, $\mathbf{n}$ is the unit vector normal to the interface directed into medium 2 and ρ_s is the surface density of free charge. The first two boundary conditions state that the tangential components of **E** and **H** are continuous across the interface. The last two conditions give the transition of the normal components of the field vectors **D** and **B** across the interface. Equation (1.62) states that the presence of a layer of surface charge on the interface results in a discontinuity of the normal component of **D** equal to the surface charge density, while (1.63) indicates that the normal component of **B** is continuous across the boundary.

If either or both of the media have conductivities that are finite and nonzero, the flow of charge across the boundary must satisfy (1.12). This results in the boundary condition

$$\mathbf{n}\cdot(\mathbf{J}_2 - \mathbf{J}_1) = -j\omega\rho_s \tag{1.64}$$

Using Ohm's law and the constitutive relation $\mathbf{D} = \epsilon\mathbf{E}$, (1.62) and (1.64) may be written

$$\epsilon_2 E_{2n} - \epsilon_1 E_{1n} = \rho_s \tag{1.65}$$

$$\sigma_2 E_{2n} - \sigma_1 E_{1n} = -j\omega\rho_s \tag{1.66}$$

where the subscript n denotes the normal component. The surface charge density ρ_s will be zero whenever

*The quantity $j\omega\mu\mathbf{M}_0$ in (1.59) is equivalent to $\mathbf{J}_m$, the fictitious magnetic current density used by many authors.

$$\sigma_1\epsilon_2 - \sigma_2\epsilon_1 = 0 \tag{1.67}$$

In general, however, ρ_s is not zero and may be eliminated from (1.65) and (1.66) to yield an alternative boundary condition on the normal component of **E**:

$$\mu_1\gamma_2^2 E_{2n} - \mu_2\gamma_1^2 E_{1n} = 0 \tag{1.68}$$

where

$$\gamma^2 = -\omega^2\mu\epsilon + j\omega\mu\sigma$$

At the surface of a perfect conductor the boundary conditions above are modified somewhat. Because all the field vectors vanish inside the perfect conductor, the boundary conditions (1.60) to (1.63) may be written

$$\mathbf{n} \times \mathbf{E} = 0 \tag{1.69}$$

$$\mathbf{n} \times \mathbf{H} = \mathbf{K} \tag{1.70}$$

$$\mathbf{n} \cdot \mathbf{D} = \rho_s \tag{1.71}$$

$$\mathbf{n} \cdot \mathbf{B} = 0 \tag{1.72}$$

where **n** is directed out of the perfect conductor and **K** is the surface current density.

Approximate boundary conditions will be discussed in chapter 2.

1.8 UNIQUENESS THEOREMS

Both the electromagnetic field and the potentials within a closed region containing sources satisfy an inhomogeneous equation such as (1.35). The complete solution within the region consists of the sum of a particular solution of the type (1.49), which gives the contribution of the enclosed sources, and a complementary solution, or solution to the homogeneous equation, which gives the contribution of the sources outside the region. A uniqueness theorem stipulates boundary conditions which assure that there is only one complementary solution for the contribution of external sources and thereby establishes a one-to-one correspondence of a field to its sources.

A uniqueness theorem for a time-harmonic electromagnetic field in a closed lossy region is given in Harrington.[1] This theorem states that *a field in a conducting region is uniquely specified by the sources within the region plus the tangential components of* **E** *over the boundary, or the tangential components of* **H** *over the boundary, or the former over part of the boundary and the latter over the rest of the boundary.* The field in a nonconducting medium is taken to be the limit of the corresponding field in a conducting medium as the conductivity goes to zero. For the stationary regime ($\omega = 0$), this statement of the theorem constitutes a sufficient condition for the uniqueness of the field. Because the stationary regime is described by Poisson's (or Laplace's) equation, a unique solution may alternately be obtained by specifying the normal component of the field on the bounding surface.[2]

If all the sources are a finite distance from a coordinate origin and the bounding surface s of the region is a sphere of radius $r \to \infty$, the solution to the homogeneous equation must vanish. This in effect assures that there are no sources at infinity and that the fields in the unbounded

---

[1] Harrington, R. F.: Time-Harmonic Electromagnetic Fields. McGraw-Hill Book Co., Inc., 1961, pp. 100–103.

[2] Jackson, J. D.: Classical Electrodynamics. John Wiley & Sons, Inc., 1962, pp. 15–17.

region represent diverging waves. The condition that assures the vanishing of the solutions to the homogeneous equation is stated in Harrington[1] as

$$\lim_{r\to\infty} \int (\mathbf{E} \times \tilde{\mathbf{H}}) \cdot \mathbf{n}\, dS = 0 \tag{1.73}$$

If the medium has a nonzero conductivity, the condition in (1.73) is satisfied automatically because the dissipation will result in the vanishing of the fields as $r \to \infty$. A nonconducting medium can be treated as the limit of the lossy case as the conductivity vanishes.

A more extensive treatment of the problems of the uniqueness of both the electromagnetic field and the auxiliary potentials may be found in Jones.[3]

1.9 TM AND TE FIELD ANALYSIS

It is shown in Morse and Feshbach[4] that *an arbitrary electromagnetic field in a homogeneous, source-free region can be expressed as the sum of a transverse magnetic (TM) field and a transverse electric (TE) field.* These partial fields can be made transverse to some particular coordinate direction, which is an advantage in the enforcement of boundary conditions. Moreover, each partial field is formed from one of the two scalars that together determine the complete electromagnetic field. The particular coordinate directions in the six coordinate systems for which such a decomposition can be effected are given in Morse and Feshbach.[4] In the common coordinate systems they are x, y, or z in rectangular coordinates; z in cylindrical coordinates; and the radius vector in spherical coordinates.

Using (1.21) and (1.22), the TM field is expressed in terms of the magnetic vector potential as

$$\mathbf{E} = -j\omega \mathbf{A} \tag{1.74}$$

$$\mathbf{B} = \nabla \times \mathbf{A} \tag{1.75}$$

where

$$\mathbf{A} = \mathbf{u}\psi^a \tag{1.76}$$

in either rectangular or cylindrical coordinates, and

$$\mathbf{A} = \mathbf{r}\psi^a \tag{1.77}$$

in spherical coordinates. In (1.76) $\mathbf{u}$ is a unit vector in one of the permissible coordinate directions mentioned above, while in (1.77) $\mathbf{r}$ is the radius vector. The function ψ^a is a solution of the scalar wave equation

$$\nabla^2\psi^a - \gamma^2\psi^a = 0 \tag{1.78}$$

Similarly, using (1.26) and (1.27), the TE field is written in terms of the electric vector potential as

$$\mathbf{D} = -\nabla \times \mathbf{F} \tag{1.79}$$

---

[3] Jones, D. S.: The Theory of Electromagnetism. Pergamon Press, Ltd. book, Macmillan Co., 1964, pp. 56-57 and 562-569.

[4] Morse, P. M.; and Feshbach, H.: Methods of Theoretical Physics. McGraw-Hill Book Co., Inc., 1953, ch. 13, pp. 1759-1767.

$$\mathbf{H} = -j\omega\mathbf{F} - \frac{\sigma}{\epsilon}\mathbf{F} \tag{1.80}$$

where

$$\mathbf{F} = \mathbf{u}\psi^f \tag{1.81}$$

in rectangular or cylindrical coordinates, and

$$\mathbf{F} = \mathbf{r}\psi^f \tag{1.82}$$

in spherical coordinates. The function ψ^f is likewise a solution of the scalar wave equation

$$\nabla^2\psi^f - \gamma^2\psi^f = 0 \tag{1.83}$$

The complete electromagnetic field in a homogeneous, source-free region is then given by

$$\mathbf{E} = -j\omega\mathbf{A} - \frac{1}{\epsilon}\nabla \times \mathbf{F} \tag{1.84}$$

$$\mathbf{H} = \frac{1}{\mu}\nabla \times \mathbf{A} - j\omega\mathbf{F} - \frac{\sigma}{\epsilon}\mathbf{F} \tag{1.85}$$

Chapter 2

PLANE WAVES

2.1 UNIFORM PLANE WAVES

From Maxwell's equations, the electric and magnetic fields in a source-free region are seen to satisfy the homogeneous wave equations

$$\nabla^2\mathbf{E} - \gamma^2\mathbf{E} = 0$$
$$\nabla^2\mathbf{H} - \gamma^2\mathbf{H} = 0 \tag{2.1}$$

where $\gamma^2 = -\omega^2\mu\epsilon + j\omega\mu\sigma$. The simplest and most fundamental electromagnetic wave solutions to (2.1) are linearly polarized uniform plane waves of the form

$$\mathbf{E} = \mathbf{E}_0 e^{j\omega t - \gamma\xi}$$
$$\mathbf{H} = \mathbf{H}_0 e^{j\omega t - \gamma\xi} \tag{2.2}$$

where $\mathbf{E}_0$ and $\mathbf{H}_0$ are vector amplitudes that may be complex, and ξ is the distance from the coordinate origin along the direction of propagation, i.e., the normal to the wave front. The plane waves are called uniform or homogeneous because the planes of constant amplitude coincide with the planes of constant phase. By substituting (2.2) into Maxwell's equations, it may be seen that the field vectors **E** and **H** are perpendicular and lie in a plane that is transverse to the direction of propagation. Plane waves with circular or elliptical polarization may be formed by combining two linearly polarized waves with the proper phasing.

It is useful to find the real and imaginary parts of the complex propagation constant γ. Thus, if

$$\gamma = a + j\beta = \left(-\omega^2\mu\epsilon + j\omega\mu\sigma\right)^{1/2} \tag{2.3}$$

then

$$a = \omega\left(\frac{\mu\epsilon}{2}\right)^{1/2}\left[\left(1 + \frac{\sigma^2}{\omega^2\epsilon^2}\right)^{1/2} - 1\right]^{1/2} \tag{2.4}$$

and

$$\beta = \omega\left(\frac{\mu\epsilon}{2}\right)^{1/2}\left[\left(1 + \frac{\sigma^2}{\omega^2\epsilon^2}\right)^{1/2} + 1\right]^{1/2} \tag{2.5}$$

The plane wave solutions (2.2) may then be written

$$\mathbf{E} = \mathbf{E}_0 e^{-\alpha\xi} e^{j(\omega t - \beta\xi)}$$

$$\mathbf{H} = \mathbf{H}_0 e^{-\alpha\xi} e^{j(\omega t - \beta\xi)} \tag{2.6}$$

This form of the solution shows that as the plane wave progresses with a phase velocity $v = \omega/\beta$, it suffers an exponential attenuation with an attenuation constant α as well as a phase change with a phase constant β. The ratio of the electric to the magnetic field amplitude is called the intrinsic impedance of the medium and may be obtained by substituting (2.6) into (1.29). This results in an expression for the intrinsic impedance η given by

$$\eta = \frac{E_0}{H_0} = \left\{ \frac{\mu}{\epsilon[1 + (\sigma/j\omega\epsilon)]} \right\}^{1/2} = \frac{j\omega\mu}{\gamma} \tag{2.7}$$

The relative magnitude of conduction current to displacement current in the medium is given by the ratio $\sigma/\omega\epsilon$. There are two cases of special interest.

Case 1. $\sigma/\omega\epsilon \ll 1$. Here the medium is only slightly lossy with the displacement current predominating. The binomial expansion of (2.4) and (2.5) yields

$$\alpha = \frac{1}{2}\sigma\left(\frac{\mu}{\epsilon}\right)^{1/2}\left[1 - \frac{1}{8}\left(\frac{\sigma}{\omega\epsilon}\right)^2 + \ldots\right] \approx \frac{1}{2}\sigma\left(\frac{\mu}{\epsilon}\right)^{1/2} \tag{2.8}$$

$$\beta = \omega(\mu\epsilon)^{1/2}\left[1 + \frac{1}{8}\left(\frac{\sigma}{\omega\epsilon}\right)^2 + \ldots\right] \approx \omega(\mu\epsilon)^{1/2} \tag{2.9}$$

which indicates that to a first approximation, the attenuation constant is independent of the frequency and the phase constant is the same as that for a perfect dielectric.

Case 2. $\sigma/\omega\epsilon \gg 1$. In this case, the conduction current predominates and the medium is very lossy. The binomial expansion then yields

$$\alpha = \left(\frac{\omega\mu\sigma}{2}\right)^{1/2}\left[1 + \frac{\omega\epsilon}{2\sigma} + \ldots\right] \approx \left(\frac{\omega\mu\sigma}{2}\right)^{1/2} \tag{2.10}$$

$$\beta = \left(\frac{\omega\mu\sigma}{2}\right)^{1/2}\left[1 - \frac{\omega\epsilon}{2\sigma} + \ldots\right] \approx \left(\frac{\omega\mu\sigma}{2}\right)^{1/2} \tag{2.11}$$

which shows that to a first approximation the attenuation and phase constants are equal.

These results in the first approximation are summarized in table 2.1.

In the case of a good conductor, it is of interest to express the attenuation over a wavelength λ in terms of decibels. Thus, since $\alpha = \beta = 2\pi/\lambda$, it follows that the attenuation over a wavelength in the medium equals $e^{-2\pi}$ or approximately 55 dB.

From the expression for the intrinsic impedance of a good conductor it may be seen that **E** and **H** differ in phase by $\pi/4$ radians. This contrasts with the in-phase relationship between **E** and **H** in a lossless medium.

The skin depth or depth of penetration in a homogeneous, conductive half space is the depth in the conductor over which the field amplitudes of a normally propagating plane wave decrease by a factor of $1/e$ or approximately 37 percent. From (2.6) it follows that the skin depth δ is the reciprocal of the attenuation constant α which is defined by (2.4). For a good conductor,

$$\delta = \frac{1}{\alpha} = \frac{1}{\beta} = \frac{\sqrt{2}}{|\gamma|} = \frac{\lambda}{2\pi} = \left(\frac{2}{\omega\mu\sigma}\right)^{1/2} \tag{2.12}$$

Plots of the skin depth for various conductivities are given as a function of frequency in appendix A.

Table 2.1.–*Approximate propagation constants and intrinsic impedance*

	$\frac{\sigma}{\omega\epsilon} \ll 1$, poor conductor	$\frac{\sigma}{\omega\epsilon} \gg 1$, good conductor
Attenuation constant α	$\frac{\sigma}{2}\left(\frac{\mu}{\epsilon}\right)^{1/2}$	$\left(\frac{\omega\mu\sigma}{2}\right)^{1/2}$
Phase constant β	$\omega(\mu\epsilon)^{1/2}$	$\left(\frac{\omega\mu\sigma}{2}\right)^{1/2}$
Propagation constant $\gamma = \alpha + j\beta$	$j\omega(\mu\epsilon)^{1/2}$	$(j\omega\mu\sigma)^{1/2}$
Intrinsic impedance η	$\left(\frac{\mu}{\epsilon}\right)^{1/2}$	$\left(\frac{j\omega\mu}{\sigma}\right)^{1/2}$

2.2 NONUNIFORM PLANE WAVES

In addition to uniform plane waves, it is possible to find more general plane wave solutions to Maxwell's equations called nonuniform or inhomogeneous plane waves. Both types of plane waves may be combined in the space domain using Fourier integral methods to meet a wide variety of boundary conditions. Nonuniform plane waves have planes of constant amplitude that are not parallel to the planes of constant phase. Thus, unlike uniform plane waves that have one and the same direction for both the normal vector to the planes of constant phase and amplitude and the Poynting vector, nonuniform plane waves in general have different directions for each of these three vectors. In a dissipative medium, the field vectors **E** and **H** of a linearly polarized nonuniform plane wave are required by Maxwell's equations to be normal. For elliptical polarization, however, the space angle between **E** and **H** is not in general $\pi/2$ but actually varies with time. The magnitude of **E** for either polarization is no longer simply related to that of **H** solely in terms of the parameters of the medium as in (2.7). Nonuniform plane waves generally exist in discontinuous media or are established by sources that are not remote.

In a lossless medium, nonuniform plane waves are sometimes called evanescent waves and have planes of constant amplitude that are normal to the planes of constant phase. In this case, the real power flow is normal to the planes of constant phase and the reactive power flows normal to the planes of constant amplitude. If the medium is conducting, the angle between these two sets of planes must be less than $\pi/2$ radians.

A detailed discussion of the properties of both uniform and nonuniform plane waves in lossless and dissipative media may be found in Adler, Chu, and Fano.[5]

[5] Adler, R. B.; Chu, L. J.; and Fano, R. M.: Electromagnetic Energy Transmission and Radiation. John Wiley & Sons, Inc., 1960, chs. 7 and 8, pp. 304–492.

2.3 REFLECTION AND REFRACTION AT A PLANE SURFACE

This section treats the interaction of a plane wave with a plane boundary between two homogeneous media. In such an interaction, the reflected and transmitted waves are also taken to be plane waves because such an assumption leads to a unique solution of Maxwell's equations. The uniqueness condition for plane waves which applies to both uniform and nonuniform plane waves is discussed in Williams.[6] Essentially, this condition states that the *transmitted field must support an energy flow from the boundary rather than toward the boundary.**

2.3.1 Snell's Laws

Two homogeneous media have a common plane boundary (fig. 2.1) that is normal to the plane of the page and contains the origin of the coordinates. A plane wave $\mathbf{E}_i, \mathbf{H}_i$ in medium 2 is incident on the boundary at an angle θ_0 and results in reflected and refracted plane waves at angles θ_2 and θ_1 respectively. The unit vector $\mathbf{n}$ is normal to the boundary plane and the unit vectors $\mathbf{n}_0$, $\mathbf{n}_1$, and $\mathbf{n}_2$ are respectively the normals to the planes of constant phase of the incident, refracted, and reflected waves. The incident plane wave may be written as

$$\mathbf{E}_i = \mathbf{E}_0 e^{j\omega t - \gamma_2 \mathbf{n}_0 \cdot \mathbf{r}}$$
$$\mathbf{H}_i = \frac{\gamma_2}{j\omega\mu_2} \mathbf{n}_0 \times \mathbf{E}_i \tag{2.13}$$

the transmitted plane wave $\mathbf{E}_t, \mathbf{H}_t$ as

$$\mathbf{E}_t = \mathbf{E}_1 e^{j\omega t - \gamma_1 \mathbf{n}_1 \cdot \mathbf{r}}$$
$$\mathbf{H}_t = \frac{\gamma_1}{j\omega\mu_1} \mathbf{n}_1 \times \mathbf{E}_t \tag{2.14}$$

and the reflected plane wave $\mathbf{E}_r, \mathbf{H}_r$ as

$$\mathbf{E}_r = \mathbf{E}_2 e^{j\omega t - \gamma_2 \mathbf{n}_2 \cdot \mathbf{r}}$$
$$\mathbf{H}_r = \frac{\gamma_2}{j\omega\mu_2} \mathbf{n}_2 \times \mathbf{E}_r \tag{2.15}$$

where $\mathbf{E}_0$, $\mathbf{E}_1$, and $\mathbf{E}_2$ are complex amplitudes independent of the coordinates. If the tangential components of the resultant field vectors are to be continuous across the boundary, it is necessary that the phases of these field vectors in (2.13), (2.14), and (2.15) be identical over the boundary. It follows that $\mathbf{n}$, $\mathbf{n}_0$, $\mathbf{n}_1$, and $\mathbf{n}_2$ are all coplanar and that

$$\sin\theta_2 = \sin\theta_0 \tag{2.16}$$

---

[6] Williams, R. H.: Propagation Between Conducting Media. Tech. Rept. EE-50, Eng. Exp. Station, Univ. of N. Mex., Albuquerque, Sept. 1961, pp. 58–59.

*In the case of layered media this must be true of the transmitted field in the last layer.

and

$$\gamma_2 \sin \theta_0 = \gamma_1 \sin \theta_1 \tag{2.17}$$

The expressions in (2.16) and (2.17) are known respectively as *Snell's laws* of reflection and refraction. It should be noted that since in general both γ_1 and γ_2 may be complex, the angles of incidence and refraction may be complex. Physically, a complex angle implies a phase change and an attenuation such that the planes of constant phase are not parallel to the planes of constant amplitude, thereby indicating the presence of nonuniform plane waves. In the event that θ_1 is complex, for example, the vector $\mathbf{n}_1$ can no longer be interpreted as a real vector normal to the planes of constant phase of the transmitted wave. For θ_1 complex, the quantity $\mathbf{n}_1 \cdot \mathbf{r}$ is complex and the vector $\mathbf{n}_1$ has complex components and can no longer be pictured. Such complex vectors are discussed in Adler, Chu, and Fano.[5]

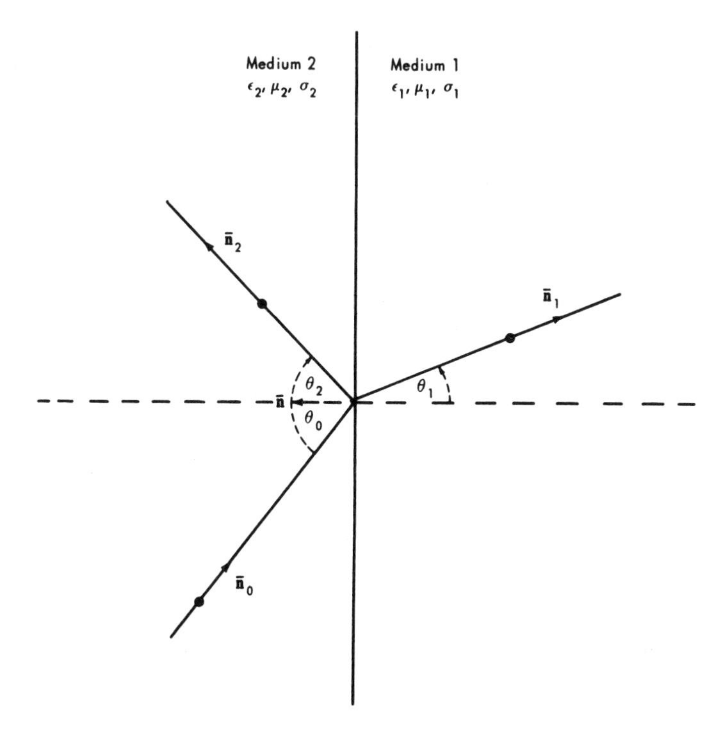

Figure 2.1.–*Reflection and refraction at a plane boundary.*

2.3.2 Fresnel's Equations

The boundary conditions on the tangential components of the field vectors may be used to determine the relation between the amplitudes $\mathbf{E}_0$, $\mathbf{E}_1$, and $\mathbf{E}_2$. Although the orientation of the primary vector $\mathbf{E}_0$ is quite arbitrary, it can always be resolved into a component normal to the plane of incidence (plane containing $\mathbf{n}_0$ and $\mathbf{n}$) and a component in the plane of incidence. This resolution is equivalent to TE and TM plane waves respectively.

Case 1. $\mathbf{E}_0$ *normal to the plane of incidence* (TE)

$$\mathbf{E}_1 = \left[\frac{2\mu_1\gamma_2 \cos \theta_0}{\mu_1\gamma_2 \cos \theta_0 + \mu_2\left(\gamma_1^2 - \gamma_2^2 \sin^2 \theta_0\right)^{1/2}}\right]\mathbf{E}_0 \tag{2.18}$$

$$\mathbf{E}_2 = \left[\frac{\mu_1\gamma_2 \cos\theta_0 - \mu_2\left(\gamma_1^2 - \gamma_2^2 \sin^2\theta_0\right)^{1/2}}{\mu_1\gamma_2 \cos\theta_0 + \mu_2\left(\gamma_1^2 - \gamma_2^2 \sin^2\theta_0\right)^{1/2}}\right]\mathbf{E}_0 \tag{2.19}$$

The magnetic field may then be obtained from (2.13), (2.14), and (2.15).

Case 2. $\mathbf{E}_0$ *in the plane of incidence* (TM)

In this case, the magnetic field vectors are normal to the plane of incidence and in a manner similar to Case 1, it follows that

$$\mathbf{H}_1 = \left[\frac{2\mu_2\gamma_1^2 \cos\theta_0}{\mu_2\gamma_1^2 \cos\theta_0 + \mu_1\gamma_2\left(\gamma_1^2 - \gamma_2^2 \sin^2\theta_0\right)^{1/2}}\right]\mathbf{H}_0 \tag{2.20}$$

$$\mathbf{H}_2 = \left[\frac{\mu_2\gamma_1^2 \cos\theta_0 - \mu_1\gamma_2\left(\gamma_1^2 - \gamma_2^2 \sin^2\theta_0\right)^{1/2}}{\mu_2\gamma_1^2 \cos\theta_0 + \mu_1\gamma_2\left(\gamma_1^2 - \gamma_2^2 \sin^2\theta_0\right)^{1/2}}\right]\mathbf{H}_0 \tag{2.21}$$

From Maxwell's equations, the electric field vectors are given by

$$\mathbf{E}_0 = -\frac{j\omega\mu_2}{\gamma_2}\mathbf{n}_0 \times \mathbf{H}_0 \tag{2.22}$$

$$\mathbf{E}_1 = -\frac{j\omega\mu_1}{\gamma_1}\mathbf{n}_1 \times \mathbf{H}_1 \tag{2.23}$$

$$\mathbf{E}_2 = -\frac{j\omega\mu_2}{\gamma_2}\mathbf{n}_2 \times \mathbf{H}_2 \tag{2.24}$$

At normal incidence, $\theta_0 = 0$, and the two cases cannot be distinguished. The electric field then reduces to

$$\mathbf{E}_1 = \frac{2\mu_1\gamma_2}{\mu_1\gamma_2 + \mu_2\gamma_1}\mathbf{E}_0 \tag{2.25}$$

$$\mathbf{E}_2 = \frac{\mu_1\gamma_2 - \mu_2\gamma_1}{\mu_1\gamma_2 + \mu_2\gamma_1}\mathbf{E}_0 \tag{2.26}$$

2.3.3 Field Reflection and Transmission Coefficients

If a field transmission coefficient τ and a field reflection coefficient ρ are defined for the TE case by the bracketed quantities in (2.18) and (2.19) respectively and for the TM case by the bracketed quantities in (2.20) and (2.21) respectively, then it may be seen that for either case

$$\rho + 1 = \tau \tag{2.27}$$

Thus, if either the field reflection coefficient or the field transmission coefficient is known, the other may readily be calculated.

The expressions for the field reflection and transmission coefficients given in (2.18) through (2.21) can also be written in terms of the intrinsic impedances of the two media referred to the normal to the interface by using (2.7). The normal impedance Z_{2n} of the incident plane wave is the ratio of the tangential component of $\mathbf{E}_0$ to that of $\mathbf{H}_0$ at the boundary. Similarly, the normal impedance Z_{1n} of the refracted wave is the ratio of the tangential component of $\mathbf{E}_1$ to $\mathbf{H}_1$ at the boundary. Thus, ρ and τ may be written

$$\rho=\frac{Z_{1n}-Z_{2n}}{Z_{1n}+Z_{2n}} \tag{2.28}$$

$$\tau=\frac{2Z_{1n}}{Z_{1n}+Z_{2n}} \tag{2.29}$$

where

$$Z_{2n}=\frac{\eta_2}{\cos\theta_0} \tag{2.30}$$

$$Z_{1n}=\frac{\eta_1}{\cos\theta_1} \tag{2.31}$$

for the TE case, and

$$Z_{2n}=\eta_2\cos\theta_0 \tag{2.32}$$

$$Z_{1n}=\eta_1\cos\theta_1 \tag{2.33}$$

for the TM case.

Simplified expressions for the transmission coefficients in (2.18) and (2.20) may be obtained for the case of a uniform plane wave that travels in a dielectric and refracts into a good conductor. For $\mu_2=\mu_1$, it follows that

$$\left|\frac{\gamma_2}{\gamma_1}\right|\ll 1$$

and the transmission coefficients become

$$\tau_{TE}\approx 2\cos\theta_0\frac{\gamma_2}{\gamma_1} \tag{2.34}$$

$$\tau_{TM}\approx\frac{2\cos\theta_0}{\cos\theta_0+\dfrac{\gamma_2}{\gamma_1}} \tag{2.35}$$

Since the transmission coefficients τ_{TE} and τ_{TM} are different functions of the media constants and the angle of incidence, the reflection coefficients ρ_{TE} and ρ_{TM} will also be different by (2.27). This means that in general an incident linearly polarized plane wave will be transmitted and reflected with an elliptical polarization. Plots of the real and imaginary parts of τ_{TE} and τ_{TM} as a function of frequency and the angle of incidence are presented in appendix A for a plane wave going from free space into sea water.

2.3.4 Power Reflection and Transmission

The average power transmitted or reflected may be determined from the condition that the normal component of the resultant Poynting vector in (1.55) must be continuous across the interface. Applying this condition to a normally incident plane wave that passes from a dielectric into a good conductor yields

$$\frac{S_i}{S_t} = \frac{1}{4}\left(\frac{2\sigma_1}{\omega\epsilon_2}\right)^{1/2} \tag{2.36}$$

where S_i = incident average power
S_t = transmitted average power

For a plane wave passing at normal incidence from a good conductor to a dielectric

$$\frac{S_i}{S_t} = \frac{1}{4}\left(\frac{\sigma_2}{2\omega\epsilon_1}\right)^{1/2} \tag{2.37}$$

2.4 REFRACTION IN A CONDUCTING MEDIUM

The field refracted into a semi-infinite conducting medium from a dielectric half space in which a uniform plane wave is propagating exhibits properties that are of great interest.

Referring to figure 2.1, a uniform plane wave propagates in medium 2, which is taken to be a perfect dielectric, $(\sigma_2 = 0)$, and is transmitted into medium 1, which is conducting. The propagation constants for the two media are defined by

$$\gamma_2 = j\omega\left(\mu_2\epsilon_2\right)^{1/2} = j\beta_2 \tag{2.38}$$

$$\gamma_1 = \left(-\omega^2\mu_1\epsilon_1 + j\omega\mu_1\sigma_1\right)^{1/2} = \alpha_1 + j\beta_1 \tag{2.39}$$

where α_1 and β_1 are given by (2.4) and (2.5), respectively. The ratio γ_1/γ_2 is called the complex index of refraction of the conductor.

From Snell's law it follows that

$$\sin\theta_1 = \frac{\gamma_2}{\gamma_1}\sin\theta_0 \tag{2.40}$$

and that θ_1 is a complex angle. This implies that the transmitted wave is nonuniform. The complex quantity $\cos\theta_1$ may be expressed as

$$\cos\theta_1 = \left(1 - \sin^2\theta_1\right)^{1/2} = ae^{j\phi} \tag{2.41}$$

Making use of figure 2.2 and (2.14), the transmitted wave may be written

$$\mathbf{E}_t = \mathbf{E}_1 e^{j\omega t - \gamma_1(z\cos\theta_1 + y\sin\theta_1)} \tag{2.42}$$

where $\mathbf{n}_1 \cdot \mathbf{r}$ has been expanded in a formal way. From (2.40) and (2.41), it follows that

$$\mathbf{E}_t = \mathbf{E}_1 e^{pz + j(\omega t + qz - \beta_2 y \sin \theta_0)} \quad (z < 0) \tag{2.43}$$

where

$$\left. \begin{aligned} p &= a(\beta_1 \cos \phi + a_1 \sin \phi) \\ q &= a(a_1 \cos \phi - \beta_1 \sin \phi) \end{aligned} \right\} \tag{2.44}$$

The expressions for p and q in terms of the constants of the media and the angle of incidence are

$$\begin{aligned} p^2 &= \frac{1}{2}\left\{-a_1^2 + \beta_1^2 + \beta_2^2 \sin^2 \theta_0 + \left[4a_1^2\beta_1^2 + \left(a_1^2 - \beta_1^2 - \beta_2^2 \sin^2 \theta_0\right)^2\right]^{1/2}\right\} \\ q^2 &= \frac{1}{2}\left\{a_1^2 - \beta_1^2 - \beta_2^2 \sin^2 \theta_0 + \left[4a_1^2\beta_1^2 + \left(a_1^2 - \beta_1^2 - \beta_2^2 \sin^2 \theta_0\right)^2\right]^{1/2}\right\} \end{aligned} \tag{2.45}$$

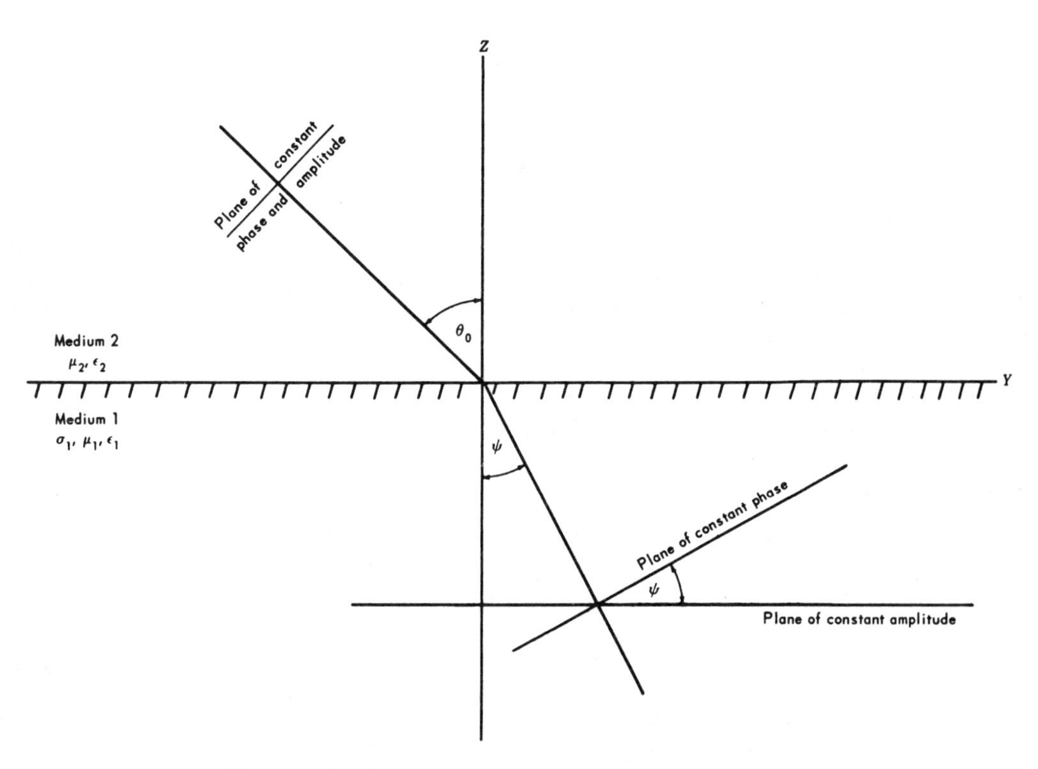

Figure 2.2.—*Refraction at a plane conducting surface.*

From (2.43), the surfaces of constant amplitude are the planes $pz = \text{constant}$, and the surfaces of constant phase are the planes $qz - \beta_2 \sin \theta_0 y = \text{constant}$. In general, these two families of planes do not coincide. It should be noted that *the planes of constant amplitude are parallel to the interface.* The angle ψ between a plane of constant amplitude and a plane of constant phase is also the angle that the normal to the planes of constant phase makes with the z-axis, and represents the real angle of refraction. This angle is defined by $-z \cos \psi + y \sin \psi = \text{constant}$. By comparison with the expression for the planes of constant phase from (2.43), it follows that

$$\cos \psi = \frac{q}{\left(q^2 + \beta_2^2 \sin^2 \theta_0\right)^{1/2}}$$
$$\sin \psi = \frac{\beta_2 \sin \theta_0}{\left(q^2 + \beta_2^2 \sin^2 \theta_0\right)^{1/2}} \tag{2.46}$$

A modified Snell's law for real angles then becomes

$$\sin \psi = \frac{1}{\beta_2}\left(\beta_2^2 \sin^2 \theta_0 + q^2\right)^{1/2} \sin \theta_0 \tag{2.47}$$

For the case where $\sigma_1 \gg \omega\epsilon_1$,

$$p \approx q \approx \left(\frac{\omega\mu_1\sigma_1}{2}\right)^{1/2} \tag{2.48}$$

and

$$\sin \psi \approx \left(\frac{2\omega\mu_2\epsilon_2}{\mu_1\sigma_1}\right)^{1/2} \sin \theta_0 \tag{2.49}$$

It follows from (2.49) that whenever $\left(2\omega\mu_2\epsilon_2/\mu_1\sigma_1\right)^{1/2} \ll 1$, the angle ψ is very nearly zero. This means that *the planes of constant phase are essentially parallel to the planes of constant amplitude and that the propagation into the conductor is in a direction normal to the surface.* Thus, the wave in the conductor has the characteristics of a uniform plane wave with a direction of propagation that is insensitive to the angle of incidence θ_0. This occurs, for example, in going from free space to metal or sea water at radio frequencies and below, and in going from free space to moderately conducting earth at VLF and below.

2.5 SURFACE WAVES

If the angle of incidence θ_0 in (2.19) or (2.21) is allowed to be complex, then it is possible to eliminate the reflected wave. The resulting plane wave will then propagate along the interface with a loss of energy into the dissipative medium. Although a pure surface wave of this type with a constant amplitude factor cannot be set up by a finite physical structure, it does describe in a local sense both the wave tilt and the energy loss into the dissipative medium of realizable surface waves that are launched over good conductors by finite antennas. The details of the solution for complex angles of incidence are given in Stratton.[7] For the common case of a TM plane wave propagating along the interface between a dielectric and a good conductor $(\sigma_1 \gg \omega\epsilon_1)$ of the same permeability, the results referring to figure 2.3 are:

In the dielectric $(z > 0)$:

$$H_x = ce^{-h_2 z} e^{j\omega t - gy} \tag{2.50}$$

[7] Stratton, J. A.: Electromagnetic Theory. McGraw-Hill Book Co., Inc., 1941, pp. 516-524.

$$E_z \approx -\left(\frac{\mu}{\epsilon_2}\right)^{1/2} H_x \tag{2.51}$$

$$E_y \approx -\left(\frac{j\omega\mu}{\sigma_1}\right)^{1/2} H_x \tag{2.52}$$

$$\frac{E_y}{E_z} \approx \left(\frac{j\omega\epsilon_2}{\sigma_1}\right)^{1/2} \tag{2.53}$$

where c is an arbitrary constant and

$$h_2 \approx -j\omega\left(\mu\epsilon_2\right)^{1/2}\left(\frac{j\omega\epsilon_2}{\sigma_1}\right)^{1/2} \tag{2.54}$$

$$g \approx j\omega\left(\mu\epsilon_2\right)^{1/2}\left(1 - \frac{j\omega\epsilon_2}{\sigma_1}\right)^{1/2} \tag{2.55}$$

$$\tan \psi_2 \approx \left(\frac{2\sigma_1}{\omega\epsilon_2}\right)^{1/2} \tag{2.56}$$

$$\tan \psi_2' \approx \left(\frac{\omega\epsilon_2}{2\sigma_1}\right)^{1/2} \tag{2.57}$$

In the conductor ($z<0$):

$$H_x = ce^{h_1 z}e^{j\omega t - gy} \tag{2.58}$$

$$E_z \approx \frac{-j\omega\epsilon_2}{\sigma_1}\left(\frac{\mu}{\epsilon_2}\right)^{1/2} H_x \tag{2.59}$$

$$E_y \approx -\left(\frac{j\omega\mu}{\sigma_1}\right)^{1/2} H_x \tag{2.60}$$

$$\frac{E_y}{E_z} \approx \left(\frac{\sigma_1}{j\omega\epsilon_2}\right)^{1/2} \tag{2.61}$$

$$h_1 \approx \left(j\omega\mu\sigma_1\right)^{1/2} \tag{2.62}$$

$$\tan \psi_1 \approx \left(\frac{2\omega\epsilon_2}{\sigma_1}\right)^{1/2} \tag{2.63}$$

$$\tan \psi'_1 \approx \frac{\omega\epsilon_2}{\sigma_1}\left(\frac{\omega\epsilon_2}{2\sigma_1}\right)^{1/2} \tag{2.64}$$

Thus, an elliptically polarized plane wave propagates along the surface in the dielectric with a wave tilt

$$\frac{E_y}{E_z} = \left(\frac{j\omega\epsilon_2}{\sigma_1}\right)^{1/2}$$

and travels nearly vertically in the conductor as an almost uniform plane wave with a propagation constant $h_1 \approx \left(j\omega\mu\sigma_1\right)^{1/2}$.

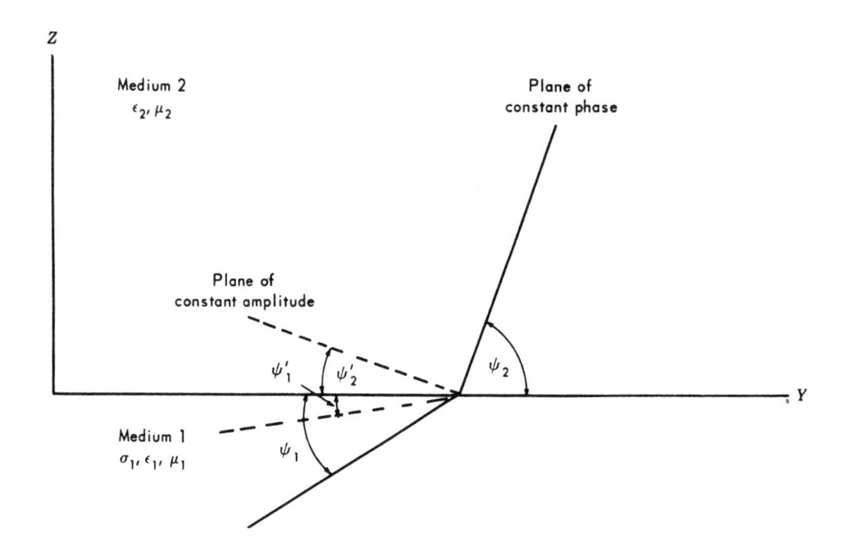

Figure 2.3.–*Surface wave along the interface between a dielectric and a conductor.*

2.6 PLANE WAVES IN LAYERED MEDIA

A detailed treatment of electromagnetic waves in stratified media is available in Brekhovskikh[8] and Wait.[9] For the purposes of this chapter, plane wave incidence on a stratified medium composed of M homogeneous layers will be discussed. Referring to figure 2.4, a plane wave is incident at an angle θ_0 on a stratified medium. The electrical constants for the mth layer are σ_m, ϵ_m, and μ_m.

For the TM case, the magnetic field has only a y-component and is a solution of

$$\nabla^2 H_{my} - \gamma_m^2 H_{my} = 0 \tag{2.65}$$

[8] Brekhovskikh, L. M.: Waves in Layered Media. Academic Press, Inc., 1960.

[9] Wait, J. R.: Electromagnetic Waves in Stratified Media. Pergamon Press, Ltd. book, Macmillan Co., 1962.

where $\gamma_m^2 = -\omega^2 \mu_m \epsilon_m + j\omega \mu_m \sigma_m$ with the real part of $\gamma_m > 0$. The solution desired is of the form

$$H_{my} = \left(a_m e^{-u_m z} + b_m e^{u_m z}\right) e^{-j\nu x} \tag{2.66}$$

where $u_m^2 = \nu^2 + \gamma_m^2$, ν can take any real value, and the real part of $u_m > 0$. The incident magnetic field H_{0y}^{inc} can be written

$$H_{0y}^{\mathrm{inc}} = H_0 e^{-z\gamma_\circ \cos\theta_\circ - x\gamma_\circ \sin\theta_\circ} \tag{2.67}$$

and therefore in (2.66) $a_0 e^{-u_0 z - j\nu x}$ can be identified with H_{0y}^{inc} if $a_0 = H_0$ and $j\nu = \gamma_0 \sin\theta_0$. It follows that $b_0 e^{u_0 z - j\nu x}$ is the reflected wave. The boundary conditions at the interface $z=0$, $z=z_1, \ldots, z=z_{m-1}$ are that the tangential magnetic and electric fields be continuous. Therefore

$$\left[\begin{array}{l} H_{m-1,y} = H_{my} \\ \\ \left(\sigma_{m-1} + j\omega\epsilon_{m-1}\right)^{-1} \dfrac{\partial h_{m-1,y}}{\partial z} = \left(\sigma_m + j\omega\epsilon_m\right)^{-1} \dfrac{\partial H_{my}}{\partial z} \end{array}\right]_{z=z_{m-1}} \tag{2.68}$$

where $m = 1, 2, \ldots,$ M. The above boundary conditions at the M interfaces result in $2M$ equations, which are linear in a_m and b_m. Because only outgoing waves are permissible in the lowest layer, which is semi-infinite, it follows that $b_M = 0$. The $2M$ equations may then be solved for the remaining $2M+1$ unknowns in terms of a_0, which is presumed to be known.

The extension of the above discussion to the TE case is straightforward. In this case,

$$E_{my} = \left(a_m e^{-u_m z} + b_m e^{u_m z}\right) e^{-j\nu x} \tag{2.69}$$

and the boundary conditions become

$$\left[\begin{array}{l} E_{m-1,y} = E_{my} \\ \\ \left(j\omega\mu_{m-1}\right)^{-1} \dfrac{\partial E_{m-1,y}}{\partial z} = \left(j\omega\mu_m\right)^{-1} \dfrac{\partial E_{my}}{\partial z} \end{array}\right]_{z=z_{m-1}} \tag{2.70}$$

where $m = 1, 2, \ldots, M$.

In calculating the surface impedance (the constant ratio of the resultant tangential electric and magnetic fields at the surface of the layered medium), it is convenient to use the well-known transmission line analogy of wave propagation, which is based on the one-to-one correspondence between reflection coefficient and impedance mismatch ratio. This analogy is discussed in Ramo and Whinnery.[10] For a medium consisting of an upper layer of thickness z_1 with an intrinsic impedance $\eta_1 = j\omega\mu_1/\gamma_1$ and a lower layer of infinite thickness with an intrinsic impedance $\eta_2 = j\omega\mu_2/\gamma_2$, the surface impedance η_s is given by

[10] Ramo, S.; and Whinnery, J. R.: Fields and Waves in Modern Radio. Second ed., John Wiley & Sons, Inc., 1953, chs. 1 and 7.

$$\eta_s = \eta_1 \frac{\eta_2 + \eta_1 \tanh \gamma_1 z_1}{\eta_1 + \eta_2 \tanh \gamma_1 z_1} \tag{2.71}$$

For an m-layered structure

$$\eta_s = \eta_1 \frac{\hat{\eta}_2 + \eta_1 \tanh \gamma_1 z_1}{\eta_1 + \hat{\eta}_2 \tanh \gamma_1 z_1} \tag{2.72}$$

where

$$\hat{\eta}_2 = \eta_2 \frac{\hat{\eta}_3 + \eta_2 \tanh \gamma_2 z_2}{\eta_2 + \hat{\eta}_3 \tanh \gamma_2 z_2} \tag{2.73}$$

$$\vdots \qquad \vdots \qquad \vdots$$

$$\hat{\eta}_{m-1} = \eta_{m-1} \frac{\hat{\eta}_m + \eta_{m-1} \tanh \gamma_{m-1} z_{m-1}}{\eta_{m-1} + \hat{\eta}_m \tanh \gamma_{m-1} z_{m-1}} \tag{2.74}$$

with $\eta_i = j\omega\mu_i/\gamma_i$ = intrinsic impedance of the ith layer
and $\hat{\eta}_i$ = impedance at the top of the ith layer

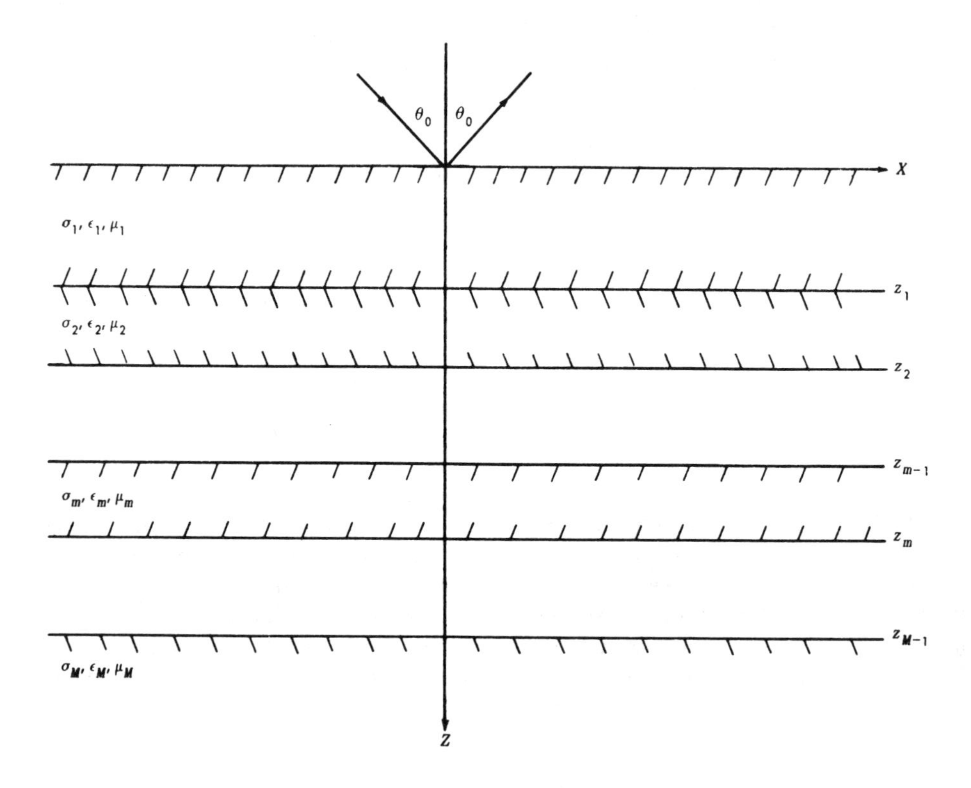

Figure 2.4.—*A stratified medium of* M *layers.*

Formulas (2.71) and (2.72) to (2.74) are exact for plane waves at normal incidence. They also hold to a good degree of approximation for arbitrary θ_0 whenever $|\gamma_0/\gamma_1| \ll 1$ because then the wave travels almost normal to the layers.

When a plane wave propagates along the normal to the interfaces of a two-layer medium, for example a sea-water layer over a semi-infinite earth bottom, it is useful to know the ratio of the resultant electric (or magnetic) field at a given depth in the upper layer to the resultant electric (or magnetic) field at the surface of that layer. This ratio at a depth z in an upper layer of thickness z_1 where $0 \leqq z \leqq z_1$ is given by

$$\frac{E_1(z)}{E_1(0)} = \frac{[\eta_2 + \eta_1 \tanh \gamma_1(z_1 - z)] \cosh \gamma_1(z_1 - z)}{[\eta_2 + \eta_1 \tanh \gamma_1 z_1] \cosh \gamma_1 z_1} \tag{2.75}$$

for the electric field, and by

$$\frac{H_1(z)}{H_1(0)} = \frac{[\eta_1 + \eta_2 \tanh \gamma_1(z_1 - z)] \cosh \gamma_1(z_1 - z)}{[\eta_1 + \eta_2 \tanh \gamma_1 z_1] \cosh \gamma_1 z_1} \tag{2.76}$$

for the magnetic field.

Numerical results in graphical form for the surface impedance of a stratified conductor of two and three layers are presented in Wait.[9] (See also ch. 3, sec. 3.3.1.) Graphs of the amplitude and phase of the components of the resultant magnetic field at the surface of a conductor of two layers are given by Dosso[11] for various angles of incidence, frequencies, conductivities, and surface layer depths. Expressions for the amplitude and phase of the components of the electric and magnetic fields in the top layer of a three-layer conductor are derived and evaluated by Dosso[12] for various frequencies, angles of incidence, conductivities, layer thicknesses, and depths in the top layer.

The two-dimensional problem of a low-frequency plane wave in free space incident on a semi-infinite conductor is treated by Weaver[13] for the case where the conductor is divided into two regions of different conductivity with a plane interface which is normal to the surface of the conducting half space. Formal solutions in terms of integrals are obtained for the electromagnetic field in the two conducting regions and numerical results for 0.1 Hz and 1 Hz are presented for the variation of the vertical magnetic field at the surface of the half space as a function of the distance from the discontinuity between the two conducting regions.

2.7 IMPEDANCE BOUNDARY CONDITIONS

If the components of the electric and magnetic field tangential to a surface S at a point P are perpendicular, the ratio of the tangential electric field to the tangential magnetic field is called the field impedance normal to S at P. In a general electromagnetic field configuration, the tangential electric and magnetic fields are not perpendicular so that the field impedance does not exist; even if it does exist, it may depend upon the field structure at the surface S.

---

[11] Dosso, H. W.: The Magnetic Field at the Surface of a Stratified Flat Conductor in the Field of Plane Waves with Application to Geophysics. Can. J. Phys., vol. 40, 1962, pp. 1583-1592.

[12] Dosso, H. W.: The Electric and Magnetic Fields in a Stratified Flat Conductor for Incident Plane Waves. Can. J. Phys., vol. 43, 1965, pp. 898-909.

[13] Weaver, J. T.: The Electromagnetic Field Within a Discontinuous Conductor With Reference to Geomagnetic Micropulsations Near a Coastline. Can. J. Phys., vol. 41, 1963, pp. 484-495.

Let the x, y-plane be the boundary between two semi-infinite homogeneous media, the fields being set up by sources in medium 2. It is shown by Wait[14] that in this plane a general wave front in medium 2 has tangential electric fields E_x and E_y related to the tangential magnetic fields by

$$E_x = \eta_1 H_y + \frac{\eta_1}{2\gamma_1^2}\left(\frac{\partial^2 H_y}{\partial y^2} - \frac{\partial^2 H_y}{\partial x^2} + 2\frac{\partial^2 H_x}{\partial x \partial y}\right) + \text{terms in } \gamma^{-4}, \text{ etc.}$$

$$-E_y = \eta_1 H_x + \frac{\eta_1}{2\gamma_1^2}\left(\frac{\partial^2 H_x}{\partial x^2} - \frac{\partial^2 H_x}{\partial y^2} + 2\frac{\partial^2 H_y}{\partial x \partial y}\right) + \text{terms in } \gamma^{-4}, \text{ etc.} \tag{2.77}$$

If H_x and H_y vary sufficiently slowly over the x, y-plane, or if γ_1 is sufficiently large, only the first term in each of the above expressions need be retained. Under these conditions

$$\left.\begin{aligned} E_x &\approx \eta_1 H_y \\ E_y &\approx \eta_1 H_x \end{aligned}\right\} \tag{2.78}$$

The field impedance normal to the surface exists and is equal to the intrinsic impedance of medium 1. Because this field impedance is independent of the field structure at the boundary, the term surface impedance is applicable.

The conditions implied by (2.78) are met at the boundary between a dielectric (medium 2) and a conductor (medium 1) whenever $|\gamma_2/\gamma_1| \ll 1$. In this case, the boundary is approximately an equiphase surface, and the wave propagates into the conductor along the inward normal to the surface in the manner of a uniform plane wave. In scattering problems, for example, the approximate boundary conditions (2.78) may be used instead of the exact ones (1.60) and (1.61) in solving for the field outside the conductor. This removes the need for considering the field inside the conductor.

Physically the impedance boundary conditions (2.78) imply that the structure of the field at some point inside the conductor is determined practically by the field distribution over the adjoining part of the surface of the conductor with dimensions of the order of a wavelength in the conductor. As the wavelength in the conductor is very much smaller than that in the dielectric, the changes of the field in the dielectric over a conductor wavelength are very small. Thus when considering the field inside the conductor, it may be assumed that the source field over the adjoining part of the conductor's surface is approximately constant and that the surface is approximately an equiphase surface.

Although the impedance boundary conditions (2.78) are given for a flat boundary, they can be applied to a curved surface if certain restrictions are placed upon the shape or curvature of the boundary. These restrictions are given by Leontovich[15] as

$$|\gamma_1 \rho| \gg 1 \tag{2.79}$$

$$\delta \ll \rho \tag{2.80}$$

---

[14] Wait, J. R.: On the Relation Between Telluric Currents and the Earth's Magnetic Field. Geophysics, vol. 19, Apr. 1954, pp. 281-289.

[15] Leontovich, M. A.: On the Approximate Boundary Conditions for an Electromagnetic Field on the Surface of Well-Conducting Bodies. Investigations of Propagation of Radio Waves, B. A. Vvedensky, ed., Academy of Sciences, Moscow, U.S.S.R., 1948.

where ρ is the smallest radius of curvature of the conducting surface and δ is the skin depth in the conductor. Condition (2.79) is sufficient for an open surface, i.e., a conductor of infinite extent, while condition (2.80) is additionally necessary for a closed surface to assure that the inward traveling field is sufficiently attenuated so that it does not appear as an outward traveling field on the far side of the conductor. For good conductors the condition (2.80) is implied by (2.79) because $|\gamma_1| = \sqrt{2}/\delta$.

Summarizing, the impedance boundary conditions, or the Leontovich boundary conditions as they are sometimes called, may be applied to the surface of a conducting body whenever the following conditions are met:

(1) $|\gamma_2/\gamma_1| \ll 1$, the wavelength in conductor is much smaller than that in dielectric
(2) $|\gamma_1 \rho| \gg 1$, the smallest radius of curvature of the conductor surface is much larger than a wavelength in the conductor
(3) $\delta \ll \rho$, the skin depth in conductor is much smaller than the smallest radius of curvature

Further discussions of impedance boundary conditions are presented by Senior[16] and Godzinski.[17]

2.8 PROPAGATION INTO A CONDUCTOR WITH A ROUGH SURFACE

A low-frequency electromagnetic plane wave propagating in free space over a homogeneous, semi-infinite, conducting medium with a rough surface will experience perturbations both above and below the interface. The field above the surface will be perturbed very little because the free-space wavelength is very much larger than the dimensions of the surface ripples usually encountered. Of concern in this section is the perturbation of the exponential attenuation suffered by the electromagnetic wave below the rippled surface. Water waves on the surface of the ocean may be considered as stationary because their velocity is very much less than that of the electromagnetic waves involved.

Wait[18] calculates the field in a good conductor with a wavy surface sinusoidal in form by assuming the Leontovich boundary conditions and a depth of observation that is small compared with the radius of curvature of the ripples. Using a rectangular coordinate system with the mean level of the conducting surface in the x, y-plane, the two-dimensional sinusoidal ripples are described by

$$z = D \cos\left(\frac{2\pi x}{L}\right) = D \cos \hat{\beta} x \tag{2.81}$$

where D is the amplitude, L is the wavelength, and $\hat{\beta} = 2\pi/L$. The plane wave field in free space above the conductor is assumed to have a wavelength that is much larger than that of the ripples and in addition it is assumed that $D \ll L$. Under these conditions, the perturbation of the free-space field is negligible and the field $\psi(x, y, z)$ in the conductor has been given by Wait as

$$\psi(x, y, z) \approx \psi_0(x, y)\left[e^{-\gamma z} + \gamma D e^{-\gamma D} e^{-(\gamma^2 + \hat{\beta}^2)^{1/2}(z-D)} \cos \hat{\beta} x\right] \tag{2.82}$$

[16] Senior, T. B. A.: Impedance Boundary Conditions for Imperfectly Conducting Surfaces. Appl. Sci. Res. Sec. B, vol. 8, 1960, pp. 418-436.

[17] Godzinski, Z.: The Surface Impedance Concept and the Structure of Radio Waves Over Real Earth. I.E.E. Monograph No. 434E, Mar. 1961.

[18] Wait, J. R.: The Calculation of the Field in a Homogeneous Conductor With a Wavy Interface. Proc. I.R.E., vol. 47, no. 6, June 1959, pp. 1155-1156.

for $|\gamma D| \ll 1$

where $\psi_0(x, y)$ = free-space field component in the plane $z = 0$ and

$\gamma \approx (j\omega\mu\sigma)^{1/2}$ = propagation constant in the conducting half space

This expression for the field shows that the perturbation is proportional to the amplitude D of the ripples and varies sinusoidally with x, thus following the form of the wavy interface. It also follows from (2.82) that whenever $\hat{\beta}$ can be neglected with respect to $|\gamma|$, i.e., the field wavelength in the conductor is much smaller than the ripple wavelength, the attenuation is the same as if a flat surface were moving up and down with an amplitude D over the point of observation.

Lerner and Max[19] derive expressions for a low-frequency field propagating above and into an ocean with a surface described by infinite trochoidal cylinders. The expressions also hold for cases in which the Leontovich boundary conditions are not valid. The above authors present plots of 20-kHz fields at a horizontal plane 25 feet under the troughs of 2-foot and 25-foot waves that have a height to water wavelength ratio of 0.05. These plots are reproduced below in figures 2.5 and 2.6 respectively and represent two extreme situations. In the case of the 2-foot waves, the skin depth is large compared with the wave height and the depths of observation are comparable with the radius of curvature of the waves. In the case of the 25-foot waves, the skin depth is less than the wave height and the depths of observation are small compared with the radius of curvature of the waves.

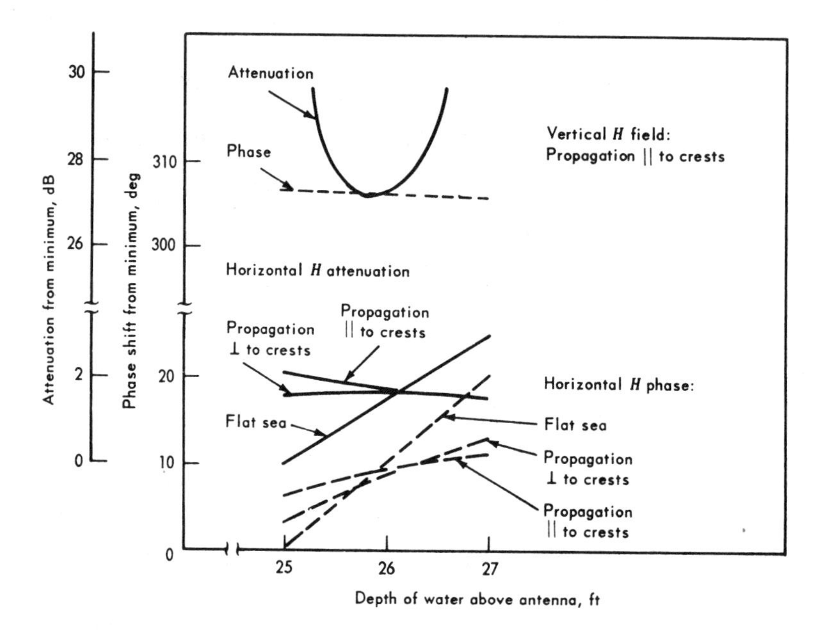

Figure 2.5.–*Fields at a horizontal plane 25 feet under troughs of 2-foot waves.*

The effect of sea surface roughness that is statistically distributed along one direction is investigated by Winter[20] who obtains expressions for the mean square deviation of the subsurface

[19] Lerner, R.M.; and Max, J.: VLF and LF Fields Propagating Near and Into a Rough Sea. Radio Sci., vol. 69, no. 2, Feb. 1965, pp. 273-286.

[20] Winter, D.F.: Low-Frequency Radio Propagation Into a Moderately Rough Sea. J. Res. Nat. Bur. Std., D. Radio Propagation, vol. 67, no. 5, 1963, pp. 551-562.

electric fields from a distantly located vertical dipole radiating at the sea surface. Selected numerical results are presented for 2 kHz.

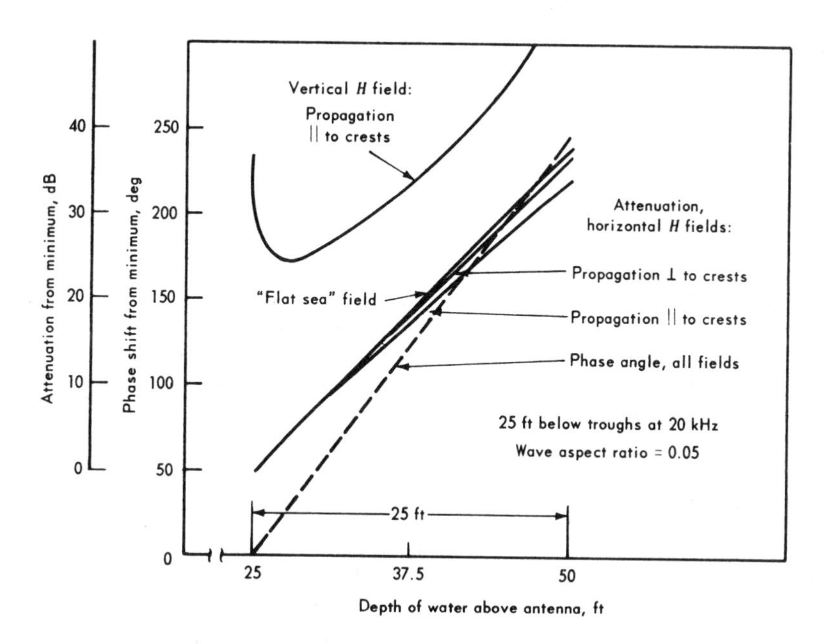

Figure 2.6.–*Fields at a horizontal plane 25 feet under troughs of 25-foot storm waves.*

A discussion and review of the effects of the rough air-sea interface on VLF and ELF propagation are presented by Weiner.[21] Expressions are derived for the perturbed plane wave fields in air and sea water for sinusoidal surface ripples. Both two-dimensional and three-dimensional surface roughness is treated. Selected numerical results are presented for the two-dimensional ripples at 3 Hz and 18.6 kHz.

[21] Weiner, K. R.: The Effect of the Rough Air-Sea Interface on VLF and ELF Propagation. Tech. Rept. EE-107, Eng. Exp. Station, Univ. of N. Mex., Albuquerque, May 1964.

Chapter 3

ELECTROMAGNETIC FIELD OF DIPOLE SOURCES

The electromagnetic field of a point dipole source may in general be determined by solving an equation of the type (1.35), (1.42), or (1.47), subject to the appropriate boundary conditions. In these equations, the impressed dipole moment distribution is of infinitesimal extent and may be expressed as a delta function of position. The solution will also apply to a finite antenna with a uniform current whenever the observation distance is much larger than the extent of the antenna.

When the fields of electric dipoles are determined, the source strength is usually expressed as a current moment with the units of ampere-meter and the electric dipole may properly be called a current element. This current moment p is related to the electric dipole moment $\hat{p}$ by $p = j\omega\hat{p}$. The fields of magnetic dipoles are expressed in terms of the loop magnetic moment m in units of ampere-meter^2. For a finite electric antenna of length L, $p = IL$, where I is the uniform current in the antenna. Correspondingly, for a finite magnetic loop antenna of cross-section A, $m = IA$ where I is the uniform current in the loop.

3.1 INFINITE HOMOGENEOUS CONDUCTING MEDIUM

The expressions for the electromagnetic fields of electric and magnetic dipoles in an infinite, homogeneous, conducting medium may be derived from the Green's function solution (1.48) for the point dipole. Using spherical coordinates with the dipole along the polar axis, the following expressions are obtained:

Electric dipole:

$$E_r = \frac{j\omega\mu}{\gamma}\frac{p\cos\theta}{2\pi r^2}\left(1+\frac{1}{\gamma r}\right)e^{-\gamma r} \tag{3.1}$$

$$E_\theta = j\omega\mu\frac{p\sin\theta}{4\pi r}\left(1+\frac{1}{\gamma r}+\frac{1}{\gamma^2 r^2}\right)e^{-\gamma r} \tag{3.2}$$

$$H_\phi = \frac{\gamma p\sin\theta}{4\pi r}\left(1+\frac{1}{\gamma r}\right)e^{-\gamma r} \tag{3.3}$$

Magnetic dipole:

$$E_\phi = -j\omega\mu\frac{\gamma m\sin\theta}{4\pi r}\left(1+\frac{1}{\gamma r}\right)e^{-\gamma r} \tag{3.4}$$

$$H_r = \frac{\gamma m\cos\theta}{2\pi r^2}\left(1+\frac{1}{\gamma r}\right)e^{-\gamma r} \tag{3.5}$$

$$H_\theta = \frac{\gamma^2 m \sin\theta}{4\pi r}\left(1 + \frac{1}{\gamma r} + \frac{1}{\gamma^2 r^2}\right)e^{-\gamma r} \tag{3.6}$$

If the medium is a good conductor so that $\sigma \gg \omega\epsilon$, then $\gamma = \beta + j\beta$ where $\beta = (\omega\mu\sigma/2)^{1/2}$. The field equations then become

Electric dipole ($\sigma \gg \omega\epsilon$):

$$E_r = \frac{p \cos\theta}{2\pi\sigma r^3}(1 + \beta r + j\beta r)e^{-\beta r}e^{-j\beta r} \tag{3.7}$$

$$E_\theta = \frac{p \sin\theta}{4\pi\sigma r^3}\left[1 + \beta r + j(\beta r + 2\beta^2 r^2)\right]e^{-\beta r}e^{-j\beta r} \tag{3.8}$$

$$H_\phi = \frac{p \sin\theta}{4\pi r^2}(1 + \beta r + j\beta r)e^{-\beta r}e^{-j\beta r} \tag{3.9}$$

Magnetic dipole ($\sigma \gg \omega\epsilon$):

$$E_\phi = \frac{-j\omega\mu m \sin\theta}{4\pi r^2}(1 + \beta r + j\beta r)e^{-\beta r}e^{-j\beta r} \tag{3.10}$$

$$H_r = \frac{m \cos\theta}{2\pi r^3}(1 + \beta r + j\beta r)e^{-\beta r}e^{-j\beta r} \tag{3.11}$$

$$H_\theta = \frac{m \sin\theta}{4\pi r^3}\left[1 + \beta r + j(\beta r + 2\beta^2 r^2)\right]e^{-\beta r}e^{-j\beta r} \tag{3.12}$$

The expressions for the field in the static case ($\omega = 0$) may be obtained from (3.7) to (3.12) by setting $\beta = 0$.

3.2 SEMI-INFINITE HOMOGENEOUS CONDUCTING MEDIUM

The following sections give the electromagnetic fields of electric and magnetic dipoles in a semi-infinite conducting medium with a plane boundary. Radiation from dipoles in air above a plane, conducting, homogeneous earth is considered by Sommerfeld[22] and Norton,[23] reviewed by Jordan,[24] and is very well known. The emphasis in the present chapter is on those cases in which either the source point or the field point or both are in the conducting medium. Applications to geophysical prospecting are omitted and may be found in texts on this subject such as Keller and Frischknecht.[25]

---

[22] Sommerfeld, A.: On the Propagation of Waves in Wireless Telegraphy. Ann. Physik, vol. 81, no. 25, 11 Dec. 1926, pp. 1135-1153.

[23] Norton, K. A.: The Propagation of Radio Waves Over the Surface of the Earth and in the Upper Atmosphere. Proc. I.R.E., vol. 25, no. 9, Sept. 1937, pp. 1203-1236.

[24] Jordan, E. C.: Electromagnetic Waves and Radiating Systems. Prentice-Hall, Inc., 1950, ch. 16.

[25] Keller, G. V.; and Frischknecht, F. C.: Electrical Methods in Geophysical Prospecting. Pergamon Press, Ltd., 1966.

3.2.1 Static Electric Dipole ($\omega = 0$)

This section gives the electric and magnetic fields of both a horizontal and a vertical dc electric dipole or current element positioned in the lower of two semi-infinite conducting media as shown in figure 3.1. Both the horizontal and the vertical dipoles are positioned on the z-axis at a depth h below the interface. The vertical current element is in the $+z$-direction, and the horizontal current element is parallel to the x-axis and is positively directed. Certain terms of the solution to the boundary value problem for both the vertical and the horizontal dipoles may be identified with an image dipole located a distance h above the interface. The distance R_1 extends from the source dipole to the field point (ρ, ϕ, z) and the distance R_2 goes from the image dipole to the field point. Using cylindrical coordinates it may be seen that

$$R_1 = \left[\rho^2 + (z-h)^2\right]^{1/2} \tag{3.13}$$

$$R_2 = \left[\rho^2 + (z+h)^2\right]^{1/2} \tag{3.14}$$

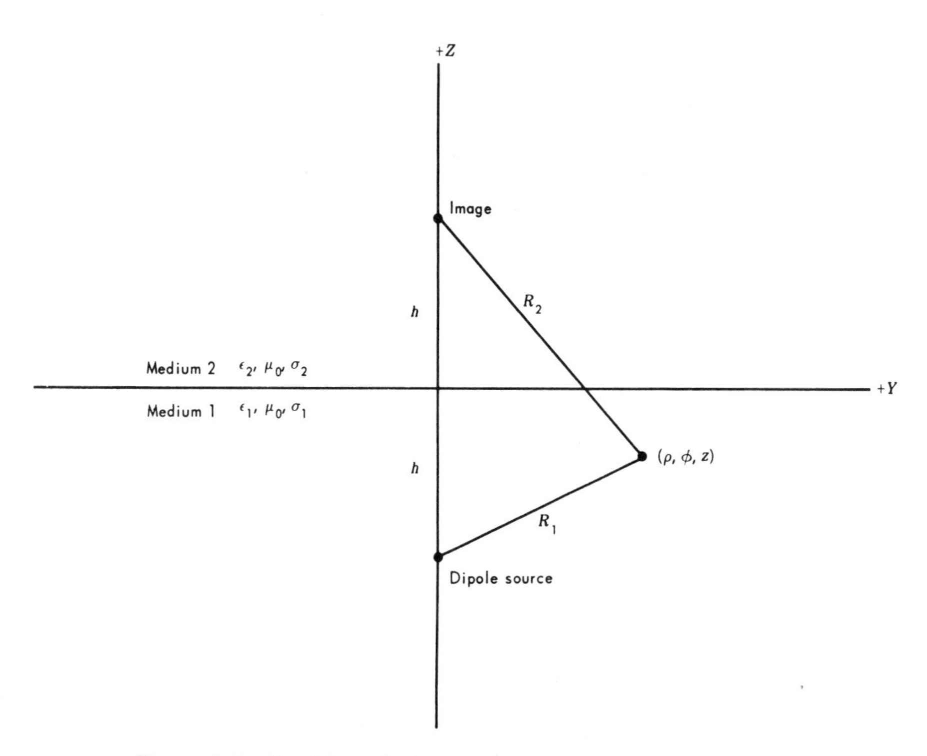

Figure 3.1.–*Position of electric dipole source and its image.*

3.2.1.1 Horizontal electric dipole

The electric and magnetic fields for the dc horizontal current element are derived by Baños and Wesley.[26] The field components are

[26] Baños, A.; and Wesley, J. R.: The Horizontal Electric Dipole in a Conducting Half-Space. Rept. No. 53-33, Scripps Inst. Oceanogr., Marine Physical Laboratory, Univ. of Cal., Sept. 1953, ch. 4.

$$E_{\rho_1} = \frac{-p \cos \phi}{4\pi\sigma_1}\left[\frac{1}{R_1^3} - \frac{3\rho^2}{R_1^5} + \frac{\sigma_1 - \sigma_2}{\sigma_1 + \sigma_2}\left(\frac{1}{R_2^3} - \frac{3\rho^2}{R_2^5}\right)\right] \tag{3.15}$$

$$E_{\phi_1} = \frac{p \sin \phi}{4\pi\sigma_1}\left[\frac{1}{R_1^3} + \left(\frac{\sigma_1 - \sigma_2}{\sigma_1 + \sigma_2}\right)\frac{1}{R_2^3}\right] \tag{3.16}$$

$$E_{z_1} = \frac{p \cos \phi}{4\pi\sigma_1}\left[\frac{3\rho(z-h)}{R_1^5} + \left(\frac{\sigma_1 - \sigma_2}{\sigma_1 + \sigma_2}\right)\frac{3\rho(z+h)}{R_2^5}\right] \tag{3.17}$$

$$H_{\rho_1} = \frac{-p \sin \phi}{4\pi}\left[\frac{z-h}{R_1^3} - \left(\frac{\sigma_1 - \sigma_2}{\sigma_1 + \sigma_2}\right)\frac{1}{\rho^2}\left(\frac{z+h}{R_2} + 1\right)\right] \tag{3.18}$$

$$H_{\phi_1} = \frac{-p \cos \phi}{4\pi}\left[\frac{z-h}{R_1^3} + \left(\frac{\sigma_1 - \sigma_2}{\sigma_1 + \sigma_2}\right)\frac{z+h}{R_2^3} + \left(\frac{\sigma_1 - \sigma_2}{\sigma_1 + \sigma_2}\right)\frac{1}{\rho^2}\left(\frac{z+h}{R_2} + 1\right)\right] \tag{3.19}$$

$$H_{z_1} = \frac{p \sin \phi}{4\pi}\left(\frac{\rho}{R_1^3}\right) \tag{3.20}$$

$$E_{\rho_2} = \frac{-p \cos \phi}{2\pi(\sigma_1 + \sigma_2)}\left[\frac{1}{R_1^3} - \frac{3\rho^2}{R_1^5}\right] \tag{3.21}$$

$$E_{\phi_2} = \frac{p \sin \phi}{2\pi(\sigma_1 + \sigma_2)}\left(\frac{1}{R_1^3}\right) \tag{3.22}$$

$$E_{z_2} = \frac{p \cos \phi}{2\pi(\sigma_1 + \sigma_2)}\left[\frac{3\rho(z-h)}{R_1^5}\right] \tag{3.23}$$

$$H_{\rho_2} = \frac{-p \sin \phi}{4\pi}\left[\frac{z-h}{R_1^3} + \left(\frac{\sigma_1 - \sigma_2}{\sigma_1 + \sigma_2}\right)\frac{1}{\rho^2}\left(\frac{z-h}{R_1} - 1\right)\right] \tag{3.24}$$

$$H_{\phi_2} = \frac{-p \cos \phi}{4\pi}\left[\frac{z-h}{R_1^3} - \left(\frac{\sigma_1 - \sigma_2}{\sigma_1 + \sigma_2}\right)\frac{z-h}{R_1^3} - \left(\frac{\sigma_1 - \sigma_2}{\sigma_1 + \sigma_2}\right)\frac{1}{\rho^2}\left(\frac{z-h}{R_1} - 1\right)\right] \tag{3.25}$$

$$H_{z_2} = \frac{p \sin \phi}{4\pi}\left(\frac{\rho}{R_1^3}\right) \tag{3.26}$$

The expressions for the fields when the upper medium is nonconducting may be obtained by setting $\sigma_2 = 0$ in (3.15) to (3.26).

3.2.1.2 Vertical electric dipole

The electric and the magnetic fields of the vertical current element may be derived from the static limit ($\omega \to 0$) of the Hertz potentials for the vertical antenna over an arbitrary earth given by Sommerfeld.[27] This results in

$$E_{\rho_1} = \frac{p}{4\pi\sigma_1}\left[\frac{3\rho(z-h)}{R_1^5} - \left(\frac{\sigma_1 - \sigma_2}{\sigma_1 + \sigma_2}\right)\frac{3\rho(z+h)}{R_2^5}\right] \tag{3.27}$$

$$E_{z_1} = \frac{-p}{4\pi\sigma_1}\left[\frac{1}{R_1^3} - \left(\frac{\sigma_1 - \sigma_2}{\sigma_1 + \sigma_2}\right)\frac{1}{R_2^3} - \frac{3(z-h)^2}{R_1^5} + \left(\frac{\sigma_1 - \sigma_2}{\sigma_1 + \sigma_2}\right)\frac{3(z+h)^2}{R_2^5}\right] \tag{3.28}$$

$$H_{\phi_1} = \frac{p}{4\pi}\left[\frac{\rho}{R_1^3} - \left(\frac{\sigma_1 - \sigma_2}{\sigma_1 + \sigma_2}\right)\frac{\rho}{R_2^3}\right] \tag{3.29}$$

$$E_{\rho_2} = \frac{p}{2\pi(\sigma_1 + \sigma_2)}\left[\frac{3\rho(z-h)}{R_1^5}\right] \tag{3.30}$$

$$E_{z_2} = \frac{-p}{2\pi(\sigma_1 + \sigma_2)}\left[\frac{1}{R_1^3} - \frac{3(z-h)^2}{R_1^5}\right] \tag{3.31}$$

$$H_{\phi_2} = \frac{p\sigma_2}{2\pi(\sigma_1 + \sigma_2)}\left(\frac{\rho}{R_1^3}\right) \tag{3.32}$$

The expressions for the fields when the upper medium is nonconducting may be obtained by setting $\sigma_2 = 0$ in (3.27) to (3.32).

3.2.2 Harmonic Dipole Sources

Many papers in recent years have considered particular aspects of the problem of dipoles in the presence of a conducting half space in which conduction current predominates. The dipole fields are presented for the far-field, near-field and quasi-static field ranges. These ranges are treated separately and reflect both the different physical nature of the field and the different mathematical approximations used in evaluating the fundamental integrals in the solutions to the boundary-value problem.

Notably useful treatments of the field expressions for the far-field, near-field, and quasi-static field ranges are presented by Wait,[28] Moore and Blair,[29] and Baños.[30] Comprehensive

[27] Sommerfeld, A.: Partial Differential Equations in Physics. Academic Press, Inc., 1949, p. 249.

[28] Wait, J. R.: The Electromagnetic Fields of a Horizontal Dipole in the Presence of a Conducting Half-Space. Can. J. Phys., vol. 39, 1961, pp. 1017-1028.

[29] Moore, R. K.; and Blair, W. E.: Dipole Radiation in a Conducting Half-Space. J. Res. Nat. Bur. Std. D Radio Propagation, vol. 65, no. 6, Nov.-Dec. 1961, pp. 547-563.

[30] Baños, A., Jr.: Dipole Radiation in the Presence of a Conducting Half-Space. Pergamon Press, Ltd., 1966.

summaries of the near-field expressions are given by Bannister[31,32] and similar summaries for the quasi-static field are presented by Bannister[33,34,35] and Bannister and Hart.[36] Although the ranges used by the above authors differ somewhat in name and overlap when compared with one another, the field expressions agree when compared with the same degree of approximation in common regions of validity.

Experimental verification of various features of the propagation equations for the conducting half space is reported by Kraichman,[37] Saran and Held,[38] and Blair.[39]

The present section reviews the expressions for the electromagnetic fields of both electric and magnetic dipole sources in and above a homogeneous, conducting half space in which conduction current predominates. The conducting half space is bounded by a horizontal plane interface separating the dielectric above (free space) from the conducting medium below. The expressions for the electric and magnetic fields are given in cylindrical coordinates (ρ, ϕ, z) with the origin in the interface and the z-axis normal to it. The positive direction of z is into the upper medium (free space). Four types of dipole sources will be considered: the vertical electric dipole (VED); vertical magnetic dipole (VMD); horizontal electric dipole (HED); and horizontal magnetic dipole (HMD). The dipoles are situated on the z-axis at a height (or depth) h with respect to the interface. *The VED and VMD are oriented in the* z*-direction and the HED and HMD are oriented in the* x*- and* y*-directions respectively.* Free-space parameters are designated ϵ_0, μ_0 with a propagation constant γ_0, while those of the conducting half space are ϵ_1, μ_0, σ_1 and γ_1. It is assumed throughout this section that $\sigma_1 \gg \omega\epsilon_1$.

3.2.2.1 Far field

The far field in the present dipole problem is characterized by the condition that $\rho \gg \lambda_0$, which states that the horizontal range from the source is much greater than a free-space wavelength. As indicated previously, the far-field radiation from dipoles in free space $(h \geqq 0^+)$ to field points in free space $(z \geqq 0^+)$ is well known and will not be discussed here. It should be noted, however, that the subsurface fields $(h \geqq 0^+$ and $z \leqq 0^-)$ may be obtained from the TM plane wave surface fields by the previous formulas (2.58) to (2.60) times an additional factor $e^{\gamma_1 z}$.

The far-field expressions for the propagation paths subsurface to free space $(h \leqq 0^-$ and $z \geqq 0^+)$ and subsurface to subsurface $(h \leqq 0^-$ and $z \leqq 0^-)$ are given below in tables 3.1 and 3.2, respectively. These formulas are valid subject to the following conditions:

(1) $|\gamma_0 \rho| \gg 1$ (far field)

[31] Bannister, P. R.: Utilization of the Reciprocity Theorem to Determine the Near Field Air-to-Submarine Propagation Formulas. USL Rept. No. 786, U.S. Navy Underwater Sound Laboratory, New London, Conn., 22 Nov. 1966.

[32] Bannister, P. R.: Surface-to-Surface and Subsurface-to-Air Propagation of Electromagnetic Waves. USL Rept. No. 761, U.S. Navy Underwater Sound Laboratory, New London, Conn., 17 Feb. 1967.

[33] Bannister, P. R.: The Quasi-Near Fields of Dipole Antennas. IEEE Trans. Antennas Propagation, AP-15, no. 5, Sept. 1967, pp. 618–626.

[34] Bannister, P. R.: Quasi-Static Fields of Dipole Antennas at the Earth's Surface. Radio Sci., vol. 1, no. 11, Nov. 1966, pp. 1321–1330.

[35] Bannister, P. R.: Quasi-Static Fields of Dipole Antennas Located Above the Earth's Surface. Radio Sci., vol. 2, no. 9, Sept. 1967, pp. 1093–1103.

[36] Bannister, P. R.; and Hart, W. C.: Quasi-Static Fields of Dipole Antennas Below the Earth's Surface. USL Rept. No. 870, U.S. Navy Underwater Sound Laboratory, New London, Conn., 11 Apr. 1968.

[37] Kraichman, M. B.: Basic Experimental Studies of the Magnetic Field From Electromagnetic Sources Immersed in a Semi-infinite Conducting Medium. J. Res. Nat. Bur. Std. D Radio Propagation, vol. 64, no. 1, Jan.-Feb., 1960, pp. 21–25.

[38] Saran, G. S.; and Held, G.: Field Strength Measurements in Fresh Water. J. Res. Nat. Bur. Std. D Radio Propagation, vol. 64, no. 5, Sept.-Oct. 1960, pp. 435–437.

[39] Blair, W. E.: Experimental Verification of Dipole Radiation in a Conducting Half-Space. IEEE Trans. Antennas Propagation, vol. AP-11, no. 3, May 1963, pp. 269–275.

(2) $|\gamma_1| \gg |\gamma_0|$ (highly conductive medium)

(3) $\rho \gg |h|$ and $|z|$ (small depths)

(4) field points in free space are very close to interface

It should be noted that in subsurface to subsurface propagation, the energy travels via an up-over-and-down mode in which the wave suffers an exponential attenuation in the conductor only in the vertically up and down portions of the path.

The surface wave attenuation factor $F(n)$, referred to in tables 3.1 and 3.2, is given by

$$F(n) = 1 - j(\pi n)^{1/2} e^{-n} \text{erfc} \ (jn^{1/2}) \tag{3.33}$$

with the so-called Sommerfeld numerical distance

$$n = -\left(\frac{\gamma_0 \rho}{2}\right)\left(\frac{\gamma_0^2}{\gamma_1^2}\right) \tag{3.34}$$

and

$$\text{erfc} \ (jn^{1/2}) = \frac{2}{\pi^{1/2}} \int_{jn^{1/2}}^{\infty} e^{-u^2} du \tag{3.35}$$

For very large values of $|n|$, $F(n) \sim -(1/2n)$, and all the fields fall off as $1/\rho^2$. Plots of the magnitude and phase of $F(n)$ are shown in figures 3.2 and 3.3.

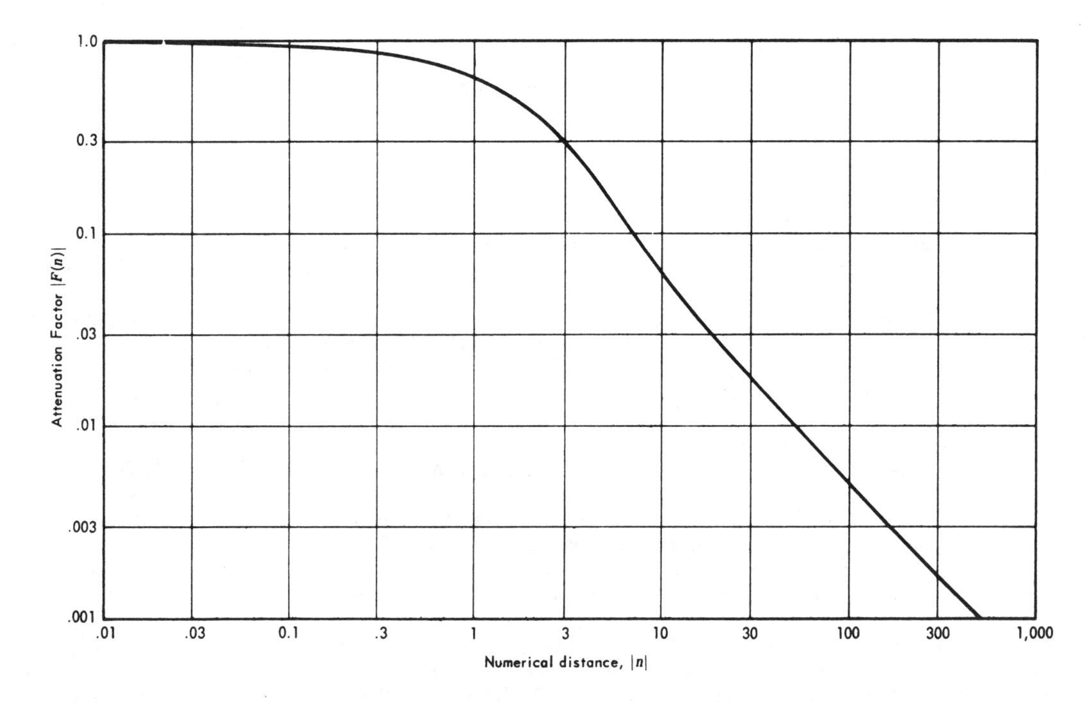

Figure 3.2.—*Magnitude of surface wave attenuation factor* F(n).

Table 3.1.–*Subsurface to free-space propagation formulas for the far-field range*

Dipole type	E_ρ	E_ϕ	E_z	H_ρ	H_ϕ	H_z
VED	$-\frac{p}{2\pi\sigma_1\rho}\frac{\gamma_0^3}{\gamma_1}F(n)e^{\gamma_1 h}e^{-\gamma_0\rho}$	0	$-\frac{p\gamma_0^2}{2\pi\sigma_1\rho}F(n)e^{\gamma_1 h}e^{-\gamma_0\rho}$	0	$\frac{p}{2\pi\rho}\frac{\gamma_0^3}{\gamma_1^2}F(n)e^{\gamma_1 h}e^{-\gamma_0\rho}$	0
VMD	0	$-\frac{m\gamma_0^2}{2\pi\sigma_1\rho^2}e^{\gamma_1 h}e^{-\gamma_0\rho}$	0	$-\frac{m}{2\pi\rho^2}\frac{\gamma_0^2}{\gamma_1}e^{\gamma_1 h}e^{-\gamma_0\rho}$	0	$-\frac{m}{2\pi\rho^2}\frac{\gamma_0^3}{\gamma_1^2}e^{\gamma_1 h}e^{-\gamma_0\rho}$
HED	$\frac{p\cos\phi}{2\pi\sigma_1\rho}\gamma_0^2F(n)e^{\gamma_1 h}e^{-\gamma_0\rho}$	$\frac{p\sin\phi}{\pi\sigma_1\rho^2}\gamma_0e^{\gamma_1 h}e^{-\gamma_0\rho}$	$\frac{p\cos\phi}{2\pi\sigma_1\rho}\gamma_1\gamma_0F(n)e^{\gamma_1 h}e^{-\gamma_0\rho}$	$\frac{p\sin\phi}{\pi\rho^2}\frac{\gamma_0}{\gamma_1}e^{\gamma_1 h}e^{-\gamma_0\rho}$	$-\frac{p\cos\phi}{2\pi\rho}\frac{\gamma_0^2}{\gamma_1}F(n)e^{\gamma_1 h}e^{-\gamma_0\rho}$	$\frac{p\sin\phi}{2\pi\rho^2}\frac{\gamma_0^2}{\gamma_1^2}e^{\gamma_1 h}e^{-\gamma_0\rho}$
HMD	$\frac{m\cos\phi}{2\pi\sigma_1\rho}\gamma_1\gamma_0^2F(n)e^{\gamma_1 h}e^{-\gamma_0\rho}$	$\frac{m\sin\phi}{\pi\sigma_1\rho^2}\gamma_1\gamma_0e^{\gamma_1 h}e^{-\gamma_0\rho}$	$\frac{m\cos\phi}{2\pi\sigma_1\rho}\gamma_1^2\gamma_0F(n)e^{\gamma_1 h}e^{-\gamma_0\rho}$	$\frac{m\sin\phi}{\pi\rho^2}\gamma_0e^{\gamma_1 h}e^{-\gamma_0\rho}$	$-\frac{m\cos\phi}{2\pi\rho}\gamma_0^2F(n)e^{\gamma_1 h}e^{-\gamma_0\rho}$	$\frac{m\sin\phi}{2\pi\rho^2}\frac{\gamma_0^2}{\gamma_1}e^{\gamma_1 h}e^{-\gamma_0\rho}$

Table 3.2.–*Subsurface to subsurface propagation formulas for the far-field range*

Dipole type	E_ρ	E_ϕ	E_z	H_ρ	H_ϕ	H_z
VED	$-\frac{p}{2\pi\sigma_1\rho}\frac{\gamma_0^3}{\gamma_1}F(n)e^{\gamma_1(h+z)}e^{-\gamma_0\rho}$	0	$-\frac{p}{2\pi\sigma_1\rho}\frac{\gamma_0^4}{\gamma_1^2}F(n)e^{\gamma_1(h+z)}e^{-\gamma_0\rho}$	0	$\frac{p}{2\pi\rho}\frac{\gamma_0^3}{\gamma_1^2}F(n)e^{\gamma_1(h+z)}e^{-\gamma_0\rho}$	0
VMD	0	$-\frac{m}{2\pi\sigma_1\rho^2}\gamma_0^2e^{\gamma_1(h+z)}e^{-\gamma_0\rho}$	0	$-\frac{m}{2\pi\rho^2}\frac{\gamma_0^2}{\gamma_1}e^{\gamma_1(h+z)}e^{-\gamma_0\rho}$	0	$-\frac{m}{2\pi\rho^2}\frac{\gamma_0^3}{\gamma_1^2}e^{\gamma_1(h+z)}e^{-\gamma_0\rho}$
HED	$\frac{p\cos\phi}{2\pi\sigma_1\rho}\gamma_0^2F(n)e^{\gamma_1(h+z)}e^{-\gamma_0\rho}$	$\frac{p\sin\phi}{\pi\sigma_1\rho^2}\gamma_0e^{\gamma_1(h+z)}e^{-\gamma_0\rho}$	$\frac{p\cos\phi}{2\pi\sigma_1\rho}\frac{\gamma_0^3}{\gamma_1}F(n)e^{\gamma_1(h+z)}e^{-\gamma_0\rho}$	$\frac{p\sin\phi}{\pi\rho^2}\frac{\gamma_0}{\gamma_1}e^{\gamma_1(h+z)}e^{-\gamma_0\rho}$	$-\frac{p\cos\phi}{2\pi\rho}\frac{\gamma_0^2}{\gamma_1}F(n)e^{\gamma_1(h+z)}e^{-\gamma_0\rho}$	$\frac{p\sin\phi}{2\pi\rho^2}\frac{\gamma_0^2}{\gamma_1^2}e^{\gamma_1(h+z)}e^{-\gamma_0\rho}$
HMD	$\frac{m\cos\phi}{2\pi\sigma_1\rho}\gamma_1\gamma_0^2F(n)$ $e^{\gamma_1(h+z)}e^{-\gamma_0\rho}$	$\frac{m\sin\phi}{\pi\sigma_1\rho^2}\gamma_1\gamma_0e^{\gamma_1(h+z)}e^{-\gamma_0\rho}$	$\frac{m\cos\phi}{2\pi\sigma_1\rho}\gamma_0^3F(n)e^{\gamma_1(h+z)}e^{-\gamma_0\rho}$	$\frac{m\sin\phi}{\pi\rho^2}\gamma_0e^{\gamma_1(h+z)}e^{-\gamma_0\rho}$	$-\frac{m\cos\phi}{2\pi\rho}\gamma_0^2F(n)e^{\gamma_1(h+z)}e^{-\gamma_0\rho}$	$\frac{m\sin\phi}{2\pi\rho^2}\frac{\gamma_0^2}{\gamma_1}e^{\gamma_1(h+z)}e^{-\gamma_0\rho}$

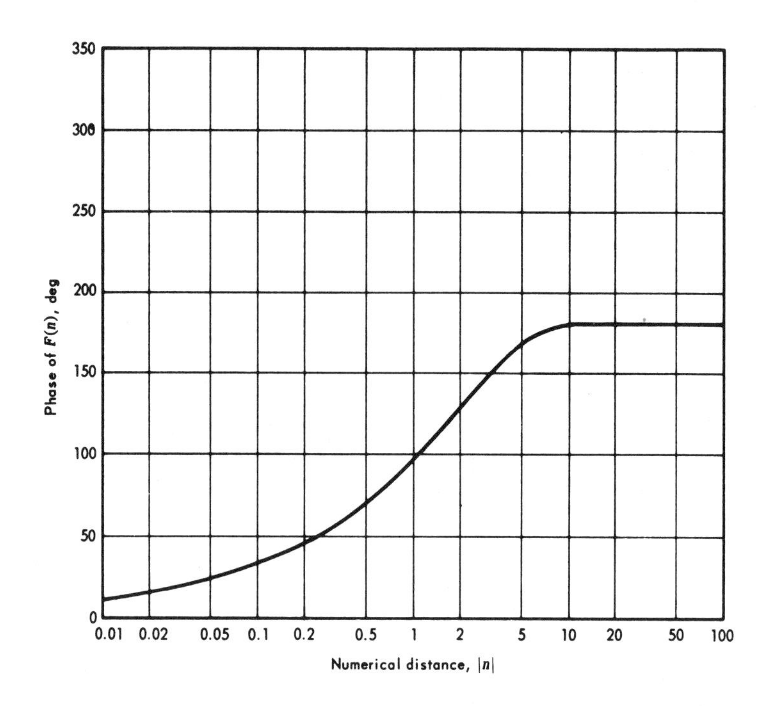

Figure 3.3.–*Phase of surface wave attenuation factor* F(n).

3.2.2.2 Near field

The near-field range is characterized in general by the condition that the distance from the source to the field point is comparable to a free-space wavelength.

3.2.2.2.1 Subsurface to Free Space ($h \leqq 0^-$ and $z \geqq 0^+$)

The near-field expressions for subsurface to free-space propagation are given in table 3.3 and are subject to the following conditions:

(1) $R=\left(\rho^2+z^2\right)^{1/2}$ is comparable to a free-space wavelength although $|\gamma_0\rho|$ may exceed unity

(2) $|\gamma_1| \gg |\gamma_0|$ (highly conducting half space)

(3) $|(\gamma_0^2/\gamma_1)\rho| \ll 1$ implying $|n| \ll 1$ (small Sommerfeld numerical distances)

(4) $|\gamma_1 R| \gg 1$ (distances much greater than a skin depth in the conductor)

(5) $R \gg |h|$

If $\rho \gg z$, the formulas in table 3.3 reduce to those in table 3.4.

In the vicinity of the z-axis, the field expressions may be obtained from those in table 3.3 by letting $\rho \to 0$. The resulting formulas in rectangular coordinates follow:

VED

$$E_z = \frac{p e^{\gamma_1 h}}{\pi \sigma_1 z^3}(1+\gamma_0 z)e^{-\gamma_0 z} \tag{3.36}$$

$$H_z = \frac{m e^{\gamma_1 h}}{\pi \gamma_1 z^4}\left(3+3\gamma_0 z+\gamma_0^2 z^2\right)e^{-\gamma_0 z} \tag{3.37}$$

Table 3.3.–*Subsurface to free-space propagation formulas for the near-field range*

Dipole type	E_ρ	E_ϕ	E_z	H_ρ	H_ϕ	H_z
VED	$\frac{pe^{\gamma_1 h}}{2\pi\sigma_1}\frac{\rho e^{-\gamma_0 R}}{R^5}\left[z\left(3+3\gamma_0 R+\gamma_0^2 R^2\right)-\frac{\gamma_0^2}{\gamma_1}R^2(1+\gamma_0 R)\right]$	0	$\frac{-pe^{\gamma_1 h}}{2\pi\sigma_1}\frac{e^{-\gamma_0 R}}{R^3}\left[1+\gamma_0 R+\gamma_0^2\rho^2-\frac{3z^2}{R^2}(1+\gamma_0 R)\right]$	0	$\frac{p}{2\pi}\frac{\gamma_0^2}{\gamma_1^2}e^{\gamma_1 h}\frac{\rho}{R^3}(1+\gamma_0 R)e^{-\gamma_0 R}$	0
VMD	0	$-\frac{me^{\gamma_1 h}}{2\pi\sigma_1}\frac{\rho e^{-\gamma_0 R}}{R^5}\left[\left(3+3\gamma_1 z-\frac{15z^2}{R^2}\right)(1+\gamma_0 R)+\left(1+\gamma_1 z-\frac{6z^2}{R^2}\right)(\gamma_0 R)^2\right]$	0	$-\frac{me^{\gamma_1 h}}{2\pi\gamma_1}\frac{\rho e^{-\gamma_0 R}}{R^5}\left[\left(3-\frac{15z^2}{R^2}\right)(1+\gamma_0 R)+\left(1-\frac{6z^2}{R^2}\right)(\gamma_0 R)^2-\frac{z^2}{R^2}(\gamma_0 R)^3\right]$	0	$-\frac{me^{\gamma_1 h}}{2\pi\gamma_1^2}\frac{e^{-\gamma_0 R}}{R^5}\left\{\left[9(1+\gamma_1 z)-\frac{15z^2}{R^2}(6+\gamma_1 z)+\frac{105z^4}{R^4}\right](1+\gamma_0 R)+\left[4(1+\gamma_1 z)-\frac{6z^2}{R^2}\left(\frac{39}{6}+\gamma_1 z\right)+\frac{45z^4}{R^4}\right](\gamma_0 R)^2+\left[(1+\gamma_1 z)-\frac{z^2}{R^2}(9+\gamma_1 z)+\frac{10z^4}{R^4}\right](\gamma_0 R)^3\right\}$
HED	$\frac{p\cos\phi e^{\gamma_1 h}}{2\pi\sigma_1}\frac{e^{-\gamma_0 R}}{R^3}\left[(1-\gamma_1 z)(1+\gamma_0 R)+\gamma_0^2 R^2\right]$	$\frac{p\sin\phi e^{\gamma_1 h}}{2\pi\sigma_1}\frac{e^{-\gamma_0 R}}{R^3}\left[\left(2+\gamma_1 z-\frac{3z^2}{R^2}\right)(1+\gamma_0 R)\right]$	$\frac{p\gamma_1\cos\phi e^{\gamma_1 h}}{2\pi\sigma_1}\frac{\rho e^{-\gamma_0 R}}{R^3}(1+\gamma_0 R)$	$\frac{p\sin\phi e^{\gamma_1 h}}{2\pi\gamma_1}\frac{e^{-\gamma_0 R}}{R^3}\left[\left(2-\frac{3z^2}{R^2}\right)(1+\gamma_0 R)-\gamma_0^2 z^2\right]$	$-\frac{p\cos\phi e^{\gamma_1 h}}{2\pi\gamma_1}\frac{e^{-\gamma_0 R}}{R^3}\left[1+\gamma_0 R+\gamma_0^2 R^2\right]$	$\frac{p\sin\phi e^{\gamma_1 h}}{2\pi\gamma_1^2}\frac{e^{-\gamma_0 R}}{R^4}\left(\frac{\rho}{R}\right)\left[\left(3+3\gamma_1 z-\frac{15z^2}{R^2}\right)(1+\gamma_0 R)+\left(1+\gamma_1 z-\frac{6z^2}{R^2}\right)(\gamma_0 R)^2\right]$
HMD	$\frac{m\gamma_1\cos\phi e^{\gamma_1 h}}{2\pi\sigma_1}\frac{e^{-\gamma_0 R}}{R^3}\left[(1-\gamma_1 z)(1+\gamma_0 R)+\gamma_0^2 R^2\right]$	$\frac{m\gamma_1\sin\phi e^{\gamma_1 h}}{2\pi\sigma_1}\frac{e^{-\gamma_0 R}}{R^3}\left[\left(2+\gamma_1 z-\frac{3z^2}{R^2}\right)(1+\gamma_0 R)\right]$	$\frac{m\gamma_1^2\cos\phi e^{\gamma_1 h}}{2\pi\sigma_1}\frac{e^{-\gamma_0 R}}{R^3}\rho(1+\gamma_0 R)$	$\frac{m\sin\phi e^{\gamma_1 h}}{2\pi}\frac{e^{-\gamma_0 R}}{R^3}\left[\left(2-\frac{3z^2}{R^2}\right)(1+\gamma_0 R)-\gamma_0^2 z^2\right]$	$-\frac{m\cos\phi e^{\gamma_1 h}}{2\pi}\frac{e^{-\gamma_0 R}}{R^3}\left(1+\gamma_0 R+\gamma_0^2 R^2\right)$	$\frac{m\sin\phi\rho e^{\gamma_1 h}}{2\pi\gamma_1}\frac{e^{-\gamma_0 R}}{R^5}\left[\left(3+3\gamma_1 z-\frac{15z^2}{R^2}\right)(1+\gamma_0 R)+\left(1+\gamma_1 z-\frac{6z^2}{R^2}\right)(\gamma_0 R)^2\right]$

Table 3.4.–*Subsurface to free-space propagation formulas for the near-field range if* $\rho \gg z$

Dipole type	E_ρ	E_ϕ	E_z	H_ρ	H_ϕ	H_z
VED	$-\frac{p}{2\pi\sigma_1}\frac{e^{\gamma_1 h}}{\rho^4}\left[\left(\frac{\gamma_0^2}{\gamma_1}\right)(1+\gamma_0\rho)\rho^2 - z\left(3+3\gamma_0\rho+\gamma_0^2\rho^2\right)\right]e^{-\gamma_0\rho}$	0	$-\frac{p}{2\pi\sigma_1}e^{\gamma_1 h}\frac{e^{-\gamma_0\rho}}{\rho^3}\left(1+\gamma_0\rho+\gamma_0^2\rho^2\right)$	0	$\frac{p}{2\pi\rho^2}\frac{\gamma_0^2}{\gamma_1^2}(1+\gamma_0\rho)e^{\gamma_1 h}e^{-\gamma_0\rho}$	0
VMD	0	$-\frac{me^{\gamma_1 h}}{2\pi\sigma_1}\frac{e^{-\gamma_0\rho}}{\rho^4}\left(3+3\gamma_0\rho+\gamma_0^2\rho^2\right)(1+\gamma_1 z)$	0	$-\frac{me^{\gamma_1 h}}{2\pi\gamma_1}\frac{e^{-\gamma_0\rho}}{\rho^4}\left(3+3\gamma_0\rho+\gamma_0^2\rho^2\right)$	0	$-\frac{me^{\gamma_1 h}}{2\pi\gamma_1^2}\frac{e^{-\gamma_0\rho}}{\rho^5}\left(9+9\gamma_0\rho+4\gamma_0^2\rho^2+\gamma_0^3\rho^3\right)(1+\gamma_1 z)$
HED	$\frac{p\cos\phi e^{\gamma_1 h}}{2\pi\sigma_1}\frac{e^{-\gamma_0\rho}}{\rho^3}\left[(1-\gamma_1 z)(1+\gamma_0\rho)+\gamma_0^2\rho^2\right]$	$\frac{p\sin\phi e^{\gamma_1 h}}{2\pi\sigma_1}\frac{e^{-\gamma_0\rho}}{\rho^3}\left[(2+\gamma_1 z)(1+\gamma_0\rho)\right]$	$\frac{p\gamma_1\cos\phi e^{\gamma_1 h}}{2\pi\sigma_1}\frac{e^{-\gamma_0\rho}}{\rho^2}(1+\gamma_0\rho)$	$\frac{p\sin\phi e^{\gamma_1 h}}{\pi\gamma_1}\frac{e^{-\gamma_0\rho}}{\rho^3}(1+\gamma_0\rho)$	$-\frac{p\cos\phi e^{\gamma_1 h}}{2\pi\gamma_1}\frac{e^{-\gamma_0\rho}}{\rho^3}\left[1+\gamma_0\rho+\gamma_0^2\rho^2\right]$	$\frac{p\sin\phi e^{\gamma_1 h}}{2\pi\gamma_1^2}\frac{e^{-\gamma_0\rho}}{\rho^4}\left[(1+\gamma_1 z)(3+3\gamma_0\rho+\gamma_0^2\rho^2)\right]$
HMD	$\frac{m\gamma_1\cos\phi e^{\gamma_1 h}}{2\pi\sigma_1}\frac{e^{-\gamma_0\rho}}{\rho^3}\left[(1-\gamma_1 z)(1+\gamma_0\rho)+\gamma_0^2\rho^2\right]$	$\frac{m\gamma_1\sin\phi e^{\gamma_1 h}}{2\pi\sigma_1}\frac{e^{-\gamma_0\rho}}{\rho^3}\left[(2+\gamma_1 z)(1+\gamma_0\rho)\right]$	$\frac{m\gamma_1^2\cos\phi e^{\gamma_1 h}}{2\pi\sigma_1}\frac{e^{-\gamma_0\rho}}{\rho^2}(1+\gamma_0\rho)$	$\frac{m\sin\phi e^{\gamma_1 h}}{2\pi}\frac{e^{-\gamma_0\rho}}{\rho^3}[2(1+\gamma_0\rho)]$	$-\frac{m\cos\phi e^{\gamma_1 h}}{2\pi}\frac{e^{-\gamma_0\rho}}{\rho^3}\left[1+\gamma_0\rho+\gamma_0^2\rho^2\right]$	$\frac{m\sin\phi e^{\gamma_1 h}}{2\pi\gamma_1}\frac{e^{-\gamma_0\rho}}{\rho^4}\left[(3+3\gamma_0\rho+\gamma_0^2\rho^2)(1+\gamma_1 z)\right]$

HED

$$E_x = -\frac{p\gamma_1 e^{\gamma_1 h}}{2\pi\sigma_1 z^2}(1+\gamma_0 z)e^{-\gamma_0 z} \tag{3.38}$$

$$H_y = -\frac{p e^{\gamma_1 h}}{2\pi\gamma_1 z^3}\left(1+\gamma_0 z+\gamma_0^2 z^2\right)e^{-\gamma_0 z} \tag{3.39}$$

HMD

$$E_x = -\frac{m\gamma_1^2 e^{\gamma_1 h}}{2\pi\sigma_1 z^2}(1+\gamma_0 z)e^{-\gamma_0 z} \tag{3.40}$$

$$H_y = \frac{-m e^{\gamma_1 h}}{2\pi z^3}\left(1+\gamma_0 z+\gamma_0^2 z^2\right)e^{-\gamma_0 z} \tag{3.41}$$

3.2.2.2.2 Free Space to Subsurface ($h \geqq 0^+$ and $z \leqq 0^-$)

The near-field expressions for free-space to subsurface propagation are given in table 3.5 and are subject to the following conditions:

(1) $R' = (\rho^2 + h^2)^{1/2}$ is comparable to a free-space wavelength although $|\gamma_0 \rho|$ may exceed unity

(2) $|\gamma_1| \gg |\gamma_0|$

(3) $|(\gamma_0^2/\gamma_1)\rho| \ll 1$

(4) $|\gamma_1 R'| \gg 1$

(5) $R' \gg |z|$

If $\rho \gg h$, the formulas in table 3.5 reduce to those in table 3.6.

The expressions for the field in the vicinity of the z-axis may be obtained from those in table 3.5 by letting $\rho \to 0$. The formulas, in rectangular coordinates, follow:

VED

$$E_z = \frac{p e^{\gamma_1 z}}{\pi\sigma_1 h^3}(1+\gamma_0 h)e^{-\gamma_0 h} \tag{3.42}$$

VMD

$$H_z = \frac{m e^{\gamma_1 z}}{\pi\gamma_1 h^4}\left(3+\gamma_0 h+\gamma_0^2 h^2\right)e^{-\gamma_0 h} \tag{3.43}$$

HED

$$E_x = -\frac{p\gamma_1 e^{\gamma_1 z}}{2\pi\sigma_1 h^2}(1+\gamma_0 h)e^{-\gamma_0 h} \tag{3.44}$$

$$H_y = \frac{p e^{\gamma_1 z}}{2\pi h^2}(1+\gamma_0 h)e^{-\gamma_0 h} \tag{3.45}$$

HMD

$$E_x = \frac{m\gamma_1 e^{\gamma_1 z}}{2\pi\sigma_1 h^3}\left(1+\gamma_0 h+\gamma_0^2 h^2\right)e^{-\gamma_0 h} \tag{3.46}$$

$$H_y = -\frac{m e^{\gamma_1 z}}{2\pi h^3}\left(1+\gamma_0 h+\gamma_0^2 h^2\right)e^{-\gamma_0 h} \tag{3.47}$$

Table 3.5.–*Free-space to subsurface propagation formulas for the near-field range*

Dipole type	E_ρ	E_ϕ	E_z	H_ρ	H_ϕ	H_z
VED	$-\frac{p\gamma_1 e^{\gamma_1 z}}{2\pi\sigma_1}\frac{\rho e^{-\gamma_0 R'}}{(R')^3}(1+\gamma_0 R')$	0	$-\frac{pe^{\gamma_1 z}}{2\pi\sigma_1}\frac{e^{-\gamma_0 R'}}{(R')^3}\left[1+\gamma_0 R' + \gamma_0^2\rho^2 - \frac{3h^2}{(R')^2}(1+\gamma_0 R')\right]$	0	$\frac{pe^{\gamma_1 z}}{2\pi}\frac{\rho e^{-\gamma_0 R'}}{(R')^3}(1+\gamma_0 R')$	0
VMD	0	$-\frac{me^{\gamma_1 z}}{2\pi\sigma_1}\frac{e^{-\gamma_0 R'}}{(R')^4}\left(\frac{\rho}{R'}\right)\left[\left(3+3\gamma_1 h-\frac{15h^2}{(R')^2}\right)(1+\gamma_0 R') + \left(1+\gamma_1 h-\frac{6h^2}{(R')^2}\right)(\gamma_0 R')^2\right]$	0	$-\frac{me^{\gamma_1 z}}{2\pi\gamma_1}\frac{\rho e^{-\gamma_0 R'}}{(R')^5}\left[\left(3+3\gamma_1 h-\frac{15h^2}{(R')^2}\right)(1+\gamma_0 R') + \left(1+\gamma_1 h-\frac{6h^2}{(R')^2}\right)(\gamma_0 R')^2\right]$	0	$-\frac{me^{\gamma_1 z}}{2\pi\gamma_1^2}\frac{e^{-\gamma_0 R'}}{(R')^5}\left\{\left[9(1+\gamma_1 h) - \frac{15h^2}{(R')^2}(6+\gamma_1 h)+\frac{105h^4}{(R')^4}\right]\left(1+\gamma_0 R'\right) + \left[4(1+\gamma_1 h) - \frac{6h^2}{(R')^2}\left(\frac{39}{6}+\gamma_1 h\right)+\frac{45h^4}{(R')^4}\right](\gamma_0 R')^2 + \left[1+\gamma_1 h - \frac{h^2}{(R')^2}(9+\gamma_1 h)+\frac{10h^4}{(R')^4}\right](\gamma_0 R')^3\right\}$
HED	$\frac{p\cos\phi e^{\gamma_1 z}}{2\pi\sigma_1}\frac{e^{-\gamma_0 R'}}{(R')^3}\left[(1-\gamma_1 h)(1+\gamma_0 R')+(\gamma_0 R')^2\right]$	$\frac{p\sin\phi e^{\gamma_1 z}}{2\pi\sigma_1}\frac{e^{-\gamma_0 R'}}{(R')^3}\left[\left(2+\gamma_1 h-\frac{3h^2}{(R')^2}\right)(1+\gamma_0 R')\right]$	$-\frac{p\cos\phi e^{\gamma_1 z}}{2\pi\sigma_1}\frac{\rho e^{-\gamma_0 R'}}{(R')^5}\left\{h\left[3+3\gamma_0 R'+(\gamma_0 R')^2\right] - \frac{\gamma_0^2(R')^2}{\gamma_1}(1+\gamma_0 R')\right\}$	$\frac{p\sin\phi e^{\gamma_1 z}}{2\pi\gamma_1}\frac{e^{-\gamma_0 R'}}{(R')^3}\left[\left(2+\gamma_1 h-\frac{3h^2}{(R')^2}\right)(1+\gamma_0 R')\right]$	$-\frac{p\cos\phi e^{\gamma_1 z}}{2\pi\gamma_1}\frac{e^{-\gamma_0 R'}}{(R')^3}\left[(1-\gamma_1 h)(1+\gamma_0 R')+(\gamma_0 R')^2\right]$	$\frac{p\sin\phi e^{\gamma_1 z}}{2\pi\gamma_1^2}\frac{\rho e^{-\gamma_0 R'}}{(R')^5}\left[\left(3+3\gamma_1 h-\frac{15h^2}{(R')^2}\right)(1+\gamma_0 R') + \left(1+\gamma_1 h-\frac{6h^2}{(R')^2}\right)(\gamma_0 R')^2\right]$
HMD	$\frac{m\gamma_1\cos\phi e^{\gamma_1 z}}{2\pi\sigma_1}\frac{e^{-\gamma_0 R'}}{(R')^3}\left[1+\gamma_0 R'+(\gamma_0 R')^2\right]$	$\frac{m\gamma_1\sin\phi e^{\gamma_1 z}}{2\pi\sigma_1}\frac{e^{-\gamma_0 R'}}{(R')^3}\left[\left(2-\frac{3h^2}{(R')^2}\right)(1+\gamma_0 R')-\gamma_0^2 h^2\right]$	$\frac{m\gamma_1\cos\phi e^{\gamma_1 z}}{2\pi\sigma_1}\left(\frac{\gamma_0^2\rho}{\gamma_1}\right)\frac{e^{-\gamma_0 R'}}{(R')^3}(1+\gamma_0 R')$	$\frac{m\sin\phi e^{\gamma_1 z}}{2\pi}\frac{e^{-\gamma_0 R'}}{(R')^3}\left[\left(2-\frac{3h^2}{(R')^2}\right)(1+\gamma_0 R')-\gamma_0^2 h^2\right]$	$-\frac{m\cos\phi e^{\gamma_1 z}}{2\pi}\frac{e^{-\gamma_0 R'}}{(R')^3}\left[1+\gamma_0 R'+(\gamma_0 R')^2\right]$	$\frac{m\sin\phi e^{\gamma_1 z}}{2\pi\gamma_1}\frac{\rho e^{-\gamma_0 R'}}{(R')^5}\left[\left(3-\frac{15h^2}{(R')^2}\right)(1+\gamma_0 R')+\left(1-\frac{6h^2}{(R')^2}\right)(\gamma_0 R')^2 - \frac{h^2}{(R')^2}(\gamma_0 R')^3\right]$

Table 3.6.–*Free-space to subsurface propagation formulas for the near-field range if* $\rho \gg h$

Dipole type	E_ρ	E_ϕ	E_z	H_ρ	H_ϕ	H_z
VED	$-\frac{p\gamma_1 e^{\gamma_1 z}}{2\pi\sigma_1}\frac{e^{-\gamma_0\rho}}{\rho^2}(1+\gamma_0\rho)$	0	$-\frac{pe^{\gamma_1 z}}{2\pi\sigma_1}\frac{e^{-\gamma_0\rho}}{\rho^3}\left(1+\gamma_0\rho+\gamma_0^2\rho^2\right)$	0	$\frac{pe^{\gamma_1 z}}{2\pi}\frac{e^{-\gamma_0\rho}}{\rho^2}(1+\gamma_0\rho)$	0
VMD	0	$-\frac{me^{\gamma_1 z}}{2\pi\sigma_1}\frac{e^{-\gamma_0\rho}}{\rho^4}\left[(1+\gamma_1 h)\left(3+3\gamma_0\rho+\gamma_0^2\rho^2\right)\right]$	0	$-\frac{me^{\gamma_1 z}}{2\pi\gamma_1}\frac{e^{-\gamma_0\rho}}{\rho^4}\left[(1+\gamma_1 h)\left(3+3\gamma_0\rho+\gamma_0^2\rho^2\right)\right]$	0	$-\frac{me^{\gamma_1 z}}{2\pi\gamma_1^2}\frac{e^{-\gamma_0\rho}}{\rho^5}\left[(1+\gamma_1 h)\left(9+9\gamma_0\rho+4\gamma_0^2\rho^2+\gamma_0^3\rho^3\right)\right]$
HED	$\frac{p\cos\phi e^{\gamma_1 z}}{2\pi\sigma_1}\frac{e^{-\gamma_0\rho}}{\rho^3}\left[(1-\gamma_1 h)(1+\gamma_0\rho)+\gamma_0^2\rho^2\right]$	$\frac{p\sin\phi e^{\gamma_1 z}}{2\pi\sigma_1}\frac{e^{-\gamma_0\rho}}{\rho^3}\left[(2+\gamma_1 h)(1+\gamma_0\rho)\right]$	$-\frac{p\cos\phi e^{\gamma_1 z}}{2\pi\sigma_1}\frac{e^{-\gamma_0\rho}}{\rho^4}\left[h\left(3+3\gamma_0\rho+\gamma_0^2\rho^2\right)-\frac{\gamma_0^2\rho^2}{\gamma_1}(1+\gamma_0\rho)\right]$	$\frac{p\sin\phi e^{\gamma_1 z}}{2\pi\gamma_1}\frac{e^{-\gamma_0\rho}}{\rho^3}\left[(2+\gamma_1 h)(1+\gamma_0\rho)\right]$	$-\frac{p\cos\phi e^{\gamma_1 z}}{2\pi\gamma_1}\frac{e^{-\gamma_0\rho}}{\rho^3}\left[(1-\gamma_1 h)(1+\gamma_0\rho)+\gamma_0^2\rho^2\right]$	$\frac{p\sin\phi e^{\gamma_1 z}}{2\pi\gamma_1^2}\frac{e^{-\gamma_0\rho}}{\rho^4}\left[(1+\gamma_1 h)\left(3+3\gamma_0\rho+\gamma_0^2\rho^2\right)\right]$
HMD	$\frac{m\gamma_1\cos\phi e^{\gamma_1 z}}{2\pi\sigma_1}\frac{e^{-\gamma_0\rho}}{\rho^3}\left(1+\gamma_0\rho+\gamma_0^2\rho^2\right)$	$\frac{m\gamma_1\sin\phi e^{\gamma_1 z}}{\pi\sigma_1}\frac{e^{-\gamma_0\rho}}{\rho^3}(1+\gamma_0\rho)$	$\frac{m\cos\phi e^{\gamma_1 z}}{2\pi\sigma_1}\frac{e^{-\gamma_0\rho}}{\rho^2}\gamma_0^2(1+\gamma_0\rho)$	$\frac{m\sin\phi e^{\gamma_1 z}}{\pi}\frac{e^{-\gamma_0\rho}}{\rho^3}(1+\gamma_0\rho)$	$-\frac{m\cos\phi e^{\gamma_1 z}}{2\pi}\frac{e^{-\gamma_0\rho}}{\rho^3}\left(1+\gamma_0\rho+\gamma_0^2\rho^2\right)$	$\frac{m\sin\phi e^{\gamma_1 z}}{2\pi\gamma_1}\frac{e^{-\gamma_0\rho}}{\rho^4}\left(3+3\gamma_0\rho+\gamma_0^2\rho^2\right)$

3.2.2.2.3 Subsurface to Subsurface ($h \leqq 0^-$ and $z \leqq 0^-$)

The near-field formulas for subsurface to subsurface propagation are given in table 3.7 and are subject to the following conditions:

(1) ρ is comparable to a free-space wavelength although $|\gamma_0 \rho|$ may exceed unity

(2) $|\gamma_1| \gg |\gamma_0|$

(3) $|(\gamma_0^2/\gamma_1)\rho| \ll 1$

(4) $|\gamma_1 \rho| \gg 1$

(5) $\rho \gg |h|$ and $|z|$

As noted in the far-field range, the energy propagates along the up-over-and-down path.

3.2.2.3 Quasi-static field

The quasi-static field range is characterized by the condition that the distance from the source point to the field point is very much smaller than a free-space wavelength. Except for certain special cases, the general field expressions are complicated and numerical computation is difficult. Fortunately, these special cases cover many practical situations, and the relevant formulas for these special cases are presented in detail.

3.2.2.3.1 Surface to Surface ($h = z = 0^+$)

The field expressions for both dipole source and field point at the free-space side of the interface of a highly conducting medium ($|\gamma_1| \gg |\gamma_0|$) are given in this section. It should be noted that the expressions will also hold at the conducting side of the interface ($h = z = 0^-$) if the following modifications are made:

(1) Multiply all field expressions for the VED by γ_0^2/γ_1^2 except E_z, which should be multiplied by γ_0^4/γ_1^4

(2) Multiply the E_z component for the VMD, HED, and HMD by γ_0^2/γ_1^2

The modifications are needed to satisfy the boundary conditions at the interface.

3.2.2.3.1.1 General quasi-static range, $|\gamma_0 \rho| \ll 1$

The expressions for the general quasi-static range are presented in table 3.8 where I_0 and I_1 are the modified Bessel functions of the first kind, of orders zero and one, respectively, with argument $\gamma_1 \rho/2$, and K_0 and K_1 are the modified Bessel functions of the second kind, of orders zero and one, respectively, with argument $\gamma_1 \rho/2$. For a highly conducting medium, $\gamma_1 \approx (j\omega\mu_0\sigma_1)^{1/2}$. The magnitudes of the quantities $I_1 K_1$, $3I_1 K_1 - (\gamma_1 \rho/2)(I_0 K_1 - I_1 K_0)$, and $T = 16 I_1 K_1 + \gamma_1^2 \rho^2 (I_1 K_1 - I_0 K_0) + 4\gamma_1 \rho (I_1 K_0 - I_0 K_1)$ are plotted by Bannister[34] as a function of $|\gamma_1 \rho/2|$ and are reproduced here in figures 3.4, 3.5 and 3.6, respectively.

3.2.2.3.1.2 The range $|\gamma_0 \rho| \ll 1 \ll |\gamma_1 \rho|$

In this range, which is sometimes called the quasi-near range, the observation distance is much smaller than a free-space wavelength but much larger than a skin depth $\delta = (2/\omega\mu_0\sigma_1)^{1/2}$ in the conducting medium. The formulas for this range are presented in table 3.9.

3.2.2.3.1.3 The range $|\gamma_0 \rho| \ll 1$, $|\gamma_1 \rho| \ll 1$

The observation distance in this range is much smaller than a skin depth in the conducting medium. Formulas for the field components are given in table 3.10.

Table 3.7.–*Subsurface to subsurface propagation formulas for the near-field range*

Dipole type	E_ρ	E_ϕ	E_z	H_ρ	H_ϕ	H_z
VED	$-\frac{pe^{\gamma_1(h+z)}}{2\pi\sigma_1}\frac{e^{-\gamma_0\rho}}{\rho^2}\left[\frac{\gamma_0^2}{\gamma_1}(1+\gamma_0\rho)\right]$	0	$-\frac{pe^{\gamma_1(h+z)}}{2\pi\sigma_1}\frac{e^{-\gamma_0\rho}}{\rho^3}\left[\frac{\gamma_0^2}{\gamma_1^2}\left(1+\gamma_0\rho+\gamma_0^2\rho^2\right)\right]$	0	$\frac{pe^{\gamma_1(h+z)}}{2\pi}\frac{e^{-\gamma_0\rho}}{\rho^2}\left[\frac{\gamma_0^2}{\gamma_1^2}(1+\gamma_0\rho)\right]$	0
VMD	0	$-\frac{me^{\gamma_1(h+z)}}{2\pi\sigma_1}\frac{e^{-\gamma_0\rho}}{\rho^4}\left(3+3\gamma_0\rho+\gamma_0^2\rho^2\right)$	0	$-\frac{me^{\gamma_1(h+z)}}{2\pi\gamma_1}\frac{e^{-\gamma_0\rho}}{\rho^4}\left(3+3\gamma_0\rho+\gamma_0^2\rho^2\right)$	0	$-\frac{me^{\gamma_1(h+z)}}{2\pi\gamma_1^2}\frac{e^{-\gamma_0\rho}}{\rho^5}\left(9+9\gamma_0\rho+4\gamma_0^2\rho^2+\gamma_0^3\rho^3\right)$
HED	$\frac{p\cos\phi e^{\gamma_1(h+z)}}{2\pi\sigma_1}\frac{e^{-\gamma_0\rho}}{\rho^3}\left(1+\gamma_0\rho+\gamma_0^2\rho^2\right)$	$\frac{p\sin\phi e^{\gamma_1(h+z)}}{\pi\sigma_1}\frac{e^{-\gamma_0\rho}}{\rho^3}(1+\gamma_0\rho)$	$\frac{p\cos\phi e^{\gamma_1(h+z)}}{2\pi\sigma_1}\frac{e^{-\gamma_0\rho}}{\rho^2}\left[\frac{\gamma_0^2}{\gamma_1}(1+\gamma_0\rho)\right]$	$\frac{p\sin\phi e^{\gamma_1(h+z)}}{\pi\gamma_1}\frac{e^{-\gamma_0\rho}}{\rho^3}(1+\gamma_0\rho)$	$-\frac{p\cos\phi e^{\gamma_1(h+z)}}{2\pi\gamma_1}\frac{e^{-\gamma_0\rho}}{\rho^3}\left(1+\gamma_0\rho+\gamma_0^2\rho^2\right)$	$\frac{p\sin\phi e^{\gamma_1(h+z)}}{2\pi\gamma_1^2}\frac{e^{-\gamma_0\rho}}{\rho^4}\left(3+3\gamma_0\rho+\gamma_0^2\rho^2\right)$
HMD	$\frac{m\cos\phi e^{\gamma_1(h+z)}}{2\pi\sigma_1}\frac{e^{-\gamma_0\rho}}{\rho^3}\left[\gamma_1\left(1+\gamma_0\rho+\gamma_0^2\rho^2\right)\right]$	$\frac{m\sin\phi e^{\gamma_1(h+z)}}{\pi\sigma_1}\frac{e^{-\gamma_0\rho}}{\rho^3}\left[\gamma_1(1+\gamma_0\rho)\right]$	$\frac{m\cos\phi e^{\gamma_1(h+z)}}{2\pi\sigma_1}\frac{e^{-\gamma_0\rho}}{\rho^2}\left[\gamma_0^2(1+\gamma_0\rho)\right]$	$\frac{m\sin\phi e^{\gamma_1(h+z)}}{\pi}\frac{e^{-\gamma_0\rho}}{\rho^3}(1+\gamma_0\rho)$	$-\frac{m\cos\phi e^{\gamma_1(h+z)}}{2\pi}\frac{e^{-\gamma_0\rho}}{\rho^3}\left(1+\gamma_0\rho+\gamma_0^2\rho^2\right)$	$\frac{m\sin\phi e^{\gamma_1(h+z)}}{2\pi\gamma_1}\frac{e^{-\gamma_0\rho}}{\rho^4}\left(3+3\gamma_0\rho+\gamma_0^2\rho^2\right)$

Table 3.8.–*The general quasi-static range,* $|\gamma_0 \rho| \ll 1$ for surface to surface propagation

Dipole type	E_ρ	E_ϕ	E_z	H_ρ	H_ϕ	H_z
VED	$-\frac{\gamma_1^2 p}{2\pi\sigma_1\rho} I_1\left(\frac{\gamma_1\rho}{2}\right) K_1\left(\frac{\gamma_1\rho}{2}\right)$	0	$-\frac{p}{2\pi\sigma_1\rho^3}\frac{\gamma_1^2}{\gamma_0^2}$	0	$\frac{p}{2\pi\rho^2}$	0
VMD	0	$-\frac{m}{2\pi\sigma_1\rho^4}\left[3-\left(3+3\gamma_1\rho+\gamma_1^2\rho^2\right)e^{-\gamma_1\rho}\right]$	0	$-\frac{m}{4\pi\rho^3}\left[16I_1K_1+\gamma_1^2\rho^2(I_1K_1 - I_0K_0)+4\gamma_1\rho(I_1K_0-I_0K_1)\right]$	0	$-\frac{m}{2\pi\gamma_1^2\rho^5}\left[9-\left(9+9\gamma_1\rho + 4\gamma_1^2\rho^2+\gamma_1^3\rho^3\right)e^{-\gamma_1\rho}\right]$
HED	$\frac{p}{2\pi\sigma_1\rho^3}\left[1+(1+\gamma_1\rho) e^{-\gamma_1\rho}\right]\cos\phi$	$\frac{p}{2\pi\sigma_1\rho^3}\left[2-(1+\gamma_1\rho) e^{-\gamma_1\rho}\right]\sin\phi$	$\frac{\gamma_1^2 p}{2\pi\sigma_1\rho} I_1K_1 \cos\phi$	$\frac{p\sin\phi}{2\pi\rho^2}\left[3I_1K_1 - \frac{\gamma_1\rho}{2}(I_0K_1-I_1K_0)\right]$	$-\frac{p\cos\phi}{2\pi\rho^2}I_1K_1$	$\frac{p}{2\pi\gamma_1^2\rho^4}\left[3-\left(3+3\gamma_1\rho+\gamma_1^2\rho^2\right) e^{-\gamma_1\rho}\right]\sin\phi$
HMD	$\frac{\gamma_1^2 m}{2\pi\sigma_1\rho^2} I_1\left(\frac{\gamma_1\rho}{2}\right) K_1\left(\frac{\gamma_1\rho}{2}\right)\cos\phi$	$\frac{\gamma_1^2 m}{2\pi\sigma_1\rho^2}\sin\phi\left[3I_1K_1 - \frac{\gamma_1\rho}{2}(I_0K_1-I_1K_0)\right]$	$\frac{\gamma_1^2 m}{2\pi\sigma_1\rho^2}\cos\phi$	$\frac{m}{2\pi\gamma_1^2}\frac{\sin\phi}{\rho^5}\left[\left(\gamma_1^3\rho^3+5\gamma_1^2\rho^2 + 12\gamma_1\rho+12\right)e^{-\gamma_1\rho}-12+2\gamma_1^2\rho^2\right]$	$-\frac{m\cos\phi}{2\pi\gamma_1^2\rho^5}\left[\left(\gamma_1^2\rho^2+3\gamma_1\rho+3\right) e^{-\gamma_1\rho}+\gamma_1^2\rho^2-3\right]$	$\frac{m\sin\phi}{4\pi\rho^3}\left[\gamma_1^2\rho^2(I_1K_1-I_0K_0) +4\gamma_1\rho(I_1K_0-I_0K_1)+16I_1K_1\right]$

Table 3.9.–*The quasi-static range,* $|\gamma_0\rho| \ll 1 \ll |\gamma_1\rho|$ for surface to surface propagation

Dipole type	E_ρ	E_ϕ	E_z	H_ρ	H_ϕ	H_z
VED	$-\frac{\gamma_1 p}{2\pi\sigma_1\rho^2}$	0	$-\frac{p}{2\pi\sigma_1\rho^3}\frac{\gamma_1^2}{\gamma_0^2}$	0	$\frac{p}{2\pi\rho^2}$	0
VMD	0	$-\frac{3m}{2\pi\sigma_1\rho^4}$	0	$-\frac{3m}{2\pi\gamma_1\rho^4}$	0	$-\frac{9m}{2\pi\gamma_1^2\rho^5}$
HED	$\frac{p}{2\pi\sigma_1\rho^3}\cos\phi$	$\frac{p}{\pi\sigma_1\rho^3}\sin\phi$	$\frac{\gamma_1 p\cos\phi}{2\pi\sigma_1\rho^2}$	$\frac{p\sin\phi}{\pi\gamma_1\rho^3}$	$-\frac{p\cos\phi}{2\pi\gamma_1\rho^3}$	$\frac{3p}{2\pi\gamma_1^2\rho^4}\sin\phi$
HMD	$\frac{\gamma_1 m}{2\pi\sigma_1\rho^3}\cos\phi$	$\frac{\gamma_1 m}{\pi\sigma_1\rho^3}\sin\phi$	$\frac{\gamma_1^2 m}{2\pi\sigma_1\rho^2}\cos\phi$	$\frac{m\sin\phi}{\pi\rho^3}$	$-\frac{m\cos\phi}{2\pi\rho^3}$	$\frac{3m\sin\phi}{2\pi\gamma_1\rho^4}$

Table 3.10.–*The quasi-static range,* $|\gamma_0\rho| \ll 1$, $|\gamma_1\rho| \ll 1$ for surface to surface propagation

Dipole type	E_ρ	E_ϕ	E_z	H_ρ	H_ϕ	H_z
VED	$-\frac{\gamma_1^2 p}{4\pi\sigma_1\rho}$	0	$-\frac{p}{2\pi\sigma_1\rho^3}\frac{\gamma_1^2}{\gamma_0^2}$	0	$\frac{p}{2\pi\rho^2}$	0
VMD	0	$-\frac{m\gamma_1^2}{4\pi\sigma_1\rho^2}$	0	$-\frac{m\gamma_1^2}{16\pi\rho}$	0	$-\frac{m}{4\pi\rho^3}$
HED	$\frac{p}{\pi\sigma_1\rho^3}\cos\phi$	$\frac{p}{2\pi\sigma_1\rho^3}\sin\phi$	$\frac{\gamma_1^2 p\cos\phi}{4\pi\sigma_1\rho}$	$\frac{p}{4\pi\rho^2}\sin\phi$	$-\frac{p\cos\phi}{4\pi\rho^2}$	$\frac{p}{4\pi\rho^2}\sin\phi$
HMD	$\frac{\gamma_1^2 m}{4\pi\sigma_1\rho^2}\cos\phi$	$\frac{\gamma_1^2 m\sin\phi}{4\pi\sigma_1\rho^2}$	$\frac{\gamma_1^2 m\cos\phi}{2\pi\sigma_1\rho^2}$	$\frac{m}{2\pi\rho^3}\sin\phi$	$-\frac{m}{4\pi\rho^3}\cos\phi$	$\frac{m\gamma_1^2\sin\phi}{16\pi\rho}$

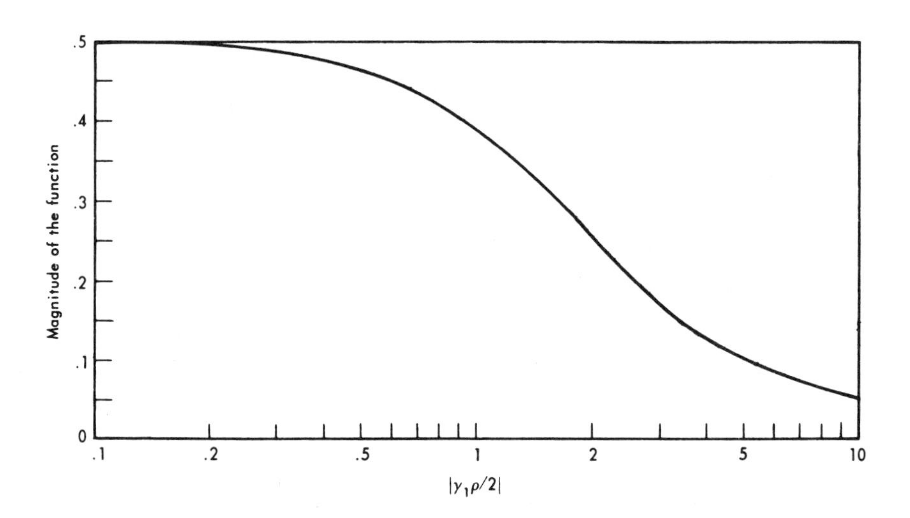

Figure 3.4.–$|I_1K_1|$ *versus* $|\gamma_1\rho/2|$.

3.2.2.3.2 Free Space to Subsurface ($h \geqq 0^+$, $z \leqq 0^-$)

The free-space to subsurface propagation formulas for the quasi-static range are presented in table 3.11 and are subject to the following conditions:

(1) $|\gamma_0 R'| \ll 1$

(2) $|\gamma_1 R'| \gg 1$

(3) $R' \gg |z|$

where $R' = \left(\rho^2 + h^2\right)^{1/2}$.

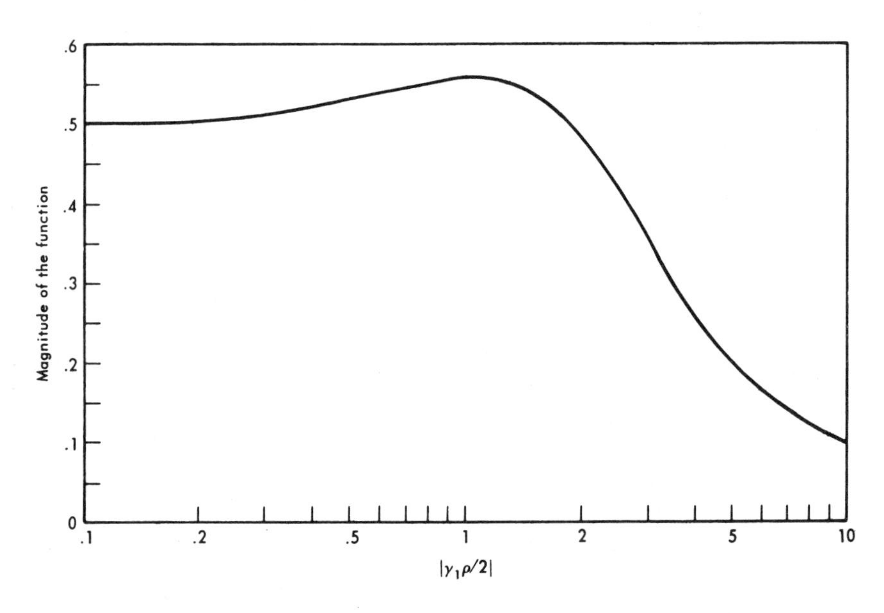

Figure 3.5.–$|3I_1K_1 - (\gamma_1\rho/2)(I_0K_1 - I_1K_0)|$ *versus* $|\gamma_1\rho/2|$

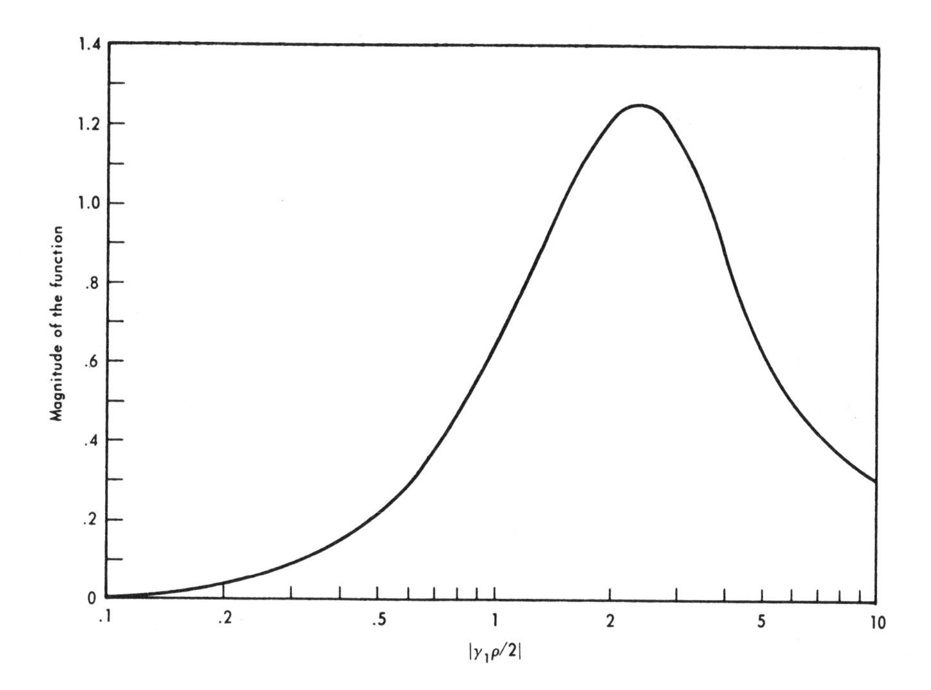

Figure 3.6.–$|T|$ *versus* $|\gamma_1\rho/2|$.

When $\rho \to 0$, $R' \to h$ and the field component expressions in the vicinity of the z-axis may be obtained from those in table 3.11. The resulting formulas are given in table 3.12.

3.2.2.3.3 Subsurface to Free Space ($h \leqq 0^-$, $z \geqq 0^+$)

The subsurface to free-space propagation formulas for the quasi-static range are presented in table 3.13 and are subject to the following conditions:

(1) $|\gamma_0 R| \ll 1$

(2) $|\gamma_1 R| \gg 1$

(3) $R \gg |h|$

where $R = (\rho^2 + z^2)^{1/2}$.

When $\rho \to 0$, $R \to z$ and the field component expressions in the vicinity of the z-axis may be obtained from those in table 3.13. The resulting formulas are given in table 3.14.

The field expressions in free space near the interface such that $\rho \gg |h|$ and z, and $|\gamma_0 \rho| \ll 1 \ll |\gamma_1 \rho|$ are given in table 3.15.

3.2.2.3.4 Subsurface to Subsurface ($h \leqq 0^-$ and $z \leqq 0^-$)

The subsurface to subsurface propagation formulas for the general quasi-static range $|\gamma_0 R_1| \ll 1$ where $R_1 = [\rho^2 + (z+h)^2]^{1/2}$ are lengthy, complicated, and difficult to use for numerical computation. A review and partial listing of these expressions together with selected numerical results are found in Bannister and Hart.[40] General formulas and selected numerical results for

[40] Bannister, P. R.; and Hart, W. C.: Quasi-Static Fields of Dipole Antennas Below the Earth's Surface. USL Rept. No. 870, U.S. Navy Underwater Sound Laboratory, New London, Conn., 11 Apr. 1968.

Table 3.11.–*Free-space to subsurface propagation formulas for the quasi-static range, with* $|\gamma_1 R'| \gg 1$ *and* $R' \gg |z|$

Dipole type	E_ρ	E_ϕ	E_z	H_ρ	H_ϕ	H_z
VED	$-\frac{\gamma_1 p}{2\pi\sigma_1}\frac{\rho e^{\gamma_1 z}}{(R')^3}$	0	$-\frac{p}{2\pi\sigma_1}\frac{e^{\gamma_1 z}}{(R')^3}\left(1-\frac{3h^2}{(R')^2}\right)$	0	$\frac{p}{2\pi}e^{\gamma_1 z}\frac{\rho}{(R')^3}$	0
VMD	0	$-\frac{3me^{\gamma_1 z}}{2\pi\sigma_1}\frac{\rho}{(R')^5}\left(1+\gamma_1 h-\frac{5h^2}{(R')^2}\right)$	0	$-\frac{3me^{\gamma_1 z}}{2\pi\gamma_1}\frac{\rho}{(R')^5}\left(1+\gamma_1 h-\frac{5h^2}{(R')^2}\right)$	0	$-\frac{3m}{2\pi\gamma_1^2}\frac{e^{\gamma_1 z}}{(R')^5}\left[3(1+\gamma_1 h)-\frac{5h^2}{(R')^2}(6+\gamma_1 h)+\frac{35h^4}{(R')^4}\right]$
HED	$\frac{p\cos\phi}{2\pi\sigma_1}\frac{e^{\gamma_1 z}}{(R')^3}(1-\gamma_1 h)$	$\frac{p\sin\phi}{2\pi\sigma_1}\frac{e^{\gamma_1 z}}{(R')^3}\left(2+\gamma_1 h-\frac{3h^2}{(R')^2}\right)$	$\frac{p}{2\pi\sigma_1}\frac{e^{\gamma_1 z}}{(R')^3}\left(-\frac{3h}{(R')^2}+\frac{\gamma_0^2}{\gamma_1}\right)\rho\cos\phi$	$\frac{p\sin\phi}{2\pi\gamma_1}\frac{e^{\gamma_1 z}}{(R')^3}\left(2+\gamma_1 h-\frac{3h^2}{(R')^2}\right)$	$-\frac{p\cos\phi}{2\pi\gamma_1}\frac{e^{\gamma_1 z}}{(R')^3}(1-\gamma_1 h)$	$\frac{3p\sin\phi}{2\pi\gamma_1^2}\frac{\rho e^{\gamma_1 z}}{(R')^5}\left(1+\gamma_1 h-\frac{5h^2}{(R')^2}\right)$
HMD	$\frac{\gamma_1 m\cos\phi e^{\gamma_1 z}}{2\pi\sigma_1(R')^3}$	$\frac{\gamma_1 m\sin\phi e^{\gamma_1 z}}{2\pi\sigma_1(R')^3}\left(2-\frac{3h^2}{(R')^2}\right)$	$\frac{m\cos\phi\rho e^{\gamma_1 z}\gamma_0^2}{2\pi(R')^3\sigma_1}$	$\frac{m\sin\phi e^{\gamma_1 z}}{2\pi(R')^3}\left(2-\frac{3h^2}{(R')^2}\right)$	$-\frac{m\cos\phi e^{\gamma_1 z}}{2\pi(R')^3}$	$\frac{3m\sin\phi\rho e^{\gamma_1 z}}{2\pi\gamma_1(R')^5}\left(1-\frac{5h^2}{(R')^2}\right)$

Table 3.12.–*Free-space to subsurface propagation formulas for the quasi-static range, when* $\rho \to 0$

Dipole type	E_ρ	E_ϕ	E_z	H_ρ	H_ϕ	H_z
VED	$\frac{\gamma_1\rho p e^{\gamma_1 z}}{2\pi\sigma_1 h^3}$	0	$\frac{pe^{\gamma_1 z}}{\pi\sigma_1 h^3}$	0	$\frac{p\rho e^{\gamma_1 z}}{2\pi h^3}$	0
VMD	0	$-\frac{3m\gamma_1\rho e^{\gamma_1 z}}{2\pi\sigma_1 h^4}$	0	$-\frac{3m\rho e^{\gamma_1 z}}{2\pi h^4}$	0	$\frac{3me^{\gamma_1 z}}{\pi\gamma_1 h^4}$
HED	$-\frac{\gamma_1 p}{2\pi\sigma_1 h^2}\cos\phi e^{\gamma_1 z}$	$\frac{\gamma_1 p}{2\pi\sigma_1 h^2}\sin\phi e^{\gamma_1 z}$	$-\frac{3\rho p\cos\phi e^{\gamma_1 z}}{2\pi\sigma_1 h^4}$	$\frac{p\sin\phi e^{\gamma_1 z}}{2\pi h^2}$	$\frac{p\cos\phi e^{\gamma_1 z}}{2\pi h^2}$	$\frac{3p\rho}{2\pi\gamma_1 h^4}\sin\phi e^{\gamma_1 z}$
HMD	$\frac{\gamma_1 m}{2\pi\sigma_1 h^3}\cos\phi e^{\gamma_1 z}$	$-\frac{\gamma_1 m}{2\pi\sigma_1 h^3}\sin\phi e^{\gamma_1 z}$	$\frac{\gamma_0^2\rho m}{2\pi\sigma_1 h^3}\cos\phi e^{\gamma_1 z}$	$-\frac{m\sin\phi e^{\gamma_1 z}}{2\pi h^3}$	$-\frac{m\cos\phi e^{\gamma_1 z}}{2\pi h^3}$	$-\frac{6m\rho\sin\phi e^{\gamma_1 z}}{\pi\gamma_1 h^5}$

Table 3.13.–*Subsurface to free-space propagation formulas for the quasi-static range, with* $|\gamma_1 R| \gg 1$ *and* $R \gg |h|$

Dipole type	E_ρ	E_ϕ	E_z	H_ρ	H_ϕ	H_z
VED	$\frac{pe^{\gamma_1 h}}{2\pi\sigma_1}\frac{\rho}{R^5}\left[3z - \frac{\gamma_0^2}{\gamma_1}R^2\right]$	0	$-\frac{p}{2\pi\sigma_1}\frac{e^{\gamma_1 h}}{R^3}\left(1-\frac{3z^2}{R^2}\right)$	0	$\frac{p}{2\pi}\frac{\gamma_0^2}{\gamma_1^2}e^{\gamma_1 h}\frac{\rho}{R^3}$	0
VMD	0	$-\frac{3me^{\gamma_1 h}}{2\pi\sigma_1}\frac{\rho}{R^5}\left(1+\gamma_1 z-\frac{5z^2}{R^5}\right)$	0	$-\frac{3me^{\gamma_1 h}}{2\pi\gamma_1}\frac{\rho}{R^5}\left(1-\frac{5z^2}{R^2}\right)$	0	$-\frac{3m}{2\pi\gamma_1^2}\frac{e^{\gamma_1 h}}{R^5}\left[3(1+\gamma_1 z) - \frac{5z^2}{R^2}(6+\gamma_1 z)+\frac{35z^4}{R^4}\right]$
HED	$\frac{p\cos\phi\, e^{\gamma_1 h}}{2\pi\sigma_1}\frac{}{R^3}(1-\gamma_1 z)$	$\frac{p\sin\phi\, e^{\gamma_1 h}}{2\pi\sigma_1\, R^3}\left(2+\gamma_1 z-\frac{3z^2}{R^2}\right)$	$\frac{\gamma_1 p\cos\phi}{2\pi\sigma_1}\frac{\rho e^{\gamma_1 h}}{R^3}$	$\frac{p\sin\phi}{2\pi\gamma_1}\frac{e^{\gamma_1 h}}{R^3}\left[2-\frac{3z^2}{R^2}\right]$	$-\frac{p\cos\phi}{2\pi\gamma_1}\frac{e^{\gamma_1 h}}{R^3}$	$\frac{3p\sin\phi}{2\pi\gamma_1^2}\frac{\rho e^{\gamma_1 h}}{R^5}\left[1+\gamma_1 z-\frac{5z^2}{R^2}\right]$
HMD	$\frac{\gamma_1 m\cos\phi e^{\gamma_1 h}}{2\pi\sigma_1 R^3}(1-\gamma_1 z)$	$\frac{\gamma_1 m\sin\phi e^{\gamma_1 h}}{2\pi\sigma_1 R^3}\left(2+\gamma_1 z-\frac{3z^2}{R^2}\right)$	$\frac{\gamma_1^2 m\cos\phi\rho e^{\gamma_1 h}}{2\pi R^3\sigma_1}$	$\frac{m\sin\phi e^{\gamma_1 h}}{2\pi R^3}\left[2-\frac{3z^2}{R^2}\right]$	$-\frac{m\cos\phi e^{\gamma_1 h}}{2\pi R^3}$	$\frac{3m\sin\phi\rho e^{\gamma_1 h}}{2\pi\gamma_1 R^5}\left(1+\gamma_1 z-\frac{5z^2}{R^2}\right)$

Table 3.14.–*Subsurface to free-space propagation formulas for the quasi-near range, when* $\rho \to 0$

Dipole type	E_ρ	E_ϕ	E_z	H_ρ	H_ϕ	H_z
VED	$\frac{3pe^{\gamma_1 h}}{2\pi\sigma_1}\frac{\rho}{z^4}$	0	$\frac{pe^{\gamma_1 h}}{\pi\sigma_1 z^3}$	0	$\frac{p}{2\pi}\frac{\gamma_0^2}{\gamma_1^2}e^{\gamma_1 h}\frac{\rho}{z^3}$	0
VMD	0	$-\frac{3me^{\gamma_1 h}}{2\pi\sigma_1}\frac{\gamma_1\rho}{z^4}$	0	$\frac{6me^{\gamma_1 h}}{\pi\gamma_1}\frac{\rho}{z^5}$	0	$\frac{3m}{\pi\gamma_1}\frac{e^{\gamma_1 h}}{z^4}$
HED	$-\frac{\gamma_1 pe^{\gamma_1 h}}{2\pi\sigma_1 z^2}\cos\phi$	$\frac{\gamma_1 p\sin\phi e^{\gamma_1 h}}{2\pi\sigma_1 z^2}$	$\frac{\gamma_1\rho p\cos\phi e^{\gamma_1 h}}{2\pi\sigma_1 z^3}$	$-\frac{p\sin\phi e^{\gamma_1 h}}{2\pi\gamma_1 z^3}$	$-\frac{p\cos\phi e^{\gamma_1 h}}{2\pi\gamma_1 z^3}$	$\frac{3\rho p\sin\phi e^{\gamma_1 h}}{2\pi\gamma_1 z^4}$
HMD	$-\frac{\gamma_1^2 m\cos\phi e^{\gamma_1 h}}{2\pi\sigma_1 z^2}$	$\frac{\gamma_1^2 m\sin\phi e^{\gamma_1 h}}{2\pi\sigma_1 z^2}$	$\frac{\gamma_1^2 m\cos\phi\rho e^{\gamma_1 h}}{2\pi\sigma_1 z^3}$	$-\frac{m\sin\phi e^{\gamma_1 h}}{2\pi z^3}$	$-\frac{m\cos\phi e^{\gamma_1 h}}{2\pi z^3}$	$\frac{3\rho m\sin\phi e^{\gamma_1 h}}{2\pi z^4}$

the VMD are given in Sinha and Bhattacharya.[41] General expressions and selected numerical results are given by Wait and Campbell[42] for the VED and by Wait and Campbell[43] for the HMD. A method of obtaining numerical results from the general expressions for the HMD is presented by Atzinger, Pensa, and Pigott.[44]

3.2.2.3.4.1 The range $|\gamma_0\rho| \ll 1 \ll |\gamma_1\rho|$

Formulas for the case where $|\gamma_0\rho| \ll 1 \ll |\gamma_1\rho|$ with $\rho \gg |h|$ and $|z|$ are given in table 3.16. As noted in the far- and near-field ranges, the energy propagates in the up-over-and-down path.

3.2.2.3.4.2 Field points on the z-axis ($\rho = 0$)

For field points on the z-axis ($\rho = 0$), the general quasi-static field expressions, which follow, simplify considerably:

VED ($\rho = 0$, $h \leqq 0^-$, $z \leqq 0^-$)

$$E_z = \frac{p}{2\pi\sigma_1 |z+h|^3} C \tag{3.48}$$

where

$$C = r^3 e^{-\gamma_1|z-h|}(1 + \gamma_1|z-h|) - e^{-\gamma_1|z+h|}(1 + \gamma_1|z+h|)$$
$$+ \frac{\gamma_0^2}{\gamma_1^2}(\gamma_1|z+h|)(2K_1 + \gamma_1|z+h|K_0) \tag{3.49}$$

with $r = |z+h|/|z-h|$. The argument of the modified Bessel functions K_0 and K_1 is $\gamma_1|z+h|$. It should be noted that the results for $h = 0^-$ may be converted to those for $h = 0^+$ by simply multiplying by γ_1^2/γ_0^2.

VMD ($\rho = 0$, $h \leqq 0^-$, $z \leqq 0^-$)

$$H_z = \frac{m}{2\pi|z+h|^3} G \tag{3.50}$$

where

$$G = r^3 e^{-\gamma_1|z-h|} - e^{-\gamma_1|z+h|} + \gamma_1|z+h|\left[r^2 e^{-\gamma_1|z-h|} - e^{-\gamma_1|z+h|}\right]$$
$$+ \frac{2}{\gamma_1^2|z+h|^2}\left[12 + 12\gamma_1|z+h| + 5\gamma_1^2|z+h|^2 + \gamma_1^3|z+h|^3\right]e^{-\gamma_1|z+h|}$$
$$- 3\left[4K_0 + \left(\gamma_1|z+h| + \frac{8}{\gamma_1|z+h|}\right)K_1\right] \tag{3.51}$$

---

[41] Sinha, A. K.; and Bhattacharya, P. K.: Vertical Magnetic Dipole Buried Inside a Homogeneous Earth. Radio Sci., vol. 1, no. 3, Mar. 1966, pp. 379–395.

[42] Wait, J. R.; and Campbell, L. L.: The Fields of an Electric Dipole in a Semi-Infinite Conducting Medium. J. Geophys. Res., vol. 58, no. 1, Mar. 1953, pp. 21–28.

[43] Wait, J. R.; and Campbell, L. L.: The Fields of an Oscillating Magnetic Dipole Immersed in a Semi-Infinite Conducting Medium. J. Geophys. Res., vol. 58, no. 4, June 1953, pp. 167–178.

[44] Atzinger, E. M.; Pensa, A. F.; and Pigott, M. T.: J. Geophys. Res., vol. 71, no. 23, Dec. 1966, pp. 5765–5769.

Table 3.15.–*Subsurface to free-space propagation formulas for the quasi-static range, when* $|\gamma_1\rho| \gg 1$ *and* $\rho \gg |h|$ *and z*

Dipole type	E_ρ	E_ϕ	E_z	H_ρ	H_ϕ	H_z
VED	$\frac{pe^{\gamma_1 h}}{2\pi\sigma_1\rho^4}\left[3z - \frac{\gamma_0^2}{\gamma_1}\rho^2\right]$	0	$-\frac{pe^{\gamma_1 h}}{2\pi\sigma_1\rho^3}$	0	$\frac{p\gamma_0^2 e^{\gamma_1 h}}{2\pi\gamma_1^2\rho^2}$	0
VMD	0	$-\frac{3me^{\gamma_1 h}}{2\pi\sigma_1\rho^4}(1+\gamma_1 z)$	0	$-\frac{3me^{\gamma_1 h}}{2\pi\gamma_1\rho^4}$	0	$-\frac{9me^{\gamma_1 h}}{2\pi\gamma_1^2\rho^5}(1+\gamma_1 z)$
HED	$\frac{p\cos\phi e^{\gamma_1 h}}{2\pi\sigma_1\rho^3}(1-\gamma_1 z)$	$\frac{p\sin\phi e^{\gamma_1 h}}{2\pi\sigma_1\rho^3}(2+\gamma_1 z)$	$\frac{p\cos\phi\gamma_1 e^{\gamma_1 h}}{2\pi\sigma_1\rho^2}$	$\frac{p\sin\phi e^{\gamma_1 h}}{\pi\gamma_1\rho^3}$	$-\frac{p\cos\phi e^{\gamma_1 h}}{2\pi\gamma_1\rho^3}$	$\frac{3p\sin\phi e^{\gamma_1 h}}{2\pi\gamma_1^2\rho^4}(1+\gamma_1 z)$
HMD	$\frac{m\cos\phi\gamma_1 e^{\gamma_1 h}}{2\pi\sigma_1\rho^3}(1-\gamma_1 z)$	$\frac{m\sin\phi\gamma_1 e^{\gamma_1 h}}{2\pi\sigma_1\rho^3}(2+\gamma_1 z)$	$\frac{m\cos\phi\gamma_1^2 e^{\gamma_1 h}}{2\pi\sigma_1\rho^2}$	$\frac{m\sin\phi e^{\gamma_1 h}}{\pi\rho^3}$	$-\frac{m\cos\phi e^{\gamma_1 h}}{2\pi\rho^3}$	$\frac{3m\sin\phi e^{\gamma_1 h}}{2\pi\gamma_1\rho^4}-(1+\gamma_1 z)$

Table 3.16.–*Subsurface to subsurface propagation formulas for the quasi-static range, when* $|\gamma_0\rho| \ll 1 \ll |\gamma_1\rho|$ *with* $\rho \gg |h|$ *and* $|z|$

Dipole type	E_ρ	E_ϕ	E_z	H_ρ	H_ϕ	H_z
VED	$-\frac{p\gamma_0^2}{2\pi\sigma_1\gamma_1\rho^2}e^{\gamma_1(h+z)}$	0	$-\frac{p}{2\pi\sigma_1\rho^3}\frac{\gamma_0^2}{\gamma_1^2}e^{\gamma_1(h+z)}$	0	$\frac{p}{2\pi\rho^2}\frac{\gamma_0^2}{\gamma_1^2}e^{\gamma_1(h+z)}$	0
VMD	0	$-\frac{3m}{2\pi\sigma_1\rho^4}e^{\gamma_1(h+z)}$	0	$-\frac{3m}{2\pi\gamma_1\rho^4}e^{\gamma_1(h+z)}$	0	$-\frac{9m}{2\pi\gamma_1^2\rho^5}e^{\gamma_1(h+z)}$
HED	$\frac{p\cos\phi}{2\pi\sigma_1\rho^3}e^{\gamma_1(h+z)}$	$\frac{p\sin\phi}{\pi\sigma_1\rho^3}e^{\gamma_1(h+z)}$	$\frac{p\cos\phi}{2\pi\sigma_1\rho^2}\frac{\gamma_0^2}{\gamma_1}e^{\gamma_1(h+z)}$	$\frac{p\sin\phi}{\pi\gamma_1\rho^3}e^{\gamma_1(h+z)}$	$-\frac{p\cos\phi}{2\pi\gamma_1\rho^3}e^{\gamma_1(h+z)}$	$\frac{3p\sin\phi}{2\pi\gamma_1^2\rho^4}e^{\gamma_1(h+z)}$
HMD	$\frac{m\cos\phi}{2\pi\sigma_1\rho^3}\gamma_1 e^{\gamma_1(h+z)}$	$\frac{m\sin\phi}{\pi\sigma_1\rho^3}\gamma_1 e^{\gamma_1(h+z)}$	$\frac{m\cos\phi}{2\pi\sigma_1\rho^2}\gamma_0^2 e^{\gamma_1(h+z)}$	$\frac{m\sin\phi}{\pi\rho^3}e^{\gamma_1(h+z)}$	$-\frac{m\cos\phi}{2\pi\rho^3}e^{\gamma_1(h+z)}$	$\frac{3m\sin\phi}{2\pi\gamma_1\rho^4}e^{\gamma_1(h+z)}$

Values of the magnitude and phase of G are plotted by Sinha and Bhattacharya[41] and are shown here in figures 3.7 and 3.8, respectively, as a function of $|\gamma_1(z+h)/2|$.

HED $(\rho=0,\ h\leqq 0^-,\ z\leqq 0^-)$

$$E_x = \frac{p}{2\pi\sigma_1|z+h|^3}M \tag{3.52}$$

where

$$\begin{aligned} M = {} & \gamma_1|z+h|\left(K_1 + \frac{\gamma_1}{2}|z+h|K_0\right) \\ & - \frac{r^3 e^{-\gamma_1|z-h|}}{2}\left(1+\gamma_1|z-h| + \gamma_1^2|z-h|^2\right) \\ & - \frac{e^{-\gamma_1|z+h|}}{2}\left(3+3\gamma_1|z+h| + \gamma_1^2|z+h|^2\right) \end{aligned} \tag{3.53}$$

and

$$H_y = \frac{-p}{2\pi|z+h|^2}N \tag{3.54}$$

where

$$\begin{aligned} N = {} & \frac{3K_0}{2} + \frac{K_1}{2\gamma_1|z+h|}\left(6+\gamma_1^2|z+h|^2\right) \\ & + \frac{r^2(z-h)e^{-\gamma_1|z-h|}}{2|z-h|}(1+\gamma_1|z-h|) \\ & - \frac{e^{-\gamma_1|z+h|}}{2\gamma_1^2|z+h|^2}\left(6+6\gamma_1|z+h|+3\gamma_1^2|z+h|^2+\gamma_1^3|z+h|^3\right) \end{aligned} \tag{3.55}$$

Values of the magnitude and phase of M and N are plotted by Bannister[40] and are shown here in figures 3.9 to 3.13 as a function of $|\gamma_1(z+h)|$.

HMD $(\rho=0,\ h\leqq 0^-,\ z\leqq 0^-)$

$$E_x = \frac{m\gamma_1^2}{2\pi\sigma_1|z+h|^2}P \tag{3.56}$$

where

$$\begin{aligned} P = {} & \frac{3}{2}K_0 + \frac{K_1}{2\gamma_1|z+h|}\left(6+\gamma_1^2|z+h|^2\right) \\ & - \frac{r^2(z-h)e^{-\gamma_1|z-h|}}{2|z-h|}(1+\gamma_1|z-h|) \\ & - \frac{e^{-\gamma_1|z+h|}}{2\gamma_1^2|z+h|^2}\left(6+6\gamma_1|z+h|+3\gamma_1^2|z+h|^2+\gamma_1^3|z+h|^3\right) \end{aligned} \tag{3.57}$$

Because of the reciprocity theorem, values of the magnitude and phase of P may be obtained from figures 3.11 to 3.13 for N. However, the curves for $|z|>|h|$ must be replaced by the curves for $|h|>|z|$ (and vice versa) and 90 degrees must be subtracted from the phase curves.

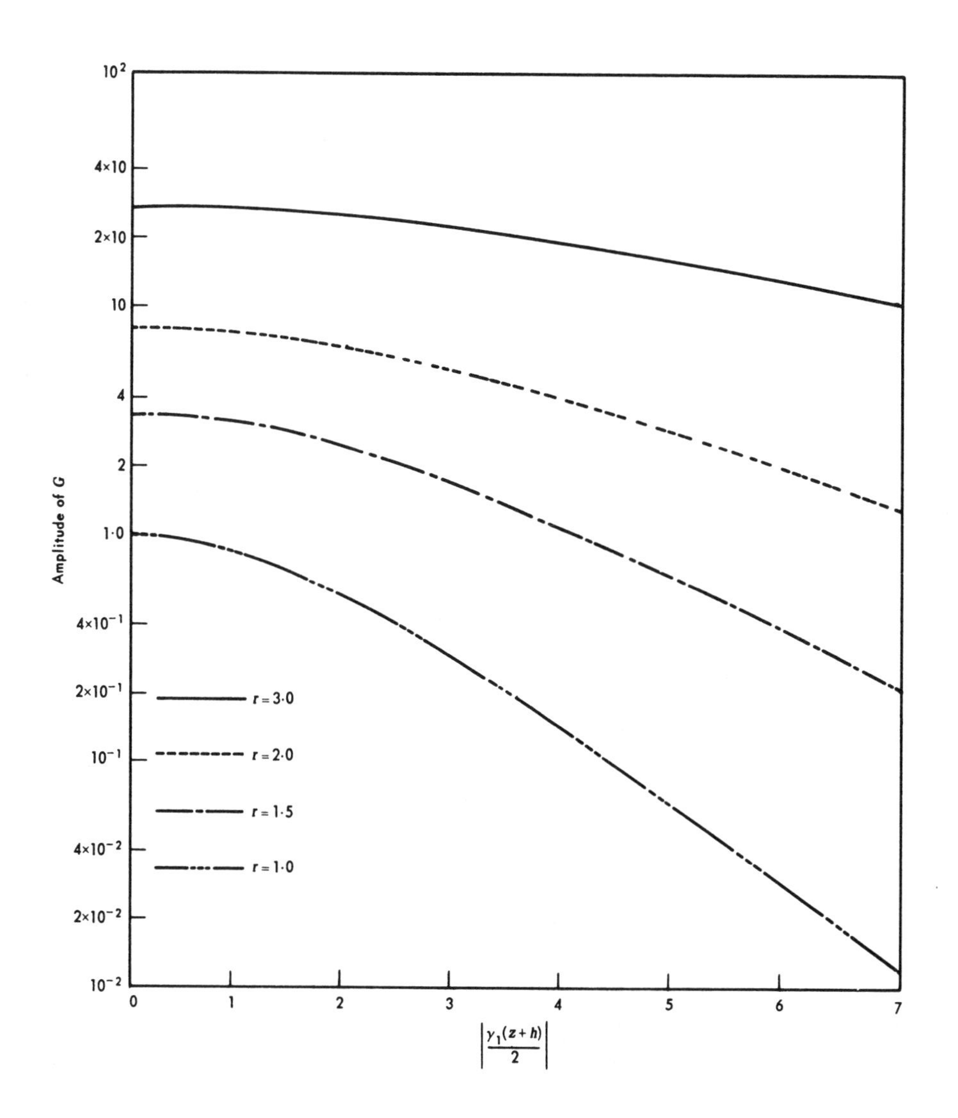

Figure 3.7.–*Variation of the amplitude of* G *with* $|[\gamma_1(z+h)]/2|$.

$$H_y = \frac{m}{4\pi|z+h|^3} U \tag{3.58}$$

where

$$U = \frac{e^{-\gamma_1|z+h|}}{\gamma_1^2|z+h|^2}\left(24 + 24\gamma_1|z+h| + 11\gamma_1^2|z+h|^2 + 3\gamma_1^3|z+h|^3 + \gamma_1^4|z+h|^4\right)$$

$$-\frac{r^3 e^{-\gamma_1|z-h|}}{\gamma_1^2|z-h|^2}\left(\gamma_1^2|z-h|^2 + \gamma_1^3|z-h|^3 + \gamma_1^4|z-h|^4\right)$$

$$-\left(12 + \gamma_1^2|z+h|^2\right)K_0 - \frac{1}{\gamma_1|z+h|}\left(24 + 5\gamma_1^2|z+h|^2\right)K_1 \tag{3.59}$$

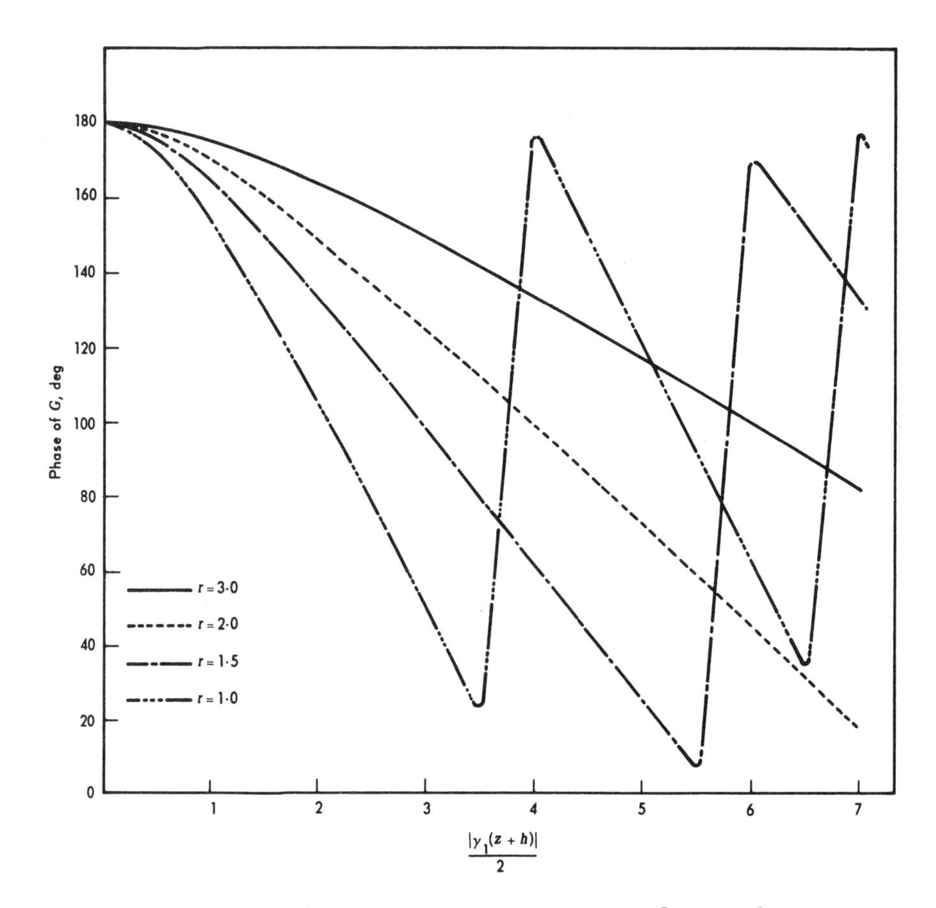

Figure 3.8.–*Variation of the phase of* G *with* $|[\gamma_1(z+h)]/2|$.

Wait and Campbell[43] plot the magnitude of $U/2$ as a function of $|\gamma_1 z|$ for the case where $h=0^-$. These numerical results are shown here in figure 3.14.

3.3 LAYERED CONDUCTING HALF SPACE

The propagation of radio waves between points over and on the surface of a stratified conducting half space is discussed by Wait.[9] Geophysical applications of dipole field propagation over a layered conductor are treated by Keller and Frischknecht[25] and Hansen, et al.[45] The present section reviews the field component expressions for those cases in which both the dipole source and the field point are in the top layer of the stratified conductor.

Because any dipole field may be expressed as a superposition of plane waves by using Fourier methods, the results of section 2.6 for plane waves over and in a layered half space may be used. Under certain conditions, however, a single plane wave may be adequate to describe the dipole field at a receiving point in the top layer. These conditions are:

(1) $|\gamma_i| \gg |\gamma_0|$, i.e., the magnitude of the propagation constant in each layer is much greater than the magnitude of the free-space propagation constant

[45] Hansen, D. A.; et al.: Mining Geophysics. Vol. II. Theory. Society of Exploration Geophysicists, Tulsa, Okla., 1967.

(2) $|\gamma_i \rho| \gg 1$, i.e., the horizontal separation between the source and receiving dipoles is much greater than the skin depth δ_i in the layers

(3) $\rho \gg h$ and z

As a consequence of these conditions, the field in the vicinity of the receiving point varies only slightly in the horizontal direction in the distance δ_i and propagates essentially as a plane wave in the direction vertically downward from the surface. The surface impedance concept of section 2.7 is therefore applicable and the results of the transmission line analogy in that section may be used.

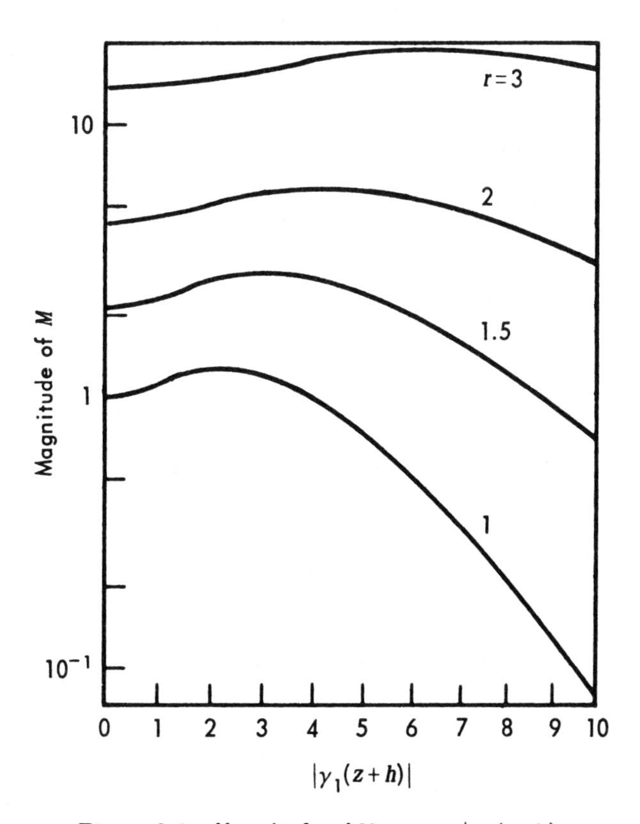

Figure 3.9.—*Magnitude of* M *versus* $|\gamma_1(z+h)$

3.3.1 Surface to Surface Propagation

For propagation between points on the surface of the layered conductor, the effect of the stratification may be accounted for by introducing an effective surface impedance η_e with a corresponding propagation $\gamma_e = (j\omega\mu_0\sigma_e)^{1/2}$. (It is assumed that all the layers have the permeability of free space, μ_0.) The problem is thus reduced to the previously known results for the homogeneous conducting half space. In terms of the parameters for the top layer,

$$\gamma_e = \frac{\gamma_1}{Q} \tag{3.60}$$

$$\sigma_e = \frac{\sigma_1}{Q^2} \tag{3.61}$$

where Q is obtained from (2.72) to (2.74). For a two-layer medium ($z_2 \to \infty$)

$$Q = \frac{\gamma_1 + \gamma_2 \tanh \gamma_1 z_1}{\gamma_2 + \gamma_1 \tanh \gamma_1 z_1} \tag{3.62}$$

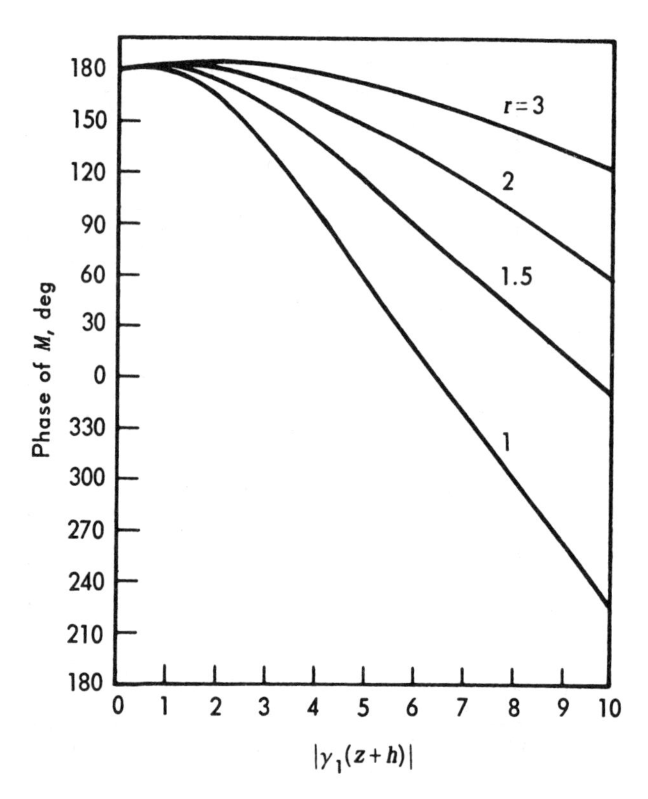

Figure 3.10–*Phase of* M *versus* $|\gamma_1(z+h)$

For a three-layer medium ($z_3 \to \infty$)

$$Q = \frac{\gamma_1 \hat{Q} + \gamma_2 \tanh \gamma_1 z_1}{\gamma_2 + \gamma_1 \hat{Q} \tanh \gamma_1 z_1} \tag{3.63}$$

$$\hat{Q} = \frac{\gamma_2 + \gamma_3 \tanh \gamma_2 z_2}{\gamma_3 + \gamma_2 \tanh \gamma_2 z_2} \tag{3.64}$$

The factor Q can be interpreted as a correction factor to account for the presence of stratification. The amplitude and phase of Q have been extensively tabulated for two- and three-layer

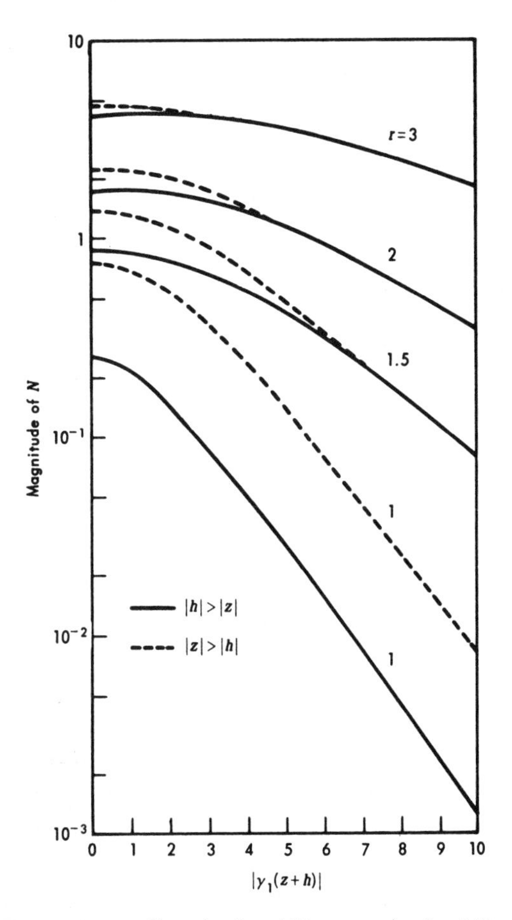

Figure 3.11.–*Magnitude of* N *versus* $|\gamma_1(z+h)|$.

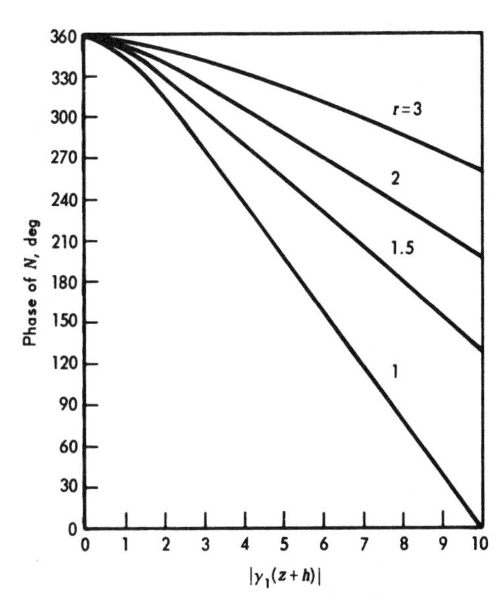

Figure 3.12.–*Phase of* N *versus* $|\gamma_1(z+h)|$ *when* $|z| > |h|$.

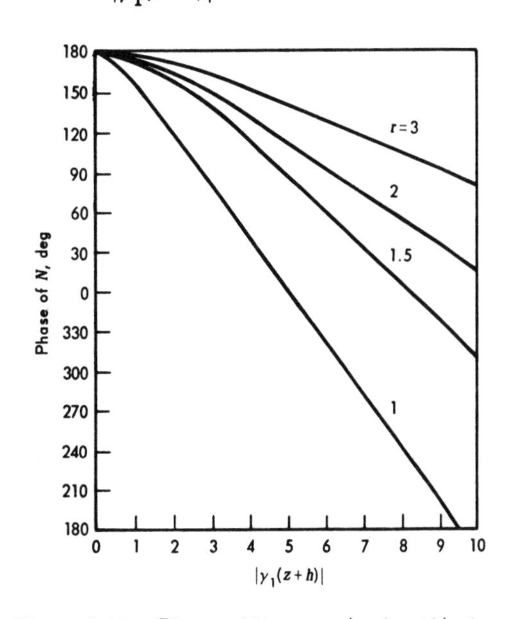

Figure 3.13.–*Phase of* N *versus* $|\gamma_1(z+h)|$ *when* $|h| > |z|$.

stratifications by Jackson, Wait, and Walters[46] and also appear in Wait.[9] It follows from this discussion that the previous results for the propagation in the far-field, near-field, and quasi-static field (when $|\gamma_1\rho| \gg 1$) regions between points on the surface of a homogeneous conducting half space may thus be amended as indicated in (3.60) and (3.61) to account for stratification under the conditions 1 and 2 above.

A further discussion of the quasi-static field components of a VMD over a layered conductor is presented by Wait[47] and Bannister.[48]

3.3.2 Surface to Subsurface Propagation

The plane wave results expressed in (2.75) and (2.76) may be used to find the dipole field in the top layer of a stratified conductor when that field is known at the surface of the layered medium. Thus for $0 \le z \le z_1$ (see fig. 2.4)

$$E(z) = \frac{E(0)}{Q_E} \tag{3.65}$$

$$H(z) = \frac{H(0)}{Q_H} \tag{3.66}$$

where Q_E^{-1} and Q_H^{-1} are defined for a two-layer medium by (2.75) and (2.76) respectively. If there are more than two layers, then the impedance η_2 in (2.75) and (2.76) is to be interpreted as the effective surface impedance of the layers below the first and may be determined from (2.72) to (2.74). Plots of the magnitude and phase of Q_E^{-1} and Q_H^{-1} for $z=z_1$ as a function of z_1/δ_1 are presented by Bannister[49] for various layer conductivity ratios and are shown here in figures 3.15 and 3.16.

As an example, the horizontal quasi-static fields in the top layer of a stratified conductor from an HED at the surface of the layered medium are obtained from table 3.9 and written

$$E_\rho = \frac{p \cos \phi}{2\pi\sigma_e \rho^3}\left(\frac{1}{Q_E}\right) = \frac{p \cos \phi}{2\pi\sigma_1 \rho^3}\left(\frac{Q^2}{Q_E}\right) \tag{3.67}$$

$$E_\phi = \frac{p \sin \phi}{2\pi\sigma_e \rho^3}\left(\frac{1}{Q_E}\right) = \frac{p \sin \phi}{\pi\sigma_1 \rho^3}\left(\frac{Q^2}{Q_E}\right) \tag{3.68}$$

$$H_\rho = \frac{p \sin \phi}{\pi\gamma_e \rho^3}\left(\frac{1}{Q_H}\right) = \frac{p \sin \phi}{\pi\gamma_1 \rho^3}\left(\frac{Q}{Q_H}\right) \tag{3.69}$$

---

[46] Jackson, C. M.; Wait, J. R.; and Walters, L. C.: Numerical Results for the Surface Impedance of a Stratified Conductor. Nat. Bur. Std. Tech. Note No. 143, 19 Mar. 1962.

[47] Wait, J. R.: A Note on the Electromagnetic Response of a Stratified Earth. Geophysics, vol. 27, no. 3, June 1962, pp. 382-385.

[48] Bannister, P. R.: Some Notes on the Utilization of a Vertical Magnetic Dipole to Determine the Properties of a Stratified Earth. USL Rept. No. 695, U.S. Navy Underwater Sound Laboratory, New London, Conn., 5 Oct. 1965.

[49] Bannister, P. R.: Determination of the Electrical Conductivity of the Sea Bed in Shallow Waters. Geophysics, vol. 33, no. 6, Dec. 1968, pp. 995-1003.

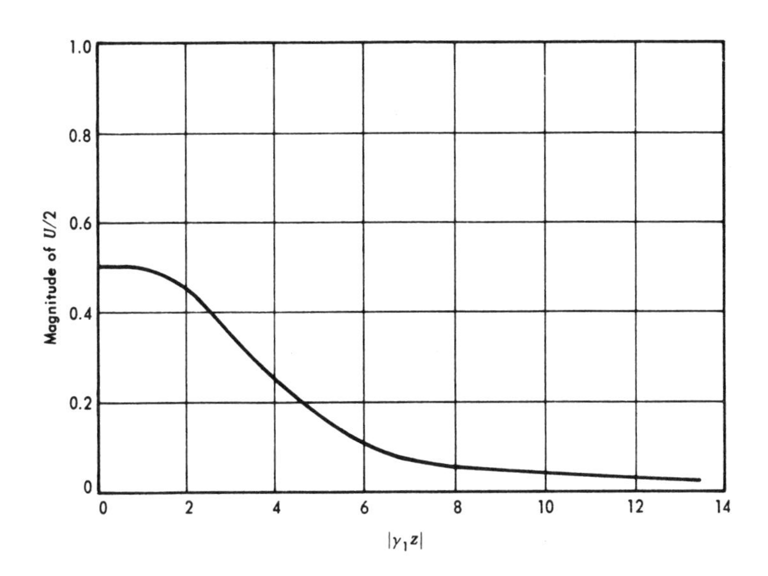

Figure 3.14.–*Magnitude of* U/2 *as a function of* $|\gamma_1 z|$.

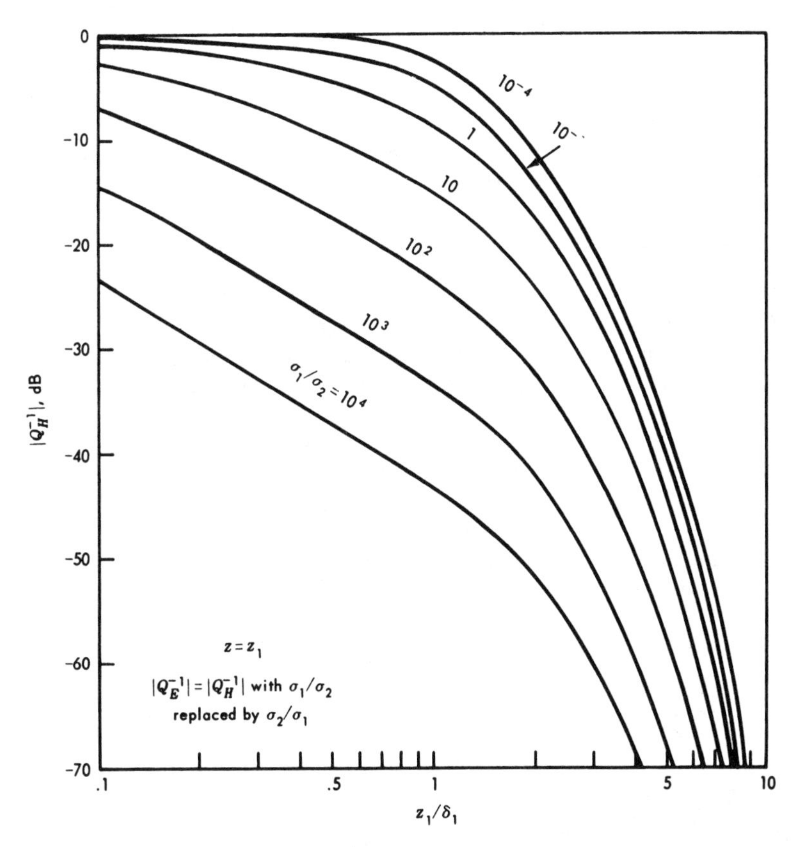

Figure 3.15.–$|Q_H^{-1}|$ *versus* z_1/δ_1 *when* $z = z_1$.

$$H_\phi = \frac{-p \cos \phi}{2\pi \gamma_e \rho^3}\left(\frac{1}{Q_H}\right) = \frac{-p \cos \phi}{2\pi \gamma_1 \rho^3}\left(\frac{Q}{Q_H}\right) \tag{3.70}$$

Similarly for the HMD,

$$E_\rho = \frac{m\gamma_1 \cos \phi}{2\pi \sigma_1 \rho^3}\left(\frac{Q}{Q_E}\right) \tag{3.71}$$

$$E_\phi = \frac{m\gamma_1 \sin \phi}{\pi \sigma_1 \rho^3}\left(\frac{Q}{Q_E}\right) \tag{3.72}$$

$$H_\rho = \frac{m \sin \phi}{\pi \rho^3}\left(\frac{1}{Q_H}\right) \tag{3.73}$$

$$H_\phi = \frac{-m \cos \phi}{2\pi \rho^3}\left(\frac{1}{Q_H}\right) \tag{3.74}$$

Plots of the magnitude and phase of Q/Q_E and Q/Q_H for $z = z_1$ are presented by Bannister[49] as a function of z_1/δ_1 for various layer conductivity ratios and are shown here in figures 3.16 through 3.19. A plot of the magnitude Q^2/Q_E versus z_1/δ_1 is given in figure 4.6.

The vertical fields in the top layer may be found from the horizontal components by applying Maxwell's equations; however, these fields are much smaller than the horizontal fields and may be neglected for most purposes.

A further discussion of the quasi-static field components at the interface between the layers of a two-layer conducting half space is presented by Bannister[50] for a HED source on the surface.

3.3.3 Subsurface to Surface Propagation

The subsurface to surface dipole field propagation formulas may be obtained by applying the reciprocity theorem to the surface to subsurface formulas. This theorem is stated in section 1.6 and in the present case applies to dipole antennas in the presence of a linear, inhomogeneous medium. Application of the reciprocity theorem results in the following relationships:

$$E_z^{HM}[z, 0] = j\omega\mu_0 \cos \phi \frac{m^{HM}}{p^{VE}} H_\phi^{VE}[0, z] \tag{3.75}$$

$$H_z^{HM}[z, 0] = -\sin \phi \frac{m^{HM}}{m^{VM}} H_\rho^{VM}[0, z] \tag{3.76}$$

$$E_z^{HE}[z, 0] = -\cos \phi \frac{p^{HE}}{p^{VE}} E_\rho^{VE}[0, z] \tag{3.77}$$

[50] Bannister, P. R.: Quasi-Static Fields Within a Two-Layer Stratified Earth Produced by a Horizontal Dipole Antenna. USL Rept. No. 956, U.S. Navy Underwater Sound Laboratory, New London, Conn., 23 Sept. 1968.

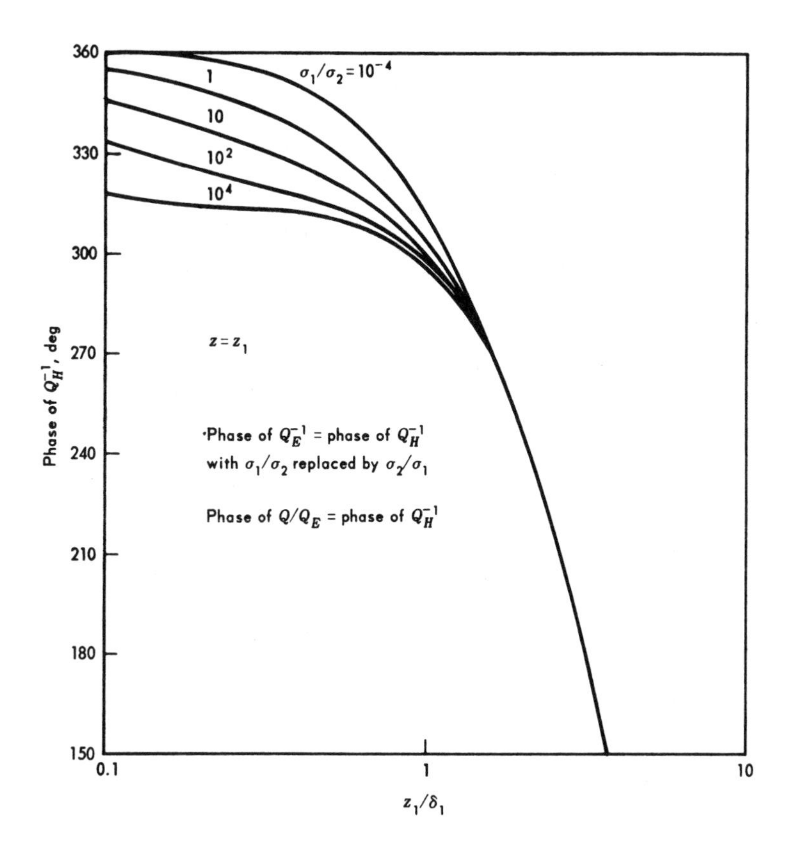

Figure 3.16.–*Phase of* Q_H^{-1} *versus* z_1/δ_1 *for* $z = z_1$.

$$H_z^{HE}[z, 0] = -\frac{\sin\phi}{j\omega\mu_0} \frac{p^{HE}}{m^{VM}} E_\phi^{VM}[0, z] \tag{3.78}$$

$$H_\rho^{HE}[z, 0] = \frac{1}{j\omega\mu_0} \frac{p^{HE}}{m^{HM}} E_\phi^{HM}[0, z] \tag{3.79}$$

$$H_\phi^{HE}[z, 0] = \frac{-1}{j\omega\mu_0} \frac{p^{HE}}{m^{HM}} E_\rho^{HM}[0, z] \tag{3.80}$$

and

$$E_z^{VE},\ H_z^{VM},\ H_\rho^{HM},\ H_\phi^{HM},\ E_\rho^{HE},\ E_\phi^{HE}[z, 0] = E_z^{VE},\ H_z^{VM},\ H_\rho^{HM},\ H_\phi^{HM},\ E_\rho^{HE},\ E_\phi^{HE}[0, z] \tag{3.81}$$

where the first quantity in the bracket refers to the depth of the dipole source for the particular field and the second quantity to the depth of the receiving dipole. In all of the above relationships, the reciprocity theorem requires that the currents in the two particular dipole antennas be the same. Thus, for example, for an electric dipole of length L and a magnetic loop dipole of area A,

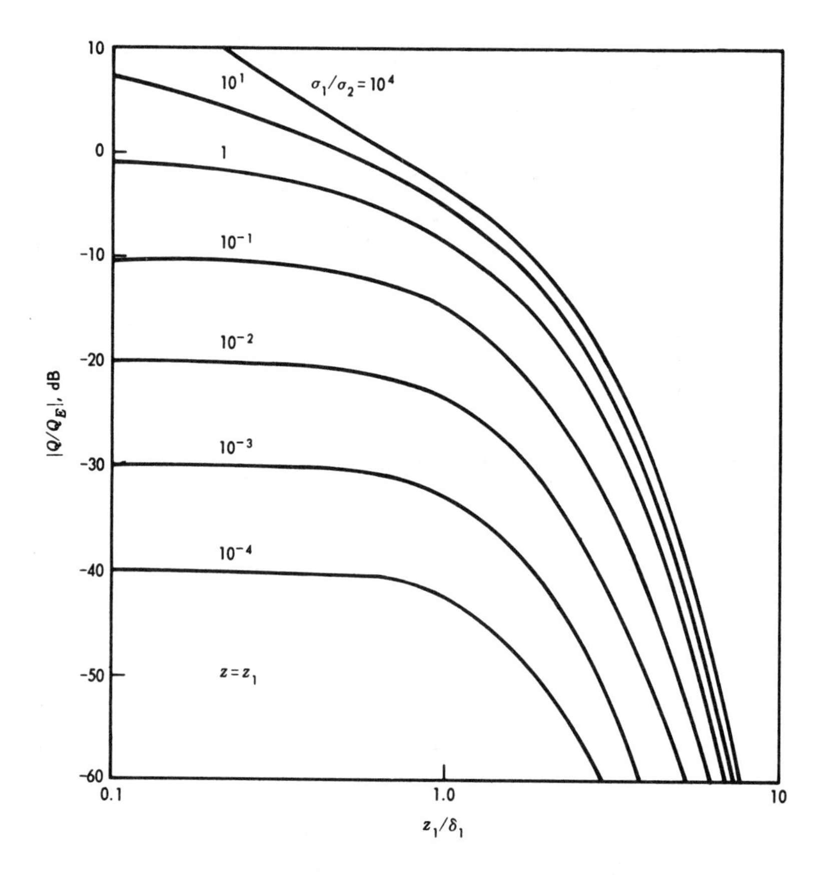

Figure 3.17.–$|Q/Q_E|$ *versus* z_1/δ_1 *for* $z = z_1$.

$$\frac{p}{m} = \frac{IL}{IA} = \frac{L}{A}$$

The resulting HED formulas for subsurface to surface propagation become

$$E_\rho = \frac{p \cos \phi}{2\pi\sigma_1 \rho^3} \left(\frac{Q^2}{Q'_E} \right) \tag{3.82}$$

$$E_\phi = \frac{p \sin \phi}{\pi\sigma_1 \rho^3} \left(\frac{Q^2}{Q'_E} \right) \tag{3.83}$$

$$H_\rho = \frac{p \sin \phi}{\pi\gamma_1 \rho^3} \left(\frac{Q}{Q'_E} \right) \tag{3.84}$$

$$H_\phi = \frac{-p \cos \phi}{2\pi\gamma_1 \rho^3} \left(\frac{Q}{Q'_E} \right) \tag{3.85}$$

where Q'_E and Q'_H have the same form as Q_E and Q_H, respectively, but with z, the depth of the receiving dipole in the first layer, replaced by h, the depth of the source dipole in that layer.

Similarly, the HMD expressions become

$$E_\rho = \frac{m\gamma_1 \cos\phi}{2\pi\sigma_1\rho^3}\left(\frac{Q}{Q'_H}\right) \tag{3.86}$$

$$E_\phi = \frac{m\gamma_1 \sin\phi}{\pi\sigma_1\rho^3}\left(\frac{Q}{Q'_H}\right) \tag{3.87}$$

$$H_\rho = \frac{m \sin\phi}{\pi\rho^3}\left(\frac{1}{Q'_H}\right) \tag{3.88}$$

$$H_\phi = \frac{-m \cos\phi}{2\pi\rho^3}\left(\frac{1}{Q'_H}\right) \tag{3.89}$$

3.3.4 Subsurface to Subsurface Propagation

The subsurface to subsurface horizontal field components may be obtained from the subsurface to surface expressions by applying the method used to account for the depth of the receiving dipole in the case of surface to subsurface propagation. Thus, for the HED, the subsurface to subsurface formulas become

$$E_\rho = \frac{p \cos\phi}{2\pi\sigma_1\rho^3}\left(\frac{Q^2}{Q'_E Q_E}\right) \tag{3.90}$$

$$E_\phi = \frac{p \sin\phi}{\pi\sigma_1\rho^3}\left(\frac{Q^2}{Q'_E Q_H}\right) \tag{3.91}$$

$$H_\rho = \frac{p \sin\phi}{\pi\gamma_1\rho^3}\left(\frac{Q}{Q'_E Q_H}\right) \tag{3.92}$$

$$H_\phi = \frac{-p \cos\phi}{2\pi\gamma_1\rho^3}\left(\frac{Q}{Q'_E Q_H}\right) \tag{3.93}$$

Similarly, the resulting HMD expressions become

$$E_\rho = \frac{m\gamma_1 \cos\phi}{2\pi\sigma_1\rho^3}\left(\frac{Q}{Q'_H Q_E}\right) \tag{3.94}$$

$$E_\phi = \frac{m\gamma_1 \sin\phi}{\pi\sigma_1\rho^3}\left(\frac{Q}{Q'_H Q_E}\right) \tag{3.95}$$

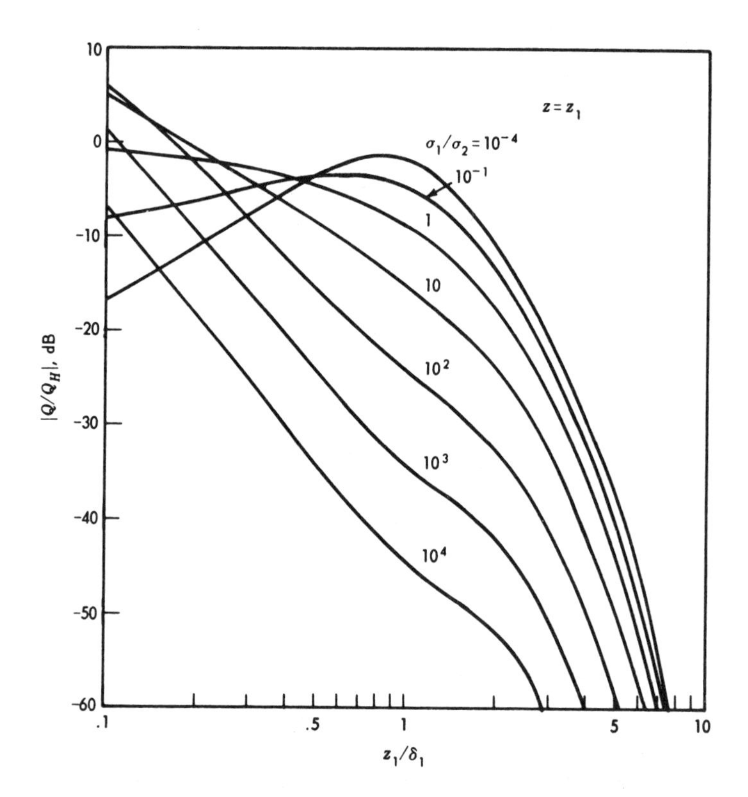

Figure 3.18.–$|Q/Q_H|$ *versus* z_1/δ_1 *for* $z=z_1$.

$$H_\rho = \frac{m \sin \phi}{\pi \rho^3}\left(\frac{1}{Q_H' Q_H}\right) \tag{3.96}$$

$$H_\phi = \frac{-m \cos \phi}{2\pi \rho^3}\left(\frac{1}{Q_H' Q_H}\right) \tag{3.97}$$

Plots of the magnitude and phase of $Q/Q_H Q_E$ as a function of z_1/δ_1 are presented by Bannister[49] for the case where $h=z=z_1$ for various layer conductivity ratios and are shown here in figures 3.20 and 3.21.

Weaver[51] solves the general problem of the quasi-static field produced by a horizontal and vertical electric dipole located in the upper layer of a two-layer conducting half space. Some numerical results are presented for particular cases for the electric and magnetic fields in the upper layer.

Another case in which both the source point and the field point are in the conductor is discussed by Wait,[52] who calculates the E_ϕ and H_z field components in the plane of separation

---

[51] Weaver, J. T.: The Quasi-Static Field of an Electric Dipole Embedded in a Two-Layer Conducting Half-Space. Can. J. Phys., vol. 45, 1967, pp. 1981–2002.

[52] Wait, J. R.: Current-Carrying Wire Loop in a Simple Inhomogeneous Region. J. Appl. Phys., vol. 23, no. 4, Apr. 1952, pp. 497–498.

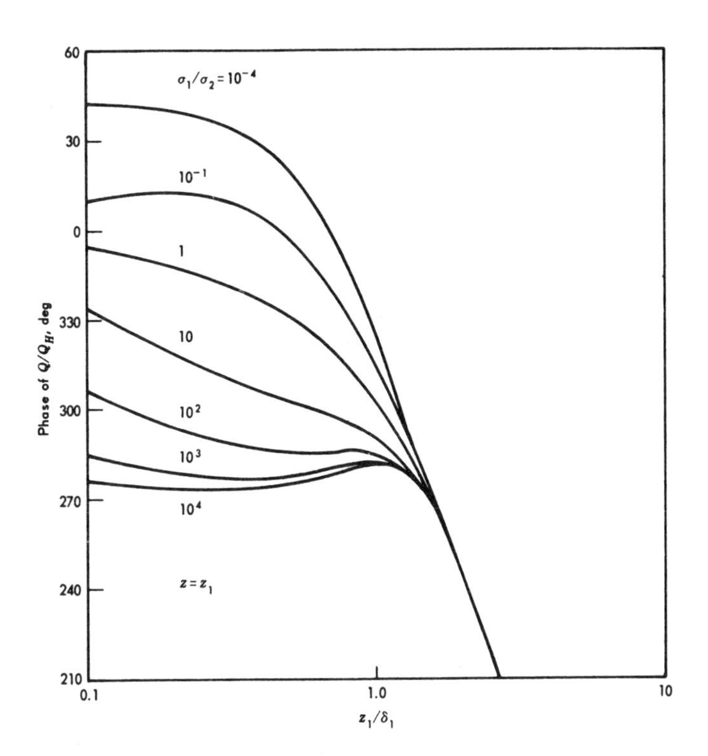

Figure 3.19.–*Phase of* Q/Q_H *versus* z_1/δ_1 *for* $z = z_1$.

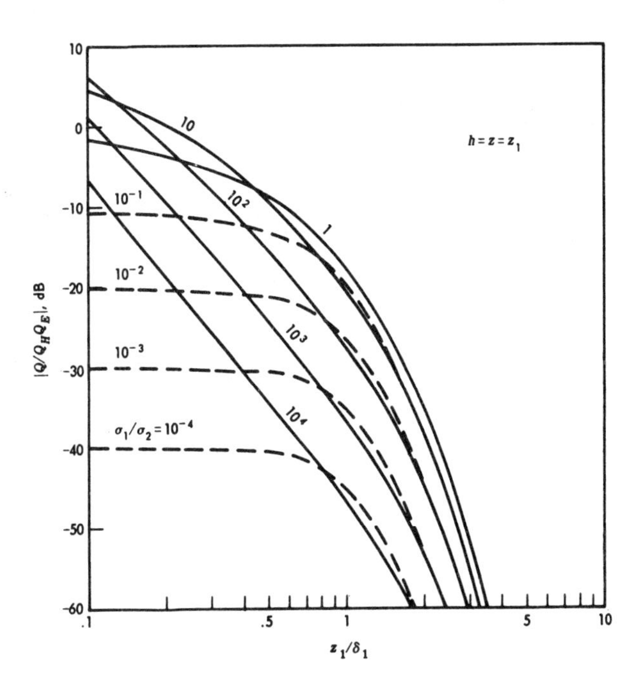

Figure 3.20.–$|Q/Q_E Q_H|$ *versus* z_1/δ_1 *for* $h = z = z_1$.

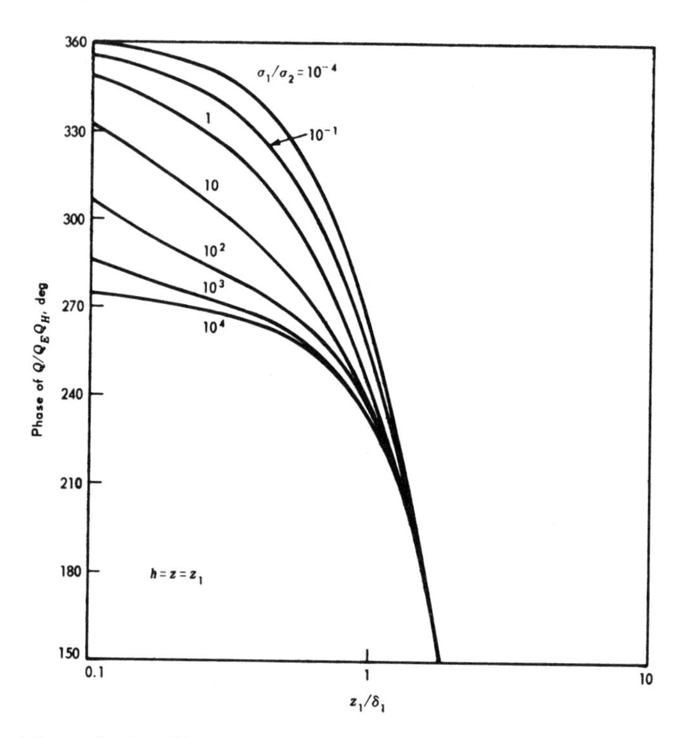

Figure 3.21.–*Phase of* $Q/Q_E Q_H$ *versus* z_1/δ_1 *for* $h = z = z_1$.

between two conducting half spaces of common permeability for a VMD also located in the interface. The expressions for these field components hold without approximation for all ranges and are given by

$$E_\phi = \frac{j\omega\mu m}{2\pi\left(\gamma_1^2 - \gamma_2^2\right)\rho^4}\left[\left(3 + 3\gamma_1\rho + \gamma_1^2\rho^2\right)e^{-\gamma_1\rho} - \left(3 + 3\gamma_2\rho + \gamma_2^2\rho^2\right)e^{-\gamma_2\rho}\right] \tag{3.98}$$

$$H_z = \frac{m}{2\pi\left(\gamma_1^2 - \gamma_2^2\right)\rho^5}\left[\left(9 + 9\gamma_1\rho + 4\gamma_1^2\rho^2 + \gamma_1^3\rho^3\right)e^{-\gamma_1\rho} - \left(9 + 9\gamma_2\rho + 4\gamma_2^2\rho^2 + \gamma_2^3\rho^3\right)e^{-\gamma_2\rho}\right] \tag{3.99}$$

Chapter 4

ELECTROMAGNETIC FIELD OF LONG LINE SOURCES AND FINITE LENGTH ELECTRIC ANTENNAS

The expressions in chapter 3 for the electric and magnetic fields of electric current elements in a well conducting medium may be integrated to derive the fields of finite electric (or loop) antennas in such a medium. For low frequencies in which the wavelength of the current in the antenna is very much larger than the antenna length, the integration is simplified because the current may be considered uniform and taken outside the integral. (In practice, the antenna wire should be insulated to maintain a uniform current that at low frequencies flows into the surrounding medium by conduction paths alone from the ends of the wire.) If, in addition, the observation distance is much smaller than the antenna length and is far from its ends, the antenna assumes the character of a long line source and the problem becomes two dimensional. In contrast, when the observation distance is much larger than the antenna length, the antenna assumes the character of an electric dipole or current element.

4.1 INFINITE HOMOGENEOUS CONDUCTING MEDIUM

4.1.1 Long Line Source

The solution for the fields of a line source of infinite length carrying a uniform current $Ie^{j\omega t}$ along the z-axis in an infinite, homogeneous, conducting medium may be obtained from (1.35) to (1.38) by evaluating the integral of the Green's function solution (1.48) over the line source. This results in the uniform cylindrical wave components

$$E_z = -\frac{\gamma^2 I}{2\pi\sigma} K_0(\gamma\rho) \tag{4.1}$$

$$H_\phi = \frac{\gamma I}{2\pi} K_1(\gamma\rho) \tag{4.2}$$

where $K_0(\gamma\rho)$ and $K_1(\gamma\rho)$ are the modified Bessel functions of the second kind, of orders zero and one, respectively. Plots, taken from Von Aulock,[53] of the magnitudes of E_z and H_ϕ in sea water for various frequencies are presented in figures 4.1 and 4.2, respectively, as a function of the distance from the line source.

It is shown analytically by Wait[54] that the fields given by (4.1) and (4.2) are not affected by insulation on the long wire (of negligible wire thickness) so long as the thickness of the insulation is very much smaller than a wavelength in the surrounding conducting medium.

---

[53] Von Aulock, W.: Propagation of Electromagnetic Fields in Sea Water. TM No. 140, Minesweeping Sec., Bur. of Ships, Navy Dept., May 1948.

[54] Wait, J. R.: Electromagnetic Fields of Current-Carrying Wires in a Conducting Medium. Can. J. Phys., vol. 30, Sept. 1952, pp. 512-523.

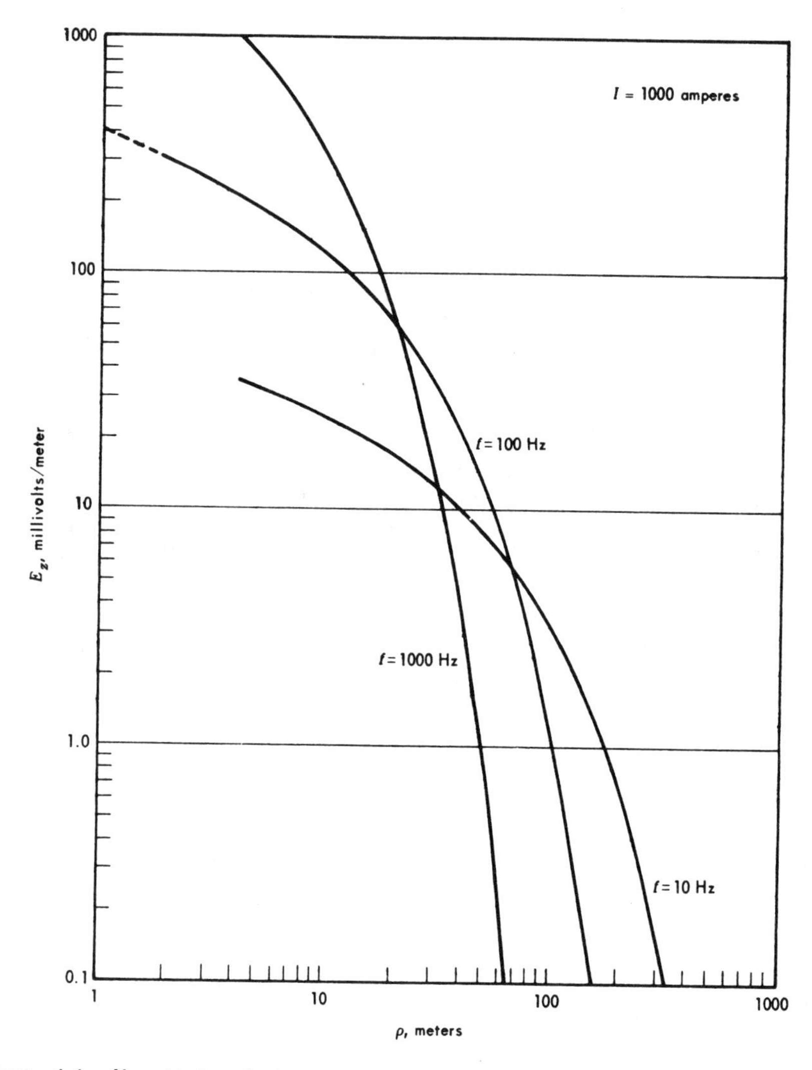

Figure 4.1.–*Magnitude of the electric field intensity in sea water of a long line source carrying 1000 amperes.*

4.1.2 Finite Length Electric Antenna

The general expression for the electric field of a linear, insulated, electric antenna of finite length that carries a uniform, harmonically time varying current is given by Wait.[54] In practice, an insulated electric antenna with electrodes at its ends carries a uniform current at low frequencies at which the length of the antenna is much shorter than a wavelength in the insulation.

Using a cylindrical coordinate system with the antenna wire extending from point A to point B on the z-axis, the distances from the ends of the antenna to the field point are given by

$$r_1 = \left[(z - A)^2 + \rho^2\right]^{1/2} \tag{4.3}$$

$$r_2 = \left[(z - B)^2 + \rho^2\right]^{1/2} \tag{4.4}$$

The somewhat complicated general formulas may be simplified for the low-frequency case $|\gamma r_1| \ll 1$ and $|\gamma r_2| \ll 1$ to

$$E_z = \frac{\gamma^2 I}{4\pi\sigma}\left(\sinh^{-1}\frac{A-z}{\rho} - \sinh^{-1}\frac{B-z}{\rho}\right) + \frac{I}{4\pi\sigma}\left(\frac{z-A}{r_1^3} - \frac{z-B}{r_2^3}\right) \tag{4.5}$$

$$E_\rho = \frac{I\rho}{4\pi\sigma}\left(\frac{1}{r_1^3} - \frac{1}{r_2^3}\right) \tag{4.6}$$

$$H_\phi = \frac{I}{4\pi\rho}\left(\frac{z-A}{r_1} - \frac{z-B}{r_2}\right) \tag{4.7}$$

where $\gamma \approx (j\omega\mu\sigma)^{1/2}$.

4.2 SEMI-INFINITE HOMOGENEOUS CONDUCTING MEDIUM

4.2.1 Long Line Source

The usual approach for determining the electromagnetic field of a long line source of uniform current $Ie^{j\omega t}$ either over or in a homogeneous (or layered) half space is to consider the electric and magnetic field as a superposition of plane waves. The mathematical details of this approach are found in Carson,[55] Von Aulock,[56] and Wait.[9]

In the present section only a long line source that carries a uniform current either on or below the surface of a conducting half space with a propagation constant $\gamma = (j\omega\mu_0\sigma)^{1/2}$ is considered. Furthermore, because in practice the condition of uniform current flow implies that the length of the line source is much smaller than a free-space wavelength, only the *quasi-static* field components are discussed. These components are expressed in rectangular coordinates with the x, y-plane coinciding with the surface of the conducting half space. The current line is on or parallel to the x-axis, and the positive z-direction points into free space above the semi-infinite conductor.

4.2.1.1 Surface to surface ($h = 0$, $z = 0$)

From Wait[57] and Bannister,[58] the general expressions for the quasi-static field components are

$$E_x = -\frac{I}{\pi\sigma y^2}\left[1 - \gamma y K_1(\gamma y)\right] \tag{4.8}$$

[55] Carson, J. R.: Wave Propagation in Overhead Wires With Ground Return. Bell System Tech. J., vol. 5, 1926, pp. 539-554.

[56] Von Aulock, W.: The Electromagnetic Field of an Infinite Cable in Sea Water. Tech. Rept. No. 106, Minesweeping Branch, Bur. of Ships, Navy Dept., 1 Sept. 1953.

[57] Wait, J. R.: Electromagnetic Waves in Stratified Media. Pergamon Press, Ltd. book, MacMillan Co., 1962, ch. 2.

[58] Bannister, P. R.: The Image Theory Quasi-Static Magnetic Fields of a Finite Length Horizontal Electric Antenna Located Near the Earth's Surface. USL Rept. No. 1011, U.S. Navy Underwater Sound Laboratory, New London, Conn., July 1969.

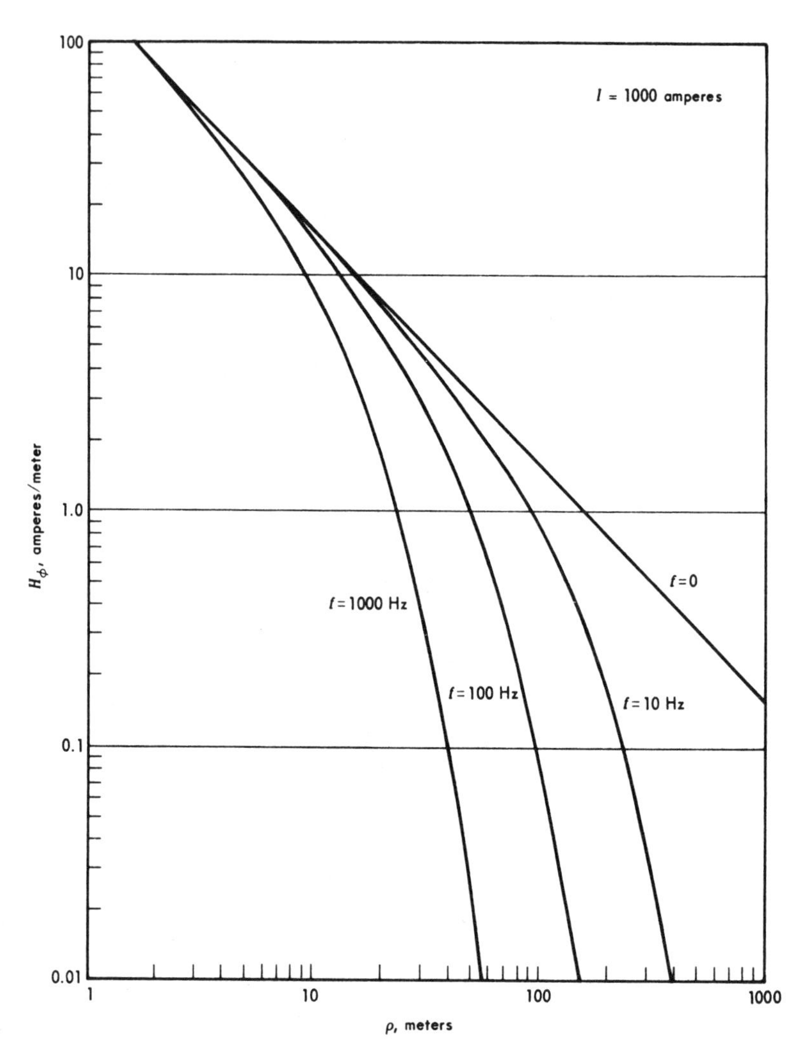

Figure 4.2.–*Magnitude of the magnetic field intensity in sea water of a long line source carrying 1000 amperes.*

$$H_y = \frac{I\gamma}{\pi}\left\{\frac{1}{3} + \frac{\pi}{2}\left[\frac{J_2(-j\gamma y)}{\gamma y} - \frac{H_2(-j\gamma y)}{(-j\gamma y)}\right]\right\} \tag{4.9}$$

$$H_z = \frac{I}{\pi y}\left[\frac{2 - 2\gamma y K_1(\gamma y) - (\gamma y)^2 K_0(\gamma y)}{(\gamma y)^2}\right] \tag{4.10}$$

where $K_0(\gamma y)$, $K_1(\gamma y)$ = modified Bessel functions of the second kind, orders zero and one, respectively

$J_2(-j\gamma y)$ = Bessel function of the first kind, order two

$H_2(-j\gamma y)$ = Struve function of the second order

4.2.1.2 Surface to free space ($h=0,\ z\geqq 0$)

Formulas for the field components of a long line source of length much greater than a skin depth in the conducting half space and with $|\gamma r| \ll 1$ are given by Bannister[59] as

$$E_x = -\frac{\gamma^2 I}{2\pi\sigma}\left(\ln\frac{2}{\gamma r}\right) \tag{4.11}$$

$$H_y = -\frac{I}{2\pi}\left(\frac{z}{r^2} - \frac{2}{3}\gamma\right) \tag{4.12}$$

$$H_z = \frac{Iy}{2\pi r^2} \tag{4.13}$$

where $r = \left(y^2 + z^2\right)^{1/2}$

If $|\gamma r| \gg 1$, the field components given by Bannister[60] are

$$E_x = -\frac{I}{\pi\sigma r^2}\left(1 + \gamma z - \frac{2z^2}{r^2}\right) \tag{4.14}$$

$$H_y = \frac{I}{\pi\gamma r^2}\left(1 - \frac{2z^2}{r^2}\right) \tag{4.15}$$

$$H_z = \frac{2Iy}{\pi\gamma^2 r^4}\left(1 + \gamma z - \frac{4z^2}{r^2}\right) \tag{4.16}$$

It should be noted that the ratio of the electric field to the tangential magnetic field at the surface of the conducting medium is equal to the intrinsic impedance of that medium.

When $|\gamma r| \approx 1$, simple approximate expressions are obtained by Bannister[61] using image theory. The image theory results are

$$E_x = -\frac{\gamma^2 I}{2\pi\sigma}\ln\frac{R_0}{r} \tag{4.17}$$

$$H_y = \frac{I}{2\pi}\left\{\frac{[(2/\gamma) + z]}{R_0^2} - \frac{z}{r^2}\right\} \tag{4.18}$$

$$H_z = -\frac{Iy}{2\pi}\left(\frac{1}{R_0^2} - \frac{1}{r^2}\right) \tag{4.19}$$

---

[59] Bannister, P. R.: Electric and Magnetic Fields Near a Long Horizontal Line Source Above the Ground. Radio Sci., vol. 3, no. 2, Feb. 1968.

[60] Bannister, P. R.: The Quasi-Near Fields of a Long Horizontal Line Source. USL Rept. No. 918, U.S. Navy Underwater Sound Laboratory, New London, Conn., 17 May 1968.

[61] Bannister, P. R.: Impedance Per Unit Length Between Two Long Horizontal Wires Above the Earth's Surface. USL Rept. No. 1028, U.S. Navy Underwater Sound Laboratory, New London, Conn., Apr. 1969.

where $r=\left(y^2+z^2\right)^{1/2}$ and

$R_0=\left\{y^2+[(2/\gamma)+z]^2\right\}^{1/2}$

4.2.1.3 Subsurface to free space ($h\leqq 0,\ z\geqq 0$)

If $|\gamma R_1|\ll 1$, where $R_1=\left[y^2+(z-h)^2\right]^{1/2}$, the expressions for the field components are given by Bannister[62] as

$$E_x=-\frac{\gamma^2 I}{2\pi\sigma}\ln\frac{2}{\gamma R_1} \tag{4.20}$$

$$H_y=-\frac{I(z-h)}{2\pi R_1^2} \tag{4.21}$$

$$H_z=\frac{Iy}{2\pi R_1^2} \tag{4.22}$$

For $|\gamma r|\gg 1$ and $r\gg|h|$, where $r=\left(y^2+z^2\right)^{1/2}$, the field components are the same as those for the surface to free-space case with $|\gamma r|\gg 1$ but with the additional factor $e^{\gamma h}$. Thus,

$$E_x=-\frac{I}{\pi\sigma r^2}\left(1+\gamma z-\frac{2z^2}{r^2}\right)e^{\gamma h} \tag{4.23}$$

$$H_y=\frac{I}{\pi\gamma r^2}\left(1-\frac{2z^2}{r^2}\right)e^{\gamma h} \tag{4.24}$$

$$H_z=\frac{2Iy}{\pi\gamma^2 r^4}\left(1+\gamma z-\frac{4z^2}{r^2}\right)e^{\gamma h} \tag{4.25}$$

4.2.1.4 Subsurface to subsurface ($h\leqq 0,\ z\leqq 0$)

For $|\gamma R_1|\ll 1$ and $|\gamma R_2|\ll 1$, where $R_1=\left[y^2+(z-h)^2\right]^{1/2}$ and $R_2=\left[y^2+(z+h)^2\right]^{1/2}$, the expressions for the field components are given by Von Aulock[56] and Bannister[62] as

$$E_x=-\frac{\gamma^2 I}{2\pi\sigma}\ln\frac{2}{\gamma R_1} \tag{4.26}$$

$$H_y=\frac{I\gamma y^2}{3\pi R_2^2}-\frac{I(z-h)}{2\pi R_1^2} \tag{4.27}$$

[62] Bannister, P. R.: The Electric and Magnetic Fields Near a Buried Long Horizontal Line Source. USL Rept. No. 991, U.S. Navy Underwater Sound Laboratory, New London, Conn., 4 June 1969.

$$H_z = \frac{Iy}{2\pi R_1^2} \tag{4.28}$$

If $|\gamma r| \gg 1$ and $r \gg |h|$ and $|z|$, where $r = \left(y^2 + z^2\right)^{1/2}$, the formulas for the field components are the same as those for the surface to surface case with $|\gamma r| \gg 1$ but with the additional factor $e^{\gamma(h+z)}$. Thus,

$$E_x = -\frac{I}{\pi \sigma r^2} e^{\gamma(h+z)} \tag{4.29}$$

$$H_y = \frac{I}{\pi \gamma r^2} e^{\gamma(h+z)} \tag{4.30}$$

$$H_z = \frac{2Iy}{\pi \gamma^2 r^4} e^{\gamma(h+z)} \tag{4.31}$$

Numerical results giving the magnitude of the field components for the surface to subsurface case under general quasi-static conditions are given in graphical form by Von Aulock.[56] These results for E_x, H_y, and H_z are presented here in figures 4.3, 4.4, and 4.5, respectively.

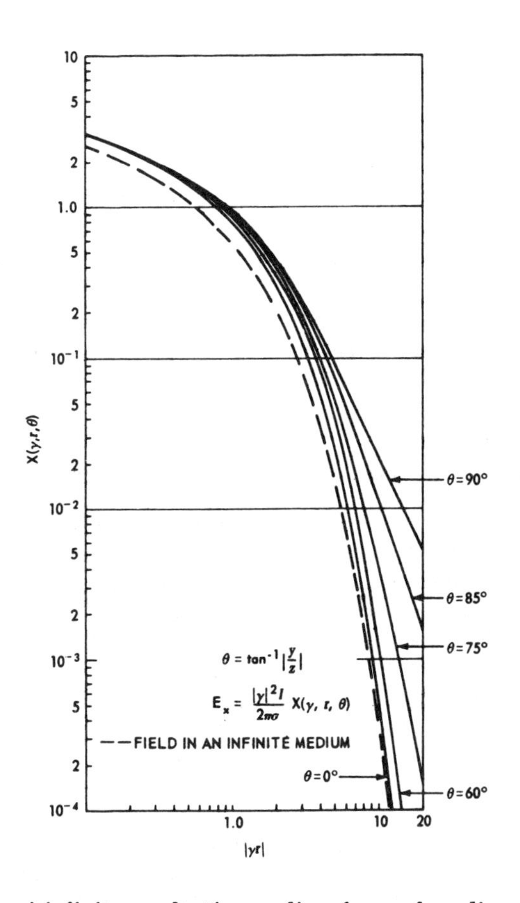

Figure 4.3.–E_x *in a semi-infinite conducting medium from a long line source at the surface.*

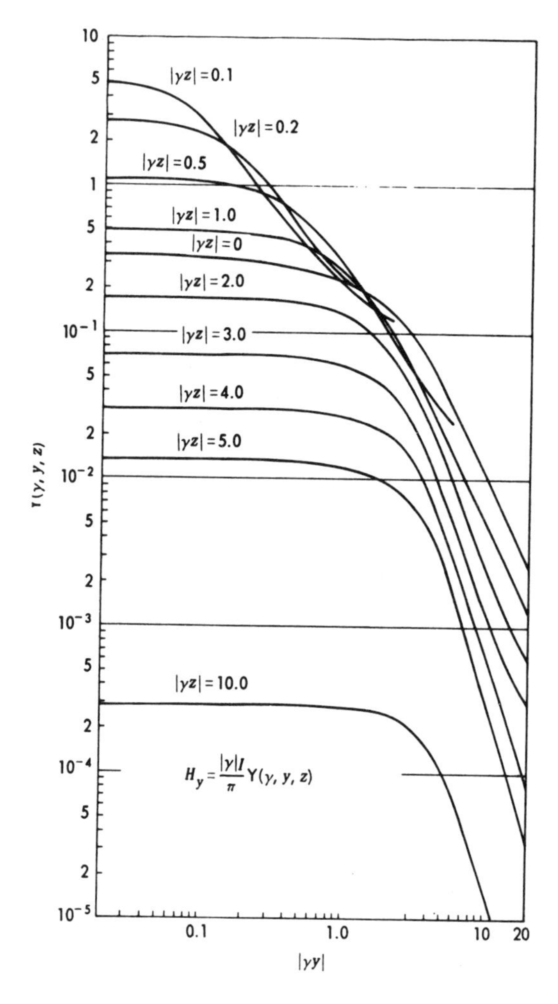

Figure 4.4.–H_y *in a semi-infinite conducting medium from a long line source at the surface.*

4.2.2 Finite Length Electric Antenna

This section presents the quasi-static field components of a finite length horizontal electric antenna that carries a uniform current either on or in a conducting half space. The electric and magnetic field may be calculated by integrating the known expressions for the horizontal electric current element over the length of the antenna. However, some of the integrals for certain cases in the general quasi-static range cannot be expressed conveniently in closed form, and image theory is used in these cases to obtain simple, approximate, closed form results that are valid throughout the range. An examination of the analytical validity of the image theory technique is given by Wait and Spies[63] and Bannister.[64]

---

[63] Wait, J. R.; and Spies, K. P.: On the Image Representation of the Quasi-Static Fields of a Line Current Source Above the Ground. Can. J. Phys., vol. 47, 1969, pp. 2731-2733.

[64] Bannister, P.: The Image Theory Quasi-Static Fields of Antennas Above the Earth's Surface. USL Rept. No. 1061, U.S. Navy Underwater Sound Laboratory, New London, Conn., 1969.

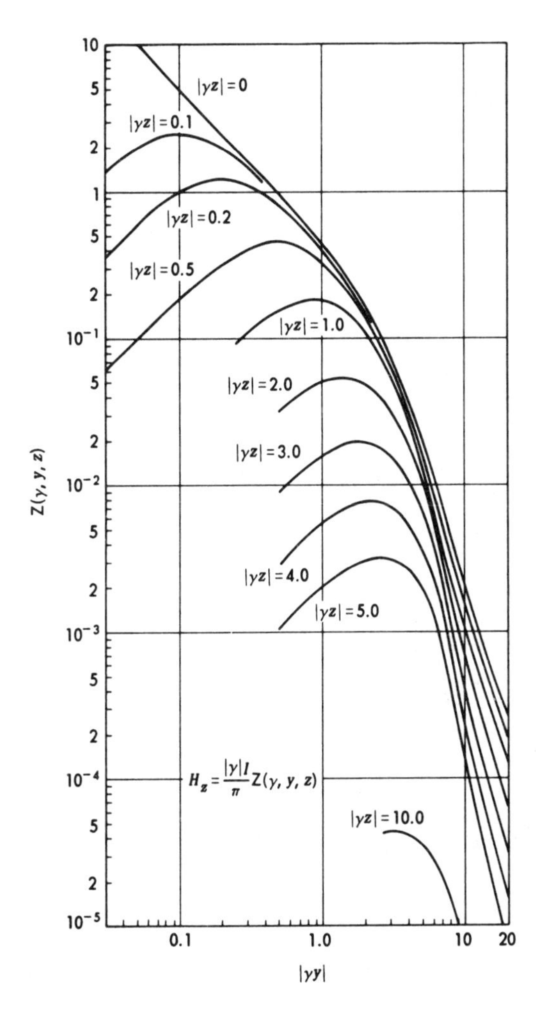

Figure 4.5.–H_z *in a semi-infinite conducting medium from a long line source at the surface.*

The field components are given in rectangular coordinates where the x, y-plane coincides with the surface of a semi-infinite conductor with a propagation constant $\gamma = (j\omega\mu_0\sigma)^{1/2}$. The antenna of length L is parallel to the x-axis with the center of the antenna at the point $(0, 0, h)$ and with the positive z-direction pointing into free space.

4.2.2.1 Surface to free space ($h = 0,\ z \geqq 0^+$)

The expressions for the field components for the general quasi-static range are derived by Bannister[64] using image theory. These components are

$$E_x = -\frac{\gamma^2 I}{4\pi\sigma}\left\{\ln\left[\frac{R_{11} - \left(x + \frac{L}{2}\right)}{R_{21} - \left(x + \frac{L}{2}\right)}\right] - \ln\left[\frac{R_{12} - \left(x - \frac{L}{2}\right)}{R_{22} - \left(x - \frac{L}{2}\right)}\right] + \frac{d^2}{2}\left[\frac{\left(x + \frac{L}{2}\right)}{R_{21}^3} - \frac{\left(x - \frac{L}{2}\right)}{R_{22}^3}\right]\right\} \tag{4.32}$$

$$E_y = -\frac{Iy}{2\pi\sigma}\left[\frac{1}{R_{21}^3} - \frac{1}{R_{22}^3}\right] \tag{4.33}$$

$$E_z = \frac{\gamma^2 I}{4\pi\sigma}\left\{\ln\left[\frac{R_{11}-(z+d)}{R_{21}-z}\right] - \ln\left[\frac{R_{12}-(z+d)}{R_{22}-z}\right] - \frac{d^2 z}{2}\left[\frac{1}{R_{21}^3} - \frac{1}{R_{22}^3}\right]\right\} \tag{4.34}$$

$$H_x = \frac{Iy}{4\pi}\left\{\frac{1}{\left[\left(x-\frac{L}{2}\right)^2 + y^2\right]}\left[\frac{z+d}{R_{12}} - \frac{z}{R_{22}}\right] - \frac{1}{\left[\left(x+\frac{L}{2}\right)^2 + y^2\right]}\left[\frac{z+d}{R_{11}} - \frac{z}{R_{21}}\right]\right\} \tag{4.35}$$

$$H_y = \frac{I}{4\pi}\left\{\frac{(z+d)}{[y^2+(z+d)^2]}\left[\frac{\left(x+\frac{L}{2}\right)}{R_{11}} - \frac{\left(x-\frac{L}{2}\right)}{R_{12}}\right] - \frac{z}{(y^2+z^2)}\left[\frac{\left(x+\frac{L}{2}\right)}{R_{21}} - \frac{\left(x-\frac{L}{2}\right)}{R_{22}}\right]\right.$$

$$\left. - \frac{\left(x-\frac{L}{2}\right)}{\left[y^2+\left(x-\frac{L}{2}\right)^2\right]}\left[\frac{z+d}{R_{12}} - \frac{z}{R_{22}}\right] + \frac{\left(x+\frac{L}{2}\right)}{\left[y^2+\left(x+\frac{L}{2}\right)^2\right]}\left[\frac{z+d}{R_{11}} - \frac{z}{R_{21}}\right]\right\} \tag{4.36}$$

$$H_z = -\frac{Iy}{4\pi}\left\{\frac{1}{[y^2+(z+d)^2]}\left[\frac{\left(x+\frac{L}{2}\right)}{R_{11}} - \frac{\left(x-\frac{L}{2}\right)}{R_{12}}\right] - \frac{1}{(y^2+z^2)}\left[\frac{\left(x+\frac{L}{2}\right)}{R_{21}} - \frac{\left(x-\frac{L}{2}\right)}{R_{22}}\right]\right\} \tag{4.37}$$

where $R_{11} = \left[\left(x+\frac{L}{2}\right)^2 + y^2 + (z+d)^2\right]^{1/2}$

$$R_{12} = \left[\left(x-\frac{L}{2}\right)^2 + y^2 + (z+d)^2\right]^{1/2}$$

$$R_{21} = \left[\left(x+\frac{L}{2}\right)^2 + y^2 + z^2\right]^{1/2}$$

$$R_{22} = \left[\left(x-\frac{L}{2}\right)^2 + y^2 + z^2\right]^{1/2}$$

$$d = \frac{2}{\gamma}$$

If each element of the finite length antenna is much greater than a skin depth away from the field point, then the simple expressions in table 3.13 for the HED may readily be integrated along the length of the antenna to obtain

$$E_x = \frac{I}{2\pi\sigma r^2} \left\langle (1-\gamma z)\left\{\frac{\left(x+\frac{L}{2}\right)}{R_{21}} - \frac{\left(x-\frac{L}{2}\right)}{R_{22}}\right\} \right.$$

$$\left. - \frac{y^2}{r^2}\left\{\frac{\left(x+\frac{L}{2}\right)\left[2\left(x+\frac{L}{2}\right)^2 + 3r^2\right]}{R_{21}^3} - \frac{\left(x-\frac{L}{2}\right)\left[2\left(x-\frac{L}{2}\right)^2 + 3r^2\right]}{R_{22}^3}\right\} \right\rangle \quad (4.38)$$

$$E_y = -\frac{Iy}{2\pi\sigma}\left[\frac{1}{R_{21}^3} - \frac{1}{R_{22}^3}\right] \quad (4.39)$$

$$E_z = -\frac{\gamma I}{2\pi\sigma}\left[\frac{1}{R_{21}} - \frac{1}{R_{22}}\right] \quad (4.40)$$

$$H_x = -\frac{Iy}{2\pi\gamma}\left[\frac{1}{R_{21}^3} - \frac{1}{R_{22}^3}\right] \quad (4.41)$$

$$H_y = -\frac{I}{2\pi\gamma r^2} \left\langle \frac{\left(x+\frac{L}{2}\right)}{R_{21}} - \frac{\left(x-\frac{L}{2}\right)}{R_{22}} \right.$$

$$\left. - \frac{y^2}{r^2}\left\{\frac{\left(x+\frac{L}{2}\right)\left[2\left(x+\frac{L}{2}\right)^2 + 3r^2\right]}{R_{21}^3} - \frac{\left(x-\frac{L}{2}\right)\left[2\left(x-\frac{L}{2}\right)^2 + 3r^2\right]}{R_{22}^3}\right\} \right\rangle \quad (4.42)$$

$$H_z = \frac{Iy}{2\pi\gamma^2 r^4} \left\langle \left\{1 + \gamma z - \frac{4z^2}{r^2}\right\} \left\{\frac{\left(x+\frac{L}{2}\right)\left[2\left(x+\frac{L}{2}\right)^2 + 3r^2\right]}{R_{21}^3} - \frac{\left(x-\frac{L}{2}\right)\left[2\left(x-\frac{L}{2}\right)^2 + 3r^2\right]}{R_{22}^3}\right\} \right.$$

$$\left. - 3z^2 r^2 \left\{\frac{\left(x+\frac{L}{2}\right)}{R_{21}^5} - \frac{\left(x-\frac{L}{2}\right)}{R_{22}^5}\right\} \right\rangle \quad (4.43)$$

where $r^2 = y^2 + z^2$.

It should be noted that the expressions (4.34) and (4.40) may be modified to hold for field points just below the surface ($z=0^-$) by multiplying by γ_0^2/γ^2 where $\gamma_0 = j\omega(\mu_0\epsilon_0)^{1/2}$ is the propagation constant of free space.

4.2.2.2 Subsurface to free space ($h \leqq 0$, $z \geqq 0^+$)

The approximate expressions (4.32) to (4.37) derived from image theory for the general quasi-static range are also valid for the subsurface to free-space case when the source is at a shallow depth ($|\gamma h| \ll 1$) in the conductor. In these expressions z must be replaced by $z+|h|$ wherever z does not appear as the combination $z+d$.

If each element of the finite length antenna is much greater than a skin depth away from the field point and also much greater than $|h|$, then the expressions for the field components are the same as those for the surface to free-space case (4.38) to (4.43), but with the additional factor $e^{\gamma h}$ for each component.

4.2.2.3 Subsurface to subsurface ($h \leqq 0$, $z \leqq 0^-$)

If each element of the finite length antenna is much greater than a skin depth away from the field point and also much greater than $|h+z|$, then the expressions for the field components are the same as those for the surface to free-space case (4.38) to (4.43) for $z=0^+$, but with an additional factor $e^{\gamma(h+z)}$ for all the components and a second factor γ_0^2/γ^2 for the E_z component.

4.3 LAYERED CONDUCTING HALF SPACE

4.3.1 Long Line Source

Wait[9] and Hansen, et al.[45] discuss the fields of a long line source for those cases where the source and the field point are over or on the surface of a stratified conducting half space.

The present section reviews the quasi-static field expressions for those cases in which the long line source and the field point are both in the top layer of a stratified conducting half space. Derivations of the expressions for the field components are given by Von Aulock[56] and Bannister.[65,66] The components are expressed in rectangular coordinates with the x, y-plane coinciding with the surface of the layered conducting half space. The line source carrying a uniform current $Ie^{j\omega t}$ is on or parallel to the x-axis and the positive z-direction points into the layered conductor (see fig. 2.4). Displacement currents are neglected in all media.

4.3.1.1 Surface to surface ($h=0$, $z=0$)

When the horizontal distance y is much greater than the skin depth δ_i in each of the layers, $|\gamma_i y| \gg 1$, and the field components are

$$E_x = -\frac{I}{\pi\sigma_1 y^2} Q^2 \tag{4.44}$$

$$H_y = -\frac{I}{\pi\gamma_1 y^2} Q \tag{4.45}$$

---

[65] Bannister, P. R.: Electromagnetic Fields Within a Stratified Earth Produced by a Long Horizontal Line Source. Radio Sci., vol. 3, no. 4, Apr. 1968, pp. 387-390.

[66] Bannister, P. R.: Further Notes on Electromagnetic Fields Within a Stratified Earth Produced by a Long, Horizontal Line Source. USL Rept. No. 908, U.S. Navy Underwater Sound Laboratory, New London, Conn., 6 May 1968.

$$H_z = \frac{2I}{\pi \gamma_1^2 y^3} Q^2 \tag{4.46}$$

where Q is defined by (3.62) for the two-layer medium, by (3.63) and (3.64) for the three-layer medium and by (2.71) through (2.74) for the m-layer medium.

4.3.1.2 Surface to subsurface ($h=0$, $0 \leqq z \leqq z_1$)

For $|\gamma_i y| \gg 1$ and $y \gg z$,

$$E_x = -\frac{I}{\pi \sigma_1 y^2} \left(\frac{Q^2}{Q_E} \right) \tag{4.47}$$

$$H_y = -\frac{I}{\pi \gamma_1 y^2} \left(\frac{Q}{Q_H} \right) \tag{4.48}$$

$$H_z = \frac{2I}{\pi \gamma_1^2 y^3} \left(\frac{Q^2}{Q_E} \right) \tag{4.49}$$

where Q_E^{-1} and Q_H^{-1} are defined for a two-layer medium by (2.75) and (2.76), respectively. If there are more than two layers, then the impedance η_2 in (2.75) and (2.76) is to be interpreted as the effective surface impedance of the layers below the first and may be determined from (2.72) to (2.74).

Plots of the magnitude and phase of Q/Q_H for $z=z_1$ are shown in figures 3.18 and 3.19 as a function of z_1/δ_1. The magnitude of Q^2/Q_E for $z=z_1$ is plotted by Bannister[66] and is shown here in figure 4.6 as a function of z_1/δ_1.

Two-layer medium with $\sigma_1 \gg \sigma_2$, for $|\gamma_1 z_1| > 2\sqrt{2}$ or $z_1 \gg 2\delta_1$, $y \gg z_1$, $|\gamma_1 y| \gg 1$, $|\gamma_2 y| \ll 1$

$$E_x = \frac{2I}{\pi \sigma_1 y^2} e^{-\gamma_1 z_1} \left[\sinh \gamma_1 (z_1 - z) - 2 \cosh \gamma_1 (z_1 - z) \right] \tag{4.50}$$

$$H_y = \frac{2I}{\pi \gamma_1 y^2} e^{-\gamma_1 z_1} \left[\cosh \gamma_1 (z_1 - z) - 2 \sinh \gamma_1 (z_1 - z) \right] \tag{4.51}$$

$$H_z = \frac{4I}{\pi \gamma_1^2 y^3} e^{-\gamma_1 z_1} \left[2 \cosh \gamma_1 (z_1 - z) - \sinh \gamma_1 (z_1 - z) \right] \tag{4.52}$$

Vertical plane ($y=0$) below the source for a two-layer medium.

The somewhat complicated analytical expressions for the field components E_x and H_y in the plane $y=0$ directly below the long line source are given by Von Aulock[56] who also presents graphs of these components. These graphs are shown here in figures 4.7 and 4.8, respectively.

4.3.1.3 Surface to bottom ($h=0$, $z=z_1$) for a two-layer medium

Case 1: $\sigma_1 \gg \sigma_2$

For $|\gamma_1 y| \gg 1$, $y \gg z_1$, and for all $|\gamma_2 y|$,

$$E_x = -\frac{I}{\pi\sigma_1 y^2}\left(\frac{\coth \gamma_1 z_1}{\sinh \gamma_1 z_1}\right)\left[1+\gamma_2 y K_1(\gamma_2 y)\right] \tag{4.53}$$

$$H_y = -\frac{I}{\pi\gamma_1 y^2}\left(\frac{1}{\sinh \gamma_1 z_1}\right)\left\langle \gamma_2 y K_1(\gamma_2 y) + \left(\frac{\sigma_2}{\sigma_1}\right)^{1/2}(\coth \gamma_1 z_1)(\gamma_2 y)^2\left\{\frac{1}{3}+\frac{\pi}{2}\left[\frac{J_2(-j\gamma_2 y)}{\gamma_2 y} - \frac{H_2(-j\gamma_2 y)}{-j\gamma_2 y}\right]\right\}\right\rangle \tag{4.54}$$

$$H_z = \frac{2I}{\pi\gamma_1^2 y^3}\left(\frac{\coth \gamma_1 z_1}{\sinh \gamma_1 z_1}\right)\left[1+\gamma_2 y K_1(\gamma_2 y)+\frac{(\gamma_2 y)^2}{2}K_0(\gamma_2 y)\right] \tag{4.55}$$

where the functions K_0, K_1, J_2, and H_2 are those defined in (4.8) through (4.10).

For $|\gamma_1 y| \gg 1$, $y \gg z_1$, but $|\gamma_2 y| \ll 1$,

$$E_x = -\frac{2I}{\pi\sigma_1 y^2}\left(\frac{\coth \gamma_1 z_1}{\sinh \gamma_1 z_1}\right) \tag{4.56}$$

$$H_y = -\frac{I}{\pi\gamma_1 y^2}\left(\frac{1}{\sinh \gamma_1 z_1}\right) \tag{4.57}$$

$$H_z = \frac{4I}{\pi\gamma_1^2 y^3}\left(\frac{\coth \gamma_1 z_1}{\sinh \gamma_1 z_1}\right) \tag{4.58}$$

For $|\gamma_1 y|$ and $|\gamma_2 y| \gg 1$, $y \gg z_1$, and $|(\sigma_1/\sigma_2)^{1/2} \tanh \gamma_1 z_1| \gg 1$

$$E_x = -\frac{I}{\pi\sigma_1 y^2}\left(\frac{\coth \gamma_1 z_1}{\sinh \gamma_1 z_1}\right) \tag{4.59}$$

$$H_y = -\frac{I}{\pi\gamma_1 y^2}\left(\frac{\sigma_2}{\sigma_1}\right)^{1/2}\left(\frac{\coth \gamma_1 z_1}{\sinh \gamma_1 z_1}\right) \tag{4.60}$$

$$H_z = \frac{2I}{\pi\gamma_1^2 y^3}\left(\frac{\coth \gamma_1 z_1}{\sinh \gamma_1 z_1}\right) \tag{4.61}$$

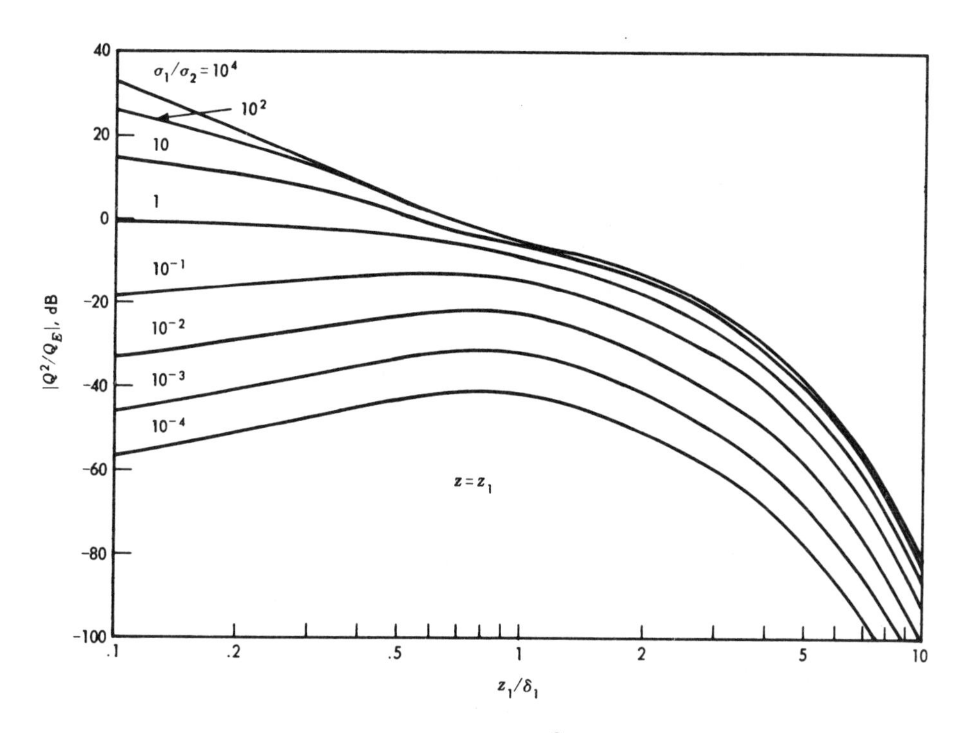

Figure 4.6.–*The magnitude of* Q^2/Q_E *versus* z_1/δ_1.

Case 2: $\sigma_2 \gg \sigma_1$

For $|\gamma_1 y|$ and $|\gamma_2 y| \gg 1$, $y \gg z_1$, and $|(\sigma_2/\sigma_1)^{1/2} \tanh \gamma_1 z_1| \gg 1$

$$E_x = -\frac{I}{\pi\sigma_1 y^2}\left(\frac{\sigma_1}{\sigma_2}\right)^{1/2}\left(\frac{\tanh \gamma_1 z_1}{\cosh \gamma_1 z_1}\right) \tag{4.62}$$

$$H_y = -\frac{I}{\pi\gamma_1 y^2}\left(\frac{\tanh \gamma_1 z_1}{\cosh \gamma_1 z_1}\right) \tag{4.63}$$

$$H_z = \frac{2I}{\pi\gamma_1^2 y^3}\left(\frac{\sigma_1}{\sigma_2}\right)^{1/2}\left(\frac{\tanh \gamma_1 z_1}{\cosh \gamma_1 z_1}\right) \tag{4.64}$$

4.3.1.4 Subsurface to subsurface ($h \geqq 0$, $0 \leqq z \leqq z_1$)

For $|\gamma_i y| \gg 1$ and $y \gg h$ and z

$$E_x = -\frac{I}{\pi\sigma_1 y^2}\left(\frac{Q^2}{Q_E' Q_E}\right) \tag{4.65}$$

$$H_y = -\frac{I}{\pi\gamma_1 y^2}\left(\frac{Q}{Q_E' Q_H}\right) \tag{4.66}$$

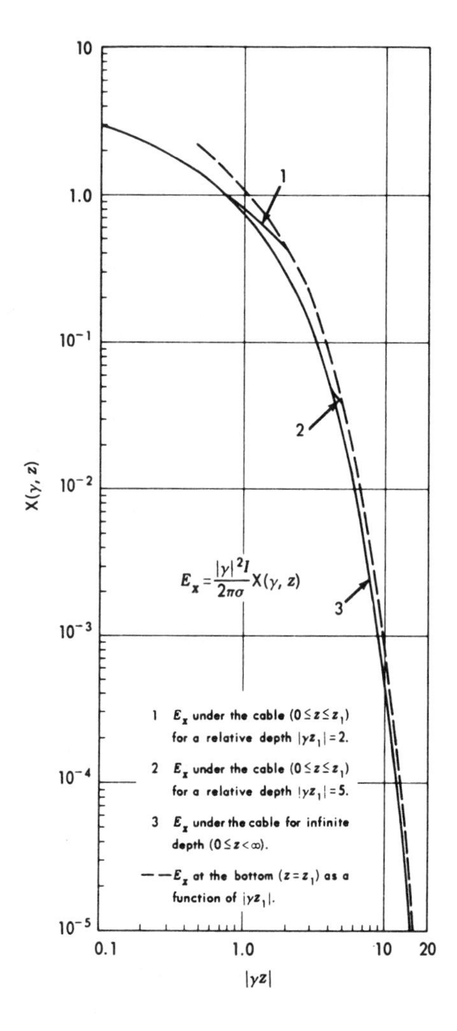

Figure 4.7.–E_x *directly under* (y = 0) *a long line source at the surface of a two-layer conducting medium* where the second medium is nonconducting.

$$H_z = \frac{2I}{\pi\gamma_1^2 y^3}\left(\frac{Q^2}{Q_E' Q_E}\right) \tag{4.67}$$

where Q_E' and Q_H' have the same form as Q_E and Q_H, respectively, but with z replaced by h. Plots of the magnitude and phase of the stratification factors above are given in section 3.3 in the treatment of dipole sources in a layered conducting half space.

4.3.2 Finite Length Electric Antenna

If each element of the horizontal finite length electric antenna is governed by the three conditions stated in section 3.3, the expressions (4.38), (4.39), (4.41), and (4.42) for the horizontal field components for the surface to surface case ($z = 0$) may be modified to account for the stratification when h or z is in the top layer in exactly the same manner as that previously used

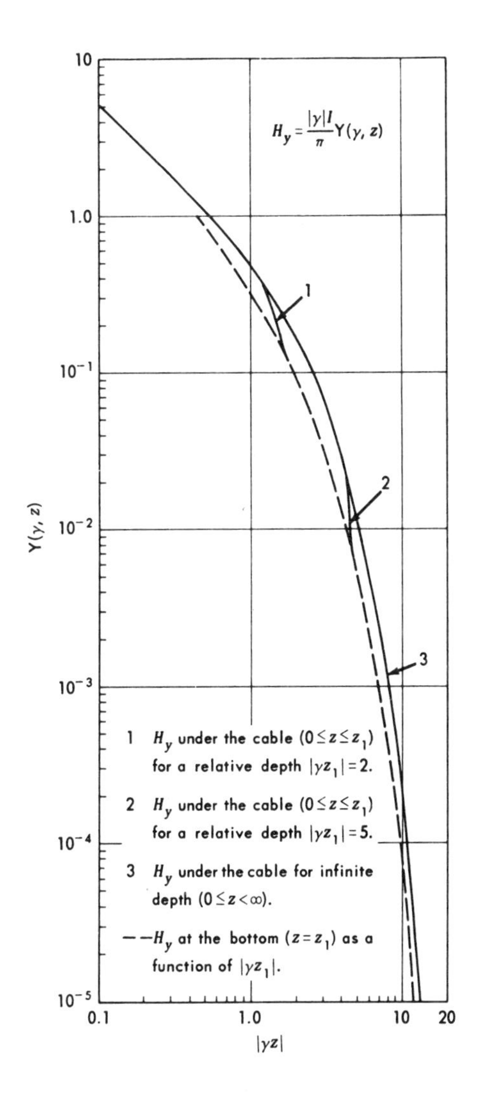

Figure 4.8.–H_y *directly under* (y = 0) *a long line source at the surface of a two-layer conducting medium* where the second medium is nonconducting.

for the horizontal field components of the HED. This procedure is justified because the stratification factors are functions of *z* only and are independent of the integration of the current elements over the length of the antenna.

The vertical fields in the top layer may be found from the horizontal components by applying Maxwell's equations; however, these fields are much smaller than the horizontal fields and may be neglected for most purposes.

Appendix A

PARAMETERS OF CONDUCTING MEDIA

Various parameters of conducting media are presented in tables A-1 and A-2 and figures A-1 through A-3. Emphasis is placed on those materials that are of interest in naval applications.

The electrical properties of sea ice are complicated and are reported in Wentworth and Cohen[67] and Luchininov.[68]

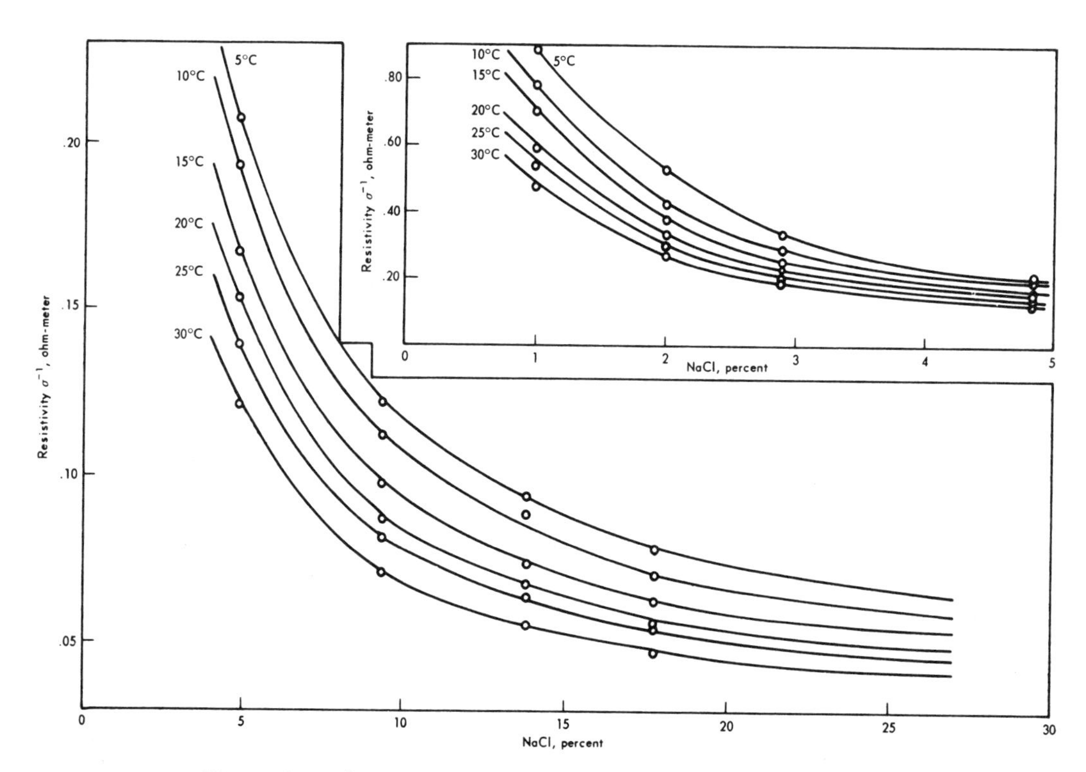

Figure A.1.–*Resistivity of aqueous sodium chloride solutions.*

---

[67] Wentworth, F. L.; and Cohen, M.: Electrical Properties of Sea Ice at 0.1 to 30 Mc/s. Radio Sci., vol. 68, no. 6, June 1964, pp. 681-691.

[68] Luchininov, V. S.: Electrical Characteristics of Ice. Soviet Phys. Tech. Phys., vol. 13, no. 3, Sept. 1968, pp. 418-423.

Table A.1.—*Conductivities and skin depths of metals at 20°C*

Metal	σ (megamhos/meter)	δ (millimeters) at 1 kHz
Silver	61	2.0
Copper	58	2.1
Aluminum	35	2.7
Brass (yellow)	15	4.1
Phosphor bronze	12	4.4
Steel		
HTS, $\mu_r = 180$	4.8	0.54
HY-80, $\mu_r = 90$	3.5	0.90

Table A.2.—*Conductivities of earth materials at normal temperatures and pressures*

Material	Approximate σ (mho/m)
Soils ($\epsilon_r \approx 10$)	
good	10^{-2} to 10^{-1}
average	10^{-3} to 10^{-2}
poor	10^{-4} to 10^{-3}
Water ($\epsilon_r = 81$)	
sea	4 to 5
fresh	10^{-4} to 10^{-3}
Snow (drifted, wet)*	10^{-6} to 10^{-4}
Ice (glacial)	10^{-6} to 10^{-4}
Permafrost	10^{-5} to 10^{-4}
Marine sands and shales	10^{-1} to 1
Marine sandstones	10^{-2} to 1
Clay	10^{-2} to 10^{-1}
Sandstone (wet)	10^{-4} to 10^{-2}
Granite**	10^{-9} to 10^{-3}

* Very dependent upon temperature, frequency, and impurities.
** Very dependent upon water content.

Graphs of the real and imaginary parts of the plane wave transmission coefficients τ_{TE} and τ_{TM} defined in (2.34) and (2.35), respectively, are presented by Simmons[69] for a plane wave going from free space to sea water ($\sigma = 4$ mhos/meter). These graphs are shown here in figures A-4 through A-7.

[69] Simmons, E. B.: Transmission Coefficients for Electromagnetic Plane Wave Radiation into a Conducting Half-Space. Tech. Rept. EE-60, Eng. Exp. Station, Univ. of N. Mex., Albuquerque, 1961.

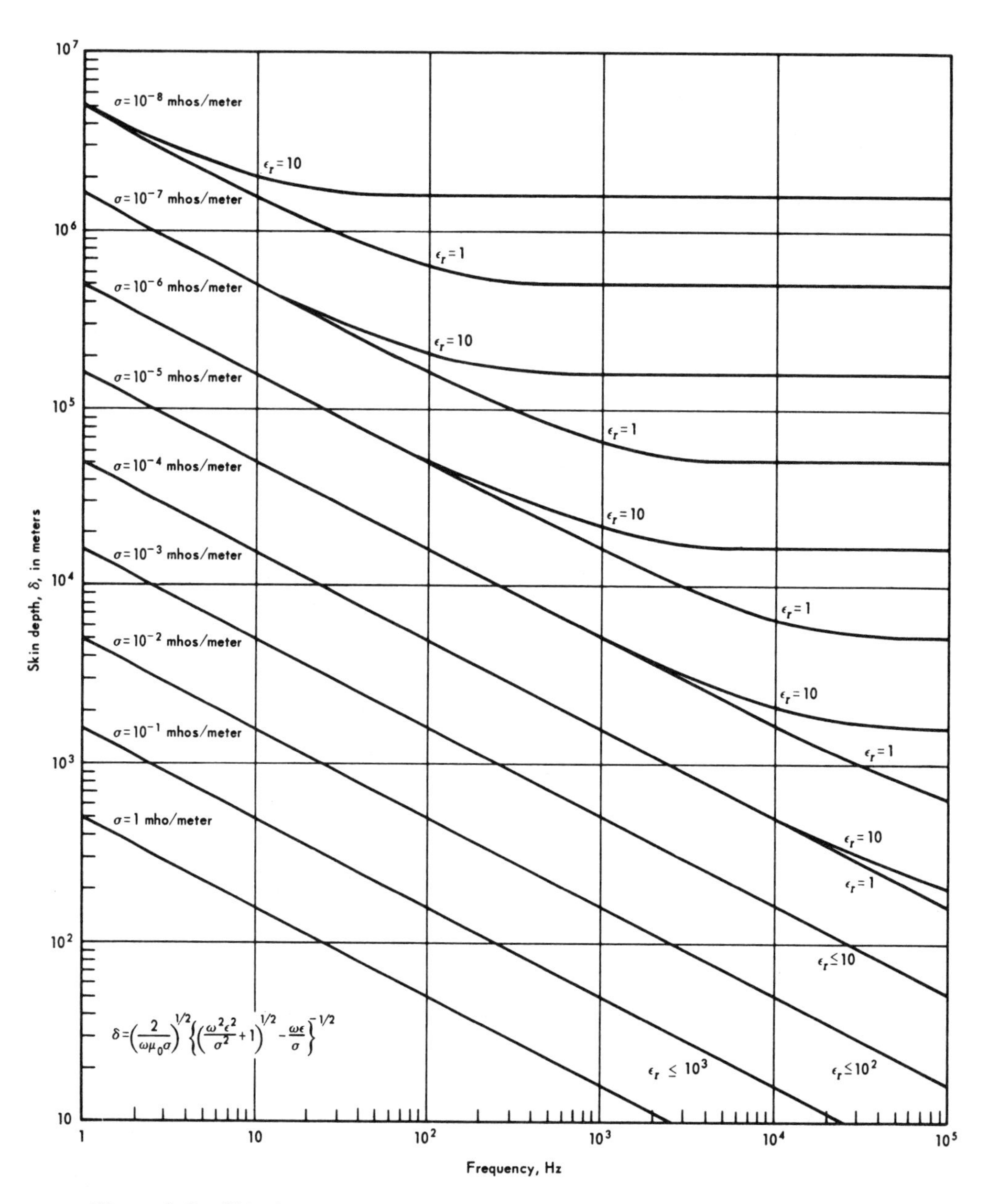

Figure A.2.–*Skin depth as a function of frequency for various conductivities and relative permittivities.*

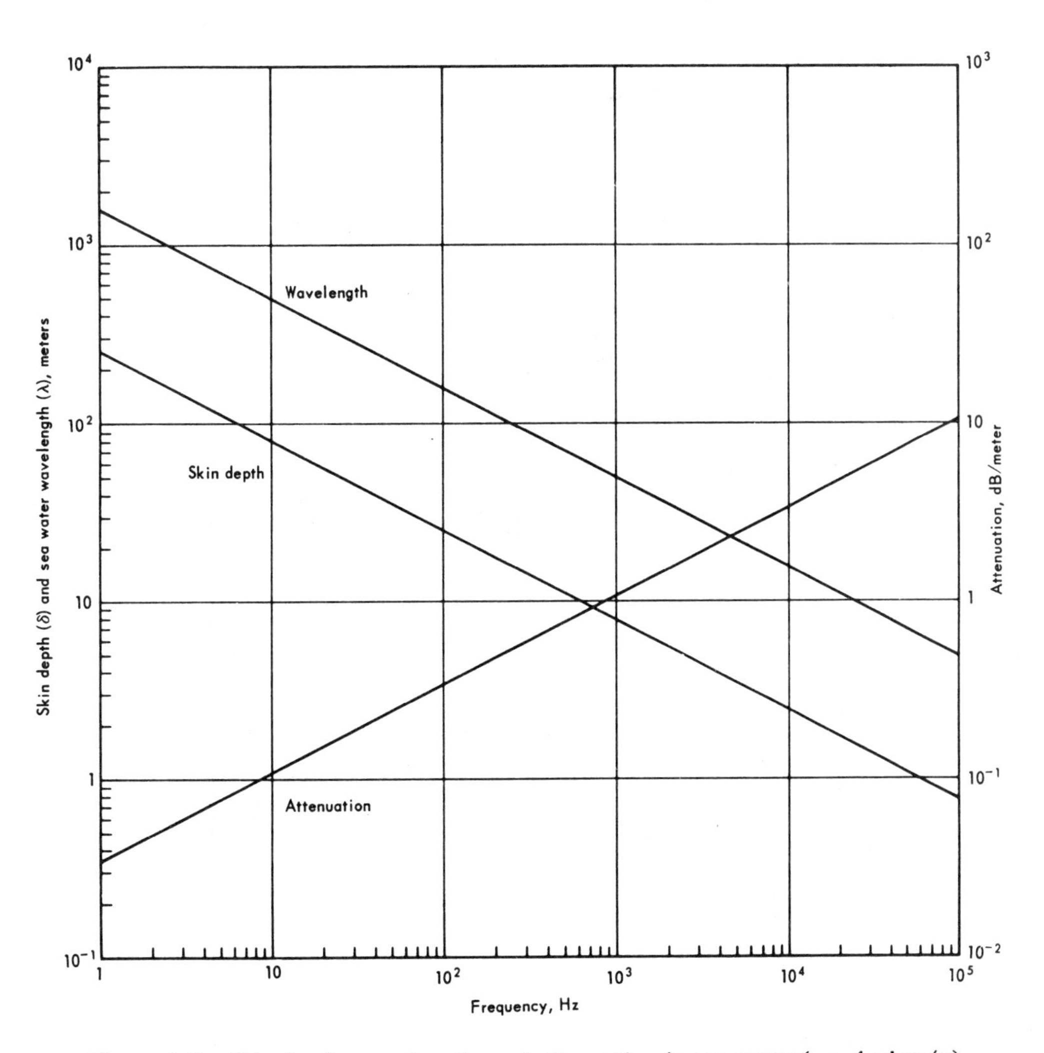

Figure A.3.—*Skin depth, wavelength, and attenuation in sea water* (σ = *4 mhos/m*).

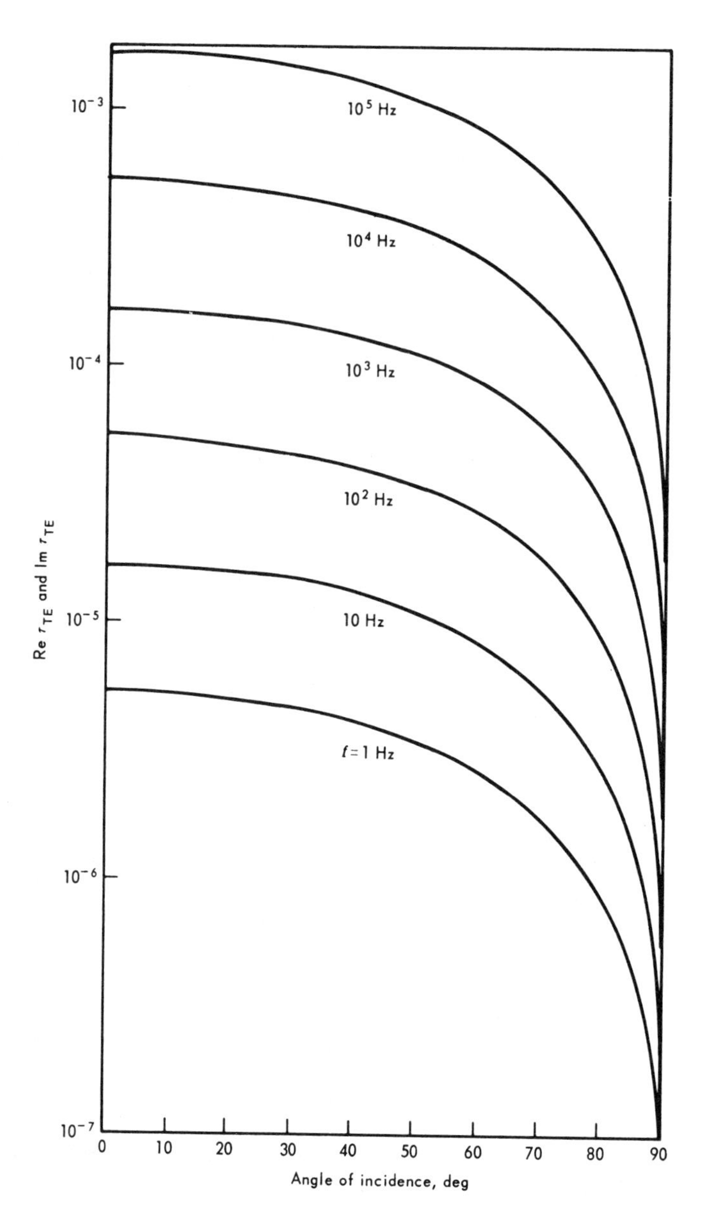

Figure A.4.–*The real and imaginary parts of the free-space to sea water transmission coefficient* τ_{TE}.

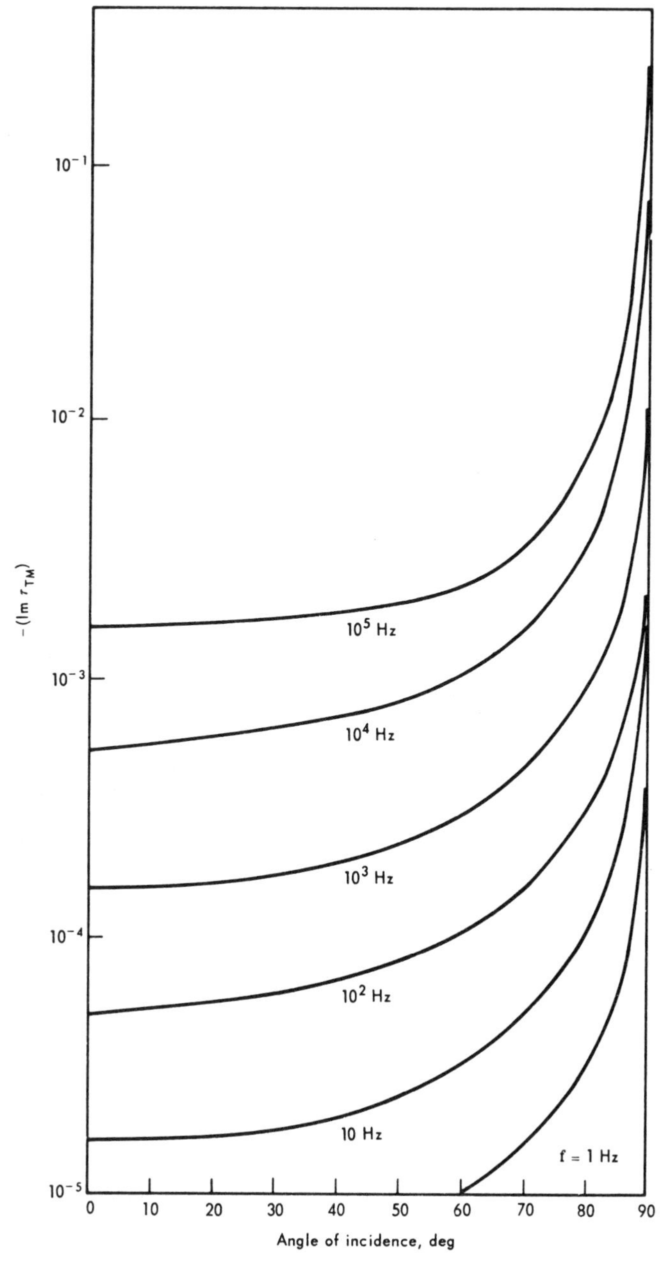

Figure A.5.–*The imaginary part of the free-space to sea water transmission coefficient* τ_{TM}.

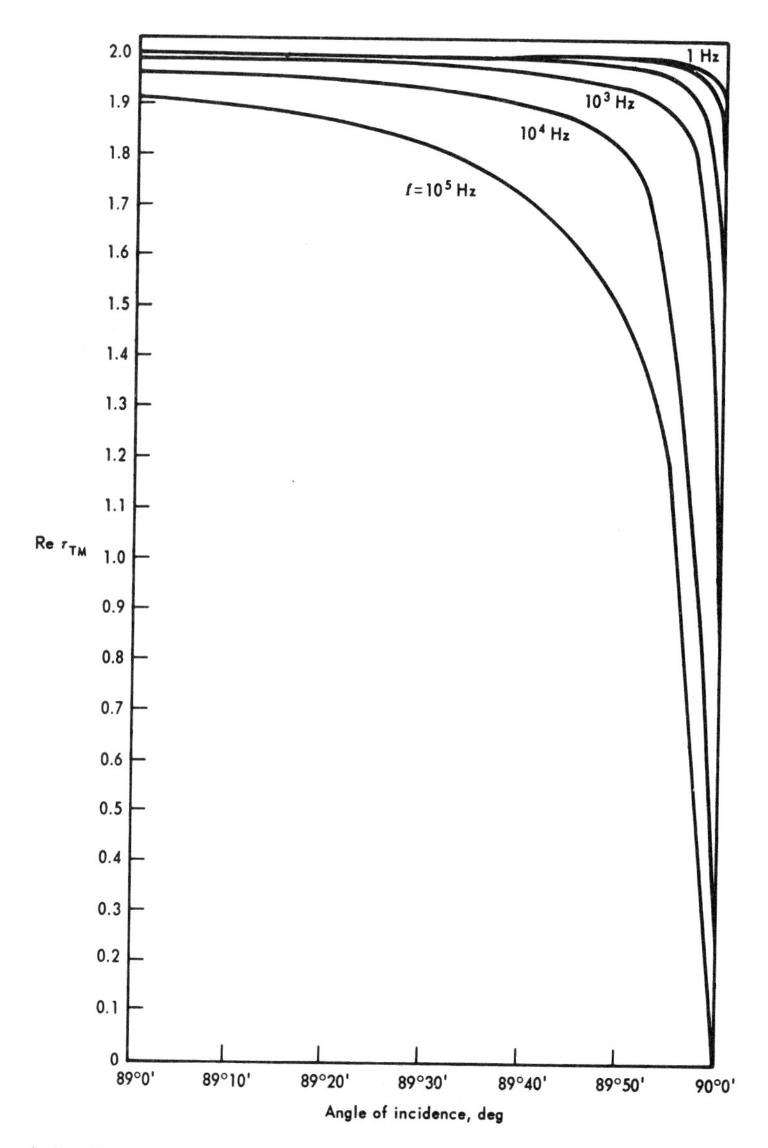

Figure A.6.–*The real part of* τ_{TM} *for angles of incidence between 89° and 90°.*

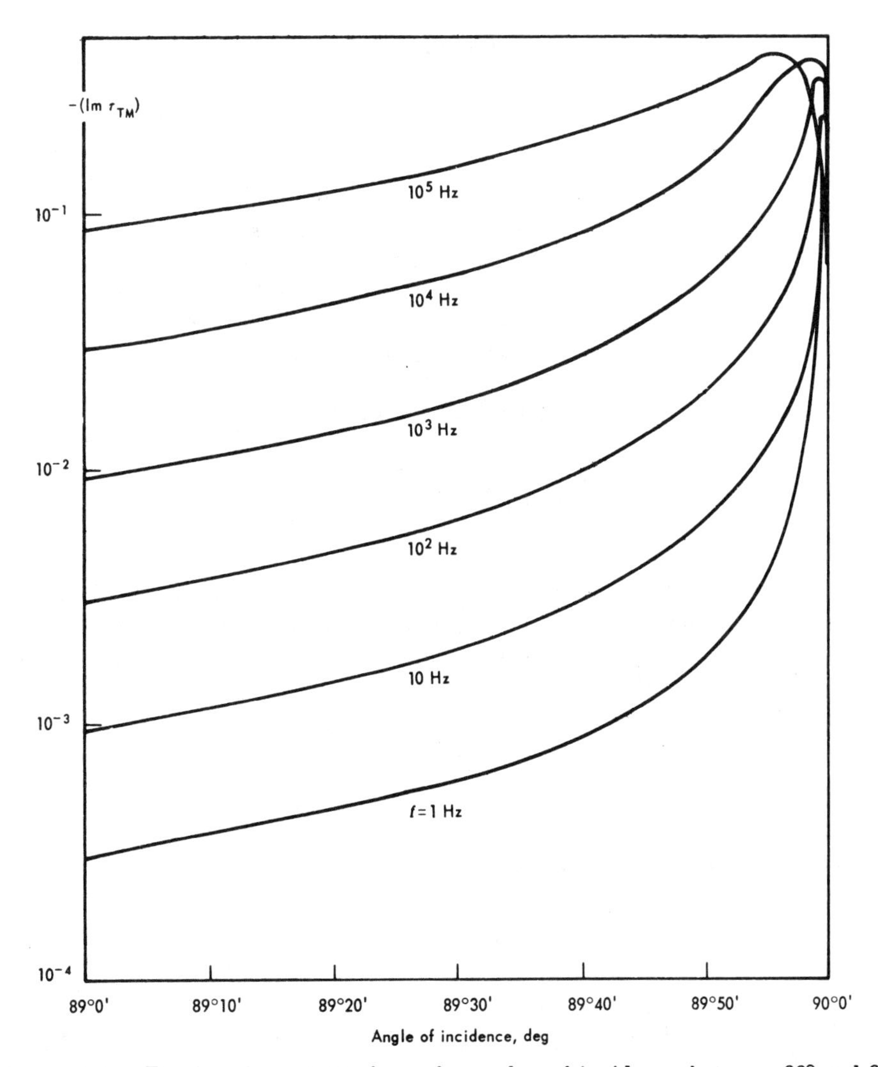

Figure A.7.–*The imaginary part of* τ_{TM} *for angles of incidence between 89° and 90°.*

Appendix B

DIPOLE APPROXIMATION SCATTERING

A perfectly conducting body (or one with negligible field penetration) has a surface current induced on it by an incident electromagnetic field. If the dimensions of the body are small compared with a wavelength in the surrounding medium, the incident field is substantially uniform over the conducting object. The resulting surface current gives rise to predominately electric and magnetic dipole moments from which the scattered field may be determined by using the appropriate dipole expressions in the previous chapters. Reflections between the scattering object and the boundaries of the conducting medium should be accounted for.

The electric and magnetic dipole moments may be calculated from the electric and magnetic polarizabilities respectively. Thus, for a perfectly conducting body embedded in a lossy medium, the x-component of the electric dipole moment $\hat{p}_x$, is expressed as

$$\hat{p}_x = \left[\epsilon\left(1 + \frac{\sigma}{j\omega\epsilon}\right)\right]\alpha_{ex} E_x \qquad \text{(B-1)}$$

where the bracketed quantity represents the complex permittivity of the surrounding lossy medium, and α_{ex} is the electric polarizability of the perfectly conducting body in the direction of the incident electric field component E_x. If the displacement current in the dissipative medium is neglected ($\sigma \gg \omega\epsilon$), then (B-1) reduces to

$$\hat{p}_x = \frac{\sigma}{j\omega}\alpha_{ex} E_x \qquad \text{(B-2)}$$

and the current moment $p_x = j\omega\hat{p}_x$ is given by

$$p_x = \sigma\alpha_{ex} E_x \qquad \text{(B-3)}$$

An incident magnetic field H_y will induce an oppositely directed magnetic dipole moment m_y, which is expressed as

$$m_y = \alpha_{my} H_y \qquad \text{(B-4)}$$

where α_{my} is the magnetic polarizability of the perfectly conducting body along the y-direction. Since m_y is opposite in direction to H_y, the polarizability α_{my} will be a negative quantity.

The polarizabilities α_e and α_m are functions of the volume and shape of the conductor and the relative orientation of the conductor with the inducing field.

Exact expressions for the polarizabilities of prolate and oblate spheroids are well known and are given below in terms of the half lengths a and b of the major axis and minor axis respectively and the semifocal distance $l = (a^2 - b^2)^{1/2}$. (When $a = b$, the spheroid becomes a sphere for which $\alpha_e = 4\pi a^3$ and $\alpha_m = -2\pi a^3$.)

Prolate Spheroid

In a field parallel to the major axis,

$$\alpha_e = \frac{4\pi a l^2}{3\left[\frac{a}{l} \ln\left(\frac{a+l}{b}\right) - 1\right]} \tag{B-5}$$

$$\alpha_m = -\frac{4\pi b^2 l^3}{3\left(al - b^2 \ln \frac{a+l}{b}\right)} \tag{B-6}$$

Similarly in a field parallel to the minor axis,

$$\alpha_e = \frac{8\pi b^2 l^3}{3\left(al - b^2 \ln \frac{a+l}{b}\right)} \tag{B-7}$$

$$\alpha_m = -\frac{8\pi a b^2}{3\left(1 - \frac{b^2}{l^2} + \frac{ab^2}{l^3} + \ln \frac{a+l}{b}\right)} \tag{B-8}$$

Oblate Spheroid

In a field parallel to the major axis,

$$\alpha_e = \frac{8\pi a^2 l^3}{3\left[a^2 \cot^{-1}\left(\frac{b}{l}\right) - bl\right]} \tag{B-9}$$

$$\alpha_m = -\frac{8\pi a^2 b l^3}{3\left(l^3 + a^2 l - a^2 b \cot^{-1} \frac{b}{l}\right)} \tag{B-10}$$

Similarly in a field parallel to the minor axis,

$$\alpha_e = \frac{4\pi b l^2}{3\left(1 - \frac{b}{l} \cot^{-1} \frac{b}{l}\right)} \tag{B-11}$$

$$\alpha_m = -\frac{4\pi l^3}{3\left[\cot^{-1}\left(\frac{b}{l}\right) - \frac{bl}{a^2}\right]} \tag{B-12}$$

The electric and magnetic polarizabilities of circular cylinders cannot be expressed exactly, but approximate values for various length to diameter ratios are presented in table B-1. These values are obtained from Taylor.[70,71]

Table B.1.–*Longitudinal and transverse electric and magnetic polarizabilities of a circular cylinder of diameter* D, *length* L, *and volume* V

$\frac{L}{D}$	$\frac{\alpha_{el}}{V}$	$\frac{\alpha_{et}}{V}$	$\frac{\alpha_{ml}}{V}$	$\frac{\alpha_{mt}}{V}$
∞	∞	2.00	-1.00	-2.00
10	58.1	2.13	-1.06	-1.94
4	15.1	2.32	-1.16	-1.85
2	7.10	2.61	-1.31	-1.74
1	3.86	3.17	-1.59	-1.58
1/2	2.43	4.22	-2.11	-1.41
1/4	1.75	6.18	-3.09	-1.27
0	1.00	∞	-∞	-1.00

Further discussions of dipole scattering by conducting objects in lossy media are found in Galejs[72] and Kraichman.[73]

[70] Taylor, T. T.: Electric Polarizability of a Short Right Circular Conducting Cylinder. J. Res. Nat. Bur. Std. B, vol. 64, no. 3, July-Sept. 1960, pp, 135-143.

[71] Taylor, T. T.: Magnetic Polarizability of a Short Right Circular Conducting Cylinder. J. Res. Nat. Bur. Std. B, vol. 64, no. 4, Oct.-Dec. 1960, pp. 199-210.

[72] Galejs, J.: Scattering from a Conducting Sphere Embedded in a Semi-Infinite Dissipative Medium. J. Res. Nat. Bur. Std. D, vol. 66, no. 5, Sept.-Oct. 1962, pp. 607-612.

[73] Kraichman, M. B.: A Dipole Approximation of the Backscattering from a Conductor in a Semi-Infinite Dissipative Medium. J. Res. Nat. Bur. Std. D, vol. 67, no. 4, July-Aug. 1963, pp. 433-443.

Appendix C

ANTENNA IMPEDANCE

The two types of antennas most commonly used inside a highly conducting medium ($\sigma \gg \omega\epsilon$) are the insulated wire electric antenna with the ends shorted to the lossy medium by means of electrodes and the insulated wire loop antenna.

INSULATED WIRE ELECTRIC ANTENNA (ELECTRODE TERMINATION)

The characteristics of this type of electric antenna may be derived by considering the insulated wire and the infinite, homogeneous, highly conducting medium to constitute a lossy coaxial transmission line that supports the principal symmetric TM mode and has a zero terminating impedance.

Von Aulock[74] derives the transmission line parameters and the input impedance for such a lossy coaxial cable under the assumption that the outer radius of the insulation is much smaller than a skin depth in the surrounding conducting medium. The input impedance Z_i of a transmission line of length d for a zero load impedance is given by the well-known expression

$$Z_i = \frac{R + j\omega L}{\Gamma} \tanh \Gamma d \tag{C-1}$$

where the line propagation constant Γ is defined as

$$\Gamma = [(R + j\omega L)(G + j\omega C)]^{1/2} \tag{C-2}$$

in terms of the per unit length resistance R, inductance L, capacitance C, and leakage conductance G. The per unit length line parameters R, L, C, and G follow:

Resistance

$$R = R_w + \frac{\omega\mu_0}{8} \tag{C-3}$$

where the first term is the internal resistance of the wire conductor and the second term is the resistance of the current return path through the surrounding conducting medium.

Inductance

$$L = L_w + \frac{\mu_0}{2\pi} \ln\left(\frac{b}{a}\right) + \frac{\mu_0}{2\pi} \ln \frac{2}{1.781 b(\omega\mu_0\sigma)^{1/2}} \tag{C-4}$$

[74] Von Aulock, W.: The Electrical Characteristics of a Laminated Coaxial Cable With Sea Water Return.

where the first term is the internal inductance of the wire conductor, the second term is the external inductance between the wire conductor and the surrounding conducting medium, and the third term is the internal inductance of the surrounding conducting medium.

Capacitance

$$C=\frac{2\pi\epsilon_I}{\ln (b/a)} \tag{C-5}$$

Conductance

$$G=\frac{2\pi\sigma_I}{\ln (b/a)} \tag{C-6}$$

In these equations,

σ = conductivity of the surrounding lossy medium
σ_I = conductivity of the insulating material
ϵ_I = permittivity of the insulating material
μ_0 = permeability of free space
a = radius of the wire conductor
b = outer radius of the insulation

The internal resistance R_w and the internal inductance L_w of the wire conductor may be calculated from the well-known expressions given in Ramo and Whinnery.[75]

At VLF and below, most practical antennas are electrically short ($|\Gamma d| \ll 1$) so that $\tanh \Gamma d \approx \Gamma d$ and the input impedance

$$Z_i=(R+j\omega L)d \tag{C-7}$$

It should be noted that unless the insulated wire antenna is very much longer than a skin depth in the lossy medium, the electrodes will have to be made extremely large to bring about good agreement between the measured value of antenna impedance and that value calculated from the transmission line formulas above.

Guy and Hasserjian[76] consider the effects of burying the insulated wire in a semi-infinite conducting medium and point out that the transmission line properties very close to the surface of the semi-infinite medium are nearly identical with those at great depths. They also show that the attenuation constant of the line (real part of Γ) reaches a maximum value at a burial depth of 0.3 of a medium skin depth. However, even at this depth the effect on the input impedance is small.

INSULATED LOOP ANTENNA

Expressions for the impedance of a thinly insulated circular loop of wire that carries a uniform current and is surrounded by an infinite, homogeneous, conducting medium are derived by Kraichman[77] by substituting a complex propagation constant in the well-known expression for the

---

TM No. 151, Minesweeping Sec., Bur. of Ships, Navy Dept., 29 Sept. 1950.

[75] Ramo, S.; and Whinnery, J. R.: Fields and Waves in Modern Radio. Second ed., John Wiley & Sons, Inc., 1953, ch. 6.

[76] Guy, A. W.; and Hasserjian, G.: Impedance Properties of Large Subsurface Antenna Arrays. IEEE Trans. Antennas Propagation, AP-11, no. 3, May 1963, pp. 232-240.

[77] Kraichman, M. B.: Impedance of a Circular Loop in an Infinite Conducting Medium. J. Res. Nat. Bur. Std. D, vol. 66, no. 4, July-Aug. 1962, pp. 499-503.

self-impedance of a loop in free space. The thin layer of insulation helps to maintain a uniform current flow around the loop but may be neglected when computing the loop impedance for an assumed uniform current flow. The validity of this approach is verified by Galejs[78] using a different analytical formulation.

Under the assumption that the radius of the loop is much smaller than a skin depth in the lossy medium with negligible displacement current, the external resistance R and the external inductive reactance X of an insulated single turn circular loop are given by

$$R = \frac{8}{3\sigma a}\left(\frac{a}{\delta}\right)^4 \tag{C-8}$$

$$X = X_0 - \frac{2\pi}{3\sigma a}\left(\frac{a}{\delta}\right)^5 \tag{C-9}$$

where a = radius of the loop
σ = conductivity of the lossy medium
δ = skin depth in the lossy medium
X_0 = inductive reactance of the loop in free space

For a flat loop of N turns, the resistance and reactance given above should be multiplied by N^2.

If the loop is symmetrically placed in a free-space spherical cavity of radius $r_0 \ll \delta$ and the cavity embedded in the conducting medium, the impedance as derived by Wait[79] is given by

$$Z = Z_0 + \frac{2\pi}{3\sigma r_0}\left(\frac{a}{\delta}\right)^4 \tag{C-10}$$

where Z_0 is the impedance of the loop in free space. The ratio of the second term in (C-10) to the resistance in (C-8) is equal to $\pi/4\,(a/r_0)$ and clearly shows the effectiveness of the spherical cavity in lowering the external resistance of the loop.

[78] Galejs, J.: Admittance of Insulated Loop Antennas in a Dissipative Medium. IEEE Trans. Antennas Propagation, vol. AP-3, no. 2, Mar. 1965, pp. 229–235.

[79] Wait, J. R.: Insulated Loop Antenna Immersed in a Conducting Medium. J. Res. Nat. Bur. Std., vol. 59, no. 2, Aug. 1957, pp. 133–137.

Appendix D

ELF AND VLF ATMOSPHERIC NOISE

The mean values of the free-space vertical electric field noise density during the summer at various locations on the surface of the earth in the northern hemisphere are presented by Maxwell.[80] These values in a 1 Hz bandwidth are shown in figure D-1. The values at other bandwidths may be obtained from the fact that the rms atmospheric noise field is proportional to the square root of the bandwidth.

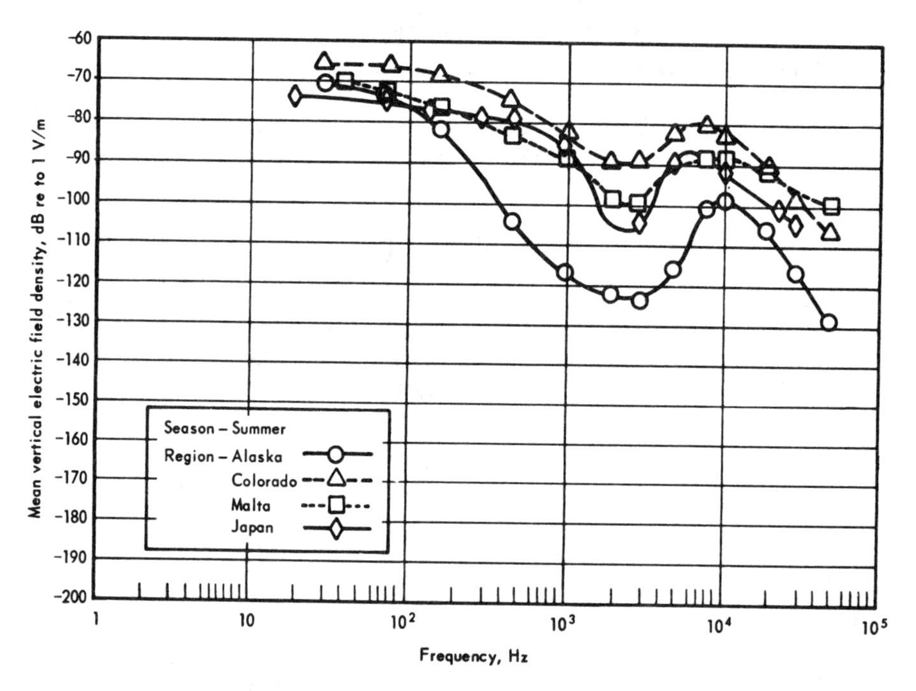

Figure D.1.–*Mean vertical electric field noise density at various surface locations of the earth in 1 Hz bandwidth.*

Surface values of the mean horizontal magnetic field noise density may be calculated from the vertical electric field values by using the free-space plane wave relationship

[80]Maxwell, E. L.: Atmospheric Noise from 20 Hz to 30 kHz. Radio Sci., vol. 2, no. 6, June 1967, pp. 637–644.

$$\frac{E}{H}=\left(\frac{\mu_0}{\epsilon_0}\right)^{1/2}=377\text{ ohms} \tag{D-1}$$

for all frequencies shown except those from 1 to 5 kHz. In this region of the first-order cutoff of the earth-ionosphere waveguide, measurements by Maxwell and Stone[81] indicate that the horizontal magnetic field values may be as much as 12 dB lower than those obtained from the free-space impedance (D-1).

For frequencies outside the 1 to 5 kHz region, the mean horizontal electric field noise density at the surface of a homogeneous, semi-infinite conductor is related to the vertical values shown in figure D-1 by the relationship given in (2.53). Both the horizontal electric and magnetic fields are exponentially attenuated with depth below the surface of the earth and are related by the expression (2.60).

[81]Maxwell, E. L.; and Stone, D. L.: ELF and VLF Atmospheric Noise. Rept. No. 30-P-6, DECO Electronics, 10 Jan. 1964.

Appendix E

SUPPLEMENTARY RESULTS

The supplementary results in this additional appendix to the second printing are listed under the pertinent section number and heading that remain unchanged and common to both printings.

3.2.1.1 Horizontal electric dipole

$$\lim_{\rho \to o} H_{\rho_2} = \frac{-p \sin \phi}{8\pi(z\text{-}h)^2} \left[2 - \left(\frac{\sigma_1 - \sigma_2}{\sigma_1 + \sigma_2} \right) \right] \tag{E-1}$$

$$\lim_{\rho \to o} H_{\phi_2} = \frac{-p \cos \phi}{8\pi(z-h)^2} \left[2 - \left(\frac{\sigma_1 - \sigma_2}{\sigma_1 + \sigma_2} \right) \right] \tag{E-2}$$

3.3.1 Surface to Surface Propagation

The magnitude of the stratification correction factor Q for a two-layer medium is plotted in figure E.1.

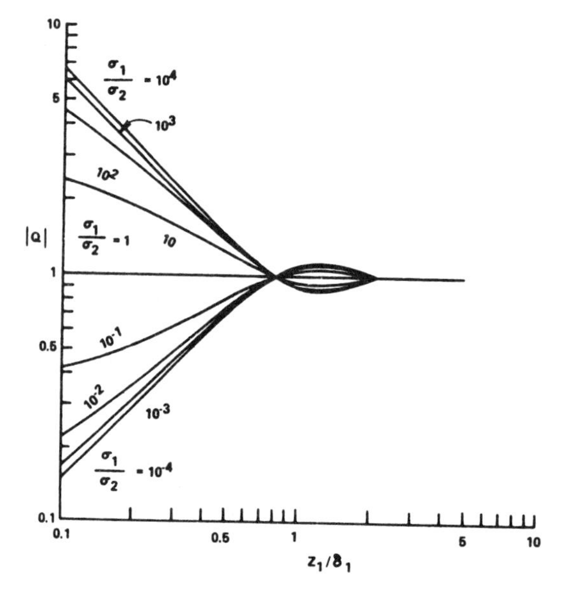

Figure E.1. – Magnitude of Q versus z_1/δ_1 for a two-layer medium.

3.3.2 Surface to Surface Propagation

To a first approximation, the vertical quasi-static fields in the top layer of a stratified conductor from an HED at the surface of the layered medium are:

$$E_z \approx o \tag{E-3}$$

$$H_z = \frac{3p \sin \phi}{2\pi\gamma_1^2 \rho^4} \left(\frac{Q^2}{Q_E} \right) \qquad \text{(E-4)}$$

Similarly for the HMD,

$$E_z \approx o \qquad \text{(E-5)}$$

$$H_z = \frac{3m \sin \phi}{2\pi\gamma_1 \rho^4} \left(\frac{Q}{Q_E} \right) \qquad \text{(E-6)}$$

Bannister[50] derives results for the quasi-static field components at the interface between the layers of a two-layer conducting half-space for an HED source on the surface. Some of these results for the case where $\rho \gg z_1$, δ_1, and δ_2 are given in (3.67)-(3.70) and in (E-3) and (E-6) above. Presented here are Bannister's expressions for the case where $\rho \gg z_1$ and δ_1 and $\sigma_1 \gg \sigma_2$ but in which (1) $\rho \sim \delta_2$ and (2) $\rho \ll \delta_2$.

(1) When ρ is comparable to δ_2

$$E_\rho = \frac{p \cos \phi}{2\pi\sigma_1 \rho^3} \, \frac{\coth \gamma_1 z_1}{\sinh \gamma_1 z_1} \left[1 + (1+\gamma_2\rho+\gamma_2^2 \rho^2)\, e^{-\gamma_2\rho} \right] \qquad \text{(E-7)}$$

$$E_\phi = \frac{p \sin \phi}{\pi\sigma_1\rho^3} \, \frac{\coth \gamma_1 z_1}{\sinh \gamma_1 z_1} \left[1 + (1+\gamma_2\rho) e^{-\gamma_2\rho} \right] \qquad \text{(E-8)}$$

$$E_z \approx o \qquad \text{(E-9)}$$

$$H_\rho = \frac{p \sin \phi}{\pi\gamma_1\rho^3 \sinh \gamma_1 z_1} \left[(1+\gamma_2\rho)\, e^{-\gamma_2\rho} - \left(\frac{\sigma_2}{\sigma_1} \right)^{1/2} \frac{\gamma_2 \rho W}{2} \coth \gamma_1 z_1 \right] \qquad \text{(E-10)}$$

$$H_\phi \; \frac{-p \cos \phi}{2\pi\gamma_1\rho^3 \sinh \gamma_1 z_1} \left[(1+\gamma_2\rho+\gamma_2^2\rho^2)\, e^{-\gamma_2\rho} - \left(\frac{\sigma_2}{\sigma_1} \right)^{1/2} \gamma_2\rho I_1 K_1 \coth \gamma_1 z_1 \right] \qquad \text{(E-11)}$$

$$H_z \; \frac{p \sin \phi \;\; \coth \gamma_1 z_1}{2\pi\gamma_1^2 \rho^4 \sinh \gamma_1 z_1} \left[3 + (3+3\gamma_2\rho+\gamma_2^2\rho^2)\, e^{-\gamma_2\rho} \right] \qquad \text{(E-12)}$$

where the functions W and $I_1 K_1$ are both of argument $\gamma_2\rho/2$ and are defined on page 3-15.

(2) When $\rho \ll \delta_2$, and

(a) $|\gamma_1 z_1| < \frac{1}{\sqrt{2}}$ (electrically thin first layer)

$$E_\rho \; \frac{p \cos \phi}{2\pi\sigma_1\rho^3} \, \frac{2}{(\gamma_1 z_1)^2} \qquad \text{(E-13)}$$

$$E_\phi = \frac{p \sin \phi}{\pi \sigma_1 \rho^3} \quad \frac{2}{(\gamma_1 z_1)^2} \qquad \text{(E-14)}$$

$$E_z \approx o \qquad \text{(E-15)}$$

$$H_\rho = \frac{p \sin \phi}{\pi \gamma_1 \rho^3} \quad \frac{1}{(\gamma_1 z_1)} \qquad \text{(E-16)}$$

$$H_\phi = \frac{-p \cos \phi}{2\pi \gamma_1 \rho^3} \quad \frac{1}{(\gamma_1 z_1)} \qquad \text{(E-17)}$$

$$H_z = \frac{3p \sin \phi}{2\pi \gamma_1^2 \rho^4} \quad \frac{2}{(\gamma_1 z_1)^2} \qquad \text{(E-18)}$$

(b) $\left|\gamma_1 z_1\right| > 2\sqrt{2}$ (electrically thick first layer)

$$E_\rho = \frac{p \cos \phi}{2\pi \sigma_1 \rho^3} \left(4e^{-\gamma_1 z_1}\right) \qquad \text{(E-19)}$$

$$E_\phi = \frac{p \sin \phi}{\pi \sigma_1 \rho^3} \left(4e^{-\gamma_1 z_1}\right) \qquad \text{(E-20)}$$

$$E_z \approx o \qquad \text{(E-21)}$$

$$H_\rho = \frac{p \sin \phi}{\pi \gamma_1 \rho^3} \left(2e^{-\gamma_1 z_1}\right) \qquad \text{(E-22)}$$

$$H_\phi = \frac{-p \cos \phi}{2\pi \gamma_1 \rho^3} \left(2e^{-\gamma_1 z_1}\right) \qquad \text{(E-23)}$$

$$H_z = \frac{3p \sin \phi}{2\pi \gamma_1^2 \rho^4} \left(4e^{-\gamma_1 z_1}\right) \qquad \text{(E-24)}$$

Appendix D – elf and vlf atmospheric noise

The average rms amplitudes of the horizontal north-south magnetic background noise during the winter at Palo Alto, California are given in table 1 for frequencies from 0.5Hz to 14Hz. These values are reported by Fraser-Smith and Buxton[82] and are typical of measurements made by other investigators in the same frequency region.

Table E.1. – Average rms amplitudes of horizontal north-south magnetic background noise during the winter at Palo Alto, California

Frequency in Hz	Amplitude in $A/m\sqrt{Hz}$
0.5	8×10^{-6}
1	3.2×10^{-6}
2	1.6×10^{-6}
4	8×10^{-7}
6	6.4×10^{-7}
8	8×10^{-7}
10	6.4×10^{-7}
14	5.6×10^{-7}

The horizontal magnetic field background noise from 14Hz to 20Hz as reported by other investigators differs little on the average from that given above for 14Hz.

[82] Fraser-Smith, A.C.; and Buxton, J.L.: Superconducting Magnetometer Measurements of Geomagnetic Activity in the 0.1 to 14Hz Frequency Region. Radioscience Laboratory Report, Stanford Electronics Laboratories, Stanford University, Stanford, CA., Sept. 1974, revised 24 Jan. 1975.

INDEX

Numbers on this page refer to PROBLEM NUMBERS, not page numbers

Absorbed power in waveguides, 14-30
Acceleration vector, 1-7 to 1-10
Admittance:
  normalized, 13-9, 13-10
  normalized load, 13-33
Air dielectric, 14-10, 14-14, 14-29
Air-filled:
  rectangular waveguide, 14-17
  waveguide, 14-19, 14-24
  X-band rectangular waveguide, 14-27
Ampere's law, 8-3 to 8-8, 8-14, 9-11, 9-18, 9-19, 9-25, 10-27, 10-29, 10-31, 10-40, 12-6
Ampere-turns, 10-9 to 10-26
  determination of, 10-9 to 10-18
Amplitude ratio, 14-42
Angle of transmission, 12-15
Angular velocity vector, 1-10
Antenna, 13-31, 14-32 to 14-46
  absorbed power, 14-43, 14-44
  dipole, 14-44
  directivity, 14-37
  gain, 14-37, 14-38
  impedance, 13-26
  radiated power, 14-43, 14-44
  radiation intensity, 14-36, 14-37
  radiation resistance, 14-34, 14-35
Attenuation, 14-13, 14-14, 14-16 14-18, 14-19
  constant, 13-2, 13-12, 13-15, 13-41, 14-18, 14-19, 14-25
  curves, 14-19
  factor, 14-17
Avogadro number, 2-2

Base vector, 1-1, 1-10
B-field:
  between parallel plates carrying current, 8-5
  due to coaxial cable carrying current, 8-14, 8-15
  due to cylindrical conductor carrying current, 8-4, 8-18, 8-28
  due to hollow cylinder carrying current, 8-22
  due to infinite planes carrying current, 8-16, 8-17
  due to long wire carrying current, 8-1 to 8-3, 8-26, 8-32
  due to loop carrying current, 8-20, 8-21, 8-23, 8-31
  due to parallel wires carrying current, 8-10
  due to ribbon carrying current, 8-12
  due to short wire carrying current, 8-13
  due to slab of volume current, 8-19
  due to solenoid, 8-7, 8-27
  due to toroid, 8-8
  due to two coils carrying current, 8-24
  in air gaps, 9-24
  near a north pole, 8-11
B-field, variation of:
  around cylinder carrying

Numbers on this page refer to PROBLEM NUMBERS, not page numbers

current, 9-23
in a solenoid, 9-22
in coaxial conductors, 8-25
in the gap in a toroid, 9-18 to 9-21
Biot-Savart law, 8-1, 8-2, 8-36
Bohr radius, 2-3

Capacitance, 3-27, 6-1, 6-2, 6-8, 6-10 to 6-12, 6-27, 6-38
Capacitor:
coaxial, 6-14 to 6-17
concentric, 6-18 to 6-21
multiple dielectric, 6-22 to 6-25
parallel-plate, 6-3 to 6-13
series and parallel combinations, 6-26 to 6-28
Cavity resonator, 14-6
Cell method, 14-47
Center of mass, 1-9
Characteristic impedance, 13-1, 13-2, 13-4, 13-5, 13-7, 13-8, 13-10, 13-12, 13-14, 13-16 to 13-19, 13-26, 13-27, 13-29, 13-39, 13-41
Charge density, 2-1, 2-3, 4-7
Charge distribution, 2-1, 2-3
Charges, 2-1 to 2-10
force between, 2-4 to 2-10
in electric and magnetic fields, 9-1 to 9-5
line, 2-3
point, 2-4 to 2-9
surface, 2-3
volume, 2-3
Circular wave guide, 14-1
Coaxial capacitor, 6-14 to 6-17
Coaxial transmission line, 13-26, 14-47
capacitance, 14-47
inductance, 14-47
Coefficient:
reflection, 12-20, 13-3 to 13-5, 13-8, 13-26, 13-29, 13-30, 13-38 to 13-41, 14-12, 14-31 14-46
transmission, 12-20, 14-12
Concentric capacitor, 6-18 to 6-21
Conservation of charge, law of, 3-29
Constant VSWR circle, 13-29
Continuous function, 1-11
Coordinate system:
fixed, 1-10
moving, 1-10
Critical angle, 12-17
Curl, 1-19, 1-20
Current density, 8-13, 8-15 to 8-19, 8-22, 8-29, 8-30, 8-33, 8-40, 8-41, 10-21, 10-27, 12-53, 12-54
Cut-off frequency, 14-1 to 14-15, 14-17, 14-20, 14-24
dominant mode, 14-17
TE and TM modes, 14-1 to 14-15
Cut-off wavelength, 14-1, 14-4, 14-7, 14-24
Cylindrical waveguide, 14-19

Density:
charge, 2-1, 2-3, 4-7
current, 8-13, 8-15 to 8-19, 8-22, 8-29, 8-30, 8-33, 8-40, 8-41, 10-21, 10-27, 12-53, 12-54
electric flux, 3-1 to 3-5, 3-7, 3-9, 3-10
Depth of penetration, 12-21, 12-22, 12-52
Dielectric constant, 6-11, 14-22
Diode, vacuum tube, 4-40
Dipole, 2-10, 4-24, 14-34, 14-36, 14-40, 14-41
antenna, 14-44
radiation resistance, 14-40, 14-41
Directivity antenna, 14-37
Displacement current, 11-29, 11-30
Divergence, 1-15, 1-18, 1-20
theorem, 1-21 to 1-23, 1-29, 3-7
Double-stub, 13-10

matching, 13-33, 13-34
tuner, 13-10

Efficiency of transmission, 13-2
Electric field:
lines of force in, 3-32
motion in, 4-35 to 4-37
Electric field due to:
charged coaxial cylinders, 3-26
charged concentric spheres, 3-8, 3-27
charged parallel plates, 3-13, 3-17, 3-29
dipole, 2-17
electron beam, 3-20
line charge, 2-19, 3-6, 3-11
point charges, 2-11, 2-15, 2-16, 3-28
ring of charge, 2-10, 2-12, 2-14
surface charge on a cylinder, 2-18, 3-19
surface charge on a disk, 2-20
surface charge on a plane, 2-13, 2-21, 3-12, 3-13, 3-15, 3-16
surface charge on a sphere, 3-8, 3-18
volume charge in a cube, 2-22
volume charge in a cylinder, 3-24
volume charge in a sphere, 2-23, 3-25
Electric flux, 3-1 to 3-5, 3-7, 3-9, 3-10
density, 3-1 to 3-5, 3-6, 3-8, 3-21, 3-22, 3-28, 3-29, 3-31
Electric potential, 4-31, 4-32
intensity, 4-32
Electromagnetic waves, field components in, 11-22
Electrostatic potential, 4-1 to 4-44
energy, 4-1 to 4-8, 4-10, 4-18, 4-21, 4-38 to 4-44
gradient of, 4-28 to 4-34
Elliptically polarized wave, 12-26
Energy:
density, 6-37, 12-1, 12-9
electrostatic, 4-38 to 4-44
in magnetic circuits, 10-44, 10-45, 10-49
plane wave, 12-9, 12-10
self, 4-4
stored, 6-33 to 6-37
Equipotential:
lines, 4-22
surfaces, 4-11, 4-14, 4-22, 4-27

Faraday's law, 11-1 to 11-20, 11-23, 12-6
Field components in waveguides, 14-20 to 14-22
Force, 6-6
between parallel planes carrying currents, 9-11
on conductors carrying currents, 9-9, 9-10
on current elements, 9-6 to 9-8
on loops carrying currents, 9-12, 9-13
on moving charges, 9-1 to 9-5
Force, magnetic, 10-43 to 10-48
between coaxial coils, 10-48
between coaxial solenoids, 10-46
between pole-pieces of a magnet, 10-43
in linear-stroke motors, 10-47
on rod inserted into a solenoid, 10-44, 10-45
Free space:
dielectric, rectangular waveguide, 14-21
transmission, 14-45
Fundamental lemma, 1-13

Numbers on this page refer to PROBLEM NUMBERS, not page numbers

Gain, antenna, 14-37, 14-38
Gauss's law, 3-4, 3-6, 3-11 to 3-27, 3-31, 3-32, 4-4, 4-7, 4-12, 4-18, 4-21, 6-1
Generator, 11-31 to 11-36
  homopolar, 11-34
  work done in a, 11-33
Gradient, 1-14, 1-16, 1-20
  of potential, 4-28 to 4-34
Green's reciprocal theorem, 4-13
Group velocity, 14-24

Hagen-Rubens relation, 12-31
Half-wave dipole, 14-38
Helmholtz coil, 8-24
H-field:
  around antennas, 8-33
  due to coaxial cable carrying current, 8-6, 8-14, 8-15, 8-40
  due to current element, 8-36
  due to cylindrical conductor carrying current, 8-4
  due to line source of current, 8-41
  due to loop carrying current, 8-37, 8-38
  due to parallel wires carrying current, 8-29
  due to a solenoid, 8-39
  in air gaps, 9-24
H-field, variation of, 8-25, 9-18, 9-22, 9-23
  around a cylinder carrying current, 9-23
  in a solenoid, 9-22
  in coaxial conductors, 8-25
  in the gap in a toroid, 9-18
Higher-order mode, 14-17

Impedance:
  characteristic, 13-1, 13-2, 13-4, 13-5, 13-7, 13-8, 13-10, 13-12, 13-14, 13-16 to 13-19, 13-26, 13-27, 13-29, 13-39, 13-41
  input, 13-24, 13-26, 13-36, 13-41
  intrinsic, 12-24, 12-33, 13-15, 14-18
  intrinsic wave, 14-15, 14-27
  line, 13-17, 13-18, 13-29
  load, 13-19, 13-21, 13-22, 13-24, 13-25, 13-28, 13-29
  normalized, 13-9, 13-10, 13-21, 13-22
  normalized load, 13-32, 13-33
  transverse, 14-18, 14-19
Incident power, 3-30
Induced charge, 6-11
Induced electric field due to time-varying magnetic field, 11-2, 11-11, 11-17, 11-18, 11-21, 11-25 to 11-27
Induced emf, 11-3 to 11-10, 11-12 to 11-16, 11-19, 11-20, 11-23, 11-31, 11-33 to 11-36
  due to time-varying magnetic field, 11-3 to 11-10, 11-12, 11-20
  motional, 11-6, 11-8, 11-13 to 11-16, 11-19, 11-23, 11-31, 11-32, 11-33 to 11-36
Induced magnetic field due to time-varying electric field, 11-29
Infinite parallel-plane transmission line, 14-18
Input impedance, 13-24, 13-26, 13-36, 13-41
Intrinsic impedance, 12-24, 12-33, 13-15, 14-18
Intrinsic wave impedance, 14-15, 14-27
Inward power flux, 12-5
Iterative method to calculate electrostatic potential, 7-23

Laplacian, 1-15
Laplace's equation, 7-1 to 7-15
  calculation of capacitance using, 7-4, 7-7, 7-11

calculation of electric field using, 7-5, 7-6
calculation of electrostatic potential using, 7-1 to 7-15
in cylindrical coordinates, 7-4, 7-5, 7-8, 7-10, 7-12
in prolate spheroidal coordinates, 7-2
in spherical coordinates, 7-6, 7-7
Left circularly polarized wave, 12-41
Left elliptically polarized wave, 12-41
Lenz's law, 11-15
Line admittance, 13-29
Linear:
arrays, 14-42
quadrupole, 4-26
Line impedance, 13-17, 13-18, 13-29
Line integral, 1-25 to 1-27, 1-30 to 1-32
Lines of force in an electric field, 3-32
Load:
admittance, 13-21, 13-25
impedance, 13-19, 13-21, 13-22, 13-24, 13-25, 13-28, 13-29
Longitudinal mass, 4-35
Lossy lines, 13-24

Magnetic:
boundary conditions, 9-24 to 9-28
dipole moment, 10-20
field intensity (see H-field)
flux, 8-9, 8-14, 8-25, 8-26
flux density (see B-field)
flux due to applied sinusoidal voltage, 11-21
Magnetic circuits, 10-1 to 10-50
equivalent electric circuit of, 10-16, 10-35
flux density in, 10-9 to 10-26, 10-31
Magnetization, 9-4 to 9-23, 9-25
curves, 10-11, 10-13 to 10-15, 10-19, 10-25, 10-26
surface current density, 9-14 to 9-17, 9-19 to 9-23
Magnetomotive force, 10-9 to 10-26
Matching stub, 13-35
Maximum effective aperture, 14-41
atenna, 14-39
dipole, 14-40
Maxwell-Faraday law, 11-21, 11-22
Maxwell's equations, 11-24 to 11-28, 13-13
Methods of images, 7-24 to 7-29
to calculate capacitance, 7-25, 7-29
to calculate electric field, 7-27
to calculate potential, 7-24 to 7-29
to calculate surface charge density, 7-26, 7-27
Multiple dielectric capacitor, 6-22 to 6-25
Mutual inductance, 10-35 to 10-42
between circular loops, 10-38
between coaxial lines, 10-36
between coaxial solenoids, 10-37
between inductively coupled thin wire rectangular circuits, 10-41
between toroid and square loop, 10-14
Neumann formula for, 10-42

Neumann formula for mutual inductance, 10-42
Newton's:
second law, 1-8, 1-9
third law, 1-9
Normal incidence, 12-11 to 12-14
Normalized admittance, 13-9, 13-10
Normalized impedance, 13-9,

Numbers on this page refer to PROBLEM NUMBERS, not page numbers

13-10, 13-21, 13-22
Normalized load:
admittance, 13-33
impedance, 13-32, 13-33

Ohm's law, 12-8

Parallel-plane:
transmission line, 14-18
waveguide, 14-13
Parallel-plate:
capacitor, 4-16, 6-3 to 6-13, 6-24
waveguide, 14-7, 14-12
Partial derivative, 1-12, 1-13, 1-17
Permeance, 10-1, 10-2, 10-8
Permittivity, 6-11
Phase constant, 13-41, 14-19, 14-25, 14-27
Phase-shift constant, 13-12, 14-20
Phase velocity, 13-1, 14-11, 14-15, 14-19, 14-20, 14-24, 14-27
Pinch effect, 9-9
Plane-parallel wave, 14-30
Plane wave energy, 12-9
P-N junction, 4-20
Poisson's equation, 7-16 to 7-22
calculation of capacitance using, 7-17, 7-22
calculation of electric field using, 7-17, 7-20, 7-22
calculation of electrostatic potential using, 7-16 to 7-22
in cylindrical coordinates, 7-16
in spherical coordinates, 7-18, 7-20
Polarized wave, 12-13
Polarizing angle, 12-38
Position vector, 1-6 to 1-10
Potential:
electrostatic, 4-1 to 4-44
energy in electrostatic fields, 4-1 to 4-8, 4-10, 4-18, 4-21, 4-38 to 4-44
energy in magnetic fields, 9-29, 9-30
function, 4-7, 4-11, 4-24, 4-44
gradient of, 4-28 to 4-34
Potential difference:
coaxial capacitor, 6-16
spherical capacitor, 6-20
Power, 12-4
absorbed, 14-30
density, 14-43, 14-45
flux, 12-10
reflected, 13-30
Poynting's theorem, 13-3
Poynting vector, 12-1 to 12-3, 12-5, 12-6 to 12-8, 12-29, 12-47
Principal maximum of array, 14-42
Propagation constant, 12-49, 13-15, 14-13 to 14-15, 14-17

Radiation:
intensity, antenna, 14-36, 14-37
resistance, antenna, 14-34, 14-35
Receiving-end impedance, 13-23, 13-27
Rectangular:
cavity resonator oscillation, 14-6
guide, dominant mode, 14-3, 14-4
impedance chart, 13-9
resonator, field components, 14-26
Rectangular waveguide, 14-3 to 14-5, 14-8 to 14-11, 14-14, 14-15, 14-19
air-filled, 14-20
dimensions, 14-3, 14-10
Reflected power, 13-30
Reflection coefficient, 12-15, 12-20, 13-3 to 13-5, 13-8,

Numbers on this page refer to PROBLEM NUMBERS, not page numbers

13-26, 13-29, 13-30, 13-38 to 13-41, 14-12, 14-31, 14-46
Refraction, 12-16
Refractive law for B- and H-fields, 9-26 to 9-28
Reluctance, 10-1 to 10-7, 10-11, 10-12, 10-15 to 10-17, 10-35
Right circularly polarized wave, 12-41

Self energy, 4-4
Self inductance, 10-27 to 10-34
of a long wire, 10-27
of a solenoid, 10-28
of a toroid, 10-29 to 10-31
of coaxial cable, 10-33, 10-34
Sending-end impedance, 13-23, 13-27
Single-stub matching, 13-32
Skin depth, 12-45, 13-15
Skin effect, 11-30, 13-12
Smith chart, 13-10, 13-17 to 13-29, 13-32, 13-33, 13-35, 14-31
Snell's law, 14-46
Square waveguide, 14-1
Standing wave, 12-32
Stokes' theorem, 1-30 to 1-32
Stored energy, 14-26
Stub:
double, 13-10
length, 13-9, 13-10
single, 13-10
Surface:
integral, 1-23, 1-24, 1-29, 1-30 to 1-32
wave, 12-17
Susceptibility, 10-21

TE and TM modes of propagation, 14-5 to 14-15
TE_{10}:
mode, 14-24, 14-28, 14-29
wave, 14-25
TEM:
mode, 14-18
wave, 14-18, 14-23
Theorem:
Gauss', 4-12
Green's reciprocal, 4-13
Poynting's, 13-3
Stoke's, 1-30 to 1-32
TM wave, 14-5, 14-9
Torque angle, 10-50
Torgue in magnetic circuits, 10-49, 10-50
Transmission coefficient, 12-20, 14-12
Transmission line(s), 13-1 to 13-41
equations of, 13-1 to 13-12
loss, 14-45
terminated, 13-9, 13-10
Transmitted and absorbed power, 14-27 to 14-31
Transverse impedance, 14-18, 14-19
Traveling wave, 12-33

Unit conductance circle, 13-29

Vector:
acceleration, 1-7 to 1-10
angular velocity, 1-10
base, 1-1, 1-10
differential equation, 1-5
position, 1-6 to 1-10
Poynting, 12-1 to 12-3, 12-5 to 12-8, 12-29, 12-47
space, 1-1
triple product, 1-2
velocity, 1-6 to 1-10
Vector function, 1-1 to 1-10
derivative of, 1-3, 1-4, 1-6, 1-10
Vector magnetic potential, 8-13 to 8-35, 10-42
around antennas, 8-33
due to circular loop, 8-31
due to coaxial cable carrying current, 8-30

Numbers on this page refer to PROBLEM NUMBERS, not page numbers

due to long cylinder carrying current, 8-28
due to parallel wires carrying current, 8-29
due to short wire carrying current, 8-13, 8-32
due to time-varrying magnetic fields, 8-34
Volume integral, 1-23, 1-28
VSWR (voltage standing wave ratio), 13-29, 13-35, 13-39
circle, 13-26

Wave:
impedance, 13-13, 14-31
propagation, 14-8
plane-parallel, 14-30
polarized, 12-13
standing, 12-32
surface, 12-17
traveling, 12-33
velocity, 14-22
Waveguide, 14-1 to 14-4, 14-16
absorbed power, 14-30
characteristics, 14-16 to 14-19
circular, 14-1
cylindrical, 14-19
parallel-plate, 14-7, 14-12
square, 14-1
Wavelength, 13-26, 13-35, 14-15, 14-16, 14-23, 14-27
dielectric side, 14-12
free space side, 14-12
Work, 4-1 to 4-7
due to moving charges, 4-5

X-band:
guide, 14-15
waveguide system, 14-31